# COFFEE TECHNOLOGY

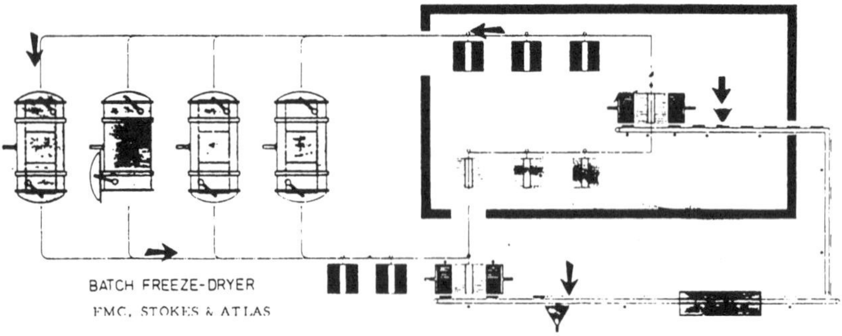
BATCH FREEZE-DRYER
FMC, STOKES & ATLAS

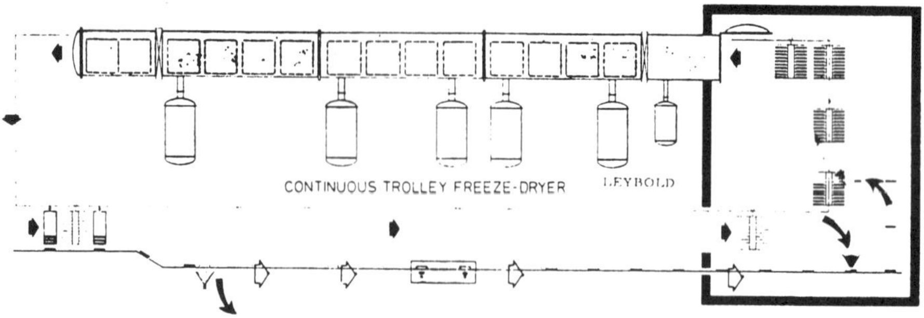
CONTINUOUS TROLLEY FREEZE-DRYER
LEYBOLD

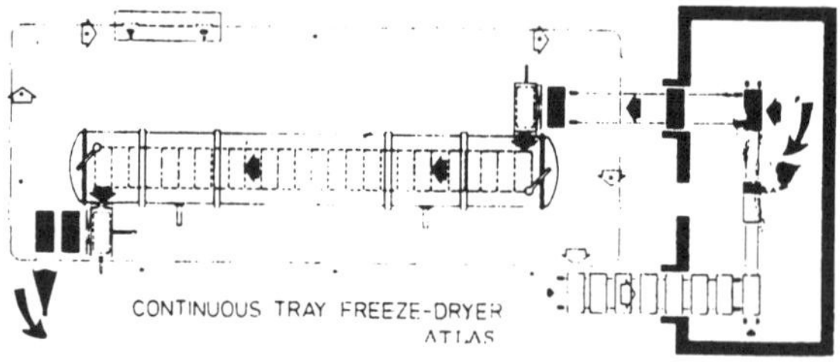
CONTINUOUS TRAY FREEZE-DRYER
ATLAS

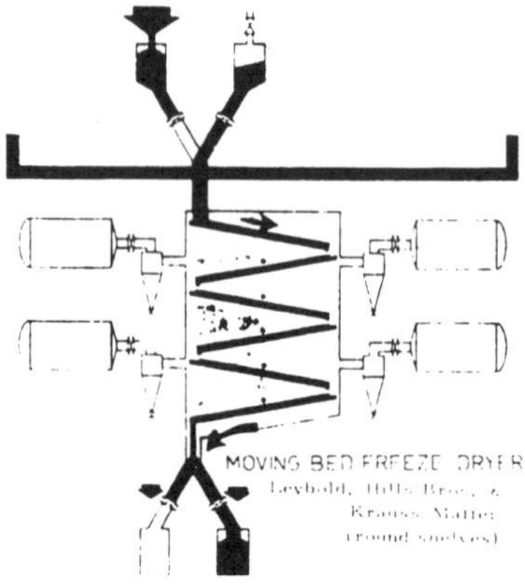
MOVING BED FREEZE DRYER
Leybold, Hills Bros. &
Krauss Maffei
(round shelves)

# COFFEE TECHNOLOGY

Michael Sivetz, Ch. E.
Coffee Industry Consultant
Corvallis, Oregon

Norman W. Desrosier, Ph.D.
The AVI Publishing Company

AVI PUBLISHING COMPANY, INC.
Westport, Connecticut

Library of Congress Cataloging in Publication Data

Sivetz, Michael.
Coffee technology.

Edition of 1963 published under title: Coffee processing technology.
Includes bibliographies and index.
1. Coffee processing. I. Desrosier, Norman W., joint author. II. Title.
TP645.S5 1979 663'.93 79-10538
ISBN: 0-87055-269-4

Printed in the United States of America

# Preface

There has been a need to revise and update the volumes dealing with Coffee Processing Technology, written over 15 years ago by the late H. Elliott Foote and Michael Sivetz, since significant growth has occurred in the field during that time. This book is an attempt to summarize present knowledge of coffee in one volume.

One object of this book is thus to organize and develop the important technical, analytical, engineering and practical aspects of the coffee industry. It endeavors to provide the reader with a central source of historical, theoretical, and currently practical information on coffee. An industry as large as that of green, roast and soluble coffees has need for a definitive technical book.

Training new personnel and correcting the misguided are two services the book can render. Better operating practice results in better use of natural resources, better processing and better value for the consumer. Coffee technology is more important today than it has ever been. The book in its own way endeavors to help the progress of the coffee industry.

The book therefore sets out to: (1) describe the coffee industry; (2) guide the reader through the process steps; (3) collect, organize, and correlate data on coffee and its properties; and (4) describe coffee flavors and aromas as to their natural origin, chemical composition and chemical changes. For the experienced coffee processor, the book relates and organizes the loose bits of information that continually come forward in numerous publications, points out accepted practices, tries to separate error from fact and to furnish thereby some enlightment on a scientific basis.

Coffee may be purchased by packet, pound or cup, but weight (the quantity measure) has value only insofar as it has acceptable flavor (the quality measure). Coffee has one basic value: it gives pleasure and satisfaction through flavor, aroma and desirable physiological and psychological effects.

We cannot explain coffee preparation in a completely scientific manner. Neither the chemical composition of the coffee bean nor the processed coffee is fully known. The influences of aging, roasting, processing and staling have not been clearly elucidated. The subjective evaluation of coffee makes the interpretation of these complex interrelationships difficult. There is much in coffee processing that is empirical and remains to be explained. In this book we have attempted to apply scientific principles as far as possible. We have included descriptive material of regions, varieties and processes where it seems to us that these influence the product quality.

The authors will appreciate criticism and the pointing out of errors and omissions of fact, data, and opinion. They fully realize that a book of this type will inevitably have defects, and they would like to enlist the help of their readers in improving the quality of a future edition.

Michael Sivetz
Corvallis, Oregon

Norman W. Desrosier
Westport, Conn.

*February 1979*

# Dedication

To the late Dr. H. Elliott Foote, Chappaqua, New York.

# Contents

# Part I
# History of Coffee

# 1

# Development of Coffee Plantations

## INTRODUCTION

Only an overall view of the coffee industry since the middle of the 19th century offers the proper perspective on past developments as well as a guide to current conditions and an insight into its future potential.

Coffee trading throughout the world, especially with Europe and the United States, was not a significant factor in world commerce until after 1850 when world shipments totaled about three million bags.

The original architects of the coffee industry are said to be the Ethiopians. Before 1200, its consumption had spread along the Red Sea to Aden, Mecca, and Cairo. By 1300, coffee was known to Persia, and by 1500 to Turkey. Shortly thereafter, coffee was being sold in Venice across the Mediterranean Sea. Rauwolf, the famous German botanist and physician, was the first European to mention coffee on his trip to the Levant. Moslem pilgrims had started to cultivate coffee in India about 1600. Coffee was first sold in Amsterdam in 1640 with regular Mocha coffee deliveries by 1663. Meanwhile, commerce in coffee spread by coastal routes to Italy, France, and England and was introduced into North America in 1668. By 1715–1730 coffee was under cultivation in Haiti, Jamaica, and Santo Domingo, and shortly thereafter in Cuba. At that time these island sources, especially Haiti, produced about 500,000 bags of green coffee each year. but the rate of production fell off markedly in subsequent years. Today Haiti produces annually about 600,000 bags. The Dutch East Indies cultivated coffee in 1718 and have had a small share of world coffee commerce ever since.

With coffee under cultivation for the first time in 1729, Brazil was producing 200,000 bags annually by 1825 and three million bags per year by 1850. Since then, Brazil has been the major supplier of coffee to the world. The first significant exports of coffee from Brazil were made about 1809, 80 years after the first plantings. It took another 50 years for Brazil's coffee exports to reach four million bags. A vignette of coffee production is shown in Fig. 1.1. Parallel development occurred after 1850 in most Latin American countries and as late as 1900 in Colombia.

the story of COFFEE
SEEDLINGS — from special nurseries are the start of the coffee we use ...from 2¾ billion trees a year.
TRANSPLANTED — after a year, the young trees need 4 more years of constant care before they bear.
BLOSSOMS — are particularly fragile, and bad weather at any time can ruin an entire year's crop.
CHERRIES — must be hand-picked just when ripe, and each cherry holds only two coffee beans.
PICKING — is arduous: the average tree yields a pound a year, which takes 2000 cherries.
TESTING — and blending are final steps before coffee is roasted here in the U.S. to your taste.


*Courtesy of Pan American Coffee Bureau*

FIG. 1.1 THE STORY OF COFFEE

Although primary emphasis is given in this text to roasted and ground (R & G) and instant coffees, the study of coffee horticulture and the processing of coffee fruit has been very actively pursued since 1900 and especially in the last two decades. Because there is a 5-yr span between seed planting and crop maturation, gathering of significant scientific information is slow. Results are more likely to be linked to general experience and observation than to systematic collection and interpretation of data. For example, the widespread use of insecticides, fertilizers, and pruning methods is often based on their reported success in similar plants, long before first-hand experience with coffee is obtained. Even today there is still a surprising lack of knowledge concerning green coffee bean quality not only in the literature but in the field. Most available information relates to achieving greater production per tree. The point is that advances made in raising other crops have been successfully applied to coffee so that productivity of the land has risen several-fold in the more advanced cultivating areas. At the same time, during the various periods of surplus production of coffee, some countries have eliminated unproductive areas of cultivation. Most coffee-growing countries employ scientific groups to evaluate soil deficiencies, fertilizer applications, water needs, mulching, pruning, diseases, yields per tree, shade factors and financial aid to farmers. Fig. 1.2 shows important aspects of the coffee cherry.

## COFFEE PRODUCTION

In the past 30 years, the coffee trade from Africa has grown more rapidly than elsewhere and has contributed at times to the world surpluses. Coffee growth and cultivation in Africa are not yet controlled enough to affect exports. Hardly any coffee was grown in Africa before 1900, and even before 1940 less than one million bags were exported annually. Current exports total about 19 million bags per year. The major producers are Ivory Coast, Angola, Uganda, Congo, Ethiopia, and Malagasy followed by the Cameroon, Kenya, Tanganyika, and Rwanda-Brundi. Only about 15 percent of all these coffees are of the Arabica species and these come mostly from parts of the Congo, Ethiopia, Kenya, and Tanganyika. Robusta (Canephora) coffees of varying types and flavor constitute about 85 percent of all African coffees. Currently, African coffees represent about 25 percent of world exports. Robusta coffees from Africa are often picked from wild trees and dried in the cherry form. This procedure makes for some difficulty in grading the types. The harsh Robusta flavor is not so acceptable to the consumer which means that Robustas sell for about half the price of good mild coffees and less than two-thirds the price of good Brazilian coffees. However, the low price naturally attracts the price conscious buyer so that Robustas have found a broad use in soluble coffee processing as well as in roast and ground (R & G) form in European countries, such as France, Portugal, England, and Belgium that have close economic relations with African nations.

The growth of mild coffees, while not quite as startling as that of the African Robustas, has been increasing through the years. In 1900, Colombian coffees represented only 2 percent of world commerce but by 1925 this had increased to 10 percent and, today, is near 15 percent. The Central American countries have increased their share of world coffee exports from about 5 percent in 1900 to about 15 percent today. This growth has occurred with mild coffees which command, as a rule, much higher prices than Brazilian coffees as they are more aromatic and flavorful.

A study of coffee usage in the past 50 years shows that on a percentage basis more mild and less Brazilian coffees have been imported into the United States

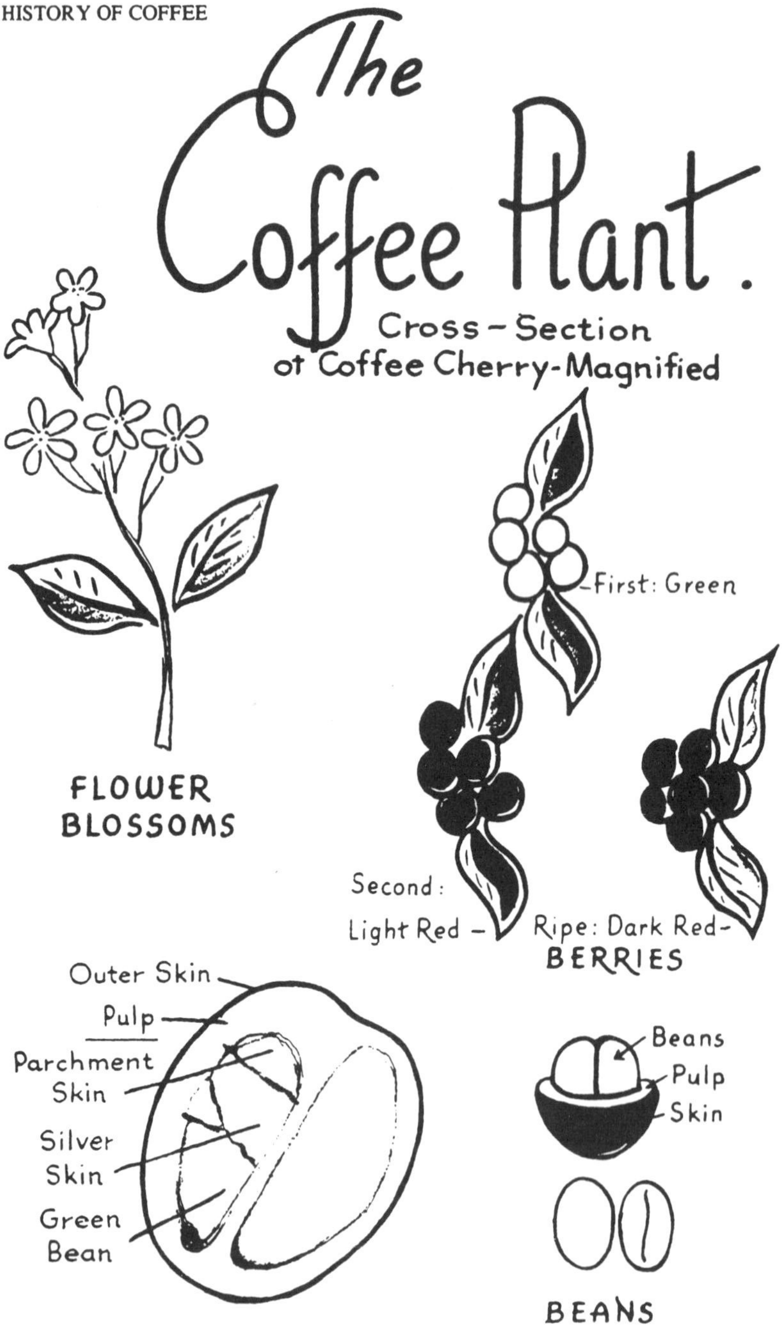


*Courtesy of Coffee and Tea Industries, New York*

FIG. 1.2. IMPORTANT ASPECTS OF COFFEE CHERRY

and, except for the large imports of Robusta coffees since 1950, better flavored blends have been available. Robusta coffees have been used for the most part in soluble coffee manufacture, although many Robustas have been used in roast coffees. This produces a quality that is not very acceptable in the United States where good blends are freely available. However, Robustas are the major coffees used in France, Portugal, and the United Kingdom as well as in Holland and Belgium.

Figure 1.3 is a map of the coffee growing world. World coffee production is shown in Table 1.1 while the exports are shown in Table 1.2.

## Brazil vs. Africa Production Levels

After the excessive productions of coffee in Brazil in the early 1960's and Brazil's dependence on its stock piles in the latter 1960's, we have seen Brazil's contribution to world supplies diminishing percentage-wise, even though exports remain substantial.

Meanwhile, Africa's production (primarily Robustas) since the end of World War II has risen to about 19,000,000 bags per year and now represents over 25 percent of all coffee used in the U.S. and Europe.

## Changing Coffee Markets

Note in the coffee export Table 1.3 that since 1920 the U.S. and world population has more than doubled, but world use of coffee has tripled. Per capita use in the U.S.A., Sweden, and other heavy coffee-drinking countries has declined noticeably in recent years. The additional coffee being produced is going into new markets like England and Japan as well as to the U.S.S.R. and its satellites. Other countries are consuming more. Since 1970 we have passed from a period of surplus coffee production into a period of deficit production.

There are still some surplus stocks in the Ivory Coast, Colombia, and Salvador, but these are relatively small when considering the 56 million bag world use in 1969–70. Stocks today are only about a six months supply. The U.S.D.A. estimated that exports in 1976–77 were about 42,000,000 bags of coffee.

Stored coffee is not always of first grade quality. Hence reported stock levels may be higher than are actually salable. Brazil is still the world's major producer. But in the past, Brazil has been repeatedly plagued with droughts, floods, and frosts. Now a serious leaf rust (fungus) disease threatens. The consequence of all these natural disasters has been a gross production decline with exports less than ten million bags in 1976 after the 1975 frost.

The sensitivity of this supply situation is well illustrated by the sharp price rise for coffees in the fall of 1969 and 1972. When coffee supply is on a "hand to mouth" basis, any major agricultural failure or political change can alter prices quickly. In 1976 green prices rose 200 to 300 percent over those of 1975.

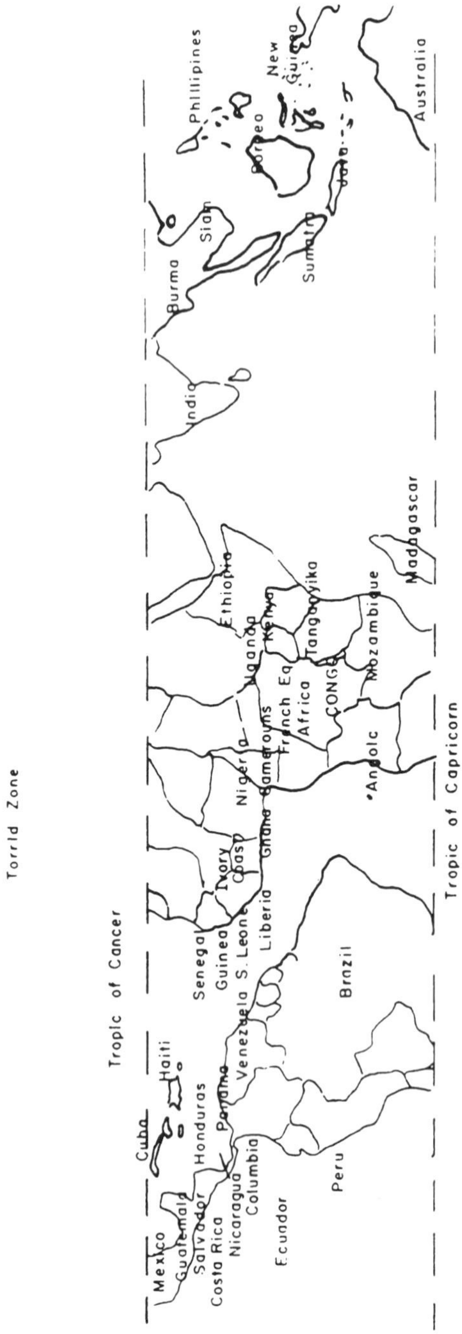


FIG. 1.3. MAP OF COFFEE GROWING ZONES OF THE WORLD

TABLE 1.1. COFFEE, GREEN: TOTAL PRODUCTION IN SPECIFIED COUNTRIES — AVERAGE 1967/68-1971/72, ANNUAL 1972/73-1976/77[1]
(IN THOUSANDS OF 60 KILOGRAM BAGS)[2]

| Region and Country | Average 1967/68-1971/72 | 1972/73 | 1973/74 | 1974/75 | 1975/76 | 1976/77 |
|---|---|---|---|---|---|---|
| North America: | | | | | | |
| Costa Rica | 1,322 | 1,335 | 1,570 | 1,390 | 1,305 | 1,500 |
| Cuba | 477 | 475 | 500 | 450 | 415 | 415 |
| Dominican Republic | 646 | 750 | 845 | 880 | 1,020 | 800 |
| El Salvador | 2,314 | 2,100 | 2,378 | 3,300 | 2,010 | 3,200 |
| Guatemala | 1,856 | 2,250 | 2,200 | 2,540 | 2,150 | 2,550 |
| Haiti | 514 | 525 | 550 | 520 | 660 | 600 |
| Honduras | 550 | 850 | 775 | 815 | 830 | 950 |
| Jamaica | 20 | 22 | 30 | 21 | 31 | 18 |
| Mexico | 3,085 | 3,700 | 3,300 | 3,900 | 4,100 | 4,300 |
| Nicaragua | 601 | 570 | 610 | 700 | 810 | 850 |
| Panama | 81 | 82 | 72 | 75 | 75 | 75 |
| Trinidad-Tobago | 60 | 50 | 30 | 65 | 42 | 57 |
| US-Hawaii | 31 | 22 | 19 | 17 | 11 | 10 |
| US-Puerto Rico | 233 | 200 | 220 | 200 | 200 | 200 |
| Total | 11,790 | 12,931 | 13,099 | 14,873 | 13,659 | 15,525 |
| South America: | | | | | | |
| Bolivia | 127 | 95 | 95 | 90 | 100 | 105 |
| Brazil | 18,370 | 24,000 | 14,500 | 27,500 | 23,000 | 9,500 |
| Colombia | 7,870 | 8,800 | 7,800 | 9,000 | 8,700 | 9,000 |
| Ecuador[3] | 1,047 | 1,100 | 870 | 1,270 | 1,170 | 1,200 |
| Guyana | 16 | 12 | 10 | 15 | 15 | 15 |
| Paraguay | 49 | 50 | 50 | 42 | 40 | 20 |
| Peru | 940 | 1,030 | 1,000 | 900 | 900 | 1,000 |
| Venezuela | 872 | 1,100 | 960 | 765 | 1,075 | 835 |
| Total | 29,291 | 36,187 | 25,285 | 39,582 | 35,000 | 21,675 |
| Africa: | | | | | | |
| Angola | 3,300 | 3,500 | 3,200 | 3,000 | 1,200 | 1,200 |
| Benin[4] | 16 | 15 | 13 | 14 | 14 | 14 |
| Burundi | 316 | 355 | 350 | 450 | 285 | 350 |
| Cameroon | 1,160 | 1,378 | 1,260 | 1,816 | 1,332 | 1,580 |
| Cent African Rep | 174 | 180 | 190 | 175 | 165 | 165 |
| Congo, Brazzaville | 15 | 14 | 10 | 10 | 10 | 10 |
| Equatorial Guinea | 125 | 115 | 105 | 110 | 90 | 90 |
| Ethiopia | 2,009 | 2,100 | 1,700 | 2,050 | 2,100 | 2,100 |
| Gabon | 16 | 15 | 9 | 10 | 10 | 10 |
| Ghana | 85 | 80 | 45 | 50 | 65 | 50 |
| Guinea | 160 | 125 | 105 | 100 | 90 | 90 |
| Ivory Coast | 4,195 | 5,050 | 3,285 | 4,500 | 5,080 | 5,300 |
| Kenya | 870 | 1,265 | 1,100 | 1,100 | 1,240 | 1,135 |
| Liberia | 71 | 85 | 65 | 75 | 75 | 75 |
| Malagasy Republic | 1,019 | 1,000 | 1,000 | 1,300 | 1,200 | 1,200 |
| Nigeria | 63 | 70 | 38 | 40 | 65 | 50 |
| Rwanda | 209 | 186 | 266 | 256 | 235 | 235 |
| Sierra Leone | 97 | 135 | 67 | 125 | 75 | 120 |
| Tanzania | 853 | 800 | 700 | 865 | 900 | 900 |
| Togo | 215 | 200 | 180 | 200 | 195 | 190 |
| Uganda | 3,047 | 3,300 | 3,100 | 3,000 | 2,800 | 2,700 |
| Zaire (Congo, K) | 1,150 | 1,380 | 1,317 | 1,150 | 1,383 | 1,433 |
| Total | 19,164 | 21,548 | 18,105 | 20,396 | 18,609 | 18,997 |

TABLE 1.1 Continued

| Region and Country | Average 1967/68-1971/72 | 1972/73 | 1973/74 | 1974/75 | 1975/76 | 1976/77 |
|---|---|---|---|---|---|---|
| Asia: | | | | | | |
| India | 1,320 | 1,580 | 1.535 | 1,630 | 1,480 | 1,715 |
| Indonesia | 2,190 | 2,700 | 2,750 | 2,675 | 2,700 | 2,800 |
| Malaysia | 91 | 65 | 67 | 70 | 100 | 100 |
| Philippines | 785 | 850 | 865 | 1,035 | 1,080 | 1,150 |
| Portuguese Timor | 54 | 65 | 60 | 75 | 75 | 65 |
| Vietnam | 52 | 55 | 55 | 60 | 60 | 60 |
| Yemen | 57 | 45 | 25 | 35 | 35 | 35 |
| Total | 4,549 | 5,360 | 5,357 | 5,580 | 5,530 | 5,925 |
| Oceania: | | | | | | |
| New Caledonia | 28 | 25 | 25 | 25 | 25 | 25 |
| Papua New Guinea | 389 | 560 | 588 | 633 | 667 | 600 |
| Total | 418 | 585 | 613 | 658 | 692 | 625 |
| World Total | 65,212 | 76,611 | 62,459 | 81,089 | 73,490 | 62,747 |

[1]Coffee marketing year begins about July in some countries and in others about October.
[2]132.276 pounds.
[3]As indicated in footnote 1, the coffee marketing year begins in some countries as early as July. Ecuador is one of these countries. Hence, the crop harvested principally during June-October 1976 in that country is shown as production for the 1976/77 marketing year. In Ecuador, however, this is referred to as the 1975/76 crop.
[4]Formerly Dahomey.

Source: Foreign Agricultural Service.

TABLE 1.2. COFFEE, GREEN: EXPORTABLE PRODUCTION IN SPECIFIED COUNTRIES — AVERAGE 1967/68-1971/72, ANNUAL 1972/73-1976/77[1] (IN THOUSANDS OF 60 KILOGRAM BAGS)[2]

| Region and Country | Average 1967/68-1971/72 | 1972/73 | 1973/74 | 1974/75 | 1975/76 | 1976/77 |
|---|---|---|---|---|---|---|
| North America: | | | | | | |
| Costa Rica | 1,172 | 1,160 | 1,400 | 1,237 | 1,152 | 1,350 |
| Cuba | 51 | 20 | 50 | — | — | — |
| Dominican Republic | 455 | 490 | 600 | 600 | 760 | 550 |
| El Salvador | 2,164 | 1,935 | 2,203 | 3,130 | 1,840 | 3,025 |
| Guatemala | 1,615 | 1,990 | 1,925 | 2,255 | 1,860 | 2,252 |
| Haiti | 319 | 325 | 340 | 305 | 440 | 380 |
| Honduras | 444 | 720 | 635 | 705 | 725 | 840 |
| Jamaica | 6 | 6 | 14 | 4 | 12 | — |
| Mexico | 1,598 | 2,100 | 1,690 | 2,156 | 2,600 | 2,800 |
| Nicaragua | 533 | 480 | 496 | 608 | 734 | 770 |
| Panama | 17 | 20 | — | — | 25 | 25 |
| Trinidad-Tobago | 45 | 40 | 13 | 55 | 29 | 42 |
| US-Hawaii | 9 | 2 | — | — | — | — |
| US-Puerto Rico | — | — | — | — | — | — |
| Total | 8,428 | 9,288 | 9,366 | 11,055 | 10,177 | 12,034 |
| South America: | | | | | | |
| Bolivia | 58 | 65 | 73 | 68 | 78 | 83 |
| Brazil | 9,869 | 15,000 | 6,370 | 19,500 | 15,000 | 2,500 |
| Colombia | 6,498 | 7,430 | 6,250 | 7,400 | 7,300 | 7,600 |
| Ecuador [3] | 835 | 860 | 625 | 1,113 | 1,005 | 1,030 |
| Guyana | 1 | — | — | — | — | — |
| Paraguay | 30 | 35 | 28 | 19 | 19 | — |

TABLE 1.2 Continued

| Region and Country | Average 1967/68-1971/72 | 1972/73 | 1973/74 | 1974/75 | 1975/76 | 1976/77 |
|---|---|---|---|---|---|---|
| Peru | 719 | 800 | 750 | 640 | 640 | 750 |
| Venezuela | 260 | 430 | 255 | 95 | 375 | 105 |
| Total | 18,270 | 24,620 | 14,351 | 28,835 | 24,417 | 12,068 |
| Africa: | | | | | | |
| Angola | 3,216 | 3,400 | 3,095 | 2,895 | 1,140 | 1,140 |
| Benin[4] | 14 | 14 | 12 | 13 | 13 | 13 |
| Burundi | 311 | 350 | 345 | 445 | 280 | 345 |
| Cameroon | 1,120 | 1,533 | 1,213 | 1,766 | 1,279 | 1,525 |
| Cent African Rep | 167 | 170 | 179 | 164 | 155 | 155 |
| Congo, Brazzaville | 14 | 13 | 9 | 9 | 9 | 9 |
| Equatorial Guinea | 117 | 110 | 100 | 105 | 85 | 85 |
| Ethiopia | 1,423 | 1,410 | 1,005 | 1,340 | 1,375 | 1,375 |
| Gabon | 14 | 14 | 8 | 9 | 9 | 9 |
| Ghana | 74 | 67 | 31 | 36 | 50 | 35 |
| Guinea | 152 | 120 | 100 | 94 | 84 | 84 |
| Ivory Coast | 4,136 | 4,985 | 3,219 | 4,432 | 5,020 | 5,240 |
| Kenya | 850 | 1,240 | 1,073 | 1,082 | 1,223 | 1,113 |
| Liberia | 67 | 80 | 60 | 69 | 70 | 70 |
| Malagasy Republic | 899 | 865 | 865 | 1,160 | 1,060 | 1,060 |
| Nigeria | 47 | 40 | 11 | 12 | 10 | — |
| Rwanda | 204 | 180 | 260 | 250 | 230 | 230 |
| Sierra Leone | 90 | 130 | 62 | 120 | 70 | 115 |
| Tanzania | 836 | 780 | 678 | 845 | 880 | 880 |
| Togo | 211 | 197 | 177 | 197 | 192 | 187 |
| Uganda | 3,031 | 3,280 | 3,078 | 2,978 | 2,778 | 2,678 |
| Zaire (Congo, K) | 1,055 | 1,265 | 1,184 | 1,017 | 1,266 | 1,316 |
| Total | 18,049 | 20,243 | 16,764 | 19,038 | 17,278 | 17,664 |
| Asia: | | | | | | |
| India | 581 | 820 | 730 | 970 | 730 | 965 |
| Indonesia | 1,484 | 1,650 | 1,795 | 1,700 | 1,800 | 1,900 |
| Malaysia | — | — | — | — | — | — |
| Philippines | 4 | | — | 100 | 110 | 150 |
| Portuguese Timor | 48 | 60 | 50 | 64 | 64 | 55 |
| Vietnam | — | — | — | — | — | — |
| Yemen | 47 | 35 | 20 | 30 | 30 | 30 |
| Total | 2,164 | 2,565 | 2,595 | 2,864 | 2,734 | 3,100 |
| Oceania: | | | | | | |
| New Caledonia | 18 | 15 | 14 | 14 | 14 | 14 |
| Papua New Guinea | 378 | 554 | 584 | 628 | 662 | 595 |
| Total | 396 | 569 | 598 | 642 | 676 | 609 |
| World Total | 47,308 | 57,285 | 43,674 | 62,434 | 55,282 | 45,475 |

[1]Coffee marketing year begins about July in some countries and in others about October. Exportable production represents total harvested production minus estimated domestic consumption.
[2]132.276 pounds.
[3]As indicated in footnote 1, the coffee marketing year begins in some countries as early as July. Ecuador is one of these countries. Hence, the crop harvested principally during June-October 1976 in that country is shown as production for the 1976/77 marketing year. In Ecuador, however, this is referred to as the 1975/76 crop.
[4]Formerly Dahomey.

Source: Foreign Agricultural Service.

Table 1.3 illustrates the rate of coffee growth and commerce in the world since 1800, in reference to the top 20 coffee growing countries that export coffee beans. Brazil's production and exports are in decline.

## United States Coffee Market

Reviewing the magnitude of the coffee industry in the United States, it is worth noting that coffee use has tripled since 1920 while population has doubled. About 19 million bags of coffee are imported annually at a value of about six billion dollars. The three major coffee processors in the United States are General Foods Corporation, Atlantic & Pacific Tea Company (A & P) chain stores, and J. A. Folger & Company which account for almost half of nationwide coffee sales. Other processors are much smaller and usually more regional. Roast coffee sells at retail at about twice its raw green coffee value; soluble coffee retails at about three times its green value. Thus, the gross retail value of coffee sold in the United States is about 10 billion dollars per year or half the amount used by the world market. The dollar value of soluble coffee sales in the United States is about 25 percent of total coffee sales.

The United States market has about 200 green coffee importers and distributors. There are about 400 roasting firms, most of which are long established and financially sound. Physical coffee plant investments are in excess of 2½ billion dollars. There are about 10,000 coffee salesmen serving about 300,000 retail stores and 300,000 restaurants. There are many regionally strong-selling coffee brands that frequently sell better than competing national brands in their own areas. The coffees sold are influenced by local tastes, buying and drinking traditions, water quality, and brewing tastes. Also, there is some general public indifference to the preparation of a good brew and recognition of good, fresh coffee flavor. The uniform quality of average soluble coffees as well as their convenience have displaced sales of some poor quality roast coffees.

In 1975, 66% of U.S. coffee imports were from Latin America (Arabicas), 29% from Africa (mostly Robustas) and 4% from Asia and Oceania (mostly Robustas). However, in 1950, African and Robustas were only about 6% of U.S. imports. This shows the marked change to lower grade quality being used in the USA and it is to be noted that this period has been accompanied by a continual per capita decline in coffee use. In 1950 Robustas were primarily used for instant coffees, but now they constitute a significant part of R&G coffees. The increased use of Robusta has also involved the increased use of decaffeinated coffees, since such processing removes the harsh character of Robustas, leaving a mild (non aromatic) low-flavored beverage.

## European Coffee Market

Although coffee houses have existed in Europe since 1700, shipments of coffee from the East Indies and Ethiopia at first were very small. By 1770, Brazil began a

TABLE 1.3. APPROXIMATE WORLD GREEN COFFEE BEAN EXPORTS SINCE 1800 (MILLIONS 60 KG BAGS).

| | Years | 1975 | 1970 | 1965 | 1960 | 1955 | 1950 | 1940 | 1930 | 1920 | 1900 | 1880 | 1860 | 1840 | 1820 | 1800 |
|---|---|---|---|---|---|---|---|---|---|---|---|---|---|---|---|---|
| | Population (Millions) | | | | | | | | | | | | | | | |
| | World | 4,000 | 3,600 | — | 3,000 | — | 2,800 | 2,400 | 2,100 | 1,900 | 1,600 | 1,400 | 1,250 | 1,100 | 1,000 | 900 |
| | U.S.A. | 220 | 205 | — | 180 | — | 150 | 133 | 122 | 106 | 75 | 50 | 34 | 18 | 12 | 6 |
| | Coffee (Million Bags) | | | | | | | | | | | | | | | |
| | U.S.A. | 19 | 22 | 22 | 22 | 20 | 19 | 16 | 12 | 10 | 7.5 | 6 | 3 | 0.6 | 0.1 | 0.06 |
| | World | 55 | 56 | 50 | 44 | 35 | 30 | 25 | 24 | 18 | 15 | 9 | 4 | 2 | 0.8 | 0.60 |
| 1. | Brazil | 13 | 18 | 16 | 16 | — | 16 | 7 | 16 | 13 | 8 | 6 | 3 | 1 | 0.1 | nil |
| | (Percent of World Crop) | | (23%) | (28%) | (42%) | | (50%) | | | (80%) | (85%) | | | | | |
| 2. | Colombia | 7 | 6.5 | 5.6 | 6 | 5 | 4.5 | 3.7 | 3.3 | 2 | 0.4 | 0.1 | — | — | — | — |
| | | | (14%) | (14%) | | | (12%) | | | (10%) | | | | | | |
| 3. | Ivory Coast (R)* | | 4.0 | 3.1 | 2.7 | 2 | 1 | 0.2 | 0.1 | — | — | start | — | — | — | — |
| | | | (9%) | (6%) | | | (3%) | | | | | | | | | |
| 4. | Angola (R) | 3.3 | 3.25 | 3 | 2.5 | 2 | 1 | 0.2 | 0.15 | 0.1 | — | — | — | — | — | — |
| | | | (7%) | (7%) | | | (3%) | | | | | | | | | |
| 5. | Uganda (R) | 3 | 2.75 | 2.5 | 2.0 | 1.5 | 1 | 0.2 | 0.1 | 0.1 | — | — | — | — | — | — |
| | | | (6%) | (5.5%) | | | (3%) | | | | | | | | | |
| 6. | Salvador | 2.5 | 2.25 | 1.67 | 1.2 | 1.1 | 1.1 | 1.0 | 0.9 | 0.7 | 0.3 | 0.1 | start | — | — | — |
| | | | (5%) | (3.7%) | | | (3%) | | | | | | | | | |
| 7. | Mexico | 2.0 | 1.6 | 1.5 | 1.4 | 1.2 | 1.1 | 0.8 | 0.7 | 0.8 | 0.4 | 0.1 | start | — | — | — |
| | | | (3.5%) | (3.2%) | | | (3%) | | | | | | | | | |
| 8. | Guatemala | 2.2 | 1.7 | 1.5 | 1.3 | 1.2 | 1.1 | 0.9 | 0.9 | 0.6 | 0.4 | 0.2 | start | — | — | — |
| | | | (3.3%) | | | | (3%) | | | | | | | | | |
| 9. | Indonesia (R) | 2.0 | 1.6 | 1.5 | 1.2 | 0.9 | 0.7 | 1.3 | 1.5 | 1.2 | 1.0 | 1.2 | 1.2 | 0.8 | 0.3 | 0.1 |
| | | | (3.3%) | (3.5%) | | | (2%) | | | | | | | | | |
| 10. | Ethiopia | 1.0 | 1.7 | 1.6 | 1.4 | 0.8 | 0.5 | 0.3 | 0.2 | 0.1 | nil | — | — | — | — | — |
| | | | (3.0%) | | | | (1.5%) | | | | | | | | | |
| 11. | Cameroon (R) | | 1.1 | 1.0 | 0.8 | 0.6 | 0.4 | 0.1 | — | — | — | — | — | — | — | — |
| 12. | Costa Rica (R) | 1.5 | 1.2 | 1.1 | 1.0 | 0.8 | 0.6 | 0.4 | 0.3 | 0.1 | nil | — | — | — | — | — |
| 13. | Zaire (R) | 1.2 | 1.0 | 1.0 | 1.5 | 0.6 | 0.6 | nil | — | — | — | — | — | — | — | — |
| 14. | Tanzania | 0.7 | 0.75 | 0.6 | 0.5 | 0.3 | 0.2 | 0.6 | 0.2 | 0.2 | nil | — | — | — | — | — |
| 15. | Peru | 0.5 | 0.71 | 0.55 | 0.5 | 0.3 | 0.20 | 0.06 | 0.02 | 0.01 | nil | — | — | — | — | — |
| 16. | Malagasy (R) | | 0.8 | 0.8 | 0.67 | 0.7 | 0.6 | 0.2 | 0.2 | 0.2 | 0.1 | start | — | — | — | — |
| 17. | Nicaragua | 0.6 | 0.5 | 0.45 | 0.4 | 0.35 | 0.35 | 0.2 | 0.2 | 0.2 | 0.1 | start | — | — | — | — |
| 18. | Ecuador | 1.0 | 0.8 | 0.6 | 0.5 | 0.4 | 0.35 | 0.25 | 0.15 | 0.06 | 0.04 | 0.04 | 0.01 | — | — | — |
| 19. | Dominican Republic | 0.6 | 0.45 | 0.40 | 0.48 | 0.4 | 0.3 | 0.25 | 0.2 | 0.1 | — | — | — | — | — | — |
| 20. | Honduras | 0.5 | 0.4 | 0.4 | 0.26 | 0.25 | 0.2 | 0.10 | 0.03 | — | | | | | | |

*R = Robusta types primarily.

TABLE 1.4. GREEN COFFEE: EXPORTS FROM SPECIFIED COUNTRIES, BY COUNTRY OF DESTINATION, CALENDAR YEAR 1975
(IN BAGS OF 60 KILOGRAMS)[1]

| Country of Destination | Angola | Brazil[2] | Colombia[3] | Guatemala | Ivory Coast | Mexico[4] |
|---|---|---|---|---|---|---|
| Europe: | | | | | | |
| Austria | — | 132,499 | — | 1,150 | — | — |
| Belgium-Luxembourg | 17,662 | 176,579 | 149,856 | 101,433 | — | 81,937 |
| Bulgaria | — | 5,833 | — | — | — | — |
| Czechoslovakia | — | 187,124 | 23,666 | 7,500 | — | 10,000 |
| Denmark | 1,979 | 533,286 | 105,617 | 18,233 | — | 3,439 |
| Finland | — | 180,727 | 324,126 | 87,117 | — | 5,169 |
| France | 65,307 | 819,176 | 201,775 | 38,050 | 1,435,517 | 77,577 |
| German Democratic Rep. | — | 446,531 | 64,876 | 42,667 | — | — |
| Germany, West | 56,862 | 553,951 | 1,842,781 | 543,033 | 98,467 | 187,354 |
| Greece | — | 186,499 | — | — | 13,400 | — |
| Hungary | — | 95,885 | 6,664 | — | 50,250 | — |
| Iceland | — | 25,225 | — | — | — | — |
| Italy | 30,841 | 1,438,750 | 33,624 | 20,350 | 102,100 | 28,827 |
| Netherlands | 227,443 | 160,821 | 757,440 | 229,917 | 331,450 | 93,132 |
| Norway | — | 334,874 | 61,376 | 24,000 | — | 17,908 |
| Poland | — | 344,884 | 122,444 | — | — | — |
| Portugal | — | — | — | — | — | — |
| Romania | — | 21,236 | 16,665 | — | — | — |
| Spain | 188,125 | 557,633 | 273,444 | 9,433 | 200,433 | 75,826 |
| Sweden | 2,200 | 837,117 | 480,222 | 39,767 | — | 1,725 |
| Switzerland | 14,419 | 181,965 | 25,417 | 11,133 | — | 3,466 |
| United Kingdom | 27,822 | 942,081 | 32,527 | 6,817 | 140,500 | 3,277 |
| USSR | — | 439,498 | — | — | — | — |
| Yugoslavia | — | 249,953 | 33,371 | 233 | — | — |
| Other | — | 6,333 | — | — | — | — |
| Total | 632,660 | 8,858,460 | 4,555,891 | 1,180,833 | 2,372,117 | 589,637 |

TABLE 1.4. CONTINUED

| | | | | | | |
|---|---|---|---|---|---|---|
| North America: | | | | | | |
| Canada | — | 139,742 | 71,162 | 8,533 | — | 2,588 |
| United States | 1,101,050 | 4,264,493 | 3,172,397 | 945,567 | 768,767 | 1,691,056 |
| Other | — | — | 1,419 | — | — | 16,666 |
| Total | 1,101,050 | 4,404,235 | 3,244,978 | 954,100 | 768,767 | 1,710,310 |
| South America | — | 378,115 | 189,975 | — | — | 29,109 |
| Africa | 120,784 | 284,994 | 6,569 | 902 | 476,783 | — |
| Asia and Oceania | 75,391 | 677,855 | 177,354 | 126,362 | 263,617 | 62,878 |
| Not specified | — | — | — | — | 353,166 | — |
| Grand Total | 1,929,885 | 14,603,659 | 8,174,767 | 2,262,197 | 4,234,450 | 2,391,934 |

[1]132.276 pounds.
[2]Includes 1,568,498 bags soluble coffee in green coffee equivalent.
[3]Includes 46,922 bags soluble coffee.
[4]Includes 4,355 bags soluble coffee and 96,923 bags roasted coffee, in green coffee equivalent.

Source: Foreign Agricultural Service.

small export trade with Portugal. Formal recording of these shipments was not begun until 1800 at which time exports were only about 1,700 lb. By 1825, Brazil's exports totaled 200,000 bags, and by 1850 some 3 million bags were exported. The distance from Santos, Brazil to Portugal is about 4,500 miles while to New York it is about 5,000 miles. Brazil was politically bound to Portugal as a colony until 1822 and remained bound thereafter in language and culture. Brazil exports were about 60 percent to Europe and 40 percent to the United States at that time. In subsequent years, Brazil coffee exports to Europe and the United States were as follows: (stated in millions of 60-kg bags) 1893—6.3/4.3; 1900—8.5/5.8; 1910–10.5/7.0; 1920–7.6/9.7; 1930–11.3/11.1. Brazil's exports reached a low of 7.3 million bags in 1942 during the war. In 1945, the United States took 83 percent of Brazil's exports, in 1946, 67 percent, and in 1958, 55 percent. In 1958, Europe took 34 percent of Brazil's exports amounting to 4.4 million bags, while in 1975 Europe took 60 percent totaling 8.8 million bags (see Table 1.4).

Historically, the consumption of coffee in Europe has increased at a similar rate to that in the United States, influenced both by increased population and higher per capita use. Table 1.5 lists an approximate per capita consumption of coffee by country as well as gross consumption in bags.

From a war period (1942–1945) when hardly any coffee entered Europe, the continent recovered its pre-war rate of use by 1958. Coffee use increased at such a rapid rate that more than 17 million bags were imported in 1960. About 20 million bags are now anticipated for annual use. Some reasons for this upsurge in coffee consumption are (1) the desire for good living after years of privation, (2) increased populations, (3) increased standards of living, and (4) common markets with enlightened civil servants.

The use of coffees in Europe is strongly related to an individual country's

TABLE 1.5. EUROPEAN COUNTRIES: POPULATION, GROSS COFFEE USE, AND PER CAPITA USE

| Country | Population, millions | Average Coffee Imports Annually, millions 60 kg bags | Per Capita Annual Coffee Use, lb |
|---|---|---|---|
| Sweden | 7.5 | 1.1 | 20 |
| Norway | 3.5 | 0.3 | 16 |
| Finland | 4.5 | 0.5 | 16 |
| Denmark | 4.5 | 0.6 | 18 |
| Holland | 11.0 | 0.9 | 10 |
| France | 55.0 | 3.0 | 10 |
| Germany | 60.0 | 3.0 | 9 |
| Bene-Lux | 10.0 | 1.0 | 13 |
| United Kingdom | 60.0 | 1.0 | 2 |
| Switzerland | 7.5 | 0.5 | 10 |
| Italy | 50.0 | 1.3 | 4 |
| Austria | 7.5 | 0.2 | 3 |
| Spain | 30.1 | 0.2 | 1 |
| Portugal | 9.0 | 0.2 | 3 |

population and culture. The finest quality coffees enter Germany; these are about two-thirds mild coffees from Central America and Colombia, about one-fifth Brazilian coffees, and the rest assorted mild coffees. Germany imports only about 2 percent Robustas while France imports more than 90 percent Robustas from the African countries. Scandinavian blends are all about three-fourths Brazilian while the United Kingdom uses about two-thirds Robustas and the balance, Brazilian and mild coffees. Holland uses about one-half Robustas, one-fourth Brazilian, and one-fourth mild coffees. Italy uses about one-half Robustas and a great variety of other types.

## BIBLIOGRAPHY

ANON. Annual. Imports of all coffees. U.S. Bureau of Customs. Superintendent of Documents, Washington, D.C.

ANON. Annual. Coffee Statistics. Pan-American Coffee Bureau. 120 Wall St., New York.

ANON. Annual. Exports of coffee. U.S. Bureau of Census, Department of Commerce, Washington, D.C.

ANON. 1958. Coffee in Latin America. 1. Colombia and El Salvador. Productivity problems and future prospects. Economic Commission for Latin America (ECLA) and Food and Agriculture Organization (FAO) of the United Nations, New York.

ANON. 1958. Origin and development of Nescafé. Tea and Coffee Trade J. *115*, No. 2, 56–58.

ANON. 1960. Coffee consumption in the U.S., 1920 to date. U.S. Department of Commerce, Washington, D.C.

ANON. 1960. Coffee in Latin America. II. Brazil, State of Sao Paulo. Productivity problems and future prospects. Parts I and II. ECLA and FAO of theUnited Nations. New York.

ANON. 1976. Coffee drinking in the U.S. Pan-American Coffee Bureau, N.Y., N.Y.

ANON. 1977. World coffee production estimates, 1976/77. FCOF-1. Washington, D.C.

BARNARD, A. 1951. Nescafé since World War II. Tea and Coffee Trade J. *101*, No. 1, Golden Anniversary Section, 50, 52.

BATEY, R. W. 1960. Upgrading soluble coffee. Coffee and Tea Inds. *83*, No. 1, 67, 145.

BATEY, R. W. 1961. How instant coffee evolved (patents). World Coffee and Tea *1*, No. 9, 57, 61.

COOK, C. W. 1959. Soluble coffee status report. Coffee and Tea Inds. *82*, No. 12, 24, 39.

COSTE, R. 1969. Coffee. Vol. 1. Green and Roast; Vol. 2. World. G. P. Maisonneuve et La Rose. Rue Victor Cousin, Paris Ve.

DIXON, R. A. 1947. Coffee essences and powders. Food *16*, 19–22.

HAARER, A. E. 1962. Modern Coffee Production, 2nd Edition. Leonard Hill (Books) Ltd., London.

HEYMAN, W. A. 1960. Flavor and soluble coffee. Tea and Coffee Trade J. *118*, No. 4, 16, 57–59.

HEYMAN, W. A. 1961. Development of instant coffee. Coffee and Tea Inds. *84*, No. 4, 26–30.

HINKS, E. 1936. Coffee essences. Food Manuf. *11*, 222.

KANNINNEN, W. H., and TAUB, S. E. 1955. Process for soluble coffee. Food Trade Review *25*, March, 16, 17.

LAWRENCE, N. A., PHILLIPS, W. H., RIFFKIN, A. H., and SALEH, A. A. 1977. U.S. Coffee Consumption. 1946–1976. F.A.S. M-275, Washington, D.C.

LINDNER, M. W. 1955. Investigations on Coffee, Coffee Substitutes and Additives. Verlag A. W. Hayn's Erben, Berlin SO 36.

MARTINEZ, A., and JAMES, C. N. 1959. Coffee–Bibliography of the Publications Found in the Library of the Institute. Instituto Inter-Americana de Ciencias Agricolas, Turrialba, Costa Rica.

PUNNETT, P. W., and EDDY, W. H. 1930. What flavor measurement reveals about keeping coffee fresh. Food Inds. *2*, 401–404. Tea and Coffee Trade J., *59*, No. 4, 554–557.

QUIMME, P. 1976. Coffee and Tea. Signet Books, N.Y., N.Y.

RICHMOND, F. W. 1977. Rising Coffee Prices. House of Representatives, U.S. Congress, Wash., D.C.

SCHAPIRA, J. D. 1975. Coffee and Tea. St. Martin Press, Washington, D.C.

SIVETZ, M. 1977. Coffee Origin and Uses. Coffee Publications, Corvallis, Oregon.

UKERS, W. A. 1922, 1935. All about Coffee. Tea and Coffee Trade J., 79 Wall St., New York.

UKERS, W. A. 1948. The Romance of Coffee. Tea and Coffee Trade J., 79 Wall St., New York.

WELLMAN, F. L. 1961. Coffee. Botany, Cultivation, and Utilization. Interscience Div. John Wiley and Sons, New York.

WILBAUX, R. 1956. Coffee Trees of the Belgian Congo, Technology of Coffee. Publication de la direction L'Agriculture des Forets et de L'Elevage. Brussels.

# 2

# Development of Coffee Industry

Green coffee beans were roasted in the home in the United States until the turn of the century, although this practice is still common in some parts of the world. Thus, in 1865 in Pittsburgh, Pa., Arbuckle's introduction of roasted beans for sale in a paper bag was a tremendous departure. The idea of buying roasted coffee beans spread, however, and necessitated a larger roasting machine at the coffee plant. It was at this time that Jabez Burns in New York and von Gimborn in Germany initiated their designs and fabrication of roasters, which subsequently led to the manufacture of commercial grinding machines when the consuming public began to buy ground coffee many years later. Special linings were also developed for the coffee bags so that the coffee oil would not penetrate to discolor the bag and soil the hands.

Shortly after 1900, there were several very important innovations: natural gas was used for roasting, the evacuated tinned can was manufactured for R&G coffee, Katyo developed an instant coffee in Chicago, and the French drip pot was invented (1906). The French drip pot's invention was the first indication that the brewing of coffee was considered an important preparative step for obtaining a flavorful beverage. Electricity was as yet far from developed but the electric coffee roaster was patented. In Italy, Pavoni took out patents for his espresso coffee machine, another indication that thought was being given to standardizing and controlling the brewing of coffee.

From 1906 to 1909, G. Washington prepared instant coffee for sale from his plant in Brooklyn, N.Y. In Elgin, Ill., Gail Borden prepared and sold a soluble coffee-milk product from 1867 until 1926. During World War I, G. Washington in Brooklyn and Barrington Hall in Minneapolis, Minn., prepared soluble coffee and continued their sales to the public after the war. Kaffee HAG of Bremen, Germany, manufacturers of decaffeinated coffee, began business in New York in 1914 but were disposed of by the U.S. Alien Property Custodian in 1919.

In 1918 Arbuckle Brothers was requested by the U.S. Quartermaster Corps to conduct much research and pilot plant manufacture of instant coffee. U.S. patent No. 1,393,045 resulted on October 11, 1921 with J. W. Scott, superintendent of Arbuckle Brothers, named as inventor. M. J. Quad and W. A. Heyman were chief and assistant chemists, respectively. The roaster coffee aroma was used in the spray drier air; vacuum drying was also done on shelves. At that time, Arbuckle Brothers owned the name Yuban and marketed a liquid coffee concentrate of that name as well as a roasted coffee. At the end of the war, the project was dissolved, the head of the company died, and the business and trademark were acquired by General Foods.

In 1926, G. Washington moved his soluble coffee plant from Brooklyn to Morris Plains, N. J., where it operated under that trade name until the 1950's. In the late 1920's and early 1930's, the soluble coffee product Fine Arts, which was drum dried, was sold in friction top cans. In 1932 General Foods Corporation acquired ownership of the Sanka trademark and the decaffeination plant in Brooklyn, N. Y., from Kaffee HAG of Bremen, Germany. In the same year, Max Morganthaler of Societé Nestlé in Vevey, Switzerland developed a process for preparing soluble coffee in percolators. From 1934 until 1940, W. A. Heyman developed a 50/50 soluble coffee-carbohydrate product which was acceptable to the U.S. Quartermaster Corps. Manufacture was carried out from 1941 through 1945. Percolator batch extractors and a box-type milk drier were used for the processing carried out at 601 West 26th Street, New York City. In 1945, this plant was sold to Hygrade Foods Inc., owners of Barrington Hall at Minneapolis. George Harrison bought the New York plant in 1948 and later sold it to Nathor. Forbes used batch extractors in St. Louis, Mo. during World War II.

In 1939 Nescafé started processing soluble coffee at Sunbury, Ohio. Percolation and spray drying of a 50 percent carbohydrate product was substantially the process technique, while Borden's process in 1941 was batch extraction followed by extract concentration in a milk evaporator and then by vacuum drum drying. Pre-war instant coffee sales were estimated at about six million pounds of the 50/50 product annually. The product was of considerably lower quality than that commonly obtainable today.

It is generally agreed that the U.S. Army's use of soluble coffee during World War II was an important factor in popularizing this beverage. From 1942 to 1945 the U.S. Quartermaster Corps bought almost 25 million pounds of soluble coffee product, half of which was carbohydrate. This represented an annual use of about three million pounds of pure coffee solubles. During the war, coffee solubles yields from roast coffee were quite low; 20 percent was common and Nescafé was probably technically ahead of the other processors in obtaining higher solubles yields from their percolator operations at higher temperature. Nescafé started processing at its Granite City, Ill., plant (near St. Louis, Mo.) in 1944 and sued Heyman at that time for infringement on its percolation patents. However, a similar suit in 1949 against Standard Brands was lost by Nescafé, allowing other firms to use percolation without fear of legal action.

General Foods Corporation placed Instant Sanka on sale shortly after 1946, and also began to utilize its patented water decaffeination process for green beans (U.S. patent 2,309,092). In about 1950, General Foods' Maxwell House instant coffee was placed on the market as a 100 percent coffee product (i.e., without addition of carbohydrates). This was subsequently followed by improved percolation and spray drying of hollow, beady particles in a tall spray drier. Thus in the early 1950's, the pioneering efforts of Maxwell House with pure coffee and "flavor-buds" allowed the company to gain consumer acceptance rapidly within a few years until half of the consumers of instant coffee were buying the Maxwell House brand. Other firms quickly changed their instant coffees to 100 percent pure coffee and also dried, large, beady particles. Nescafé was the last company to abandon the carbohydrate addition several years later, and never processed a large beady particle, but developed an agglomerate. The wartime spray dried powders containing carbohydrate were very dusty, light colored, and had the consistency of talcum powder. About 1950, the name was changed from "soluble" to "instant" coffee. The war product was difficult to dissolve; the post-war product was instantly soluble in boiling water.

At the close of World War II, Nescafé started numerous soluble coffee plants throughout the world in conjunction with their milk drying facilities. In 1948 and 1949, plants were erected in Freehold, N.J.; Hayes, England; Bugalagrande, Colombia; and Chesterville, Ontario, Canada. In the following years new soluble coffee plants also started at Orbe, Switzerland; Marseille, France; Dennington, Australia; Auckland, New Zealand; Norway; Denmark; Brazil; Peru; Chile; South Africa; Spain; Japan; and elsewhere.

By the close of World War II the National Research Corporation in Cambridge, Mass. had developed a soluble coffee process using the Kennedy extractor, with acid hydrolysis of spent coffee grounds followed by pseudo freeze-drying under vacuum on a metal belt with preliminary extract concentration in an evaporator. This process was used by some of the same development personnel to prepare Holiday instant coffee near Boston, Mass. This firm ceased functioning after about 1957. A similar process had been operated from 1952 to 1974 by Penndale near Philadelphia, Pa.

Freeze drying was commercialized in the mid 1960's when General Foods introduced MAXIM. This was followed by Tasters Choice from Nestles. The early production qualities were superior tasting to the spray dried products of commerce. However, progressively higher soluble yields and increasing use of Robustas have modified the quality over time, and consumer demand has leveled.

Most freeze dried coffee users came from the spray dried instant user groups. Agglomerated spray dried coffee became widely established in the late 1960's, almost eliminating all spray dried retail products.

## CHRONOLOGY OF DEVELOPMENTS

Soluble coffee plant developments and construction were carried out with considerable rapidity in the 1950's and early 1960's as follows:

1952 Tenco, Folger, and Beechnut build soluble coffee plants.
Nescafé abandons carbohydrate product.
1953 Hills Brothers pioneers soluble coffee process.
Swenson Evaporator Company and Folger improve spray drying.
Nescafé introduces decaffeinated soluble coffee.
1954 J.F.G. Company and George Harrison build plants. Salvador produces Tenco process soluble coffee.
The use of Jabez Burns' continuous roaster and blender is extended.
1955 Maxwell House and Chase and Sanborn erect new large soluble coffee plants.
Philippine plant designed by Heyman is started.
1956 Percolation for improved quality is started by Folger.
German development of soluble coffee begins.
1957 Package plants for percolator-spray drier process are sold in Europe.
Maxwell House, using coffee oil, adds aroma to soluble coffee in jar. Recovery and add-back of other aroma constituents is started by other firms.
Tenco process plants are started in Canada, San Francisco, Germany, and Mexico.
Jewel Tea, Kroger, and Nescafé build plants in the United States.
Borden changes from drum to spray drying.
The G. Washington Company, originators of soluble coffee in the early 1920's, sell their pioneer plant to American Home Products in 1943 who sell it to Otis McAllister in 1957, and later to Tenco.
1958 Tenco plant starts in Guatemala.
Savarin starts roast and soluble coffee plant at Palisades Park, New Jersey.
Electronic sorting of roast beans in Europe and green beans in Latin America and Africa begins.
1959 Soluble chicory-coffee plant is started in New Orleans by Wm. B. Reily.
Building of many new plants and formation of private label soluble coffee companies take place, such as Coffee Instants, Sol Cafe, and United Instant Coffee Corporation.
1960 Safeway Stores build soluble coffee plant in San Francisco.
Nescafé introduces aroma into instant coffee.
1961 Many new soluble coffee plants are started in many parts of the world, such as Nicaraguan plant for MJB Company, plants in Monterrey, Mexico; Costa Rica; Jamaica; West Indies; Caracas, Venezuela; Sydney, Australia; Moscow, Russia; Frankfurt-am-Main and Mannheim, Germany; Tokyo, Japan; Florence, Italy; and Paris, France.
1962 Soluble coffee plants are started in France, South Africa, Abidjan, Ivory Coast, Brazil, Angola, Japan, Jamaica, Philippines and Vancouver, Canada.
Aromatized instant coffees appear in Europe.

1965 Freeze dried coffee appears.
1967 Agglomeration.
1970 Brazilian instant coffee production reaches 80,000,000 lbs/yr.
1972 Antipollution enforcements increase.
1975 Continuous freeze drying systems.

## COFFEE ASSOCIATIONS

The New York Coffee Exchange was founded in 1882. For such an exchange to operate effectively, it was necessary to establish progressively better standards of coffee quality for contract descriptions. Although coffee roasters and green coffee brokers were loosely organized about 1900, the National Coffee Association (NCA) was formally founded in 1911. The Pan-American Coffee Bureau (PACB), representing Latin American coffee-growing countries, was established in 1937.

The Coffee Brewing Institute (C.B.I.) was founded in 1953 to promote the proper brewing of coffee and was a non-profit group supported jointly by the NCA and PACB. It was closed in 1976.

There are several green coffee associations in New York, New Orleans and on the Pacific Coast.

In the early 1970's, the National Coffee Service Association, with offices in Chicago, was formed and now has about 500 members. They operate office coffee systems around the country.

A scientific group was formed out of the I.F.C.C. (Institute Francaise du Cafe et Cacao) called A.S.I.C. (Association Scientific International du Cafe) which sponsors technical meetings every two years in various countries.

The I.C.O. (International Coffee Organization) is located in London and has been an instrument of the coffee growing countries involved in controlling exports and prices.

## COFFEE QUALITY

Even the most preliminary scientific studies of coffee were not undertaken until about 1920, starting with Professor Prescott's work at the Massachusetts Institute of Technology. Staling of coffee was later studied at Columbia University in New York City by Professors P. W. Punnett and W. H. Eddy (1930).

The tin can filled with roasted and ground coffee and sealed under vacuum was used for about 25 years before its effectiveness was scientifically studied. A Tea and Coffee Trade Journal survey of coffee freshness conducted in 1932 showed that 60 percent of the trade samples were over 10 days old; the control standard was ground coffee 10 days old. Today such a standard would hardly be adequate. The acceptance of the evacuated can of ground coffee was slow, partly due to the extra cost of the can over bagged coffee. Even until 1940 most of the roasted and ground coffee sold in the United States was not vacuum packed. The use of stale coffee was widespread and staleness was accepted as a part of coffee flavor. Even today only a small fraction of coffee sold in Europe is packed in vacuum cans. Further, Brazilian coffees dominate the most popular blends as they did in the earlier coffee trade. Today the average United States coffee blend is very commercial and there is much room for improvement. A barely acceptable flavor is always related to brand cost. The finest coffee blends are not sold in the most mass produced coffee brands, but can be found in gourmet shops.

In the United States there are some premium priced coffee blends that are vacuum packed. However, for the gourmet, coffee specialty shops are available in the larger cities to offer special blends and roasts at special prices. Green coffee quality changes in time and place; consumer tastes and standards also change. Public dislike of the lower grade and lower price roast coffees has contributed to rapid and broad acceptance by the consumer of instant coffees. In the United States since 1962 one cup in three is instant coffee. Many instant coffees are better flavored than the lowest grade roast coffees, and invariably are less costly.

The real bulk of scientific coffee knowledge has been evolved since 1950 under sponsorship of the Coffee Brewing Institute, the Quartermaster Corps, and private soluble coffee firms studying the chemistry and flavor origins of coffee. The coffee industry is now strongly influenced by firms and personnel who were not connected with the regular coffee business before 1940 or even 1950.

## SOLUBLE COFFEES

A significant development since about 1955 has been the increasing number of soluble coffee plants constructed in the coffee-growing countries. Whereas imports of soluble coffee into the United States has totaled only a few million pounds per year in the past or only a few percent of soluble coffee use, this new development is likely to mean that soluble coffee imports into the United States and other consuming countries will rise considerably provided tariff barriers are not raised in the consuming countries.

About 10 to 15 percent of any country's coffee crop is marginal quality coffee; this is high in black, nipped, discolored, and otherwise abnormal beans. None of the poor appearance and little of the bad flavor come through the soluble coffee process. The result, especially for mild coffee, is a very pleasant flavored instant coffee when produced in a properly designed, equipped, and operated soluble coffee plant. It is a useful outlet for low grade coffee beans.

Many of today's operating soluble coffee plants have percolation cycles lasting more than 1 hour and fine particle spray drying which contribute to marked flavor loss in the final instant coffee powder. In the growing as well as the consuming countries, enterprising men recognize the advantage of soluble coffee processing of marginal quality green coffees. These coffees are purchased locally and inexpensively and thus are exported at a relatively good price. More and more of these coffees are being processed as soluble coffees. By coordinating green coffee growing, processing, and bean grading with the soluble coffee processing, a better grade of green coffee can be exported at a better price.

Consuming countries may raise import duties on soluble coffee. However, living standards are improved all around by free trade because the coffee-growing countries may become good customers for the manufactured items of the coffee-consuming countries. Hence, some agreement that is reasonable for both must be reached. More soluble coffee will eventually be processed in the producing countries. It is inconsistent for the United States to raise duties on such imports while giving money to the same countries to help raise living standards. In addition, the European common markets and Central American common markets, for example, make for stronger and more realistic bargaining agreements. The continuing objectives of the European common market have been to effect a more equitable distribution of low priced goods among themselves while eliminating marginal or inefficient producers.

Whether the controlling factors in the coffee economy and technology of the world are surpluses and withholding agreements, shortages and panic prices (1976), higher living standards for the people in the coffee-growing country, import tariffs, export taxes, or coffee substitutes, each will play a part in the often dramatic changes of the multi-billion dollar coffee industry.

## Basis of Soluble Coffee Acceptance

Soluble coffee has been on sale to the public in the United States since 1909, but sales for many years were small. Nescafé enjoyed some success in promoting the sale of instant coffee about 1940 when workers in war plants in the United States were anxious to save time. Even after 1946 there was grave uncertainty as to the salability of soluble coffee to the public. It is now agreed that one major factor in public acceptance of soluble coffee was the exposure of 15 million United States military personnel to such a beverage over a four-year period. Further, four years of operating experience in a dozen firms built up a reservoir of processing "know-how." The worst methods were abandoned and intermediate methods were continued, but the best methods were still being evolved. In brief, percolation, large particle spray drying, and a 100 percent coffee solubles process were evolved in about 1950 to produce a truly acceptable instant coffee flavor and product. The four-year period prior to that was one of increasing instant coffee sales associated with considerable uncertainty about continuing wartime processing and post-war promotion. The big increase occurred after 1950. This was a period of technical shifts toward the processing methods of the best selling brand.

Too much credit has been given to the conditioning of wartime personnel to instant coffee quality; not enough has been said about the achievement of instant coffee after the war when the paying public judged the value of its purchase.

Acceptable instant coffee was commercially developed primarily in the United States, which is still the largest consuming country. Current use is at the rate of about 180 million pounds of instant coffee per year in the United States, with less than one-fourth of this in the rest of the world.

Although Nescafé has more soluble coffee processing plants throughout the world than any other firm, Maxwell House instant coffee is probably the largest selling brand, holding over half the United States instant coffee sales market. Both technical and marketing factors have contributed to shaping sales distribution of each firm.

An effective tool applied in this period to measure product acceptance by the public was the tasting poll on unmarked samples in both public places and in the home. Before 1940, it was common for the president of a coffee firm to decide what blend and roast he would sell. After 1946, a marked shift away from such personal and arbitrary decisions was arrived at by using independent and objective consumer panels, preferably among the lay public. Thus, the instant coffee product was designed to suit the purchasing consumer. Another tool of the time was the application of marketing studies to determine the amount of competitors' shelf space, competitive prices, product turnover, competitive discounts, advertising, and sales methods. The A. C. Nielsen firm in Chicago developed its services particularly in these areas. Reports were subscribed to by individual brand companies which received monthly or quarterly verbal and written presentations comparing conditions in the industry on a chronological as well as a regional basis. Such feedback of marketing knowledge was very effective in offering management the facts regarding its investments in advertising, product improvements, and pricing. In other words, as soon as instant coffee was accepted by the public, it became important not only to enter the market vigorously but to use all known means to present a more economical and appealing instant coffee product to the public to gain wider product acceptance with the least investment.

The coffee trade journals publish the annual advertising expenses of major instant coffee and regular coffee firms. In 1975, television advertisements by coffee firms in the United States cost about 40 million dollars; newspaper advertising cost about six million dollars; and radio advertising cost less than one million dollars. Almost half of the television advertising expenditures was underwritten by one company, which is in proportion to their instant and regular coffee sales nationally. Most soluble coffee firms spend a maximum of perhaps 5 percent of their advertising budgets for research and development in product improvements, and a number of firms engage in practically no research and development outside the bounds of normal plant process control.

Instant coffee is a convenience food compared with brewed roast coffee. It minimizes consumer attention to preparation, eliminates cleaning after preparation, reduces product waste, and does away with investment in and maintenance of a brewing device. Some instant coffees do not become stale as do roasted and ground coffee. The fact that a cup of instant coffee is about half the price of brewed coffee is also an important factor in public acceptance.

## Technical Progress

Before 1940 the availability, cost, and quality of stainless steel and its fabricated equipment, process instruments, and spray-drying systems and techniques were highly inadequate in their application to soluble coffee processing. Great technical strides were made during World War II. Wartime processing of soluble coffee as well as spray-drying techniques for detergents and other substances made great progress as a direct result of accelerated process developments from parallel fields. Since 1955, improved analytical techniques have revealed important knowledge about coffee chemistry and flavor, of which a good deal has been promptly applied to practice. New knowledge has pointed up the need for further research. For example, ash composition of wood or coffee beans is known to change the course of the chemical flavor substances in roasting. This type of knowledge, carefully applied, accelerates the rate of industrial progress and the improvement of product quality.

Whereas before WW II there were few engineers and scientists working in the coffee field, the manufacture of instant coffees oriented hundreds of professionally trained personnel to the production of better quality beverages at lower costs.

## Relative Costs for Roasted and Instant Coffees

Marketing reports and surveys concerning roast and instant coffees have usually used low yields of soluble coffee when converting to roast or green coffee equivalents. Not all soluble coffee plants achieve the same yield, but a fair figure for Brazilian and mild blends is about 0.375 lb of instant coffee per pound of roast coffee or 0.33 lb of instant coffee per pound of green coffee at medium roast. Darker roasts, imperfect beans, and other factors may contribute to somewhat lower yields, whereas Robusta blends may easily produce 40 percent yields.

In comparing roast and instant coffee costs, the cost of the green coffee for roast coffee use is invariably higher than for instant coffee blends. This means that the profit on the soluble coffee product is accordingly higher. Instant coffee blends are usually one-half to two-thirds the cost of the blends used for roasted coffee sale.

However, capital investment and operating costs are considerably higher for instant coffee manufacture than for roasting, grinding and packaging. Higher skilled labor is also required in operating instant coffee processing facilities.

## Coffee Supply Increase via Instants and "New Markets"

A pound of R & G coffee extracted on the basis of 0.4 lb of solubles per pound of green beans (which is what is being taken) yields 180 g of instant coffee. If spray dried, it makes 90 cups of coffee; if freeze dried, it makes over 100 cups of coffee beverage.

The same pound of R & G coffee brewed at home from regular grind makes about 50 cups of coffee beverage. Hence instant coffees produce twice as many cups of coffee from the same raw material as compared to home brewing. With increased coffee prices in 1976, office coffee service users have been producing 1.75 oz and 2 oz pouches of R & G coffee, which produces 12 cups, or the equivalent of 96 cups per lb R & G coffee. A 1.5 oz pouch yields 120 cups per lb.

## BIBLIOGRAPHY

ANON. Annual. Exports of Coffee. U.S. Bureau of Census, Department of Commerce. Washington, D.C.

ANON. Annual. Imports of all coffees. U.S. Bureau of Customs. Superintendent of Documents, Washington, D.C.

ANON. Annual. Unilac Corporate Reports. Vevey, Switzerland.

ANON. 1960. Coffee consumption in the U.S., 1920 to date. U.S. Department of Commerce, Washington, D.C.

ANON. 1960. Coffee in Latin America. II. Brazil, State of Sao Paulo, Productivity problems and future prospects. Parts I and II. ECLA and FAO of the United Nations. New York.

ANON. 1960. New Safeway instant coffee plant. World Coffee and Tea *1*, No. 1, 31.

ANON. 1959. The Tenco Story. Tenco, Linden, N.J.

ANON. 1958. Kroger develops instant coffee process. Food Eng. *30*, No. 4, 57–58.

ANON. 1958. Nestlé's Freehold, N.J., instant coffee plant. Coffee and Tea Inds. *81*, No. 12, 49, 52.

ANON. 1958. Origin and development of Nescafé. Tea and Coffee Trade J. *115*, No. 2, 56–58.

ANON. 1939. Coffee extract. Food Inds. *11*, 91.

BARNARD, A. 1951. Nescafé since World War II. Tea and Coffee Trade J. *101*, No. 1, Golden Anniversary Section, 50, 52.

BATEY, R. W. 1961. How instant coffee evolved (patents). World Coffee and Tea *1*, No. 9, 57, 61.

BATEY, R. W. 1960. Upgrading soluble coffee. Coffee and Tea Inds. *83*, No. 1, 67, 145.

COLODNEY, B. 1959. High finance in coffee. The Analysts Journal, November, 69–78.

COSTE, R. 1955. Coffee. Vol. 1. Green and Roast; Vol. 2. World. G. P. Maisonneuve et La Rose. Rue Victor Cousin, Paris Ve.

FOSTER, A. C. 1960. Instant coffee in the 60's. Coffee and Tea Inds. *83*, No. 1, 62–63.

HARDY, W. L. 1955. Soluble coffee—product of tomorrow. Tea and Coffee Trade J. *79*, No. 1, 50–51.

HARRISON, G. 1960. Military, civilians switching to solubles. Tea and Coffee Trade J. *118*, No. 6, 26, 28, 64.
HEYMAN, W. A. 1961. Development of instant coffee. Coffee and Tea Inds. *84*, No. 4, 26–30.
HEYMAN, W. A. 1961. Soluble coffee's role in foreign policy. Tea and Coffee Trade J. *121*, No. 3, 20, 36, 38.
LEE, S. 1961. Monterrey, Mexico soluble coffee plant. Tea and Coffee Trade J. *121*, No. 3, 22, 23.
LEE, S. 1962. Soluble coffee's changing profile. Tea and Coffee Trade J. *122*, No. 4, 24, 78, 82–87.
PUNNETT, P. W., and EDDY, W. H. 1930. What flavor measurement reveals about keeping coffee fresh. Food Inds. *2*, 401–404. Tea and Coffee Trade J. *59*, No. 4, 554–557.
QUIMME, P. 1976. Coffee and Tea. Signet Books, N.Y., N.Y.
SHAPIRA, R. G. 1975. Coffee and Tea. St. Martin Press, N.Y., N.Y.
SIVETZ, M. 1976. History of soluble coffee. Tea and Coffee Trade J. Aug. 1976.
SIVETZ, M. 1977. Coffee Origin and Uses Coffee Publications, Corvallis, Oregon.
WELLMAN, F. L. 1961. Coffee. Botany, Cultivation, and Utilization. Interscience Div. John Wiley and Sons, New York.
WILBAUX, R. 1956. Coffee Trees of the Belgian Congo, Technology of Coffee. Publication de la direction L'Agriculture des Forets et de L'Elevage. Brussels.
WOOD, J. H. 1959. Growth trends in soluble coffee. Tea and Coffee Trade J. *117*, No. 1, 22–24, 71 72.

# 3

# Development of Coffee Uses

*Michael Sivetz*

## CULTURAL ASPECTS

A cup of coffee to an Arab, a Brazilian, a Colombian, an Italian, a German or an American is different. A cup of coffee here is not the same as a cup of coffee there. Local customs and sentiment prevail. The cup of coffee is different in many ways, not just in one way. The blend is different, the roast is different, the grind is different, the method of beverage preparation is different. The Italian or Scandinavian is unlikely to take instant coffee, whereas the English-man and Japanese find instant coffee as normal, especially with plenty of hot milk. Americans are more likely to accept instant coffee as normal, also.

In some parts of Belgium, Holland, France and Switzerland, the use of chicory with coffee is normal. To the Italian and Turk, the dark roast bean, finely ground, is normal. To the Brazilian, a dark roast, fine grind, strong tasting beverage loaded with sugar is normal.

And so when a Latin American visits the U.S.A., he is dismayed when he is served a cup of tea-looking beverage, called coffee, and he can see the bottom of the cup. There is nothing so familiar to a person as his food, and when this is "foreign," he becomes uncomfortable and unhappy. Similarly, when a U.S. citizen visits New Orleans, he may be surprised with the black, strong demi-tasse served as "cafe."

Americans (U.S.) drink coffee with their main course in a meal and this is certainly uniquely cultural. The Italians have their Espresso beverage preparation machines, and the Brazilian his Quador. The Americans continue to distill their coffee aromas with percolators, while Europe tends to use Drip brewers. The U.S. market is mainly for vacuum packed R&G coffee of unknown blend, whereas, in Europe, bag-pack roast beans of known origin are much more normal.

## PERSPECTIVE IN COFFEE USE[1]

Only a small percentage of the world's population drinks coffee beverage. Of the 5 billion world population, only 1 billion are in the developed nations, where

[1]Source: Coffee Origins and Uses by M. Sivetz, 1977.

over 95 percent of exported coffees are drunk. And 10 percent of the world population is in the U.S. and Europe, where two-thirds of the world's coffee is consumed. We can further say that coffee is drunk primarily in the higher living standard countries; hence, coffee is very much a priced item if not a luxury item. As the world population explodes, growing at almost 100 million persons per year currently, coffee, a non-essential food requiring large acreages and much labor for growth and processing, will become more expensive and more luxurious to use.

Of the one-third of world coffee exports, say 63 million bags, going to the U.S.A., a country of Anglo-Saxon heritage (tea drinkers), 20 percent goes to instant coffees constituting about one-third of consumed cups of beverage. A good deal of Robusta imports go into instant coffee manufacture, but still we can say 25 percent of roast and ground coffees have Robusta in their blends. Another 25 percent are unwashed milds (Brazils), not truly top quality beans. When further considering the U.S. market of branded merchandise (without real origin identity), light roasts, coarse grinds, weakly prepared and filtered beverages, plus the damage of heating, reheating, urn holding, etc., it is unlikely that 140,000 bags of coffee entering the U.S.A. are properly prepared and served for full fresh flavor. Along with the United Kingdom, Canada and Australia, the U.S. uses milk with the beverage (75 percent do). This means that one-sixtieth or less than 2 percent of U.S. imports are likely to be used to make a flavorful fresh cup of coffee. In Europe, the percentage of ultimate good coffee consumed may be five-fold greater, but yet a very small percentage of the total.

A major reason for these circumstances is that coffee, carrying caffeine, is used more for its "lift" or stimulating effects, than for its satisfaction in flavor. Very few people, an infinitesimal number, really know how to make a fresh, good-tasting beverage. The variables involved are numerous, and the pitfalls many. Interwoven into this complex fabric are the social factors of politics among nations, geographical proximities to supplies, religious and spiritual relations, historical, colonial, neo-colonial, and purely commercial factors. Also, the consumer lacks education and significant experience with coffee quality, as compared to commercial sellers. History has laid down traditions that the average coffee consumer barely realizes exist.

## Middle East and Mediterranean Areas

Ethiopia was the geographical and original agricultural source of mild coffee beans. However, it was in Arabia along the Red Sea near Aden, where coffee shrubs were commercially cultivated with irrigation for over a thousand years. The coffee bean consequently was called *Coffea arabica,* ignoring its agricultural origin in Ethiopia, but recognizing its growth and use throughout the Arabian countries. The coffee bean and its beverage were especially noted by pilgrims visiting Mecca each year. Indeed, these berries were precious, as the Arabs truly appreciated the taste as well as the exhiliration of a good cup of coffee. The green beans were roasted just before grinding and beverage preparation. The coffee was fresh, a quality which has been largely lost in today's commercialization. Since the beans were costly, they were ground very fine after roasting so as to produce the highest yield of solubles, oils and flavors. Although the Arab custom was to heat the grounds in water in an Ibrik pot over a fire, boiling was not allowed. Separation of insoluble grounds was attained mostly by keeping these insolubles in the pot or cup. Being valuable, the coffee beverage was consumed from a demi-tasse sized cup containing 2 to 3 fl oz (60-90 ml). It had a solubles concentration of 3 to 4 percent; in other words, the coffee beverage was very strong-tasting and black. It was made strong enough for one to appreciate the coffee flavor, not weak and watery as is now practiced in the U.S.A. Colloidal oils and fatty acids that carried much aroma and flavor were allowed to enter the beverage to be consumed and were not filtered out. These properties gave the beverage a burnt and oily taste.

It was only a small expansion in the use of coffee beverage when its use spread to Cairo, the Levant and Turkey a thousand years ago. All these places were on or close to the Mediterranean Sea. Small wonder then that coffee beans and their use quickly found their way to Venice, Italy and Greece, and later to Spanish, French and African Mediterranean seaports. Hence, Mediterranean people learned to prepare and drink their coffee beverage in a similar way. Further, most of these Mediterranean people liked to use much sugar in their black coffee beverage and a few used a twist of lemon peel. Lemons grow profusely along the Mediterranean Sea, and sugar is a major crop in Egypt.

Quality of taste and aroma were a conscious part of the appreciation of the beverage. Consequently, by 1900 Italy developed the Espresso machine for individual fresh beverage preparation, thereby controlling portions, time, water temperatures, and extraction method—a scientific control on preparation. To this day, this part of the world uses negligible amounts of instant coffee, except in the tourist trade.

The Portuguese and Spaniards relished the same dark French/Italian roast. They carried their tastes to the new world—the Caribbean, Central and South America. Since about 1905, the Italian espresso extraction machine has facilitated brewing this strong beverage from dark roast, fine grind coffee. The lower level of aromatics in this dark roast bean and the greater release of oil from it (due to greater cell wall destruction), allows the volatile aromatics to be better protected from evaporation and oxidation when they are dissolved in the oils. The net result is a reduced rate of oxidation or staling, which would otherwise be accentuated with lighter roasts in these warm countries. One reason why dark colored bean roasts were used here is because elemental roasting utensils, like a metal globe or fry pan over a charcoal fire, were used. Also, when nonuniform quality green coffee beans are used, some beans are nearly always burnt before most of the coffee beans have the desired dark roast color and flavor development. An important reason for the dark color of French/Italian type roasts is that hundreds of years ago grinding of the roasted coffee beans was done in a mortar and pestle (made of stone or brass). A medium to lightly roasted coffee bean has a very tough texture; it cannot be readily fractured like a dark roast, brittle coffee bean. Also, a dark roast coffee bean, with more cell destruction, is easier to extract solubles from than a lightly roasted coffee bean. The dark, brittle coffee beans were literally pulverized in a mortar and pestle. This great exposure of cell surface allowed faster and easier extraction of solubles. It also allowed considerably more oil and fatty acids to be liberated from the bean.

## EUROPEAN TRANSFORMATION OF COFFEE USE

The purist traditions of the Arabian and middle east-Mediterranean areas underwent many and varied changes when in the 18th and 19th centuries, European wealthy classes accepted coffee and demanded it from the Americas, Dutch East Indies, Ceylon, Africa (after 1900, and especially after 1930). Different and larger roasting machines were devised, burr-disc grinders became larger, and then roller grinders were introduced. Commercialization grew. Large volume production and distribution were introduced, invariably inconsistent with delivering quality or freshness. Politics infused primarily Brazilian coffee use in Scandinavia, and high Robusta use in France, Belgium, United Kingdom, Netherlands, etc. where they had African colonies. New type brewing devices were evolved in France, U.K. and elsewhere in the 1800's. New sources of coffees came and went, motivated by colonial profiteering and later destroyed by natural diseases or political repressions. Brazil's early 1800's thru 1930's domination of world exports yielded to production increases from Colombia and other Central and South American milds producers. But after 1946, African production grew rapidly—especially Robusta production, which readily found purchasers both in Europe and the U.S.A. The Anglo-Saxon influence in the U.S.A., Canada, and Australia made for a markedly different type of coffee use than in other European countries.

## MILK PRODUCT ADDITIVES TO COFFEE BEVERAGE (NEUTRALIZERS)

Since "black" coffee beverage is often bitter to harsh tasting (the usual improper preparation), it was natural for the European drinkers to add milk or cream to the coffee to smooth out its harshness. Milk and its derivatives were the basic agricultural activity in all the cooler, mountainous areas, such as Switzerland, Scandinavia, Germany, United Kingdom, France, Netherlands and Belgium. Cream, with its butter fat content, smooths the taste of the coffee. Dairy products also have nutrition. Milk and milk products were not that plentiful nor that much consumed along the Mediterranean Sea, especially hundreds of years ago. Even today and especially before World War II, milk was a rarity for children in many parts of the world. Without refrigeration dairy products are very perishable. Milk was also expensive in tropical and subtropical areas. In many world areas coffee and milk both were and are luxuries, and certainly not an item for everyday consumption. However, in the cooler European climates, cow's milk was more common, not so expensive and perishable.

Hence, the fashion of half coffee beverage with half boiled milk (boiling killed bacteria) came in to use. "Cafe au lait" is common all over Latin America for breakfast, in France, Spain, Portugal and England. Where safe to use, additions of small quantities of cold cream or milk to a cup of hot coffee beverage (as in the U.S.A.) became the cultural pattern. The protein of milk reacts with the tannin-like acids in the coffee beverage to give a smoother tasting beverage, but a definitely dairy-tasting beverage. If a coffee beverage has become more acid from being held hot too long before serving, the milk can neutralize this acidity. Coffee acidity is irritating to many people's stomachs. In fact, the Borden Food Co. in 1970, introduced their brand, KAVA, which is a neutralized instant coffee for those people with sensitive stomachs. Instant coffee beverages are now more acid than freshly brewed coffees, so they need a dairy neutralizer to make the coffee beverage more palatable.

Further, young people in their late teens or early twenties, when first tasting coffee beverage, do not find it naturally appealing. Most young people in the U.S.A. and northern Europe have been brought up on milk. It is an easier transition for them to add a little coffee beverage to milk at first. This coffee flavors the milk and makes it hot. Later as a taste for coffee is acquired, less milk and more coffee beverage is consumed. Many people, as they grow older, never stop adding milk to coffee beverage. The average coffee beverage they usually drink is so bitter that the milk or cream buffer is needed so that they can consume the beverage. For example, in the U.S.A. two-thirds of the coffee drinkers use dairy additives, and two-thirds of these use sweeteners in addition. Only 10 percent of the coffee drinkers use sweetener only. The use of milk with coffee beverage, however, is a common practice all over the world. This is especially so with the morning coffee. In Latin America and many Mediterranean countries, milk products are not added to coffee beverage when consumed in the late afternoon or evening. In parts of Switzerland, France, Germany and some other European countries where chicory is added to coffee beverage, milk is usually added. Otherwise the chicory-coffee beverage alone is too strong and unpalatable to enjoy.

## NON-DAIRY ADDITIVES

In recent years there has been widespread use of whiteners with coffee beverage instead of real milk or cream. This has happened because whiteners are cheaper, stable, and relatively uniform. Whiteners have been accepted for vending machines, in restaurants and institutions, and even in the home.

In the U.S.A. cream consumption has declined 50 percent per capita from 1950 to 1970. Mostly it is due to increased use of non-dairy coffee whiteners. In 1971, dry coffee whitener sales were 100 million lbs per year, as were frozen coffee whiteners. By 1971, 25 percent of the natural cream market was whiteners.

## SWEETENER

Sugar was a relatively rare commodity until after 1600, the beginnings of colonialism. Sugar cane and, later, beets were its main source. In the U.S.A. half the coffee drinkers add sugar or sweetener. A heaping teaspoon of sugar is about ten grams and 40 calories. Sugar additions may vary from a half a teaspoon to two heaping teaspoons in the U.S.A. In Brazil and some other countries, coffee drinkers are accustomed to add 3 heaping teaspoons of sugar to their demitasse of coffee beverage. If we assume 10 grams of sugar are used per cup of coffee, there are a billion pounds of sugar used per year in half the cups of coffee beverage consumed in the U.S.A.

### Unrefined Sugar

In many parts of the world different quality sugars are used with coffee. The sugar qualities can vary from those that are dark brown and carrying a heavy molasses flavor, to those that are light brown with a slight molasses flavor. The latter are referred to as single crystallization sugars, and are more common in Latin America. Double crystallized white-colored sugars with no molasses taste are more commonly used in the U.S.A. and Europe. On the other hand, there are single crystallized brown sugars (large ⅛ in. size). These are specialty items for use in gourmet shops and restaurants. They have a molasses flavor but are not as hygroscopic as ordinary single crystallized, fine grained sugars.

The molasses taste from unrefined sugar can be very prominent in a coffee beverage. Some people acquire a taste for this combination of flavors. On the other hand, people accustomed to using refined sugars do not like to have the molasses taste introduced into their coffee beverage.

### Chicory *(Cichorium intybus).*

This hardy rooted plant is related to the dandelion. It also resembles parsnip and sugar beet. The ideal root size is 3.5 in. diameter by 13 in. long. It is better known for its leafy, above the soil (second) growth, called endive. This is used for salads with or in place of lettuce. It usually has a blue flower, grows wild, and is often seen along roadsides. Its leaf was used for salads since Roman times. The root is cultivated in France, Belgium and Holland. It is harvested in September. Process: Roots are trimmed, washed, cut into 1 in.-sized pieces, and air-dried overnight.

**Chicory composition.**—It is about 75% carbohydrates, of which 70% is water soluble. The insoluble 25% is as follows: 2.5% moisture, 4.7% ash, 3.6% oils, 7.5% protein, 7.7% crude fiber.

There are about 8.6 percent reducing sugars and 0.4 percent sucrose. Inulin, a starchy substance of the root, can yield levulose sugar which is almost twice as sweet as sucrose. The dried cuttings of chicory root are roasted much like coffee, but much more slowly and at lower warm air temperatures so that drying, hydrolysis of sugars, and color development are uniform. The roasted pieces are ground in a roller mill and are screened to 6 mesh for uniformity. The roast and ground chicory must be packaged in moisture tight containers. Roast or instant chicory is shipped in bulk to coffee roasters in the U.S.A. for blending with coffee or for separate retail packaging.

**History.**—Chicory use was launched after 1789. At that time, the French Antilles supplied two-thirds of the coffee beans drunk in Europe. The revolution in France sparked revolutions in Santo Domingo and Haiti. These major sources of coffee beans were lost to France. Consequently, France had to depend on British ships to bring coffee beans from the Dutch East Indies, which was much more expensive. In 1799 Napoleon became ruler of France. In November 1806, Britain blockaded all ship traffic to France. Exclusion of foreign shipments required France to become more self-sufficient in every way. Although some coffee beans were smuggled into France, it was during this period that chicory substitute for coffee came into widespread use.

**Use.**—Chicory root was grown in Michigan, U.S.A. until World War II when it became uneconomical to produce. Annual U.S. use is about ten million pounds of chicory per year, whereas in Europe it is 100 million pounds per year. For the U.S.A. at a 33 percent weight level in coffee mixtures, this is less than 1 percent of U.S. coffee use. The level of chicory blended with roast coffees depends sometimes on where the blend is shipped. For example, there may be only 10 percent chicory in chicory-coffee blends shipped to the northern states, whereas 35 or 40 percent chicory may be used in the south. Bulk chicory costs about 25¢ per lb compared to coffee at $2.50 per lb or $^1/_{10}$ as costly as coffee. In England, Belgium, Switzerland and other European countries instant coffees with chicory are widely sold.

**Taste.**—Chicory does darken cup color, add body and add a spicy-peppery tang to the coffee beverage. In this respect chicory is a better substitute than roasted cereal grains, peas or acorns.

## CHOCOLATE FLAVORED COFFEE

Chocolate has a highly compatible flavor with coffee. In Mexico, pastillas or discs of sugar, coffee, chocolate, cinnamon, and sometimes nutmeg are commonly sold in local "cantinas." These discs, 2 in. diameter by 0.5 in. high, are broken up and dissolved in boiling water. Usually a cocoa sludge remains at the bottom of the cup. For a more elegant preparation, the consumer can use chocolate syrup in strong coffee and top the spiced beverage with whipped cream. Adding instantly dissolving chocolate/sugar powder to strong coffee beverage gives another chocolate coffee flavor variation.

Another tasty after dinner drink is to add about 1 fl oz of creme de Cacao (alcoholic extract of cocoa with added sugar) to 2 fl oz of espresso or strong coffee beverage; top with whipped cream. This is sometimes called Cafe Cacao; when cold and served in a wine glass, called Coffee Alexander.

Caffe Borgia is the Italian version of half espresso coffee and half hot chocolate, topped with whipped cream and some grated orange peel. It somewhat resembles Orange Cappuccino. In the early 1970's, a few firms introduced chocolate flavored instant coffee products, called Cafe Vienna or Cafe Viennese, the latter with cinnamon. Suisse Mocha had powdered milk and sugar with the instant coffee; Cafe Mocha was chocolate flavored. Bavarian Mint was instant coffee-chocolate flavored; while Capri was chocolate-coconut flavored.

All the above products contain non-dairy creamer. All these products are sweet, milky beverages. Some of these dry mixes make a pleasant drink, but are in no way coffee beverage, only coffee flavored. The weight percent true coffee solubles in these products is usually very small, and much higher in lower priced ingredients. The non-alcoholic flavors often have alcoholic counterparts.

## ALCOHOL AND SUGAR CONTAINING COFFEE BEVERAGES

### Creme de Cafe

This liqueur is a potent after-dinner drink. It carries the stimulation of caffeine and coffee roasted products. The increased blood circulation helps to distribute the alcohol and sugar throughout the body. Normally the 1 fl oz consumed after a heavy dinner does no more than settle the stomach and mind. The liqueur can be prepared in several ways commercially or domestically. Usually about 35 percent grain alcohol (70 proof vodka will do) or brandy is used to extract good quality R&G coffee, in about the same proportions of alcohol to coffee that one would use for brewing coffee. Steeping the coffee grounds in the alcohol at room temperature is sufficient to thoroughly withdraw the coffee's aromatic and flavor factors. The mixture of alcohol and R&G coffee after intermittent agitation may be filtered through paper or fine cloth to yield a clear coffee-red colored extract. Then enough sugar is added to sweeten the coffee-alcohol mixture to taste, but not to saturation. The sugar does have the effect of smoothing out the taste and reducing the harshness of the alcohol. Too much sugar can make the beverage taste objectionably sweet. The water binds with sugar and alcohol to preserve the liqueur flavor constituents. The coffee aroma and flavor constituents are preserved for many months, even years, when the liqueur is stored at room or lower temperatures. Notable brands of cream de cafe liqueur are Tia Maria from Jamaica, Kahlua from Mexico, Illy's espresso from Italy, and others from Europe, U.S.A., Latin America, Russia, and elsewhere. Coffee liqueur can also be prepared by mixing (1) a strong coffee beverage (with 10 to 20 percent coffee solubles) with (2) simple syrup of 80 percent sucrose, and then adding (3) high (over 100) proof alcohol. The quality of the liqueur flavor resulting can be no better than the quality of the coffee and brandy used.

### Other Liqueur Additives

Some of the more common coffee drinks mixed with other liqueurs are: creme de cacao, Cointreau, Curacao or Le Mariner (orange liqueurs), Anisette or even Kummel (anise-like), and creme de menthe (mint). Care must be taken in most additions, because liqueur flavors are so strong that even a half fluid ounce of liqueur per 2.5 fl oz demitasse coffee beverage may be too strong to give a properly balanced blend. Topping the coffee-liqueur mixture with whipped cream smooths out the taste delightfully. Brandy or whiskey (0.5 oz) may be added to a demitasse strong beverage.

## SPECIALTY COFFEES

### Irish Coffee

Irish coffee (½ fl oz Irish whiskey per 2 fl oz strong coffee beverage) benefits from the use of a small amount of sugar and a topping of cold cream, whipped or unwhipped.

### Cafe Royal

Cafe Royal is made up similarly but with brandy, with or without whipped cream topping.

### Cafe Brulot or Cafe Diablo

Cafe Brulot or Cafe Diablo is made by placing sugar, cloves and lemon rind in a dish of warm brandy and igniting the delectable beverage with a lighted match. As the brandy-spice mixture burns, blue flames dance in a darkened room. The fire makes a dramatic and memorable occasion prior to serving. Cover the flaming bowl with a plate to cut off air and the dancing blue flames will cease. Then (with a large ¾ fl oz tablespoon) the spice flavored warm brandy is spooned into a demitasse cup or even regular sized (6 fl oz) cup. This is followed by 3 fl oz strong hot coffee beverage. Then a generous dab of whipped cream is floated on the beverage. This delightful drink should be served promptly.

## Arabian Coffee

The green coffee beans are roasted in a metal globe over a charcoal fire just before use. The darkly roasted beans are finely ground in a mortar and pestle. Cardamom is usually mixed with the ground coffee to spice it, although sometimes cloves or saffron are used for a different flavor effect. The "3 time boil-up" method may be used, or the coffee beverage may be served from a brass pot with long spout (stuffed with fiber to strain out grounds). Tiny porcelain cups without handles are only one quarter filled each time. Persons are served in accordance with their rank, and rarely refuse coffee since they consider its service high hospitality.

## Turkish Coffee

This coffee beverage is served in tiny porcelain cups about the size of egg shells. The beverage is prepared in an Ibrik, a cone shaped pot with flared neck, and with a handle, which is held over the fire. The beans are darkly roasted and pulverized. Four tablespoons of coffee are added to 1.5 cups of water. After the water has boiled, 4 teaspoons of sugar are dissolved in it. The coffee is brought to a boil, and immediately removed from the heat—a process that is repeated three times. The neck of the Ibrik is wider to allow the froth or foam to accumulate without boiling over. A fraction of an ounce of cold water is added. The coffee with froth is added to each of four demitasse cups. It is important that the froth be present and in each cup; as the Arabics say, "Host loses face, if the froth is absent from the face of the cup."

In Indonesia, a largely Moslem country, the coffee beans are roasted lighter than in the U.S.A., and are then disc ground to something like vend grind, that is, less than 10 percent on 20 mesh, 45 percent on 30 mesh and 45 percent thru 45 mesh, particles ranging 0.2 to 0.6 mm. A heaping teaspoon of this R&G is added to a cup followed by boiling water and slight stirring. The wetted fines settle to the bottom of the cup and the coffee beverage is consumed with a gentle overflow past the lips, hopefully with few grounds.

## Brazilian "Cafezinho"

Cafezinho is really the diminutive word for cup of coffee. It means demitasse of strong coffee beverage prepared in a Brazilian style. The Brazilians use a darkly roasted, finely ground coffee. For every 4 fl oz water per demitasse cup, a heaping tablespoon of coffee is used. The water is brought to a boil and then is immediately taken off the heat. The coffee grounds are added and stirred for a few seconds. After steeping a minute or two, the mixture is stirred again. Then it is poured through a flannel sac called a quador, which is held on a wire ring. The beverage, free of grounds, filters through rapidly. It is promptly served. One adds sugar as he prefers. In Brazil, it can be a heaping teaspoon or two of sugar. This makes a delightful, if strong, beverage. The beverage is well suited to the Brazilian coffee beans. Demitasse porcelain cups are usually kept warm and sterile in hot water baths. In Brazilian cities, cafezinhos are often taken at stand-up bars on the street, producing a quick "drink-and-run" atmosphere in larger cities like Sao Paulo. In Colombia, for example, the local coffee beverage is called tinto.

### Espresso Type Coffee

This coffee is made in an Italian espresso machine. Similar coffee beverages can be made in domestic Italian drip pots (machinettas), or domestic upward steam-water pressured brewers. The latter brewer and the commercial espresso machine are similar in performance. A porous cartridge is filled with R&G coffee, and a measured volume of pressured super-heated water is pushed through the R&G coffee in the cartridge. The coffee beans are darkly roasted; the oil seeps over the bean surfaces to give the beans a glistening appearance. The beans are finely ground just before brewing. Beverage is prepared one cup at a time and is served immediately for drinking. The time and temperature factors are closely controlled. A valve on the side of the espresso machine allows steam to be bubbled through milk to heat the milk quickly for those who want their coffee beverage capped with milk. Hence, the term Cafe Cappuccino. Italians, Argentinians, Brazilians, and other Latins using the espresso coffee usually add copious amounts of sugar to the small 2 fl oz beverage serving. Others prefer a twist of lemon rind to flavor the espresso coffee.

### Vienna Coffee

After the Turks fled Vienna, about 1683, Kolschitzky founded his coffee house on the 100 bags of coffee beans the Turks left behind. Since then Vienna has had a reputation for its coffee beverage and pastries. Today coffee served in Vienna is often from espresso machines. The term "Turkish" tradition of froth on the cup has persisted, so Robusta coffee beans, which foam excessively are used in some Vienna coffee shops. The term "Vienna" means coffee beverage with a helping of whipped cream on top. Germans call this "mit schlag."

## COFFEE AROUND THE WORLD

Reference to tables in Chapter 1 shows the quantities and origins of coffee imported by the major consuming countries. These tables reveal a good deal about the quality of imports and hence, coffee. By gathering additional data on how these statistics have varied through the years, one can gain further insights on coffee-use habits. Other significant data are the current and past custom duties on coffee, or non-duties on favored nation coffees, as well as historical facts in colonialism and post colonial treaties. Also, the cost of the coffee to the purchaser in terms of hours worked reflects standard of living. Then, are the coffees sold as beans, ground or by bagged or canned brand names? What kind of roasting and grinding are used? What kind of brewing methods are used, and what sorts of brewing equipment? How is the coffee served; when and where is it drunk mostly; what sized cups are used; and what water quality exists? Altogether, these variables contribute to markedly different tasting coffees around the world.

## Coffee Drinking Habits

Coffee drinking habits in the major consuming areas will be discussed briefly in what the author considers to be generally the best to worst quality preparations. It is notable that one can take fine quality green coffee beans, and blend, roast, grind, brew and serve them in a manner that destroys their good flavors. It is being done all the time virtually everywhere, but especially in institutional service areas. On the other hand, the best care and knowledge in preparation will not upgrade poor quality, stale coffees. Obviously both conditions exist all over the world. But basically, the consumer's knowledge and diligence in purchasing (what is made available commercially), storage, and preparation are important. In some countries like the U.S.A. where public acceptance of branded vacuum canned or pouched semi-stale coffees of light roast, coarse grind predominates; the public is to blame for not educating itself above the service of a hot black cup of tasteless if not bitter-harsh beverage.

The nations to be discussed are Latin America, Italy, Spain, Germany, Scandinavia, Switzerland, France, Holland-Benelux, U.S.A., Canada, Australia, U.K., Greece, Turkey, and Japan.

## Latin America

In the growing countries of Latin America, coffee beverage preparation is rather uniform and is mostly done with dark roasts and fine grinds. It produces a strong-tasting beverage (lots of coffee to water), which is freshly prepared when requested and served in 3 to 4 fl oz (100 to 120 ml) portions in small porcelain cups. Espresso machines are widely used in Argentina, Colombia and Brazil. In the home, a heaping tablespoon of R&G coffee is used per cup, to which just boiled water is added. This is steeped and mixed to assure thorough wetting and solubilization. After 2 to 3 minutes, the slurry is poured through a cotton or flannel strainer to separate out the insoluble grounds. The beverage is served immediately. In cafes and restaurants a larger sock may be used to produce more beverage that is held in a hot water jacket. Where U.S. type urns and filter half gallon systems have been adopted, the resulting beverage is invariably poorer; but this system is used often to accommodate American tourists. Low grade beans are often used, which produce a low grade beverage.

## Italy and Spain

The quality of coffee beans used in Italy and Spain vary widely from good to low grades. If one takes urn-prepared coffee in hotels, it is undrinkable with or without milk. But if one asks for espresso, especially if one sits at a bar in front of an espresso station, then one will get a drinkable beverage. This is for several reasons: the espresso station stocks whole bean roast coffees, and only grinds them finely on request or when they will be used up in a few hours. The finely ground coffee, even if not top grade, is measured in the espresso cartridge, inserted into the head, and is extracted in seconds under pressurized water conditions. The R&G coffee actually never attains the boiler water temperature (10 psig), because it is cooled by the interconnecting piping, valves and chamber as well as R&G coffee. Proof of this is that the porcelain demitasse cups must be kept warm on top of the machine; if the cups were not kept warm, the resulting beverage in a cup held at room temperature would not be at the proper drinking temperature. The whole method of espresso brewing is complementary to dealing with a fresh R&G coffee—not exhaustively extracting it, and promptly serving it.

These machines can pressure extract a pound of finely ground R&G coffee in 5 minutes to produce a good-tasting beverage. Pressurized water is necessary to force the hot water through the resistant granular mass in a short time. The shortness of brewing time, as well as the non-exhaustive extraction, reduces bitterness and harshness. These types of institutional machines are considerably more expensive than the U.S. mass produced half gallon brewer, but pressurized extraction of fine grinds in two minutes yields a superior tasting beverage.

An interesting and noteworthy point is that Italy consumes hardly any instant coffee, and although Spain's use of instant coffee is growing, it is still a small amount. Spain uses mostly decaffeinated Robustas which take on a mild taste after decaffeination.

## Germany (with similarities in Austria)

Historically, Germany has always been a large importer of coffee beans, usually of good quality. German immigrants and merchants who settled in Mexico, Central America and South America were able to carry on a strong trade with Germany. Costa Rica grows fine coffees and sends most of them to Germany. Also, there are many German settlers in Costa Rica. Brazil also has a number of German families in coffee agriculture and commerce. The Germans had occupied

Kenya and Tanzania, as well as Camerouns, so they had a good knowledge of East Africa and the fine, mild coffees grown there. Many Germans are involved in the coffee trade in Colombia. So coffee is a dear subject to the Germans, from whom we get the expression *kaffee klatch*.

Since coffee beans imported into Germany carry a very high customs duty, retailed coffee is a very expensive if not luxurious product, drunk with special care on special occasions. For this reason, Robustas were slowly imported to Germany. Even in 1961, many retail coffee roasters were selling whole bean freshly roasted coffees. But in the 18 years since then, there has been a marked decline in the quality of German coffees. The retail roasters have gone out of business, while two large chains, have proliferated to over 1,000 retail shops each. By 1976 whole roast beans of identified origin were no longer seen, but prebagged whole beans by brand name are sold in the indicated chains and in large department stores. It seems that with the industrial growth of Germany after World War II, and with vast numbers of immigrant labor there, the average German has stopped taking the time to pick out freshly roasted, identified coffees. Or perhaps he feels that lower price is now more important.

Instant coffee is only one-sixth of all coffee processed; however, this is about 25% of all cups consumed, although its use appears to be leveling out, if not actually declining. Vacuum packed R&G coffee in cans is approaching 40%, while decaffeinated and health coffees are about 14% of the coffee market.

## Scandinavia

Through long term treaties, Brazil has been the major supplier of coffees to all four Scandinavian countries. Hence, these nations are "conditioned" to the taste of Brazilian beans, and use them as strongly in beverage as they do in Brazil. Brewing has traditionally been by steeping in a pot, but since about 1970 automatic drip coffee brewers using filter papers have come to dominate the market.

## Finland

Finland is worthy to discuss first because it is somewhat apart from the other 3 Scandinavian countries in its language and cultural habits, including coffee. Since about the mid 1960's, Finland has progressively used less Brazilian and more and more top quality milds from Colombia, Central America and Kenya. Very little Robusta type coffees are processed or sold in Finland. Heavy duties and sales taxes are placed on coffee which make it somewhat of a luxury item; however, Finland has the highest per capita consumption, exceeding 10 kg green coffee per person per year. Finns living near the Swedish border buy their coffee in Sweden because it is relatively cheaper. Historically, it is important to note that until the early 1960's 95 percent of the coffee sold was sold in roast bean form, being ground at the retail store "point of sale," usually in ¼ kg units. Roast color is light, almost as light as in the U.S.A. Before World War II, 55 percent of the coffee beans were sold green, and were roasted at home. By 1970, major coffee sales were predominantly from four major roasters, which used vacuum packed (Hesser type) foil bags in cartons, as well as vacuum packed R&G coffee in "tins."

## Sweden

Sweden imports close to 1.8 million bags of green coffee per year. In the mid 1950's through the 1960's Brazil supplied almost 70 percent of the imports, Colombia 15 percent, and Central America 10 percent. Brazil supplied less than 50 percent in 1975 after having yielded 55 to 60 percent in 1968. Colombia's share is up to almost 20 percent as are African milds, especially Kenya. Central America still furnishes 8 to 10 percent of all imports. These statistics over 20 years indicate a great deal of stability in the coffee types being purchased, as well as the emphasis on quality. Sweden does have, after all, one of the highest living standards in the world, and so their coffee selections are in accord, except for the heavy Brazilian portion that tradition has established. About two-thirds the coffee sold is in vacuum packed tins. Sweden, like the other Scandinavian countries, has a high per capita coffee use, exceeding 10 kg per person per year. Flexible pouch packs that are gas flushed are commonly sold in 500 gram units. Granulations are very fine, and automatic drip brewers are widely popular. The roast is medium dark and 25 percent of the coffee market is now institutional. Instant coffee is only 2 percent of the coffee market. An interesting aspect of the coffee market is that imported brands of canned R&G coffees or instants from the U.S.A. and elsewhere are readily available at competitive prices.

## Denmark

Denmark imports almost 1,000,000 bags of green coffee per year, and per capita use also exceeds 10 kg per yr. Imports are still 70 to 80 percent Brazilian, and in recent years more Robusta types have been imported. Vacuum packed cans are available, but most sales are in gas flushed foil pouches of 500 gram size. There is an import duty on green coffee. Little instant coffee is sold, but various brands from Germany and other countries are available. In this dairy and cheese country the heavy use of milk is common with coffee beverage, and helps to make Robusta blends tolerable. There are about 25 commercial roasting firms.

## Norway

Norway imports a bit over 600,000 bags of green coffee per year. Brazil supplied 80 to 90 percent from 1945–1957 under barter agreements, but since then

Brazilian bean imports have still been near 75 percent. In 1960 the government ceased its involvement with imports, so there is no duty on green coffee imports, only on roasted. Instant coffee sales are less than 1 percent of coffee imports. There are about 20 coffee roasting firms, located mostly in Oslo, but also in Bergen and Tronheim. Norway uses a darker roast, and imports very little Robusta.

## Switzerland

Coffee use in Switzerland is rather unusual. Of the million bags of coffee imported each year, only about two-thirds is consumed in that country; the rest is exported or transported as roasted, instant or decaffeinated coffee. The green coffee brokers working out of primarily Lausanne actually buy and sell contracts for considerably more than that. About 40 to 50% of the coffee consumed in Switzerland is instant coffee, and no doubt this is tied to the fact that being a dairy country, the beverage is consumed with considerable milk or cream. Many Swiss use up to 20% chicory in their coffee beverage, which requires milk or cream to be drinkable. The use of chicory is a frugal habit that has persisted in spite of Swiss affluence. In January, 1977 it was not unusual to pay 75 cents for a cup of coffee in a Swiss restaurant. Swiss industrial operations are highly mechanized, sanitary, and impressive. Equipment manufacturers say that the Swiss are difficult to please and are more fussy about technicalities than are the Germans. Haco supplies Migros, the largest Swiss supermarket chain, with its coffees which are branded Mocafino. Today Haco uses the best coffee beans available, takes reasonable solubles yields and produces the best tasting freeze-dried instant coffees in the world—by some accounts. The entire controlling factors seem to be the use of good raw material plus good technology and machinery. Swiss instant coffee, like Swiss chocolate, is expensive, because there is a ten cent per lb. import duty. Most Swiss coffee comes from Brazil, with notable amounts from Central America and Kenya-Tanzania; as much as 25% Robustas are imported also. Green coffee comes on barges up the Rhine, e.g. Migros-owned barges, from ports like Rotterdam.

## France

France imports 4.5 million bags of green coffee each year; 40% is Robustas from Ivory Coast, 10% are Robustas from Cameroon, and altogether over 60% was Robustas in 1976 (see T&CTJ Dec. '76). Brazil in the early 1970's was shipping over 1,000,000 bags per year to France, but this fell off after the 1975 frost. Most of France's imports of coffee were from the Franc zone, a very political influence on France's coffee sources and quality, due to preferential import taxes. Five percent mild coffees came from Latin America, raised to 10 percent in 1976 due to shortages in Brazilian beans. The price increases in 1976 were reacted to by the French consumer not by reducing consumption but by buying lower grades at lower prices. With government control over retail coffee prices, business can be sustained only by using lowest grades of green coffee at the lowest prices. France has been a major chicory grower and user, and high prices of coffee can easily contribute to increased use of chicory to ameliorate the coffee price increases of 1976–1977. It is reported that soluble coffee use in France in 1976 was 22% of the market, which is the result of a slow but continual acceptance of instants.

It seems that the French consumer prefers to buy from his local roaster, so national or regional brands do not succeed very well. Espresso machines are widely used in France and Bene-lux, and no doubt help to deliver a better cup of coffee from Robustas than would otherwise occur. In some regions, e.g. Belgium-Holland border, some Frenchmen actually prefer to drink Rioy Brazilian coffees. Per capita consumption is only about five kg green coffee beans per person per year, and invariably the coffee is consumed with milk or cream. The roasts are dark brown. About 12% of the coffee used is decaffeinated.

## Netherlands

The per capita consumption is about seven kg per year, and close to six million bags of green coffee are imported per year. This is almost a doubling of coffee imports since 1960. Brazil has been supplying about 25%; Angola, 25%; and altogether about 35% Robustas with contributions from Indonesia, Uganda and Oamcaf. About 25% of milds come from Latin America.

Instant coffees represent about 8% of the market. Sixty percent of instant imports (20% duty ad valorem) have been from Brazil. Green or roast bean duties are 10 percent, and 16 and 23 percent respectively if decaffeinated. About 16 million lbs instant coffee per year are exported, about half to Germany and 25% to France and 25% to Belgium. Douwe-Egberts exports liquid coffee extracts to Germany for the vending machine.

## Belgium-Luxembourg

Antwerp is the principal coffee trading center not just for Belgium but for western Europe. Douwe-Egberts has a roasting plant here as does Rombouts. Jacmotte in Brussels is another large roaster. Chicory and powdered caramel extenders are sold for addition to coffee, and are more used in rural areas. The Rombouts plastic stacked cup with R&G coffee, already in a porous tissue paper layer, constitutes a prepared fast beverage preparation and is widely used. Belgian imports are about 1.3 million bags per year, $^1/_5$ of which come via Holland, and Robustas are only about 15 percent of imports.

## United States of America

The statistics show that U.S. per capita use of coffee has fallen 35% in the past 15 years (1960 to 1975). Since 1970, there has been a rash of retail shops that have opened that specialize in selling whole roasted beans, at least of identified and better quality and fresher than what the vacuum canned mass producer is manufacturing. U.S. green coffee imports have even fallen off from the 1968 peak of 25 million bags to 19 million bags, whereas soluble coffee use has been maintained at 25% of green use, but represents 35% of cup use. Before 1968, Brazil contributed about 7 to 9 million bags per yr to U.S. imports, but in 1974 this fell below 3 million bags. Whereas the Robusta producing countries increased their share to U.S. imports, this happened because the large roasters depended on competitive pricing. The poorer tasting beans were used, reaching 35% of U.S. imports in the '70's. A good deal of Robustas go into instant coffee manufacture, but not all. Angola Robustas reached 2.4 million bags in 1974, and low grade Indonesians were 1 million bags per year. Other Robustas came from Uganda and Ivory Coast, together constituting 1.5 million bags. In 1974 Colombia shipped 3 million bags to the U.S.A.

It is well known in the coffee trade and in coffee growing countries that U.S. roasters as a rule will not pay better prices for better quality coffee. So we have the terminology *European preparation* of green beans at a few cents per lb more cost, but with considerably more clean flavor. There are a few stars in quality in the U.S.A., and all the roasters have not succumbed to lower price-lower quality status. The worst facets of the U.S. coffee trade quality are in institutional sales.

Part of the problem commercially is that proper selection, roasting, brewing, etc. fragments the mass production system, hence works against the large roasters who control most of the sales. In the face of declining coffee use in the U.S.A., countries like Italy, Japan, U.K., Germany, etc. have been increasing their coffee consumption in total and sometimes per capita.

**Modes of Coffee Use.**—Coffee markets are quite fragmented. In the U.S.A. the retail vacuum packed R&G coffee brands, and the instant coffees constitute the two largest selling markets. Then, there are regional brands and regional coffee type preferences. There are various water qualities, type brewers, additives (sugar, cream, chicory) etc. The Pan American Coffee Bureau (PACB) publishes annually a breakdown on types of coffee usage in the U.S.A. This information is useful in understanding marketing mechanisms, and in identifying sales intensive areas. These PACB studies cover pricing, drinking habits by age and ethnics, summer/winter coffee usage, regional coffee usage, use with sweeteners and/or creamers vs age, per capita use of R&G coffee and instants, time of day and place of use.

**U.S. Coffee Sales by Original Bean Varieties.**—*100% Colombian.*—The Colombian Federation of Coffee Growers has for twenty years aggressively promoted its 100% Colombian coffees. There have been and are many coffee roasters that put out such a Colombian vacuum pack brand. Colombians are a premium quality and priced coffee, for which there seems to be a sizeable market. Such premium priced sales are often influenced by the difference in price between Colombian beans and other green coffee beans at that specific time.

*100% Hawaiian Kona.*—Another coffee sold by origin, especially on the U.S. west coast, and in specialty shops all over the U.S.A. is Kona coffee. Kona is the only commercial coffee grown in the U.S.A., on the island of Hawaii. Due to the high cost of U.S. labor, Kona coffee production has fallen from 150,000 bags in 1957 (a peak year) to 45,000 bags in 1969. Kona coffee is very strong and rich in flavor. It actually can be used to give 20–25% more cups beverage per pound, than average commercial coffee blends because of this extra strength. Due to Kona coffee's intense flavor properties, it makes an excellent beverage from instant or freeze-dried coffee. Freeze-dried Kona is sold at premium prices in some specialty shops.

*100% Brazils.*—These R&G coffees have been marketed in the New York City area from time to time by private roasters. The Brazilian Coffee Institute does not promote its coffee like the Colombians do. Brazilian beans lack the aroma of Colombian milds of equal grades.

*Other 100% Coffee Bean Types.*—In retail shops in the larger cities one can buy freshly roasted coffee beans from Jamaica, Costa Rica, Guatemala, Mexico, Salvador, Ethiopia, Kenya, Mocha, etc. at about 50 to 100% over the cost of average priced vacuum packed brands.

## U.S. Coffee Drinking

Since the nation consumes almost half the world's green coffee bean production, U.S. coffee drinking habits have been studied. Some findings include:

1. Drinking coffee beverage with food. This is seldom done elsewhere in the world except during breakfast.
2. Drinking weakly prepared beverage, lacking flavor and aroma, watery. Can see bottom of cup through beverage. People drink many cups of weak coffee per day, up to 60 cups per lb R&G.
3. Two thirds use milk, cream or whitener, consuming coffee flavored milk.
4. Reheating coffee beverage is a hard dying habit in the U.S.A., encouraged by false economy and advertisements.
5. Urns hold deteriorating hot beverage for hours.
6. Large volumes of coffee beverage served per cup (6 or 7 fl oz). This is common, especially in vending type paper or plastic cups.
7. U.S. brand use since World War I obscures coffee origins and qualities. There is an increased use of poor tasting Robusta coffees. The U.S. retail coffee

buyer is not knowledgeable about coffee identities, tastes and original bean sources.

8. Stale coffee flavor is considered part of the normal coffee flavor. Ground coffee purchases are mostly in evacuated cans and not truly roaster fresh. R&G coffee overbought in excessively large cans requires over a week to consume and staling sets in.

9. Water impurities like chlorine, alkalinity and organics downgrade brewed coffees.

## U.K.—UNITED KINGDOM

The U.K. is still pretty much a tea drinking population, with per capita tea use of 4 kg per yr. London has the principal Robusta trading market. Annual imports were about 2.25 million bags in 1974 and 2.8 million bags in 1973. The origin of the major portions of green coffees are from Uganda (35%), other African Robusta producers (18%), Brazil (25%), and Latin America (10%). In the period 1963–1973 the imports of green coffee virtually doubled. Former colonial and protectorate countries are major sources of imports, since they are not taxed: Uganda, Kenya, Tanzania, Ghana, Sierra Leone, and Papau.

Instant coffee constitutes 80% of the coffee consumption, and the U.K. has the dubious distinction of being the highest user of instant coffees in the world.

Britons take their coffee half and half: half hot beverage (strong) and half hot milk. For most Britons, coffee is a quick, satisfying milk drink. Of 78% of adult Britons who drink coffee, 80% use instant. The majority then adds hot milk. When one takes coffee in this manner, instants and Robustas are palatable.

## JAPAN—FAR EAST

An interesting parallel case in coffee use to the U.K. is Japan. Both peoples are traditionally tea beverage drinkers, but since World War II they have become progressively greater coffee drinkers. Both countries are islands and are oriented to seafaring and world commerce. Both are manufacturers and exporters of machinery. Both nations import close to 2 million bags of coffee beans each year. Both drink lots of instant coffee; at the 80% level in the U.K. and 50% level in Japan. Both possibly were introduced to coffee during and after World War II. Today instants are used mostly in homes, and R&G is sold in Kissatens (coffee shops).

Since the early 1960's many Japanese trading firms have established themselves around the world, and many of them buy and sell green and instant coffee. In the early 1970's one bought controlling interest in a large instant coffee processor near Londrina in Brazil, that operates a plant producing ten million pounds of instant coffee per year. There is a large Japanese agricultural group in Brazil, and one of them grows substantial amounts of green coffee. The dramatic increase in coffee consumption in Japan is clear. In 1960, Japan only imported 178,000 bags of coffee, and by 1975, the imports had increased ten fold.

## SUMMARY

There are numerous differences in coffee preparation, use, and buying habits in Europe, the United States, and elsewhere.

1. Former colonial affiliations of the United Kingdom, France, Portugal, Belgium, Holland, and Italy bring much Robusta coffee into these respective countries and these cups have strong Robusta flavors.
2. Scandinavian blends are three-fourths Brazilian coffees, and these cups taste accordingly.
3. High import duties on coffee make it a luxury commodity.
4. With high German coffee duties, only better quality coffees are marketable.
5. Cartel market controls on coffee as well as higher sales markups differentiate coffee marketing in Europe compared with the United States.
6. Coffee in Europe is more commonly sold from specialty shops from single variety roast bean bins; grinding is at point of purchase or at home.
7. Roast coffee packed in evacuated cans represents less than 10 percent of European retail sales. Vacuum packing of roast coffee in foil bags has been growing fast in the last several years. This delays staling more than six weeks.
8. Due to common use of custom roasts and blends, small batch roasting is prevalent in many countries.
9. Possibly due to the broader use of Robustas and Brazilian coffees with fewer mild coffees, uniformity and taste controls appear to be more lax than in the United States.
10. In Europe there appear to be more ways of brewing coffee than in the United States. For example, the espresso method in Italy is "the way" to prepare coffee beverage there.
11. The best quality coffee blends are found in northern Germany. In southern Germany the cup flavors are heavier and less aromatic; use of chicory in this area is also common. The coffee beverage in Switzerland is similar to that in southern Germany. In France, the coffee roast is darker than in Scandinavia; the Scandinavian degree of roast and bean color approximate those of the United States. In Italy, coffee beans are dark roasted. Drinking coffee with hot milk, often on an equal volume basis, is quite prevalent.

Soluble coffee sales in Europe by 1960 had made little progress and represented only a few percent of total coffee sales especially in Italy and the Scandinavian countries. Both the United Kingdom and Japan were exposed to United States Army roast and instant coffee and since 1960 have experienced a great increase in coffee drinking.

Soluble coffee in Europe is sold mostly in 50 and 250-gm tin cans, but other sizes are also available. Glass jars have been used only to a limited extent because they are more expensive. Currently, there is little uniformity in the quality of instant coffees, which vary from country to country. Some of these variations are due to selective import restrictions on green coffee types as well as varied processing facilities.

# Part II
# Green Coffee Technology

# 4

# Coffee Horticulture

It is the purpose of this chapter to review and summarize the principal problems encountered by the average coffee grower, and to furnish a guide to the latest literature in the field.

## LITERATURE ON GROWING COFFEE

To begin, there is the readily available and valuable collection of coffee references: "Coffee—Bibliography of the Publications Found in the Library of the Institute." The institute is the Inter-American Institute of Agricultural Sciences (IAIAS), Orton Memorial Library, at Turrialba, Costa Rica, and the collection was compiled by Angelina Martinez and C. Noel James. All the references can be obtained from the Orton Library, where photostatic copies may be made promptly at moderate charges. The records are kept up-to-date by the monthly publication of new current additions in preparation for future editions. Among other things, it lists 25 different bibliographies on coffee.

Although IAIAS in Costa Rica did intensive agricultural work in the early 1960's, the most current leads to new developments are now coming from several other sources as well. Namely, in Brazil, where rust, the fungi attacking, damaging and killing coffee shrubs, became prominent in the mid 1970's. An intensive effort has been made by Brazilian schools under the sponsorship of the IBC (Instituto Braseilero do Cafe) to take broad steps to stop the spread of this devastating disease. Extreme measures were taken, such as destroying millions of trees in and around Victoria, wide spread use of anti-fungal chemical sprays and for a while packaging suspect green coffee beans in paper bags for shipment as a means of quarantine. This disease prompted co-operative efforts from scientists in Colombia, Central America and elsewhere. Consequently, useful information was disseminated to all these countries and can be found in their central agricultural laboratories. Rust also appeared in Nicaragua in 1977, and created terror among the local farmers, as well as the governments themselves that depend on coffee export incomes. In Colombia, the FNC (Federacion Nacional do Cafeteros) has a central research station in Chinchina, where many aspects of coffee agriculture are studied and field tested. In fact it was here that Colombia initiated its program for planting the Caturra shrub variety that can produce heavily and does not require shade trees.

There are active research groups in Kenya and the Ivory Coast. The latter is supported by the I.F.C.C. (Institute Francaise du cafe e cacau) which has laboratories near Paris. However, the most up to date literature comes from the papers of A.S.I.C. each 2 years, and copies can be obtained from the I.F.C.C.

One of several large research projects on coffee was the program of the IBEC Research Institute of New York, carried out in Brazil in the States of São Paulo and Parana, for more than ten years. IBEC published ten bulletins about coffee. These are listed in the bibliography and are available to the public. Excellent information is also found in the Pan Am Coffee Institute, N.Y., N.Y., and the Foreign Agricultural Services, U.S. Department of State, Washington, D.C.

The International Food Information Service at Reading, U.K., also publishes abstracts, some of which pertain to coffee shrub propagation.

## CULTURE OF THE EVERGREEN COFFEE PLANT

### Location

Geographically coffee grows only between the Tropic of Cancer and the Tropic of Capricorn. Outside this area there is opportunity for frost, from which the coffee plant cannot survive. Small islands near the perimeter of the Torrid Zone are less likely to be exposed to frost than hilly-to-mountainous areas nearer the equator. In choosing a location, the chief considerations are altitude, soil, climate, disease and economy.

**Altitude.**—Normally one finds that the higher the altitude is, the better the quality of the coffee. The limiting altitude is the frost danger zone. Coffee beans grown in Kona, Hawaii can be at 1,000 feet above sea level. Kona is close to the Tropic of Cancer, yet does not suffer from frost exposure because it is near the sea.

However, in Parana, Brazil, which borders the Tropic of Capricorn, each year winter threatens severe frost damage to coffee shrubs. Here the elevation is close to 2,000 feet above sea level. In 1973 Brazilian plantings were shifted to Minas Gerais where there was virtually no danger of frost. Frost damage turns the leaves of the coffee tree brown, killing them, much as the frosts do to deciduous trees in the fall in temperate climates. In Parana soybean plantings in '74 displaced coffee shrubs. The coffee shrub move to Minas was a good move, because better quality coffee beans are produced there. Although coffee trees grow at 6,000 to 7,000 feet

TABLE 4.1. COFFEE FRUIT HARVESTING TIME ABOUT THE WORLD—LISTED NORTH TO SOUTH OF EQUATOR IN TORRID ZONE.

| Location Relative to Equator: Degrees North or South | Altitude Above Sea Level: Feet | Political Name | Country | Trade Names of Coffees: Examples | Harvest Months |
|---|---|---|---|---|---|
| 18–20° North | 3,000 | Haiti | isle | Haiti | Nov.–Dec. |
| 19.5° | 2,000 | Hawaii | isle | Kona | Nov.–Dec. |
| 18° | 3,000 | Dom. Rep. | isle | — | Nov.–Dec. |
| 16–22° | 4,000 | Mexico | N.A. | Tapachula | Nov.–Feb. |
| 15° | 3,000 | Arabia | Near E. | Mocha | Oct.–Dec. |
| 15° | 5,000 | Guatemala | C.A. | Antigua | Nov.–Dec. |
| 14° | 2,000 | Salvador | C.A. | Salvador | Nov.–Dec. |
| 14° | 2,000 | Honduras | C.A. | Honduras | Nov.–Dec. |
| 13° | 1,500 | Viet Nam | Asia | Viet Nam | Nov.–Dec. |
| 12° | 2,000 | Nicaragua | C.A. | Matagalpa | Nov.–Dec. |
| 12° | 3,000 | India | Asia | Arabica/Robusta | Nov.–Jan.–Mar. |
| 10° | 4,000 | Costa Rica | C.A. | Cartago San Jose | Nov.–Dec. |
| 10° to 13° | 4,000 | Venezuela | So. Amer. | Tachira | Nov.–Jan. |
| 9° | 3,000 | Panama | C.A. | — | Nov.–Dec. |
| 8° | 5,000 | Ethiopia | Africa | Djimmas Harrars | Nov.–Jan. |
| 7 to 1° | 5,000 | Colombia | So. Amer. | MAM's | All year |
| 6° | 1,500 | Ivory Coast | Africa | | Nov.–Jan. |
| 5° | 1,500 | Cameroun | Africa | Arabica | Nov.–Dec. |
| | 3,000 | Cameroun | Africa | Robusta | Jan.–Feb. |
| 4°N to 4°S | 3,000 | Zaire(Congo) | Africa | Robusta | Jan.–Feb. |
| | 5,000 | Zaire(Congo) | Africa | Arabica | Jan.–Feb. |
| Equator | 4,000 up 9,500* | Equador | So. Amer. | Robusta/Arabica | June–Aug. |
| Equator | 4,000 | Kenya | Africa | Robusta/Arabica | Sept.–Nov.–Dec. |
| Equator | 4,000 | Uganda | Africa | Robusta | Sept.–Nov.–Dec. |
| 2° to 4° South | 4,000 | Rwanda | Africa | OCIR | |
| | | Burundi | Africa | OCIRU | |
| 5° | 2,000 | Sumatra | isle | Ankola | all year |
| 7° to 8° | 2,000 | Indonesia | isle | Java | all year |
| 7° to 12° | 2,500 | Angola | Africa | Ambriz Amboim | 3% Arab. Apr.–June 97% Rob. June–Aug. |
| 5° to 15° | 4,000 | Peru | So.Amer. | — — | wash–May–Aug. "nat"–June–Dec. |
| 18° to 22° | 2,500 | Brazil | So.Amer. | Santos, Parana | May–Sept. |
| 20° | 1,500 | Malagasy (Madagascar) | isle | Kouilov | June–Oct. |

*Highest grown coffee in world.

above sea level in Costa Rica the rate of bean growth is slow (dense). Bean yields per tree are lower at these cooler temperatures. Coffee grows up to 9,500 feet above sea level in Ecuador, highest level in the world. Hence, it is less economical to grow coffee at these higher altitudes. Growing coffee at 6,000 to 7,000 feet in Colombia or Ecuador (which is on the equator) is common, and is practical.

Growing Arabica coffee shrubs at much below 2,000 feet (above sea level) in Latin America is not feasible because of the excessive heat, weeding, leaching of soils by heavy tropical rains, erosion, infestation, etc. Robusta shrubs thrive in more humid, warmer zones, hence, generally grow better at lower altitudes than Arabicas.

## Soil

The usual criteria of fertility, i.e., presence of an adequate supply of essential mineral elements, temperature, moisture, friability, pH, drainage, and degree and orientation of slope are important. The soil for coffee trees must not only be porous to allow heavy rains to drain off, but organic enough to hold moisture between rains. Soil must not be dry and hard so that the feeder roots and deeper water tapping root growths are inhibited.

Interestingly enough, most of the world's coffee is grown on volcanic soils in areas near extinct and live volcanoes. These soils are very rich in minerals and slightly acidic (pH 4.5-5.0), taking up iron. The "terra roxa" or red earth of Brazil, on which coffee shrubs thrive, was formerly virgin forest. It is also rich in iron. Potassium is needed for healthy growth of coffee shrubs.

## Climate

Climate includes mean temperature, diurnal as well as seasonal variations in the temperature; the possibility of frost; rainfall and its distribution throughout the year, its variations from year to year, and the possibilities of water and power supplies for irrigation, if necessary; sunshine and cloud patterns.

Annual average temperatures are 68° F with usually only plus or minus 10° F variations at most, yet infrequently. Between 60 and 80 inches of rainfall annually is ideal with not more than 3 to 4 consecutive months with less than 2 inches of rain.

The first rain after a dry season in Central America, e.g., in May, triggers flowering of the coffee shrubs and six months later, almost to the day, the fruit will be ripe for harvest. Weather can also be damaging to production. In recent years, drought in Brazil and excessive rain in Colombia have caused significant production losses.

**Effects of Clouds**.—There are daily, weekly, monthly and seasonal variations to consider. Light intensity is also influenced by haze. For example, it is common practice to burn fields prior to planting new grain crops each year in Central America and in Brazil.

This burning produces smog and haze that reduces sunshine considerably. Such smog or haze conditions can be so intense that airplane flying visibility is markedly impaired.

Afternoon cloud accumulations such as occur in Parana and Sao Paulo in Brazil markedly reduce direct sunshine allowing plants to grow without the need for shade trees. Clouds in the high mountains of Colombia have, since 1972, encouraged the Federacion to make massive plantings of the Caturra without shade.

Steep slopes as in Colombia contribute to erosion and even landslides. On the other hand, irrigation water is used to sustain coffee shrubs in Yemen (Biennial cycle production). (Fig. 4.1.)

When temperatures are too high and rainfall is inadequate, trees can bear a bumper crop, followed by a year of markedly reduced production. This situation is accepted as normal in all growing countries.

**Shade Protection**.—Shade has an equilibrating effect on the tree. It seems to keep the flowering in balance with the trees' strength and environmental capabilities. Shade in Brazil in Parana, São Paulo and Minas States, is provided by the afternoon columnas clouds, in Colombia by the clouds at high altitudes.

Shaded trees are protected from extremes in wind and dryness. The ambient air is quieter about the coffee trees making it cooler than in general in the day and warmer in the evening. Thus shade offers a more uniform and equitable environment for the shrubs to flourish. (Fig. 4.2, 4.3)

The coolness, due to shade in the day, reduces transpiration both from the leaves and the roots. Direct sunlight on the base of the plant can dry out and inhibit root growth, hence mulching and elimination of weeds is important.

FIG. 4.1. STEEP MOUNTAINOUS CLOUDY AREAS IN COLOMBIA

FIG. 4.2. CENTRAL AMERICAN PLANTING WITH SHADE TREES

FIG. 4.3. BRAZILIAN PLANTING WITHOUT SHADE TREES

Shade is particularly more effective in keeping the plant cooler during the drier season. Shade minimizes weed growth. Uniformity of growth uniformizes the use of labor during growth and during picking and processing.

Today, much of the coffee of the world is still grown under shade trees. It was long thought that the coffee tree could not stand full sun without damage. In Brazil, no shade trees are used, but the trees are planted in close groups of 4 to 6 individual trees, and each group is counted as one tree or "foot" (*pé* or *cova*) as it is called in Brazil. This furnishes self-shading, and the individual trees do not receive full sun.

The trend now is to use less and less shade but to supply the tree with greatly increased amounts of both fertilizer and water. Under these conditions, they bear more heavily and still keep healthy. The coffee grower must determine the right amount of shading for his particular conditions. Shade trees may be useful for erosion control on steep slopes. If rainfall is deficient, it may be supplemented by irrigation if feasible.

## Economic Factors

Figures 4.3 and 4.4 show the picking of the ripe berries by women into hand baskets. The persons doing the picking also include whole families, children and aged persons (cheap labor). They often average one U.S. dollar equivalent per day (8 hours) without any fringe benefits, except possibly the plantation owner may supply meals during a one-to-two month picking campaign. Many of these people live on the edge of starvation, certainly nutritional starvation. Their clothes are tattered and torn.

Many governments and coffee producing countries do little to upgrade the lives of these poor people, who are often afflicted with sickness and disease. Payments for picking are often related to the weights picked per person. These wandering people usually earn their highest income during coffee picking time. The weight picked in L.A. is the "arroba" 12.5 kg. There are no unions. Most "cherry" pickers have no shoes or worldly goods.

FIG. 4.4. HAND PICKING CHERRIES

The price of land and the cost of preparing it, availability and costs of labor, methods, costs, and distances of transportation and communication—these factors and many others must be weighed in deciding on a location for a coffee plantation.

## Productivity and Shrub Spacing

Spacing is a very important question, and it has direct bearing on capital cost for land, yields per unit of area, and harvesting costs. Should the planting be single or multiple? Should it be contoured or rectangular? The latter depends on the steepness of slopes and the character of the soil. Is it firm or easily eroded? The closeness or density of planting depends largely on how much shading or exposure to sunshine is needed and what the cloud pattern may be. In some localities, it is quite regular; in others, it varies greatly. Records of past conditions are desirable. Results and recommendations of nearby experiment stations can be helpful in determining the best spacing for local conditions. (See Fig. 4.3, 4.4)

Productivity per acre or hectare (2.25 acres) is directly related to number of trees per unit area.

Coffee trees are usually laid out in a square pattern. At 11 foot spacings, there are 350 trees per acre or 800 trees per hectare. At 9 foot spacings, there are 537 trees per acre; at 8 foot, 680 trees/acre.

Hence, productivities are about two to three (60 Kg) bags green coffee per acre or one to two pounds of Arabica beans per tree. Plantations that produce say 20,000 bags of coffee per year are about 10,000 acres in size or about 16 square miles requiring perhaps 500 residents as families. This is almost 5,000 hectares and 50 sq km area.

In many areas and countries like Colombia or Guatemala, which are quite mountainous, fewer trees per hectare are planted, and one family may care for its own two or three acres. Bean yields will be influenced by botanical varieties, quality of soil, use of fertilizer, disease, moisture, sun, etc.

If trees are to mature fully, the eight foot spacing is too close. Average yields, by example, are as follows: Brazil 400 lbs/acre or 450 Kg/hectare; Colombia 480 lbs/acre or 525 Kg/hectare; and Salvador 600 lbs/acre or 660 Kg/hectare.

These yields are doubled and quadrupled with higher density intensive cultivation which is now being pursued in Colombia. These new cropping procedures work on a five-year growth cycle. They use no shade trees, but rather use Caturra variety closely planted within a few feet of each other (see photo) for self shading. This can give high yields of green dry coffee beans like 3,000 lbs Bourbon variety or 20 to 30 sixty Kg bags per acre (60 bags/hectare), and 50% more yield from Caturra. This sort of intensive cultivation requires fertilization to sustain the fullest bean yields. However, the cup quality of the coffee seems to suffer from such mineral fertilization and intensification of agricultural production. It remains to be seen how extensively these newer methods will be used.

## Growth Cycles

Although coffee trees can be 20 or 30 feet high, depending on species and environment, commercial coffee shrubs are usually not much larger than about eight feet high and ten feet in diameter. The shrubs are pruned so as to facilitate the picking of the fruit.

Recent methods of planting that eliminate shrubs after 5 to 8 years, do not allow the shrubs to reach their full size. Rates of growth are slower in shaded areas and/or higher elevations that are cooler. Bean seeds become denser with slower growth rates, flavor quality rises while productivity falls.

In suitable environments, coffee shrubs will bear fruit for 40 or more years, although productivity is the greatest between 5 and 15 years of age. It normally takes about 3 to 5 years from seedling to full producing shrub.

In some countries, coconut or banana trees are interdispersed with coffee trees for shading as well as for two croppings of land.

## Species

The Arabica species of coffee plant must have come from Ethiopia originally. Since it was cultivated in Arabia for 1,000 years, it naturally would be recognized as Arabian. The variations in a species are considerable. There are at least 12 distinct variations in Ethiopia where the plant has grown wild for at least 2,000 years.

The species most used for commercial coffee production are *Coffea arabica,* about 75 percent of world production; *C. canephora,* about 24 percent; and *C. liberica,* about 1 percent or less. *Canephora* is increasingly used because it has been found to be valuable in the manufacture of soluble coffee. It is probable that the use of *liberica* will gradually disappear as it is distinctly larger in size which makes it difficult to handle in standard machinery, and it seems to have no outstanding advantages except that it is highly resistant to disease. The relative sizes of the green coffee beans are: *arabica,* 1,200; *canephora,* 1,600; and *liberica,* 800 per lb.

## Variety

Many varieties of *C. arabica* such as *tchertcher, sidama, gimma, local bronze,* and *bronze tip* in Africa, *kents* in India, *blawan pasoemah* in Java, *bourbon vermelho, bourbon amarello* (yellow-colored when ripe) *commum* or *typica* in Brazil, *mundo novo* and *caturra* throughout Latin America and others have been used commercially on a wide scale. Today's favorites in Latin America seem to be *mundo novo, bourbon vermelho,* and *caturra* although several others are still in use.

Several varieties of *canephora* are used, but the most common variety is *robusta*. Extensive study of this subject has been made, and much information is available in the literature.

## Propagation

This too is an important phase of coffee growing. Having decided what species and variety to use, it is then desirable to secure high quality selected seed carefully prepared from high yielding trees. Seeds in the form of parchment coffee are sometimes upgraded by flotation in water to eliminate poor, shriveled, or diseased beans.

When seedlings appear in the nursery, the best specimens are transplanted and the poor ones discarded. If seeds are grown on the plantation, they are taken from the best trees. Much study has been given to propagation, and the Instituto Agronomico in Campinas, Brazil, has been one of the leading institutions working on the genetic improvement of coffee.

It is important to make note of the asexual or vegetative propagation of coffee. The chief advantage is that genetically uniform stock is obtained by this method. Figure 4.5 shows the Brazilian method of transplanting nursery seedlings. The sticks are arranged to give partial shade at first.

The accompanying figure illustrates how a coffee shrub is propagated from seed, although it may also be propagated by rootings of branch cuttings.

Figure 4.6 shows the seeds in pergamino being set out in rows. Figure 4.7 shows

FIG. 4.5. Brazilian method of transplanting nursery seedlings.

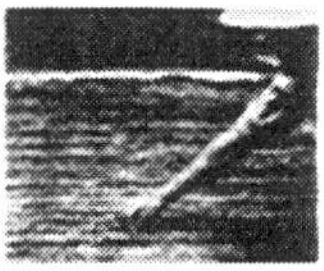

Fig. 4.6. Seeds in Pergamino Being Set Out in Rows

Fig. 4.7. Seedlings Popping Up After a Few Weeks

Fig. 4.8. Young Plants 6-8 in. Tall

Fig. 4.9. Plants Individually Potted in Shaded Area and About One Foot Tall

Fig. 4.10 Shows Foot Tall Plants Being Transplanted to the Field

Fig. 4.11. Pruning Three Year Old Plants

the seedlings popping up after a few weeks. Figure 4.8 shows the young plants after they are about 6-8 inches tall.

Figure 4.9 shows the young plants individually "potted" in a shaded area and about a foot tall. Figure 4.10 shows these one foot tall plants being transplanted to the field. Figure 4.11 shows pruning on a 3-year-old plant. Figure 4.12 shows a flowering plant. Figure 4.13 shows the berries fully developed on the tree. Figure 4.14 shows a close-in view of branch with ripe berries and flowers. Figure 4.15 shows seedlings in plastic bags under lath shade.

FIG. 4.12. FLOWERING PLANT

FIG. 4.13. BERRIES FULLY DEVELOPED ON TREE

## Flowering

The time of flowering determines the time of maturing of the fruit. It is desirable to control the flowering so that the harvest may be gathered most efficiently and at the lowest cost. Careful studies have been made on the physiology and the mechanism of flowering, particularly those factors which trigger the opening of the mature buds. However, despite a substantial amount of research, there is not yet a practical method of controlling flowering under commercial field conditions. Depending on the supply of labor and other factors, it may be of advantage to stretch out the period of flowering or to make it as brief as possible in each tree, but to accelerate some sections of the plantation and retard others so that the pickers will encounter high concentrations of ripe cherries and also have time to cover the whole plantation while ripe cherries are available. It would seem likely that, in the future, one or more control methods will be effective for the timing of this natural process. The use of timed irrigation has been utilized to a limited extent as one method of controlling flowering. Hormone and chemical sprays are possibilities, and the timing of fertilizer application has also been suggested as another method.

## Pruning

Some planters prune coffee trees, sometimes using intricate and highly developed methods; others do no pruning except for what occurs accidentally during the harvesting. In all probability, pruning is beneficial for increasing yields, but

FIG. 4.14. CLOSE-IN VIEW OF SHRUB BRANCH WITH RIPE FRUIT AND FLOWER IN COLOMBIA

sometimes little response is obtained because some other factor limits the yield. Pruning will become increasingly important as labor costs rise to keep man-hours highly productive. A pruned tree is easier to harvest.

## Weed Control—Use of Mulch

In order that the coffee trees may not be subjected to competition with weeds for nutrient material and water which might be so severe as to limit yields, weeds must be eliminated as nearly as possible. The old method was hand cultivation with a

FIG. 4.15. COLOMBIA, SEEDLINGS IN PLASTIC BAGS UNDER LATH SHADE

hoe or a similar tool, and this is still used; but as the cost of labor rises, it is generally found less costly to use herbicides of which there are many on the market. Two general types are recognized: (1) those effective on broad leaf weeds and (2) those used to kill the narrow leaf grasses. They must be used carefully and in the right concentrations in order not to damage the roots and foliage of the coffee trees. The U.S. FDA analyzes imported coffee beans for chemical residues.

Another line of attack is the use of mulch—usually dead vegetation, sometimes plastic or paper strips. Mulch has several advantages. It suppresses weeds and conserves soil moisture. It keeps soil temperatures several degrees cooler, which may be of great advantage to the health and yield of the trees in excessively hot weather. It may increase the efficiency of fertilizer assimilation, especially of phosphorus. Vegetation mulch has two disadvantages: it may be a fire risk, and it may exaggerate frost damage if cold weather occurs. This results because some of the heat of the sun during the day is prevented from reaching and warming the soil. The fire hazard may be minimized by the use of fire breaks to localize a fire. Mulch should probably not be used at any time or place where frost is a threat.

## Diseases

The further environmental conditions are from ideal for a given species or variety of species, the more likely disease and insect infestation will occur. Drought, lack of nutrients, injury, overbearing, dampness, and coolness can be contributing factors.

Aggressive levels of diseases will attack both healthy and unhealthy plants. Potential diseases attack when environmental conditions are conducive to its spread. Diseases can flourish on abandoned plantations and carry over to nearby healthy ones. Localized and incidental attacks are likely to occur any time. Diseases are caused by fungi, viruses and bacteria. Unless they are understood and recognized, they cannot be fought.

Bourbon variety of Arabica coffee is particularly susceptible to leaf spot disease, a fungus carried by spores. In 1972–1973 the IBC in Brazil destroyed many fungi-infested trees and had a general quarantine of many zones, especially the region around Espirito de Santos. Leaf spot disease is encouraged by drought and over-bearing of trees. In 1969 it contributed to marked losses in production for the first time in Brazil. Commonly called rust disease *(Hemileia vastatrix),* it has been known since 1869 in Ceylon, when all coffee trees were destroyed.

In India and in East Africa an Arabica strain of seed called Kent's coffee is rather resistant to this fungus. Spraying the shrubs with fungicides containing copper salts helps to reduce disease activity.

## Insects

The coffee bean borer attacks Robusta coffees and the antestia bug prefers the Arabica coffees. There are bean, branch and stem borers—a highly specialized group of insects.

The antestia bug bores directly into the berry before the fourth month of development while the berry is still soft. This results in spotted and irregular shaped beans, often with considerable bean loss since the damage externally is not very evident. Also, flowers and young leaves are attacked. Arsenite and pyrethrum sprays are effective against this bug; DDT has been used reluctantly in severe situations.

Thrips in East Africa defoliate plants rapidly and are associated with poor environmental conditions for the coffee plant. Paris green, copper-aceto-arsenite, is an effective spray for thrips.

There is also a green scaling insect, and borers—stem, branch, berry and bean.

The mealy bug, protected by ants, attacks coffee plants in unsuitable environments. Grease banding trunks and the introduction of parasites have kept the mealy bug under control. The reader is referred to Haarer's and Wellman's writings for additional information.

## Fertilizer

This is perhaps the most important problem of coffee horticulture. The first principle is to supply the major and minor elements to the coffee tree in adequate, but not excessive amounts, and under conditions in which the food material will be available to the tree to a maximum degree and not tied up in the soil complex.

To maximize yields and profits, the concept should be that of removing all limiting factors so that the tree may be allowed to give its maximum performance. This requires that consideration be given to all the mineral elements required for plant growth (both macro- and micro-elements). Also the possibility of excessive and harmful quantities of certain elements must be avoided, as well as the more obvious possibilities of nutritional deficiencies. Heavily fertilized shrubs tend to produce inferior tasting beans.

The soil is one of the principal factors to consider when evaluating the fertility requirements of coffee. The various soils have different capacities to supply the essential nutrient elements. Also some soils have unusually high fixation capacities, and they may "fix" elements such as phosphorus so that they are not available to the coffee plant. In addition to the soil there are a number of other factors which should be considered when developing fertility recommendations for coffee. These include (a) the species and variety of coffee that are used. There is some evidence that certain varieties such as *mundo novo* are more responsive to fertilizers than others. (b) Climate (including temperature and rainfall) is important as these factors influence the uptake and utilization of fertilizers by the coffee plant. (c) Sunshine and shade must be right—not too much and not too little. This may be controlled by spacing. (d) Pruning methods must be used which are consistent with the overall cultural practices being utilized in the locale. (e) Finally, good disease, pest, and weed control are essential. The importance of good weed control cannot be overemphasized as it will be of little value to spend money for fertilizers if they are to be used mainly to increase the size of the weed crop. (f) Soil acidity helps solubilize some minerals.

Assuming that the other factors are optimal or nearly so, one should apply as much fertilizer as can be economically and efficiently used. The problem of choosing the source of fertilizer material should be resolved on the basis of cost per pound of the various essential elements which field experiments indicate are required for maximum yields and maximum economic returns to the grower. The source may be partly organic, such as compost of vegetation and coffee fruit pulps or the rich fertilizer produced from the raising of poultry on a large scale. The mainstay, however, is usually chemical fertilizer because of low cost per pound of plant nutrient (NPK, etc.) and because it is usually more available and standardized in its nutrient content which allows it to be used without waste in quantities precisely matched to the needs of the trees. This presumes, of course, that the essential research has been conducted to provide the basic information regarding deficiencies, excesses, etc. In general, the elements may be supplied equally well from organic or inorganic sources. The principal advantage of the organic material is that it may improve the physical characteristics of the soil and often supply needed trace elements. However, this is usually offset by the problem of handling a large amount of bulk material and of insufficient supply for large-scale coffee plantations.

The major elements which are needed are N, P, K, Ca, S, and Mg, and the minor ones are Fe, Mn, B, Cu, Zn, and Mo. The minor ones are known as trace elements because only very small quantities are needed. The quantities, however small, are absolutely essential, and the lack or shortage of any one of them results in loss of yield and damage to the health of the tree. Present day developments tend toward the use of chemical fertilizers not only balanced as to the major elements, but also containing the proper amounts of the trace elements in order to guard against the possibility of production-limiting trace element deficiencies.

Radioactive tracer compounds have been used successfully in penetrating minutely into nutritional processes in order to pinpoint exact deficiencies and sources of trouble. Chemical analyses of the leaves has also been useful in establishing nutritional levels.

Also, more and more deficiencies are being recognized by specific symptoms in the appearance of the trees. Potassium deficiency kills the oldest leaves and gives heavy defoliation. Phosphorus deficiency is uncommon; but when it occurs, it causes the oldest leaves to mottle producing yellow spots and a reddish tint. Defoliation occurs with fungus attack increased. Nitrogen deficiency is exhibited by yellowing of the leaves.

The plant needs nitrogen as nitrates, phosphorus as phosphates and potassium. Nitrogen uptake increases with the rainy season and decreases after bearing fruit (in the dry season). Urea is the ideal way to supply nitrogen to the coffee tree. Ammonium sulfate, due to its acid contribution, is the least desirable fertilizer. Manganese, while essential, becomes toxic to the plant if present in excessive quantities in which case it may prevent the assimilation of other elements such as iron. Iron, which is often abundant in many soils, may be unavailable because it exists in almost absolutely insoluble form. Iron and other elements may be held in assimilable form as chelates which are metallic salts of ethylenediamine-tetracetic acid. A rough guide to the proper amount of fertilizer to be used is the amount of each element removed in the harvesting and pruning. Usually, because of certain extraneous losses, larger amounts than this must be supplied in order to keep the fertility of the soil unimpaired. Chemical fertilizers, like 100% soluble fish "non-burning" organic fertilizers, are almost 100% available to the plant.

## BIBLIOGRAPHY

ANON. 1957. Coffee in Paraguay. Report 18E. IAIAS.

ANON. 1957. Report on coffee cultivation in Costa Rica. Report No. 20. IAIAS.

ANON. 1957. Report on coffee cultivation in Ecuador. Report No. 24. IAIAS.
ANON. 1957. Report on coffee cultivation in Nicaragua. Report No. 21, IAIAS. No. 21, IAIAS.
ANON. 1957 Report on coffee cultivation in Panama. Report No. 22. IAIAS.
ANON. 1957. Report on coffee cultivation in Peru. Report No. 23. IAIAS.
ANON. 1957. Report and recommendations on coffee cultivation in Guatemala. Report No. 17. IAIAS.
ANON. 1957. Technical Progress in Coffee Culture. *In* Coffee, Cocoa, Tea *1*, No. 3, 111–116.
ANON. 1958. Coffee growing in Hawaii and Guatemala. Caribbean Commission Publications Exchange Service No. 12. Trinidad, S.A.
ANON. 1958. Report and recommendations on coffee cultivation in El Salvador. IAIAS.
ANON. 1962. Report on the "Marly Bean" Condition of Market Coffee in Jamaica due to Iron Deficiency. Report No. 49. IAIAS.
BEAUMONT, J. H., and FUKUNAGA, E. T. 1958. Factors affecting the growth and yield of coffee in Kona, Hawaii. Hawaii Agricultural Experiment Station, Bulletin 113, June, 29–37.
BEAUMONT, J. H., FUKUNAGA, E. T., and LANGE, A. H. 1956. Initial growth and yield response of coffee trees to a new system of pruning. *In* American Society for Horticultural Science, Proc. *67*, 270–278.
COOIL, B. J., and FUKUNAGA, E. T. 1958. Mineral nutrition: high fertilizer applications and their effects on coffee yields. Coffee and Tea Inds. *81*, No. 11, 68–69; *also as* Hawaiian Agricultural Experiment Station. Miscellaneous Paper No. 100.
DE GIALLULY, M. 1958. Factors which affect the intrinsic quality of green coffee. Instituto Interamericano de Ciencias Agricolas, Turrialba, Costa Rica; *also* Coffee and Tea Inds. and the Flavor Field *81*, No. 11.
FUKUNAGA, E. T. 1955. Chemical control of noxious weeds in the coffee plantations of Kona, Hawaii. Inter-American Institute of Agricultural Sciences (IAIAS).
FUKUNAGA, E. T. 1959. A new system of pruning coffee trees. Hawaii Farm Science *7*, No. 3, 1–3.
HAARER, A. E. 1962. Modern Coffee Production. 2nd Edition. Leonard Hill (Books) Ltd., London.
IBEC Research Institute, New York, (now IRI Research Institute, Inc.)
FRANCO, C. M. 1958. Bull. No. 16. Influence of Temperature on Growth of Coffee Plant.
LOTT, W. L., MCCLUNG, A. C., DE VITA, R., and GALLO, J. R. 1961. Bull. No. 26. A Survey of Coffee Fields in São Paulo and Parana by Foliar Analysis. IAIAS.
LOTT, W. L., MCCLUNG, A. C., and MEDCALF, J. C. 1960. Bull. No. 22. Sulfur Deficiency in Coffee. IAIAS.
LOTT, W. L., NERY, J. R., GALLO, J. R., and MEDCALF, J. C. 1956. Bull. No. 9. Leaf Analysis Technique in Coffee Research. IAIAS.
MEDCALF, J. C. 1956. Bull. No. 12. Preliminary Study in Mulching Young Coffee in Brazil. IAIAS.
MEDCALF, J. C., BONTEMPO, A., and FAVRE, G. 1961. Bull. No. 25. The Use of Pre-Emerge Herbicides for Weed Control in Young Coffee. IAIAS.
MEDCALF, J. C., and DE VITA, R. 1960. Bull. No. 19. The Use of Pre-Energy Herbicides for Weed Control. IAIAS.

MEDCALF, J. C., and LOTT, W. L. 1956. Bull. No. 11. Metal Chelates in Coffee. IAIAS.
MEDCALF, J. C., LOTT, W. L., TEETER, P. B., and QUINN, L. R. 1955. Bull. No. 6. Experimental Programs in Brazil. IAIAS.
MES, M. G. 1956–1957. Bull. No. 14. Studies on the Flowering of Coffee Arabica L. IAIAS.
INSTITUTO AGRONOMICO DE SÃO PAULO. 1956. Literature on Coffee. List No. 1. Campinas, São Paulo.
INSTITUTO INTERAMERICANO DE CIENCIAS AGRICOLAS DE LA OEA. 1962. List of Publications. Turrialba, Costa Rica.
INTER-AMERICAN INSTITUTE OF AGRICULTURAL SCIENCES. 1960. Coffee—Bibliography of the Publications available in the Library of the Institute. Bibliographical List, No. 1, Revised. Turrialba, Costa Rica.
JOHNSTON, W. R., and FOOTE, H. E. 1951. Development of a New Process for Curing Coffee. Food Technol. *5*, No. 11, 464–467; *and in* Coffee and Tea Inds. 1952, *75*, No. 1, 12–13, 40, 42; *also in* Indian Coffee Board Monthly Bull., 1952, *16*. No. 4, 69–73; *and in* Coffee Board of Kenya, Monthly Bull., 1952, *17*, No. 201, 204–206.
MARTINEZ, A., and JAMES, C. N. 1960. Coffee—Bibliography of the Publications found in the Library, Inter-American Institute of the Agricultural Sciences. Turrialba, Costa Rica.
QUIMME, P. 1976. Coffee and Tea. Signet Books, N.Y., N.Y.
SHAPIRA, R. G. 1975. Coffee and Tea. St. Martin Press, N.Y., N.Y.
SIVETZ, M. 1976. History of soluble coffee. Tea and Coffee Trade J. Aug. 1976.
SIVETZ, M. 1977. Coffee Origin and Uses. Coffee Publications, Corvallis, Oregon.
SYLVAIN, P. G. 1957. Report on a Program on Coffee Research and Promotion in Nicaragua. Report No. 16. IAIAS.
WALLACE, T. (Editor). 1960. Trace Elements in Plant Physiology. Chronica Botanica Company, Waltham, Mass.
WELLMAN, F. L. 1961. Coffee—Botany, Cultivation, and Utilization. Interscience Division, John Wiley and Sons, New York, and Leonard Hill (Books) Ltd., London.

# 5

# Harvesting and Handling Green Coffee Beans

The technology to be discussed here and in Chapters 6 and 7 will begin with the harvesting and handling of the fruit and proceed to the "green" coffee stage, that is, the dried seed free from external layers (skin, pulp, mucilage, parchment, and silver skin).

## HARVESTING PROBLEMS

The ripe coffee fruit, called cherries or grapes, resemble cranberries quite closely in external appearance (size and color). The part that is used commercially is the seed; one, two, or three seeds may occur in a single fruit. Seed maturity occurs 6 to 9 months after blooming which is triggered by a rainfall. The fruit matures more slowly under shade. In all probability, the triggering of the flowering is a result of a combination of factors including (1) rain which relieves a moisture deficit in the plant, and (2) a reduction in temperature which usually accompanies the rain. Under experimental conditions hormones have been used to trigger the flowering.

Sometimes the entire flowering occurs at one time, but more often there are two, three, or four independent flowerings when successive rains occur. Triggering depends upon the stage of maturity of the flower buds. The subsequent fruit crop matures according to the flowerings that have occurred from three weeks (a rather rare condition) to about nine months earlier, depending on the timing of the flowering as well as on the local climate, and the physiological state of the tree subsequent to flowering. Perhaps the average span is about three months. Within a few days after the fruit turns from green to yellow, the fruit turns red and may be picked. Full maturity develops about eight days later. The fruit stays at its prime for about a week or more depending on weather conditions; then it gradually turns soft and darkens in color until it is almost black. It may be picked slightly into the overripe stage without serious loss of quality, but this is undesirable because the fruit is hard to handle and transport in this soft condition.

Sometimes ripe fruit and flowers appear simultaneously (Fig. 5.1). This varies from country to country, from place to place, and in any one place according to seasonal weather variations in a given year. For this reason, and because coffee harvesting has not yet been mechanized, it is difficult to have the right amount of man power available for hand picking at the time and place of ripening. If the period is long, as in Colombia, a relatively small labor force may harvest the coffee continually as it ripens. During a short off-season the workers may be employed advantageously to cultivate the plantation. Since a large area must be covered repeatedly for a small amount of coffee each time, the cost of labor is rather high. If the period is short, e.g., 2 to 3 months, as in Brazil and many parts of Central America, the lowest cost might be less than 15 percent of the total cost of production. However, a large labor force, which cannot always be employed economically in the long period between harvests, is, nevertheless, necessary. Labor costs may be extremely high if nearly all the coffee is to be picked as ripe fruit. If this is impracticable, part of the crop is strip-picked, i.e., all the coffee is taken from each tree after most of the fruit has matured. The mixture may be as high as 15 percent green fruit at the beginning of the harvest, and 15 percent mostly overripe and partially dried fruit on the tree or on the ground beneath the tree at the

FIG. 5.1. ARABICA SHRUB IN FLOWERING

end. In Brazil the usual system is to make one or two sweepings under the tree to collect the fruit from the earliest flowerings that have matured and fallen to the ground. These sweepings are made prior to the final stripping mentioned earlier. Figure 5.3 showed a method of selective picking.

Sometimes a compromise is made whereby 2 to 5 selective pickings of 90 to 95 percent ripe coffee cherries are made which may total 5 to 50 percent of the whole crop; then the remainder is stripped off in all stages of ripeness.

In any case, the ripe cherries may be separated from the rest by water flotation and screening to be described below. They are next pulped, fermented, washed, and dried. The green coffee is then safe from deterioration if it is stored and handled with reasonable care. On the other hand, cherries that are green, or overripe and dried to the point of 45 percent moisture content or less, cannot be pulped because the skin is hard; hence, they must be dried as whole fruit either in mechanical driers or on the sun-drying terrace to make *natural* coffee. Sometimes no separation is made, and the entire strip-picked heterogeneous mixture is made into *natural* coffee. It is obvious that economics, specifically the supply, quality, and cost of labor in connection with the span of the harvest are important in determining how coffee is to be harvested and dried. The ideal method of harvesting only ripe cherries and making washed coffee is naturally more expensive than merely drying the whole fruit; yet it is common practice in many coffee countries to produce 90 percent or more of the entire crop as washed coffee. In other countries, e.g. Brazil, as little as 3 to 5 percent is fermented and washed.

Figure 5.5 shows the cross section of a ripe coffee cherry.

## Coffee Bean Quality

Brazil, which produces about one-third of the world's coffee, has a wide variety of climate and local rainfall distribution. It also has a wide variety of curing practices ranging from the best to the worst. Ninety-five percent or more of Brazil's coffee is natural. Certain regions are noted for hard, soft, or Rioy characteristics.

In large areas of Brazil, mainly in the states of São Paulo and Parana, a long dry period usually coincides with the harvest season. This causes a very rapid passage from the ripe to the partially dry stage, thus rendering it impossible to harvest a large portion as ripe fruit. The dryness of the weather tends to minimize deterioration, hence the natural harvesting system is almost necessary under these conditions. In other areas of intermittent wet weather during the harvest season with prolonged periods of ripeness and threatened damage to partially dried fruit, the fermented and washed coffee treatment is more suitable.

The factors determining which method is to be used is governed by tradition and economics. Complications involve the amount of rainfall during the harvest season, types of microorganisms present, dirt, elapsed time, etc. Much of Brazil's

FIG. 5.2. CLOSE-UP VIEW

FIG. 5.3. ARABICA SHRUB WITH RIPE AND NEAR RIPE CHERRIES

coffee, as well as that of many other coffee-producing countries, might be greatly improved by applying the principles of (1) picking more of the fruit when it is ripe in order to make more washed coffee, (2) speeding up the processing, (3) using better methods of sanitation, and (4) avoiding rain damage by using mechanical driers.

An economic factor in determining coffee quality is the well-known law of supply and demand. If production outruns demand and prices fall, buyers of coffee become more quality-conscious. They are then in a position to insist on the highest grades only. In times of scarcity, the reverse is true and quality tends to decrease. In a buyers' market, there is also a tendency to produce more washed coffee rather than natural coffee.

FIG. 5.4. UGANDA ROBUSTA SHRUB WITH FRUIT.

## FERMENTED & WASHED VS. NATURAL CHERRY DRYING

The two main methods used are the fermentation of mucilage followed by washing (wet) and the natural (dry) processes. In the washed or wet process, the ripe fruit is squeezed in a pulping machine which removes most of the soft outer pulp or fibrous fruit flesh, leaving a slippery exposed layer of mucilage. Since the layer of mucilage cannot be readily dispersed in water, it is removed by one of several methods to be described below leaving the clean parchment layer. The product is called washed coffee because the mucilage is finally removed by washing with water. Next, the coffee bean is dried to about 12 percent moisture. The thin yellowish parchment layer, separated from the seed by shrinking of the latter, is removed by a hulling machine, and commercial green coffee results. Figure 5.6 shows the different appearances of natural and washed coffee on the drying terrace.

In the natural or dry method, the fruit is allowed to remain on the tree past the fully ripe stage and is partially dried before harvesting. After harvesting, it is dried to about 12 percent moisture in the whole fruit form after which all the outer layers are removed at one time by hulling, and the commercial bean is obtained directly. The latter method is simpler and cheaper, but the resultant natural coffee product is usually of poorer quality than washed coffee. For combination processes suitable for mixtures of cherries in various stages of ripeness, see the section on classification.

In order to obtain the highest quality of washed coffee beans, it is only necessary to use the following simple principles. (1) The fruit must be picked while it is in prime ripe condition; green and overripe cherries must be excluded. (2) The fruit must be processed as quickly as possible. (3) Any contamination by foreign bodies, especially microorganisms, must be avoided. To apply these principles may sometimes be difficult, but, in general, it is well worth the effort to pulp, ferment and wash as high a percentage of the crop as is economically possible.

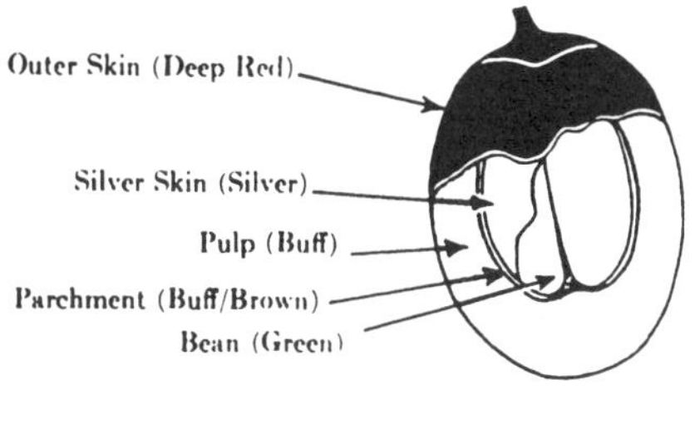


FIG. 5.5. CROSS SECTION OF COFFEE CHERRY

A. Natural Coffee

B. Washed Coffee

Fig. 5.6. Sun Drying of Coffee

## Speed of Processing

For a well-organized operation, the sequence is: (1) picking from early morning to late afternoon; (2) prompt transportation to the central processing plant; (3) pulping; (4) mucilage removal in not more than eight hours (overnight); and (5) then washing and drying the next day in 6 to 20 hr in mechanical driers or in 5 to 7 days by sun drying. Washed coffee may be stored in parchment or it may be hulled when ready for classification for export.

Natural coffee is merely dried—in about three days in mechanical driers or in 3 to 4 weeks in the sun. Normally, cherries are first dried on the ground or on patios for days to weeks, then final drying is done with machines. Speedy processing is also desirable for natural coffee.

Natural coffee is merely dried—in about three days in mechanical driers or in 3 to 4 weeks in the sun. Speedy processing is also desirable for natural coffee.

For washed coffee, if not more than 36 hr elapse between picking and the beginning of drying, no noticeable deterioration in cup quality takes place. In practice, delays may occur due to poor organization of the operation or because of rainy weather if sun drying is used. Assuming the absence of infection, the rate of deterioration is a function of time, temperature, and natural biological changes such as metabolism, respiration, and enzyme activity in and around the living seed (coffee bean) as with practically all fresh fruits. Blanching to inactivate the enzymes has been tried but it was found that minimum effective blanching conditions damaged coffee quality. Hence, speedy processing is highly desirable in order to obtain maximum quality. Speed is accomplished by organization and planning, training and discipline in the labor personnel, and by providing reliable and well maintained equipment of ample capacity to meet peak loads.

Drying should be done as far as possible in mechanical driers in order to avoid costly and quality-damaging delays by rain during the harvesting. Delays may be an important source of quality loss because both internal biological deterioration and growth of infecting microorganisms are encouraged if the storage of wet coffee, either washed or natural, is prolonged.

## Sanitation

The third factor in securing high quality is sanitation. As long as coffee has a high moisture content (above 18 to 20 percent), it provides a fertile medium for the growth of microorganisms such as molds, fungi, and bacteria. Their effects on the coffee are very diverse, and specific for each organism; some have much worse effects than others, but practically all are harmful to quality. The foreign flavors, once implanted in the coffee, cannot be removed. This is the source of such low quality characteristics as "Rioy" (from the city of Rio de Janeiro)—hard, musty, sour, fermented, earthy, rain-damaged, etc.

The obvious countermeasures are to keep all processing equipment clean by thorough washing and, at critical points, to use mild disinfectants (steam, detergents, lime) and since complete sterilization is neither practical nor necessary, to minimize the microorganism population by processing the coffee quickly. By far the most important factor in safeguarding the high quality of the coffee, once it has ripened on the tree, is to process quickly and cleanly.

## Comparison of Washed and Natural Coffee

The next point to be considered, since perhaps more than half of the world's production is natural coffee, is the difference in quality between washed and natural coffee. In general, washed coffee, carefully prepared and handled, is clean in flavor and free from undesirable elements. Yet it is sometimes lacking in body or full flavored richness compared to well-prepared natural coffee.

Some experiments carried out in order to learn how to prepare pulped coffee with a minimum of water in regions where water is in short supply showed that pulped coffee treated with mucilage-digesting enzymes (a commercial product) and then dried *without washing* had an improved body when cup tested. It is not practical to dry pulped coffee in mucilage directly without enzyme treatment because it takes too long and quality is damaged. Natural coffee, since it is always dried in contact with its mucilage, has a better body than it otherwise would because of this fact. Natural coffee must be dried much more slowly than pulped coffee for the following reasons: (1) there is much more water to remove (about four times as much as from washed coffee per pound of dry green coffee); (2) natural coffee is harvested with a wide range of moisture content, usually from about 30 to 65 percent, and hence it must be dried gently to avoid overdrying of the driest portions; and (3) the extra layers of the whole fruit offer more resistance to the loss of water. Since the natural coffee remains moist for a much longer period than washed coffee, the microorganisms which are always present have a much better chance to flourish. If the skin is intact, infection penetration becomes more difficult. Because much natural coffee falls from the tree before harvesting, it may become infected by soil organisms, especially if rain occurs before it is harvested and produces musty, groundy, or rain-damaged flavor.

If natural coffee is dried in mechanical driers, low temperatures favor the growth of microorganisms while higher temperatures produce cooked and sour flavors in the product. (In the trade, the term *acid* has a good connotation, and *sour*, a bad one.) Both of these causes of damage occur mainly while the coffee is very wet, i.e., above 30 percent moisture. The conditions of drying most favorable to good quality natural coffee during the first or wettest stages are sun drying (1) on the tree, (2) on the *dry* ground under the tree, or (3) on a clean brick or concrete terrace in rainless weather, preferably in bright sunshine with protection against night dampness in covered heaps. If rain occurs during the drying, or if the natural coffee is stored in large masses in a moist condition for more than overnight periods, intense microbiological activity may take place even to the extent of producing a considerable temperature rise and serious quality damage. The delay promotes natural changes in the coffee independently of the problems of infection.

Under ideal conditions of dry weather, absence of green cherries in the picking, freedom from infection, sun drying for the first stages, and machine drying for the last stages, natural coffee may be of excellent quality, clean tasting and full bodied and, while different, fully as desirable as washed coffee. But the ideal conditions for drying natural coffee are nearly always difficult or impossible to realize, and most natural coffee is, to varying degrees, inferior to its washed coffee counterpart. An important element in the production of high grade natural coffee is the exclusion of green fruit which, when very immature, imparts an intolerable flavor (described as foul). Well-disciplined pickers keep green berries to a minimum. Sometimes they are removed by hand before processing. Flotation (some greens float but the separation is not sharp) and screening (the greens are small) will remove more of them before pulping. In addition, some greens are removed by screening immediately after pulping, and of the few that are left after drying, some may be removed by hand- or machine-picking. The green coffee beans from green fruit have a whitish color rather than the normal gray-green.

## Relative Processing Costs

The natural coffee process is inherently cheaper than the washing process because it is simpler, and thus, less labor and machinery are required. It has one notable cost disadvantage, however; much more water has to be evaporated, and this takes heat energy (partially supplied without cost by the sun), time, drier capacity, and extra labor on the drying terrace.

Many of the same factors damaging to quality apply also to washed coffee. It, too, may be damaged by unduly long periods in a wet condition, i.e., by natural deteriorative changes and by infection from microorganisms. It may lack a full-bodied flavor because it is dried in the absence of its mucilage but, if processed properly, it will always have a clean characteristic coffee flavor free from contaminating elements.

## Drying Washed Coffee

The drying of washed coffee is simpler because (1) the coffee has a very uniform moisture content of about 53 percent at the start; (2) much less moisture has to be evaporated; (3) higher drying temperatures and rates may be used; and finally, (4) machine driers may be used all the way from start to finish which means complete independence of the effects of possible (in many cases probable) unfavorable weather. For these reasons, damaging delays are minimized.

To summarize: the *best* natural coffee is of a different character, but, by and large, about equal in quality to the *best* washed coffee grown under the same conditions. In general, it is harder to obtain the best quality of natural coffee. Too many conditions, difficult to control, tend to result in mediocre grades, and there is always a serious risk of getting a very poor quality. Usually, natural coffee costs less to produce.

On the other hand, the production of washed coffee from ripe cherries may easily be kept under tight quality control in a properly organized operation. The most difficult step is obtaining a high quality of ripe fruit at the start. Poor processing methods will also result in poor washed coffee.

Depending on local conditions, any one plantation may produce anywhere from 100 percent natural coffee, in which case no equipment for pulping, fermentation and washing is needed, to 90 to 95 percent washed coffee. Even under the best conditions for washed coffee, small amounts of green and overripe fruit are unavoidably harvested. When separated, these are dried without pulping to form natural coffee which is marketed separately from the washed coffee.

Bulk densities for washed coffee are shown in Table 5.1.

TABLE 5.1. BULK DENSITIES OF WASHED COFFEE IN VARIOUS STAGES OF PROCESSING

| | Kg per cu m | Lb per cu ft |
|---|---|---|
| Ripe cherries | 616 | 38.4 |
| Pulped beans | 846 | 52.8 |
| Washed beans | 665 | 41.5 |
| Dry parchment | 352 | 22.0 |
| Green coffee | 650 | 40.6 |

## HARVESTING

A series of studies by the Economic Commission for Latin America (ECLA) and the Food and Agriculture Organization (FAO) of the United Nations on "Coffee in Latin America" covering Colombia, Salvador, and the state of São Paulo, Brazil, presents some interesting facts on relative labor costs in terms of man-hours necessary to harvest coffee.

In order to afford a perspective on the industry in the above three countries, a few basic statistics are given in Table 5.2.

TABLE 5.2. RELATIVE STATISTICS FOR SALVADOR, COLOMBIA AND SÃO PAULO, BRAZIL

| | Salvador[1] | Colombia[2] | São Paulo |
|---|---|---|---|
| Crop of → | | | |
| Production, 60 kg bags | 1,417,000 | 6,137,000 | 14,880,000 |
| Total area in coffee, hectares | 137,000 | 777,000 | 1,699,400 |
| Number of plantations | 20,000 | 234,700 | 104,800 |
| Average area plantation, ha | 6.9 | 3.3 | 16.2 |
| Average yield, kg per ha | 659 | 523 | 446 |

One hectare = 2.471 acres.

Table 5.3 shows relative labor costs in man-hours for Salvador, Colombia, and São Paulo, Brazil.

It has been reported that harvesting costs in Brazil (strip-picking) are 13.5 percent of total production cost, in Guatemala (selective) 41.4 percent, in Hawaii (selective) 54.5 to 74 percent, and in Costa Rica (selective) 60 percent. It is probable that the basis of comparison for these latter figures is not exactly the same as for those in Table 5.3.

Salvador's product is about 85 percent washed and 15 percent natural coffee; Colombia's is nearly 100 percent washed coffee; while Brazil's is probably 95 to 97 percent natural coffee. Coffee in Salvador and Colombia is grown under shade trees; Brazil uses no shade but grows 3 to 5 individual trees together in one spot

TABLE 5.3. LABOR COSTS IN PRODUCING AND HARVESTING COFFEE
*Man-hours per 100 kilograms of green coffee*

| | Salvador | | Colombia | | São Paulo | |
|---|---|---|---|---|---|---|
| % | m-h | % | m-h | % | m-h | % |
| Picking | 78.8 | 36.5 | 80.0 | 46.8 | 70.0 | 51.5 |
| Processing | 11.4 | 5.3 | 19.8 | 11.6 | 6.7 | 4.9 |
| Cultivation | 125.9 | 58.2 | 71.0 | 41.6 | 59.3 | 43.6 |
| Total | 216.1 | 100.0 | 170.8 | 100.0 | 136.0 | 100.0 |

which results in a certain degree of self-shading. Each group is counted as one tree in rating the size of a plantation in Brazil.

In most coffee-producing countries, the question as to whether natural or washed coffee is to be produced is settled; the pattern is fixed on one method or the other. In some countries, notably Brazil, where the choice is debatable, and where either method or a combination of varying proportions of washed and natural coffee processing may be used, the factors influencing the choice of method in a given situation are complex. The dominant considerations are economic. Will the increased market value of the higher quality washed coffee justify its extra cost? The level of coffee prices, the climate of the plantation, the cost, supply, and quality of the labor, the availability and cost of power, water, and processing machinery, the character and temperament of the grower, and many other minor circumstances, all influence the choice.

## Harvesting for Washed Coffee

To obtain the full advantage of the wet processing of coffee, it is necessary to obtain as high a percentage of ripe fruit as possible. To be pulpable, it must be soft, which means slightly underripe (yellow), ripe (red), and slightly overripe and soft (black). This requires selective picking in order to avoid hard green and hard partially dried fruit. Since the coffee on any one tree and in any one area ripens successively, this means starting the picking when the first ripe fruit appears and making repeated pickings. If this is done well, little or no fruit is allowed to reach the hard, partially dried stage, i.e., beyond pulping. But, at the end, diminishing returns make selective picking too costly, and it is necessary to accept some green and some overripe cherries in the final picking.

Although perhaps only 3 to 5 percent of Brazil's coffee is pulped and washed, the organization of a typical washed coffee operation is interesting. Because the harvest season is usually short, the common arrangement is a compromise in which both washed and natural coffees are produced. A typical modern, progressive plantation is organized with a sufficient labor force to pick ripe fruit selectively for washed coffee from 40 to 50 percent of the crop. To extend the harvest season and to reduce the number of workers required, the remainder of the crop is harvested after the selective picking is finished. Thus, more time is afforded for the natural coffee to dry partially on the trees.

During the selective picking, in order to keep the ripe fruit clean, much of it is picked into baskets supported from the belt of the picker, and cloths are spread on the ground under the trees to catch that which falls. The baskets may be funnel-shaped. Their range is limited from a waist-high level to the height which the picker can reach. Low cherries are picked onto the cloths and high cherries are reached from ladders or by bending the high branches down.

It requires patience and persistence on the part of management and the careful working out of inspection and incentives, rewards and penalties.

In order to be able to pick a maximum of ripe fruit, a skillful manager observes his local conditions minutely so that he knows which plots on his plantation ripen first as well as the complete chronological succession of ripening in all areas. Sometimes, rather large retardation or accelerations occur because of the orientation of slopes and sun exposure, air drainage, prevailing winds, variations of elevations, and other factors. There are also possibilities of extending the harvest period by deliberately accelerating or holding back the time of ripening by the timing of irrigation or by pruning methods.

It is also possible that chemical or hormone spraying may be able to accomplish similar results in the future.

## Natural Coffee (Strip-Picking)

As pointed out before, strip-picking for natural coffee starts late when most of the coffee has passed through the ripe and soft overripe stages and has dried from the 65 to 70 percent moisture content of the ripe cherry down to 30 to 40 percent moisture. At this point, the cherry that is wanted is relatively hard and dry on the surface and easy to handle. Varying portions, depending on the weather, have, by this time, fallen from the tree, and usually the ground beneath the tree has been previously prepared by clearing away weeds and debris and by leveling so that the fruit will not be lost. While most of the coffee is partially dried, a substantial amount is green, ripe, and soft overripe. The proportions vary as the harvest season progresses. Usually only one, infrequently two, and rarely three passes are made. The pickers try to leave the more immature fruit for the later passes, but this is difficult, and separation during the picking is very imperfect. The rather heterogeneous product is dried in the form of the whole fruit. The cost economies of this method are obvious.

If facilities are available (ample water supply, canals, tanks, and pulping machinery), a crude separation is made by flotation in water. This serves to wash dirt from the coffee and to separate the part which floats (most of the partially dried coffee) from that which sinks (most of the soft fruit which can be pulped). If the soft pulpable fruit is not removed, it tends to prolong the drying and to harbor microorganisms which may raise the incidence of infection throughout the whole mass.

Thus, coffee processing may be classified as (1) 100 percent natural, (2) natural with about 5 percent washed, (3) roughly equal quantities of natural and washed, and, (4) 90 to 95 percent washed.

In Colombia, where conditions are particularly favorable for washed coffee, practically 100 percent washed coffee is produced because the harvest season extends over about nine months of the year, and the relatively small labor force which is adequate for harvesting may be usefully employed in the off-season for the cultivation of the trees. The longer Colombian harvest season is due to the fact that the coffee areas are near the equator where there is little seasonal temperature change and an equable distribution of rainfall. These factors cause the trees to bloom over a long period and the fruit to mature correspondingly about six or more months after blooming. These conditions, combined with relatively cheap labor, make it easy and economical to pulp and wash all the coffee using mostly small scale, hand methods without the need for power machinery.

A very important influence on the efficiency of a coffee picker is the density of the coffee cherries, i.e., the yield per tree. The picker can fill his container much more easily and quickly if the fruit is plentiful. This is why it is so desirable to bring yields up to the maximum potential by sound horticultural methods.

In strip-picking, as practiced in Brazil, the picker runs his closed hand along the branch trying to avoid stripping off the leaves. Some leaves are unavoidably knocked off. Sometimes the tree is beaten with wooden sticks to knock down any fruit that is loose enough to be dislodged. The cherries fall to the ground which has usually been prepared beforehand or to a cloth spread under the tree as in selective picking. If the coffee is admixed with dirt, leaves, and twigs, as it usually is, it is winnowed in a light breeze as the picker throws it high in the air from a flat basket or tray. The impurities are blown out and away as the coffee falls back into the tray. A high degree of skill is often developed by the picker in this operation.

It is important to gather up all the coffee because fruit that is allowed to remain on the ground until the next season forms a breeding place for diseases and pests, particularly the coffee borer.

## TRANSPORT OF CHERRIES

The selectively- or strip-picked coffee is finally placed in bags and transported to the central processing plant by horse or mule-back, horse- or ox-drawn carts or wagons, motor trucks or railroad according to the topography of the plantation and the condition of its roads. The main consideration is to transport all ripe or soft cherries to the plant the day they are picked. If they are partially dried to 45 percent moisture or less, speed is not so imperative, although even this coffee should not be exposed to rain.

## STORAGE OF HARVESTED CHERRIES

### Washed Coffee

Ripe or soft cherries may be dumped into a funnel-shaped dry tank or into an ordinary tank partly filled with water. Where water is used, the cherries which float are drawn off the top and those which sink are removed from the bottom. The mass is stirred so that the lighter fruit may float free from the heavier. The floats are considered inferior and consist of any partially dried hard cherries and some green cherries (not all green cherries float), and usually substantial quantities (5 percent or more) of red ripe cherries containing only one seed or underdeveloped small seeds. If, for any reason, the freshly picked ripe fruit must be stored for a few hours before processing it is better that it be in water to prevent any heating due to microorganism activity. If a dry tank is used, the cherries are drawn out of the bottom into a stream of water flowing in a canal, and then are conveyed to the classifying system which delivers them to the pulpers. Storing of ripe cherries under water is shown in Fig. 5.7.

### Natural Coffee

On arrival at the plant, natural coffee may or may not be subjected to flotation to remove soft cherries (which sink). This step has the advantage of eliminating a high-moisture component allowing more even drying of the natural coffee, and the disadvantage of wetting the partially dried coffee and thus retarding its drying. The natural coffee, however, proceeds at once to the drying operation on the terrace or in the mechanical drier as the case may be (see Fig. 5.8).

## CLASSIFICATION OF CHERRIES

Broadly speaking, this step applies to any mixture of harvested coffee. Although there are many variations in the layouts of classification systems, they all make use of the very simple principles of flotation, and screen separation by size. Systems of this type require rather large amounts of water. It may be that, in the future, means will be developed to classify coffee fruit into separate grades by color electronically as is now done in the case of green and roasted coffee beans. If

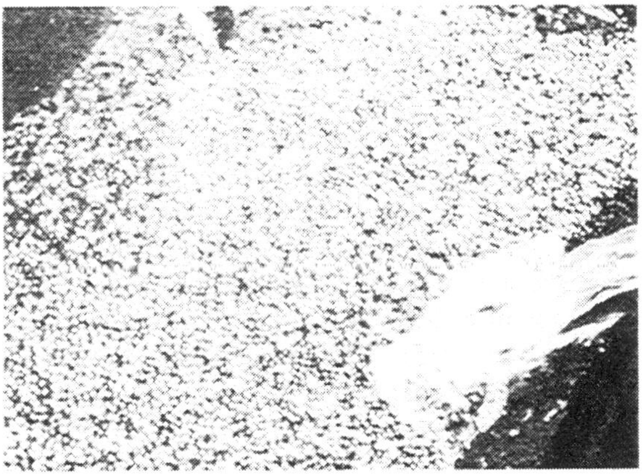

FIG. 5.7. STORING RIPE CHERRIES UNDER WATER BEFORE PULPING

*Courtesy of Pan American Coffee Bureau*

FIG. 5.8. STORING FRESHLY HARVESTED NATURAL COFFEE IN DRY BIN BEFORE DRYING

this were practical, a considerable improvement in quality might be possible because the methods now in use are not clear-cut. The system shown in Fig. 5.11 is typical for washed coffee; it is well-designed and works fairly satisfactorily for any mixture of coffee fruit in various stages of ripeness. The general aims of the system are (1) to get rid of sticks, stones, dirt, and any other foreign matter, (2) to pulp all of the pulpable coffee and separate this coffee into two sizes, large and small, before pulping, as well as to remove the pulps, (3) to separate the hard, partially dried cherries for the production of one grade of natural coffee, and (4) to separate the green cherries to produce a second grade of natural coffee.

It will be noted that the system as a whole is crude and, to a degree, unsatisfactory because coffee as harvested is such a variable material. In addition to its variations in maturity, coffee varies in size which complicates the screening problem, and in buoyancy which makes it difficult to separate sharply the different types by flotation. Even a careful picking will always contain a substantial quantity of imperfect fruit. It may have been damaged by insects, microorganisms, malnutrition, crushing or other obscure causes; parchment shells of normal size and appearance may contain only a vestige of a normal coffee seed. These factors add to the difficulty of removing bad fruit, and of classifying and upgrading the sound fruit. However, such a system does greatly improve the best grades by eliminating *most* of the undesirable lower grades.

The picked fruit is weighed or measured (Fig. 5.9) in order to determine the picker's compensation. Berries may be inspected at this point to see whether or not they meet the standards set by management. The fruit is placed in the receiving tank which, in the system shown in Fig. 5.11, is dry and funnel-shaped. The

FIG. 5.9. MEASURING THE DAY'S PICKING

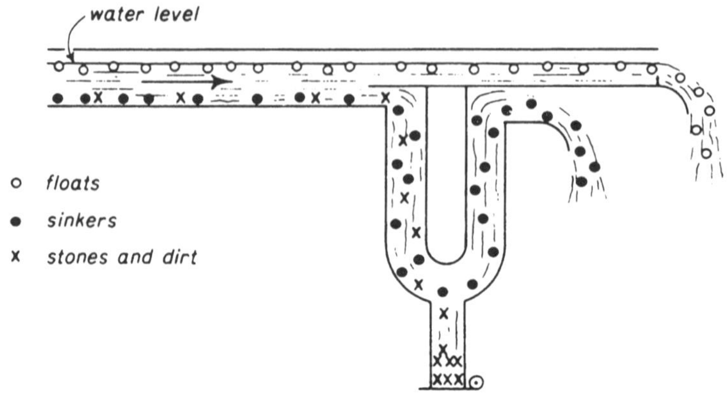


FIG. 5.10. DIAGRAM OF STONE AND DIRT REMOVER

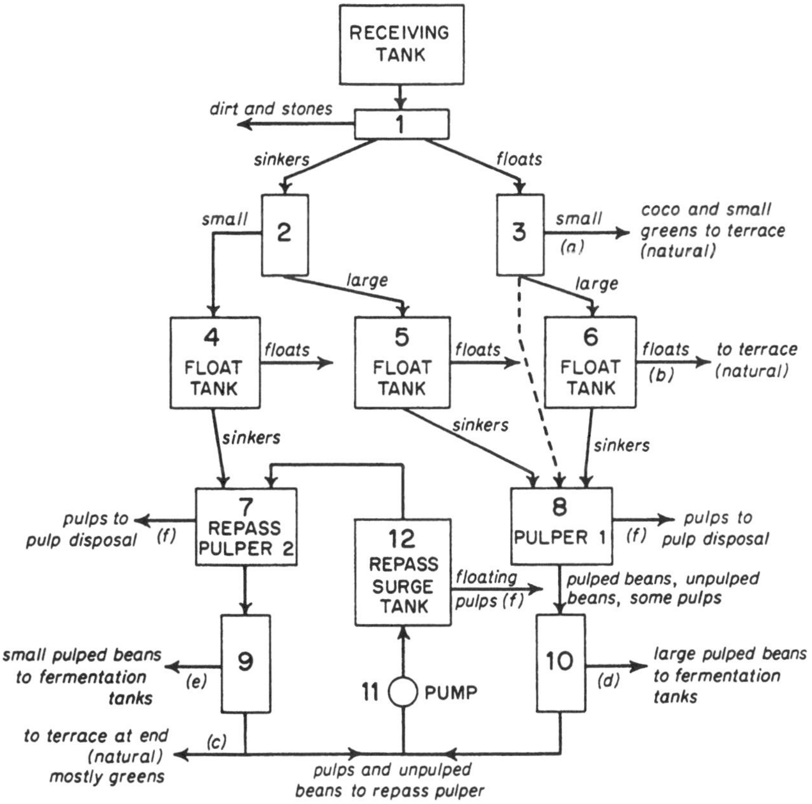


FIG. 5.11. CLASSIFICATION FLOW DIAGRAM

cherries are drawn off from the bottom and fall into a canal with running water supplied from a large reservoir or diverted from a stream which conveys them to a combination float separator and stone and dirt remover (Fig. 5.10). The berries that sink (sinkers) with the stones and dirt pass down the U-shaped member; the stones and dirt that collect at the bottom are much heavier than the coffee. They are removed at intervals by opening the bottom of the trap. Since the cherry is only a little heavier than water, the upward velocity of the water carries the cherry up the U-shaped member and over the top. The floats by-pass the U-shaped trap. The floats and sinkers are led by separate canals into the two rotating screens, 2 and 3 (Fig. 5.11).

The separator, unit 1, accomplished only a crude separation, and some floats go with the sinkers and vice versa. The sinkers are mostly ripe cherries and the floats are mainly hard, partially dried fruit (25 to 50 percent moisture), but the floats also include some ripe cherries containing only one bean (peaberries) instead of two or three. The green cherries are divided between floats and sinkers. The floats and sinkers each pass through the rotating screens operated in a water stream for a size separation. In the float separator 3, the small sized coffee, which passes through the screen, consists mostly of partially dried and small green cherries. This portion is sent directly to the sun-drying terrace and is dried as natural coffee, product a. The larger sized coffee, which cannot pass through the screen, contains most of the one-bean ripe cherries and the larger greens. This portion is passed to a concrete float tank 6 with one exception which will be explained below.

In the similar separator 2, the heavy coffee sinkers, mostly ripe cherries with some greens, are separated into two sizes so that when they are pulped, the more nearly uniform large size beans may be sent to a pulper adjusted to that size, and the more nearly uniform small size beans may be sent to a pulper adjusted to that size. In this way, the pulpers do a cleaner job of separating the coffee beans from the pulps. The two fractions are discharged from the rotating screens into float tanks 4 and 5 according to their size.

Tanks 4 and 5 will contain mostly sinkers and a few floats, while tank 6 will contain mostly floats and a few sinkers. In the three tanks 4, 5, and 6, a clean flotation separation is made by stirring and standing, thus correcting the errors of the crude separation at 1.

From tank 6, the floats are drawn off from the top surface at an overflow point and are carried directly to the drying terrace and dried as natural coffee, product *b*. Products *a* and *b* have very nearly the same composition and are united on the terrace. The main purpose of tank 6 is to reclaim any ripe sinkers which may have slipped through unit 1 with the floats.

If the quality of the picking is high so that there are very few overripe and partially dried cherries in the floats from unit 1, i.e., if the large sized coffee which will not pass through screen 3 consists mostly of one-bean cherries, then float tank 6 is by-passed, and this coffee goes directly to pulper 8 which is adjusted for large sized cherries. Usually one-bean ripe coffee is of as high quality as regular two- or three-bean cherries, and the yield of washed coffee is increased by this amount. If float tank 6 is used, these one-bean floats are lost for pulping and they become natural coffee—a defect in this form of classification system. The reason this occurs is that the system is designed primarily for all mixtures of coffee in all stages of ripeness ranging from nearly 100 percent ripe cherries to 5 or 10 percent ripe cherries, and is designed to handle any type of coffee mixture that may be harvested. If tank 6 cannot be by-passed because of considerable amounts of partially dried cherries, then the percentage of ripe floats is usually quite low, and much potential washed coffee is not made into natural coffee.

The sinkers in tanks 5 and 6 are of the larger size. They are drawn out from the bottoms of the tanks and pulped in pulper 8 which is properly adjusted for that size. The sinkers of tank 4 are smaller and are pulped in pulper 7 which is adjusted to a closer clearance to fit the smaller size.

The pulpers, to be described in more detail below, pulp the sinkers of tanks 4, 5, and 6 and separate the greater part of the pulps as product *e* from the pulped coffee. The separation is imperfect, and some pulps are found along with the pulped beans; also, some cherries escape pulping. The three products (pulped beans, pulps, and unpulped cherries) are next passed through rotating screens, 9 and 10, also operating in water streams. The pulped beans, a few small unpulped red and green cherries, and a few pulps, go through the screens. Most of the pulps and large size unpulped cherries do not go through the screens and are led to the repass storage tank 12 for temporary storage until they can be passed through pulper 9, having the small clearance. Most of the conveying of coffee is done by gravity in water streams in canals. The repass coffee must be raised by means of a pump to tank 12 where it is fed to pulper 7 by gravity.

Pulper 8 is operated until all of the day's picked coffee from float tanks 5 and 6 has passed through it. Repass coffee is recirculated continuously through pulper 7, screen 9, pump 11, and repass tank 12 as long as appreciable amounts of pulped beans (product *e*) appear.

When the pulped beans cease to flow, the repass pump is turned off, and the remaining coffee, in the system which contains a high percentage of green cherries, is passed from the repass tank through the pulper and screen and then diverted to the drying terrace to form product *c* to be dried as natural coffee. This grade is inferior to products *a* and *b*, and is dried separately.

If high grade ripe cherries are picked specifically for washed coffee, product *d* may comprise 90 percent or better of all the products and is, of course, the highest grade of coffee produced. The next highest grade is product *e* which is also mainly washed coffee but contains small amounts of small whole cherries. The best grades of natural coffee are *a* and *b* which are blended and dried together. The lowest grade is product *c*. Product *f* is the pulp discharged from the two pulpers which may contain small amounts of pulped beans. These represent a loss of coffee which is kept to a minimum by careful maintenance and precise adjustment of the pulpers. Even with the best possible adjustment, the pulped beans may contain 10 to 15 percent of pulps which constitutes a maximum of about 25 percent of the total weight of pulp in the original cherries. This amount causes little difficulty in the pulped coffee beyond adding slightly to the drying load. It disappears during the subsequent hulling.

The coffee as picked is thus separated into four useful products: large pulped beans, product *d;* small pulped beans, product *e;* repass residue, mostly greens, product *c;* and whole fruit floats dried as natural coffee, products *a* and *b*.

The pulp from the pulpers is disposed of in various ways. It may be dumped into a stream and wasted where it may also cause water pollution problems. It may be returned to the soil as mulch or compost where at least part of its valuable fertilizing elements is reclaimed. It may be sun-dried on a terrace and used as fuel. The dry pulp containing approximately 5 percent moisture has a heating value of about 6,000 BTU per lb (3,300 kg-cal per kg). Together with the dry parchment from washed coffee and the dry hulls from natural coffee, this fuel might be used to generate steam to furnish heat in the mechanical driers for the coffee. Roughly half of the total fuel requirements would be supplied in this way.

The fresh pulp contains about 9 percent of sugar, and proposals have been made to extract it and make alcohol and vinegar by fermentation. This has met with little or no commercial success. The dried material has been mixed with molasses and used as a cattle feed supplement. Cattle will eat it but cannot tolerate high percentages of it in the diet. Apparently, the best use for it is as a fertilizer and soil conditioner.

### Simple Classification

If the harvested fruit has not been previously classified according to the elaborate system described above, the entering cherries are passed first over a vibrating screen with smaller openings than the cherries to remove small impurities such as dirt and gravel, and then over another vibrating screen with openings large enough to allow the cherries to pass through leaving behind the larger stones, as well as twigs, leaves, and other debris. The cherries fall into a receiving tank of water. The sinkers are drained from the bottom of the tank or siphoned through a pipe from the bottom up and over the top edge and delivered to the pulpers. The floats are processed separately.

## PULPING

The pulping operation has two steps. First, the fruit is squeezed between the roughened surface of either a rotating cylinder in one type of pulper or a disk in another type and a stationary member called a breast with a smooth, channeled, or slotted surface sometimes lined with rubber (see Fig. 5.12, 5.13, 5.14 and 5.15). The distance between the moving member and the breast is carefully adjusted so that the space narrows as the fruit is carried through. The breast is often held in position by springs that permit adjustment of tension. This passage produces the squeezing action which detaches the skin and flesh, i.e., the pulp, from the seed. The final minimum clearance is made just large enough to allow the seed to pass through without being crushed or bruised.

The drum which squeezes the fruit against the breast is covered with small projecting mounds each having a cutting lip to help drag the pulps through the slot to be described below.

In the second step of the operation, the seeds are separated from the skins. This is accomplished by means of a plate with a carefully ground straight, sharp edge. This is fixed at right angles to the cylinder or disk with a precisely regulated

FIG. 5.12. PULPING MACHINE WITH OSCILLATING SCREEN

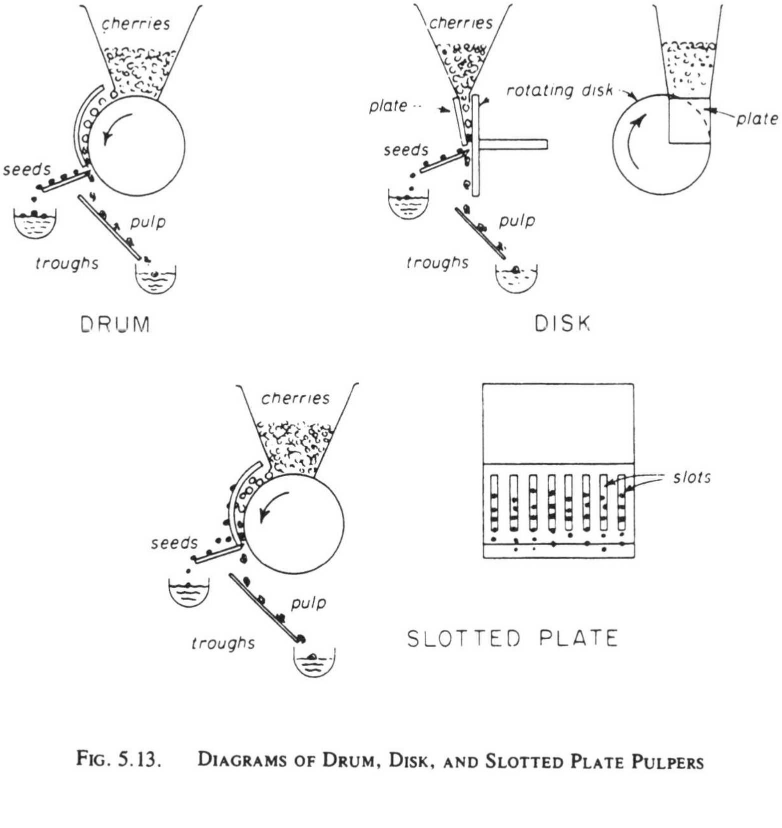


FIG. 5.13. DIAGRAMS OF DRUM, DISK, AND SLOTTED PLATE PULPERS

FIG. 5.14. OPERATING PULPER. CHERRY FEED AND PULPED COFFEE DISCHARGE

FIG. 5.15. BATTERY OF PULPERS SHOWING PULP DISCHARGE SIDE

clearance of about one-sixteenth of an inch. The rough surface of the moving drum or disk forces most of the flexible fibrous pulp through the crack between plate and drum which is too small to allow the seed to pass through. The seed is deflected onto and over the plate and falls into a conveyor trough which carries the coffee in a stream of water to the next operation. The pulping is carried out with the help of a small stream of water which lubricates the squeezing and separating processes and helps to carry the coffee to the next screening step (see Fig. 5.14).

Pulping is inexpensive. The machines are low in cost and have high capacity and low power requirements. A medium sized machine requiring about ½ hp to drive it is rated at 3,000 lb per hr of skinned coffee beans.

A third type of pulper removes the pulp by cast dimples (one-quarter spheres) on a drum turning against a breast in the form of a slotted plate (Fig. 5.13). When the pulp has been torn or rubbed off, the bean is small enough to pass through the slots while the skin is carried downward through the clearance between the plate and drum that is too small for the coffee beans. Figure 5.15 shows the pulp discharge side of a battery of drum-type pulpers.

In all types of pulpers, there is a chance that some beans will be injured and their parchment coating cut and nicked. This is kept to a minimum by carefully adjusting clearances to suit the size of the bean being handled and by regulating the speed of rotation of the drum or disk for best results. To some extent, the fruit is previously classified by size as described above so that a nearly uniform berry size is fed to any one pulper. If the parchment is damaged, the exposed seed may later show discoloration, and microorganisms may enter and impair the flavor.

It is important to use clean water in all stages of the processing to prevent infection. Use of dirty river or swamp water as well as too much reuse of water may spread organisms which will contribute to coffee bean flavor contamination. Imperfect cherries which are infested with insects, or crushed, moldy, fermented or otherwise damaged should be eliminated as far as possible in the picking and handling of the harvested fruit.

## REMOVAL OF MUCILAGE

### Composition of Mucilage

The freshly pulped coffee seeds are covered with a slippery mucilaginous layer approximately $^1/_{16}$ of an inch (1.50 mm) thick but somewhat variable, which, when the coffee is freshly pulped, is translucent and colorless; but, on exposure to air, turns brown probably through enzymic oxidation resembling that in freshly cut apples or peaches. Figure 5.16 shows pulped coffee before and after mucilage removal. It is said to consist chemically of protopectin, pectin, pectin esters, and small amounts of sugars along with the naturally occurring enzymes pectase, pectinase, pectinesterase, and protopectinase. The mucilage is insoluble in water, has no definite cellular structure but appears to be an amorphous gel. It clings to the coffee too tenaciously to be dispersed and removed by simple washing.

The presence or absence of coffee mucilage has an influence on pulping. Mucilage does not form in the cherry until the fruit is nearly ripe; hence, a green

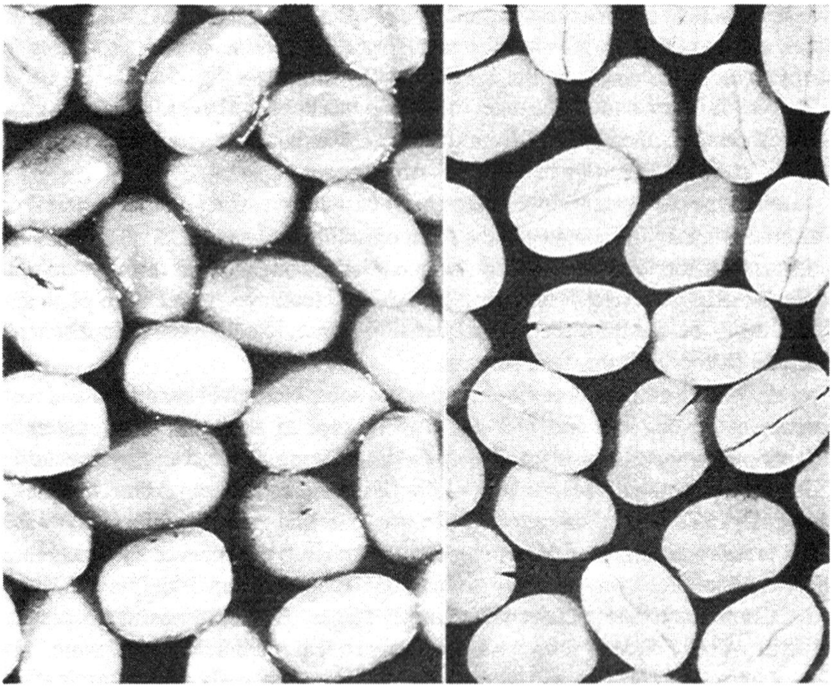

FIG. 5.16. PULPED COFFEE BEFORE AND AFTER MUCILAGE REMOVAL

cherry lacks the cushion and lubrication necessary in pulping, and the seed may be crushed or cut which, in all probability, will eventually impair its flavor. If the coffee is very overripe but still soft, natural fermentation of the mucilage has taken place within the cherry, the mucilage has disappeared as a slippery gel, and again, the seed may be damaged in pulping through lack of lubrication.

In making natural coffee, the whole fruit is dried. As the drying proceeds, the mucilage is digested and liquified and constitutes food material for the seed. It allows metabolism and respiration to continue. This modifies the flavor and may damage or improve it, according to the presence or absence of contaminating organisms. This depends on how carefully the natural coffee is handled during the drying.

## Methods of Mucilage Removal

There are five general methods of removing the mucilage:

Natural fermentation
Fermentation with added enzymes
Chemical methods
Warm water
Attrition

**Natural Fermentation**.—This is the time-honored, simple process which has been in use since coffee was first pulped, probably at least 100 years, and it is still used for 90 to 95 percent of present washed coffee production.

The pulped coffee from the pulper is placed in concrete or, in a few cases, in wooden tanks. The water which carries the coffee to the tanks is drained away, and the coffee is held until the mucilage is dispersible. The natural enzymes present in the mucilage itself accomplish the digestion at first; later, yeasts and bacteria from the tank walls, the water used, and the outer surface of the fruit skins begin to grow in quantities large enough to form similar pectic enzymes which continue the process. The quantities and importance of the secondary enzymes will depend on the sanitation practices, i.e., the care with which coffee and tanks have been washed. Sometimes, but not usually, the tanks are treated with a lime wash to destroy microorganisms and to furnish calcium ions which are favorable to the natural fermentation.

In any case, the pulped coffee is allowed to remain in the tanks until the mucilage has become completely dispersible. Then it should be washed as soon as possible after this point has been reached because excessive numbers of microorganisms may introduce damaging amounts of undesirable flavors. The simple and satisfactory test universally used is to rub a few beans in the hand while washing them under a small stream of water. After the slimy digested mucilage has been washed away, the clean beans feel gritty (like pebbles) when the digestion is complete.

The size of the fermenting tanks ranges from a maximum of about 3-m (10-ft) cubes (30 cu m or 1,000 cu ft) (see Fig. 5.17) to about ½ cu m (15 cu ft) of any convenient shape. Sometimes they are in the shape of long troughs, e.g., 2 ft wide,

*Courtesy of Pan American Coffee Bureau*

FIG. 5.17. LARGE FERMENTATION TANKS

2 ft deep, and 150 ft long, with the proper gradient for washing the coffee when fermentation is complete. This saves one step in transferring the coffee from one place to another. Figure 5.18 shows this operation.

When mucilage has been digested and rendered dispersible by natural or accelerated fermentation or by chemical treatment, the coffee is washed (1) in washing machines, (2) in concrete tanks fitted with paddles to give violent agitation, or (3) by hand or mechanical paddles operating in long canals in which water is flowing. A size and quality classification is often made at the same time in canal washing with the lighter, smaller, and poorer grades being carried the greatest distance and the best coffee settling out first. The different grades may then be dried separately.

The time required for the digestion varies over wide limits from about 6 hr to as much as 60 to 72 hr depending mainly on three factors: (1) temperature, (2) thickness of the mucilage layer, and (3) the concentration of the pectic enzymes. If the mucilage is thin and the temperature around 86 F (30 C), which usually means altitudes at or near sea level, only minimum time is required, while maximum time is necessary for thick mucilage with temperatures of around 50 F (10 C) or lower

*Courtesy of Pan American Coffee Bureau*

FIG. 5.18. MANUAL WASHING IN CANAL

found at altitudes of 6,000 ft (1,800 m) or higher. For a typical, average coffee plantation at altitudes between 2,000 and 4,000 ft (600 and 1,200 m), the natural fermentation time is usually about 24 hr. Often the fermentation is complete in less time, but since the process is usually started at the end of the day on which the coffee is picked, it is inconvenient to wash it before the second morning after picking it. Sometimes it is ready in 24 hr or less in which case it may be washed late in the day following the picking. Figure 5.19 shows medium sized fermentation tanks.

**Chemistry of Fermentation**.—The known chemistry of coffee mucilage fermentation is quite complex; thus, only the main reactions will be shown. Coffee mucilage, the mesocarp of the fruit, contains about 85 percent of loosely bound water and 15 percent of solids in the form of an insoluble, colloidal hydrogel without cellular structure at the time of fruit maturity. The solids consist of about 80 percent pectinic acids (12 percent of the mucilage) and 20 percent sugars (3 percent of the mucilage). The pectinic acids are composed of linear polymerized chains of acids formed mainly from hexose hydroxyl sugars, chiefly galactose with a little arabinose and other sugars. Esters of these acids and acid chains are also formed. The pectinic acids are further classified as protopectins (with high molecular weight), pectins, and pectic acid (with decreasing molecular weight ranges). Viscosimetric methods indicate molecular weights up to 70,000 for protopectins. During the development of the fruit, the carboxylic acid is formed from the sugar (alcohol) by oxidation. The acid anhydride is formed by the loss of

*Courtesy of Pan American Coffee Bureau*

FIG. 5.19. MEDIUM SIZED FERMENTATION TANKS

one molecule of water, then two anhydride molecules form the dimer with the release of another molecule of water. The trimer is formed by a similar reaction, and this continues until several hundred anhydride molecules have combined to form the protopectin chain.

In addition to the pectinic acids and sugars, several enzymes are formed in the mucilage and in the pulp surrounding the mucilage as the fruit develops: protopectinase, pectinase, pectinesterase, and pectase. If the fruit is not harvested but remains intact until it falls from the tree and finally dries, these enzymes cause fermentation of the mucilage until it is gradually broken down in steps to the simple components from which it was formed, each enzyme acting specifically on the component indicated by its name. This takes place within the skin until the mucilage disappears as such. It is transformed from the hydrogel to the hydrosol condition. The digested mucilage becomes food material for the metabolism of the seed.

The enzymes break down the large molecules by breaking the polymer bonds, gradually producing shorter chains until the simple monomolecular acids and esters remain.

If the fruit is pulped, many of the same changes go on; but in this case, the beans are exposed to a diversity of microorganisms such as yeasts, molds, fungi, and bacteria which find here a favorable medium for growth. In their development, they produce their own enzymes which, when the organisms become significantly numerous, may act on the coffee mucilage. The usual sequence is that the yeasts find favorable conditions first and produce an alcoholic fermentation from the sugars present. Then bacteria, using the alcohols as food material, form acetic, lactic, butyric, and higher carboxylic acids more or less in that sequence by means of oxidative fermentations. When butyric acid begins to form, coffee cup quality begins to suffer. Bacteria, molds, and fungi may produce undesirable flavors and aromas which sometimes develop in the coffee when fermentation is prolonged.

**Chemical Removal of Mucilage**.—When pulped coffee is treated with alkalies, neutralization of the carboxyl groups in the chain takes place with the formation of soluble salts along with hydrolysis which results in depolymerization as a secondary effect. Treatment with dilute acid causes hydrolysis and depolymerization. The chemical reactions are very fast compared with the action of enzymes, usually taking place in less than an hour.

The fermentation reactions are mildly exothermic as a whole, and the temperature may rise, usually not more than 5 to 8 F (3 to 5 C), or it may remain about constant where ambient temperatures are lower than those of the coffee. On one plantation in Salvador, the coffee is heated by blowing hot air through it in order to shorten the fermentation time. Temperatures above 100 F (38 C) are dangerous in that some enzymes and microorganisms may be destroyed and the digestion retarded. Also, the coffee itself may be damaged in quality by long periods in a wet condition at elevated temperatures.

Sometimes, but not usually, the fermentation is carried out under water. It is hard to see any advantage in this although the digestion will go on under water. When no water is present, occasional stirring is feasible if the tanks are small; this is desirable in that it equalizes temperature, but it is not considered necessary.

**Fermentation with Added Enzymes**.—In order to accelerate the digestion of the mucilage, it has been found advantageous to use small amounts of pectic enzymes probably containing pectase, protopectinase, pectinase, and pectinesterase as the chief active ingredients. This material is produced from molds and is a carefully standardized product sold for this purpose under the trade name of Benefax. In large doses at around 100 F (38 C), it will digest the mucilage in 5 min, but this procedure is not economical. Used in small quantities and at ambient temperatures, it digests the mucilage in 5 to 8 hr at low cost.

In practice, the cherries are pulped at the end of the day, and the enzyme is mixed with the pulped coffee which is allowed to stand overnight. It is washed the next morning and is ready for drying early in the day. The commercial enzyme contains a filler such as starch or any inert, dry, finely ground material in order to make it easier to distribute the enzyme evenly throughout the pulped coffee. In this form, one-quarter of 1 percent of the enzyme-filler mixture is used, based on the weight of the wet pulped coffee. The weight is usually determined by measuring the volume. Pulped beans weigh 53 lb per cu ft or 846 kg per cu m. The yield of finished green coffee would be about 131 lb per lb of commercial Benefax (10 percent active pectic enzyme, 90 percent starch or other filler). The cost of the enzyme would be approximately a few tenths of a cent per pound of dry green coffee.

In this way the digestion time is shortened, which tends to improve quality, eliminate undesirable fermentations, allow the maintenance of regular production schedules, and greatly reduce the required capacity and cost of fermenting tanks. The shortening of the time from perhaps 36 to 8 hr means that much less fermentation tank capacity is required. These pectic enzymes have been in commercial use in Brazil and Central America for two decades although their use is declining in favor of attrition methods.

Benefax as such, and sometimes in the 100 percent active enzyme form, has been used to raise the quality of partially dry and immature fruit which passes through the pulpers intact. If this fraction is passed through a scarifying machine like a corn grinder which is adjusted so that the berries are cut and bruised to allow access to enzyme, the mucilage is digested in a few hours, which loosens the pulp. The fruit is kneaded, whereupon the pulps are easily removed. It may now be handled as washed coffee with a very marked improvement in quality as compared with that obtained by drying it as natural coffee without enzyme treatment. The reason for the improvement is mainly that processing time is shortened, thus suppressing the action of harmful organisms.

**Alkaline Digestion of Mucilage**.—Sodium hydroxide solution dissolves coffee mucilage rather rapidly, and it has been proposed for commercial use. However, the time and alkali concentrations in this coffee treatment must be closely controlled or damage may result. To date, it has not been used commercially to any great extent. Sodium carbonate may also be used in the same way.

An approximately 3 to 5 percent caustic soda or 6 to 8 percent sodium carbonate solution is used, and the time of contact ranges between 30 and 60 min. The pulped coffee is placed in the concrete fermenting tank in the usual way. The water is drained out, and the dilute alkali solution is added in just sufficient quantity to cover the coffee. When digestion is complete, the coffee is washed in the usual way. As with pectic enzymes, the coffee must be well mixed with the digesting agent for best results. It is desirable to do this in relatively small tanks for ease in handling. The use of wetting agents was found to accelerate the digestion in this process. The operation is shown in Fig. 5.20.

*Cafepro*.—A method for using lime in the form of a calcium hydroxide suspension either alone or mixed with sodium or potassium hydroxide or sodium carbonate was developed in 1953. The process is continuous; the pulped coffee from the pulpers is fed into a horizontal mixer with a screw at one end to regulate the flow followed by paddles on the same shaft to ensure uniform contact between

FIG. 5.20. ADDING SODIUM HYDROXIDE SOLUTION TO PULPED COFFEE

the coffee and the alkali. Wash water is added near the discharge end. When discharged from the mixer, the coffee is completely washed and proceeds at once to the drying operation. The capacity of the machine is matched to that of the pulper. If water for the washing is in short supply, a lime and calcium chloride solution is used after which the coffee is sufficiently clean to be dried without washing. This machine is shown in Figs. 5.21A and 5.21B.

A standard model of the machine operates at 40 rpm and requires about 2 hp. Its capacity is 5,700 lb of cherries (3,250 lb of pulped coffee or 1,066 lb of dry green coffee) per hr. A 2 percent alkali solution is used, and about 120 gal per hr of this solution are required.

Mucilage was removed in 20 min with 1 percent sulfuric acid solution. The coffee was poor in appearance, but had good cup quality.

Fig. 5.21A. Cafepro

Fig. 5.21B. Cafepro Closed

The main elements of cost in the accelerated digestion methods are those of (1) the agent, (2) power, and (3) extra labor as contrasted with the cost of storage tanks for the slow, natural fermentation process. Two other important questions must be answered in deciding which method to adopt in any given case: (1) How is coffee quality affected? (2) Are differences in the yield of coffee obtained by the different methods?

As to the first question, many cup test comparisons have been made between coffee fermented naturally and coffee prepared by quick methods of mucilage removal (added enzymes, alkali treatment, and attrition). In no case has any marked difference been found, provided an alkali treatment was not too severe. Very minor differences have been reported, however, and, if any generalization can be made, it may be said that the fast processes tend to produce coffee which, while clean and bland in taste, has a little less body than coffee which has been in contact with its mucilage for longer periods.

Regarding the second question, statements have been made by several workers, and careful quantitative experiments have been carried out to support them, that the solids in moist pulped coffee may disappear slowly due to processes of metabolism and respiration as time goes on until the coffee is dry. The chemical mechanism is probably the conversion of carbohydrates into carbon dioxide and water, and possibly, protein into ammonia. Laboratory experiments have actually recovered carbon dioxide from fermenting coffee. If this is true, any quick method of mucilage removal will result in a greater yield of dry green coffee as compared with that obtained by natural fermentation. The advantage becomes greater the longer the time required for natural fermentation in a given locality.

**Warm Water.**—A simple method for removing coffee mucilage quickly is to mix equal weights of pulped coffee and water and heat to 122 F (50 C) plus or minus 5 F (3 C) as quickly as possible. The warm water alone is able to break down the structure of the pectic materials of the mucilage gel, and the coffee may be washed after as little as 3 min contact with the water at this temperature. The coffee may be added to water heated as much above the required temperature as is necessary to result in a mixture at the right temperature.

The chief expense in this very simple method would be the cost of fuel. With heating oil at 140,000 BTU per gal, and a 50 percent efficiency factor, about 680 lb (308 kg) of pulped coffee equivalent to 230 lb (104 kg) of finished green coffee could be produced from 1 gal of heating oil assuming that the temperature of the water and coffee were raised from about 68 to 122 F (20 to 50 C).

When the mucilage has been removed by washing, the coffee may be spread on the drying terrace as shown in Fig. 5.22.

**Attrition Methods of Removing Mucilage.**—Several machines have been designed to rub off the mucilage from pulped coffee by attrition or scrubbing. In one type, the pulped beans are pressed against each other and against the

*Courtesy of Pan American Coffee Bureau*

FIG. 5.22. SPREADING WET WASHED COFFEE ON DRYING TERRACE

roughened lining of the machine while being forcibly fed through the machine by a screw against resistance generated by a partially throttled discharge. Clearances are carefully adjusted so that the parchment layer is neither crushed nor broken and the corners of ribs or projections are rounded to avoid cutting the beans. Since the exact clearance is rather critical, beans of different sizes cause difficulty since the larger ones may be damaged and the very small ones may escape thorough treatment.

*Aquapulper.*—Such a machine is the *Raoeng* pulper developed in Java and once manufactured by Krupp in Germany, later by Bentall in England as the Aquapulper, and more recently by Hentschel in Germany. It was originally designed to accomplish both pulping and mucilage removal in one operation, i.e., whole cherries were fed in, and clean, washed coffee ready for drying was discharged.

The machine consists essentially of a rotating horizontal cylinder inside a fixed cylinder with a slightly ribbed surface. The inner cylinder is fitted with cleats, both longitudinal and transverse, to assure the positive partial rotation of the coffee. At the feed end, a few turns of screw conveyor ribs feed the coffee through the machine at a uniform rate. Some resistance and pressure is built up within the machine by a partially throttled and adjustable outlet. Water is fed in along with the coffee and also about midway in the outer cylinder. Small slots are provided in the bottom for the escape of water, mucilage, and finely ground pulps with retention of the coffee. The slots are too small to allow the passage of the wet parchment coffee. Clearances may be adjusted by moving the rotating cylinder laterally within the fixed cylinder. Since they are both slightly cone shaped, their relative position determines the exact clearance.

A standard model is equipped with a 30 to 40 hp motor and its capacity is about the same as three regular pulpers requiring less than 1 hp each. The extra power is expended in grinding up the pulps. The machine is usually set with just enough clearance to avoid bruising the coffee beans; a small amount of mucilage is left on most of the coffee which causes little or no difficulty on drying.

The machine has usually been used as a simple scrubber for coffee that has already been pulped in the regular way in standard pulpers. In this case, the power requirements are much less and the capacity is greatly increased. The protective cushioning effect of the pulps is absent. The machine is shown in Fig. 5.23A and 5.23B.

*Courtesy of E. H. Bentall & Company*

FIG. 5.23A. AQUAPULPER CLOSED

*Courtesy of E. H. Bentall & Company*

FIG. 5.23B. AQUAPULPER OPEN

*The Hess Washer.*—This one is another low pressure, low power machine which removes mucilage by attrition (Hess, U.S. patent 2,722,226, 1955). It is essentially a trough which vibrates longitudinally and contains fixed and moving baffles. The coffee beans are rubbed against each other and on the baffles, and, by the time the coffee has traveled from one end of the trough to the other, the mucilage has been rubbed off. The trough is square with about 1 sq ft cross-section area, about 10 ft long and is driven by a 3 hp motor; however, it is available in several sizes. The vibration is generated by an eccentric or crankshaft. It has an amplitude of an inch or two and a frequency of about 300 cycles per min. The trough itself oscillates and is supported on small wheels running in tracks. It has vertical baffles attached to its bottom and sides which move with it. Other fixed baffles are suspended from above.

Wash water is led in at intervals over the open top. The pulped coffee is fed in by gravity at one end along with part of the wash water and passes in a devious path under the fixed baffles and over the moving ones. The washer may be fed directly from the discharge of a pulper. The coffee overflows at the other end free from mucilage. The water-mucilage suspension is separated from the clean coffee by a screen as it leaves the washer. The forward movement through the trough is accomplished by having a higher head of coffee at the feed end; the reciprocating motion causes the coffee and water mixture to seek its own level like a liquid.

In another version of the same machine, twin screw conveyors are substituted for the baffles in order to produce a more thorough mixing and scrubbing action. The forward motion of one screw is opposed and neutralized by the reverse motion of the other so that the net result is only mixing. The coffee is propelled forward by the hydraulic head at the feed end as before. The vibratory motion is the same in both versions.

A standard model has a capacity of about 1,500 lb of washed coffee (from 1,800 lb of pulped coffee) and yields the equivalent of about 620 lb of dry green coffee per hour with a water consumption of 7.5 gal per 100 pounds of dry green coffee. This machine is free from the defects that cause damage to the coffee parchment and excessive power consumption. No critical adjustments of clearance are necessary. The mechanism must be anchored very firmly to an exceptionally solid base since the inertial forces tending to loosen its mountings are rather large. Balancing springs of the correct natural frequency and tension would tend to reduce these forces.

*HAES*[1] *Washer.*—Possibly the simplest coffee scrubber is one in use in the Kona coffee region on the island of Hawaii in the Hawaiian Islands. It consists of a vertical steel cylinder (or a pair) covered at the top, approximately 8 in. in diameter, and about 3.5 ft high. A vertical shaft with round horizontal arms is rotated at about 300 rpm. Round baffle arms are attached to the inside of the stationary cylinder. The pulped coffee and water are fed in at the bottom. The

[1]Hawaiian Agricultural Experiment Station, University of Hawaii, at Kealakekua, Hawaii.

coffee is subjected to a rapid and thorough but gentle rubbing action since the round arms have no sharp edges and no close clearances are needed. The mucilage is removed by the time the coffee overflows from the top outlet. The capacity is large and the power consumption is small. A 1 hp machine will remove the mucilage from about 900 lb (408 kg) of pulped beans from 1,500 lb (680 kg) of cherries per hour. If water warmed to 110 to 120 F (43 to 49 C) is used, a smaller sized ½ hp machine will produce washed coffee at the same rate. (Fig. 5.24)

**Drying Coffee Without Washing**.—It has been found that, if desired, coffee may be dried without washing (see p. 105). It is not practical to dry pulped coffee directly because the hydrophilic nature of the mucilage lowers the vapor pressure of the moisture in the coffee at the mucilaginous surface, and drying is extremely difficult and slow. Such coffee is also likely to be of poor quality. But if the mucilage is digested with pectic enzymes in a short time, e.g., overnight, the coffee may then be spread on the drying terrace and dried successfully without washing. If drying conditions are good, i.e., if favorable weather prevails, normal drying of washed coffee usually requires about 5 to 7 days. If the coffee is not washed but dried in *digested* mucilage, it passes through an initial sticky stage for the first day or two; then the surface becomes dry and the coffee may be handled in

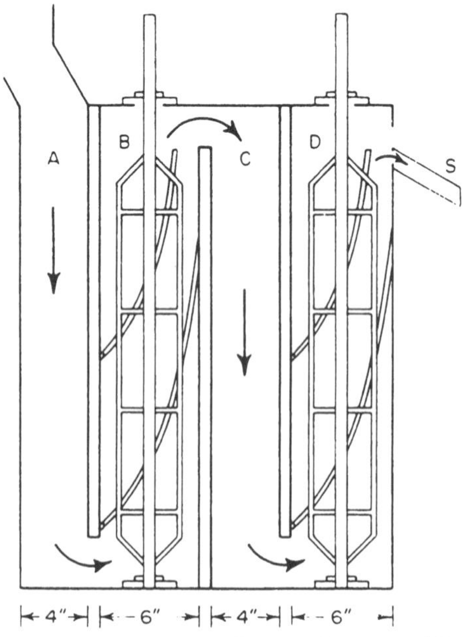


*Courtesy of University of Hawaii*

FIG. 5.24. DIAGRAM OF HAES WASHER

the usual way thereafter. The presence of the digested mucilage will require about one extra day compared with the normal drying time for clean, washed coffee.

Coffee in digested mucilage may also be dried in a mechanical drier, but the mucilage adheres to the walls of the drier and introduces a rather troublesome cleaning problem. The mucilage causes no difficulty on the terrace, and it can easily be washed away by rain or hose.

The advantages of the method are: (1) in locations where an extreme scarcity of water exists, it greatly reduces the water required to produce pulped coffee; (2) such coffee has been shown to have a somewhat improved flavor described as body or mild richness; (3) experiments on quantitative weight loss indicate that there is little or no loss of solids in coffee dried in digested mucilage. There may even be a slight gain in weight due perhaps to the diffusion of some solids from the mucilage through the parchment layer into the inner coffee seed. It was observed that green coffee produced in this way is slightly darker in color than standard coffee.

The disadvantage of the method is that coffee in digested mucilage is somewhat more difficult to handle and to dry than clean, washed parchment coffee.

**Loss of Solids During Fermentation.**—In 1950, a report was received from Colombia from an obscure, word-of-mouth source which could not be traced, that fermenting coffee loses 2 to 5 percent of its dry weight in 48 hr (see p. 99). If true, this would have an important bearing on the desirability of using quick methods of removing mucilage.

Although it is difficult to make reliable measurements on a large scale (20,000 lb of green coffee) a loss of about 2 percent in 24 hr occurs. The results of these experiments are shown in Fig. 5.25.

A similar investigation showed losses of 1.73 percent at 20 hr, 5.58 percent at 28 hr, and 9.03 percent at 44 hr. Also reported were losses of 9 to 12 percent in pulped coffee fermented under water for 48 hr which suggests that it may not be good practice to ferment coffee under water. Part of the weight loss may be due to the leaching out of soluble solids from the seed.

Although it is more difficult to make reliable measurements on the large scale (20,000 lb of green coffee), a loss of about 2 percent in 24 hr was determined. The results of these experiments are shown in Fig. 5.25.

It is very difficult to determine the magnitude of these losses quantitatively because of many variable factors not yet well understood and very hard to control. As in the case of most fruits, individual units vary in size and in imperfections of many types. There is reason to believe that enzyme activity, which is involved in the changes of weight, varies with the stage of the harvest season. The result is that many replicate experiments must be carried out, and as many variables as possible must be eliminated by the design of the experiments in order to obtain results that are statistically meaningful. At Bernadino de Campos, seven conditions were used with 30 replicates for each condition.

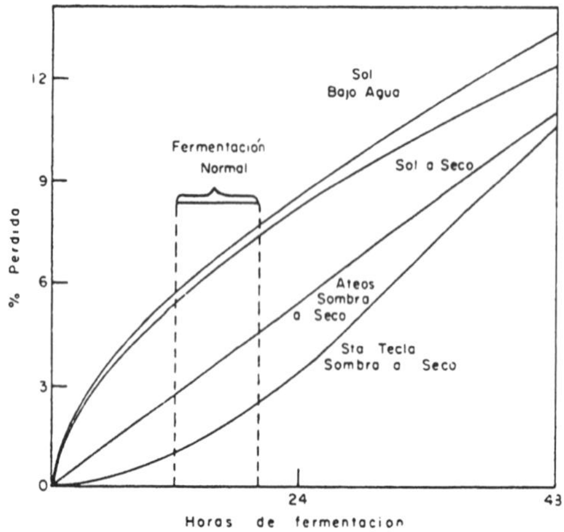


FIG. 5.25. WEIGHT LOSS IN RELATION TO NORMAL FERMENTATION

In general, these experiments were carried out by taking many small lots of pulped coffee of equal weight, removing the mucilage from some of them quickly by means of pectic enzymes or alkali, and allowing the remaining lots to undergo natural fermentation for 36 to 48 hr. The lots processed by the fast method were washed and sun-dried as quickly as possible, and the naturally fermented lots were washed and sun-dried at the end of the selected fermentation period in the same way. The dry parchment coffee was hulled by hand, with care being taken to lose none of the coffee. The whole process was carried out quantitatively with careful weighings throughout. Exact moisture contents were determined for each lot of dry green coffee and the bone dry weights calculated. In some cases, weights of 100 bean lots were determined to see whether or not the average bean weights of the fast and slow lots were different.

A summary of the data on solids loss from all sources is shown in Table 5.4 and Fig. 5.26 which show the results in graphic form.

In summary, it seems practically certain that more or less rapid losses of solids take place in pulped and washed coffee as long as it remains moist and alive. These losses are probably due to normal metabolism and respiration resulting in the formation and loss of volatile by-products such as carbon dioxide, water, and possibly ammonia, and by the leaching of soluble solids when the coffee is washed, with more solids becoming soluble as time goes on. The rate of this loss in weight varies widely with local conditions, but is probably seldom less than 0.5 percent in 36 hr, the average time required for natural fermentation. It may, under favorable conditions (for weight loss), be much higher than this. The loss is accelerated markedly as the temperature of the natural fermentation rises. The temperature is mainly determined by the prevailing weather conditions. Very little temperature rise occurs due to the heat generated by the fermentation.

TABLE 5.4. DATA FOR FIGURE 5.26

| Curve | | Hours | Percent Loss |
|---|---|---|---|
| 1 | Santa Cecilia, Brazil | 0 | 0.00 |
| | | 8 | 0.00 |
| | | 36 | 0.48 |
| | | 60 | 0.67 |
| 2 | Guatemala | 48 | 1.0 |
| 3 | Atibaia, Brazil | 8 | 0.3 |
| | | 36 | 1.6 |
| | | 48 | 2.5 |
| | | 84 | 4.9 |
| 4 | Colombia | 48 | 2–5 |
| 5 | Salvador | 24 | 2.7 |
| | | 40 | 6.1 |
| 6 | Salvador | 20 | 1.73 |
| | | 28 | 5.58 |
| | | 44 | 9.03 |

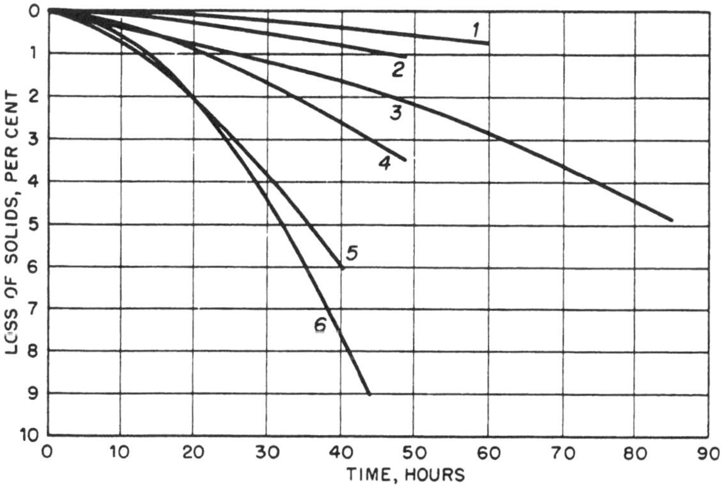


FIG. 5.26. SUMMARY OF WEIGHT LOSS EXPERIMENTS

Any method of quick removal of mucilage will avoid this weight loss, and the value of the increased yield of coffee will partially or entirely, and sometimes much more than pay for any extra cost of processing. It is probable that mucilage may be removed by an efficient and simple machine such as the HAES design used in the Hawaiian Islands, with its low power, low water consumption, and high capacity, much more cheaply than by natural fermentation with its investment in costly fermentation tanks.

**Speed, a Desirable Factor.**—It has been well established that coffee does not improve as processing time increases, especially while the coffee contains large amounts of moisture. Indeed, it is highly probable that deterioration of flavor quality starts as soon as the coffee is picked and continues until the moisture content approaches 12 percent. This figure is the average standard for dry coffee, and coffee of this moisture content is in approximate equilibrium with ambient air of average relative humidity.

The term curing for coffee is a misnomer in so far as it implies an improvement in quality, as it does with many food products such as cheese. For coffee, all efforts are directed toward conserving the original high quality which most coffee has at the time of picking if it is picked, as it should be, at peak maturity. The best way to forestall deterioration is to carry out all the steps of processing, including the drying, as rapidly as possible. Only when dry is coffee relatively safe from quality loss. The only exception to this rule seems to be that coffee may acquire a slightly more desirable body when it is dried in contact with its digested mucilage as was pointed out previously.

## BIBLIOGRAPHY

ANON. 1962A. The overflowing cup. Time Magazine *20,* No. 4, July 27, *66.*

ANON. 1962B. Soothing the coffee nerves. Time Magazine *20,* No. 9, Aug. 31, *58.*

BECKLEY, V. A. 1933. Enzymes and bacteria in coffee fermentation. 1930. Fermentation of coffee. Department of Agriculture Bull. No. 8. Kenya Colony and Protectorate. Nairobi.

CARBONELL, R. J. 1953. Recommendations on the new method of processing coffee by means of caustic soda and the results with chemicals which add to its effectiveness. Boletin Tecnico No. 14. Ministerio de Agricultura y Ganaderia Centro Nacional de Agronomia. Santa Tecla, El Salvador.

CASE, E. M. 1936. Coffee preparation. A comparison between coffee cleaned by fermentation and by the Raoeng pulper. Monthly Bull. *2,* No. 17, 93–94. The Coffee Board of Kenya.

CHOUSSY, F. 1948. Technical studies on the fermentation of coffee. Agrotecnia (Cuba) *2,* No. 8, 532–553.

COLEMAN, R. J., LENNEY, J. K., COSCIA, A. T., and DI CARLO, F. J. 1955. Pectic acid from the mucilage of coffee cherries. Arch. Biochem. Biophys. *59,* 157–164.

DAVIES, E. DE L., and JONES, M. A. 1953. Cafepro–Machine for removing the mucilage chemically. Servicio Cooperativo Interamericano de Agricultura *3,* No. 4, 151–155.

FOOTE, H. E. 1962. Factors affecting cup quality in growing and processing coffee fruit. Coffee and Tea Inds. *85,* No. 9, 11–18.

1963. Factors affecting cup quality in growing and processing coffee fruit in Latin America. Coffee and Tea Inds. *86,* No. 1, 38–42.

FRITZ, A. 1933. Processing of coffee without fermentation. Revista Agricola (Guatemala) *11*, No. 6, 285–290.

1935. Study of wet fermentation of coffee. L'Agronomie Coloniale *24*, 41–47, 72–84; *also in* Chimie et Industrie *34*, 1197.

FUKUNAGA, E. T. 1957. A new mechanical coffee demucilaging machine. Bull. 115. Hawaii Agricultural Experiment Station. University of Hawaii, Honolulu.

GOTO, Y. B., and FUKUNAGA, E. T. 1956. Harvesting and processing for top quality coffee. Extension Circular 359. University of Hawaii, Honolulu.

HAARER, A. E. 1962. Modern Coffee Production. 2nd Edition. Leonard Hill (Books) Ltd., London.

JOHNSTON, W. R., and FOOTE, H. E. 1951. Development of a new process for curing coffee. Food Technol. *5*, No. 11, 464–467; *also in* 1952. Coffee and Tea Inds., *75*, No. 1, 12–13, 40, 42; *also in* 1952. Indian Coffee Board Monthly Bull., *16*, No. 4, 69–73; *and in* Coffee Board of Kenya, Monthly Bull., 1952, *17*, 201, 204–206.

JONES, M. A., and BAYER, J. J. 1956. Cafepro–Machine for the chemical removal of mucilage from freshly pulped coffee. Servicio Cooperativo Interamericano de Agricultura. 1954. Further experience with Cafepro–A machine for the chemical removal of mucilage from freshly pulped coffee. Instituto Agropecuario Nacional. Guatemala City.

JONES, M. A., and DAVIES, E. DE L. 1953. Preliminary results on the removal of mucilage using lime and soda. Instituto Agropecuario Nacional, Guatemala.

KRUG, H. P. 1947. Causes of coffee hardness. I. Bull. 48, 397–406. (State) Secretaria da Agricultura, São Paulo, Brazil.

MARTINEZ, A., and JAMES, C. N. 1960. Fermentation *in* IAIAS Coffee Bibliography, p. 517–520, Nos. 4633 to 4664.

PEDERSON, C. S., and BREED, R. S. 1946. Coffee fermentation. Food Research *11*, No. 2, 99–106.

PERRIER, A. 1931. Investigations of the role of pectinase in the fermentation of coffee. Compt. Rend., Acad. Sci., Paris. (14) *193*, 547–549.

1932. Investigations on the fermentation of coffee. Comptes Rend. (14) *194*, 306–308.

QUIMME, P. 1976. Coffee and Tea. Signet Books, N.Y., N.Y.

SACHS, B., and SYLVAIN, P. G. (Editors), 1959. Advances in Coffee Production Technology. Coffee and Tea Inds., 106 Water St., New York.

SHAPIRA, R. G. 1975. Coffee and Tea. St. Martin Press, N.Y., N.Y.

SIVETZ, M. 1976. History of soluble coffee. Tea and Coffee Trade J. Aug. 1976.

SIVETZ, M. 1977. Coffee Origin and Uses of Coffee Publications, Corvallis, Oregon.

SMITH, F., and MONTGOMERY, R. 1959. Chemistry of Plant Gums and Mucilages. Reinhold Publishing Corp., New York.

SPRINGETT, L. B. 1940. Development of quality and color by means of fermentation. Instituto de Defensa del Cafe de Costa Rica. Revista *10*, No. 69, 38–45.

1951. Quality and cup testing of coffee. Revista de Agricultura (Costa Rica) *23*, No. 5, 133–134.

SPRINGETT, L. B. 1935. Quality of Coffee. Spice Mill Publishing Company, Water St., New York.

STERN, J. 1948. Methods of fermenting coffee. Revista Soc. Mexicana Hist. Nat. 7, Nos. 1–4, 25–34.

1944. Notes on the study of the fermentation of coffee. Variations of the pH and the temperatures in the fermenting tanks. Superintendencia dos Servicos do Cafe. Sao Paulo, Brazil. Bull. 19, No. 205, 284–292.

1945. The use of lime in the wet processing of coffee. Revista Cafetalera de Guatemala *1*, No. 6–7, 30, 45.

SUMMER, J. B., and MYRBACK, K. 1951. The Enzymes. Vol. I, Part 2, Chapter 21. Academic Press, New York.

SUMMER, J. B., and SOMERS, G. F. 1947. Chemistry and methods of enzymes. 2nd Edition. pp. 71–72, 111–112. Academic Press, New York.

# 6

# Drying Green Coffee Beans

There are no reliable figures for the percentages of coffee dried in the sun and by mechanical driers. Although much of the world's coffee is still sun-dried, it is very probable that there is a present tendency to dry more and more in mechanical driers. The main reasons for this are: (1) with rising labor costs, the relative cost of machine drying tends to become lower since much less labor is required than in sun drying; (2) the development of highway systems and public power make the use of mechanical driers practicable; and (3) unfavorable weather adds to the cost and often lowers quality in open air drying, whereas mechanical driers do not depend on weather.

The drying process requires sensible and latent heat of evaporation of the water. The value of the latter is about 585 gm-cal per gm or 1,053 BTU per lb of water at 68 F (20 C) which may be taken as an average drying terrace temperature. For 104 F (40 C), the values are 575 gm-cal per gm and 1,034 BTU per lb.

The heat efficiency of the drying operation, whether sun or mechanical, is calculated as the ratio of the latent heat of the water actually evaporated, as defined above, to the heat input into the system. If, in addition, the material being dried must be heated to a drying temperature above ambient temperature, this is included in the theoretical heat requirement. This latter factor does not apply in sun drying.

For example, in mechanical drying, if 100 lb of water are evaporated, which requires a total of 100 × 1,053 = 105,300 BTU, with the consumption of 2 gal of fuel oil rated at 140,000 BTU per gal, the heat efficiency would be (105,300 ÷ 280,000) × 100 = 37.6 percent.

The same calculation may be applied to sun drying when it is known how much sun heat is available for the evaporation of water. This will be explained below. Figure 6.1 shows a typical terrace drying layout.

## SUN AS A HEAT SOURCE

The sun radiates a definite and constant quantity of energy at all times. This is known as the solar constant, and it is measured in units of heat received by a unit

FIG. 6.1. A BENEFICIO FEATURING SUN DRYING

*Courtesy of Pan American Coffee Bureau*

FIG. 6.2. SMALL DRYING TERRACES IN COLOMBIA WITH GRAVITY ENTRANCE AND DISCHARGE

area in one minute. A British thermal unit (BTU) is defined as the quantity of heat necessary to raise the temperature of 1 lb of water 1° F and the gram-calorie (gm-cal) as the quantity of heat necessary to raise the temperature of 1 gm of water 1° C. The solar constant has a value of 7.15 BTU per sq ft per min (1.94 gm-cal per sq cm per mm) at the earth's outer atmosphere on a surface perpendicular to the sun's rays when the earth is at its mean distance from the sun. Some of the solar energy is lost in passing through the earth's atmosphere; this factor, naturally, varies with atmospheric conditions. Hence, it is necessary to measure the quantity of solar energy received at any given time and place in order to find out how much is available for useful work.

The instrument used for this purpose is called a pyroheliometer, which consists of two concentric rings, one black and one white, mounted in a large vacuum tube. In sunlight or daylight, the black ring becomes warmer than the white ring in proportion to the intensity of the radiation. The temperature difference is measured by a multiple thermocouple which produces a voltage difference measured by a sensitive potentiometer reading to millionths of a volt. The voltage readings are converted to energy received per unit area per minute by means of a constant factor. When the instrument is used, the plane of the rings is leveled by means of a spirit level, or the instrument may be pointed in the direction of the sun, making the plane of the rings perpendicular to the sun's rays. Figure 6.3 shows the observed elevation of the sun for the dates indicated.

The results of readings taken at 2-hr intervals (8:00, 10:00, 12:00, 2:00, and 4:00) for surfaces (1) level and (2) perpendicular to the sun's rays, were plotted in the form of curves for (1) clear weather, (2) thin clouds, and (3) thick rain clouds. These are shown in Fig. 6.4 and 6.5. The observations were made at Matão, in the state of São Paulo, Brazil (south latitude of 21.5°). It was found that on an average clear day in August (mid-winter), a total of about 2,120 BTU per sq ft per 11-hr day or 575 gm-cal per sq cm per 11-hr day was actually received on a level surface compared with 4,719 BTU per sq ft (1,280 gm-cal per sq cm), which is the solar constant for 11 hr on a surface perpendicular to the sun's rays. Thus, it is seen that about 45 percent of the theoretical energy is actually available for useful work. The losses are due (1) to the passage of the radiation through the atmosphere— they would be higher if clouds were present, and (2) to the average low angle of the sun.

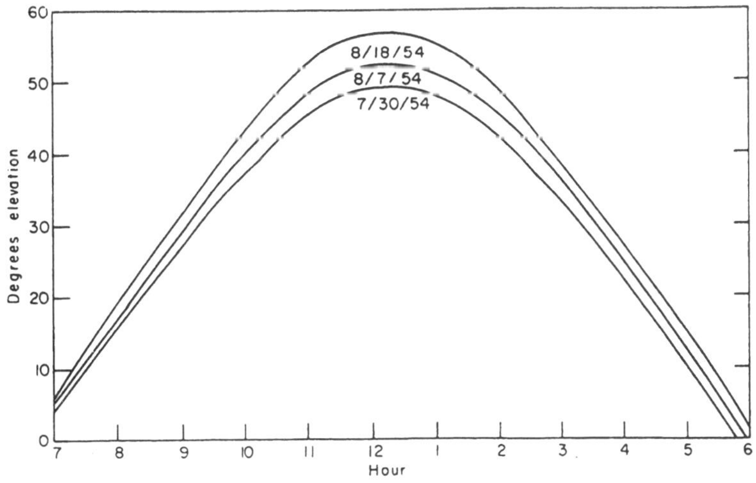


*Courtesy of I.R.I. Research Institute, Inc.*

FIG. 6.3. ELEVATION OF THE SUN

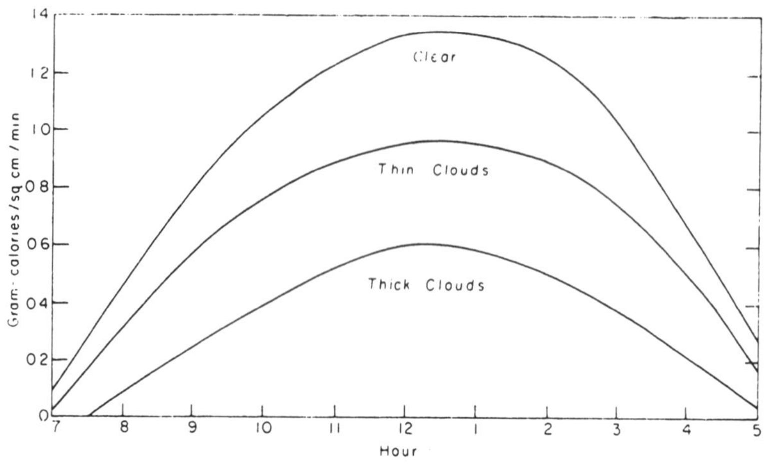


*Courtesy of I.R.I. Research Institute, Inc.*

FIG. 6.4. SOLAR RADIATION ON LEVEL SURFACE

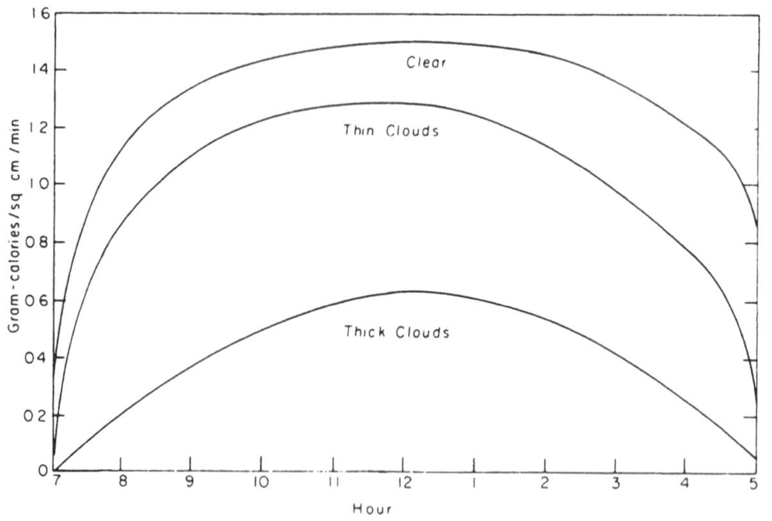


*Courtesy of I.R.I. Research Institute, Inc.*

FIG. 6.5. SOLAR RADIATION ON PERPENDICULAR SURFACE

For a level surface, the portion of total energy received on a unit area is equal to the sine of the angle of elevation of the sun above the horizon.

## Drying Efficiency

Of the available 45 percent of the total energy, only 7 to 13 percent can be used in the evaporation of water from a wet material such as coffee beans. The efficiency is highest when the coffee beans are wettest.

Not all the energy comes directly from the sun. Some comes by reflection and refraction from the bright sky as a whole. Observation shows that about 77 percent is directly from the sun and the remaining 23 percent is from sky or indirect radiation.

The available energy on a typical clear day was found to be as shown in Table 6.1 and Fig. 6.4 and 6.5.

TABLE 6.1. AVAILABLE SOLAR ENERGY

| | Level Surface | | Perpendicular Surface | |
|---|---|---|---|---|
| Time | Gm-cal per sq cm per min | % of Solar Constant | Gm-cal per sq cm per min | % of Solar Constant |
| 8:00 A.M. | 0.562 | 29.0 | 1.187 | 61.1 |
| 10:00 A.M. | 1.092 | 56.3 | 1.435 | 73.9 |
| 12:00 M. | 1.343 | 69.2 | 1.492 | 76.9 |
| 2:00 P.M. | 1.219 | 62.9 | 1.467 | 75.5 |
| 4:00 P.M. | 0.612 | 31.6 | 1.172 | 60.5 |
| Average | 0.966 | 49.8 | 1.351 | 69.6 |

The highest individual readings were 73.4 percent of the solar radiation constant for a level surface, and 81.4 percent for a perpendicular surface. These were obtained at solar noon with the sun at an elevation of about 55°.

Figures 6.4 and 6.5 show the effects of thin and thick clouds in reducing solar radiation. The thick clouds are characteristic of a rainy day, and the amount of radiation that is able to penetrate them is rather surprising. The total quantity of energy falling on a considerable area like a coffee drying terrrace is not negligible. For example, a level terrace 1,000 ft (305 m) square having an area of 1,000,000 sq ft (93,000 sq m) will, in clear weather with the sun at an average winter elevation, actually receive enough solar energy to evaporate two million pounds of water daily at 100 percent efficiency. This would be equivalent to the burning of 70 tons of coal per day. Even at the efficiencies of 7 to 13 percent actually obtained in drying coffee, 140,000 to 280,000 lb of water could be evaporated from coffee per day on the above drying terrace under good weather conditions.

## Sun Drying Technique

A study of the effects of various factors on drying time and drying efficiency led to the following conclusions:

(1) The loss of water from ripe coffee cherries at 60 to 65 percent moisture (wet basis) is about 100 times as fast as from the same coffee with 10 to 12 percent moisture content.

(2) Up to a depth of about 4 in. (10 cm) the rate of *water loss* per unit area of drying terrace surface rises with an increase of the depth of the layer although the

rate of *coffee drying* (water loss per unit weight) is retarded somewhat. This is shown in Fig. 6.6.

(3) Washed pulped coffee requires about one-third of the drying time required for natural or whole fruit coffee. Coffee, dried on screen-bottomed trays with access of air beneath, dries much faster than coffee spread on a brick or concrete drying terrace.

(4) Hourly stirring shortens drying time by about 14 percent.

(5) The drying capacity of a brick or concrete drying terrace in good weather is shown approximately in Table 6.2.

## OPERATION OF THE SUN DRYING TERRACE

### Washed Coffee Beans

Freshly pulped and washed coffee in Latin America usually contains between 52 and 54 percent moisture on the wet basis (108 to 117 percent, dry basis) after draining off excess water. Beans are spread on the drying terrace by hand carts

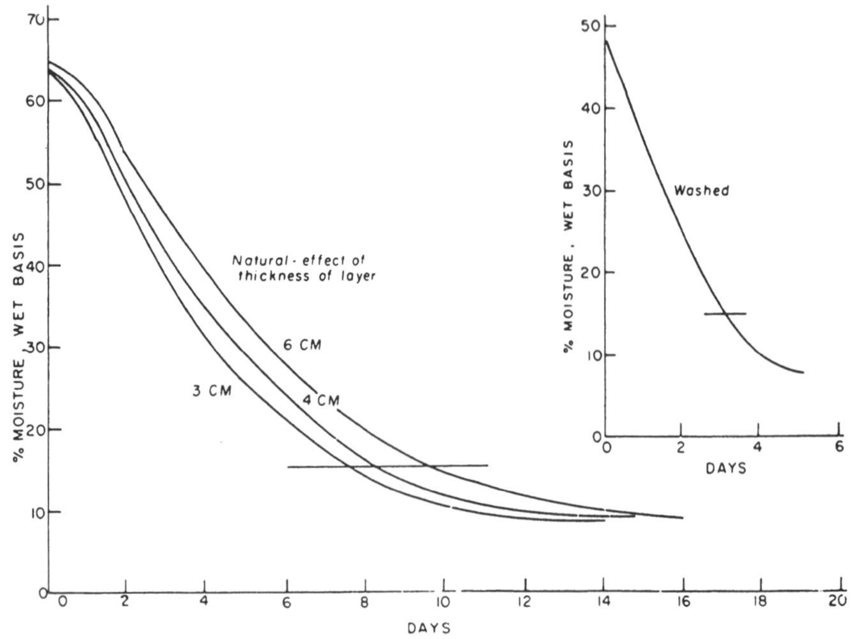


*Courtesy of I.R.I. Research Institute, Inc.*

FIG. 6.6. DRYING CURVES OF WASHED AND NATURAL COFFEE

perforated for water drainage and equipped with a door in the bottom which is opened for spreading the beans. The bottom of the cart clears the terrace by the amount of the desired thickness of coffee layer, and the wet coffee beans run out as the cart is pushed along.

TABLE 6.2. DRYING CAPACITY OF BRICK OR CONCRETE TERRACES

| Moisture and type of coffee | Weight of coffee: Kg per sq m per day | Weight of coffee: Lb per sq ft per day |
|---|---|---|
| Natural | | |
| 65% ripe cherry coffee | 3.0 | 0.62 |
| 12% dry green coffee | 0.5 | 0.10 |
| Washed | | |
| 53% wet parchment | 9.0 | 1.86 |
| 12% dry green coffee | 3.7 | 0.77 |

The thickness of the layer is usually 2 to 4 in. (5 to 10 cm), and the coffee is stirred at frequent intervals during the daytime drying period. This is done with wooden rake-like tools pushed by hand. In Salvador, this operation has been mechanized by using light pneumatic tire tractors to pull the rakes in order to save labor.

In the late afternoon the coffee is gathered into heaps and covered with tarpaulins for the night for several reasons:

1. The coffee is warm from the sun's heat, and some of this heat is conserved since the coffee can be delivered slightly warmer the next day, thus accelerating the drying.

2. The moisture distribution tends to be equalized during the night since little or no drying goes on at this time.

3. The tarpaulins also protect the coffee from rain and from taking up moisture from the cooler and damper night air. Since sun radiation is absent in the night, the air temperature may drop considerably. This sharply raises the relative humidity of the ambient air.

If, during the daytime drying, rain occurs or even threatens, the coffee is usually hastily gathered into heaps and covered to avoid reversing the drying process and prolonging the operation which in turn increases the cost and endangers the quality by favoring infection. Often the covering of the coffee develops into a race with a sudden shower which the coffee grower sometimes loses. Figure 6.7 shows hand methods of stirring coffee while it is drying. (See also Figs. 6.8, 6.9, 6.10)

The coffee is gathered into heaps by broad scrapers, usually hand drawn, but sometimes drawn by draft animals or tractors followed by hand sweeping. In a few areas, the coffee is protected against rain by movable roofs running on tracks. In this case it is not gathered into heaps. This method is fast and labor-saving but it requires a heavy investment in the construction of roofs.

*Courtesy of Pan American Coffee Bureau*

FIG. 6.7. HAND METHODS

Ambient temperature and relative humidity are important factors in determining the speed of drying on the terrace. Rains occurring during the drying may cause the beans to reabsorb moisture, which may result in loss of color and cup quality and give the dried beans a spotty appearance. If the parchments are broken before the coffee is dry, the resulting bean is bleached and poor in cup quality.

## The Determination of Moisture in Natural and Washed Coffee

In order to know when to stop the drying operation both on the drying terrace and in a mechanical drier, the moisture content must be determined. This may be done in a number of ways with varying degrees of accuracy. For this purpose it is desirable for one to follow the drying from about 20 percent moisture content down to the standard 12 percent or even below. This is important because seriously overdried coffee suffers irreversible quality damage.

The standard method usually accepted is to heat 2 to 5-gm granular samples in aluminum moisture dishes in a drying oven with circulating air at 212 to 220 F (100 to 104 C) for 24 hr, cooling the samples in a desiccator to prevent reabsorption of moisture, and weighing the samples before and after drying on an analytical balance. This method is too slow to be used in the field.

If sufficiently brittle, samples may be ground to facilitate evaporation of moisture.

Another standard method is to boil a 5 to 10-gm sample of coffee with toluene

Fig. 6.8. Colombia. De-skinned, Fermented and Washed Coffee Beans are Dried in the Sunshine from 50 to 15% Moisture. This Takes About a Week Depending on the Weather. The Coffee Bean Layers Must be Frequently Turned Over to Expose the Wetter Darker Colored Coffee Beans Underneath to Drying Air and Radiant Heat. Note That the Coffee Beans are Spread on Retractable Drawers Which Can Place the Coffee Beans Under the Roof in Case of Rain.

(boiling point 232 F (111 C)) in a flask with a reflux condenser collecting the water which distills off in a graduated tube where it may be measured volumetrically. The process is continued until water no longer is collected. This usually takes 1 to 2 hr. These methods, while reasonably reliable and reproducible, are often impractical because of the rather expensive equipment and trained personnel required, and also because they are too slow to be of use in process control where results are often needed in a few minutes.

FIG. 6.9 and 6.10. BRAZIL. SLUICING WHOLE CHERRIES INTO WHEEL BARROW, SPREADING ON PATIO.

For this purpose various electrical instruments are used. While the instruments are expensive, the results usually are available in 5 min without requiring trained personnel, and they are reasonably reliable if standardized occasionally against the oven or toluene methods. Standardization is usually done at the beginning by the instrument manufacturer who uses natural, parchment, and green coffee and furnishes a table for each type of coffee.

If instruments are not used, supervisors of the drying operation use certain rule of thumb methods which have the advantage of being fast and, with skill and experience, are sufficiently reliable to serve the purpose though usually less dependable than the instrument tests.

The coffee sample, whether natural or washed, is shelled by hand or in a small portable huller, and the green beans are examined for color and hardness (rubbery or brittle). Biting and thumbnail penetration and, for parchment coffee, the ease with which the parchment and silver skin break away, are useful tests for judging the moisture content. Coffee at 12 percent moisture has a light greenish-blue color. At higher moistures, coffee appears black and at still higher moisture white or gray.

When the coffee has reached approximately 12 percent moisture content, which may take about a week for washed coffee and about three weeks for natural coffee, it is gathered into bins where it is stored for further moisture equalization until it is to be hulled, see Fig. 6.11. Sometimes it is sold as parchment coffee without hulling. In this case, the buyer collects coffee from a large surrounding area and operates a hulling establishment from which he sells the product as finished green coffee. This system is in vogue in Colombia where the usual coffee plantation is small and does not justify the installation of hulling machinery, which is relatively elaborate, expensive, and of high capacity. The large plantations usually have their own hulling machinery.

## Natural Coffee

The problems of drying natural coffee are somewhat different from those of drying washed coffee. The latter has a very uniform initial moisture content, although even with washed coffee, size differences result in different drying rates for individual beans. Natural coffee, especially when strip-picked, may have a

*Courtesy of Pan American Coffee Bureau*

FIG. 6.11. GATHERING THE DRY COFFEE. TRACTORS ARE ALSO USED IN STIRRING DURING DRYING very wide range of initial moisture content from ripe fruit at 65 to 70 percent moisture to partially dried material at 25 to 30 percent moisture. Because of the many layers enclosing the seed, the drying is much slower than with washed coffee which is an advantage as far as moisture equalization is concerned. The drying of natural coffee is very much faster when the fruit has a high percentage of ripe cherries which allows the wettest portions to catch up partially with the drier portions. The process of moisture equalization is still an important part of the drying. This is accomplished by allowing longer rest periods, that is, by gathering the coffee into piles earlier in the afternoon and speading it out later in the morning than is the practice with parchment coffee, and occasionally leaving it in piles all day. The management of this operation is an art requiring practice and skill, and it is controlled on the spot by visual and tactile inspection of small samples. Figures 6.12 and 6.13 show mechanical methods of handling coffee on the drying terrace.

When the coffee is stored in a mass, the enclosed air reaches an equilibrium relative humidity which allows drying of the wettest units and absorption of water by the driest so that all parts tend to attain the same moisture content.

Another problem relating especially to natural coffee is the matter of infection. Since it dries slowly, it remains moist longer; there is, therefore, a longer period in which damaging microorganisms may grow. Furthermore, since the fruity outer layers offer a better medium for their growth than the clean parchment of washed coffee, natural coffee is more difficult to protect against this kind of damage. Rainy weather during the sun drying period tends to increase the risk of such damage, although there are indications that very heavy rain may be able to wash away microorganisms and thus be less damaging than light rain or dew. Nevertheless, any rain retards the drying process and coffee should be protected from it if possible.

*Courtesy of Pan American Coffee Bureau*

FIG. 6.12. BAG FILLING ON THE TERRACE

*Courtesy of Pan American Coffee Bureau*

FIG. 6.13. SUCTION AIR LIFT FOR TRANSFER TO STORAGE

Table 6.3 shows the course of a small-scale drying experiment carried out on ripe coffee cherries. The results are given on a daily basis.

It may be noted from column 3 that the rate of water loss is about 100 times as fast when the coffee is in the form of the ripe cherry as when it contains about 10 percent moisture. Column 4 gives the exact heat equivalent of the water lost each day based on a latent heat of evaporation of 580 cal per gm at 85 F (29.4 C). Column 5 gives the total solar energy that fell each day of the experiment as observed by the pyrheliometer. Column 6 is the percentage ratio of column 4 to column 5. It will be observed that, as in the case of water loss, the efficiency is much higher in the earlier than in the later stages. This fact throws light on how best to handle coffee on the drying terrace. It would indicate that the coffee beans should be dried first on the terrace and later in mechanical warm air driers.

## SOLAR STILL DRYING

Another possible application of the sun drying of coffee is the solar still designed by Telkes (1953). The method has not yet been applied to coffee, but the theory is interesting. It was originally proposed for small-scale use on life rafts to

TABLE 6.3. TERRACE DRYING OF NATURAL COFFEE BEANS[1]

| 1. Day | 2. Percent Moist. | 3. Water Loss, gm | 4. Equivalent Calories | 5. Available Calories | 6. Heat Efficiency |
|---|---|---|---|---|---|
| Start | 63.90 | — | — | — | — |
| 1 | 61.48 | 1087.0 | 630,460 | 1,033,200 | 61.02 |
| 2 | 52.25 | 592.1 | 343,418 | 1,033,200 | 33.24 |
| 3 | 45.08 | 355.5 | 206,190 | 1,033,200 | 19.96 |
| 4 | 39.64 | 399.0 | 231,420 | 725,400 | 31.90 |
| 5 | 32.09 | 177.4 | 102,892 | 990,000 | 10.39 |
| 6 | 28.09 | 186.5 | 108,170 | 954,000 | 11.34 |
| 7 | 23.34 | 212.7 | 123,366 | 1,088,000 | 12.24 |
| 8 | 17.10 | 49.0 | 28,420 | 1,033,200 | 2.87 |
| 9 | 15.49 | 32.0 | 18,560 | 1,033,200 | 1.80 |
| 10 | 14.45 | 45.7 | 26,506 | 1,008,000 | 2.57 |
| 11 | 12.88 | 34.2 | 19,836 | 1,033,200 | 1.97 |
| 12 | 11.66 | 17.7 | 10,266 | 1,033,200 | 0.99 |
| 13 | 11.02 | 29.7 | 17,226 | 1,033,200 | 1.67 |
| 14 | 9.92 | 9.6 | 5,568 | 1,033,200 | 0.54 |
| | | | | Overall efficiency | 13.80 |

[1]Initial weight, 6kg (13.23 lb). Drying area, 1,800 sq cm (1.93 sq ft). Initial moisture, 63.90%. Depth, 6 cm (2.4 in.).

produce fresh drinking water from the sea, and for large-scale use for the same purpose on dry, tropical islands where little rainfall occurs and solar energy is at a maximum.

The solar still is essentially a heat trap, an enclosure containing a material to be dried such as coffee or a liquid to be evaporated such as sea water. It has a transparent cover of glass or plastic with roof-like slopes for drainage of the condensed water on the inside. The radiant energy passes through the cover with little loss and without substantial heating of the cover, and then it is absorbed by the wet material with a rise of temperature which raises its vapor pressure and the absolute humidity of the saturated air within the enclosure. The cover, being cooler and kept so by the ambient air, condenses moisture which drains away either for use or disposal. Hence, a partial distillation is accomplished. One obvious advantage in drying coffee by this method is that the coffee is protected from rain and requires little or no daily handling.

## Available Solar Energy

The solar constant mentioned earlier is 1.94 gm-cal per sq cm per min or 429.1 BTU per sq ft per hr at normal incidence outside the earth's atmosphere. For a horizontal surface at 20° latitude, the actual energy reaching a level surface in full sunlight averages about one-half these values throughout a 12-hr day or 0.97 gm-cal per sq cm per min = 214.6 BTU per sq ft per hr. For the 12-hr day, the total energies reaching the unit areas are 698.4 gm-cal per sq cm per day = 6,984 kg-cal per sq m per day and 2,574.6 BTU per sq ft per day, respectively. The latent heat of evaporation for water at 68 F (20 C) is 585 kg-cal per kg or 1,053 BTU per lb. The efficiency of the operation has been found to be approximately 70 percent, and hence 6,984 ÷ 585 × 0.7 = 8.36 kg $H_2O$ per sq m per day and 2,575 ÷ 1,053 × 0.7 = 1.71 lb $H_2O$ per sq ft per day would represent the capacity of the solar still in full sunlight during the constant-rate period of drying.

## Water Evaporated from Four Types of Coffee Beans

The following four types of wet coffee to be considered:

| | Type Bean | % Moisture |
|---|---|---|
| A. | Coffee cherries | 65 |
| B. | Partially dried fruit | 25 |
| C. | Strip-picked fruit | 40 |
| D. | Pulped, washed coffee | 53 |

These have the following weight relations on drying calculated on the basis of a unit weight of dry, green coffee beans:

| Type Bean | A | B | C | D |
|---|---|---|---|---|
| Weight of finished green coffee (12% $H_2O$) | 1.00 | 1.00 | 1.00 | 1.00 |
| Weight of wet coffee | 5.00 | 2.34 | 2.93 | 2.34 |
| Weight of water removed | 3.00 | 0.34 | 0.93 | 1.09 |
| Weight of dry coffee before hulling | 2.00 | 2.00 | 2.00 | 1.25 |
| BTU per lb green coffee to evaporate $H_2O$, 70% efficiency | 4,513 | 511 | 1,399 | 1,640 |
| Kg-cal per kg green coffee to evaporate $H_2O$, 70% efficiency | 2,507 | 284 | 777 | 911 |
| Bulk density of wet coffee | | | | |
| lb/cu ft | 36 | 36 | 36 | 41.5 |
| kg/cu m | 577 | 577 | 577 | 665 |

For the remaining calculations, only coffees A (65%) and D (53%) will be used. The other two may be calculated similarly if desired. For brevity, the metric equivalents are omitted.

If it is assumed that a ½ in. layer of beans is to be spread on 1 sq ft of solar still, the latter will be loaded with the following quantities:

| | A | D |
|---|---|---|
| Lb per cu ft as above | 36 | 41.5 |
| Wet coffee, lb | 1.50 | 1.73 |
| Water removed, lb | 0.90 | 0.82 |
| Unhulled dry coffee obtained, lb | 0.60 | 0.94 |
| Green coffee produced from each lot, lb | 0.30 | 0.75 |
| BTU required at 70% efficiency | 1,354 | 1,236 |
| BTU available per day | 2,575 | 2,575 |
| Number of lots per day | 1.90 | 2.08 |
| Green coffee per sq ft per day, lb | 0.57 | 1.56 |

In order to obtain maximum heat efficiency, it would probably be desirable to use about 1 in. of thermal insulation under the coffee bed. Calculations show that if this much is used, with coffee temperatures ranging from 100 to 140 F (38 to 60 C) about 2 to 7 percent of the total solar energy will be lost downward through the coffee bed.

All the above figures are based on practically ideal performance. Three factors may tend to reduce this and should be investigated further. (1) The coffee will doubtless lose water more slowly as it becomes nearly dry. The fact that the temperature is higher than it would be on the open terrace would tend to sustain the drying rate. (2) The absence of sunshine will, as always, cut down efficiency and interfere with scheduling and perhaps control. (3) An important question is whether top quality may be obtained by this method considering the elevated temperatures and moisture-saturated atmosphere in which the drying is carried out.

The chief advantages to be hoped for would be: (1) reduction in the area necessary to dry a given amount of coffee; (2) efficient use of low-cost solar energy; (3) reduction of labor costs; (4) partial independence of rainy weather, though interruptions of drying would still be present; and (5) a higher temperature during drying which might be advantageous not only in speeding the drying but also in facilitating the equalization of moisture content throughout each bean.

## MACHINE DRYING OF COFFEE BEANS

It is true that sun drying of coffee is attractive if the climate is such that dependable sunshine is available during the harvest season, because a paved terrace is simple to design, construct, and operate and because the sun's energy and the winds are free of direct cost. Yet the heat efficiency is low, 7 to 13 percent, necessitating a large terrace area and much labor. For these reasons, the overall cost of sundrying may be higher than that of machine drying. In choosing between the two methods, a careful economic balance must be calculated, and the resulting choice will depend on the level and trend of labor costs as well as terrace construction, maintenance, and replacement costs as compared with the costs of machinery, power, fuel, and the operation of mechanical driers. If the weather is not dependable, it is always preferable to dry by machine because wet weather may add enormously to the cost of sundrying as well as endanger the quality of the coffee by stimulating the growth of molds. When rain, especially prolonged rain, falls, the terraces are suddenly overloaded and wet coffee must wait, sometimes for long periods, before it can be dried.

Machine drying is, in general, independent of weather. On many coffee plantations the old method was sun drying, and terraces were built. As the more modern machine processes developed, and the old terraces remained available without further investment, it was frequently convenient to use them for a preliminary drying operation before machine drying. For this step, the wet washed coffee is spread on the terrace for about one day. During this short period, since the coffee loses water at the highest rate in the beginning, roughly one-third of all the water that is to be removed is evaporated on the terrace, leaving a lighter task for the mechanical drier in terms of capacity, fuel, power, and time required. In this manner, the moisture content is reduced from about 53 percent to somewhere around 44 percent (wet basis).

Further advantages of machine drying are: (1) dependable scheduling of the whole operation is possible; (2) since the process is faster than sun drying, there is less time for deterioration in quality while the coffee remains wet; (3) the fuel efficiency of a mechanical drier may range from 40 to 60 percent (though fuel must be purchased) as compared with the 7 to 13 percent obtainable on the drying terrace; (4) much less space is necessary to house the drying machinery compared with the huge areas required for the drying terraces; and (5) the heavy capital investment in terrace construction is eliminated.

The most important advantage, however, which mechanical drying has over sundrying is the reduction in labor required. With the rising trend in labor costs found almost everywhere in the world, this becomes an important consideration. While conditions vary greatly from place to place, it may be roughly estimated that one man operating a mechanical drier of large capacity can do the work of 20 to 30 on the drying terrace, although the latter, too, may be cut down somewhat by mechanization.

## General Principles of Machine Drying

The operating conditions common to most mechanical driers are as follows. The drying air is heated and passed through a bed of coffee. Since the coffee is granular, whether natural or washed, not much resistance is offered to the passage of air, yet there is enough to ensure efficient heat and moisture exchange between air and coffee as well as even distribution of air flow in all parts of the bed. The method of through-circulation drying is much more efficient as to heat and moisture exchange than mere passing of the air over the surface of the material being dried.

The coffee mass is usually kept in motion, which favors even drying. Both air and coffee temperatures are carefully controlled. The higher the temperature, the more fuel economy is possible, but coffee has definite temperature-time limits which, if exceeded, will damage its quality. To illustrate how heat efficiency rises with temperature, a study of a continuous coffee drier showed the rather striking temperature effects reported in Table 6.4. This was a pilot plant drier tested without proper insulation and without air recirculation; hence, the efficiencies are low but comparable for the different temperatures used. So there is every reason to perform the drying at the highest possible temperature that the coffee will tolerate without damage.

TABLE 6.4. EFFECT OF DRYING TEMPERATURE ON BEANS VS. HEAT EFFICIENCY

| Air Temp. F | Air Temp. C | Dry Green Coffee Production, lb/hr | Steam Consumption, lb/hr | Lb Steam Per Lb Coffee | Heat Efficiency, Percent |
|---|---|---|---|---|---|
| 126 | 52.2 | 13.0 | 136.2 | 10.5 | 16.1 |
| 143 | 61.7 | 28.9 | 211.5 | 7.5 | 22.2 |
| 176 | 80.0 | 52.9 | 272.0 | 5.1 | 31.3 |

**Safe Temperature Limits**.—It should be clearly understood that to specify a temperature limit alone is not enough. The time during which the coffee is maintained at a given temperature is just as important in its effect on quality.

Since conditions in different types of driers vary greatly, and since coffee temperatures are constantly moving up or down, temperature programming even in the same type of drier varies with different operators. In addition, the course of the drying may be interrupted by moisture equilibrating rest and storage periods in which neither heating nor drying goes on. For these reasons, it is impossible to specify safe levels for coffee temperature or even safe combinations of time and temperature. It is even less meaningful to specify drying air temperatures because coffee temperatures may be very different from air temperature especially at the beginning of drying when the cool coffee has entered the drier and is losing moisture rapidly which results in evaporative cooling. At times, the outside of the coffee beans may be considerably warmer than their centers. Overheating during drying produces sour or cooked flavors in the brewed coffee.

The best generalizations that can be made are:

(1) The true temperature of the coffee itself, to be reliably comparable, must be determined apart from the stream of drying air. One method is to remove a sample from the drier, place it quickly in a relatively large heat insulated container such as a thermos bottle with a thermometer, and take the maximum reading shown by the thermometer a few minutes after the sample is taken. The result will be very slightly below the true average coffee temperature in the drier. Even with this method, errors due to non-uniform temperatures in different parts of the drier remain.

(2) As an approximate guide, it may be stated that coffee will tolerate temperatures of 104 F (40 C) for a day or two, 122 F (50 C) for a few hours, and 140 F (60 C) for less than an hour without damage.

However, even this must be qualified since excellent tasting dried coffee beans have been prepared using fluidized bed driers at 160° F.

Actually, dry (12 percent) green coffee suffers perceptible flavor loss in about a week at 90 F (32 C); it is best stored under refrigeration.

Air temperatures as high as 180 F (82 C) have been used on cool, wet coffee without overheating the coffee. The tolerance of coffee itself to high temperatures probably varies quite markedly according to its moisture content, and there is reason to believe that it is least tolerant in the early stages when its moisture content is highest. The figures above apply to coffee in the later stages of drying when it contains less than 20 percent moisture. This is a rather prolonged period when there is little evaporative cooling and the air and coffee temperatures are not far apart.

This leads to a paradox. It is still good practice to use drying air of relatively high temperature in the first stages of drying because the coffee enters the drier at ambient temperature, and it is desirable to warm it up to a maximum safe drying temperature as quickly as possible. Since the coffee is wettest at this time and loses moisture readily (often 20 times the rate at the end of the run), large amounts of latent heat of evaporation are absorbed from the air without raising the temperature of the coffee. The air itself is quickly cooled by this effect and tends to approach its wet bulb temperature which, in heated air, may be far below its dry bulb temperature. For example, ambient air at 68 F (20 C) and 50 percent relative humidity (rh) would have a wet bulb (wb) temperature of 57 F (14 C). When this air is heated to a coffee drying temperature of 140 F (60 C), the rh is lowered to 6 percent and the wb temperature is 80 F (27 C), a difference of 60 F (33 C) between dry bulb (db) and wb temperatures. This is clearly shown by the psychrometric chart. Since the surface of the coffee is moist for a time, it is actually equivalent to a wet bulb. Hence, it is desirable to use the hottest drying air in the early stages in order to increase drying rate and heat efficiency. Even though the coffee is most susceptible to heat damage in this period, it is protected by the cooling effect of evaporation. This is why it is important to determine the true coffee temperature as distinguished from drying air temperature.

## OPTIMUM DRYING CONDITIONS

In establishing the best drying conditions for any drier, certain general principles of drying apply. Three factors make for high efficiency and low cost together with good control of coffee quality:

(1) The greatest possible *production capacity* from the available equipment, obtained by operating at the highest temperature consistent with good cup quality. This implies good temperature controls and accurate sensing elements. The maximum speed of drying is produced by high air flow, but this is limited by maximum flow that may be used efficiently; otherwise, power fan, and duct capacity may be wasted.

(2) *High heat efficiency* which minimizes fuel consumption. This is obtained by minimizing heat losses caused by heat and air leaks and by using partial recirculation of the moist exit air if the design of the drier permits.

(3) *Labor economy,* obtained (a) by using a few drying units of large size rather than the reverse, although flexibility to meet fluctuating coffee flows must not be overlooked, (b) by making full use of material handling machinery to move the coffee, and (c) by using automatic controls for temperature and relative humidity of the drying air.

## Establishing Optimum Conditions

In order to determine the correct air flow, the properties of the drying air should be known. The best single tool for understanding these properties is the psychrometric chart based on fundamental studies made by Willis H. Carrier in the early 1900's. It gives the complete relationships between dry bulb (db), wet bulb (wb), and dew point (dp) temperatures, relative and absolute humidities (rh and ah), and density, as well as other properties of air. If any two properties are known, the others may be determined from the chart.

The chart shows that ambient air, heated to the moderate temperatures used in the drying of coffee, has its relative humidity greatly lowered and its drying capacity greatly increased. For example, if ambient air at 70 F (21 C) and 85 percent Rh and having a drying capacity of only 6 gr per lb of dry air, is heated to 158 F (70 C), its relative humidity will be lowered to the rather amazing figure of only 7 percent and its drying capacity will be raised to 142 gr per lb.

**Drying Potential**.—The first principle is that the relative humidity of the drying air is the important factor which determines the drying potential of the air. Drying potential is defined as the driving force which accomplishes drying, and it is measured by the difference between the water vapor pressure of the material to be dried and the water vapor pressure in millimeters of mercury (mm Hg) corresponding to the relative humidity of the air. The vapor pressure of water (as of all liquids) is a function of temperature as shown on the curve of Fig. 6.14. The water vapor pressures of the water in air saturated with moisture are the same as those shown on the curve, and the percentage of relative humidity expresses the fraction of this saturation water vapor pressure in air which has a moisture of less than saturation. Example: if air has a relative humidity of 60 percent, and a temperature of 113 F (45 C), then the vapor pressure is $0.60 \times 71.9$ (from Fig. 6.14) $= 43.0$ mm of mercury.

In the drying operation, the material being dried has a certain water vapor pressure. If its surface is moist, this vapor pressure will be the same as that shown on the curve for the existing temperature. If the surface is dry, the vapor pressure will be less and, as the material becomes drier and the remaining moisture is held

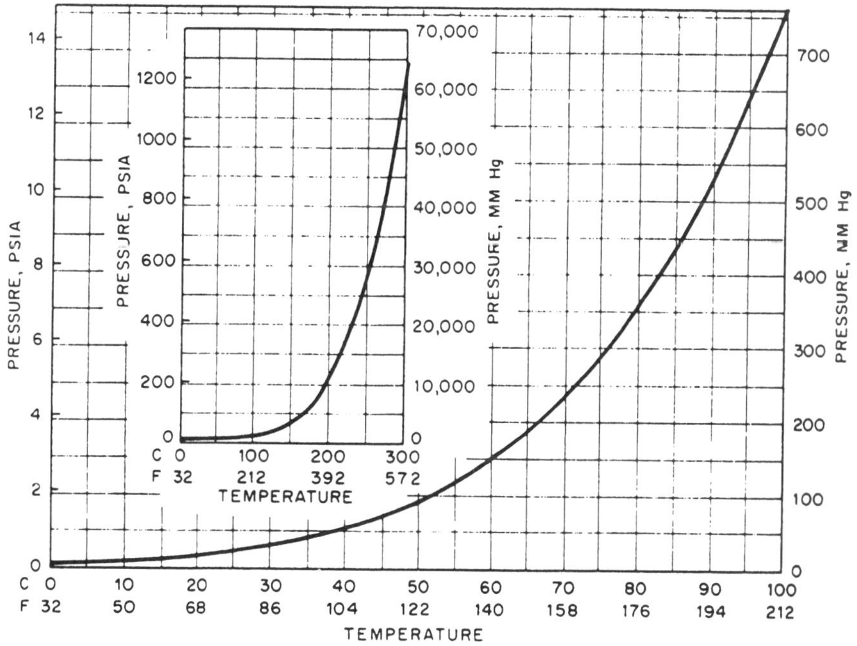


FIG. 6.14. VAPOR PRESSURE OF WATER

more tenaciously (bound water), the vapor pressure will diminish until it is near zero. This material vapor pressure constitutes the driving force *toward* loss of water, i.e., drying. On the other hand, the vapor pressure of the air as measured by the curve and the relative humidity constitutes a driving force *against* drying, and the difference, i.e., the excess of the material vapor pressure over that of the air, is the drying potential in any given circumstances. This is one factor influencing the speed of drying.

At the beginning of the drying operation, the coffee vapor pressure is high and remains constant as long as the surface is moist. After a time, the moisture at or near the surface is exhausted, and moisture must migrate from the interior of the bean to the surface. This is the so-called falling rate phase of drying. As time goes on, the water begins to be held more and more firmly by chemical forces, such as the bonds of water of hydration and of water which is part of the structure of the chemical constituents of the coffee solids. The drying process gradually becomes slower and the surface vapor pressure continually diminishes as time goes on.

A corollary to the principle that the relative humidity of the drying air should be as low as possible in order to have high drying potential and fast drying, is that, within moderate limits, drying rate depends much more on the relative humidity of the air than on the dry bulb temperature.

Heat transfer to the bean from the air actually is the controlling factor in water evaporation rates from the beans.

When material and air vapor pressures are equal, drying potential is zero, and drying ceases. At this point, the water content of the material is said to be in equilibrium with the drying air of that particular relative humidity. Raising the temperature will disturb this equilibrium. One should not overlook the fact that surface vapor pressure of the material is a dynamic concept and depends on the rate of moisture diffusion and moisture gradient between center and surface. If the movement is slow, this becomes the limiting factor, and drying cannot proceed faster than water can reach the surface no matter how low the relative humidity nor how high the velocity of the drying air—factors which, in the early stages, do accelerate drying.

**Wet Bulb Temperature in Adiabatic Drying**.—Carrier's second principle of drying clarifies the relationship between the wet bulb temperature of the drying air and the taking up of moisture by the air. This principle holds when drying air passes over or through a moist material under adiabatic conditions which means that no heat is added or taken away externally to or from the drier. The entering air has a definite and constant set of properties. It must have a relative humidity less than 100 percent or it can not do any drying. Hence, differences must exist between *dry bulb, wet bulb,* and *dew point* temperatures. Carrier stated that as the air passes over or through the moist material (assuming that its surface is moist and its vapor pressure is, therefore, as high as possible) the wet bulb temperature of the drying air remains constant while its dry bulb temperature falls as the air picks up moisture, and its dew point temperature rises. If the process continues until the air becomes saturated, the dry bulb temperature goes all the way down to the wet bulb temperature, and the dew point temperature rises all the way to the wet bulb temperature reflecting the increased moisture content of the air up to the limit of its capacity.

This is a simple concept, but it is by no means obvious. It means, in brief, that exactly the right amount of heat is furnished by the cooling of the drying air to supply the latent heat of evaporation of the moisture taken up by the air from the material being dried. The assumption is made that the initial temperature of the material is at or near the wet bulb temperature of the air. If it were cold, it would require sensible heat from the air to warm it and the wet bulb temperature would fall according to the quantity of heat necessary to warm the material. This important principle is used to calculate the drying capacity of the air.

## Drying Capacity of Air

The drying air entering a drier has a known moisture content most conveniently determined by taking its dry and wet bulb temperatures, and using the psychrometric chart to determine its other properties. This moisture content is known as the air's absolute humidity and it does not change unless moisture is added to or taken away from the air. Absolute humidity is expressed in grains of water per pound or per cubic foot of dry air. One pound equals 7,000 gr. The specific gravity of air under various conditions is also given on the psychrometric chart; hence, conversions may easily be made between pounds and cubic feet. The unit, grain, *gr* (7000 per lb) should be carefully distinguished from the unit, gram, *gm* (454 per lb).

In order to attain saturation during the drying operation, the dry bulb temperature must fall until it is equal to the wet bulb temperature which remains constant; hence, the air may take up moisture only to the extent of saturation at wet bulb temperature, not at the initial dry bulb temperature, a severe limitation. Obviously, the difference between the initial absolute humidity and the absolute humidity of saturation at wet bulb temperature is the drying capacity of the air. An example will make this clear.

Assume drying air at 180 F (82 C) dry bulb and 90 F (32 C) wet bulb. From the psychrometric chart, one can see that the dew point is 40 F (4 C), and that the absolute humidity (ah) is 36.5 gr per lb of dry air. The saturation absolute humidity at 90 F (32 C) is 218 gr per lb. Hence, the drying capacity is 218 − 36.5 = 181.5 gr per lb or, since air at 180 F (82 C) dry bulb and 90 F (32 C) wet bulb has a specific gravity of 16.4 cu ft per lb, the drying capacity is 181.5÷ 16.4 = 11.07 gr per cu ft.

**Recirculation of Drying Air**.—A third principle is useful in designing driers. In the example cited above, saturated air was discharged at 90 F (32 C). If ambient air temperature is assumed to be 60 F (16 C), the discharged air is 30 deg above the ambient air temperature. The 30 deg of sensible heat of this air might be used to help warm the incoming ambient air by means of a heat exchanger in order to save fuel in heating this air to 180 F (82 C). This is not generally done. On the other hand, part of the used exhaust air might be added to the incoming air. This would, of course, add moisture which would apparently lower the drying capacity of the mixture. The wet bulb temperature and relative humidity would also be raised, which would lower the drying potential. At first it would seem that the disadvantages outweigh or, at least, counterbalance the advantages. But a study of the actual figures involved reveals a net gain resulting from partial recirculation.

The first point that should be considered is the effect of the natural moisture content of the ambient air used for drying on the properties of the heated drying air. Since coffee is largely raised and dried in the tropics, the ambient air usually has a temperature not far from 70 F (21 C). Its relative humidity may range from 30 percent to nearly 100 percent.

Let us examine in detail what happens when air at 70 F (21 C) db temperature is heated for use in a drier. Two cases will be considered, viz., very moist air of 90 percent RH and very dry air of 10 percent RH. Using the psychrometric chart, we find the complete list of properties shown for each in Table 6.5 after the heading "Initial." If both wet air and dry air are heated to 160 F (71 C) as in drying coffee, it is found that the heated air has the two sets of properties shown after "Heated" in

Table 6.5. If the air is used to dry a material such as coffee, and the air picks up moisture to the limit of its capacity, both the wet air and dry air will be found in both cases saturated at the respective wet bulb temperatures. The further assumption is made that the coffee has a moist surface as in the initial stages of drying.

The drying potential, as explained above, is the difference between the vapor pressure of the material being dried and that of the drying air. Without specifying exact vapor pressures, the *relative* drying potential is measured by the difference in relative humidities of the material and the air. The relative humidity of the material, when it has a moist surface, is 100 percent of water vapor pressure at the given temperature (Fig. 6.14); hence, the relative drying potential are 100 − 7 = 93 percent for the moist air and 100 − 1 = 99 percent for the dry air, i.e., only a 6 percent advantage for the dry air.

To compare actual vapor pressures, again referring to Table 6.5 and Fig. 6.14, we find that the vapor pressure of the moist air heated to 160 is 244 × 0.07 = 17 mm Hg, and that of the dry air is 244 × 0.01 = 2 mm Hg, while the vapor pressure of the moist material (coffee) is 244 mm Hg. The true drying potential of the moist air is 244 − 17 = 227 mm Hg and that of the dry air, 244 − 2 = 242 mm Hg. Completely dry air would have a drying potential of 244 mm Hg. Hence, we arrive at the same result as before and find that the moist air has 93 percent and the dry air 99 percent of the theoretical drying potential for the given temperature of 160 F (71 C). We may conclude that the initial moisture content of the drying air makes little difference in its drying potential when heated.

TABLE 6.5. CHANGING PROPERTIES OF DRYING AIR
*Temperatures are Fahrenheit*

| Conditions | Dry Bulb | Wet Bulb | RH % | Dew Point | AH. gr per lb |
|---|---|---|---|---|---|
| Initial | | | | | |
| Moist air | 70 F | 68 F | 90 | 67 F | 100 |
| Dry air | 70 | 47 | 10 | 17 | 13 |
| Heated | | | | | |
| Moist air | 160 | 94 | 7 | 67 | 100 |
| Dry air | 160 | 78 | 1 | 17 | 13 |
| Saturated at W.B. temp. | | | | | |
| Moist air | 94 | 94 | 100 | 94 | 250 |
| Dry air | 78 | 78 | 100 | 78 | 145 |

The drying capacity of the air in grains per pound is the difference between its initial moisture content (ah) and its final or saturated moisture content, i.e., 250 − 100 = 150 gr per lb for the moist air and 145 − 13 = 132 gr per lb for the dry air. The surprising result is that the dry air has about 12 percent *less* drying capacity than the wet air because the final temperature and moisture content of the dry air are limited by a lower wet bulb temperature. The dry air has the further disadvantage that the temperature of the operation as a whole is kept lower. These results are shown in Table 6.6.

It may be concluded, then, that even extreme variations in the moisture content of the ambient air used for drying have only minor effects on the drying capacity of the air. Moist air (if not too moist) is, in general, more desirable than dry air. Variations in moisture content of the ambient air are disadvantageous, although only to a minor extent because of consequent variations that are necessary in the programming of the operation. For instance, the controls must often be reset in response to weather variations.

**Humidification of Drying Air**.—To go one step further, let us examine the effect of adding moisture to the drying air by recirculating some of the warm, moisture-laden air leaving the drier. By means of automatic controls, the mixture may be maintained at a selected constant level of relative humidity. In this case, two levels will be selected, 15 and 20 percent RH. For this calculation the initial conditions of the ambient air do not matter because variations in moisture content will be compensated for by adding the proper amount of moist exhaust air to bring the drying air up to a constant moisture level. The heated air and the final exit air will have the properties shown in Table 6.6.

TABLE 6.6. HUMIDIFIED DRYING AIR
*Temperatures are in Fahrenheit*

| Conditions | Dry Bulb | Wet Bulb | RH % | Dew Point | AH gr per lb | Capacity, gr per lb |
|---|---|---|---|---|---|---|
| Heated | | | | | | |
| (a) | 160 F | 116 F | 20 | 100 F | 300 | 200 |
| (b) | 160 | 109 | 15 | 90 | 223 | 179 |
| Saturated at W.B. | | | | | | |
| (a) | 116 | 116 | 100 | 116 | 500 | ... |
| (b) | 109 | 109 | 100 | 109 | 402 | ... |

The drying potentials are 80 percent for the wet air and 85 percent for the dry air. In other words, adding moisture to entering air increases the drying capacity considerably as shown in the last column and lowers drying potential moderately. The drying potential is not lowered seriously, although this factor becomes more important in the later stages of drying. At either degree of humidification (15 or 20 percent) the drying will proceed satisfactorily. Recirculation has the further advantage of maintaining constant drying conditions regardless of the moisture content of the ambient air.

If one selects air at 160 F (71 C) db and 15 percent RH having an ah of 223 gr per lb as the standard drying air, and assumes that the ambient air contains 100 gr per lb of moisture 75 F (24 C) db and 80 percent RH, the saturated return air will contain 402 gr per lb of moisture. Calculations show that one may mix about 40 percent of this return air with 60 percent of ambient air to obtain a mixture of the standard drying air specified above. The advantage is that about 40 percent of the sensible heat of the return air will be saved. The temperature of the mixture entering the heater will be about 89 F (32 C) instead of 75 F (24 C), so the heater will have less work to do with a consequent saving of fuel.

A study of the performance of a commercial coffee drier indicated that about 21 percent of the fuel required without recirculation was saved by recirculating to 15 percent RH for the drying air.

The calculation is made according to the equation

$$1 \times a + x \times b = c(1 + x)$$

or

$$x = \frac{c - a}{b + c}$$

where:

$x$ = the number of pounds of return air added to each pound of ambient air
$a$ = the ah of the ambient air in grains per pound
$b$ = the ah of the return air
$c$ = the ah of the mixed drying air

The first form of the equation states that 1 lb of ambient air containing $a$ gr per lb of moisture plus $x$ lb of return air containing $b$ gr per lb of moisture gives $1 + x$ lb of drying air containing $c$ gr per lb of moisture. The percentage of return air in the mixture is $x/(1 + x) \times 100$, and the percentage of ambient air is $1/(1 + x) \times 100$.

The conditions of the drying air may be controlled conveniently in practice by: (a) a dry bulb thermostat to control the air heater, and (b) a wet bulb thermostat to operate a set of dampers in the air duct system regulating the percentage of return air to be mixed with the ambient air to obtain the desired moisture content. The remainder of the return air escapes into the atmosphere carrying the moisture evaporated from the coffee out of the system.

**Pickup Efficiency.**—One more term which measures the performance of a drier remains to be defined. If heated air has a certain drying capacity as defined above in grains per pound of dry air, and it is passed through or over the material to be dried, it will be found (by determining the dry and wet bulb temperatures) to have certain absolute humidities also in grains per pound before and after passing through the drier. The difference between its initial and final ah is called the pickup, also measured in grains per pound. The ratio of actual pickup to drying capacity, i.e., the possible or theoretical pickup, is defined as the pickup efficiency and is expressed as a percentage. The factors affecting pickup efficiency are chiefly the thickness of the bed of material through which the drying air passes and the velocity of the drying air. A general indication of the pickup efficiency is how nearly the exit drying air approaches 100 percent RH. In order to maintain pickup efficiency at a reasonably high point, the velocity of the drying air may be reduced during the later stages when the coffee is giving off moisture slowly.

## Sonic Vibrations

It has been found that moist materials subjected to sonic vibrations in the frequency range of 8 to 16 kc may be dried more rapidly at lower temperatures and to lower moisture levels than by conventional methods. It has been proposed to dry heat-sensitive materials, textiles, paper, wood, films, and other products by this method. It is especially applicable to the falling rate period of drying in which moisture is removed slowly and inefficiently by ordinary air drying.

Sound waves of this type have been applied to homogenizers where the vibrations are set up within the liquid being treated, and to the drying of powdered and granular fluid materials as well as continuous films and sheets. For drying purposes, the sound waves are generated in the atmosphere surrounding the material and thus communicated to the material itself. The intensity may range from 130 to 170 decibels and the optimum frequency is determined by the resonance characteristic of the material. The energy attacks the force of attraction between molecules of moisture and the product to be dried by alternately raising and lowering the surrounding air pressure at the frequency employed. The ratio of compressed air used to power the whistle to the acoustic energy produced is such that 13.5 to 17.5 cfm of air at 20 to 30 psig will produce sound waves in the 160 decibel range in a laboratory rotary drier some 3 in. (7.6 cm) in diameter and 2 ft (61 cm) long. The air used to generate the sonic energy may be used to remove the water vapor given off by the material being dried.

When parchment coffee at about 18 percent moisture was subjected to this treatment, it was found that the rate of water loss in drying to about 12 percent moisture was increased 2 to 2.5 times compared with the normal rate without sonic treatment.

While complete quantitative aspects, costs of sonic energy per unit weight of product, precise methods of application on a large scale, and the overall economics of sonic treatment have not yet been determined, such a process might possibly be practical as a drying method for coffee in the future.

# DESCRIPTIONS OF DRIERS

## Hot Air and Screen Bottom Trays

One of the simplest types of driers consists of a horizontal wood or masonry tunnel, the top of which is enclosed by trays with bottoms of wire screen (Fig. 6.15) or perforated metal. The wet coffee is placed in the trays and heated air under

*Courtesy of American Drying Systems Company*

FIG. 6.15. HOT AIR AND SCREEN BOTTOMED TRAYS

sufficient pressure from a power-driven fan and heater is passed up through the coffee. The source of heat may be wood, coal, oil, steam, or even dry coffee parchment or whole fruit hulls. If the fuel produces smoke, the drying air is usually heated indirectly by means of a heat exchanger in order to avoid damage to the cup quality of the coffee. This operation is carried out under a roof, which may be movable, so that sun may be used if available. Loading, unloading and stirring the coffee during the drying are all done by hand. This method is used only for small scale operations. It may or may not have automatic temperature control. The process has rather poor fuel efficiency, and the labor cost is high in relation to production. However, it is suitable for certain conditions and may be applied to natural or washed coffee.

## Wilken Rotary Plow Drier

The Wilken drier shown in Fig. 6.16 is a simple and rather efficient development of the hot air-screen drier previously described. The distinguishing features are automatic loading and unloading and rotating power-driven rakes to keep the coffee continually stirred. The drying air may be heated indirectly by means of a heat exchanger. The model shown is 20 ft (6.1 m) in diameter and is designed for pulped and washed coffee. It has a capacity of 5,000 to 5,600 lb (2,268 to 2,540 kg) of finished green coffee per 24 hr. With an oil consumption of 3.1 gal (11.7 liters) per hr, it has a fuel efficiency of about 40 to 45 percent.

## The Guardiola Rotary Drum Drier

A successful and simply designed drier is the Guardiola, named after its inventor Don José Guardiola who lived in Guatemala. It is essentially a horizontal

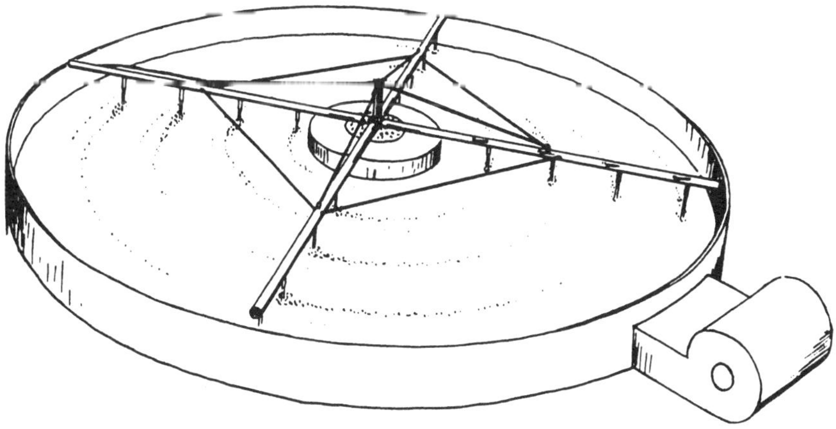

FIG. 6.16. DIAGRAM OF WILKEN DRIER

cylinder which revolves at about 4 rpm. Its cylindrical walls are fabricated from perforated sheet steel to allow free outward passage of moist air. Heated air is led in through a hollow trunion in each end and is distributed into the coffee from an axial duct with radial perforated arms. The warm air passes through the tumbling coffee and escapes through the outer cylinder wall (Fig. 6.17).

Large sizes of this machine accommodate batches of up to 12,000 lb (5,450 kg) of wet washed parchment coffee. Such a machine is approximately 16 ft (4.9 m) long and 6 ft (1.8 m) in diameter with a net volume of some 300 cu ft (8.5 cu m). Nearly 5,000 lb (2,260 kg) of dry green coffee are produced from each batch. Since wet washed coffee has a uniform moisture content of about 53 percent, it can be dried fairly uniformly (except for individual bean size variations) from start to finish without interruptions for moisture equalization, especially since the drying period is rather long, sometimes as much as 60 hr. The Guardiola drier is strictly for batch, not continuous operation. Natural coffee may be dried in this machine, but the process is slower and must be carried out at a lower temperature because natural coffee is more susceptible to heat damage than is parchment coffee. It may also take longer because more moisture must be removed depending on the state of dryness of the entering coffee (Fig. 6.18).

The standard accessory equipment for this machine consists of a wood, parchment, hull, steam, or oil fired furnace with an indirect air heater, i.e., a heat exchanger. The air is moved by a fan driven by a 10 hp motor.

The best practice is to control the air temperature by hand or thermostat according to a thermometer in the hot air duct near the drier. The air temperature should be regulated according to the coffee temperature, which is determined by another thermometer whose bulb is in the coffee itself. If the coffee temperature is kept in a safe range below 113 to 122 F (45 to 50 C), the air temperature may be as high as 194 F (90 C) in the early stages when the mass of relatively cool coffee is

*Courtesy of Geo. L. Squier*

FIG. 6.17. STEAM HEATED GUARDIOLA TYPE DRIER

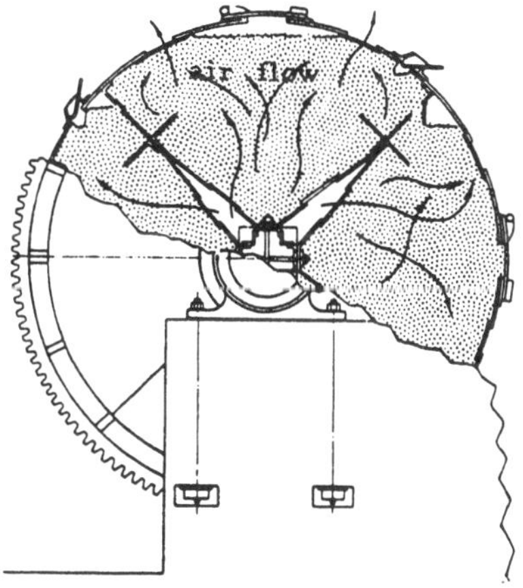


FIG. 6.18. GUARDIOLA SECTION

being warmed up and when the drying rate is high resulting in marked evaporative cooling. The air temperature is gradually lowered as the drying proceeds and the coffee temperature rises. In some cases, automatic temperature controls are used. By using as high an air temperature as possible within safe limits, the drying time for this size of batch may be shortened to 30 to 36 hr. An installation of Guardiola driers is shown in Fig. 6.19.

A very common practice is to pre-dry the wet washed parchment coffee on the drying terrace for about one day before using this or other mechanical driers. In good weather, the moisture is lowered to around 45 percent, thereby removing about one-third of the total amount of water that is to be removed. The capacity, time, power, and fuel required are accordingly reduced by about one-third.

To dry 12,000 lb (5,450 kg) of wet parchment coffee from 53 percent to 12 percent moisture in about 36 hr means the removal of 5,592 lb (2,530 kg) of water. This would require an air flow of about 3,000 cfm in the first part of the run and considerably less in the remainder.

The drier is charged through a hopper and a cylinder door. When the drying is complete, the drier door is opened; the machine is rotated continually until the coffee is discharged by gravity beneath the drier.

Some Guardiola driers are divided longitudinally into quadrants and laterally by circular partitions perpendicular to the axis so that small lots of coffee beans of different moisture may be dried altogether.

This drier has been used extensively throughout the coffee growing world for many years, and it performs well, on the whole. It is simple, trouble-free, and does not require a high degree of skill to operate. It is not designed for recirculation of the drying air; hence, heat and fuel efficiency are low. If the drying air is heated

*Courtesy of E. H. Bentall and Company*

FIG. 6.19. GUARDIOLA INSTALLATION

indirectly by means of a heat exchanger to avoid smoke contamination of the coffee, the heat efficiency is cut down somewhat more. If the available fuel is used to generate steam for heating the drying air, the cost of the heated air may be as much as 65 percent higher than with direct heating because of boiler inefficiency and heat exchange losses. In the direct method, the products of combustion are included in the drying air. If the combustion is complete and clean, the mixture may be used without damage to the coffee. Smoke, however, will impair coffee quality by introducing off-flavors.

The performance of a 12-ft model Guardiola drier in Costa Rica in November, 1958 is given in Table 6.7.

## The Torres Rotary Drier

The Torres drier, designed and manufactured in Brazil, is intended primarily for natural coffee although washed coffee may be dried in it. It embodies the principles of through circulation of air and rest periods for moisture equalization, but has no recirculation of the drying air.

Since natural coffee is usually of heterogenous moisture content, it is advantageous to interrupt the drying from time to time and store the coffee in equalizing bins. One reason for this interruption is that the different coffee berries tend to become more nearly equal in moisture content. Second, the moisture in each berry

TABLE 6.7. PERFORMANCE OF THE GUARDIOLA DRIER

| | |
|---|---|
| Wet parchment charged (52.0% moisture) | 9,150 lb |
| Dry parchment discharged (12.18% moisture) | 5,035 lb |
| Water evaporated | 4,115 lb |
| Time in drier | 33.5 hr |
| Average drying air temperature | 142 F (61 C) |
| Fuel oil used (140,000 BTU per gal) | 68 gal, 9,520,000 BTU |
| Water evaporated per gallon oil | 60.5 lb |
| Theoretical heat requirements: | |
| (a) To heat 9,150 lb 70 to 130 F, or 60° F, 0.605 BTU per lb per ° F | 332,000 BTU |
| (b) To evaporate 4,115 lb water at 1.023 BTU per lb | 4,210,000 BTU |
| | — |
| Total | 4,542,000 BTU |
| Heat efficiency | 47.7 percent |
| Drier capacity (12.8 percent parchment) | 3,600 lb per 24 hr |

tends to become more evenly distributed, i.e., it migrates from the center to the surface from which it can be evaporated more readily during the next drying period.

The main unit of the drier consists of a cylinder some 7 ft (2.1 m) in diameter and about 30 ft (9.1 m) long. It has a concentric cylinder on the inside that is about 3 ft (0.9 m) in diameter. Both cylinders are made of heavy wire screening supported in a stiff frame. The whole unit rotates on rollers at about 2½ rpm.

The coffee is charged into the annular space between the inner and outer cylinders. The drying air is blown into the inner cylinder and forced out through the coffee. The air heating and blower system is much like that of the Guardiola drier. The volume and drying capacity are also roughly the same as those of the Guardiola, but the Torres drier may be charged and discharged while running, thus allowing either batch or continuous drying. This is accomplished by an ingenious method of handling the coffee described below.

The program for drying includes filling the drier with coffee, allowing it to remain for a drying period, removing it to a bin for a rest period, returning it to the drier, and repeating the cycle until the coffee is fully dry, i.e., until it contains about 12 percent moisture. An elaborate system of bins and coffee handling machinery—conveyors, elevators, chutes, hoppers, etc.—is necessary, and the accessories of the drier occupy many times as much building space as the drier itself. The drier proper is shown in Fig. 6.20.

The periods spent in the bins are two to three times as long as those spent in the drier; hence, to keep the drier operating continuously, three or more batches are usually being handled at one time. This results in rather complicated schedules for programming the movement of coffee. The system is often a combination of batch and continuous operation.

The main advantage of this method is that a material of heterogeneous moisture may be brought gradually to a homogeneous final moisture content without overdrying the portions which were driest at the beginning. The disadvantage is

*Courtesy of Pan American Coffee Bureau*

FIG. 6.20. TORRES DRIER

that the overall drying cost is loaded with a rather top-heavy investment necessary for the accessory equipment, a large building, bins and handling machinery, and high power consumption for moving so much coffee beans so many times.

The drier may be used for parchment coffee, but the method is somewhat cumbersome and expensive for this purpose. Furthermore, the added handling may break the parchment. If parchment is removed from the bean while the coffee is wet, permanent discoloration of the final green coffee which impairs its appearance and quality may result.

The method of charging and discharging the coffee while the drier is running is interesting. At each end of the drier there are four radial ducts, one in each quadrant, in a plane at right angles to the axis revolving with the drier. They lead into or from the annular coffee space from or to the periphery. The outer ends of these ducts are U-shaped so that free coffee flow is impeded. At the charging end of the drier the U, which is about 1 ft long and 6 in. in diameter, swings downward and passes under a flow of coffee that fills the U. When the U-shaped end reaches the top point in its rotation, the coffee flows by gravity into and down the main portion of the duct and into the coffee chamber. This leaves the U-shaped end empty and ready for the next filling. This is repeated by each of the four ducts in turn; thus, there is a slow intermittent feeding of coffee into the drier. The rate of ingress is regulated by the rate of discharge at the other end of the drier. The drier remains full of coffee after it has once been filled. At this point, if charging and discharging continue, the operation becomes continuous. If charging and discharging are stopped, the drying becomes a batch operation.

At the discharge end, the operation is reversed; when a duct points downward, coffee flows from the main chamber into the U-shaped end but does not flow out because the U is turned upward. When the position is reversed, one-half turn later, the measured quantity that filled the U flows out into a hopper and the coffee is conveyed away. The amount of coffee discharged at each quarter turn of the drier may be varied by dampers or slide gates. Note the discharge ducts at right in Fig. 6.20.

The equalizing bins are filled by means of bucket elevators, belt conveyors, and their accessories together with a complex switching and dispatching system. The bins are numerous and large, usually constructed of wood. They have hopper bottoms for easy discharge and are provided with several openings for the ingress of outside air. These distribute the air at many points throughout the mass of coffee. Since the coffee is discharged warm from the drier, convection currents are set up in the ventilating air and the stored heat of the coffee is utilized to continue a slow drying while the coffee is resting, i.e., equalizing the moisture distribution.

On the whole, the Torres drier is best suited to the drying of natural coffee. Temperatures must be kept lower than for washed parchment coffee because coffee quality is much more easily damaged if the whole fruit is heated, especially while the moisture is high as in ripe and overripe cherries. As in the case of parchment coffee, it is difficult to set precise time, temperature, and moisture limits. The moisture range is greater than in parchment (for natural coffee, 60 to 15 percent), the degree of heterogeneity of moisture content is greater, the presence of the outer layers of fruit and the microorganisms they may contain have their influences on coffee flavor and color; even the fruit size variations are greater than with parchment. A rough estimate is that, in the early stages of drying at least, the temperature of the coffee itself should not be above 100 to 105 F (38 to 41 C) for any extended period. This means that the removal of water from natural coffee is slower and less efficient than from parchment—an argument for using the washed coffee process. Except for the lack of recirculation, however, the Torres drier does the job it was designed to do for natural coffee very well, and, if careful attention is given to the temperature control of both coffee and drying air, it is possible to obtain the best quality of natural coffee. The drier has had considerable commercial application in Brazil.

## The Moreira Vertical Drier

This drier was also designed, and is manufactured and used in Brazil. It is especially suited to the use of wood as fuel, and its most interesting feature is its wood burning furnace which accomplishes the smokeless combustion of wood with remarkable success. Thus, no heat exchanger is necessary and fuel efficiency is high. The products of combustion are clean and do not contaminate the coffee. Recirculation of the drying air is not used, although it would not be difficult to do so. A diagram of this drier is shown in Fig. 6.21.

The firebox is cylindrical and positioned nearly vertically, but leaning 15° toward the drier at the top. It is 30 in. (76 cm) in diameter and 40 in. (102 cm) high.

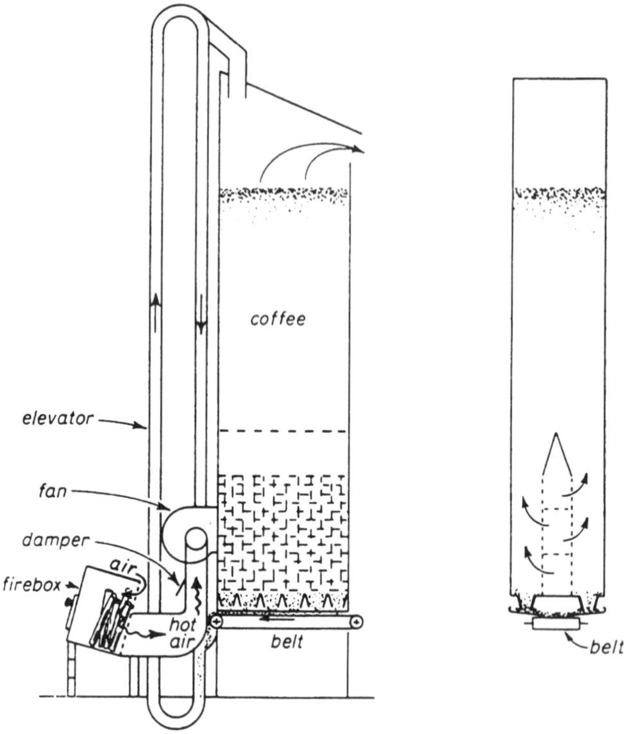


FIG. 6.21. MOREIRA DRIER

The nearly vertical side which is lower is fitted with a grate against which the sticks lean in a nearly vertical position. It is fed by hand and carefully placed in this position. The ambient air inlet is in the upper part of this grate. Here it meets the glowing charcoal face of the wood fire. Combustion gases pass downward across the face of the fire and out through the lower portion of the same grate into a suction chamber on the inlet side of a motor-driven fan which forces the combustion gases and the heated air through the coffee in the drier. As fresh wood is fed in at the back of the firebox, it is gradually heated mainly by radiation until volatile smoke-forming gases are driven off. These are drawn under suction into the fire but they cannot get into the drying air stream except through the brightly glowing coals of the face of the fire. The temperature is so high here that the volatiles are completely oxidized and become smokeless. By the time that the wood has reached the fire face, it is completely transformed into charcoal and burns without smoke. This design is very successful, and coffee bean quality is preserved.

The fan delivers drying air to a chamber whose only outlet is through screens set in vertical walls. The drying air passes into and through the coffee which is contained in a tower some 30 ft (9 m) high, 6 ft (1.83 m) long, and 4½ ft (1.37 m) wide. The coffee is charged into the tower by means of a bucket elevator. When the drier is full, the coffee is discharged slowly at the bottom of the tower to a short horizontal conveyor belt which carries it to the hopper of the same bucket elevator by which the drier was filled. Thus, the coffee is continually circulated during the run. The tower is bifurcated at the lower end, the legs straddling the drying air chamber with its screen walls. The air is blown into the two legs of the divided portion of the tower. The dimensions of each of the two screens are 6¼ ft (1.9 m) long by 5 ft (1.5 m) high. The coffee is discharged through several small funnels in the bottom of each leg of the tower onto a vibrating plate inclined toward a central trough in which the conveyor belt moves. The rate of discharge is regulated by changing the clearance between the funnels and the plates. Because of the small size of the funnel mouths, most of the drying air is forced upward through the main mass of the coffee.

Since the drying air which passes upward becomes saturated with moisture at a relatively low level in the tower and can do no further drying, most of the upper part of the tower serves as an equalizing bin. If the coffee is cooled in the elevator, as it may be slightly, some moisture is likely to be condensed on the coffee, but the quantity would be negligible. Some of the sensible heat of the air would be conserved in warming the descending coffee thus improving heat efficiency, and it is practically certain that the full drying capacity of the air is utilized by this arrangement at least until near the end of the drying. Recirculation could be accomplished quite easily by adding a return duct, but this is not usually practiced.

It is evident that the drier is designed for batch operation although, if the discharge were very slow, the operation could be made continuous by making only a single pass through the drier. A preliminary drying of wet parchment coffee could be made in this way.

The drier is suitable for both natural (whole fruit) coffee and washed coffee. The inlet air temperature is controlled from a thermometer probe and a manually adjusted damper opening into the suction side of the fan and admitting ambient air as required. The highest temperature is obtained by having the damper fully closed.

If the circumstances are such that wood is the most desirable fuel to use, the Moreira drier is well adapted to the purpose. The design is compact, and, except for the lack of air recirculation, the heat efficiency is good. The quality of the coffee is satisfactory. Repeated handling breaks some of the parchment or hulls, which then exposes the seeds and often results in discoloration and uneven drying of the beans. This defect, however, is not serious.

## American Vertical Grain Drier

This is a drier originally designed for grains such as corn, wheat, and rice. It was modified about 1952 for use with coffee, especially washed coffee. Ordinary cereal grains contain around 22 to 28 percent moisture before drying; washed coffee contains about 53 percent moisture. Hence, wet washed coffee must lose about 2½ times as much water as harvested grain.

This columnar drier has features similar to those of the Moreira drier and may be equipped with equalization bins for intermittent drying as in the Torres drier.

Figure 6.22A shows the general layout. The principal features are two columns with louvers through which the coffee passes downward. Each column is open to the inside chamber into which the heated drying air is forced from below. The only path outward for the drying air is through the coffee. At the bottom on each side, a slowly turning bar with a cross-shaped cross section discharges the coffee into a trough at a regulated rate which may be changed at will by means of a variable speed drive. At the bottom of the trough, a horizontal screw conveyor carries the coffee to one end of the trough where it is led by hopper and chute into a bucket elevator. The bucket elevator carries the coffee into the top of the drier for another cycle or it may be switched into a bin for a period of rest (see Figs. 6.22B and 6.22C).

The heated air is supplied by a fan and a direct-fired oil heater, which is controlled by a thermostat whose sensing element is located in the hot air chamber within the drier. A sufficient depth of coffee must be maintained at the top of the drier so that the resistance to air flow is equal to or greater than that through the louvers; otherwise, drying air will be wasted by a leak or short circuit without doing useful work. This type of drier easily lends itself to the use of recirculation of drying air.

The drier is manufactured in five models which dry from 2,000 to 32,000 lb of dry parchment coffee per 24 hr. The drier shown in Fig. 6.22A has a rated daily capacity of 7,500 lb. The drier proper is about 11 ft. high, the columns are 9½ in. thick from inside to outside, and the length is about 8 ft. The overall height of each louver is 1 ft. The drier, whose performance characteristics are discussed later, was designed for the partial recirculation of the drying air. It is connected with moisture equalizing bins, so the operation is carried out with alternating drying and rest periods.

The drier is operated by filling it with wet parchment either (1) directly from the washing process at about 52.5 percent moisture, (2) after pre-drying on the drying terrace for about a day which results in a moisture content of 40 to 48 percent, or (3) after pre-drying by one pass at a much lower rate through the same type of drier.

**Operation of Drier.**—When the drier is full of wet parchment coffee, the fan and air heater are turned on and the coffee is discharged at the bottom from both columns and carried to the top of the drier for a closed circuit recycling of the coffee. This is continued until the first *recycled* coffee has passed from the top to the bottom of the drier. At this point, the complete drier charge has had the same treatment and is presumably homogeneous.

If the equalizing bins are to be used, the coffee discharge from the bottom of the

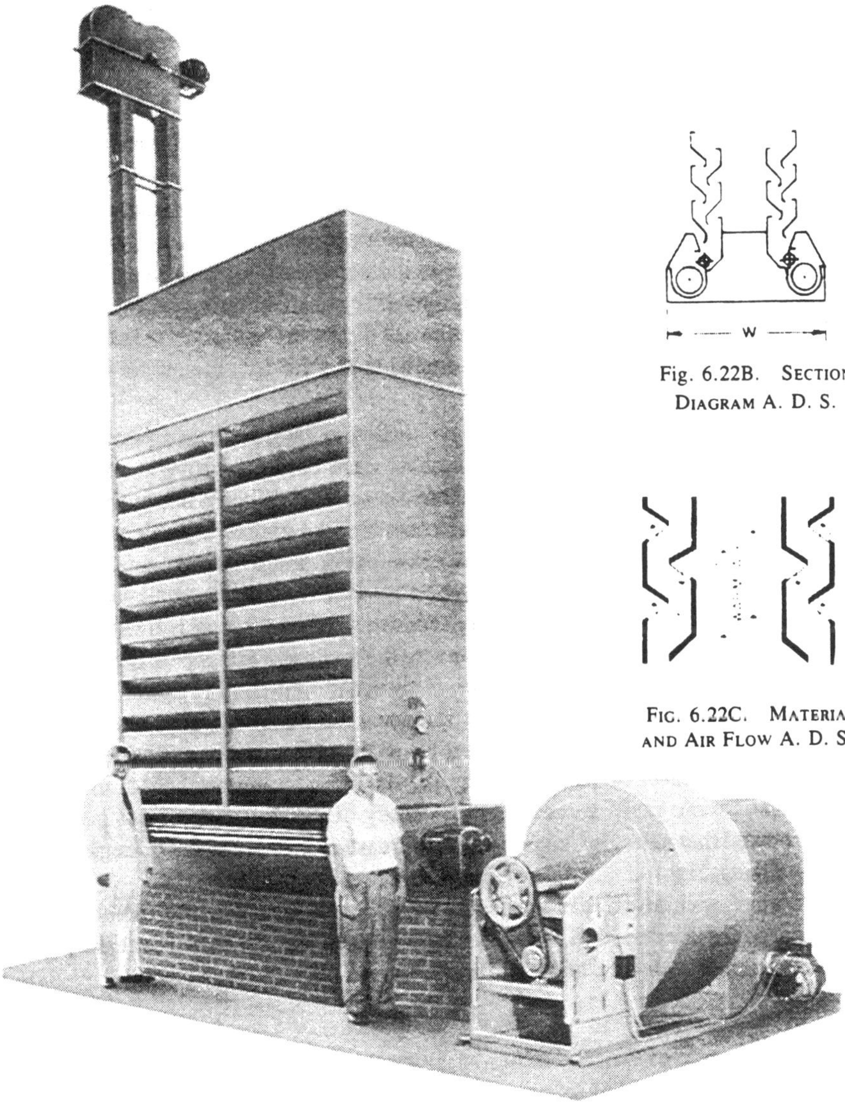


Fig. 6.22B. Section Diagram A. D. S.

Fig. 6.22C. Material and Air Flow A. D. S.

*Courtesy of American Drying Systems Company*

Fig. 6.22A. A. D. S. Drier

drier is now carried to the first bin, and the drier is replenished with more wet coffee. This process is continued until the bin is full. The next step is to feed the drier from this bin and to discharge the coffee into the second bin. This procedure is continued until the coffee is dry. When the drier discharge shows the correct moisture content, the coffee is moved to the dry coffee storage. Once the remaining equalizing bin is empty, coffee from the bottom is carried to the top of the drier until about one drierful has been recycled. Then the fan and heater are turned off and the coffee in the drier is discharged.

**Equalizing Bins.**—Usually, two equalizing bins are used with one drier. Their size is governed by such considerations as: (1) the desired relative times for active drying and rest, (2) how much coffee it is convenient to handle in one lot which may be determined by the desired relations between coffee drying cycles.

Two important processes take place in the equalizing bins. Since any granular material dries unevenly because the grains differ in size, during periods of rest all grains tend toward moisture equilibrium with the air in the mass of the coffee. This means that the wetter grains will dry somewhat and the drier grains will either gain in moisture or at least stop drying.

The other process which takes place during resting is that of equalizing the moisture content throughout the volume of each coffee bean. When a granular material dries, the water leaves only from the surface. When it leaves the surface, the outer layers become drier than the center, and a moisture gradient is formed from center to surface. The drier surface has less drying potential than the original fresh wet surface, and drying is slowed down. If the process continues without interruption, the rate is progressively reduced until the drying is finished. If the drying is stopped intermittently, however, the moisture gradient drives moisture from the center to the surface which becomes relatively wet again. When the drying is resumed, it goes faster for a time until the gradient again becomes steep. The result of alternate drying and resting is that the surface is consistently at a higher average level of wetness while the drying is going on; hence, the whole operation is more rapid and efficient. A study of the operation of this particular drier showed that its drying capacity was increased by 20.3 percent through the use of equalizing bins—a substantial improvement.

In normal operations, bins are used unless there is insufficient coffee to make a full charge for drier and bin. The drier may be used without bins if necessary, and this contributes a degree of flexibility to the plant.

**Performance of Grain Drier.**—Having described the operation of the drier, we now pass to a study of its performance. This will deal with such phases as the coffee production capacity, drying air flow, capacity, and efficiency, and the heat and fuel efficiency and power requirements.

The drier in Fig. 6.22A, having a capacity of 3,000 lb of wet (52.5 percent) parchment, is rated by the manufacturer at 7,500 lb of dry (12.5 percent) parchment daily. If predrying is not used, the quantities shown in Table 6.8 are involved:

TABLE 6.8. RATED DAILY CAPACITY OF GRAIN DRIER WITHOUT PREDRYING BEANS

| | Lb |
|---|---|
| 52.5 percent wet parchment | 13,816 |
| 00.0 percent (bone dry solids) | 6,563 |
| 12.5 percent dry parchment | 7,500 |
| Water evaporated | 6,316 |

Since one complete charge for the drier plus one bin are about 10,000 lb, one lot would be dried in 17 hr, 23 min, or 1.38 lots would be dried per day. If pre-dried coffee is used, its moisture content may range between 40 and 48 percent, which means that from 18.93 to 45.57 percent of the total water that is to be removed in drying from 52.5 to 12.5 percent moisture is removed in the pre-drying.

If it is assumed that the volumetric capacity of the drier and one bin accommodates the same weight of pre-dried coffee as wet (52.5 percent) coffee, viz., 10,000 lb, the pre-dried coffee contains more dry weight, and a correspondingly larger yield of dry coffee is obtained. Since less moisture needs to be removed, the coffee will dry in a shorter time. If the further assumption is made that moisture can be removed at the same rate as in the case of unpredried coffee, viz., 6,316 lb per 24 hr (which is not strictly true because the coffee loses moisture fastest when it is wettest, as shown in Table 6.14) we have the figures shown in Table 6.9 for the comparative daily dry parchment production capacity of the drier under discussion for coffee pre-dried to 47 percent moisture.

TABLE 6.9. RATED DAILY CAPACITY OF GRAIN DRIER WITH PRE-DRYING BEANS TO 47 PERCENT MOISTURE

| | Lb |
|---|---|
| 47.0 percent wet parchment | 16,018 |
| 00.0 percent (bone dry solids) | 8,490 |
| 12.5 percent dry parchment | 9,702 |
| Water evaporated | 6,316 |

One charge of 10,000 lb would dry in about 15 hr or 1.60 lots per day. In this case, the drier capacity would be increased by about 30 percent by pre-drying to 47 percent moisture. For other stages of pre-drying, the following calculation gives the daily capacity of the drier.

Let:

$x$ = the bone dry weight of coffee solids

$a$ = the moisture content of the pre-dried wet parchment expressed as a decimal number

$b$ = the weight of water evaporated per 24 hr

The factor 0.875 is equal to one minus the moisture of the dry parchment (12.5 percent) also expressed as a decimal.

Then

$$\frac{x}{1-a} - \frac{x}{0.875} = b$$

or

$$x = \frac{b}{\frac{1}{(1-a)} - 1.14286}$$

For everyday use, a table or graph may be constructed easily from this equation.

**Air Flow and Air-Drying Capacity**.—The conditions specified in the following discussion are those ordinarily used in the drier described above.

The standard air flow passing through the coffee is about 8,000 cu ft per min (cfm). This air is heated by a burner using Diesel oil of approximately 140,000 BTU per gal heating value. All parts of the duct work through which the air is passed have a cross section sufficiently great to keep the air velocity down to 1,500 ft per min or less. So the cross section measures 5.3 sq ft or greater. The pressure rise in the fan from inlet to outlet is about 1 in. of water. Under these conditions, the fan, which has a wheel diameter of 24½ in., is driven at about 1,039 rpm, requiring about 3 hp.

The drying air is heated to a dry bulb temperature of about 158 F (70 C) at the start and is maintained at this point until the coffee temperature has reached about 122 F (50 C). Then it is gradually lowered at such a rate that the coffee temperature remains within safe limits, usually 122 to 127 F (50 to 53 C).

Moist air from the outlet of the drier is partially recirculated to the inlet of the fan. The exact amount of recirculation is automatically controlled by the wet bulb thermostat as described on p. 154.

The coffee is relatively cool when charged, so if the drying air has a dew point temperature substantially above that of the coffee, moisture will condense from the air onto the coffee until the coffee temperature has risen above the dew point of the air. Since this tends to delay the drying, recirculation is not used until the coffee temperature has risen above 89 F (32 C), which is the dew point of air at 158 F (70 C) and 15 percent RH. It may require about 15 min after starting for the coffee to reach this temperature. After the initial warm-up period, automatic recirculation is turned on and is used for the remainder of the run. It has been found in practice that the drying air temperature should be reduced gradually during the course of the run (see p. 154) until, at the end, it is about 135 F (57 C).

To simplify the calculations, let us assume that one charge of coffee is that amount which will be dried in exactly 24 hr, e.g., 13,816 lb of 52.5 percent parchment. It has been found from experience that, on the average, the dry bulb temperature of the drying air for the first 8 hr of drying is about 154 F (68 C), for the middle 8 hr, about 147 F (64 C), and for the last 8 hr, 139 F (59 C), and that *at all times* it will have a relative humidity of 15 percent. The air will have a *theoretical* drying capacity (see Table 6.12), assuming that it becomes saturated on passing through the coffee, as shown in Table 6.10. Inlet and outlet refer to the air passing through the coffee bed.

The absolute humidity (ah.) is obtained from the psychrometric chart and is given in grains (gr) of moisture per pound of drying air. One pound equals 7,000 gr. Capacity is given in the same units and is obtained by subtracting the absolute humidity of the inlet air from that of the outlet air. The specific gravity of air at 154 F and 15 percent RH is 16.6 cu ft per lb. Calculations for the *theoretical* drying capacity of an 8,000-cfm flow of drying air are shown in Table 6.11. The actual moisture pickup is much less than this, as will be shown below.

The linear velocity of the drying air in the coffee is roughly estimated at 500 ft per min. The drier has 20 louvers each 8 ft long. The coffee bed in each one is 6 in. wide and 4 in. thick. For 20 percent open area, the total interstitial area for the 20 louvers of a cross section in the coffee at right angles to the air flow is about 16 sq ft. This means that the air would pass through the coffee in about 0.04 sec. Because of this brief time of contact, the drying air does not become saturated in its passage through the coffee. Hence, the pickup efficiencies are less than 100 percent. Records of an actual drying run made at an air flow of 5,000 cfm show the pickup efficiencies (see definition, p. 142) given in Table 6.12. The normal air flow of 8,000 cfm would show poorer results, although the coffee would be dried faster.

TABLE 6.10. THEORETICAL DRYING CAPACITIES OF Air in GRAIN DRIER

| Period | | Db, °F. | Wb, °F. | RH, % | Dp, °F | Ah, gr/lb | Capacity, gr/lb |
|---|---|---|---|---|---|---|---|
| 1st | Inlet | 154 | 103 | 15 | 86 | 190 | 143 |
| | Outlet | 103 | 103 | 100 | 103 | 333 | |
| 2nd | Inlet | 147 | 98 | 15 | 81 | 161 | 122 |
| | Outlet | 98 | 98 | 100 | 98 | 283 | |
| 3rd | Inlet | 139 | 91.5 | 15 | 74 | 128 | 102 |
| | Outlet | 91.5 | 91.5 | 100 | 91.5 | 230 | |

TABLE 6.11. DRYING CAPACITY OF AIR FLOWING AT 8,000 CFM IN GRAIN DRIER

| Period | Temp., °F | Air Flow, cfm | | Sp Gr, cu ft, lb | | Air Flow, lb/min | | Capacity, gr/lb | | Capacity, gr/min | Conversion Factor | | Capacity, lb/hr | | Hr | | Lb per 8 hr |
|---|---|---|---|---|---|---|---|---|---|---|---|---|---|---|---|---|---|
| 1st | 154 | 8,000 | ÷ | 16.3 | = | 491 | × | 143 | = | 70,213 | × 0.0085714 | = | 602 | × | 8 | = | 4,816 |
| 2nd | 147 | 8,000 | ÷ | 16.0 | = | 500 | × | 122 | = | 61,000 | × 0.0085714 | = | 523 | × | 8 | = | 4,184 |
| 3rd | 139 | 8,000 | ÷ | 15.6 | = | 513 | × | 102 | = | 52,326 | × 0.0085714 | = | 449 | × | 8 | = | 3,592 |
| | | | | | | | | | | | | | | | 24 hr | | 12,592 |

TABLE 6.12. MOISTURE PICKUP EFFICIENCIES IN GRAIN DRIER

| Time After Start, Hr | Pickup Efficiency, Percent |
|---|---|
| ½ | 70 |
| 1 | 76 |
| 2 | 81 |
| 3 | 78 |
| 5 | 76 |
| 7 | 78 |
| 9 | 78 |
| 11 | 69 |
| 13 | 64 |
| 15 | 59 |
| 17 | 54 |
| 19 | 46 |
| 21 | 37 |
| 23 | 33 |
| 24 | 22 |
| 25 | 20 |
| | Avg. 59 |

Let us assume that the average pickup efficiency for the first 8 hr period is 70 percent, making an allowance for the greater velocity of 8,000 cfm; for the second period, 60 percent; and for the third period, 30 percent. Applying these factors to the theoretical drying capacities calculated in Table 6.11, we arrive at an estimated actual moisture pickup for the 8,000 cfm air flow shown in Table 6.13.

Thus, it is seen that an air flow of 8,000 cfm heated to the temperatures specified and having a constant entering relative humidity of 15 percent, should be ample to accomplish the specified water removal of 6,316 lb per 24 hr.

*Changes in Drying Conditions*.—It will be noted that conditions in the early stages are quite different from those in the later stages of the drying cycle. At first, the grain surfaces are wet, the coffee vapor pressure is high, and the air is at a maximum temperature. Its vapor pressure is moderate, and the vapor pressure difference between coffee surface and air is high which results in a high drying potential. Coffee temperature cannot rise above the wet bulb temperature of the drying air; thereby, the coffee is protected against damage. Pickup efficiency is high, and this, combined with high drying capacity, results in a very high rate of water evaporation (see Table 6.14).

TABLE 6.13. PRACTICAL DRYING CAPACITY OF AIR

| Period | Temp., ° F. | Lb/ 8 hr | Pickup Eff., % | | Total, lb water | Lb/ hr | Per-Cent | Water to be Evaporated | Lb/ hr |
|---|---|---|---|---|---|---|---|---|---|
| 1st | 154 | 4,816 | 70 | × | 3371 | 421 | 49 | 3095 | 387 |
| 2nd | 147 | 4,184 | 60 | × | 2510 | 314 | 36 | 2274 | 284 |
| 3rd | 139 | 3,592 | 30 | × | 1078 | 135 | 15 | 947 | 118 |
| Tot. or Av. | | 12,592 | 53 | | 6959 | 290 | 100 | 6316 | 263 |

TABLE 6.14. DRYING CURVE DATA FOR GRAIN DRIER

| Time After Start, hr | Percent Moisture | Wet Weight, lb | Water Loss, lb |
|---|---|---|---|
| 0 | 52.5 | 10,715 | . . . |
| 2 | 48.8 | 9,941 | 774 |
| 4 | 44.9 | 9,237 | 704 |
| 6 | 40.8 | 8,598 | 639 |
| 8 | 36.9 | 8,066 | 532 |
| 10 | 33.1 | 7,607 | 459 |
| 12 | 29.5 | 7,219 | 388 |
| 14 | 26.5 | 6,925 | 294 |
| 16 | 23.6 | 6,662 | 263 |
| 18 | 20.9 | 6,435 | 227 |
| 20 | 18.4 | 6,237 | 198 |
| 22 | 16.8 | 6,117 | 120 |
| 24 | 15.4 | 6,016 | 101 |
| 26 | 14.0 | 5,918 | 98 |
| 28 | 13.0 | 5,850 | 68 |
| 30 | 12.5 | 5,817 | 33 |
| | | | 4,898 |

As the run proceeds, the coffee surface becomes less wet as the rate of its water supply falls off and the coffee vapor pressure diminishes. The coffee ceases to be equivalent to a wet bulb thermometer, and its temperature may now rise above the wet bulb temperature of the air. This proceeds gradually until, near the end of the run, the rate of moisture removal is so slow that there is little evaporative cooling and the temperature of the coffee becomes practically equal to the dry bulb temperature of the air. This is why the inlet air temperature must be lowered to protect the coffee from over-heating.

In the last few hours of the run, drying potential is quite low. Lowering the air flow slows down the drying to some extent, although, as may be seen from Table 6.12, at 20 percent pickup efficiency, five times the theoretically required air flow is being used. Cutting it to one-half, for example, might have very little effect on drying rate.

The data given in Table 6.14 are observations made on a typical drying run. From them a drying curve may easily be plotted. It is of interest to note how the rate of water loss falls off throughout the entire run and that drying air requirements diminish greatly in the later stages.

The effects of the use of bins and of recirculation on the capacity of the drier are shown in Table 6.15.

TABLE 6.15. EFFECTS OF BINS AND RECIRCULATION ON PRODUCTION CAPACITY
*Relative production of dry parchment per 24 hr.*

| | Bins | Recirculation |
|---|---|---|
| Not used | 100.0 | 100.0 |
| Used | 120.3 | 90.4 |

The figures of the first column of Table 6.15 show that the capacity of the drier is increased about 20 percent by using bins for moisture equalization; those of the second column show that the capacity of the drier was diminished by about 10 percent by the partial recirculation of the moist outlet drying air.

**Fuel Consumption and Heat Efficiency in Grain Drier.**—Observations on heating-oil consumption were made on four drying runs which showed the effects of the use of equalizing bins and of the partial recirculation of drying air. The results are shown in Table 6.16. This shows that fuel efficiency was increased by about 9 percent by the use of bins and by about 27 percent by the use of air recirculation. (Fig. 6.23, 6.24)

TABLE 6.16. EFFECTS OF BINS AND RECIRCULATION ON HEAT EFFICIENCY IN GRAIN DRIER
*Relative amounts of water evaporated per gallon of oil*

| | Bins | Recirculation |
|---|---|---|
| Not used | 100.0 | 100.0 |
| Used | 108.9 | 126.7 |

Thus, both equalization bins and recirculation improve heat efficiency and both are desirable even though drying is slowed down somewhat by recirculation, as shown in Table 6.15.

Experimental data show that, under normal conditions of operation, the American Drying Systems drier evaporates about 67 lb of water per gal of fuel oil. Including the heat necessary to warm the wet coffee up to the drying temperature, about 8 percent of the total, 1 gal of oil at 140,000 BTU per gal should theoretically evaporate about 127 lb of water from coffee. This performance, then, represents about 53 percent heat or fuel efficiency.

Another way of measuring the performance of the machine is to observe the production of dry (12.5 percent) parchment from wet parchment of various initial moisture contents. The experimental data showed that 1 gal of oil would, in practice, produce about

70 lb dry parchment from 52.5 percent wet parchment
87 lb dry parchment from 50.0 percent wet parchment
117 lb dry parchment from 45.0 percent wet parchment
137 lb dry parchment from 40.0 percent wet parchment

These performance figures, viz., 67 lb of water evaporated per gallon of fuel oil and 790 lb of water evaporated per 24 hr per 1,000 cfm of airflow at an average dry bulb temperature of 147 F (64 C), see Table 6.13, should apply approximately to any commercial size of any well designed coffee drier.

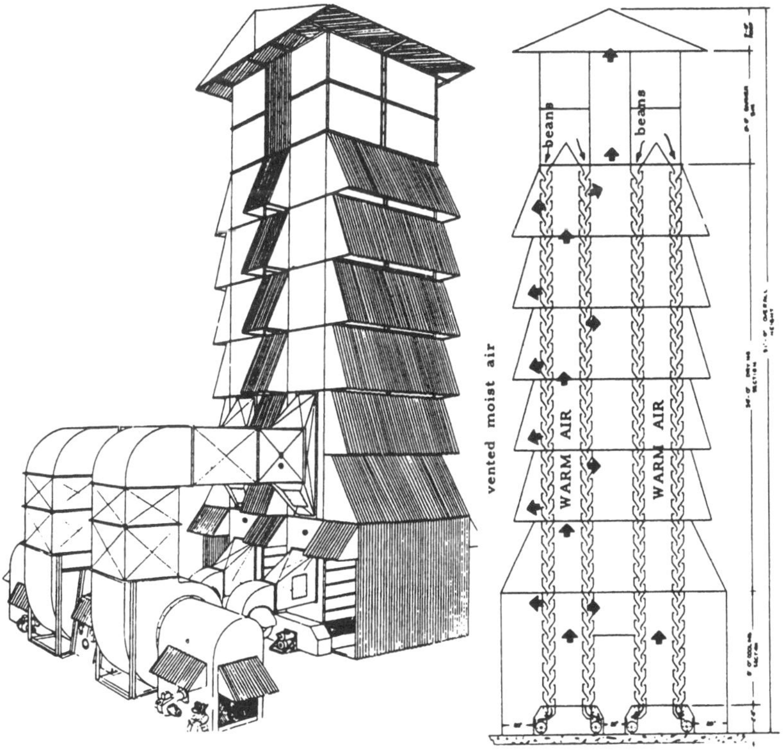


FIG. 6.23. WARM AIR GRAIN DRIER

## Fast Drying of Green Coffee Beans

It has been found possible to dry wet (52.5 percent) parchment coffee in 4 to 5 hr in a continuous horizontal rotary drier having through-circulation of the drying air. As might be expected, the coffee, although of excellent quality, was somewhat unevenly dried. The method is not in commercial use.

The conditions used in this case are given in Table 6.17.
This was a pilot plant drier and very inefficient in the use of heat. Normal efficiencies would be about 40 percent.

## Cost of Drying in Continuous Rotary Drier

The costs of drying 54 percent wet parchment coffee by means of a continuous rotary drier 6 ft 4 in. in diameter and 30 ft long and capable of producing 1,000,000 lb of finished green coffee in a four month harvest season would be approximately

FIG. 6.24. USING THE MOISTURE METER FOR RAPID RESULTS

TABLE 6.17. CONDITIONS IN CONTINUOUS ROTARY DRIER[1]

| | |
|---|---|
| Maximum coffee temperature | 140 F (60 C) |
| Air inlet temperature | 170 to 175 F (77 to 80 C) |
| Air inlet pressure | 3.5 in. of water |
| Washed coffee feed | 73.4 lb per hr |
| Dry parchment production | 38.0 lb per hr |
| Finished green coffee production | 28.9 lb per hr |
| Drier rotation speed | 1 revolution in 11 min |
| Drying time | 4.5 hr |
| Depth of coffee at discharge point | 6 in. |
| Moisture content of wet parchment | 54 percent (wet basis) |
| Moisture content of dry parchment | 11 percent (wet basis) |
| Heat required per pound green coffee | 12,000 BTU |
| Oil consumption | 2.5 gal per hr |
| Weight dry parchment per gal oil | 15 lb |
| Heat efficiency | 11 percent |

[1]Link belt rotolouver drier.

those given in Table 6.18. This estimate is based on the assumptions shown in Table 6.19 which applied in Brazil several years ago. The experimental drier used in this work is shown in Fig. 6.25.

## DESIGN FACTORS FOR DRIERS

Now that the details of operation of several types of driers have been examined, the main principles of design may be discussed.

TABLE 6.18. DRYING COSTS IN CONTINUOUS ROTARY DRIER[1]

| | Per lb Green Coffee |
|---|---|
| Capital costs | $0.00960 |
| Overhead | 0.00151 |
| Labor | 0.00067 |
| Operating costs | 0.01772 |
| Total | 0.02950 |

[1]Link belt rotolouver drier

TABLE 6.19. BASIC FACTORS IN DRYING COSTS

| | |
|---|---|
| Cost of drier delivered | $48,000 |
| Annual charges, 20 percent[1] | 9,600 |
| Common labor daily wages | 1.50 |
| Electric power | 0.025/kwhr |
| Fuel oil | 0.50/gal |

[1]Capital charges, depreciation, obsolescence, overhead, maintenance, building.

An efficient drier should have a high capacity per dollar of investment cost. It should have low operating, maintenance, and labor costs. These are obtained by having good design of details, freedom from breakdowns and lost time, and high quality of accessories, especially coffee handling machinery. The coffee should be handled gently so as not to break the parchment. The drier should make full use of automatic controls in order to maintain steady conditions and a high quality of product.

High production capacity is attained with the highest temperatures consistent with safeguarding the quality of the coffee. Air flows should be as high as possible consistent with high pickup efficiency. The latter should average better than 50 percent for the entire run (see Table 6.12). Recirculation of part of the exit drying air to a point around 15 percent relative humidity for the inlet drying air has the advantage of maintaining steady air drying capacity. While recirculation may reduce production capacity somewhat, this disadvantage is usually outweighed by the advantage gained in increased fuel efficiency. Here, economic factors must be balanced in order to determine whether to recirculate or not.

The intermittent use of equalizing bins increases both production capacity and fuel efficiency. Pickup efficiency is promoted by carefully designing the drier so that the thickness of the coffee bed through which the drying air is passed is sufficient to allow the air to become nearly saturated—at least during the first half of the run. Too thick a bed will increase the resistance of the coffee to air flow and needlessly increase air pressure and power consumption.

In order to obtain high heat efficiency, hot air ducts should be short and insulated against heat loss if necessary. For minimum power consumption, all air ducts should be large enough in cross section to reduce velocity below 1,500 ft per min, and they should avoid sharp turns in order to reduce friction and the building up of static pressure.

*Courtesy of Link Belt Co.*

FIG. 6.25. PILOT PLANT ROTOLOUVER DRIER

Given these conditions of drier design, skillful operation, and reasonably steady coffee flow, drying costs will be kept to a minimum and quality will be maintained at a maximum.

## SHIVVERS DRIER

The first coffee bean drying application on a commercial scale took place in November, 1975 near Cordoba, Mexico. An interesting variation on the drying of washed coffee beans in pergamino is the drying of coffee cherries in about 18 hours. (Fig. 6.26)

Why is the Shivvers drier more efficient? There is a mechanism, a rotating radial helical transporter that continually sweeps the driest beans from the bottom of the silo and pushes them to the center lift for spreading on top of the batch, thereby effecting mixing both at the bottom and top of the batch. This mixing is essential to minimizing overheating and damage to localized areas of beans, and results in uniformly dried batches in eight hours.

In-let air temperatures as high as 170° F were used in Colombia in March 1976 and the resulting eight hour dried 3.5 ton batch of beans were at least as good as reference standards, and better than from other driers.

The coffee beans are charged at 53 percent moisture, and in the first few hours so much water is driven off that the vented moisture steams and forms clouds as it leaves the silo drier and condenses in the outside cooler air. Indeed, the inside of the silo is near saturation temperatures almost all the time because condensation occurs on one's eye glasses and camera lenses almost immediately. The product moistures were measured at Cenicafe, the agricultural laboratory of the National Federation of Coffee Growers of Colombia at Chin-China and were near 13.5 percent prior to unloading and storage.

It is important to note that the bean temperatures during the drying period were less than 100° F at all depths. The 8.0 ton wet batch was 50 cm deep in pergamino in the 15 foot diameter silo, but 80 cm depths at 12 wet tons had been run earlier. The helix transporter rotates clockwise in about one hour when beans are wet and speeds up to about 25 minutes when drier. The location of the screw at the bottom of the silo is revealed by a decline or valley at the upper bean level, and its location can be followed.

Thermometers were inserted into the bean bed from the outer side wall at 15 cm above the base, 25 cm and 40 cm above the base. Also a thermometer was on top of the beans which was always five degrees Fahrenheit or less than several inches below the top level of beans. Bean samples were taken from different levels and their moistures were measured.

This sized drier consumes about eight gallons per hour of diesel oil fuel to heat air flowing at about 15,000 CFM, and with several inches static pressure forces the warm air up through $^{3}/_{16}''$ holes, uniformly across the silo area. Direct oil firing is used with no trace of combustion odors. Screw conveyors are recommended for loading or unloading so as to get the fullest use of the drier during the short coffee harvest season.

The fullest potential and capacities of this drier have not been realized because this work has only begun. However, it already seems clear that the Shivvers drier is faster, and more efficient than any other drier currently in use in drying coffee beans.

The "old wives' tale" that coffee beans have to be dried in the sun to get the right flavor has already been pretty well laid aside since almost all countries use mechanical drying. In 1976, El Salvador Socafe virtually abandoned patio drying for large scale mechanized drying on very tall grain driers. However, their drying conditions, starting with 53 percent moist, washed beans, was sufficiently severe, that there was noticeable bean flavor down-grading in the tall grain driers being used. Sun drying, due to its prolonged nature, especially in wet periods, when natural sun drying can take three weeks, can result in deterioration of taste quality of the coffee beans.

One of the really important benefits of the Shivvers drier is that it allows the washed pergamino coffee, or cherry, to be promptly dried. This obviates all sorts of deteriorating steps caused by intermittent drying, storage and, re-handling, with its implications of contamination, lack of uniformity, etc. Note that this helix sweep *does not* damage the pergamino as screw conveyors and elevators do when re-cycling grain drier beans.

## FLUIDIZED BED DRYING

A little over five years ago the Central American Technology and Industrial Investigations Institute (ICAITI) in Guatemala sponsored research on fluidized bed roasting of coffee beans and fluidized bed drying of parchment-covered coffee beans. To date there has been no real commercialization of either process, although ICAITI has recommended such drying to local farmers.

In the ICAITI report many dried coffee beans were said to be poor or unacceptable in taste and this implied that although the conclusions indicated that the process was acceptable and useful, it was just not practical.

Since fluidized bed drying worked so well in roasting, and roasting is 90 percent drying, then it was only logical to extend the application of the air fluidized bed to green coffee drying in the growing country. In fact, the drier system used is a partial bed fluidization with the balance being gravity movement of the beans. In any case, the beans—all the beans—are in substantial movement and acquire a uniform exposure to heated air and equilibration of temperature among themselves.

Running drying tests at a production plant revealed some anomalies of very considerable importance.

The fact is that all beans processed are not identical in quality—and for many reasons. There are cherries delivered day by day, lot by lot and week by week, variation in pulping, fermentation, washing and process times. Drying is also carried out in different ways; some beans are patio dried initially; some are subjected to direct diesel-fired heated air in the initial drying stages; some by indirect steam-heated air in the final stages of drying. There can be significant parchment stripping—perhaps 5 percent of throughput—which gives the beans a harsh taste. Results can be affected by mechanical breakdowns or changes altering routine drying operations. The actual degree of drying—that is the final moisture—may be a 10, 11 or 12 percent, and usually the lower the final moisture, the more flavor is lost. The nature of the crop itself—the range of large and small beans, defective beans, the degree of final classification, grades and so forth—produce unexpected variance. So can the degree of treatment of the beans in classification: for example overheating in parchment removal contributes a heavy, aged taste.

It is important to know of these anomalies since the test-dried coffee beans are being compared to normal production beans, and the production beans are not always either a good or a uniform control. For example, test dried beans were hand-hulled and hand sorted, production beans were machine hulled and sorted. So, even if the parchment dried beans tasted identical, the sorted polished beans would not.

Sivetz found that fluid bed drying of parchment coffee beans can be carried out successfully in two to four hours using inlet air temperatures from 160° F to 260° F (70 C to 127° C) and bean bed temperatures from 115° F to 190° F (46° C to 88° C) with effluent air temperatures about 10° F above parchment bean bed temperatures. The product beans are very uniform in appearance and color, in this case exhibiting the definite bluegreen color aspect characteristic of high quality Arabica coffee beans.

It appears that the next step is for processors in the growing countries to use, say a 200 lb parchment product fluid bed drier to prove its virtues for themselves—virtues which include simplicity, economy and performance. Uniformity of the final product without parchment damage are two big advantages over conventional, time-consuming methods, and this is a step toward real industrialization of the whole process.

## BIBLIOGRAPHY

AMERICAN SOCIETY OF HEATING AND AIR-CONDITIONING ENGINEERS. 1955. Psychrometric chart. 345 East 47th St., New York.

AMERICAN SOCIETY OF HEATING REFRIGERATING AND AIR-CONDITIONING ENGINEERS. 1939. Revised Bulkeley Chart. Based on Table 6, Chapter 1. Heating, Ventilating Air-Conditioning Guide.

CARRIER CORPORATION. 1959. Carrier Psychrometric Chart. 1. Normal Temperatures. II. High Temperatures. AC 467, Syracuse, N. Y.

FOOTE, H. E., and JOHNSTON, W. R. 1949. Pilot plant development of a new process for the curing of coffee. Unpublished Report. Standard Brands, Inc., New York.

HODGMAN, C. D., WEAST, R. C., SHANKLAND, R. S., and SELBY, S. M. 1962–63. 44th Edition. Handbook of Chemistry and Physics. Chemical Rubber Publishing Company, Cleveland, Ohio.

IVES, N. C. 1950. Coffee-Drying Project. Progress Report. Inter-American Institute of Agricultural Sciences, Turrialba, Costa Rica.

IVES, N. C. 1953. Some temperature, time, moisture relationships and processes for drying grains. Inter-American Institute of Agricultural Sciences, Turrialba, Costa Rica.

LLEWELYN, D. A. B. 1955. Coffee Drying. Preliminary report to the Coffee Board of Kenya. East African Industrial Research Board, Nairobi, Kenya.

MADISON, R. D. (Editor), and CARRIER, W. H. 1949. Fan Engineering. An Engineer's Handbook on Air, Its Movement and Distribution in Air-Conditioning, Combustion, Conveying, and other Applications employing Fans. 5th Edition, Revised. Buffalo Forge Company, Buffalo, N. Y.

MCLOY, J. F. 1959. Mechanical drying of Arabica coffee. Kenya Coffee *24*, No. 280, 117–129, 131, 137.

SIVETZ, M. 1976. Shivvers driers. Tea and Coffee. *148* (7) July.

SIVETZ, M. 1977. Coffee Origin and Uses. Coffee Publications, Corvallis, Oregon.

TELKES, M. 1953. Fresh water from sea water by solar distillation. Ind. Eng. Chem. *45*, 1108–1114.

VAN ARSDEL, W. B. 1962 Psychrometric Chart. Avi Publishing Co., Westport, Conn.

VAN ARSDEL, W. B. 1963. Food Dehydration. Vol. I. Principles. Avi Publishing Company, Westport, Conn.

VAN ARSDEL, W. B., and COPLEY, M. J. (1964). Food Dehydration. Vol. II. Products and Technology. Avi Publishing Company, Westport. Conn.

# 7

# Hulling, Classifying, Storing, and Grading Green Coffee Beans

Because of the marked difference in structure and size of wet and dry processed coffee bean hulls, slightly different machines are used for each type, but the basic principles originally conceived and applied by Smout underlie the operation. The hulling mechanism consists of a chrome iron screw with helical pitch increasing toward the discharge end. The 36-in. long screw rotates within matching concave covers. The top cover is held down with quick-opening clamps. Hulling is achieved by creating friction among the beans lying along the screw. The broken parts of the parchment, along with dust, are forced or fall through a perforated steel plate at the bottom by means of an air suction to a cyclone collector. A weight loaded discharge gate regulates the back pressure on the issuing beans so that their residence time in the cylinder is regulated in accordance with the type of hulling desired. As beans fall out of the discharge, countercurrent air flow carries away pieces of parchment. A 1,250 lb per hr pergamino feed requires a 10 hp motor. The screw turns at 200 rpm. See Figs. 7.1 and 7.2.

The machine that removes the dried skin, pulp, and parchment from the dried cherry resembles the hulling mechanism except that it has an assortment of star-shaped chrome iron screws rotating at 200 rpm between projecting steel staves from the housing. This baffling arrangement tears the husk off the dried cherries as they move through the machine. Openings in the staves allow the broken pieces to be drawn through and also help, by the movement of air, to remove the frictional heat of hulling. The screw is about twice as long as is needed for parchment removal alone so as to allow more working time and distance for full husk and parchment removal as well as polishing of the bean surfaces to improve their appearance. Machines of phosphorbronze enhance the blue-green coloration of the beans. Beans of Uganda Robusta dry cherry have heavy silverskin coats that ordinary polishing does not remove, but moistening the skins allows their removal by special machines. Dirty parchment can affect the final bean color and polish adversely. An average hulling loss for arabica parchment coffee in East Africa was

*Courtesy of G. L. Squier Company*

FIG. 7.1. COFFEE HULLER CLOSED

FIG. 7.2. COFFEE HULLER OPEN

found to be 19.76 percent. Experience in Latin America shows about the same result.

## INFLUENCE OF BEAN MOISTURE IN HULLING AND STORAGE

The drying of beans on the sun drying terrace and by machine has been described in Chapter 6. If drying is carried out too rapidly case hardening may occur, which is common in the drying of many grains. The surface is overdried and shrinks irreversibly to prevent easy movement of moisture from within the grain in an outward direction. Worse than this, the bean may become pale and bleached in appearance—signifying flavor deterioration. There is ample field evidence that when drying is done too rapidly under excessively warm temperatures, the potential cup flavor is largely lost from coffees that otherwise would have been considered excellent. Later discussion of green coffee storage will indicate that moisture and bean temperature are the key factors in green coffee flavor retention during storage. Thus the drying process not only causes some physical change during the shrinkage of the cell structure, but temperature and moisture conditions cause denaturation of proteins. There is almost a direct relationship between viability of the bean and its cup flavor. If the bean has been dried or stored so as to be able to germinate, it invariably has more cup flavor and bean color than the bean whose time-temperature-moisture exposure has destroyed its germinating power.

If the wet pergamino with 52 percent moisture enters a mechanical drier initially, then initial drying temperatures of the circulating air can be about 180 F (85 C). This moisture evaporates rapidly, causing the beans to experience self-cooling. After a few hours, the air temperatures are reduced to 167 F (75 C). Six to twelve hours later, the circulating air temperature may be reduced to 150 F (65 C) until the moisture in the beans is reduced to 12 percent. These temperature examples are only illustrative since the operator uses his discretion and judgment depending on his experience, urgency of processing, type of equipment used, and size and type of beans. The point is that air temperatures above 140 F (60 C) for very many hours at the end of the drying process can be injurious to coffee color and flavor. There are cases where drying is inadvertently carried too far and bean moistures of 8 to 10 percent result with a marked loss in coffee quality. Upon storage, such coffee beans will reestablish their equilibrium moistures with the atmosphere, but the harm done to the coffee quality will not be corrected. When the coffee is dry, it must be cooled to room temperature promptly to stop deterioration and to avoid the situation where the heat is retained in a large body of coffee upon discharge from the drier. Bringing the pergamino to the proper final moisture level facilitates removal of the hull and minimizes pressure of the huller on the bean, as well as minimizing the inclusion of hull dust into the bean surface. Also, reduced work effort means less horsepower in hulling and the beans are not subjected to localized heating and loss in shape. Where polishing practices are used, proper final moistures minimize the polishing needed to remove traces of adhering silver skin.

## Moisture Content During Transportation

Coffee bagged in the coffee-growing country should have a standard moisture content of 12 percent with a practical tolerance of about 1 percent. Variations in bag weight outside of this range may be cause for complaint, although payments are usually adjusted to net weight of coffee received at the port of entry. This makes allowances for spillage and torn bags as well as for moisture variations en route. Obviously the exporter bagging 14 percent moisture knows that moisture will drop during shipment, so this should not be a source of complaint. However, if the beans are exposed to conditions where moisture is picked up, there is a serious hazard of spoilage. Even if the shipment arrives in good order, the buyer may be poorly disposed toward the shipper since he has paid for more coffee than he actually gets. In any case, the seller runs the risk of obtaining a bad reputation.

Normally, moisture is about 12 percent when coffee is bagged and does not alter much during short storage and shipment to the consuming country. However, where longer storage occurs under high relative humidity in warm climates, moisture may be absorbed from the atmosphere and raise moistures in the bag 1 or 2 percent; on the other hand, long storage at relatively low relative humidities (35 percent) may cause bean moistures to fall to 10 percent. Moisture measurements on imported coffees often show much variation from new to old crops and allowances in roasting losses must be made accordingly.

Green coffee stored for long periods in the producing country under ambient conditions may suffer marked damage. For example, beans originally at 12 percent moisture may rise to as much as 15 percent moisture in damp storage at altitudes of several thousand feet. Even though ambient temperatures are only about 70 to 75 F (21 to 24 C) mold growth may be an imminent danger. In 85 to 90 F (29 to 32 C) seaports also having high relative humidities, beans may become bleached out in color and lose most of their desirable flavor properties within a few weeks. Since port storage often lasts for months, the amount of quality loss in the ports is great and often explains the poorer qualities received by buyers in consuming countries as compared with samples received earlier from the same export agents.

The bright greenish-blue color of new crop coffee with high moisture (14 percent) is a misleading guide to bean quality because it is not equivalent to the genuine bluish-green color and waxy appearance of high grown properly dried mid-crop coffee. The former usually gives a thin cup and the color fades rapidly. Brown coloration in beans is often caused by improper patio and/or mechanical drying. A mottled roast bean coloration is usually due to uneven drying.

The batch drying of beans of uniform size and type will be more uniform than the drying of beans of different initial moisture, size, and density.

Musty flavors may easily develop when moist coffee is held in piles during periods when drying facilities are inadequate to handle through-put, so that partially dry pergamino is held and stored under humid conditions. A green flavor may be imparted to a coffee that is not thoroughly dried.

## Viability of Green Coffee Seed in Storage

The coffee seed is the coffee bean in its parchment. It may be sown after pulping or after drying. The bean is viable even if the parchment is removed, but this must be done with care since it must not damage the plumule, minute young leaves, or the radicle, embryonic root. Germination is the growth and development of the seed. Water softens the seed coat (parchment), causes the seed to swell and burst the coat. This facilitates the passage of oxygen from the air into the seed and carbon dioxide out. This is respiration, a process of life; it is related to metabolism, the movement of food materials. Thus water, oxygen, and temperature govern the rate of respiration and, subsequently, of germination. During this growth process the food materials in the seed are consumed. The tiny leaves then begin photosynthesis.

The point of the foregoing paragraph is to show that the bean in parchment is a living seed and the means used to dry and store it may destroy its living characteristics. In the coffee-growing country, it is a matter of common knowledge that storage of dried beans in parchment results in much less deterioration than storing the same beans hulled. This is a very important point since storage in parchment is necessary to help protect the coffee flavor during long months of storage in the coffee-growing country under what often would be undesirable temperature and/or humidity conditions. The parchment and silver skin are seed coats and thus probably act to dampen the otherwise severe ambient influences that may do great damage to the seed (bean).

Figure 7.3, which shows the respiration rate of rough and milled rice vs moisture content of the grain illustrates two important points: (1) that the respira-

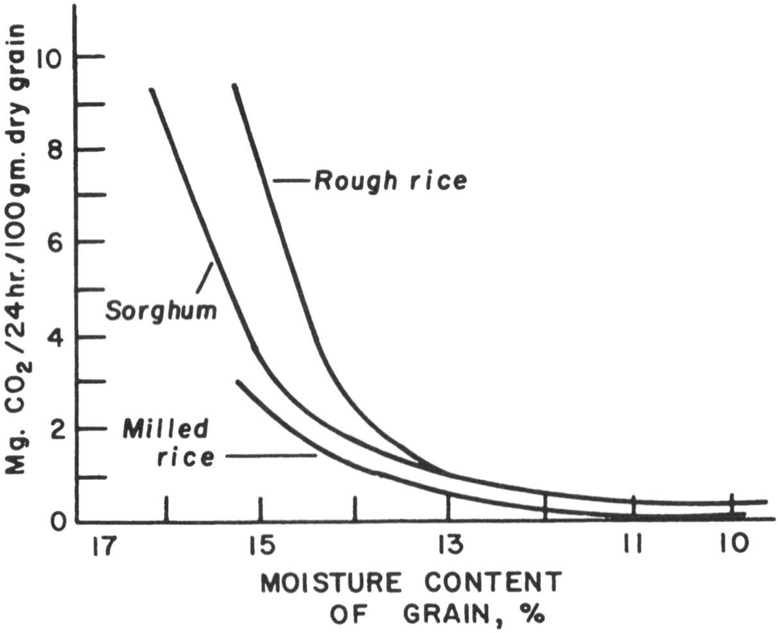


FIG. 7.3. RESPIRATION RATE VS MOISTURE CONTENT OF GRAINS

tion rate, hence metabolism, of milled rice (without seed coat) is several-fold less than the rough rice (in seed coat), and (2) that as moisture content of the grain rises above 14 percent there is a sharp increase in rate of metabolism and respiration. In other words, at moistures below 13 percent the seeds in both cases are relatively dormant.

This respiration characteristic is known to be general for all seeds, but the maximum allowable moisture for relative dormancy varies somewhat. The process of respiration is the conversion of oxygen to carbon dioxide—an exothermic reaction. An average of 4.4 mg $CO_2$ evolved per 24 hr per 100 gm grain and a heat of formation of 96 cal per 44 mg $CO_2$, results in a heat rise of 0.5 F (0.25 C) per 24 hr in the grain. Since a large mass of beans or grain is self-insulating, the temperature rise will be cumulative day after day. The rate of respiration and therefore, heat evolution, increases with rising temperatures. Thus when coffee is not fully dry, it is important to keep good air circulation about the coffee so as to maintain it at an ambient temperature which will minimize temperature buildups if they should occur. Another reason for keeping minimal storage temperatures is that lower temperatures yield lower rates of metabolism. Altogether it can be seen from this situation that high ambient port temperatures of 90 to 100 F (32 to 38 C) and high relative humidities (80 percent) may create circumstances where respiration rates are increased with local self heating and loss of vegetable food for the plant; also, the degree of viability is reduced. All this is accompanied by a chemical change, bleached colors, and a loss in desirable coffee flavors. Figure 7.4 shows the equilibrium moisture of coffee beans at various temperatures and relative humidities, and similar data on lumber, wheat flour, and other grains.

Under such circumstances, it is best to determine the moisure content of the beans quickly by means of a moisture tester (see Fig. 7.3). The important point is to keep the bean moisture sufficiently low so that respiration is relatively dormant, and conditions so dry that it tolerates higher temperatures without damage. Then, later storage conditions must be such that the bean moisture and temperature do not rise to the point of high rates of respiration.

Seeds up to 16 weeks old had 95 percent viability and germinated within 2 weeks. By 21 weeks the seed had 87 percent, by 30 weeks had only 60 percent viability. By 47 weeks only 22 percent of the beans germinated; this took 26 days from sowing. Coffee seeds will be viable for decades if properly dried and kept cool. Under warm, humid conditions all viability can be lost in a few months. The three major factors affecting the life of seeds in storage are moisture, temperature, and oxygen or air composition. Elimination of oxygen will eliminate respiration, and a reduction in oxygen will reduce the rate of respiration. Storage at 40 to 86 F (4 to 30 C) with humidities between 35 and 55 percent retains high germination rates at five months. At 76 percent relative humidity the germination rate is constant at 41 F (5 C) but falls off quickly at higher temperatures.

The influences of relative humidity and ambient temperature on bean moisture

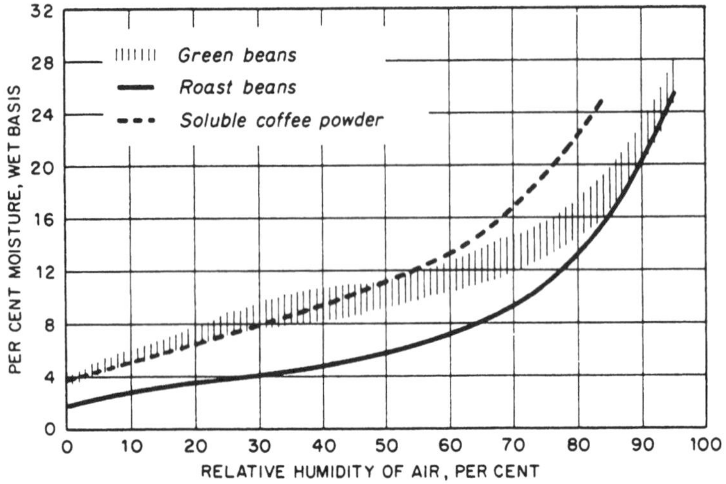


FIG. 7.4. EQUILIBRIUM MOISTURE CONTENT OF COFFEE VS RELATIVE HUMIDITY OF AIR

are important to the storage of coffee at all times but particularly when the beans are undergoing ocean shipment. Before World War II, coffee bean cargoes were protected in most cases only by natural air circulation. Damage to the beans would occur when moisture that was liberated in a warm location condensed as water droplets on coffee in cool locations. Condensation in the hold might occur when a ship passed through cold waters or during squalls. Water transport under continually humid and rainy circumstances requires dehumidification of the air circulated by forced air blowers.

Freezing should not cause a seed to lose germinating power since there is no free moisture, and proteins are not damaged. Shipment of beans in parchment with bulk density of about 25 lb per cu ft is more costly compared with hulled beans of about 45 lb per cu ft; furthermore, only 80 percent of the weight is green coffee.

The magnitude of the heating due to respiration is shown by the following example of a warehouse containing 8,000 69-kg bags of green coffee which may have a respiration rate of 4.4 mg $CO_2$ per 24 hr per 100 gm of beans. The heat of combustion of carbon is 8 kg-cal per gm C or 8 gm-cal per mg C = 96 gm-cal per 12 mg C or 96 gm-cal per 44 mg $CO_2$, hence 9.6 gm-cal per 4.4. mg $CO_2$ (as in example above). Therefore we have 9.6 gm-cal per 100 gm beans per 24 hr or 6624 gm-cal per 69 kg beans per 24 hr or 26.3 BTU per bag per 24 hr or 210,400 BTU per 8000 bags per 24 hr for the whole warehouse. Some means of cooling must be found in order to prevent a damaging rise in temperature.

## Pallet Storage of Bagged Green Coffee

In the United States where most of the coffee comes from Brazil, Colombia, and Central America, pallet sizes usually are 57 to 60 in. (1.45 to 1.52 m) by 68 to 72 in. (1.73 to 1.83 m) to accommodate 20 bags in four layers piled in alternate directions. The bag weights range from 60 to 70 kg (see Table 7.1), and the total weight of one loaded pallet is about 1.5 tons. Since bulk densities of the different coffees vary, some bags are slack and stack well and others are plump and stack with more difficulty. The moisture content also affects bulk density and absorption of moisture may cause swelling and bursting of bags.

The reasons for moving coffee in bags rather than by the cheaper bulk handling with machinery are to facilitate handling in the absence of such machinery and to allow the segregation of large and small lots. The sizes are related to the load a man can carry. Figure 7.5 illustrates a plan view of a warehouse for storing palleted coffee beans.

There is no storage plan that will fully utilize all storage space all the time. The smaller the lots of bags handled, the more space must be devoted to accessibility, or extra work is involved in removing one lot to gain access to another. Further, each pallet needs aisle space of 4 to 8 in. on each side depending on the size of

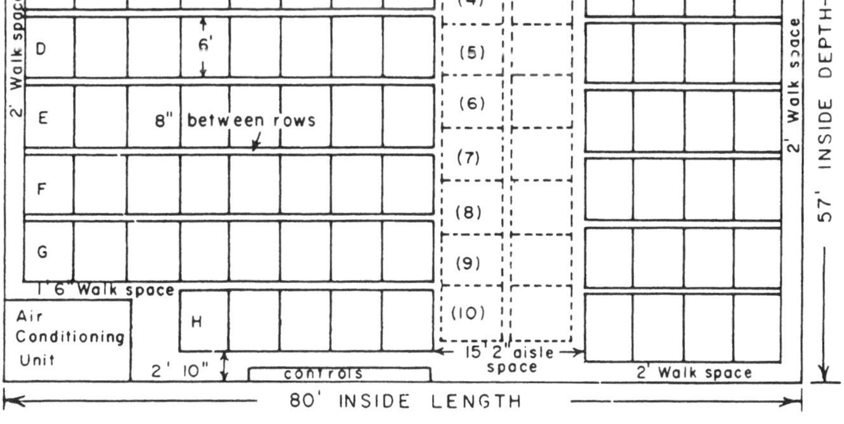


FIG. 7.5. PALLET LAYOUT

TABLE 7.1. NET BAG WEIGHTS OF GREEN COFFEE BEANS SHIPPED FROM PRODUCING COUNTRIES

| | |
|---|---|
| 60 kg or 132.3 lb | Brazil<br>Zaire<br>Kenya<br>New Guinea<br>Dominican Republic |
| 70 kg or 154 lb. | Colombia<br>El Salvador<br>Costa Rica<br>Guatemala<br>Nicaragua<br>Mexico<br>Honduras<br>Peru<br>Ecuador (up to 202 pounds)<br>Venezuela |
| 75 kg or 165.5 lb | Haiti<br>Dominican Republic<br>Java and Sumatra |
| 80 kg or 176.4 lb | Angola<br>Arabia |
| 90 kg or 198.4 lb | Cuba |

pallet, bag, and overhang of bag, otherwise the bags catch and a pallet cannot be freely moved in and out of location. Aisle space along walls for inspection, and aisle space between stacks for the fork trucks to maneuver are also needed. Thus only two-thirds to three-fourths of the floor space is really available for accessible storage. Often, even less of the floor space is available because of continual withdrawals. Stacks are usually only four high, sometimes five high, but not higher because higher stacks are unstable and fall. Another reason is that excessively high telescoping lifts on the fork truck are required which add to expense and require more skill in maneuvering. Higher stacks also increase floor loading and add to construction expense. A 5 by 6 ft pallet stacked four units high exerts a force on the floor of 600 lb per sq ft, which is a severe floor loading. This is why many older storage warehouses accommodate stackings only one or two pallets high. The higher stackings can only be made on floors designed for them and thus are more common in the newer coffee warehouses and plants.

The pallet is usually about 6 in. high, and the coffee stacked on it, depending on plumpness of bag fills and levelness of layering, will be about 3½ ft or, overall, 4 ft per pallet. A 20 ft eave will allow five stack storage and an 18 ft eave only a four stack storage at the wall.

For prolonged storage in warm, humid climate, air conditioning is recommended; in freezing climates, space heating and even humidification may be required.

## Bulk Storage of Green Coffee Beans

Bulk storage of green coffee is done in varying degrees in both the producing and consuming countries. In the coffee-growing country, bulking is usually in quantities not greater than a few hundred bags per crib, and is usually done in wood frame construction. Bulk storage of green coffee in the United States serves a different purpose. It usually is used to provide storage of blends for 24 to 48 hr for a roasting operation. For example, one large processor has a false wall around a shaft in his building for its entire height of eight stories. The open area is about 150 sq ft, so when the shaft is full it holds about 5,000 bags of coffee. A large mid-west United States roasting plant built in 1950 has a monolithic reinforced concrete tower, 45 by 25 by 125 ft, partitioned into about a dozen shaft areas. The purpose of this tower is to provide a reservoir of numerous green coffees that may be withdrawn below at set rates for purposes of blending. As one shaft empties, another one of similar quality may be substituted so as to maintain the flow of blended coffee continuously. Each shaft may hold about 1,000 bags of coffee.

Bare concrete is not a good storage wall for green coffee beans. It permits moisture penetration and dusts alkaline concrete particles throughout its life. Thus, concrete storage shafts have to be coated to eliminate moisture penetration and dusting, and the coating must be inert to the green coffee.

Actually, bulk storage of green coffee beans can only be useful to a firm handling large quantities of relatively uniform beans. In the growing country this is seldom the case except at ports or points of grading and export. Also tall shafts with residual dust are explosion hazards, and long-term storage lasting more than a few months requires other precautions to avoid compaction and deterioration.

For example, long term storage requires air circulation through the beans to control humidity and bean moistures and temperatures. If good air circulation cannot be assured, then movement of the beans out of the bin and into the same or another bin must be possible to allow sampling and full inspection of the stored beans. In the long-term storage of grain, it is necessary to incorporate a temperature monitoring system so that hot spots may be detected before any respiration and condensation damage occurs. Thus the base cost of any silo system needs to be augmented with costs of conveyors, cleaners, discharge valves, air recirculatory and conditioning systems, and air filters. In a roasting plant where green coffee is fed to a continuous roaster processing 70 bags of green coffee per hour, a series of green coffee storage bins feeding a continuous green coffee blender justifies the use of bins with immediate storage capacities of several hundred bags each. Figure 8.4, Chapter 8, illustrates such an installation which is typical of recently constructed United States roasting plants. Storage in these bins is live and the bins may be emptied every day or every week. They are usually constructed of galvanealed steel in groups so as to use the economical advantage of common walls. The zinc plated steel will neither flake as will galvanized steel, nor rust like plain, unprotected steel.

## Palletized Vs. Silo Storage

In palletized storage, pallets are normally stacked five units high (100 bags) which occupies a floor space of 5 by 6 ft or 30 sq ft and has a height of about 25 ft.

In the other case a round silo occupies about the same floor space, has the same total height, including a bottom cone section, and holds about 150 bags of bulk coffee. (Fig. 7.6)

Storage in pallets is the least expensive method. Storage in a galvanized steel silo without accessories for loading and unloading costs about 12 times as much as storage in five pallets; a glass-lined steel silo costs about 24 times as much as the pallets; and a concrete silo costs about 36 times as much.

Because of the cost, the use of palletized coffee bags has been an attractive and most popular method for storage aside from such considerations as accessibility to small lots, normal air circulation about the bags and easy inspection and sampling. A fork truck is needed to carry pallets, but invariably one is needed to handle packed stock as well, so this is not an additional investment. Large silo storage of green coffee beans results in classification of beans by size, so the handling of blends in such silos is subject to examination of the degree of segregation, and consequent variation in discharge blend as the storage level in the silo rises and falls. An objection to silos is that accurate inventories are difficult to maintain as the level of beans is not always horizontal and the height in the bin, even if scaled, can be quite a few bags in error by visual estimate. Often the decision to use large silos rests more on objectives in processing than on economy. However, silos do have the advantage of reducing labor required for handling bagged coffee especially when double handling of bags for pallet storage and for use in the process can be eliminated. This labor saving can be important where labor rates are high and union pressures can be exerted, as in the United States. If steel silos are to be used outdoors in a warm, humid climate, they must be insulated and equipped with a circulatory air system for controlling humidity and temperatures.

FIG. 7.6. BULK BAG STORAGE

Bulk storage of pergamino is attractive because it retains the good flavor properties of the bean until it is hulled. However, since its bulk density is 25 lb per cu ft, it occupies much more storage space than hulled coffee. Where deterioration of hulled beans is serious, good economics might indicate investment in larger storage facilities although this is sometimes not obvious to investors.

### Air-Conditioned Storage for Green Coffee Beans

It has already been shown that there is marked coffee flavor deterioration and loss accompanied by bleaching of bean color when good quality coffees are stored at tropical seaports with a temperature of over 80 F (27 C) and more than 50 percent relative humidity. For export, the agent usually endeavors to keep the green bagged coffee at higher altitudes and cooler temperatures until the boat for shipment is near its arrival date. This timing is not always practiced nor is it always dependable, with the result that the coffee may stay in the hot humid port for several months. In the case of a soluble coffee plant located in such a climate, it is worthwhile to compare the cost of air-conditioned storage with an estimated value of deterioration of that coffee. Because a large loss occurs every year, it is reasonable to assume that the amortization of the warehouse and air-conditioning investment may be spread over ten years, although the equipment will be used for 20 to 25 years. The warehouse should be reasonably air tight and insulated especially from the sun's direct heat rays on the roof. Assuming a basis of a 20 ft eave (five pallets high), a 5,000 sq ft floor area holding about 8,000 bags of green coffee may need 40 tons of refrigeration and associated ducting, blowers, cooling tower, and instruments. The overall construction would certainly have to be better than just a rain-protected shed as is commonly used in the producing country. The actual size of the warehouse may be five- to ten-fold greater, but for purposes of evaluating air-conditioned warehousing under the circumstances, bean storage from the end of one crop to the beginning of the new crop is what is needed.

## SIZE, SHAPE AND DENSITY OF COFFEE BEANS

Notable differences in the size and shape of coffee beans are influenced by botanical variety and environmental growth circumstances. Visual examination of green coffee beans reveals some distinct shapes and, to a lesser degree, intermediate shapes. In any given growth area these differences are readily recognizable and usually can be associated with coffee sources and properties. The most common bean is oval with one flat face. The rounded back portion may be a shallow portion of an ellipse or it may be quite rounded, and therefore this dimension or thickness, is used in grading. Three dimensional grading of flat beans is done first by width, then by thickness, and then by length. Bean shapes may be pointy, boat-shaped with tips curling in, plump to round, long and short. Aside from these variations, there is another distinctly shaped bean, the peaberry, that occurs in small percentages in every normal crop.

The peaberry contains only one seed within the coffee cherry and this is rounded, having no flat side. Therefore, it is classified only on a two-dimensional basis, length, and width. The peaberry is the same in cup quality as the paired flat-faced bean. However, in Central America, the peaberry is called caracol (snail) in Spanish, and carries the connotation that it is a premium quality bean. These rounded seeds are usually separated from the flat-sided bean and are sold separately. They bring a good price, verifying their better acceptance. Peaberries, because they are round, roll better and some believe that this better flow during roasting contributes to better heat transfer and a more even, and better developed, roast. Robusta beans are similar to peaberry but usually brownish in color.

Elephant seeds are also separated out. These are two seeds rolled into one; they have an uneven thin-walled shape and are usually large. In addition, broken and immature beans are separated out; these are called triage (rejects).

In Brazil a Bourbon bean is characteristically small and curly. Maragogripe is a variety of arabica discovered near Bahia, and has very large berries of heavy flavor. This variety is mostly sold in Europe at a premium price.

The shape of the bean has some influence on the means used for its separation. Peaberries are readily separated on a wire reel and/or vibrating table. However, when flat bean sizes are graded, the process is more complex, and the standards will vary from country to country as well as from one growth area to another. Size alone becomes, in some cases, a factor secondary to uniformity of size. For example, Brazilian coffee beans are, on the average, smaller in size than high grown mild beans while some Robusta beans are smaller than Brazilian beans. It often becomes necessary for the coffee associations and/or cooperatives to establish their own standards as a means of uniformity in sales, but these regional and national standards often have little or no relation to each other. Further, systems of grading are so arranged that a maximum number of beans are classified as top grade, which may or may not be best for the buyers' interests. Some reference will be made here to Brazilian standards, which system is similar to other Latin American exporters.

A single grading machine is not capable of dealing with all kinds of coffee from all countries; these machines must be adjusted in reel, slot, and hole sizes according to the coffee being graded and the consumer market being served. Figure 7.7. shows a reel grader* and Fig. 7.8 shows the full size of slotted and round holes used in graders. Brazilian coffee is classified and described both in Brazil and the United States by the size of beans passing over numbered screens with round holes. Figure 7.8 shows the slotted holes used to analyze and classify peaberries. When size is specified, it is customary to have over 85 percent of the beans in that size category.

*Note cylindrical brush at top, inside used to clean screens.

*Courtesy of Wm. McKinnon and Co., Ltd.*

FIG. 7.7. REEL GRADER

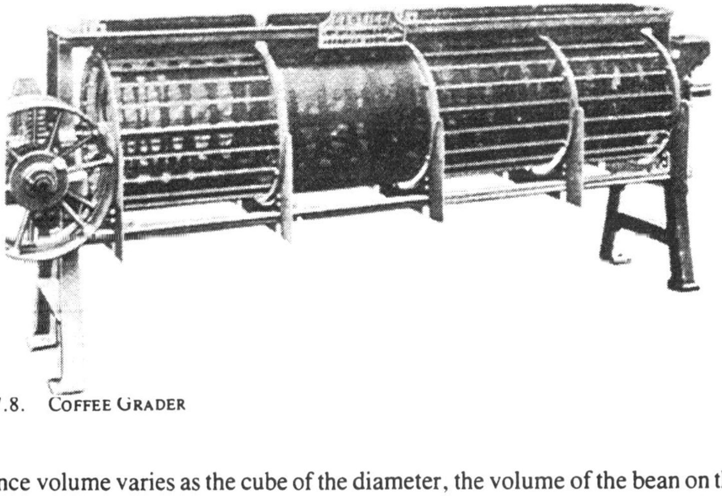

FIG. 7.8. COFFEE GRADER

Since volume varies as the cube of the diameter, the volume of the bean on the No. 19 screen, $^{19}/_{64}''$, (7.500 mm) compared with that of the bean on the No. 13 screen, $^{13}/_{64}''$, (5.000 mm) will be $(3/2)^3$ or 3.4 times the volume (and weight) of the smaller bean. This much difference in weight makes a marked difference in roasting characteristics.

Research has shown that heavy yielding trees usually produce heavy, large beans. Long beans are not necessarily flat or narrow, but are generally heavy. Wide beans are heavy and flat beans are usually light. Width and boldness seldom correlate. Additional research in Kenya-Tanzania has shown that: the weight of a coffee bean varies between 0.10 to 0.20, averaging 0.155 gm; the range of bean length from one tree varied from 5.3 to 12.0, averaging 9.8 mm. Bean width measured by shaking the beans through sheets with circular perforations showed that 70 percent of the beans were larger than 6.50 mm and 30 percent were larger than 6.88 mm. Measurements for boldness or thickness of beans were based on slotted holes and 80 percent of the beans were smaller than 4.00 mm. These beans are smaller than Brazilian beans which, on the average, are 6.0 mm; Kenya beans are at least 10 percent wider. Studies on specific gravity of beans showed a range of 1.15 to 1.40, but no correlation was made between specific gravity and cup quality. This aspect deserves further study. The size of the bean does not relate to cup quality since some of the finest Kenya coffee is a small, well made bean, as is Mocha. Size and shape of the beans seem to result from many variables of age and environment that influence the tree, and from the picking. Roasting a non-uniform screened sample revealed a very non-uniform roast and green-burnt flavor. Screen sizing after roasting showed that burnt beans swelled the most.

Table 7.3 gives the composition of fresh coffee pulp, and Table 7.7 (p. 205), that of the mucilage.

## Coffee Grading by Imperfections

Grading of green coffee has to do with the lack of uniformity of bean size and shape and with the presence of foreign matter. These, together with other characteristics which will be dealt with later, are rated as imperfections. Table 7.2 shows Brazilian coffee grades by imperfections. These standards were worked out more

TABLE 7.2. BRAZIL COFFEE BEAN GRADES BY IMPERFECTIONS PER KILOGRAM

| Grade Number | Imperfections as Black Beans or Equivalent |
|---|---|
| 1 | None |
| 2 | 6 |
| 3 | 13 |
| 4 | 29 |
| 5 | 60 |
| 6 | 115[1] |
| 7 or 8 | More than 115 |
| Lower than 8 | Not admitted into the United States |
| Equivalents to one black bean: | 3 shells |
| | 5 quakers |
| | 5 broken beans |
| | 1 small pod (large pod equals two black beans) |
| | 1 medium stone |
| | 2 small stones (large stones equal 2 or 3 black beans) |

[1]One hundred and fifteen black beans are about 3 percent of the coffee. This implies that perhaps 15 or more percent of discolored beans constitute the coffee.

TABLE 7.3. CHEMICAL COMPOSITION OF FRESH COFFEE PULP

| | Percent |
|---|---|
| Moisture | 42.5 |
| Raw fibre | 27.5 |
| Sugars | 9.5 |
| Tannins | 8.6 |
| Minerals | 3.8 |
| Waxes, fats, resins | 1.2 |
| Volatile oils | 0.1 |
| Others | 6.8 |
| Total | 100.0 |

than 35 years ago. They are not used directly on coffees from countries of origin other than Brazil although they might be so used to advantage. With the increasing use of electronic sorters, it is becoming economically practicable to obtain the equivalent of grade Nos. 1, 2, or 3 on the Brazilian imperfection table from any reasonably good lot of coffee. Much of the contaminating material (black beans, pods, and sticks), very small amounts of which may greatly damage cup quality, can be eliminated by careful processing. The machines can do the rest. Brazilian grade No. 4 is the standard of trade reference, and all lower grades are discounted. (See Fig. 7.9)

## Green Coffee Bean Grading by Color

Although beans of first class color and appearance may carry a tainted flavor, bean color is an important index in indicating off-quality. The blue to gray-green is the most desirable, on the basis of cupping experience. Bleached-out pale white to gray colorations testify to poor storage or processing history. Yellow color with multicolor patterns denotes age of one or more years—an old crop. Under-dried new crop coffee has a different bluish-green color at first, but it soon fades and the cup is weak. Brown is frequently associated with improper drying of the beans, picking of overripe cherries or overfermentation. Natural beans dried in the cherry are often brownish and mottled. Brown coloration can be caused by contact with pulpy skin during fermentation. Blotchy or mottled color patterns on the bean are often due to uneven drying. Holes and decay patches which may be black, pink, or green are due to borer insects. Pulper-nipped beans have characteristic black discolorations of small cavities. Foxy flavored beans are usually reddish, attributed to overripeness, cherry damage, absorption of ferment liquors, or improper washing. Blacks are caused by disease, and ambers, smooth and yellow, are caused by picking immature berries.

High grown mild coffees have a characteristic waxy, bluish gray-green color which is an intrinsic part of high grown, good flavored coffees. Beans from Brazil

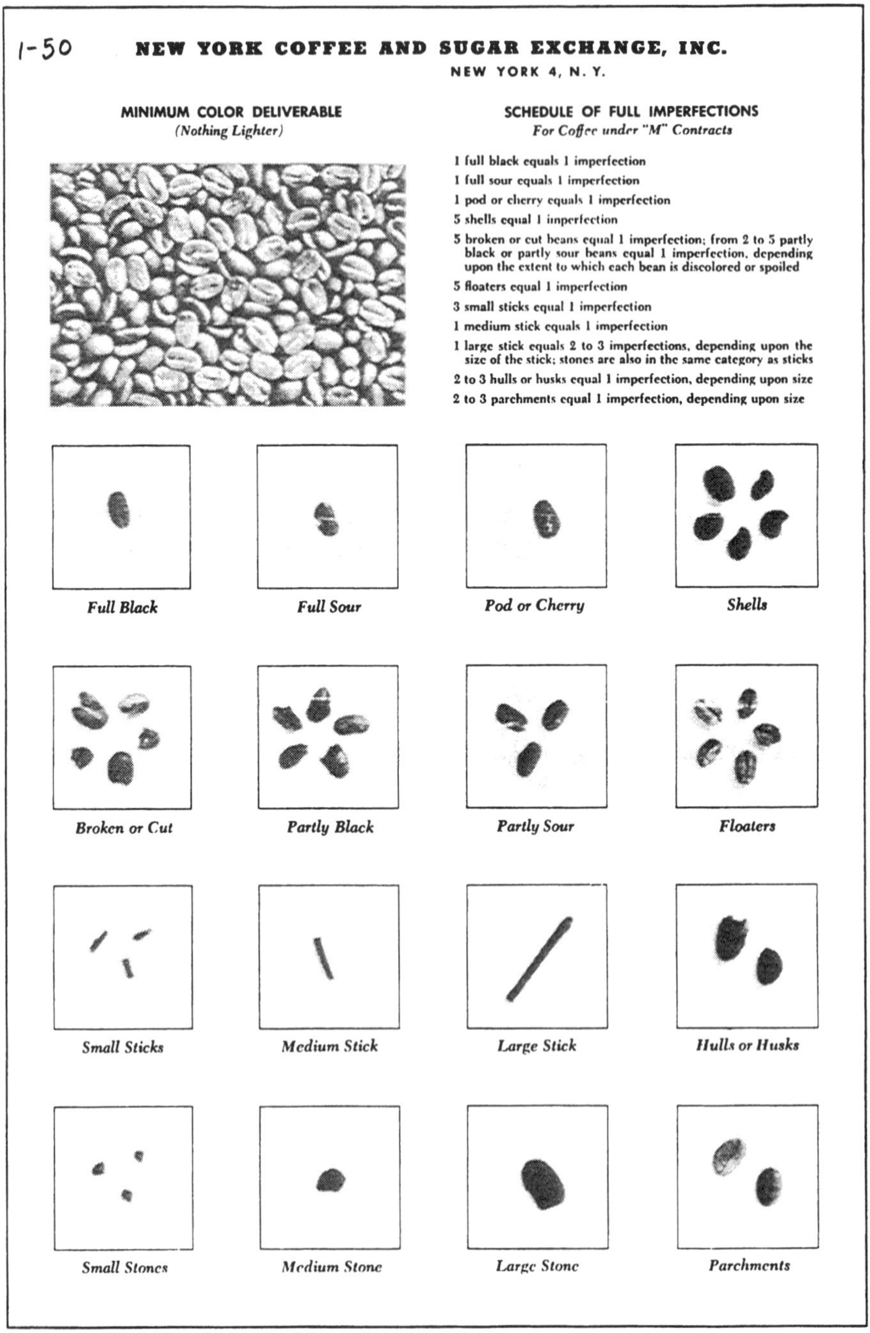

1-50
NEW YORK COFFEE AND SUGAR EXCHANGE, INC.
NEW YORK 4, N. Y.
MINIMUM COLOR DELIVERABLE
(Nothing Lighter)
SCHEDULE OF FULL IMPERFECTIONS
For Coffee under "M" Contracts
1 full black equals 1 imperfection
1 full sour equals 1 imperfection
1 pod or cherry equals 1 imperfection
5 shells equal 1 imperfection
5 broken or cut beans equal 1 imperfection; from 2 to 5 partly black or partly sour beans equal 1 imperfection, depending upon the extent to which each bean is discolored or spoiled
5 floaters equal 1 imperfection
3 small sticks equal 1 imperfection
1 medium stick equals 1 imperfection
1 large stick equals 2 to 3 imperfections, depending upon the size of the stick; stones are also in the same category as sticks
2 to 3 hulls or husks equal 1 imperfection, depending upon size
2 to 3 parchments equal 1 imperfection, depending upon size
Full Black
Full Sour
Pod or Cherry
Shells
Broken or Cut
Partly Black
Partly Sour
Floaters
Small Sticks
Medium Stick
Large Stick
Hulls or Husks
Small Stones
Medium Stone
Large Stone
Parchments


FIG. 7.9. COFFEE BEAN DEFECTS

TABLE 7.4. BRAZILIAN GREEN COFFEE BEAN GRADING SYSTEM

*Varieties:* Brazil's coffees are mostly Arabica species. Main varieties are red and yellow, bourbon, mundo nova, red and yellow caturra.
*Types:* Brazil coffees are classified by type, on defects in a 300-gram sample. Range is from Type 2, best to type 8, poorest.
*Grades:* Coffee for export is graded by size and shape of beans on No. 14 to No. 19 screens, and must be within these proportions: impurities 4 percent, peaberries 8 percent, small beans 10 percent, good beans 78 percent.
*Qualities:* Coffee for export is also classified by cup quality. Descriptions are: strictly soft, soft, softish, hard, Rioy, Rio flavor. Soft categories are sweet in the cup, Rio categories generally harsh and pungent, hard coffees clearly harsh.
*Origin classification:* in addition to Brazil's standards by type, grade, cup quality, coffees are usually also designated by port of origin: Santos, Rio, Victoria, Parana, Angra dos Reis, etc.

SCALE OF DEFECTS
(in 300 gram sample)

| | |
|---|---|
| 1 large stone | equals 5 defects |
| 1 medium stone | equals 2 defects |
| 1 small stone | equals 1 defect |
| 1 large twig | equals 5 defects |
| 1 medium twig | equals 2 defects |
| 1 small twig | equals 1 defect |
| 1 black bean | equals 1 defect |
| 1 pod bean | equals 1 defect |
| 1 large husk | equals 1 defect |
| 2 to 3 small husks | equals 1 defect |
| 2 sour beans | equals 1 defect |
| 2 beans in parchment | equals 1 defect |
| 3 shells | equals 1 defect |
| 5 unripe beans | equals 1 defect |
| 5 broken beans | equals 1 defect |
| 5 quakers | equals 1 defect |
| 2 to 5 weevils | equals 1 defect |

OFFICIAL CLASSIFICATION TABLE

| Type | | Defects |
|---|---|---|
| 2 | | 4 |
| 2 | | 5 |
| 2 | | 6 |
| 2 | | 8 |
| 3 | | 10 |
| 3 | | 11 |
| 3 | | 12 |
| 3 | | 14 |
| 3 | | 16 |
| 3 | Type 3/4 | 19 |
| 4 | | 22 |
| 4 | | 24 |
| 4 | | 26 |
| 4 | | 30 |
| 4 | | 33 |
| 4 | Type 4/5 | 36 |
| 5 | | 39 |
| 5 | | 42 |
| 5 | | 46 |
| 5 | | 52 |
| 5 | | 59 |
| 5 | Type 5/6 | 66 |
| 6 | | 73 |
| 6 | | 80 |
| 6 | | 86 |
| 6 | | 99 |
| 6 | | 111 |
| 6 | Type 6/7 | 123 |
| 7 | | 135 |
| 7 | | 147 |
| 7 | | 160 |
| 7 | | 180 |
| 7 | | 200 |
| 7 | | 220 |
| 7 | | 240 |
| 7 | Type 7/8 | 260 |
| 8 | | 280 |
| 8 | | 300 |
| 8 | | 320 |
| 8 | | 340 |
| 8 | | 360 |

are never this color but are usually pale green to tan and are often bleached when large inventories of coffee are being held. A non-uniform coloration of beans in the sample tray is a sure sign of non-uniform quality, which is more strikingly revealed after roasting. Old crop beans usually have lost their original luster. Beans exhibit distorted shapes because of drought or other growth causes and shrivelled beans give a different shading impression in the sample tray and are cause for critical evaluation of the cup. A residue of silver skin on numerous beans might mean that hulling was done before the beans were sufficiently dry or that cleaning was inadequate.

TABLE 7.5. METHODS OF CLASSIFYING GREEN COFFEE BEANS

| Item | Equipment | Principles | Objects being removed |
|---|---|---|---|
| 1. | Tank of water | Floatation | Leaves, twigs, immature cherri-stones settle. |
| 2. | Canal of water | Levitation (dynamic) | Immature cherries. Immature bean in parchment. |
| 3. | Vibrating perforated table | Limited opening | Twigs, sticks, etc. larger than normal coffee bean. |
| 4. | Vibr. solid table | Mechanical lift on sloped table | Stones, metals, heavy bad taste beans, etc. |
| 5. | Vibr. table with air lift | Air and mech. lift | Light density defective beans. Heavy density defective beans. |
| 6. | Wire reel rotating cylinder | Slotted aperature | Fines, broken beans and specific thickness of beans. |
| 7. | Inclined moving band | Flat beans lay on band | Round beans (peaberry) roll off. |
| 8. | Slotted reel or slotted vibr. table | Flat beans pass thru slots | Round beans (larger than flats). |
| 9. | CATADOR—air fluid bed | Air levitation | Defective light beans, husks, chaff and dust lifted off, heavy beans remain. |
| 10. | a) Rubber belt<br>b) Magnetic discs | Magnetic drum<br>Magnetic throw | Ferritic: steel, iron, etc. |
| 11. | Electronic sorting | Light reflection from each bean surface | Programmed to remove black, white, red, etc. colored beans. |
| 12. | Manual sorting | Eye/hand | All sorts of defective beans. |
| 13. | Roaster machine | Color uniformity<br>density uniformity | Unroasted.<br>Excessively dense or light. |
| 14. | Prepare beverage | Taste | Guide to uniformity, and detection of off flavors. |
| 15. | Mixing station | Blending | Lack of uniformity. |

TABLE 7.6. GREEN BEAN GRADES HARVESTED—BRAZIL (Type In Percent)
*Sao Paulo crops of 1967-70*

| Year of harvest | 1967 | 1968 | 1969 | 1970 | 4-year |
|---|---|---|---|---|---|
| 1,000 bags classified | 3,670 | 4,021 | 3,289 | 3,213 | average 3,548 |
| 2 | 0.08 | 0.05 | 0.01 | 0.03 | 0.04 |
| 3 | 1.55 | 3.08 | 3.27 | 1.95 | 2.46 |
| 4 | 9.83 | 12.01 | 12.09 | 8.24 | 10.54 |
| Total under 4 | 11.46 | 15.14 | 15.37 | 10.22 | 13.04 |
| 5 | 27.95 | 30.86 | 31.58 | 27.75 | 29.54 |
| 6 | 43.82 | 43.48 | 41.76 | 47.54 | 44.15 |
| 7 | 13.64 | 9.48 | 9.38 | 13.41 | 11.48 |
| 8 | 2.02 | 0.68 | 0.65 | 0.98 | 1.08 |
| Below 8 | 1.11 | 0.36 | 1.26 | 0.10 | 0.71 |
| Total over 4 | 88.54 | 84.86 | 84.63 | 89.78 | 86.96 |

Source: O Estado de S. Paulo

## Coffee Bean Grading by Roasting

The appearance of the roasted coffee bean is a further clue to its quality and uniformity. When the bean center cut opens due to swelling to reveal the tan parchment clearly, the roast is called brilliant and is most desirable. Brownish center cuts are a sign of poor bean development; they are very undesirable. Dull bean surfaces indicate mediocre or poor coffees. If all the roast beans are not close to the same shade of color, this shows lack of uniform roasting, resulting from the presence of nonuniform lots of green coffee. Pale yellow beans are immature, do not roast well and are nutty or grassy in flavor. Some pales have a rank odor when crushed, and thus may contaminate the rest of the coffee. An extreme range of roast where some beans are not developed while others are burned gives green-burnt tastes which are not found in good coffee flavor. The aromas driven off during roasting usually reveal the quality of coffee being developed. Burning or popping of some beans long before others is a sure sign of non-uniformity. Such roasts are not clean because oils are driven off from some beans and other beans burn while the roast is brought to a certain median acceptable condition. Mottled green beans often give mottled roast colorations; this type is found to deteriorate more rapidly than uniformly colored beans. Hollow or thin walled beans are revealed by burning of the thin tips.

Blends that differ sufficiently in moisture, size, and variety may not give full development to each type of bean, hence it may be necessary to roast each part of the blend separately for full development before blending. Even similar coffees from the same growth areas, if not of the same size, moisture, or age may give a very non-uniform roast color when mixed. Grinding the roast beans and comparing the colors of bean surfaces and of the grounds reveal the penetration of the roast which is not always the same for different types of beans.

## Coffee Grading by Cupping (Tasting)

The ultimate criterion of coffee quality is cupping flavor; the value of the coffee is further judged by price and delivery time. Briefly, the coffee flavor must be typical of the class it represents, e.g., high or low grown mild, Brazilian grade No. or type of Robusta coffee. No off-flavors should be recognizable from process and/or grading shortcomings. Some of the typical expressions encountered in cupping green coffee samples are listed here.

*Acidy:* desirable flavor note of high grown coffees.

*Sour:* an undesirable acid flavor.

*Harsh:* sometimes acidy, too, but notably irritating in throat after tasting.

*Dirty:* an undesirable fuzzy taste that dominates the flavor background.

*Body:* usually a desirable taste sensation imparting a feeling of heavy texture, usually associated with very high grown coffees, but some Robusta coffees also have heavy body. This is desirable in instant coffees.

*Earthy:* similar to earth and related to mold organisms; very undesirable.
*Musty:* similar to earthy and moldy and due to such causes, also called groundy.
*Fruity:* high in acetaldehyde, attributed to overripeness and late picking.
*Foxy:* from reddish beans with brown stain at split and with a sour taste.
*Fermented:* caused by enzymes changing sugars to aldehydes and acids which give a characteristic flavor often carried into the bean from fermentation in the tank, delayed drying or microorganism contamination.
*Hay-like:* a common aroma and taste from grassy foods that have been dried.
*Grassy:* a green flavor, differing from hay; often occurring in new crop coffee, and not considered a normal or desirable part of coffee flavor.
*Foul:* a very objectionable flavor related to rot and fermentation.
*Rioy:* a very objectionable medicinal taste, both harsh and heavy; occurs in many regions, especially in the state of Victoria, Brazil.
*Nutty:* tasting nut-like and without development of coffee flavor.
*Winey:* having the body and texture of wine, often desirable; occurs in naturally dried arabica coffees.
*Woody:* a wood-like taste occurring in old, deteriorated beans lacking flavor.
*Rubbery:* an undesirable taste often encountered in Robustas.
*Sweet:* denotes a smooth cup free of any foreign flavor; also called soft in Brazil.
*Hard:* strong, harsh but not Rioy, common in Parana, Brazil.
*Neutral:* a flavor usually noted as desirable, as in Angola Robusta as opposed to Robusta flavors, which can be harsh and disagreeable. *Neutral* also can be related to Brazilian coffees which are not very aromatic or acidy yet form a good base flavor on which to build a blended flavor.
*Thin:* denotes a wateriness or lack of flavor in the cup; typical of low grown mild coffees.
*Fragrant:* denotes appreciable delicate aroma.
*Rich:* more than body and aroma; a buttery, satisfying cup.
*Mellow:* means fully developed in flavor and aroma; not harsh, but balanced.

There are, of course, many other lesser terms which are often used regionally to characterize certain types of coffees from local growth areas.

## Coffee Bean Grading by Origin and Age

The most striking association and identification for a coffee is its point of origin. The more one knows about the coffee's origin, the more confident one can be about its uniformity and properties. Normally the respective details about the coffee's origin are: country of origin, state or region where grown, port of embarkation, the name of the mill or exporter, the name of the grower, and the location of the grower's plantations. The names of the mill and the grower are usually indicated on the bags (mark) unless they have contracted to provide the coffee to someone else. In Brazil the state and port must be noted; in Mexico, the state; in Colombia, the growth district and port. If there are no excessive stocks of coffee held in the producing country, the coffees will be from the most current harvesting; otherwise age should be noted. Such pinpointing of growth sources have considerable bearing on the quality, and such names as Coatepec, Antigua, Medellin, and Armenia have strong meaning.

Beans stored for several years, as in Brazil in the mid 1960's, took on a yellow to white color, with a loss of virtually all the green color. On roasting, aroma was lacking and "body" was stronger. It was common to blend old and new crop beans for use in soluble coffee manufacturing.

## Grading Beans by Levitation and Vibration

Figures 7.10 through 7.13 show an air lift classifier and Fig. 7.14, a rotary green coffee bean cleaner, used in Brazil, to remove residual parchment and debris after hulling.

After the green coffee beans are dried to 12 percent moisture, they are hulled and classified by size in a rotary wire slotted reel then classified by density in an air lift.

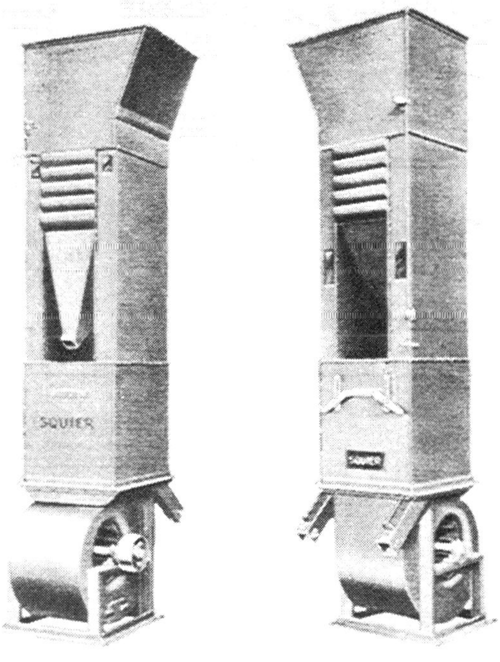

*Courtesy of G. L. Squier Co.*

FIG. 7.10. CATADOR

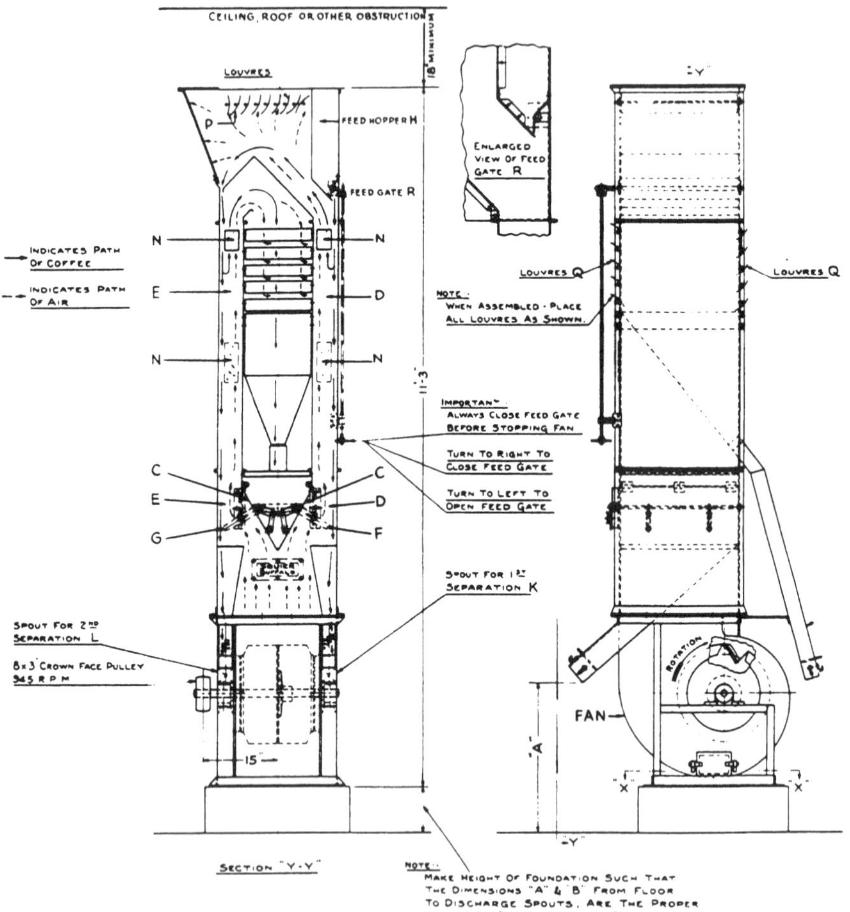


FIG. 7.11. SCHEMATIC DIAGRAMS OF SEPARATOR

FIG. 7.12. BAGGING AIR-CLASSIFIED COFFEE BEANS IN A SERIES OF SEQUENTIAL BEAN SEPARATORS IN BRAZIL

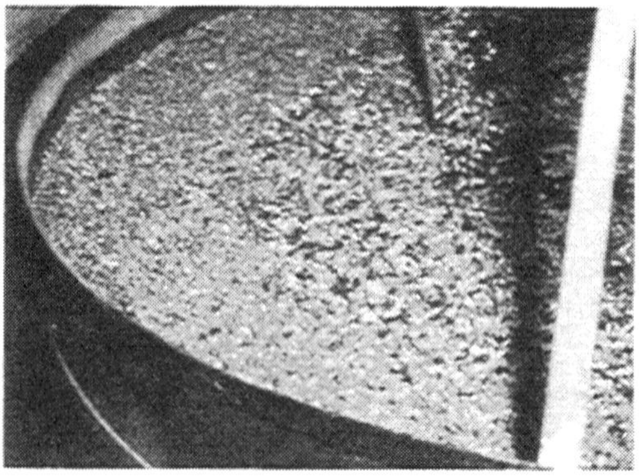

*Courtesy of Pan American Coffee Bureau*

FIG. 7.13. ROTARY GREEN COFFEE CLEANER

*Courtesy of Davidson and Co., Ltd.*

FIG. 7.14. PEABERRY SEPARATOR

After this, the hulled beans are finally classified on a vibratory wire table or air flotation table like the Oliver, see Figs. 7.15 and 7.16. These separations can be good or poor and are often governed by the consuming market. In any case, the mill operator is often not very scientific and simply does what he has been doing for years or what is most economical, not necessarily using the best classification system. For example, good classification is obtained by sizing first, and then using the air lift to obtain density separation. But in practice, the air lift is used first

FIG. 7.15. ROTARY MAGNETIC STONE SEPARATOR BY D'ANDREA, IN USE AT CIA. CACIQUE, BRAZIL

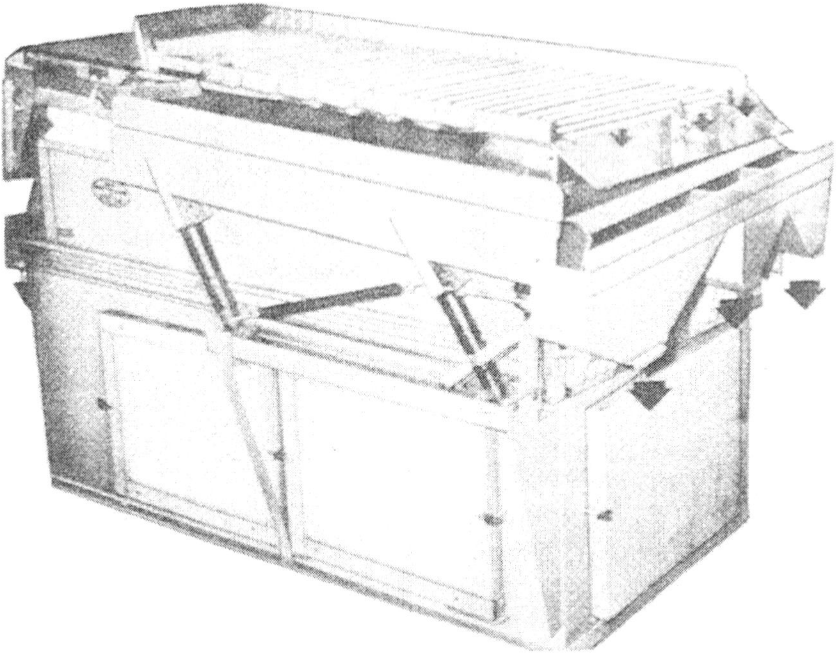

FIG. 7.16. GRAVITY SEPARATOR

*Courtesy of Pan American Coffee Bureau*

FIG. 7.17. THREE-TIERED VIBRATORY BEAN GRADER

*Courtesy of Pan American Coffee Bureau*

FIG. 7.18. AIR LIFT CLASSIFIER—NOTE ROTARY CLEANER AT LEFT

because it removes loose hull fragments and preliminarily cleans the dried beans by separating out the stones, pits, sticks, and so on.

Figures 7.19 and 7.20 show two other mechanical methods of cleaning and grading green coffee beans.

*Courtesy of Pan American Coffee Bureau*

FIG. 7.19. HAND SORTING—STATIONARY TRAYS

FIG. 7.20. THE ICORE ELECTRONIC (COLOR) BEAN SORTERS IN THE IVORY COAST

## Grading Green Coffee Beans

In any case, after the hulled green coffee has been so classified it then undergoes visual inspection for black and discolored beans which must be removed. These few black and discolored beans would otherwise not only give non-uniform appearance but would contribute disproportionate off-flavor, and thus downgrade the selling price. A classic picture from almost any coffee country is the assembly of women, who work for a pittance, picking out the off-grade beans from piles or from moving belts. Even today roasted beans are handled this way in some European countries. Where there are several percent of rejects, such picking becomes very expensive and inefficient. Part of the problem, is, of course, due to the poor preliminary separations made by machines. But as grades of coffee go down, the percentage of rejects goes up. Thus, hand-picking for both good and poor grades is being gradually supplanted by electronic bean sorters to increase quality, uniformity, and profit. The latter is illustrated by the following example. Electronic bean grading rates are 100 to 150 kg per hr or 600,000 beans per hr. Assuming a maximum human defective bean picking rate of 1 bean per sec (2 percent of throughput), 60 beans per min or 1.2 lb beans per hour can be picked manually.

An interesting application of these sorters to green coffee occurs in Africa, particularly where Robustas are being graded. The processors learned, that because the flavor of Robusta coffee was less desirable, less in demand, and thus, about half as profitable as mild coffees, something extra had to be offered to make Robusta sales more attractive. One cannot help but marvel at the physical uniformity in color, size, and density of Robustas as compared with coffees from any other part of the world. The processors in Leopoldville, Belgian Congo, were the first to apply intensively the use of electronic sorters to produce coffee with a very uniform physical appearance. Figure 7.19 shows a row of hand sorters classifying green coffee in Leopoldville.

Figure 7.25 shows the method of determining size distribution. When the various machines shown in previous figures have done their best, which is far short of perfection, green coffee still needs to be sorted either by hand (see Figs. 7.19 and 7.21) which is becoming increasingly expensive, or electronically as shown in Fig. 7.20. The beans are individually inspected by the electric eye, and accepted or rejected. The good beans are released at one point, the rejected at a different position.

Another method allows the beans to fall in a single line; beans rejected by the electric eye are blown aside and fall into a separate receiver. The electric eye's decision may be based on lightness or darkness or on color.

**Sorter Cost—Profit Example.**—About 1958 electronic sorting of roast coffee beans picked up at a fast rate in Europe, thus displacing hand picking. Germany, which traditionally uses good coffee, imposed a 5 DM per kg tax on green coffees after World War II. This, in effect, largely eliminated the importation of low grade coffees, and made the consumer doubly critical in his choice of

*Courtesy of Pan American Coffee Bureau*

FIG. 7.21. HAND SORTING—MOVABLE BELT

roast coffee beans due to their high price. Thus, with no cheap labor and often no labor at all, the German roaster turned to the electronic sorters to attain high quality coffee and yet maintain a profit. The electronic sorters first cut down the amount of rejects from about 4 to 2 percent. The rejected coffee had to be sold at less per lb.

Aside from profit, the electronically sorted beans offer quality heretofore unattainable by human picking. The electronic sorter offers uniform beans and bulk shipments identical to samples and recovers good beans often lost among a majority of poor beans. The throughput capacity of the sorters do not vary with changing percentages of rejects. Doubling rejects would require at least doubling the number of women pickers. The machine does not tire; its efficiency is the same day after day, immune to weather, lighting, illness, and so on.

The use of electronic sorters for roast and green coffee grading operations integrated with a soluble coffee manufacturing operation has some true advantages. In the coffee-growing country, centralized buying of green coffee for milling, export, and soluble coffee use as well as for local consumption allows the fullest distribution of types of coffees under a single control and achieves the highest profit. For example, the economics of operating a sorter as opposed to hand labor are already desirable on any basis. The first point of profit evolves from a higher export price on the green coffee which use of the sorter allows. The sorter further establishes a reputation for high quality coffee not heretofore attained, nor offered by competitors. Then, the rejects can be further graded for use in a soluble coffee plant where physical appearance and some off-flavor are not so critical as for coffees used in regular coffee sales. The operation of the mill or beneficio thus is geared to these two sales outlets. Whatever is left can usually be sold on the local market.

## Bulking or Blending Green Coffee

In producing green coffee any processing plant will obtain many small lots of different sizes and grades. To collect lots of marketable size and the best possible quality, processors must blend these small lots. This is done in towers of wood or metal containing baffles. The coffee is raised to the top in bucket elevators. Care must be exercised that each large lot is made up of small ones of approximately the same grade; otherwise quality may be unnecessarily lowered. Figure 7.22 shows the use of grading screens.

## Uniformizing Coffee Beans

To appreciate how a small amount of impurity (whether it is a defective coffee bean, or a foreign piece of metal) can influence the taste of the resulting coffee

FIG. 7.22. DETERMINING BEAN SIZE DISTRIBUTION WITH GRADING SCREENS

beverage it is constructive to see how many distinct classification steps are required to uniformize and elevate the grade of coffee beans:

1. Growing fruit on a healthy shrub in a suitable environment without insect or disease damage.
2. Harvesting only ripe healthy fruit from the branches of the shrub, and not picking green, overripe, dried, fallen, or otherwise abnormal fruit.
3. Density classification: water floatation of freshly picked fruit in canals (to float off immature cherries called "stinkers" which remain whitish) and to allow stones, metals and heavier to sink.
4. Pulping machine tearing off the skin of the cherry without damaging or nipping the seed.
5. Properly fermenting the fruity mucilage so that no remaining fruit on the seed can consequently cause anomolous tastes and defects in the beans.
6. Prompt drying of the fermented and washed beans in the sun. Warm air bean drying machines can be used in the final stages of drying to avoid uncontrolled fermentations, overheating, over- or under-drying, mold growth, etc.
7. Hulling: Removing the skin or parchment layer from the beans after drying.
8. Classifying the cleaned beans by size, shape and density, rejecting broken, deformed and abnormal. (Figs. 7.22, 7.23, 7.24)
9. Color sorting: Removing beans of abnormal colors (black, white, or red) from cleaned green (or roast) beans.
10. Removing foreign matter like sticks, stones, wires, dust, dirt, etc.
11. Density classification of beans in air levitation and vibrating conveyor tables.

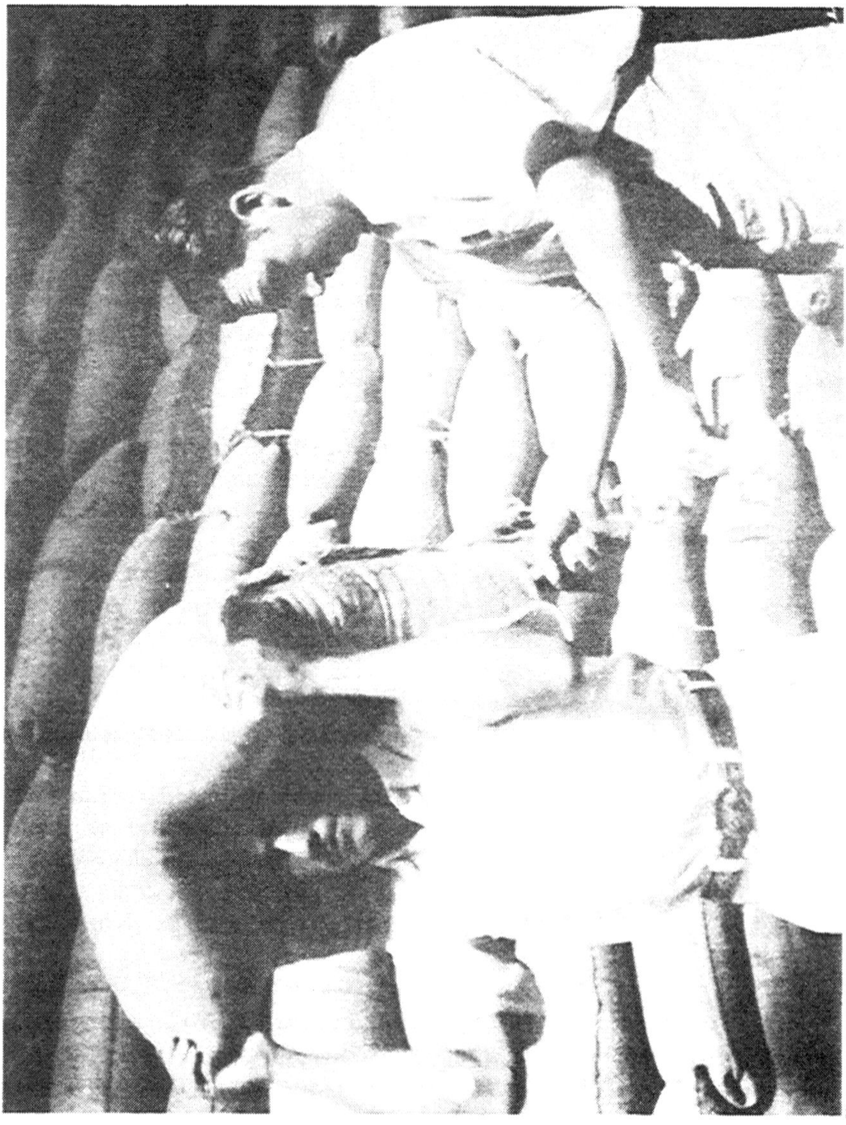

FIG. 7.23. SAMPLING BEANS

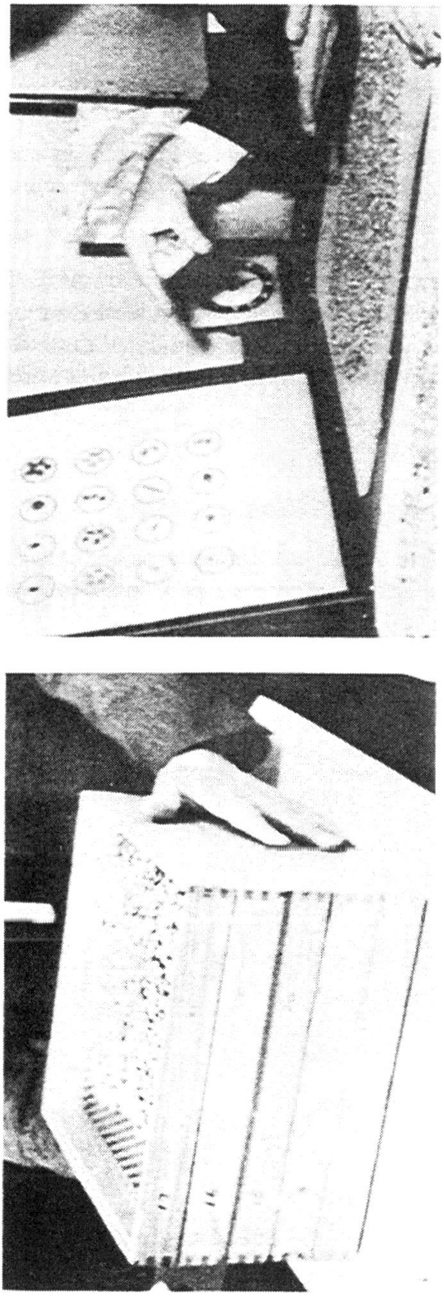

FIG. 7.24. SPREAD OUT BEAN SAMPLES ARE BEING COMPARED TO GRADING CHART NOTATIONS

12. Odor classification of green coffee beans. There must be no moldy fermented or foreign aroma. Also use of bean hand feel and bean appearance guides.
13. Classifying for uniformity of roast color development: rejecting burnt or unroasted beans. (Fig. 7.25)
14. Classifying roast coffee beans for acceptability and uniformity in taste.
15. Classifying green beans by age of crop and source, as well as time and conditions in shipment.

In the growing countries, having uniformized, purified and graded the coffee beans to the above extent, one seriously considers the mixing of lots. In blending lots of coffee beans, as with mixing food products, the addition of a poorer grade to a better one, will invariably downgrade the mixture closer to the qualities of the poorer grade.

## Relating Bean Prices to Taste Quality

Refer to the following table of coffee types and pricing.

New York coffee bean prices are a good index to the relative taste quality of many original bean varieties:

FIG. 7.25. A-F. GRADING BRAZILIAN COFFEE BEANS BY APPEARANCE, TASTE AND SMELL

COFFEE BEAN ORIGINS, TYPES, PRICES & DATES

| | Origins | Type | Prices Dec. 1971 | Prices March 1973 | Fall/ Winter '76 |
|---|---|---|---|---|---|
| a) | Colombian MAM's | | 50.0–53.0¢/lb | 72.0¢/lb | $1,80–$2.00 |
| b) | Costa Ricans | | 52.0–54.0 | — | 1.80 |
| c) | Guatemalans | prime washed | 46.0–47.0 | — | 1.78 |
| d) | Salvador | central standard | 46.0–47.0 | — | 1.75 |
| e) | Mexico | prime washed | 46.0–47.0 | — | 1.83 |
| f) | Brazil, Santos 4's | | 43.0–44.5 | 58.0¢/lb | — |
| g) | Brazil, Parana 4's | | 42.5–43.5 | — | |
| h) | Ethiopia-Djimmas | | 43.0–43.5 | — | 1.75 (poor) |
| i) | PWA-Ambriz | Robustas | 43.0–43.2 | 48.0¢/lb | 1.50 |
| j) | Uganda | Robustas | 42.5–43.0 | — | 1.45 |
| k) | Indonesians | Robustas | 37.0–37.8 | — | 1.35 |

Normally the 10 percent to 20 percent higher price between Brazils and top grade mild beans is well worth the price difference. Few commercial coffee buyers are willing to pay that small difference usually. Green bean prices are quoted to tenths of a cent per pound, because this can become a large absolute dollar difference when thousands of bags of coffee are involved in a transaction.

Selling price is not always controlling. The laws of "supply and demand" usually regulate coffee pricing. Some large sources of green coffee beans can be monopolized by the larger coffee buyers. For example, Brazils, Robustas from the Ivory Coast, Angola and Uganda may be virtually pre-sold at "arranged" prices with the large coffee roasters. This does not leave many coffee beans in quantity or quality for the smaller roasters. Possibly no coffee beans are available outside the ICA quotas.

TABLE 7.7. CHEMICAL COMPOSITION OF COFFEE MUCILAGE

| | Percent |
|---|---|
| Water | 84.2 |
| Protein (N × 6.25) | 8.9 |
| Sugars | 4.1 |
| Pectic acid | 0.9 |
| Ash | 0.7 |
| Total | 98.8 |

## Future Technology

Whereas the availability of cheap labor in the past has tended to discourage research and development in the production of green coffee, the fact that cheap labor is now rapidly disappearing tends to stimulate a search for better, cheaper, and more mechanized methods of coffee production. There is a present day need for much more research in this and indeed, in all parts of the coffee industry. Some excellent work has been carried on for many years by The Inter-American Institute Of Agricultural Sciences at Turrialba, Costa Rica. Such work should be greatly expanded.

Better horticultural methods, better processing of ripe fruit to form green coffee, and better control of flavor and aroma are certain to be developed in the future. Better sorting and the raising of grades and quality are moving ahead rapidly. Much valuable information is now available; more workers are needed to consolidate, complete, and expand it.

# Part III
# Roast Coffee Technology

# 8

# Coffee Bean Processing

In the producing country, coffee beans are usually transferred by hand from the grading mill to the truck that takes them to port. Then, the bags of coffee are moved again by hand in and out of warehouse storage. Manual labor also is used to sling the bags into ships' holds and remove the bags from the ships when they arrive. Figure 8.1 shows bags being unloaded from a ship's hold. Belt conveyors sometimes are used to help reduce manual labor in the loading and unloading of ships and trucks. The bags of coffee may have to be stored temporarily upon arrival in the consuming country, although in most cases the coffee beans are already consigned to a roasting firm by the time the coffee arrives in port. The bags of coffee may be placed on the roasting firm's pallets at the dock to facilitate subsequent movement by fork trucks and conveyors. Figure 8.2 shows a fork truck placing palleted coffee bags onto a truck. Another more recently developed means for handling the movement of palleted coffee bags is to place the pallets on special link belt tracks within a truck, so that when palleted coffee bags arrive at the roasting plant, they can be pulled off the truck and placed on the receiving track in a few seconds. Figure 8.3 shows such a conveying system with weigh-in scale.

The weighed coffee is either sent directly to bag dumping or the pallets are taken to storage. The bags of green coffee beans usually are dumped through a steel, sometimes magnetic, grill located a few inches from floor level. The released dust is taken away by air suction below or above the grill which may or may not have a hood. Figure 8.4 shows a large hooded bag dump station. The emptied burlap bags are shaken clean of any coffee beans or debris and are bundled for sale.

## CLEANING GREEN COFFEE BEANS

Green coffee beans often are not clean when sewn up in bags for export; furthermore, they may break and accumulate dust and dirt in transit.

*Courtesy The Port of New York Authority*

FIG. 8.1. BAGS OF GREEN COFFEE IN SHIP'S HOLD

*Courtesy The Hyster Company*

FIG. 8.2. FORK TRUCK MOVING PALLETED BAGS OF COFFEE

*Courtesy Link-Belt Company*

FIG. 8.3. PALLETED GREEN COFFEE BAGS BEING MOVED BY CONVEYOR FROM TRUCK INTO PLANT

*Courtesy Jabez Burns & Sons, Inc.*

FIG. 8.4. GREEN COFFEE BAG DUMP, STORAGE BINS, AND BLENDING SYSTEM

Coffee received at the roasting plant should be reasonably clean but in practice heavy foreign materials such as nails, coins, buttons, stones, and pebbles, and light contaminants such as wood splinters, strings, lint, chaff, corn, and beans other than coffee often appear and must be dealt with. Large bodies are removed in a rotary screen with holes just large enough for the coffee to pass through. Air levitation removes the light material. Since the 1970's, mostly air cleaning is used, but it is important to understand how a Gump cleaner works.

The B. F. Gump Company coffee bean cleaner shown in Fig. 8.5 removes both light and heavy objects. (No longer sold). Entering green coffee beans are distributed over a longitudinal slot over the valley of two pocketed rotating drums. The beans fit into the pockets and slip under a rubber flap as the pockets on the drum face rise up and outward. Elongated articles such as wire, string, and twigs cannot pass the rubber flaps and thus remain in the valley of the pitched drums until they are discharged at the lower sloped end. Figure 8.6 shows a detail view of this

*Courtesy B. F. Gump Company*

FIG. 8.5. GREEN COFFEE BEAN CLEANER

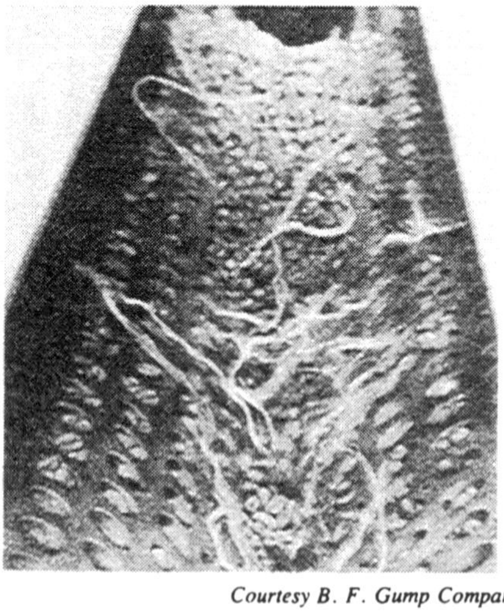

*Courtesy B. F. Gump Company*

FIG. 8.6. DETAIL OF COFFEE BEAN-STRING SEPARATOR

bean-string separation on the rising drum faces. The pocketed coffee beans passing under the wiper fall out of the pockets on the outer side of the drums and onto a vibrating gyratory screen. Fine particles and broken beans fall through the perforated holes; these holes are $^{3}/_{16}$ in. in diameter, the same as the roasting cylinder perforations. The main flow of beans leaves at the lower end of the perforated metal surface and falls into a funnel. Here, a strong blast of air draws fines and light matter up and away into a collection cyclone. The cleaned green coffee beans are then conveyed to storage or to process area.

Normally only a fraction of one percent of green bean cleaning loss occurs in the consuming country's imported coffees. When low grade beans are cleaned, several percent loss may occur. Weight loss records in green bean cleaning can point out unusually high losses in a particular lot of coffee, and identify it with its supplier.

The cyclone dust collection drums must be kept clean to avoid back-up of dust into the cyclone which will reduce its dust-removal effectiveness. The system may be so designed as to deliver the deposits to one central cyclone collector, thereby obviating manual emptying of drums. Often the air suction systems for removing green coffee dust are joined together so that one blower serves the tentacles of a multi-ducted cleaning system. The top of each green coffee bean storage bin usually has an air suction connection to remove dust released as green coffee falls into the bin. The cases that convey green coffee by means of an air blower system must have one cyclone collector for the beans followed by another for the dust. Movement of green coffee always causes dust which, aside from being unsanitary, may under special conditions of air dryness and dust density, create an explosion hazard. Airveying of coffee beans and/or ground roast coffee is used in some plants.

## Airveying Coffee Beans

Since green coffee beans are twice as dense as roast coffee beans, higher air velocities (i.e., more air in a given duct) are required to levitate green coffee beans. Horsepower needs are higher for airveyors than for screws and elevators, but airveyor initial investments are often less. The choice of system depends on the transit distance, elevation required, tortuousness of the path of coffee bean movement and available space. Airveying is preferred under crowded conditions, since tubes take little space. Figure 8.7 shows the Probat flowsheet of the Nabob plant, which has an airveyor system, used to carry: (1) green coffee from the bean cleaner to storage, and (2) weighed blends of green beans to the bins over each roasting machine via a tube switching station. The linear air flows required for levitation are about 4,000 to 6,000 ft per min. The structural supports required for airveyors are nominal. Airveying of beans is done from a volumetric metering valve set for a high velocity air flow. In the Probat system the green beans arrive over the roaster relatively unmixed, and are mixed in the roaster cylinder. If two Probat roasting machines require different blends, the console circuitry sends the blends out accordingly. Key cleanout points along the airveyor tubes must be provided in case of a plug although this seldom occurs with a properly operated system.

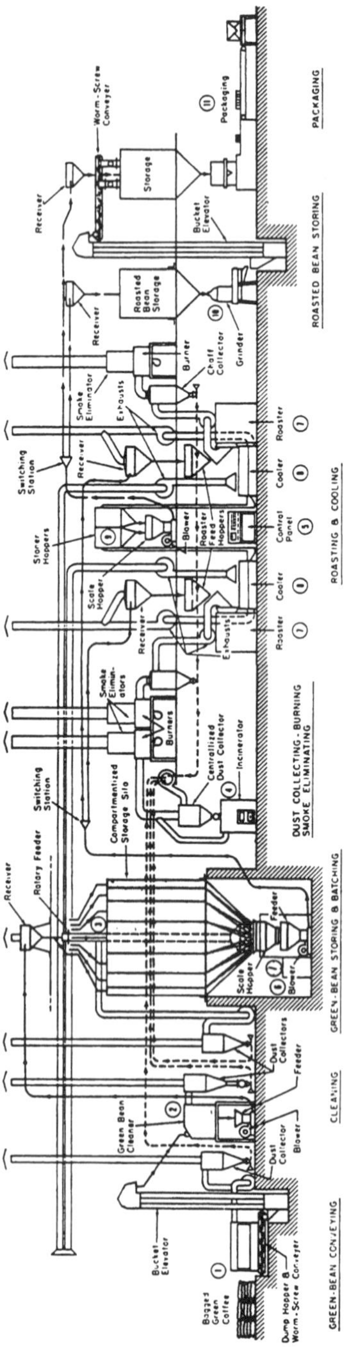


*Courtesy Food Engineering*

Fig. 8.7. Flowsheet of Probat Coffee Roasting Plant

## BULK STORAGE OF GREEN COFFEE BEANS

Individual green coffee bins hold several hundred bags of green coffee and are usually built in two rows in groups of ten to twenty over volumetric coffee bean feeders emptying into continuous mixing or blending screws. To avoid double handling of green coffee at the roasting plant, green coffee beans are dumped on arrival, cleaned and then sent to bulk bin storage until they are blended and roasted. Coffee bags are often stored at a warehouse near the docks as well as at the processing plant.

A continuous coffee roaster may produce over 10,000 lb (75 bags) of roasted coffee per hour. The green coffee bin storage may average only a few days to one week's supply. Since uniform quality green coffee lots are 250 bags, bins are designed to accommodate multiples of this amount, that is, 500, 750, or even 1,000 bags. Larger bins involve less investment per unit stored, but have the disadvantage of requiring several coffee lots that must be judged similar. Bulk storage of green coffee is more economical in use of floor space and in manual handling and blending. If green coffee beans are drawn from, say five bins in different proportions to make up a specified blend, when the supply of one type of coffee or lot is exhausted, it is a simple matter to replace it with a similar type from another bin. The blending process is related to producing a large continuous volume of uniformly blended coffee beans day after day, and month after month. With enough extra green coffee bin storage space to accommodate most incoming shipments, the bin storage minimizes the problem of searching for bagged lots of coffee on pallets which may be inaccessible.

### Design of Bean Bins

Bean storage bins have 45 degree slopes at the bottom so that all beans will run out through the volumetric feeder to the blending screw; periodically bean flow may be diverted after the feeder to check-weigh the rate of feed. Another chute above the feeder allows the coffee beans in the bin to be run out if necessary. Sight glasses in the vertical and slope sides of the bin show bean level. Bean level may also be detected by an electrically actuated probe connected to a board signal and bell citing high or low level. The sheet steel used for bins can be black but will undergo rusting in transit and even in use. Galvanized or zinc dipped sheet tends to flake after use. Electroplated zinc on steel affords protection against corrosion and does not flake; this is called galvanealed plate. Galvanealed plate should be welded with brass rod to cover any bare steel; if steel welding is done, the exposed steel surfaces need to be protected from rusting. Since the top levels of the bins support less coffee weight, the steel sheet gauges at the top bin sections may be thinner than the lower ones.

The bins should have a top and side door large enough to allow a man to enter to inspect and/or repair the bin. A dust tight light inside the bin roof allows inspection of contents. A calibrated scale inside or outside along the sight glass is useful in establishing coffee inventory. The low level bin indicator should be on the sloping side so as to minimize the time between signal and actual exhaustion of the coffee supply.

## Screw Conveyors and Bucket Elevators

These are the most common means of transporting coffee beans in coffee plants in the United States; airveyors and belts are used less frequently. Figure 8.8 shows a Probat installation in Europe using a belt conveyor and knife block diversion of beans. In the first process step, the rate at which coffee bags are emptied is governed by the removal capacity of the conveyor. For example, a bucket elevator, that has buckets spaced 1 ft apart with each bucket holding 1 lb of green coffee beans, moving at 60 ft per min carries away 3,600 lb of green coffee per hour. With 1.5 lb of green coffee per bucket, the carry rate is 5,400 lb of green coffee per hour. If this latter conveyor feeds a green coffee bean cleaner, rated at 6,000 lb of green coffee per hour, the cleaner will not be overloaded.

With bagged coffee on pallets brought to the man opening the bags and dumping the green coffee beans, a dumping rate of 40 bags per hour or about 6,000 lb of green coffee per hour is easily maintained.

There are so many screw conveyors in use that standards have been established regarding their design and application. For example, by specifying the bulk density of the coffee, the rate of weight movement desired, the angle of incline, and the moving level, a standard diameter screw conveyor at a suitable speed and motor size will be chosen by the fabricator. The pitch of the screw governs the rate of coffee bean forward movement. The shaft of the helix or screw is usually supported from Arguto bearings which may have to be replaced occasionally. Arguto is a natural oily lignite cut into a wood sleeve. Ball or roller bearings require grease lubrication and are not used normally in coffee conveyor systems. Lubricant leakage may cause serious contamination of the coffee flavor. The screw flights are normally $^1/_{16}$ to ⅛ in. above the conveyor shell so that the flights push the coffee beans but do not rub against the metal shell base. Too large a space between flights and shell may cause excessive crushing and cutting of beans without permitting them a positive movement through the screw and housing. Shaft bearings are located about every 9 ft so that the hollow shaft does not sag too much and rub and wear against the bottom metal surface. Misalignment of screw shafts may cause shell wear and excessive friction, thereby increasing the work load.

Screw drives are similar to those described for elevators in that the motor works through a belt and a speed reducing drive; the motors are also protected against drawing excessive currents by circuit breakers. By placing an overflow outlet at

*Courtesy Probat-Werke*

FIG. 8.8. COFFEE BEAN BELT CONVEYOR WITH KNIFE DIVERSION

the discharge end of a screw conveyor, end plugging is avoided. An overhanging lip on the screw conveyor cover assures that any water that falls on the conveyor will be kept out. For outdoor installation a gasket between the cover and the conveyor flange is desirable. Use of quick-opening lip clamps located at intervals of several feet keeps the conveyor cover on snugly yet allows quick disassembly when required. Jamming the coffee in a screw conveyor may cause the shaft connecting pins to shear. Detection and repair of such failures can take several hours, as they will remain unnoticed until the coffee supply is gone. The screw drive may be turning but the sheared shaft connection is not. Normally, shaft drives are located so that the screw pulls rather than pushes the coffee. The pushing pressure of a sealed-off compartment of coffee is powerful enough to push off the conveyor cover and spill out the coffee. Standard lengths of conveyor may be replaced when screws, or more usually housings, have become worn or actually worn through resulting in coffee slippage or leakage. Screw conveyors should have access all along their lengths for inspection and maintenance; when they are located in relatively inaccessible places inspection and repair may take a long time and even require temporary plant shutdown. Inaccessibility also produces a less safe installation. During repair or inspection work, it is safer to have lock-out circuits that are controlled in the vicinity of the conveyor so that no one else may start motor movement. Frequently, a group of conveyors is started automatically on signal as part of a series of processing steps. It is desirable during periodic shutdowns to clean fine dust and broken beans from conveyor bottoms and, at the same time, allow for screw and housing inspection. The wedging of a metal nut or bolt by the screw flight may jam the conveyor and at other times wear through the trough. Normally the coffee level is one-quarter to one-third the height of the screw flight, or 3 in. in a 9 in. screw.

Figure 8.9 shows one screw conveyor discharge to a crossover screw conveyor with both screw conveyors having overflows; the elevator overflow is obscured. The top duct of an elevator discharge may have a flip gate operated by chain from below so as to divert coffee flow to an alternate bin or conveyor.

Sometimes it is desirable to have a reverse coffee movement in a particular screw conveyor as, for example, the blending screw, if coffees are to be moved from one bin to another via an elevator. Wall thicknesses on screw conveyors must be specified in accordance with the severity of service.

## WEIGHING GREEN COFFEE BEANS

Green coffee beans are weighed many times between bagging in the producing country and entering the roasting machine. Normally green coffee bags are weighed when entering the consuming country as well as when entering the roasting plant warehouse as a check on shipping documents. Green coffee is weighed again upon removal from the warehouse and upon entering the green coffee cleaning process; cleaning weight losses and roast weight losses are also

*Courtesy J. A. Folger & Company*

FIG. 8.9. GREEN COFFEE BEAN BINS, CONTINUOUS BLENDER, AND ROASTER recorded. These losses vary among different green coffees, among the same coffees at different times and especially with the degree of roast. Bagged coffee is weighed on a platform scale at floor level. Palleted coffee bags may be set on the scale platform with a manual or powered fork lift truck. Gross weights of incoming truck shipments of bagged coffee are also made by weighing the truck on a truck scale before and after unloading.

Coffee beans in process may be weighed by a scale in the flow path. Figures 8.10 and 8.11 show two batch dump scales of the type used in coffee plants that, for example, dump 50 lb charges with each dump registering on a counter. This type of batch dump scale may be located between an elevator discharge duct and a receiving bin. By such a system of dump scales, accurate green and roast coffee inventories may be kept on bins. The dump batches, however, must be calibrated from time to time to assure correctness. In order that the batch scale does not become a bottleneck in the coffee flow path, the rate of coffee bean flow passable through the batch dump scale must be specified for its service. The batch dump

*Courtesy Howe-Richardson Scale Company*

FIG. 8.10. BATCH DUMP SCALE

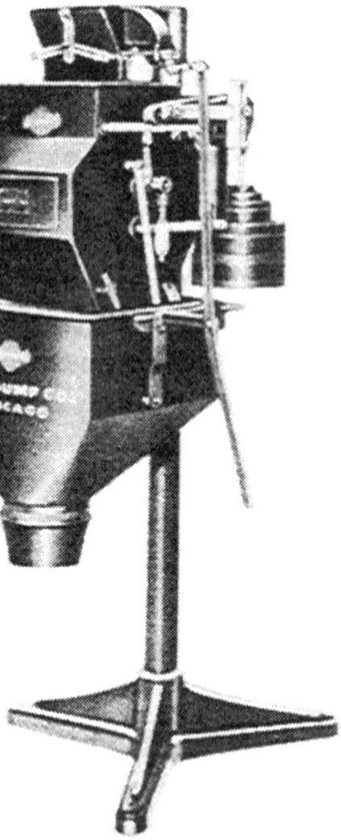

*Courtesy B. F. Gump Company*

FIG. 8.11. FIFTY POUND UNIT SCALE

scale takes the major portion of coffee bean flow at a high flow rate, and the final weight before dumping is a dribble so as not to overrun the set balance weight. Where there is considerable dust in passing green coffee or ground coffee through the scale, the scale housing will confine the dust, otherwise no housing is needed. There are conveyor belts that weigh or deliver granular material at a set rate, but this type of scale has not had much use in the coffee industry. Scales, in general, need periodic calibration and adjustment, preferably by the scale field technician, because of scale abuse, accumulation of dirt, wear of balancing knife edges, corrosion and erroneous tares.

## BLENDING GREEN COFFEE BEANS

Coffee beans purchased throughout the year may vary substantially. Yet inventories of beans must fulfill a need for uniformity from crop to crop (year to year), which is the main reason for commercial coffee bean blending. Some other reasons for blending coffee beans are as follows:

1. To control raw material prices, e.g., Robustas are more plentiful and lower in price.
2. Fermented tastes and off-flavored beans must be blended at 10–15 percent level to obscure any overall bad tastes.
3. Shift in availability or pricing of specific bean varieties, e. g. flood, drought, disease, etc. must be subsequently reflected in coffee blend composition changes.
4. A poor tasting coffee bean can sometimes be roasted more darkly to obscure its naturally poor taste.
5. Bean quality variances which are evident in whole green or roast beans, are obscure in ground coffee.
6. New crop coffees which are usually thinner in taste, need to be blended with old crop heavier tasting beans. Low acid old coffee beans must be balanced off with higher acidity new crop coffee beans.
7. Blends allow flexibility, so a large coffee buyer is not totally dependent on a single coffee bean source of supply.
8. Blending is necessary to smooth out variations in lots of bags.
9. Brazil beans which are neutral are enhanced in a blend with aromatic mild beans, e. g. U.S. military formula uses 70 percent Brazils to 30 percent MAM Colombian milds.

Secrecy in blend formulas is, therefore, usually to hide the undesirable aspects, not to hide the virtues, of the coffee composition. Blend formulas are secret. The least number of people know: a) how blend compositions are changing, b) who are the bean suppliers, c) the extent of supply, d) as well as how cheap is the coffee bean quality used.

Blending is a means to uniformize quality control while trying to control the maximum prices and to expand availability of a raw material. The big roasters must evolve blend formulas based on sources with large quantities of coffee beans available to them throughout the year at competitive prices. This is why it is commendable for a large coffee roaster to pre contract his coffee purchases, often at discounts. He later does not have to rush around to fulfill his needs to buy spot coffee beans at premium prices.

## Blending Process—Volumetric

In a small coffee roasting plant, where 1 or 2 four-bag batch roasters are used, green coffee can be blended in bag portions within the roasting cylinder. However, in larger coffee roasting plants, where many batch roasters or even continuous roasters are used, blending of green coffee beans is best done before roasting, and not in the roasting cylinder. Many large firms still batch blend their green coffees in 20-bag lots in a rotating drum with air suction to draw off dust. Recently-built coffee roasting plants use continuous green or roast coffee bean blending. To prepare the blend, selected beans are proportioned by volumetric feeders from several green coffee storage bins. Volumetric coffee bean feeders are not as accurate as weight feeders, due to variations in bean sizes and shapes, and, therefore, bulk density. But volumetric feeders are accurate enough for most blending purposes, and are less costly than weight rate feeders. Figure 8.12 shows a section through the Draver type volumetric pocket rotary bean blending feeder, with adjustable rotary speed. Larger pockets allow larger bean throughout at a given rotational speed. The speed is adjusted at the feeder by the ratchet length adjustment so that delivery rates of 5 to 100 percent of maximum feed rate are obtainable. Normally, rates between 10 and 25 percent are used. Two bean feeders set at 30 and 70 percent will deliver 100 percent of rated capacity; four bean feeders set at 10, 25, 25, and 40 percent will deliver 100 percent bean blend rate. Blending rates of 20 to 200 bags per hour are in use. Some volumetric feeders are driven from a common drive shaft while others have independent motor drives.

The volumetric bean feeders empty into a mixing screw conveyor. This type of conveyor is similar to an ordinary screw conveyor except that some flights are cut so that forward bean movement is reduced and rotary motion is increased. All the coffee beans are not mixed until they pass the last feeder. Some signal or visual control needs to be provided so that as one type of green coffee bin empties, the same portion of a similar coffee blend is substituted from another bin so as to maintain full bean flow and constant blend composition. At the mixing screw discharge, the blended beans may be elevated to a storage bin supply for the roaster, either batch or continuous. Thus adequate bin storage is maintained to effect blending 24 hr a day with a minimum of labor. Such blending systems are ideally suited to large roast or soluble coffee operations. The only precaution necessary with the volumetric feeder is to see that no wire, bolt, or other debris clogs the bean flow through the pockets. Normally such foreign matter is removed upstream of the volumetric feeders in a magnetic dump grate and the green coffee bean cleaner.

## Blending: U.S. vs. Europe

There are two ways of handling coffee: (1) large volume processing as in the United States, and (2) custom blending and roasting small batches, which is more prevalent in Europe.

The sales and processing in the United States are related to single type blends and roasts from single brand large processors with the roast and ground coffee sold mostly in evacuated cans. Some roasters pack from one hundred- to several hundred thousand pounds of roast and ground coffee each day. The coffee roasting plants in the United States usually are located at principal coffee import cities so that the packed coffee may radiate out to sales areas in the interior of the country. This seaport location between coffee supplies and distribution eliminates double

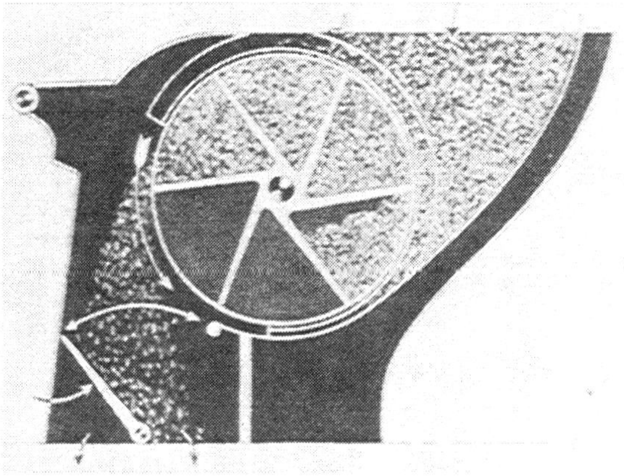

*Courtesy B. F. Gump Company*

FIG. 8.12. ROTARY POCKET FEEDER FOR COFFEE BEAN BLENDING

transit of coffee. These same seaport cities are themselves large consumers of coffee as is true, for example, of New York, Boston, Philadelphia, New Orleans, Houston, Los Angeles, San Francisco, Portland, and Seattle in the United States.

The newest roasting plants endeavor to hold manual labor and handling to a minimum. Such installations have continuous blending and roasting. Figs. 8.3 and 8.7 show some of these plants.

Coffee marketing methods in Europe differ markedly from those in the United States because of: (1) relatively higher prices of coffee, (2) traditional methods of buying coffee beans, (3) a wide choice of blends and roasts, (4) sales are usually from specialty shops, (5) the general lack of standards and (6) the absence of a mass market.

The exclusive use of batch roasting machines in Europe until about 1960 has now given way to several continuous roasting operations in Finland, Germany, Holland, and elsewhere. European processing and equipment are best typified by Probat; Fig. 8.7 shows plant flowsheet, and Fig. 8.13 shows an isometric view of another coffee plant. Probat uses sectored silos for green coffee storage, the silo sectors are often quite small, each holding as little as a dozen bags of coffee. After they leave the bean cleaner, green coffee beans are airveyed to a receiving cyclone centered over the sectored silos and via a duct positioner by means of a remote control index. Thus each sector holds an unblended green coffee type which is drawn out for blending.

## Automatic Batch Blending by Weight

Electronic consoles can give the following services: (1) directing dumped green coffee beans into sectored silo storage; (2) programming the blending of green

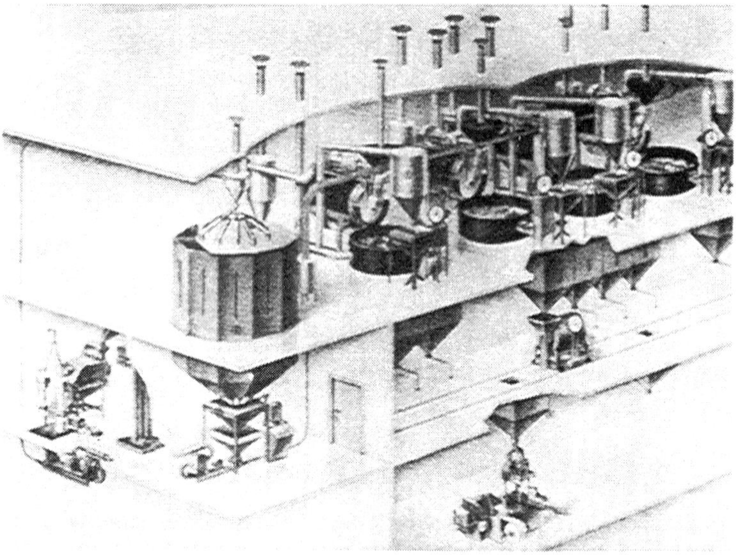

*Courtesy Probat-Werke*

FIG. 8.13. ISOMETRIC VIEW OF COMPLETE ROAST COFFEE PLANT

coffee from the sectored silos by weight into a hopper from which the unmixed batch is airveyed to the chosen roaster (Fig. 8.14); (3) programming the filling of the roaster with coffee, controlling the time/temperature performance of the roaster, and discharging roasted beans for cooling and storage.

The green coffee blending can be set with punched percentages or with an indexed coded blend card. Beans can be drawn from as many as ten bins in amounts ranging from 10 to 600 lbs to give the desired blend by weight. For the greater flexibility needed in the European market, such weighed blends may be altered as frequently as desired in only a few seconds. Stepping switches effect the sequencing of blending.

Compared to the simplicity of equipment used for continuous blending and roasting, this elaborate electronic system is effective so long as it is reliable. As long as custom blending and roasting are done, such a system saves labor. Europe, however, is beginning to convert to mass production and this trend will influence the choice of new equipment.

Blending roast coffees is also desirable. For example, soluble coffee plants in the coffee-growing countries, where poorer grades of coffees are often used, call for darker roasts for some beans than for others.

*Courtesy Probat-Werke*

FIG. 8.14. BATCH WEIGH BLENDING OF BEANS WITH AIRVEYING AWAY

## Blend Variations Thru the Years

Different botanical varieties of coffee growths have become more available in time while other growths have diminished or disappeared. Over 100 years ago Arabica coffee beans from Mocha, Arabia and Java (Indonesia) were highly prized for their taste qualities. Today Mocha coffee beans are of such small production that they are commercially unimportant. The Arabica Java coffee production has been substantially wiped out by shrub diseases. Indonesian Robustas are now commercially large exports and of a very low grade. Since 1950 Robusta bean exports from Africa have grown from a negligible level to ⅓ of world production, and are continuing to grow. Brazil, meanwhile, although producer of ⅓ of the world's coffee, has had serious coffee production problems due to drought, rain-floods, frost, and fungus disease in the past few decades. These have cut annual coffee bean production in half several times during this period.

In the consuming country to blend in a few percent Indonesian or Ecuadorian Robustas (naturals) with good quality Arabica beans will give a poor-tasting coffee blend. With 15 to 20 percent Uganda Robustas or with neutral Brazil beans, the whole bean mixture takes on a noticeable and objectionable Robusta taste. Use of Robusta beans in a coffee blend discourages the use of any good grades of mild aromatic flavored coffee beans. Putting it another way, increasing the portion of top quality beans in a blend cannot overcome the bad taste of the small portion. Where there are over-riding factors in a roaster's sales area such as poor water quality for brewing, use of stale R&G coffee, use of very dark roasts, additives like chicory, low prices are controlling, then very low grade coffee bean blends will be sold. Instant coffees are usually made from 100 percent rejected beans in the soluble coffee processing plants in the coffee growing countries. The continued use of Robusta beans in freeze-dried coffee blends has diminished the advantage that freeze-dried coffee technology offers in making better tasting coffee beverage products.

1977 GREEN COFFEE BLEND PRICE COMPARISON

| Coffee | Cost U.S. cents/lb | Cost High Robusta | Blend I High Robusta | Blend II Low Robusta |
|---|---|---|---|---|
| Robusta | 180 | 50% | 20% | |
| Brazils | 190 | 10 | 30 | |
| Centrals | 210 | 25 | 30 | |
| MAMS Colombian | 200 | 15 | 20 | |

## COFFEE ROASTING

The evolution of coffee roasting in the United States is graphically condensed in Fig. 8.15. The key dates are related, interestingly enough, to war-time periods when storages of labor required more efficient means of roasting.

Until almost 1900, coffee beans were roasted at home in Europe and in the U.S.A., in various types of frying pan or hand turned cylinders. The difficult part of roasting coffee beans is that heat must be applied quickly and uniformly, so people tried to keep the beans moving while they applied considerable heat. If the heating source were too hot, the beans would scorch; if the heating source were not hot enough, the beans will not pyrolize or roast. In either case an unpalatable product was produced, often with smoke and fumes that clung for hours, if not days, to kitchen walls, clothes, and body. Roasting was a time-consuming smoky process. If the coffee beans were not continually agitated, non-uniform roasting and some burnt beans resulted. One type of home roaster was a globe-like chamber at the end of a metal rod. The metal chamber was spun over a charcoal fire to roast a few ounces of coffee beans, just before grinding and beverage preparation. Some globe-like metal chambers were perforated to vent gases and to help improve heat transfer to the tumbling coffee beans.

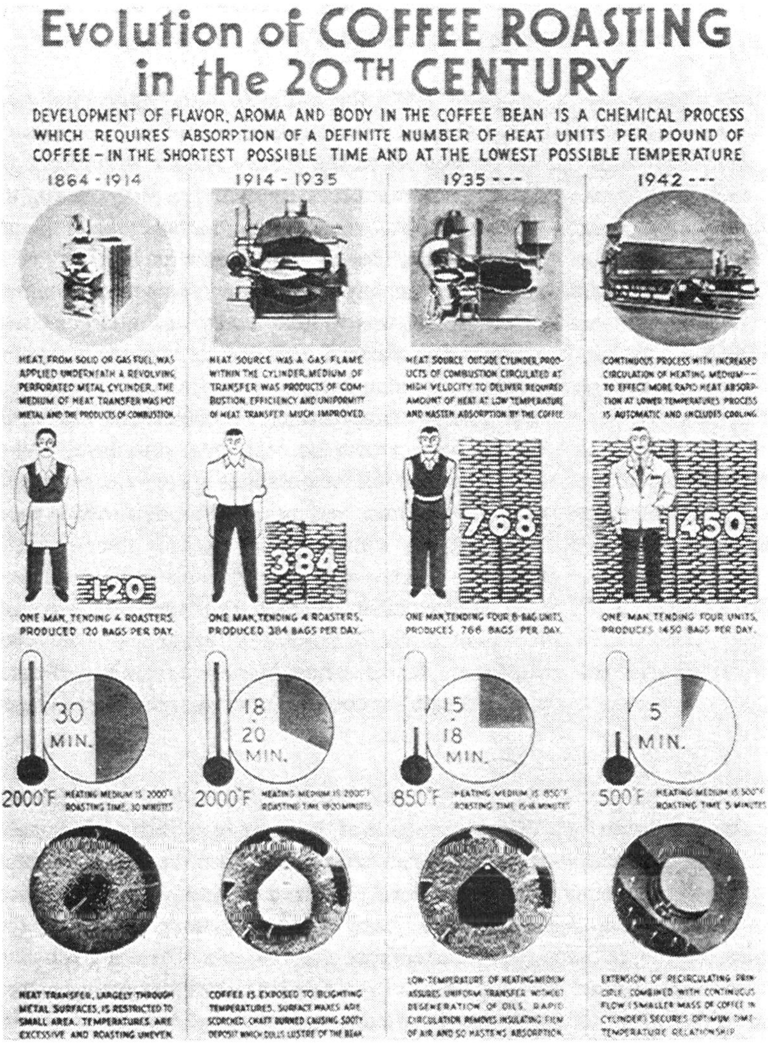


*Courtesy Coffee & Tea Industries Magazine*

FIG. 8.15. EVOLUTION OF COFFEE ROASTING

Before 1914, coffee beans were mostly heated by conduction of heat through a rotating steel wall. When the beans were fully roasted, they were dumped onto a stone floor and spread out for cooling. Gas temperatures about the steel drum were near 2,000 F (1,093 C), internal temperatures were not monitored and roasting time for the coffee beans was about ½ hr per batch. Colors of the roasted beans were uneven and it was normal to have some burnt beans. Roasting controls were manual and productivity was about 100 bags per man-day with 12-hr work days.

Working conditions and methods were unsafe, especially the open furnaces, handling of hot drums, burning coal and raking beans for cooling.

**Jubilee Roaster.**—Shortly after 1900, the availability of commercially refined mineral oil influenced the design of coffee roasting machines. During World War I, Jabez Burns & Sons, Inc. developed the Jubilee roaster with direct gas flame heating of the air within the perforated roasting cylinder, and covering the flame with a hood so that the beans thrown over from the rotating cylinder did not contact the flame. About the same time, the direct gas flame air heated roaster, Monitor, by Huntley Manufacturing Company of New York, was made commercially available. Manufactured gas fuel was available only in the larger cities; hence, fuel oil was used by many other roasters. This change in mode of heating was a significant departure from coal fired units. Direct fired oil or gas heating of roaster air allowed better control of gas temperatures in contact with the beans, effected more heat transfer by air contact with the bean than from metal wall to bean, required lower hot gas and cylinder wall temperatures, gave a cleaner roaster operation, shortened roasting time to 20 min, and increased productivity to about 400 bags of coffee roasted per man-day with one man watching three roasters. Many Jubilee roasters are still in use today with over 60 years service. The net result was that coffee beans were more uniformly roasted and fuel was used more efficiently. The manufacturer claimed that 70 percent of fuel heat was absorbed directly in the coffee roasting process. It was during this period that the perforated cooling tray with air suction underneath for cooling the discharged roasted beans evolved.

**Thermalo Roasters.**—In 1935, Jabez Burns & Sons, Inc., modified the Jubilee roaster by moving the burner outside of the roasting cylinder. Also part of the formerly exhausted hot gases were recirculated back into the burner to obtain improved heat recovery from the fuel used. The heated gases were also recirculated over the tumbling beans through the perforated cylinder wall at higher velocities resulting in better heat transfer from gases to bean. The design changes resulted in reducing batch roasting time to 15 min and in reducing gas temperatures to below 800 F (427 C), with the specified weight loading. Roasting time and temperatures could be varied by changing the weight of the coffee batch. The Thermalo coffee bean roaster doubled the productivity of the roasting operator handling three roasters, to 800 bags of coffee per man-day. It also gave a more uniform bean roast and better development of coffee flavor; a more uniform bean swelling was an indication of a rapid expansion of the bean during the exothermic period. Figure 8.16 shows a section through the Thermalo roaster, including recirculating gas ducting, blower, and damper. Figure 8.17 shows a battery of Thermalo roasters. Green coffee beans are charged from the front overhead bin into the roaster. The burner is at the rear as is the chaff and dust cyclone collector.

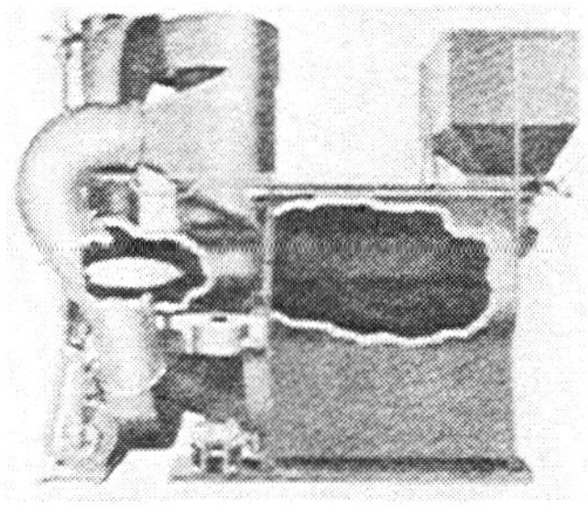

*Courtesy Jabez Burns & Sons, Inc.*

FIG. 8.16. SIDE SECTION OF THERMALO ROASTER

*Courtesy National Coffee Association—U.S.A.*

FIG. 8.17. BATTERY OF THERMALO ROASTERS

**Continuous Roaster.**—The next most significant commercial development took place about 1940 when Jabez Burns & Sons, Inc., offered for sale their continuous roaster. About this time, several large firms on the west coast of the United States had built custom designed continuous roasters which are still in use today. Figure 8.18 shows the outside view of the continuous roaster; Fig. 8.19 shows a section through the roaster. Roasting productivity was doubled to 1,600 bags of coffee per man-day with one man operating two 12,000-lb per hr continuous units. The repetitive labor of loading and unloading coffee was eliminated. The operator had only to check outcoming roast coffee color and make infrequent adjustments in air temperature controls, bean feed rates, and damper settings. The recirculating gas temperatures for roasting were reduced; 500 F (260 C) and beans

Courtesy Jabez Burns & Sons, Inc.

FIG. 8.18. OUTSIDE VIEW OF CONTINUOUS ROASTER

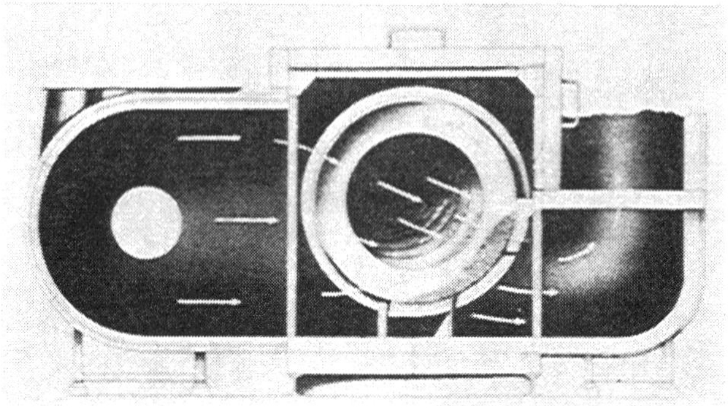

Courtesy Jabez Burns & Sons, Inc.

FIG. 8.19. SECTION ACROSS CONTINUOUS ROASTER

were roasted in 5 min. There was an improved development of coffee flavor and bean size, and slightly less weight loss. The beans were more uniformly roasted.

The continuous roaster is a long perforated cylinder ($^3/_{16}$ in. holes), several feet in diameter with an internal spiral sheet 8 in. high. The coffee beans are compartmented into 4 to 6 in. spacings of the spiral. Hot gases passing through the cylinder wall perforations are sufficiently forceful and turbulent to pass through the few inches of beans with good heat transfer. The coffee beans ride the upward movement of the perforated cylinder wall and tumble back upon each other. The roasting process occurs progressively along the 15 ft length of the cylinder; in the 3,000 lb per hr unit. The roasted beans continue to spiral through with a heat lock or center disc separating the heating from the cooling parts of the perforated cylinder. In the next 6 ft, the perforated cylinder has a hood and a suction of ambient air cools the roasted beans almost to room temperature. The passage of large volumes of heating or cooling air at high velocity through the perforated cylinder over the coffee exposes each bean to the same treatment and produces a uniform result. Water spray cooling is often used to supplement the air cooling of beans. A&P was the first to use the Burns Continuous roaster in 1942 in Brooklyn, New York. Now most major coffee roasters in the U.S.A. use continuous roasters. Many are used in Europe and in Brazil also.

**Starting and Stopping the Roaster.**—The heat load or demand of a batch roaster during warming up and initial roast is different from heat demand during roasting as well as at the end of the roast when the beans liberate heat. These conditions have been taken into account in the Probat batch roaster, and must be taken into account by every roaster operator if uniform roast coffee is to be obtained from all batch roasts. After starting a roaster, whether batch or continuous, the metal is not at thermal equilibrium with the process conditions for perhaps an hour or longer. This phenomenon is apparent with a continuous roaster as the recirculatory gas temperature must be lowered several degrees after an initial period while passing the same green coffee at the same feed rate and to the same degree of roast.

Starting and terminating the green coffee feed to a continuous roaster while obtaining the least off-color roasted coffee must be done carefully. There are several methods. One way, on starting, is to double the rate of green coffee feed for the first 20 to 30 sec so that the first surge of coffee in the first spirals is abundant enough to absorb the frontal heat load of bare metal which is not yet in thermal equilibrium with the passing coffee. Otherwise, the frontal coffee may take on a very dark roast and even burn. This depends to a large extent on the preheating time and temperature of the continuous roaster. In shutting down the continuous roast with relatively uniform roast coffee being run out, the green coffee feed is cut off and in about 4 min or 80 percent of roast time the heating of the recirculatory gases is cut off. Under these conditions there is enough heat in the cylinder and in the recirculating gases so that the last of the coffee leaves at nearly the desired roast color. During any outages of flame or heat, provisions must be made for diverting coffee beans roasted to an unacceptable degree. These underroasted beans must be subsequently brought up to the proper roast by an experienced operator. This is far more difficult to judge correctly than normal green coffee feed.

**Roasting Time.**—Operating a continuous roaster is simple, and this adds to its advantage when operating with unskilled labor. Aside from fuel combustion settings which are normally fixed, temperature control settings, damper control settings on the portion of gases being vented, and velocity of recirculatory gases, there are only two other important settings: (1) the speed of cylinder rotation set by means of a variable speed drive, which in turn determines the forward rate of coffee movement or coffee residence time, and (2) the rate at which coffee beans are fed into the machine. Of course, bean moistures, types, sizes, and so on also influence the rate of roasting. For the smallest continuous roaster, rated at 3,000 lb per hr of green coffee feed, there are the following relations between cylinder rpm and coffee holdup time:

| | | | | | | |
|---|---|---|---|---|---|---|
| Rpm of cylinder | 2 | 3 | 4 | 5 | 6 | 7 |
| Coffee hold time, min | 22 | 16 | 12 | 10 | 7½ | 6½ |

About three-fourths of the hold time is in the roasting section; the balance is in the cooling section.

The current green coffee feeder may measure either by volume or by weight. The latter means of delivery is preferred since weight is more directly related to heat load than is volume of coffee. The volumetric feeder is usually geared to the cylinder drive but may be independent. It is the same type of pocket feeder as is used in blending beans. Often it is the black, light density beans that take up the most volume and are most easily roasted and burned. Also, the flow rates of old, dry coffees can be regulated more precisely by their weight.

**Water Quench.**—Where a water quench is used to cool roast coffee more quickly, whether it is in a batch or continuous roaster, it is best delivered via a spray nozzle either directly on the beans or on the outer cylinder wall. For such thermal cycling, a stainless steel perforated cylinder offers some advantages in cleaning also. In a batch roaster a metered amount of water is sprayed. If the water spray is directly on the beans, the amount of water can be judged in terms of the residual moisture left in the roast beans. The amount of water may also be metered by an electrical timer switch controlling a solenoid valve on a water supply line having constant pressure. If a water spray is used in a continuous roaster, it may be sprayed continually on the discharging beans in the cooling air section, or the water may be sprayed on the outer perforated metal wall of the cylinder without wetting the beans.

**Dark Roasts.**—French and Italian roasts may be obtained with care in a continuous roaster, even though much coffee oil and combustible volatiles are driven off. Control of roasting is achieved by the release of the excess exothermic reaction heat via the exhaust gas damper which controls the temperature of recirculatory hot gases and the ambient fresh air intake to the burner. A roasting equilibrium has to be established so that a dark roast is achieved yet the hot, oil-laden blue gases are emitted. The darker roast beans may be held longer in the roaster cylinder while maintaining a relatively cooler recirculation gas temperature. This requires a damper that can be sensitively adjusted to control gas flow. Dark roasts can be carried out just as fast as lighter roasts, but control is more important at the faster rates.

Dark roasts leave oil and caramelized sugar deposits on the cylinder wall. These deposits under prolonged exposure to such continual roasting conditions may cause partial or complete closure of the cylinder wall perforations. Also, if these deposits carbonize, they are exceedingly difficult to remove. Closed holes contribute to localized overheating of beans and the beginning of bean fires. Thus the roaster cylinder wall perforations must be cleaned routinely. Cylinder wall tar buildup is due to neglect of roaster inspection and cleaning. Roaster doors, ports and ducting should be inspected weekly and cleaned as frequently as necessary. Chaff, fines, broken beans, and oily residues contribute to smouldering and fire which may lead to a roaster shutdown.

**Roasting Instruments.**—Before 1940, the only roasting instrument was a dial thermometer to indicate the temperature of the gases and tumbling beans. The key factor was the operator's determination of the final coffee bean roast color which marked the end of the roast by discharging the coffee to a cooling cart. Figure 8.20 shows the operator sampling roast bean color. From experience, the operator adjusted the rate of heating, damper setting, and weight of coffee batch, but no other controls were usually available. Since the degree of roast was not always uniform, these circumstances left room for improvement. In addition, roast coffee at that time was sold by appearance in the United States as it is now in Europe, and non-uniform roasting did not derive full value from the green coffee. With increased use of evacuated cans for packaging ground coffee, a variation in coffee bean development caused a variation in bean bulk density, which resulted in slack fill or a coffee bulkiness that did not fit within a standard 1 lb coffee can. The latter case was quite serious and called for better roasting controls.

After 1946 increasing numbers of roasters had thermal probes set into the tumbling coffee beans and connected to a temperature recorder controller so that the fuel supply, and consequently heating, would be discontinued when the probe indicated a set temperature and an alarm signal was given to the operator who could then discharge the roast beans into the cooling cart. This control system is widely used today. However, operator experience in starting up a cold machine as well as different coffee blends and degrees of roast are still important for such batch roasting operations and, in the author's experience, beans are never quite as uniform as those processed in a continuous roaster. In any case, extreme variations of roast have been reduced.

**Roaster Safety Features.**—The instrumentation of the batch and continuous roasters is also part of their safety features. The use of gas fuel may be hazardous. Unburned gas accidentally released into a roaster may ignite and explode. Direct gas firing to heat recirculatory gases is cleaner than fuel oil firing. Every gas burner and some fuel oil burners have gas pilot lights, and these are ignited first by an electric spark. Then the main fuel burner is ignited from the pilot flame. The electric control starting panel for such ignition is tied into a light sensitive electric

*Courtesy National Coffee Association—U.S.A.*

FIG. 8.20. SAMPLING ROAST COFFEE BEANS FROM BATCH ROASTER

eye that monitors the main flame. If delivery of the main fuel flow does not ignite within a few seconds, something is wrong, and the main fuel supply will be shut off automatically. The fault must be found and corrected before restarting the pilot-main burner cycle. A simple yet frequent fault may be low gas supply or oil pressure, or a shut off fuel valve. A time delay for air purge usually assures removal of any residual combustible gases before the firing cycle is started by the automatic controls. In the event of a flame failure, an alarm actuated by the flame monitor will alert the operator to the flame-out. Immediate remedial action is needed to avoid under-roasting or later burning of partially roasted coffee, as well as to minimize the throughput of coffee which is not properly roasted.

If there is an electric power failure, a batch roaster may be emptied of coffee beans at the critical stage by manual rotation of the cylinder. The probability of such an occurrence is small. However, in a continuous roaster some of the coffee beans are always at the exothermic point, and electrical power failure will cause fire and burning of some coffee beans unless prompt action is taken. In most installations an auxiliary drive such as a gasoline engine takes over and the beans may be discharged from the roaster. Also a bank of carbon dioxide cylinders may be used to drive out oxygen and cool the beans to below ignition temperature. Protection of the roasting cylinder from thermal damage as well as from subsequent production outage is the main consideration.

**European Coffee Roasting Equipment.**—In any discussion of roasting and roasting equipment, processors and manufacturers become extremely biased toward the merits of their methods. This is mentioned here because roasting equipment in Europe differs in one design feature from the widely used Jabez Burns & Sons roasters in the United States. The latter have perforated cylinders for air flow patterns while the former do not. The Probat roaster section is shown in Fig. 8.21; air heating is done from below the double solid walled steel cylinder holding the bean. Hot gases enter the back end of the cylinder, pass through the tumbling coffee beans, and leave via a front duct which is also used for charging green coffee beans. An air duct damper above the rotating solid walled cylinder that holds coffee beans may be opened to allow more of the heating to be done by the hotter gases outside the cylinder wall than by the hot air passing through the bean filled cylinder. There is no recycle of hot gases leaving the roaster which might increase the heat efficiency. At the end of the roast, the whole Probat roaster front slides out about one foot to allow the roast beans to discharge into a cooling cart. In Europe the most common coffee roasting machines are the Probat and the Gothot, both produced by German manufacturers. The Probat machine batch roasts in less than 10 min. All the roasting machines mentioned have been in use for decades and each is an efficient roaster when operated properly. Since about 1950 Probat has augmented the dial thermometer on the roaster face with a temperature recorder controller which not only starts and ends the roast by a set temperature, but programs two heating rates. Green beans may not enter the roasting machine until it warms up to a set temperature. A high rate of heating is used to bring the moisture out of the green beans, and a lesser rate is used until the exothermic process occurs. When the thermal probe in the beans signals the set temperature for roasting, the burner fuel is automatically cut off and the roast beans are discharged. Thereafter a new green coffee bean fill, roast, and discharge cycle is automatically initiated. Thus batch roastings may be done at a rate of 4 per hr. The standard size Probat roaster holds 240 kg or four bags of coffee. Other roaster sizes are available down to a fraction of a bag for use in retail shops and for test purposes.

**Coffee Roasting Equipment Developments.**—From the foregoing descriptions of types of roasters and the roasting process, certain generalizations may be made. There is a definite trend to larger coffee roasting plants using continuous roasters, especially in the United States. As a result, roasting time has been reduced from 20 to 5 min. Changes that have occurred include automation, better controls, less labor per unit of coffee processed and, therefore, more efficient equipment. The faster roast, through better heat transfer, gives better flavor development, but if higher temperatures are used, poorer flavor results.

**Engineering Factors in Coffee Roasting.**—A good way to see what is happening to the coffee in the roaster, the gases and the fuel is to make a material and thermal balance, and to examine the blower capacities and power loads.

Consider a continuous roaster that processes 3,000 lb per hr of green coffee with an 18 percent weight loss. Roast or shrinkage loss depends upon the coffee, the degree of roast, residual moisture, and other factors. This means that 2,400 lb per hr of roast coffee will be produced, 500 lb of water will be evaporated before pyrolysis, and 100 lb of carbon dioxide will result. Since roast coffee may have two percent carbon dioxide, we assume that half the $CO_2$, (400 cu ft) is liberated during roasting and half (50 lb) is associated with the 2400 lb of roast coffee leaving the roaster.

*Roaster Blowers.*—The recirculatory gas fan moves 14,000 cfm at 450 F (232 C) and 8 in. of water suction which corresponds to about 8,000 cfm at 80 F (27 C) and ten air horsepower. A 50 percent efficient fan needs a 20 hp motor. The cooling air fan draws 8,000 cfm of ambient air raised to about 140 F (60 C) at 6 in. of water suction without duct restriction requiring 7.5 air horsepower or, at 50 percent efficiency, a 15 hp motor. The combustion air blower delivers 16 oz per sq in. pressure with 360 cfm ambient air requiring 1.5 air horsepower or, at 50 percent efficiency a 3.0 hp motor drive. The roaster variable speed cylinder drive of the roaster uses 1.5 motor horsepower. The gear fuel oil pump delivers about 10 gal or 80 lb of oil per hour with a regulating pressure relief valve set at 50 psig using a fractional horsepower motor. Table 8.1 gives the material and heat balance around the continuous roaster, and is sufficiently accurate to show the relative importance of the contributing factors. The roaster process heat utilization is about 80 percent; the balance is heat losses through ducting. The pyrolysis heat release, depending on degree of roast, constitutes up to 15 percent of the heat input.

*Roaster Air Flow Requirements.*—The total gas flow from the roaster vent stack is about 1,000 cfm at 480 F (249 C). In addition, at this temperature: (1) Combustion blower air at 360 cmf-NTP[1] is about 600 cfm; (2) Fuel oil is converted to water vapor and $CO_2$ gas. The carbon of the fuel adds nothing to gas volume when it becomes $CO_2$, but water vapor from the fuel is developed at about 90 lb per hr or almost 3600 cfm; (3) The water vapor from the green coffee is almost 200 cfm-NTP or 350 cfm at 480 F.

The following calculations illustrate the amount of excess air used and the amount of $CO_2$ in the vented and recirculated gases. The equation shows the equivalent oxygen demand by the fuel oil for complete combustion.

| | | | | | | | |
|---|---|---|---|---|---|---|---|
| $C_{12}H_{26}$ | + | $18\frac{1}{2}O_2$ | → | $12\ CO_2$ | + | $13\ H_2O$ | moles |
| 170 | | 592 | | 528 | | 234 | weights |

[1]NTP, normal temperature and pressure, 70 F, 29.92 in Hg barometric pressure.

TABLE 8.1. MATERIAL AND HEAT BALANCE 3,000 LB/HR CONTINUOUS ROASTER[1]

| Heat Loads location | | lb/hr | Specific heat | Temperature change F | Temperature change C | BTU/hr |
|---|---|---|---|---|---|---|
| 1. Sensible | a. Bone-dry coffee | 2500 | 0.4 | 380–80 | 196–27 | 300,000 |
| | b. Water-liquid | 500 | 1.0 | 212–80 | 100–27 | 66,000 |
| | c. Water-vapor | 500 | 0.4 | 472–212 | 244–100 | 52,000 |
| 2. | Fuel oil | 60 | 0.9 | 472–80 | 249–27 | 20,000 |
| 3. | Combustion air | 1500 | 0.25 | 480–80 | 249–27 | 150,000 |
| 4. Latent | Water evaporation | 500 | | 1,000BTU/lb | | 500,000 |
| | Surface heat losses—through bare and insulated walls | | | | | |
| 5. | Surface estimated at 500 sq ft. MgO thermal conductivity, 0.3 BTU/Hr/sq ft/°F/ft. For 2 in. insulation and avg. temperature difference 330° F. See* ... | | | | | 300,000 |
| 6. | Vent gas loss** ... | | | | | 400,000 |
| Total heat requirements ... | | | | | | 1,788,000 |
| 7. | Exothermic heat gain—Pyrolysis The estimate of heat release is based on carbon to $CO_2$, but since the $CO_2$ is partly formed from thermal decomposition, half of the calculated value isused.*** Combustion of carbon to $CO_2$ yields 94,000 cal/gm mole of $CO_2$ or 169,000 BTU/lb mole of $CO_2$; 100 lb $CO_2$/hr is 2.27 lb moles × 169,000 = 384,000 BTU/hr ÷ 2 ... | | | | | 192,000 |
| Total net heat requirement ... | | | | | | 1,596,000 |
| Heat supply from 80 lb/hr of fuel oil at 20,000 BTU/lb ... | | | | | | 1,600,000 |

[1]Basis: An operating case. No fresh air is taken in at damper that vents recycled roaster gases to stack.

*Recirculating gas is usually 500° to 550° F depending on the roast time and other factors.
**60,000 CFH × 0.66 lb/cf × 400° F × ¼ = 400,000 BTU/hr vented gas heat.
***2500 lbs/hr roasted beans × 0.01 = 25 lbs/hr $CO_2$ ~ 180 CFH or 3 CFM at 80° F.

It is interesting to note that in a fluidized bed using clear hot air (e.g. Sivetz, M. U.S. Pat. No. 3,964,175) and inciting vigorous bean movement, the rough bean surfaces are cleaned smooth, thereby increasing heat transfer across the bean surface.

Each pound of fuel oil requires 592/170 or 3.5 lb of oxygen for combustion; this is 42 cu ft of oxygen or 200 cu ft of ambient air. Each pound of fuel oil gives 528/170 or 3.1 lb of $CO_2$ or 27 cu ft of $CO_2$ at NTP. From Table 8.1, the fuel flow is 60 lb per hr or 1 lb per min. The combustion air flow is about 80 percent in excess at 360 cfm. When the products of combustion have been cooled to ambient temperatures (assuming that the water vapor produced has disappeared), a gas volume balance on the operation for each minute would be approximately as follows:

| Combustion air | CFM | Product Gases | CFM |
|---|---|---|---|
| Air for combustion | 200 | $CO_2$ from combustion | 21 |
| $O_2$ for combustion | ( 42) | $CO_2$ from coffee | 3 |
| $N_2$ | (158) | $N_2$ | 158 |
| Excess air | 160 | Excess air | 160 |
| Total | 360 | Total | 352 |

With the actual quantity of air used for combustion, the Co₂ content of the gaseous products approaches 8 percent by volume in vented gases.

*TAR FORMATION in Recirculatory Gases.*—The hot gases leaving the rotating cylinder area, when the beans rise above 400° F, are high in aldehyde, ketone, alcohol, acid and other complex organic vapors. The same gases are in the roasting cylinder and bean area. These organics are very reactive and readily polymerize at temperatures over 500° F to form the smog and smoke associated with coffee roasting. These micron sized particles cannot be readily dissolved nor collected nor filtered. These are the tars from roasting, and they deposit on the surface of every bean, as evidenced by the bean surfaces being darker than the inside of the bean. (Note that in the Sivetz once through hot air roaster, the beans are always lighter on the outer surface than inside). These tars contribute a harsh raspy taste and aftertaste, as well as coalescing later in extracts and causing separation problems of extract from ice in freeze concentration operations. Further, these tars are the air pollutants, with which we will deal later. These tars also deposit on the roaster cylinder, housing, ducts, dampers, etc., taking with them oils, all of which subsequently plug roaster cylinder perforations, and their coatings form a heat resistant layer on the steel rotating cylinder. Therefore, as this coating builds up, it requires higher input heating temperatures of the recirculatory gases in order to maintain the same roasting time. But the higher operating gas temperatures trigger at some point a fire of the tarry coating and/or of the beans.

The phenomena described occurs in every commercial coffee bean roaster, and manual cleaning is very time consuming, expensive and has prompted Burns to offer its ACO 1200° F heat cleaning cycle as an alternative. These tars are considered carcinogenic in nature, typical of byproducts from pyrolytic processes.

*Rate of Heat Transfer From Air to Bean in Continuous Roaster (Heating or Cooling)* If the coffee beans traverse the roaster in 7½ min (5 min in the roaster section and 2½ min in the cooling section), there are about 250 lb (green coffee equivalent) in the roaster section and 125 lb in the cooling section. The entering air temperature differential between air and bean is about 400 F (222 C) and the exit differential is quite small, a 200 F (111 C) mean differential. Assuming a ¼ sq in. surface per bean and about 2,800 green beans per lb, there are about 5 sq ft of bean surface per pound of green coffee. By equating the useful heat of about 1,000,000 BTU per hr into 12,500 sq ft of bean surface and a 200 F (111 C) temperature differential, the heat transfer film coefficient from air to bean is ½ BTU per hr per sq ft per ° F.

*Thermal Conductivity Rate to Center of Coffee Bean.*—After the heat passes through the film resistance at the surface of the bean, it is useful to determine the rate of heat transfer through the mass of bean. Assuming that the thermal conductivity of wood is similar to green coffee, $413 \times 10^{-6}$ cal per sec per sq cm per ° C per cm, or multiplying by 2900 to obtain units of BTU per hr per sq ft per °F per in., we have about 15 BTU per hr hr per sq ft °F per bean depth.

This heat transfer rate is twice as great as that across the film coefficient. The film heat transfer rate, therefore, is controlling, and the internal bean temperature will be rather uniform.

However, at least half of the heat entering the bean is by conduction from the hot cylinder wall, plus 10–15 percent by radiation; so that the beans in a conventional rotary cylinder roaster receive heat at far lesser rates than in a roaster in which over 90% of the heat transfer is from hot air to bean.

## Types of Roasters—Based on Methods of Heat Application

**Aerotherm.**—A roaster that levitates the beans in a 392 F (200 C) turbulent flow of hot air to achieve batch roasts in 2 to 3 min will now be compared in its heat and power use to the Burns and Probat roasters. The largest type of batch roaster offered is 110 lb rated at 24 roasts per hour, or 2,640 lb of green coffee roasted per hour. This is some 25 percent less throughput capacity than the continuous roaster already discussed, so that the heat demand and air flow power use may be readily compared, with one exception. The Aerotherm has an indirect air heater which almost triples heat input and doubles air blower horsepower requirements. With direct fire heating of air and the shorter roasting time at lower inlet air temperatures, the net higher fuel use is only about 25 percent. The Aerotherm roaster is claimed to attain uniform roast color, reduced roast weight loss, and better flavor development. However, sales were discontinued about 1970 by the manufacturer.

**Fuel Cost.**—The cost of fuel in roasting has a great deal to do with selection of fuels. In some localities gas costs more than oil; in most localities electricity is the most expensive fuel by at least several-fold. Gas is the cleanest hydrocarbon fuel and is usually preferred for its complete combustion. Electric roasters, in which the air is heated by electricity, with heat transferred to the beans by convection, are not usually found in commercial use.

In 1978, fuel costs in the USA were approximately:

| SOURCE | PRICE |
|---|---|
| Electricity | 3.5¢/KWH (1,000 BTU/¢) |
| Natural Gas | 35.0¢/100,000 BTU |
| LP Gas | 50.0¢/100,000 BTU |
| Fuel Oil (Gal) | 50.0¢/140,000 BTU |

To roast, it takes 450-500 BTU/lb green coffee beans, or 200 lbs of beans can be roasted for about 40¢ worth of fuel, which is about $^1/_5$¢ of fuel per lb of coffee roasted.

**Bean Levitation**—Assuming a 0.15 gm green bean (about 3,000 beans per lb of roasted coffee), and 0.30 cm equivalent bean radius, the resulting weight distribution of bean per unit area is 0.57 gm per $cm^2$. The kinetic head of air flow upward to sustain such an average green bean particle is the above pressure divided by the air density at 400 F (204 C). This is 29 gm per 44,800 $cm^3$. The velocity head is 1100 cm air. Since air velocity squared is 2 gh, we have $v^2 = 2 \times 980$ cm per $sec^2 \times 1{,}100$ cm, or v = 1500 cm per sec; in the English system, 50 fps or 3000 fpm.

| | |
|---|---|
| *Coffee heating requirement per 110 lb batch:* | |
| Bean sensible: 100 lb × 0.4 sp ht × 400 F is | 16,000 BTU |
| Water sensible: 10 lb × 1.0 sp ht × 150 F is | 1,500 BTU |
| Water latent: 10 lb × 1,000 BTU per lb is | 10,000 BTU |
| Heat losses from system are not taken into account here | — |
| Total | 27,500 BTU |
| At 24 batches per hour (2,640 lb per hr) the heat demand is | 725,000 BTU/hr |
| At 6,000 cfm at 400 F (204 C) the air heating load is | 2,200,000 BTU/hr |

Fuel heat use in the last case is 800 BTU/lb green coffee compared with 400 BTU/lb theoretical as shown above for indirect air heat. Assuming a 12 in. water pressure loss for the air passing through heater and roaster at a 6,000 cfm, 400 F air flow and 40 percent overall fan efficiency, a 30-hp motor is needed. Motor power is also needed for the cooler and primary furnace air and the total is about 50 hp.

**Radiant Heat**—The roasters discussed so far use direct firing of the fuel into the circulating air stream heating the tumbling beans. The major heat tranfers to the beans is through gas contact, with lesser quantities by conduction through direct contact with the metal wall and other beans of dissimilar temperature. A small amount of heat is also transferred to the beans by radiation from the wall and from hot gases at 500 to 600 F (260 to 316 C) ambient temperatures. Radiant heating is dependent on the emissivity of the gas and is proportional to the fourth power of the absolute temperature. The proportionality constant derived by Stefan-Boltzman is $0.173 \times 10^{-8}$ BTU per sq ft per hr per $R^4$ (F). Only a relatively few gases radiate energy and only at certain wave lengths in the infra-red spectrum (see frequencies and wave lengths below.) The only gases that have emission bands of sufficient magnitude to consider are carbon monoxide and dioxide, hydrocarbons, water vapor, sulfur dioxide, ammonia, and hydrogen chloride (polar compounds). Symmetrical molecules like hydrogen, oxygen, nitrogen, and so on are not absorptive or emissive for practical purposes. Since the gases in the roaster are 90 percent oxygen and nitrogen, the emissivity of the roaster gases is small. Another consideration is the absorptive capacity of the beans for such radiations; this would be perhaps a 0.2 factor when green and a 0.8 factor when roasted. The geometry and wave length of radiation is also a consideration, but in the roaster cylinder this would be about a 1.0 factor (PL) but the gas emissivity would be only about 0.13.

If the roasting temperature is taken at 540 F, the absolute Fahrenheit temperature is 1000 R (Rankin), and the factors for the absorption of radiant heat by the beans during roasting would be less than 1 percent. The radiant heat contribution from higher temperatures and from red hot electric heaters can be up to 30 percent.

## Means of Radiant Heating for Roasting Coffee

The following listing shows the types of radiant energy to be discussed and their respective frequencies and wave lengths (frequency × wave length is Plank's constant):

| Radiation | Frequency, Megacycles | Wave Length |
|---|---|---|
| Infrared | 500 million to 1 million | 0.001 to 0.5mm |
| Microwave | 5,000 | 0.1 m |
| Dielectric | 10 to 5 | 50 to 100 m |

**Roasting with Infra-Red Radiation.**—Green coffee may be roasted by radiant heat. The advantages of this type of heat transfer are that the heat penetrates the bean and there is little temperature gradient from the bean surface to the interior depending on radiant wave length used. The difference in taste and aroma development may be sufficiently distinctive to appeal to some roasters. Since electricity is used, the heating cost is higher. Radiant roasters are made in cup test sizes as well as 100, 200, and 500-lb green coffee capacity, and resemble a conventional roaster in that they have a solid rotating cylinder, a cyclone for chaff removal, front door discharge and a front rotary cooler pan.

**Microwave Heating to Roast Coffee.**—In the United States the firms of Raytheon and Girdler manufacture food heating equipment using high frequency radio waves as the heating medium. The Raytheon unit operates at 2,500 megacycles and heats coffee and other substances much faster than at lower frequencies such as short wave radio and dielectric heating. Raytheon sells a large number of radar ranges per year to restaurants for quick thawing of frozen food and for preheating foods before serving. At this frequency, water is heated more rapidly whereas, at the dielectric frequencies, carbohydrate materials are heated faster.

One can roast coffee experimentally in a Pyrex tube with a rubber stopper while it is rotating in the microwave field. Rotation is to assure good mixing. The heating can be done in air at a satisfactory speed but vacuum reduces the tendency to ionize the air under higher power applications. However, at pressures below 6 mm Hg, coronas of ionized gases appear which furnish a conducting path for electricity and result in an electric discharge, overloading of the equipment, and shutdown with some coffee burned in the process. Ultra-high vacuums might be better, except for the evolving of water vapor and organic compounds from the coffee. This type of equipment shows small promise. Moisture from the green coffee does not condense in the heating tube at temperatures up to about 212 F (100 C) but once the coffee is dry and temperatures exceed 300 F (149 C), there is sufficient localized heating which progressively concentrates on the spots of least resistance. Once carbonaceous coffee is formed it is a good electrical conductor and the flow of excessive current in a localized spot causes electrical discharge. Use of microwave heating in coffee roasting would be expensive and would involve the design and operation of new equipment; the advantages of such rapid heating would have to be demonstrated to justify investment.

**Roasting Coffee with Dielectric Radiation (short-wave radio band).**—The Girdler equipment operates in the 27,000 to 2 megacycle frequency range, and thus can be used for both microwave or dielectric heating. Allis-Chalmers equipment operates at 30 megacycles, which is in the dielectric region only. Dielectric heating is used to heat the laminated glued sections of plywood uniformly and thoroughly to effect their thermal set. Higher microwave frequencies heat faster but design applications are easier at the dielectric range of wave lengths. A Pyrex tray of green coffee heated in the dielectric field leaves the Pyrex unheated while the coffee gives up its moisture and then heats to the roasting point. Roasting proceeds satisfactorily but volatiles and condensation must be removed to eliminate points of high electrical conductance, that would make for electrical discharge and termination of the heating process. This factor as well as the continual agitation of the coffee must be incorporated into any working design.

## Elimination of Roaster Smoke

**Washing Roaster Smoke.**—Some coffee plants discharge their roaster smoke through a water sprinkling duct or tower before releasing it to the atmosphere. Spray units usually are ineffective even in absorbing the water soluble volatiles although some of the volatile acids, aldehydes, and other organics are removed. The only colloidal particles that contribute to the opaqueness of the smoke are hardly affected. The use of a caustic in the recirculating water helps to neutralize the absorbed acids and to dissolve some of the fatty acids from the aerosol cloud. The lack of efficiency is due to inadequate residence time of the gases in the mixing water spray as well as to a lack of formation of enough surface to effect a high aerosol contact and transfer of droplets to the spray (or in some cases countercurrent packed tower). One way to break up the air stream effectively so that it undergoes intimate mixing and contact with an alkaline recirculating solution, is to suck the roaster fumes into a water ejector. With a 1,000 cfm exhaust as from the 3,000 lb per hr roaster, two such roasters could be handled by a 14-in. polyvinyl coated ejector pulling up to 2500 cfm using 160 gpm of solution at 80 psig or requiring about a 15 hp pump. The initial investment and maintenance will be somewhat higher for the ejector than for the burner. Water makeup, chemicals makeup, cleaning and straining the solution reservoir are not attractive. This type of scrubbing will not dissolve the pure hydrocarbons, but these are only a small fraction of the total contaminants passing out of the roaster stack. They are not very reactive chemically and are aerosols.

**Ionization-Precipitation of Aerosols in Roaster Smoke.**—Aerosols can be passed between two wires having a large voltage difference. The aerosol particles, as a rule, carry an electric charge which prevents their coalescence. They usually carry some water and often electrical conductors. The approach of these minute charged particles to a high voltage wire creates a highly ionized field where the charge of the aerosol particles is neutralized and the particles may then coalesce with other neutralized particles to become large enough to fall. The charge will also become discharged in the highly ionized gas field surrounding the wires. Deposits of particles on the wire, though not normally a problem, may be removed by cutting off the high voltage electricity while the wires are washed. If this type of system is used to precipitate the aerosols constituting the roaster smoke, it often pays to pre-wash the roaster smoke so as to remove all washable items before the ionizing conditions are applied. Sometimes, disposable air filters (fiberglass) are useful in removing gross particulate matter before the washer or ionizer operates. Ionization systems in the United States are built by Cottrell (originator), Smokatron, Minneapolis-Honeywell, Westinghouse, and others. The high voltage terminals constitute a hazard that employees need to recognize.

These systems have not provided a practical solution to eliminating roaster smoke.

In 1951, Jabez Burns & Sons, Inc., offered for sale to the coffee roasters an afterburner which substantially effected the complete combustion of roaster smoke by passing it through a gas or oil burner. This converted all organic, volatile, odorous coffee substances into odorless, clean gases containing only $CO_2$ gas and water vapor. Even fine coffee dust and chaff that carried over were burned. The reasons for this afterburner demand was that the oily smoke and odors as well as fine dust from coffee roasters were creating a public nuisance, corrosion, and/or filth damage to adjoining property. Thus, neighbors individually or collectively through their city government exercised pressure to stop the annoying dirty smoke erupting from the coffee plants. Laws to eliminate smog in many cities are now enforced. Now most roasters, even when they move to a relatively less populated area, include afterburners or equivalent smoke eliminators to maintain good public relations. With increases in population density, diminution of air and water pollution is receiving greater attention. Smoke is a general term for a visible gas leaving a stack or other roasting or incinerating equipment. The white cloud of condensable water vapor is not in any way a problem. The problem lies in odorous constituents, solid particles, and aerosols, such as colloidal droplets of oil and tars which result from the dry distillation of the coffee bean during roasting. The particulate matter forms a haze that is diluted as it blows away.

**Afterburners.**—Pollution of the atmosphere may be removed by oxidation; this does not mean that an open flame is required. It does mean that the aerosols need to be brought up to their combustion temperatures, with adequate air to cause oxidation to water, $CO_2$ and $SO_2$. Of course NO or $NO_2$ as well as $SO_2$, even when oxidized, are very acidic and irritating gases, but these may be easily dissolved by passing the bases through a water wash. This is not practiced as a rule, because sufficient improvement is achieved when the smoke is oxidized to a clear (but acid) moist discharge gas. From a chemical viewpoint some roaster gas substances are easier to oxidize than others. For example, $H_2S$, aldehydes, carbon monoxide, hydrogen, and unstable organics oxidize readily at less than 800 F (427 C), but stable organics such as methane or long chain fatty acids require oxidation temperatures of up to 1,200 F (649 C). Particulate matter that comes from carbonized chaff requires temperatures of up to 1,800 F (982 C). Carbon particles are usually not an important contaminating factor from roaster gas discharge, provided all the chaff is burned. Temperatures of 1,000 to 1,200 F (649 C) may be sufficient to burn out the roaster smoke before discharge to the atmosphere.

Referring back, for purposes of comparison, to the 3,000 lb per hr green coffee continuous roaster to determine the amount of extra fuel needed to heat the smoky exhaust gases to 1250 F (677 C) to clear combustion gases, the following figures are cited. The roaster exhaust gases are assumed to be at 450 F (232 C), passing through insulated ducting to conserve heat content, before heating to 1250 F (677 C). Figure 8.22 shows a roof afterburner connected after a cyclone from three batch roasters.

*Afterburner fuel needs for 3,000 lb per hr continuous roaster:*

| | | |
|---|---|---|
| 1. Water vapor: | 500 lb/hr evaporated from green coffee beans<br>100 lb/hr formed from fuel combustion | |
| Total | 600 lb/hr at 0.5 sp ht and 800 F (427 C) Rise | 240,000 BTU/hr |
| 2. Vented gases: | 3,600 lb/hr at 0.25 sp ht, 800 F (427 C) Rise | 700,000 BTU/hr |
| 3. Carbon dioxide: | 200 lb/hr from fuel combustion<br>50 lb/hr from coffee beans | |
| | 250 lb/hr at 0.25 sp ht, 800 F (427 C) Rise | 50,000 BTU |
| Total afterburner reheat requirements are: | | 1,000,000 BTU |

This is $^2/_3$ of the fuel requirements for roasting; hence, is an extra cost for the simple and effective service achieved. However, with increasing fuel costs, the annual costs for fuel afterburner, in large roasting operations is worth recovery.

*Courtesy Probat-Werke*

FIG. 8.21. SIDE SECTION THROUGH ROASTER

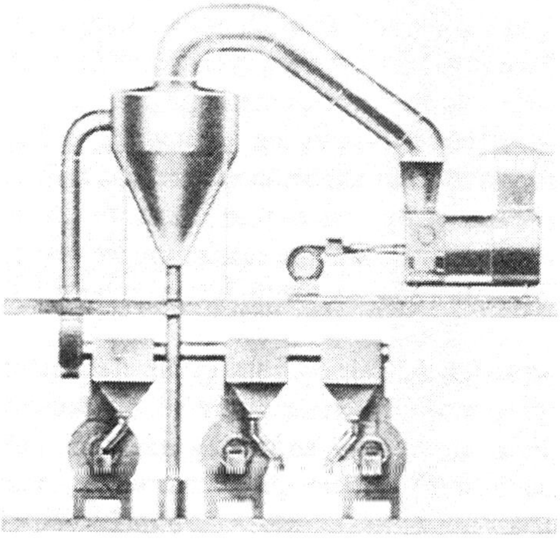

*Courtesy Jabez Burns & Sons, Inc.*

FIG. 8.22. AFTERBURNER CONNECTED TO THREE BATCH ROASTERS

Coffee roaster gases have much particulate matter, and thus the use of a burner using gas or oil is suited to burn up the chaff, dust, and oils. In the above calculation, no credit is taken for the heat liberated by the complete combustion of the gas and particulate substances; this liberated heat would be significant and would perhaps reduce the afterburner fuel needs by 10 or 15 percent.

**Catalysts.**—Since afterburners and heat recovery systems are costly in fuel and capital investment, and since considerable know-how has been gained in the 1970's with the catalytic oxidation of auto and factory exhausts, the use of platinum on ceramic pellets as a means of oxidizing smoke when exposed at 600° F has come into use. This is basically the Houdry process developed in the 1930's and largely applied to petroleum vent gases and catalytic cracking of petroleum fractions which today are more often done in fluidized ceramic beds. The operating principle is that when the combustible gas, oxygen and the platinum catalyst come together, oxidation occurs at much lower temperatures than would be possible without the catalyst. The principle is sound and has many useful applications but, depending on the amount of chaff, oils, ash, etc., fouling of the porous ceramic catalytic surfaces will occur in the course of time. This will require burning out the fouling matter to help restore catalytic efficiency, or replacement of the catalyst bed with fresh catalyst. In any case, the roaster gases must have a sufficiently high temperature when they enter the catalyst bed to effect the chemical oxidation in air. The oxidation occurring on the catalyst bed results in heating the bed as well as the gases passing through. The issuing gases often rise several hundred degrees to 1,000 F (538 C) after passing through the bed. Obviously, a portion of these gases may be recycled into the part of the gas stream entering the catalyst bed to preheat the entering air mixture, or a heat exchanger may be used. Usually these measures are not applied to batch roaster systems because of the wide variation in temperatures and gas flow. A similar catalytic muffler is used on the automobile gas engine exhaust to oxidize the incompletely burned gases. Sometimes, such catalytic oxidations are desirable to recover heat from the normal exhaust fuel gases or to eliminate combustible condensables from the exhaust duct.

Normally chaff should be collected from cyclone collectors and burned in a solid fuel type of furnace or incinerator where the residence time for burning is longer than in a fuel afterburner. Also, the lightness of chaff is such that it is quickly carried through a high velocity gas stream without complete combustion. Chaff burns completely only at temperatures over 1200 F (649 C).

## CONVEYING, STORAGE, WEIGHING, AND BLENDING OF ROAST COFFEE BEANS

The coffee leaves the roaster through the stoner and proceeds to the stoner bin, which serves as a surge bin as the roast beans pass through a continuous flow type scale (Richardson-bucket). This gives a weight balance between green coffee entering and leaving the roaster system. The roast beans may then be bucket-elevated and screw-conveyed to the appropriate storage bin. The roast coffee storage bin is similar to the green coffee storage bin. With reduced bulk density of the roast beans, the same size bin will hold only half as many pounds of roast coffee. A nominal bulk density that may be used for roast beans is 20 lb per cu ft. Sheet metal wall thicknesses of the bins are about 14 gauge backed up every few feet with channel steel; top tier sheeting, which does not require as much support, may be 16 gauge. Smaller bins are usually sufficiently strong if 18 gauge sheet steel is used. If galvanealed steel surfaces are used, brass welding rods do not destroy the protective surface, otherwise a protective coating should be used with steel welding. The level of the coffee in bins may be monitored by an instrument such as the Bindicator which has an electric motor-operated propeller that stops when the level of coffee rises above the propeller and trips an electric circuit. There are also dielectric probes, pressure switches, and some other level probes that are equally reliable. Inspection doors and dust-proof light as well as internal and/or external calibrated scales should be provided to show contents in the coffee bins. If volumetric bean feeders are located at the discharge of each bin, roast coffees may be blended via a mixing screw. The mixed roast coffees may be lifted by bucket elevator to the storage bins above the grinder. Airveying of roast coffee causes only a little particle breakage. Roast coffee blending may also be done by drawing the desired weights of each type into a scaled bin.

## Roast Coffee Stoner

The coffee leaving a roaster is usually air-lifted to an overhead storage bin as shown in Fig. 8.23. The purpose of this air lift is to convey the coffee so as to effect a separation of heavy particles from the roast coffee beans. Using a 0.16 gm bean with 0.3 cm equivalent radius, the weight per unit area is 0.6 gm per cm$^2$, and this requires an air velocity head of at least 450 cm to sustain it in mid-air flow, not counting drag or buoyancy effect. The minimal velocity of ambient air for lift is thus 950 cm per sec, 31 fps, or 2,000 fpm. If the air-flow area at the boot is 1.0 sq ft for the 3,000 lb per hr continuous roaster, then the upward airflow is 3,000 cfm. Actually, air-flow is several times higher than this so that the "lift" on the roasted beans may be regulated by flaps above the boot and before the fan. Also, since all beans are not uniform, the lift velocity must accommodate the densest beans. Adjustment of airflow by the flapper gates allows only beans to be lifted; a dancing motion of the beans is produced indicating that some beans fall back somewhat before they are lifted away. Heavier particles such as stones cannot rise in the air lift stream; hence, they accumulate in the perforated boot. This aspect gives rise to the name of the unit, "stoner." A flip gate at the bottom of the boot allows stones and heavy objects to be removed. In the event that a bean is jammed in the lift, the balanced gate may be opened to drop beans out so that the lifting action may be resumed. In order to control the flow of beans and to spread them out across the air-lifting perforated metal chute, an adjustment height on the entering spreader is provided as well as a guillotine gate. In order to pick up ferrous objects, a permanent magnet is placed below the chute but before the spreading guillotine.

*Courtesy Jabez Burns & Sons, Inc.*

FIG. 8.23. ROAST COFFEE BEAN STONER

Figures 8.24 and 8.25 show duct and belt magnets. Roast beans occupy only a small fraction of the lift duct cross-sectional area. Assuming a 9 in. water pressure drop and 5,000 cfm air flow and 50 percent efficiency, a 15 hp suction blower motor is required.

## Coffee Bean Properties During Roasting

Roasting is the step relating to coffee aroma and flavor development in the processing of green coffee beans. The aroma and flavors developed are characterized by the type of green coffee. The degree of roast is related to the type of green coffees being processed and the market or disposition of the roast coffee. The manner in which the chosen degree of roast is attained is dependent on the type

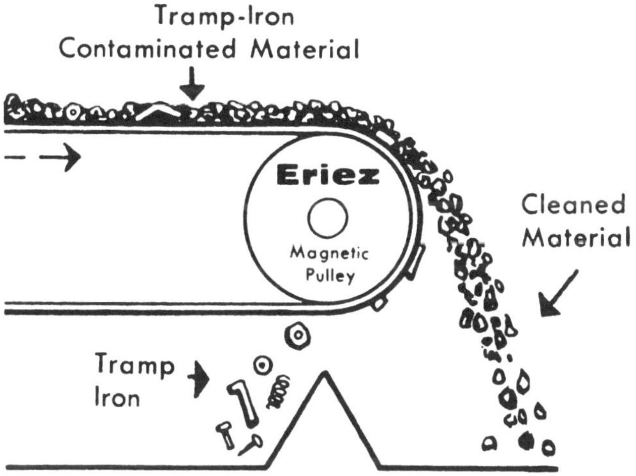


FIG. 8.24. BELT MAGNET

*Courtesy Erie Mfg. Company*

FIG. 8.25. DUCT MAGNET

of roasting equipment used. Although green coffees vary in chemical and physical properties, the chemical and physical changes they undergo during roasting are similar even though they vary in degree. Roasting of green coffee beans is essentially a process of exposing the green coffee beans to a warming process that is sufficiently fast to drive off the free and bound moisture of the bean and the dried bean residue is heated to more than 400 F (200 C). At about this temperature, pyrolysis, or thermal decomposition and chemical change, occurs within the bean.

In a fraction of a minute exothermic (heat liberating) chemical reactions occur. The bean temperatures rise to 392 to 410 F (200 to 210 C) with an accompanying dry bean weight loss rising from 4 to 6 percent. With a green bean starting moisture of 12 percent, this is equivalent to a 16 to 18 percent total roasting loss. The higher the percent loss, the darker the roast color. The brown color development of the bean occurs during this period of rapid loss in weight. Most of the sucrose is altered and most of the swelling of the bean (to almost twice its original volume) also occurs during this period, with the simultaneous revelation of the chaff at the bean crevice. The chlorogenic acid, which is at about 7½ percent in the green bean, and slightly less at the beginning of pyrolysis, falls to about 4 percent in the roasted beans. The pH of brew extractables which is at about 6.0 in green coffee, falls gradually to about 5.5 at the beginning of pyrolysis, then becomes about 4.9 in roast mild coffees. This process is substantially a dry distillation of thermally decomposed organic matter constituting the green dry coffee bean. The roasting process is terminated at the desired flavor which is equivalent to the desired degree of chemical pyrolysis, and concomitant weight loss; it is guided by the darkness of color developed at the point of roasting. The roasting process also may be terminated by the temperature of the beans as it is under automatic temperature control systems on batch roasters. The key point is that the desired flavor and aroma are attained at the desired degree of sugar pyrolysis and caramelization.

Roast weight losses occur at two rates. The first rate is slow (low slope) and is due to the evaporation of water from the bean. The second rate (steep slope) is pyrolysis. As the rate of water loss falls off, the rate of carbon dioxide gas evolved rises rapidly. This point of transition occurs at the beginning of pyrolysis, that is, at about 365 F (185 C) or at about the time of 10 percent weight loss from the green bean. Although a 2 percent $CO_2$ liberation was used in the roaster calculations, a 1 percent $CO_2$ liberation is possible at the lightest palatable roasts. The general character of the phenomenon described is quite similar in coffees that differ as much as Robustas, Medellin and Brazilian coffees. The similarity in what occurs is not to be confused with the fact that different sizes, types, moistures, and other factors cause one lot of beans to come to the desired flavor development (as well as color and weight loss) faster than another lot.

## Volatiles, Flavors and Aromas

Supplementary knowledge may be gained about the roasting process by examination of the volatile gases evolved as related to bean temperature, color, or weight loss. The rate of expulsion of volatile acids reaches a maximum during roasting which might indicate when to end the roast. This chemical method of relating the end of the roasting period to the rate of acid distillation or the rate of carbonyl evolution, or the rate of dimethyl sulfide evolution is, in fact, more closely related to taste and aroma than bean color or bean temperature. This relationship is emphasized today because current sensitive ionization equipment for gas chromatography analyses makes such a monitor possible. In any case, it is a means of relating chemical composition development within the bean to volatiles being driven from the bean. The maximum acidity curve is further instructive in showing the changing volatiles content of the bean and the fact that taking the roast to darker stages reduces the volatiles residue in the bean. In the extreme case all volatiles would be driven off and only charcoal would be left. We know that prior to pyrolysis the coffee beans have a maximum acidity but such coffee is not yet palatable. In other words, the pyrolysis must be carried to a point where there is a balance of chemical products that best pleases the senses.

Roasting beans in a closed cylinder does not allow volatiles and water to escape; hence, these alter the course of roasting as done commercially. The result is a very acid cup. Roasting correlations are best made in comparison with weight loss, not time. Different coffees give different rates of $CO_2$ evolution. Roasted proteins give carbonyls. Roasting of dry green coffee solubles under pressure develops a coffee flavor (not as good as from the bean) and a similar pH profile; several percent water yields an excessively acid cup. The smaller or the more broken beans of green coffee give a weaker and poorer coffee flavor, and less volatiles are produced. The chemistry of the roasting process is complex, but it is worthwhile to examine what is known about the changes in sugars and other carbohydrates, oils, proteins, and minerals.

*Sucrose,* which constitutes about 7 percent of green coffee, is altered to simple, carmelized, and decomposed sugar products. The sucrose is first dehydrated, then hydrolyzed to reducing sugars as the temperature rises to the pyrolysis point. Then the reducing sugars are dehydrated, polymerized, and partly degraded to volatile organics, water, and $CO_2$ gas. Some pyrolysis products react with proteins and degradation products to form other coffee substances.

*Starches and dextrins* undergo some hydrolysis at roasting temperatures in the presence of water to yield water soluble polysaccharides and polysaccharides that are later solubilized in the percolator hydrolysis step. A small portion of the starches is partially degraded, liberating water and $CO_2$. Some starches and dextrines also undergo carmelization and even carbonization, depending on degree of roast.

*Pentosans* partially decompose to yield furfural, which is at its highest level in lightly roasted coffee and may easily be identified from its characteristic cereal odor. Industrially, furfural is prepared by acid hydrolysis of pentosans from corncobs and other food wastes.

*Cellulose, hemi-cellulose, and lignins,* which respectively constitute the woody, fibrous, and binding matter of the bean cell structure, are not very much affected by the roasting and are mostly water insoluble. Shrinking and then swelling of the woody cellular structure occurs with drying and pyrolysis. But whatever hydrolysis or carbonization may occur is small compared to that occurring in the smaller carbohydrate molecules. The internal gas pressures created within the bean cells at the time of pyrolysis (with the liberated water) softens the cellular structure sufficiently to cause it to swell and release the $CO_2$ gas.

*Acids* are formed mostly from the carbohydrates as they undergo thermal decomposition to carboxylic acids and then $CO_2$ gas. The coffee solubles, before a palatable roast is achieved, show a pH change from 6.0 to 6.5; once a palatable roast is achieved, the acidity is higher—a pH of as low as 4.9 but usually about 5.1 with Arabicas (Robustas are much higher in pH, about 5.5). Darker roasts drive off volatiles and acidity and give coffee solubles with pH's several tenths higher. In general, light roasts give a more acid cup than dark roasts. The predominant acids are chlorogenic, acetic, and citric. The chlorogenic acid occurs in green coffee at about 7.0 percent but about one-third to one-half is destroyed during roasting. About 0.4 percent of roast coffee is acetic acid with several tenths of the higher homologous acids also present. Similar losses in citric and malic acids also occur but the percentages vary somewhat with coffee and roast. The simultaneous formation and decomposition of acids during roasting explain why brew pH of light (unpalatable) roasts rise above pH 6.0 before they fall to their most acid tasting cup (light roast but palatable). Medium roasts have cup pH's of 5.0 and darker roasts have cup pH's up to 5.3. Optimum flavored roasts are obtained just before the rate of volatile acid evolution starts to fall off, e.g. pH 4.90.

At the beginning of the roast, at bean temperatures of about 380 F (194 C), volatile acids that form, such as formic and acetic (and homologues), due to their low equivalent weight, far offset the loss in acidity caused by decomposition of chlorogenic acid. Then the rate of formation of volatile acids equals its release from the bean, while the chlorogenic and other non-volatile acids continue to decompose. The rate of volatile acid formation thus rises to a maximum, levels out and then falls off while at the same time the rate of acids being driven off is only partly influenced by their rate of formation. Since most of the effective acid flavor is in the volatile acids, when this acid contribution starts to fall off with the continuing fall-off in non-volatile acid content, the most acid cup point is passed. It is necessary to appreciate the fact that in the dynamics of the acid formations and their eventual reduction, each acid develops at a different rate. The formation of phenolic acids is important because of their strong flavor contribution, rather than their acid factor. A further complexity is that only about one-third of the non-volatile acids is water soluble; the rest are bound up as salts.

*Volatiles* that give aroma and flavor are formed and retained within the cell structure of the bean. They are the decomposition and reaction products that constitute only about 0.04 percent of the roast coffee. They are mostly aldehydes and ketones from protein and carbohydrate breakdown under heat with a small but odorously important amount of sulfides from certain proteins.

*Proteins* are made water insoluble (denatured) at temperatures far below pyrolysis. Hydrolysis of peptide linkages liberates some carbonyls and amines.

Hydrogen sulfide is driven off in notable quantities although hardly any remains in the roasted bean. High grown coffees liberate and retain dimethyl sulfide, a very important aroma and flavor constituent of good quality coffees. Small amounts of methyl mercaptan and dimethyl sulfide are also liberated from the sulfide type of proteins. Hydrolysis releases some amino acids. Sugars and nitrogenous substances polymerize into caramel and browning-type products. The fishy and ammoniacal odor from dark roast coffees is due to amines. The insoluble particles of protein are partly associated with fatty substances, and when roast coffee is brewed, much protein is in the colloidal brew particles that give brewed coffee its turbidity. Some of the insoluble proteins are later hydrolyzed in the percolation process to yield water soluble amino acids, as well as coagulated tars. The 13 percent protein in green coffee is sufficient to give marked flavor contributions through its decomposition and the reaction products of pyrolysis.

*Caffeine* is, for the most part, unaffected in the roasting process, except for minute amounts sublimed at 350 F (176 C) and accumulated within roaster stacks. With reduced roasting temperatures, sublimation is more vs. less with lighter roasts. Chemically the caffeine is stable at roasting temperatures.

*Trigonelline*, the N-methyl betaine of nicotinic acid which occurs in green coffee at about 1.1 percent, the same percentage as caffeine, suffers only about a 10 percent decomposition loss during roasting.

*Oils* in the green coffee are partly unsaturated, hence susceptible to breakage under thermal stress at points of double bonds. Green coffee has about 12 percent oil as determined by petroleum ether extraction. The temperatures attained in roasting are not high enough to cause much change in 95 percent of the oils. Natural vegetable oils are glycerides; thus the heating of these glycerides in the presence of water and acids causes some hydrolysis to glycerine and fatty acids. If short chain, volatile fatty acids form, they will partly volatilize. Blush smoke is the colloidal oil driven off especially during dark roasts. The presence of fatty acids is quite evident by their action in reducing the surface tension of coffee extracts, i.e., less foaming in the cup. Roasting breaks down and opens up the bean cell structure. It liberates oils that have been chemically bound so that they become free to move about in the bean. The oils coagulate, seep through the bean, and wet the bean surface in dark roasts. These dark roasts also have a characteristic oily odor that one associates with circumstances where vegetable oils have been cooked. Thus, the uniform distribution of oil throughout the cytoplasm of the green bean forms tiny droplets of coalesced oil at the cell walls in the roast bean and the oil becomes mobile through the cell walls because the cytoplasm has been destroyed. The cells in the roast bean are cavities containing $CO_2$ and other volatile gases.

*Carbon dioxide gas* which does not exist free in the green coffee bean is formed during roasting and pyrolysis. At least 1 percent of the green coffee is given off as $CO_2$ gas during roasting; darker roasts give about 2 percent. Normally $CO_2$ is a product of decomposition of carboxylic acids and is related to such sources during the roasting process. As the acids formed within the bean are driven out, they also undergo decomposition. The residual roast bean, depending on degree of roast and whether a water quench is used (water drives out $CO_2$ gas) has 1.5 to 2.0 percent $CO_2$; the fine grind coffee analyzed immediately after grinding has only half as much $CO_2$ gas content. Using a roast bean density of 0.7 gm per $cm^3$ and an equivalent diameter of 0.8 cm, the roast bean volume is ¼ $cm^3$ and its weight is 0.18 gm. At 2 percent $CO_2$, 3.6 mg $CO_2$ has a volume of 1.8 $cm^3$ $CO_2$ at NTP, or the gas occupies seven times the bean volume. At 50 percent bean wall volume, the gas, if pressurized, is at 14 atmospheres or about 200 psig. The $CO_2$ gas is partly adsorbed.

Whole bean roast coffee keeps its $CO^2$ better than ground coffee, that has so much more surface area for moisture absorption and less distance for the $CO_2$ to diffuse out. The lesser surface per unit weight of the whole roast bean retards the rate of moisture entering the bean. Moisture absorption by the roast bean whether the moisture is from the atmosphere or from liquid spray, liberates the $CO_2$ gas promptly. As the $CO_2$ leaves, so do the aromatic flavorful volatiles while, at the same time, oxygen enters. Whole roast coffee beans stale noticeably within two weeks; ground coffee stales within hours when exposed to the atmosphere.

*Minerals* which exist in the coffee during roasting are separated from their organic origins and catalyze the pyrolysis reactions. Phosphatides (lecithin-like phospholipids), which are the colloidal portion of brewed coffee, form some phosphates. The potassium and calcium alkaline ions form salts with organic acids when liberated from their natural chemical orientation.

Thus roasting changes the color, size, and shape of the bean physically. Chemically, pyrolytic changes alter the original organic compounds to decomposed, transformed, and reacted organic compounds which now include caramels, volatile acids, volatile carbonyls and sulfides. The characteristic coffee flavor has been developed in the roasted bean. Heat has been used to drive out the moisture from the green bean and to raise it in temperature until mineral-catalyzed self-sustaining exothermic reactions occur; the process is stopped abruptly by rapid cooling of the beans below pyrolysis temperatures. All the free moisture (about 12 percent) is lost from the beans plus about 4 to 6 percent by weight of the chemical substances (depending on degree of roast) of the original green coffee bean. Another one percent of $CO_2$ gas remains in the roast coffee even after grinding.

## Influences of Degree of Roast

So far, the effects of roasting to develop coffee flavor have been covered in a general way. It is desirable to view the differences resulting from light, intermediate, and dark roasts.

Light roasts have more acidity, and hence are more suited to areas where alkaline waters are used. Since lower grade coffees have their poor flavor characteristics more readily revealed in light roasts, it is usually better to roast these coffees dark so as to drive out most of the characterizing volatiles. From the foregoing discussion, the advantages of blending roasted coffees are self-evident. Habituation and local taste preferences often govern the degree of roast. In cities this is usually darker than average, as is indicated by the expression, "city roast." This type of roast, something lighter than French, and without any burnt character, is the degree of roast usually used in the preparation of roast coffees for soluble coffee manufacture. Since more carbohydrates are made water soluble in plant percolation than in brewing at home, the additional carbohydrate flavor needs to be balanced off by a deeper roast.

There is a relationship between visual color of the roast bean and the percentage of roast loss. A light to medium cinnamon-colored roast which is just within the palatable range would have about a 14 percent loss; a fully developed roast of deep brown but not blackish-dark would have about a 15½ percent loss; a high roast which is dark brown with tinges of black would have about a 17 percent loss, and is the darkest usually used in the United States. However, specialty roasts, such as the French, are very dark brown with some oil slicks on the bean surfaces, and have about an 18 percent roast loss; the Italian roast is extremely dark brown at about a 20 percent loss. In all cases, it is assumed that the green coffee contains about 12 percent moisture.

The amount of solubles extractable by water at 212 F (100 C) will vary among coffees, but darker roasting will reduce such extractables. For example, a medium roast at 15 percent roast weight loss may yield 23 percent extractable solubles; but the same coffee roasted to an 18 percent roast loss may give only 21 percent extractable solubles. In regard to hydrolyzable solubles which are of interest for soluble coffee manufacture, these also are reduced in darker roasts. Using the same case, hydrolyzable solubles may go from 25 to 22 percent. In other words, the total reduction in extractable and hydrolyzable solubles is from 48 to 43 percent. This has marked influence on the actual yield of solubles attainable from the percolators. Whereas 38 percent may have been the solubles yield under percolation conditions with the lighter roast, they may only be 35 percent from the darker roast. The darker roast beans have suffered more destruction to their cell walls and their structural strength, and hence their compressibility is greater. For this reason compaction and excess pressure drop of dark roast coffee percolator beds are more frequent than with lighter roast coffees. Dark roast beans are also more brittle, and in grinding, more fines are formed. These fines also contribute to blocking off flow paths in the ground coffee percolator bed and give higher pressure drops or more resistance to extract flow. Although the above roast weight losses were cited, it is possible to distinguish by eye about eight distinct color shades from the lightest to the darkest palatable roast; these can be referred to as cinnamon, medium high, city, full city, French, and Italian. Darker roast flavors are more faithfully carried through the soluble coffee manufacturing process and give a better tasting instant coffee powder.

As different coffees are used, or as different blends are roasted, continual tasting is essential to control the flavor development by means of a ligher or darker roast. A light roast of blended coffee usually reveals more disparity in flavor development of the individual types than the darker roasts. The liberation of more fatty acids in dark roasts and their carryover into the extract and spray dried powder help retain more coffee flavor through the spray drying process, and reduce foaming in the cup.

The use of a water quench to add several percent of moisture to the beans on discharge from the roaster darkens the bean surface and thus dry and wet bean surfaces are not comparable for color matching.

The amount of extractable and hydrolyzable solubles will depend on grind; in the example given, the grind is coarse. In pulverized coffee samples, for example, the 212 F (100 C) extractables may be 30 or more percent. However, for any given grind, the darker roast will still be a softer bean cell residue and will still have less extractables than the lighter roast.

The bulk density of the roasted beans will vary with the degree of roast. It will also vary with the speed of roast and the original moisture of the green coffee beans. Roast bulk densities for Robusta beans may range from 33 to 23 lb per cu ft, while a 50 percent mild/50 percent Brazilian blend of good coffees may vary from 23 to 18 lb per cu ft (free fill of cubic foot box). These degrees of bulk density differences are also reflected in the ground coffee used in the percolation columns.

Most bean swelling occurs at the pyrolysis temperature. Fast roasts on large beans, especially new-crop coffees with more than average moisture, may cause a 10 to 15 percent larger swelling than normal. Sometimes, this is a problem when filling a fixed volume 1 lb vacuum can. A United States west coast roaster produces such bean development intentionally and then has to force the ground coffee into the can to make it fit.

## Flavor Development

The development of the roasted flavor in the coffee bean can also be done by roasting green beans in an oil bath. It is analogous to French frying potatoes, nuts, fish, meats, and so on. Developing the roast beef flavor in the oven or the charcoal broiled flavor in ground beef is also somewhat analogous. The chemical changes in flavor development are more uniform across the coffee bean than they are in the beef (raw or burnt). Roasting coffee beans is a process that, for the most part, takes place in the absence of air. It can be compared with the dry distillation of bituminous coal in order to produce coke, or wood being converted to charcoal. Many of the volatiles derived from coffee roasting are identical with those derived from the dry distillation of wood or even tobacco. Each of the odors and flavors derived from woods: oak, hickory, bay, pine, eucalyptus, are different and enjoyed or used (e.g., for smoking meats) according to the types of flavor volatiles evolved. So it is with different coffees. Some have more volatiles than others, and some have constituents in the volatiles that others do not. Thus, each is valued for its aroma and flavor content (chemical volatiles), and the result is further shaped by the degree to which these volatiles are developed or driven out of the roasted coffee bean. Roasting time, the air temperatures used, the means of heat conveyance (radiant, conduction, or convection), the means and rate of cooling, all these will have some bearing on the final flavor (chemical) residue in the roasted bean. The intrinsic potential flavors lie within the bean; the mechanisms of roasting can only bring out what is there. The raw material can be improved by roasting e.g. under steam pressure for Robustas. An oak log will still smell of oak volatiles; and a hickory log will still smell of hickory volatiles.

Green bean properties that influence the mode and rate of roasting are specific heat of the bean, moisture, size, and shape of the bean, wholesomeness of bean, its origin, botanical variety, age, methods of curing and drying, and a few other factors. For example, peaberries roast faster than the normal hemi-ellipsoidal shaped bean due to their roundishness and better rolling ability as well as the fact that they are usually smaller. Immature berries and quakers hardly roast at all and appear as light yellow, tough beans. When smaller beans are mixed with larger beans of the same type coffee, the smaller beans will roast darker. New or recent crop coffees do not roast as deeply as aged, dry, or old-crop coffees. One reason for this is that there is less moisture in aged coffees. Since mild coffee beans are usually denser than Brazilian beans which are smaller than mild beans, Brazilian beans tend to roast darker than mild beans when such mixtures are roasted together. Robusta and mild coffees roast more uniformly but are not a usual type of blend. Measurements of the bean specific gravities show a range of 1.15 to 1.40 , although other measurements show less variation on commercial lots entering the United States. Brazilian beans are 1.20, Central American beans 1.24, Colombian beans 1.27, and Angola Robustas are 1.18. These specific gravities vary 0.01 to 0.02 plus or minus from lot to lot.

The interesting point here is the rate and degree of swelling that the beans undergo during the roast. The four above types taken to a medium and dark roast are, respectively: 0.65/0.57; 0.67/0.58; 0.75/0.65; and 0.75/0.65 specific gravities. These figures are only illustrative since it has been explained already that swelling will be influenced by rate of roast (hence, roasting machine), bean age, initial moisture, and so on. The higher green bean density of Colombian beans appears related to a longer growth period at cooler climates. Green beans that have been sorted out from better beans during grading are among the "imperfectos" which are non-uniform in density (even if the same in size), roast very unevenly, and have a low bulk density, e.g., 40 instead of 45 lb per cu ft. Brazilian beans are physically soft but have a hard or harsh (not mild) taste. A broader coffee flavor development may be achieved by evenly roasting the same type of coffee to different degrees and then blending the roasted beans. When different green beans cause poor overall flavor development when roasted together, it sometimes becomes desirable to roast the different green coffees separately before blending.

## Additional Notes About Coffee Bean Roasting

**3 Step Process: Drying, Pyrolysis & Cooling**.—Imported coffee beans usually have about 12 percent moisture, and as the beans are exposed to high roasting temperatures ranging from 500 to 800 F, the moisture is evaporated as the beans warm up. Most of the moisture is released in the first few minutes of the roasting period, which can be as short as 5 min or as long as an hour, although usually in the U.S.A. roasting time is about 15 min. In about 8 min the beans will be up to about 300 F and will have lost their green color, having passed through a yellow color, and will begin to turn tan. It is in the period from about 260 F to 370 F that the more tightly held moisture is slowly driven off. By 11 to 12 min the bean temperature has risen to 390 F and the color becomes light brown. At 14 min, the bean temperature is about 410 F, and is a medium brown color. At 16 min the bean is about 440 F and has a dark brown, almost French roast flavor; if allowed to continue another minute, the temperature will pass 450 F and become an Italian flavored and colored bean (dark dark brown) at 465 F with 20 wt % loss.

The actual time, temperature and color effects will vary somewhat with method of roasting, time cycle, type bean, initial moisture, rates of heat transfer, etc. but the general description of what occurs, holds. What is really significant in all cases of roasting is that a great deal of heat is released from the bean, being triggered at near 400 F, and is considered to be pyrolysis or chemical decomposition of sucrose. Measurements of temperatures vs time made in late 1974 by Sivetz, and reported at the Hamburg A.S.I.C. meeting in June 1975, shows that this thermal release is measureable and is directly related to the taste quality of the green coffees being roasted, that is, highest grown best tasting and most aromatic beans release the most heat, then lower grown washed milds, then Brazils, and virtually no heat release from Robusta type beans.

It is the pyrolysis reactions that form aldehydes and other appealing aromas, whose presence and stability is so fleeting. The flavors developed will depend on how far pyrolysis is allowed to proceed (that is, how dark a roast bean color is allowed to develop). Some of the pyrolysis chemical products are carbon dioxide gas, aldehydes, ketones, ethers, acetic acid, methanol, vegetable oil, mist, and glycerol, which are volatilized from the beans. The characteristic odors of coffee bean roasting are acidic, acrid coffee aromas. The bean roasting process usually takes about 7 minutes in a continuous roaster, and 12 to 15 minutes in a batch 4 bag roaster. When bean roasting time exceeds a half hour, the coffee beans tend to develop more of a flat baked flavor than a rich aromatic flavor. Roasting is often accompanied by popping sounds and oily smoke, which is mostly water vapor but with a blue color. As soon as the desired bean color is reached, the beans must be removed from the heated gases and be promptly and positively cooled by ambient air and/or a water spray. Most of the sprayed water evaporates off, cooling the beans, with hardly any water being absorbed by the beans if the water spray is light. Cooling of the roasted beans stops the pyrolysis reactions. Holding roasted beans hot downgrades their flavor.

**Results of Bean Roasting**.—About 2 percent carbon dioxide gas has been produced within the roast bean, the cell structure has been disrupted, sugars and starches have been converted to homologous acids like acetic, proprionic and butyric, as well as homologous aldehydes like acet-, propyl, butyr- and valer-. Water soluble proteins have become denatured. Many proteins have reacted with sugars to form browning products and caramel.A few percent of the coffee oils have broken down to yield fatty acids and glycerol. The softening of the coffee bean cellulose structure with heat and pressure of pyrolysis products has caused the roasted bean to swell to twice its green bean volume. In so doing the chaff has been released from the fold of the bean. The vegetable oils within the bean cells absorb and carry dissolved aromatics. Cooling makes coffee beans brittle enough for grinding. Perhaps up to 1 percent of the caffeine in the original bean is volatilized in roasting and sublimes onto the chimney walls as white crystals. In most roasting machines the vent gases, laden with reactive aldehydes, form insoluble aerosols and sooty matter that deposits within the roasting chamber and chimney, as well as oily matter. If these surfaces are not cleaned every several months, depending on the intensity of production and darkness of roasts, fires can occur in the chimney as well as within the roast chamber. If carbon dioxide gas purge and cooling and/or water cooling is not immediately available, severe and expensive damage can be done to the roaster and/or chimney system. This can result in loss of coffee as well as loss in production time until that equipment is cleaned and repaired. Chaff buildup in chimneys is not uncommon. In Brazil a refractory chimney that effectively burns up the chaff is often used. However, in the U.S.A. and elsewhere, it is common to pelletize the collected chaff and to return it to the product flow entering just before the grinders. The chaff has a bitter and unpalatable taste, and represents 0.6 percent wt or more. In Europe and outside the U.S.A. Probat roasters are used, since they have a long history of production, but since Gothot has offered 6 minute roasts. Gothot has been gaining European sales. In Europe and the U.S.A. most large roasters use the continuous roasters in sizes running at 10,000 lbs/hr made by Jabez Burns. In the U.S.A. most batch roasters are also Burns Thermalos, since Burns has been a U.S. manufacturer for almost 100 years. However, some inroads have been made by some other roaster equipment fabricators, like Presca in Europe, Wolverine in the U.S.A., and now Sivetz with smaller roasters.

*Origin aromas.*—Are strikingly revealed when the green coffee beans are being warmed up past 200 F. This of course should only be done on a single pass air flow, and is easily appreciated by wafting in the aromas from the small Sivetz fluid bed roasters described here. The odors rising from Mandehlang (Sumatra)

Arabicas, Kalosi (Sulewasi), Colombians, Brazils, etc. are those of their environments, just like one is struck with those tropical or characteristic odors when one steps off the jet plane. These odors are very characteristic and often easily associated with the environment of origin.

Examining green beans before, during and after roasting tells a great deal about their quality. Also, smelling the aromas rising from the beans during the whole roasting cycle reveals foul properties, like Rioness, without even tasting the beverage.

*Dirty tastes & tars from commercial roasters*.—Until one tastes a cleanly roasted coffee with a single pass hot air flow, one cannot appreciate how dirty the taste of commercial coffees are. Interestingly enough, Struther Wells' U.S. Patent 3,809,775 May 7, 1974 cites this contamination and alleges it to be carcinogenic in nature. Ordinarily, these observations would have been interesting only, but the binding facts were real proof of these tars. Experiments have shown that 2 to 10 percent by volume of sediment result from the liquid coffee concentrate from commercially ground vacuum-packed coffees after 1 to 3 days. However, coffees roasted on the once thru air flow yielded virtually no tars. However, persuading proof of the downgrading influence of these tars is to taste side by side the same green coffee beans roasted on a once thru fluid bed vs commercial cylinders with recycled vent gases. This has been done on clean good quality Kenya, Guatemala and Salvador coffees, and the differences were startling. The commercial coffee roasts were dirty, left a coating in one's mouth, and diminished the real flavor differences of origin type coffees. It is interesting that even though sample roasters, cylinder types, are not once thru air systems, they give cleaner tasting roasts than the commercial air recycle roasters.

*Smoke*.—At one time it was believed that the blue haze from coffee roasters was a fine oil. Normally when one does not recycle vent gases, there is no smoke at all until the bean temperature rises over 400 F. Washed clean milds even to 420 F produce hardly any smoke, but naturals like Brazils or Robustas throw off considerably more chaff and more smoke. But the greatest amount of smoke is liberated when one recyles the vent gases. During the pyrolysis of sucrose many aldehydes are given off in the vent gases. If they are simply released to atmosphere, they cool and are diluted and yield little smoke. But if these same aldehydes are recycled back into the burner or hot air recycle stream, they react and condense to form micron-sized insoluble particles, which is smoke and is difficult to eliminate except by reheating to say 1200 F with an afterburner, at which temperature they are completely oxidized to carbon dioxide. The Gothot system recycles as much of the 1200 F gases back to the roasting process as possible, by diluting the hotter gases with ambient gases. The Proctor Schwartz heat recovery system simply passes the vent hotter gases through a heat exchanger of ceramics, and then later intake ambient air picks up the heat from the ceramic hot reservoir.

*Degree of roast*.—Degree or bean color and flavor development. The degrees of roast are fundamentally tied to weight percent loss from green to roast bean. Green bean moistures are close to 12 percent but sometimes run to 14 percent, and roasters often add 3 to 4 percent moisture to help their profits. The light brown color of most American roasts is more a tribute to profit taking than to endeavoring to get the best and most flavor and aroma from the coffee beans. Good quality coffee beans have this optimum level of flavor aroma development closer to 16 percent weight loss. The loss of 19 to 20 percent wt, as in Italian roasts, results in reduced aromatics, and is more characterized as a cultural taste tradition. Several hundred thousand bags of very lightly roasted coffee beans are imported into the U.S.A. per year, mostly from Mexico but some from Colombia & elsewhere. Because export taxes are less on roasted coffee beans, they sell at lower prices, and are profitable items blended into freshly roasted coffees, often flowing out through the hotel and institutional trades, where pricing is highly competitive and qualities are a secondary factor. The degree of roast can be pretty well ascertained by measuring reflective light from a screened fraction of ground roasted coffee, as herein illustrated. Latin American roasts run more to the 17-18 percent wt loss, and are in the same class as French roasts. Some people refer to French roasts as darker than Italian, but this is not true. Some beans really taste much better in dark roasts, e.g. Brazils & New Guinea, whereas dark roasts of some central regions like Salvador, Guatemala, Costa Rica & Mexico can develop a very bitter taste, and care is needed when roasting to avoid making such a mistake.

Now with a fluid bed roaster without vent gas recycling, bean color development and pyrolysis are different. First, with a one pass air flow virtually none of the aldehyde tars deposit on the bean surface. Second, due to the higher rates of direct hot air to bean convective heating, the pyrolysis is triggered within the bean, and in fact, the most central parts of the bean are darker than the surface layer. In other words the exothermic heat from the center is going outward, since the bean surface temperature is less. Therefore the fluid bed roasted beans appear lighter in color than they really are after grinding. But, due to the lower heating temperatures used, the sectional colorations are relatively uniform by inspection. It is clear that when one uses high temperature air for 5 minute roasts, an unbalanced flavor development will occur even if the acidity is increased, as it is for the Robusta coffee beans.

*Sucrose & pyrolysis*.—Sivetz observed a "thermal bump" which was larger with better quality, harder beans, but it was not possible to correlate the larger thermal releases with original sucrose content of the bean because of the lack of published literature in this regard. However, from scattered data available this relationship appeared to hold. The best high grown coffee beans have about 7 percent sucrose, and lesser growths, 4 or 5 percent. All sucrose disappears in roasting and is converted to water, carbon dioxide and various aldehydes, caramels, etc. So the increased thermal bumps could very well be associated with increased sucrose decomposition and heat release. (See Sivetz, 1977).

*Green bean density.*—In Guatemala a relationship between higher bean density and quality was developed. In recent measurements, Sivetz found that coffee beans from Mandehlang, Sumatra and Kalosi, Sulewasi have some of the highest densities, e.g. 780 and 775 grams per liter. This is an easy measurement to make. Fill a one pound coffee can with water; the water will weigh very close to 1 liter. Then weigh vibrated green coffee beans in that same can (after drying can), and the density will result in grams per liter. Colombian Supremos were 765 gm/1; Costa Rican Naranja, 760 gm/1 (thin but clean flavor); Harrar Ethiopians of very thin quality were 700 gm/1; and good flavored Brazils were 635 gm/1. Robusta beans are smaller and denser and are not part of this discussion. According to this data and other information, bean densities are an important index to bean quality, and are easily measured before roasting.

*Defective green beans.*—will not yield a uniform color when roasted among wholesome beans, and so in Germany it has been a custom to manually or electronically sort out the lighter beans after roasting for premium priced sales of whole bean roasted coffees. Such a market is diminishing in the 1970's.

**Mexican sugared coffee**.—In Cafe El Marino in Mazatlan, Mexico, an appreciable amount of this type product is produced for low priced retailing, as with several other Mexican roasters. The method of roasting and the equipment used is quite different from conventional roasters. A heavy steel sphere about 5 feet in diameter, has a 2.5 in. D shaft with reduction gears, and gear drive to rotate the sphere at about 30 rpm. The sphere has about a 12 in. diameter access hole, into which about 500 lbs semi-roast coffee is loaded (10 wt percent loss); similar light or ''half roast'' coffee is shipped to CFS and others in the U.S.A. An equal weight of sugar (sucrose) is added, the loading port is sealed, and the sphere is allowed to rotate while a hot blast burner heats the sphere from below. There are no instruments used in this operation; it is all done by experience. The roasting cycle takes almost an hour, and the water and gases formed in the caramelization process vent out of the unit more or less. The hazardous and unpredictable nature of this operation can be appreciated when it is realized that in about 1970 the internal pressure from one of these spherical roasters got so high that it blew out the loading port, and then the jet blast from the open port levitated the 2 to 3 ton sphere up through the concrete ceiling of the building, thrusting it across the street. Miraculously, no one was killed, but damage was very extensive. The pyrolysis of the sugar in the presence of the coffee beans (usually low grade) develops much heat and liberates much steam (water) and carbon dioxide as caramelization occurs, and the molten mass coats the beans while roasting them to a black color. At the proper time the port is opened, and the contents, a viscous, lumpy mass, is hoed out of the sphere onto the concrete floor to cool, congeal, and then be broken up in a hammer mill. The operation supposedly originated in Spain perhaps 150 years ago, but is no longer used in Spain as the public prefers better quality coffee. In Mexico too, this coffee with caramel is a dying business as living standards rise. It is not a good tasting product, but a low priced one, since sugar is much cheaper than coffee beans. In 1974, there were such cylindrical roasters working on a one hour cycle in Jakarta, Indonesia.

**Fluid bed non-gas recycling roaster.**—In addition to the points already mentioned, the vigorous bean recirculation which causes bean rubbing, helps rub off dust, dirt and chaff-silver skin from the green beans while they are in the drying cycle, and immediately removes these light weight contaminants away from the beans into a collection cyclone. The significance of this action is that the bean surfaces are smoother and cleaner, which improves heat transfer and mixing. The removed fines do not have a chance to burn and smoulder as they do in conventional roasters, thereby contributing smokey, smouldering tastes. The overall effect is a cleaner operation, and a better taste.

Fluid bed roasting allows one to see the whole mass of coffee beans through a sight glass throughout the process. Simple as this is, it is not possible to see the process in a conventional rotary cylinder roaster. One sees the snow storm of chaff released when the beans expand near 400 F, as well as the dust and fines leaving the green coffee beans that have just been charged. The gradual change of the bean colors from green, to yellow to tan, then to brown shades is an exciting part of the roasting process that virtually no one ever sees, and is particularly demonstrative to gourmet shops that can show the public. Expansion of the roasted beans is dramatically viewed also through the sight glass as their level rises in the sight glass. This also confirms proper swelling for finalization of roast.

Altogether, the coffee beans undergo good development; that is they swell better than in conventional roasting and show a nice, clean bean surface, with a clear demarkation of the chaff at the center fold. The beans are not dull looking from deposited soot, but are shiney to glistening, yet no oil reaches the surface. Another advantage is in Italian roasts, where 20 wt percent loss can be taken, yet oil does not flow out immediately to the bean surface, but comes out some hours later. This phenomena helps keep the roaster clean, and minimizes any chance for fires. The fluid bed roaster is therefore a safer roaster to use, in so far as fire possibilities are concerned. The lower cost of the fluid bed roaster is because one uses a stationary box instead of a driven cyclinder. Engineering skills in the design of the shape of the box, as well as air flows and pressures, result in simplicity of design, translated into economy of construction and operation.

**Air-pollution.**—*Traps*.—Air pollution comes mostly from the coffee roaster gas effluents. After-burners which can raise roaster effluent gas temperatures to 1400 F for over ½ second to oxidize all organic matter, require several times the normal fuel use just for roasting. This is too expensive to tolerate on large installations. An article from the January 1973 issue of WC&T clearly shows how a ceramic heat recovery system conserves fuel use while eliminating organic vapor discharges. The June, 1971 issue of WC&T describes TRAPS (Thermal Regenerative Air Purification System).

In all cases the use of 1400 F for ½ sec was essential to meet effluent regulations, and the TRAPS system made it possible to do this economically. The TRAPS system is much more efficient and easier to apply on a large continuous roaster with continuous water quench than on varied numbers of batch coffee roasters. The U.S. EPA gas effluent codes require four things: 1) no visible discharge; 2) limits on particulate matter; 3) limits on odor (especially if neighbors complain); and 4) limits on unburned hydrocarbons.

As things stand for many coffee roasting operations, wet scrubbers and afterburners either do not meet effluent regulations and/or are too expensive to operate. Noise pollution can also be an objection.

Thermal recovery has become a necessity due to fuel costs. Fuel shortages and the enforcement of stringent limits on gaseous effluent discharges which calls for higher investment in such gas cleaning equipment.

*Self-sustaining chaff after-burner*.—Outside cities and in areas where the 1400 F-½ second rule is not needed or enforced, a very effective elimination of roaster smoke and chaff can be achieved in a refractory brick chimney. This "after-burner" system is used by many roasters in Brazil. It has a low initial investment, and practically no maintenance thereafter. The after-burner heat is supplied by the coffee oils and chaff themselves which burn on the insulated refractory cyclone walls and chimney to give a relatively odorless and particulate-free stack emission.

This system was first applied in 1966 by Cacique, the Brazilian instant coffee company. In the next few years Dominium, Vigor, Brasilia, and others have put these systems to use. The recovered ashes at the bottom of the self incinerating cyclones are effectively recycled to the farm for fertilizer being rich in potassium and phosphates.

**Water pollution**.—This effluent comes primarily from washing spray driers, extract tanks, pipe lines, trays, and floors. These washdowns are intermittent and voluminous. Water pollution constitutes the most difficult problem and requires the most expensive solutions.

Normally instant coffee plants run such effluents to city sewers or public creeks, and if contamination is low and the process purges small, this is an economical and easy solution. However, when neighbors complain of water color and odor and city sewage treatment capacities are overloaded, then the instant coffee plant must solve its own problems. Coffee colors are not removed by biological action in water treatment. Water effluent disposal via spray ponds and water percolation through field soils have proved to be impractical.

## GRINDING ROAST COFFEE

After developing coffee flavor by roasting, it is desirable to extract efficiently the roast coffee solubles and volatiles which are the coffee flavor and aroma. The solubles could be extracted from the whole roast bean but the yield would be low and flavor would be poor. Extraction may be made more thorough to give a higher yield of solubles faster by breaking down the whole roast bean into smaller pieces. Such breakdown exposes much greater surface for the liberation of $CO_2$ gas and absorption of hot water, while at the same time it shortens the distance from the center of each particle to the surface, thereby greatly reducing the diffusion distance for coffee solubles. Greater surface exposure also increases the amount of colloidal substances that are free to go into water solution (or suspension); most of these substances are high molecular carbohydrates and fatty substances. Thus, finer grinds not only increase extraction efficiency and yield but alter the nature of the solubles and colloids entering the brew. Taste is therefore altered accordingly.

Grinding is a general term that means reducing the size by crushing, rubbing, grating, cutting, tearing, and any other process that will cause particle-size reduction. Depending on what is being reduced in size, one mechanism may be better than another. In most grinding machines several mechanisms are at work. If the substance is tough, such as green coffee beans, cutting and tearing are a better means of size reduction. In the case of roast coffee, which is brittle, crushing, and rubbing may have some success, as may impact or shock forces, depending on the roast. Thus the character of the coffee will not only determine the best machine for size reduction, but the toughness of the coffee will determine the work required to reduce it in size. The finer the final particle size desired, the more work it takes to attain the greater surface exposure. Size reduction occurs in progressive steps, since notching out the desired size from a much larger particle is not feasible. The total surface area per unit weight of matter is in inverse proportion to the diameter of the particles. Further, in any size reduction process, complete uniformity of reduced particle size is not possible; thus, only the best possible uniformity of reduced particle sizes must be the goal. For extraction purposes, the surface of the whole bean is not as permeable as the exposed surface of the inner bean.

### Grinding Work

Table 8.2 shows the average particle size relation for whole beans for variously used ground coffees and the number of particles and the surface area per unit weight. Thus, fine grind coffee has 4,000 times the number of particles and 16 times the surface area per unit weight compared to the whole roast bean. Halving the particle size makes twice the surface and eight times the number of particles. The work required to attain this subdivision is greater as the particles become finer. In other words, fine grinds take more work to prepare, and still more work is required to make the finest fraction.

For example, if it is assumed that the number of particles produced is a very rough measure of the work required in grinding, the column in Table 8.2 showing the ratios of the number of particles indicates that some 100 times as much work is needed to produce a fine grind as is needed to produce cracked beans. Of course, the figures are to be regarded as an order of magnitude and not as a precise calculation.

There are complicating factors such as the greater difficulty to ''bite'' into the small particles than into whole or cracked beans. The difficulty in biting is partly due to the greater difficulty in holding the smaller particle. In mixtures of particle sizes the coarser particles will be subject to more shear and ''bite'' than smaller particles. The pushing together of small particles causes slippage which reduces the effectiveness of the working surfaces of the size reduction machine on the larger coffee particles. Dulling of cutting and shearing surfaces of the grinder as well as over-feeding the machine will cause size reduction variances resulting in more work for a given size reduction and/or failure to achieve the desired size reduction. Normally, the horsepower required to effect a given range of size reductions by a given machine can only be determined in practice since there are notable variances in bean properties and machine characteristics. Some of the grinding machines used in the past as well as types of coffee grinding machines that may be used will be briefly discussed below. Also, the toothed roll grinder manufactured by B. F. Gump Company will be evaluated, as it is the most commonly used coffee grinding machine in the United States, if not in the world.

TABLE 8.2. PARTICLE SIZE VERSUS NUMBER OF PARTICLES PER UNIT WEIGHT

| Particle Size Description | Size, mm | No. Particles Per Gram | Increase Particles/gm | Ratio of Increase | Total Area, sq cm/gm |
|---|---|---|---|---|---|
| Whole bean | 6.0 | 6 | — | — | 8 |
| Cracked bean | 3.0 | 48 | 42 | 1 | 16 |
| Instant R & G for percolation | 1.5 | 384 | 336 | 8 | 32 |
| Regular | 1.0 | 1,296 | 912 | 22 | 48 |
| Drip | 0.75 | 3,072 | 1,776 | 42 | 64 |
| Fine | 0.38 | 24,572 | 21,500 | 512 | 128 |

## Grinding Machines

**Mortar and Pestle.**—This was and still is in some areas the basic means of reducing roasted beans to a powder, e.g., Moslem countries, with excellent results.

**Burr-Mill.**—The original means for grinding grains was a burr-stone mill, which is still used in many industrially backward areas of the world. One large circular stone with a lower serrated (rough or burr) face is rotated over a stationary identical circular stone with upper serrated face. Grain particles are fed into the center hole of the upper stone and, as they suffer some degradation in size, they become small enough to move out to the periphery of the stone. The rotation causes shearing of particles against the burrs or face of the stone while the weight of the stone against the grains constitutes the grinding pressure. The grains also rub against each other. The issuing powder is wheat, corn, or rye flour. This principle was used for years in grinding coffee, and still is today. The stones have been replaced by cast iron serrated disks, the stone weight has been superseded by a compression spring, and the drive is now a geared down motor instead of an ox. The grinding disks are upright and the coarseness of the grind is adjusted by the spacing or pressure on the disks facing each other. This type of coffee grinder is used in markets where the consumer grinds his coffee from whole roast beans at the time of purchase. The same type of grinder may be used at home using hand or electric power to grind whole roast beans just before brewing. Such home grinding was more common before the advent of vacuum packed ground coffee.

Vacuum packing of foods by means of a can closing machine operating in a vacuum was patented in 1900 by E. Norton of Chicago. Hills Brothers of San Francisco in 1903 were the first to apply this method of packing to roast coffee commercially. As the grinding operation was assumed by the coffee roaster, the excessive heating of the disks at high rates of grinding was revealed. Particle size standards did not yet exist and each packer ground the coffee to what seemed to him to be the most acceptable particle sizes. Disk grinders at working rate of 500 to 1,000 lb per hr could make a regular type of grind in one pass. But it was more efficiently done in two passes by using a scalping screen to return the over-sizes to another disk grinder.

**LePage Cut**.—About 1914, a milling engineer, Jules LePage, employed by B. F. Gump Company, in Chicago, developed companion cutting rolls; one roll had peripheral cuts and the other roll had longitudinal cuts (Fig. 8.26). The roll cuts or teeth "bit" the coffee particles under roll pressure. This resulted in more efficient size reduction than with the disk mills. Further, the throughput capacities were greater without crushing the beans as much and with more uniform particles, less heating, and less work by the rolls. There were less fines and less oil was exuded. However, the chaff was released in its almost entire flaky form which was in appearance a departure from disk grind. By 1921, grading sieves were incorporated below the grinding rolls and an air suction removed the chaff to a cyclone collector. A chaff grinder was later added and the ground chaff was mixed back into the ground coffee. Without chaff return, the ground coffee had a clean appearance and was called "steel cut" coffee; return of the ground chaff (1 percent of roast coffee) and mixing it into the ground coffee altered the steel cut appearance. The mixing action on the ground coffee also contributed to oil release. The light chaff color overtones on the dark oily coffee grounds were visually undesirable. The chaff tastes bitter and contributes a downgrading flavor effect to the mixture. The whipping action of mixing that drives out oils also contributes to staling.

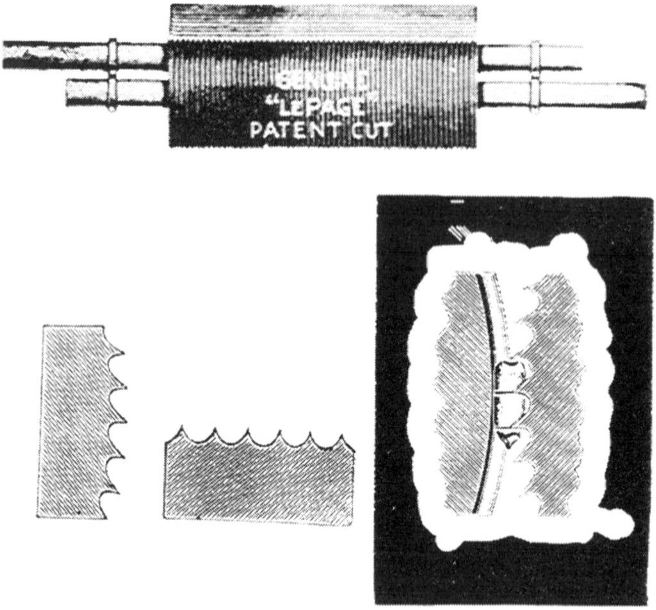


*Courtesy Jabez Burns & Sons, Inc.*

FIG. 8.26. DETAIL OF LEPAGE ROLLS AND CUTS

**Gump Grinder**.—In the late twenties, with increasing demand for finer grinds, another roll and breaker bar was added; this provided three gradual reductions. Thereafter, another pair of reduction rolls was added. The bean thus passed through two breaker and two reduction passes or a total of four gradual reductions. Figure 8.27 shows the present section through a B. F. Gump Company grinder arrangement of three cracking and one set of regrind rolls as is commonly used in soluble coffee plants. The second set of regrind rolls is needed in commercial grinding of drip and fine grind vacuum pack coffees. The chain drive from the motor is used (after gearing down speed) to drive directly the lower rolls that do the major work in size reduction; the upper trio of cracker rolls is driven from the opposite end of the driven reduction rolls since they only use a fraction of the total horsepower. The current arrangement of grinder rolls was introduced in 1930, and it eliminated the grading sieve, air aspiration of chaff (also chaff collector and grinder and mixer), oversize return and bucket elevator. This Gump grinder is so widely used that it raises the question of whether other types of grinder might not be applicable to coffee.

**Other Grinders**.—In fact, there are other types of grinders that will effect coffee bean size reduction without elaborate scalping and recycle of coarse

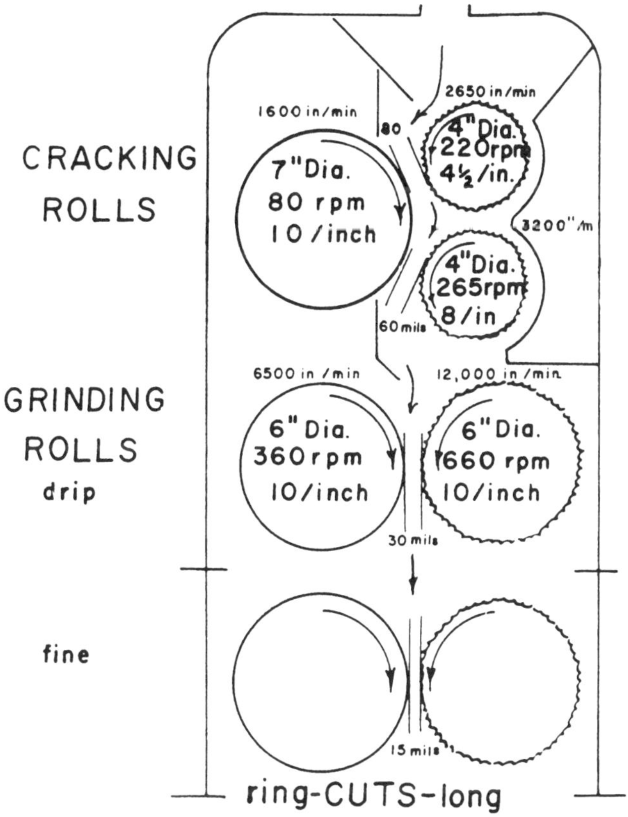


*Data Courtesy B. F. Gump Company*

FIG. 8.27. SECTION OF LEPAGE ROLL GRINDER

fractions. They can produce particle size ranges that are almost as uniform as the Gump grinder, but usually yield several percent more fines. Since the Gump grinder has been developed with the coffee industry it is natural that it would give better performance with coffee than a grinder developed for other purposes. However, there is no need to hesitate to use the Fitzpatrick hammer mill which operates with cutting blades impacting the beans and achieves good cutting action. Small bean fragments fall through a slotted opening. The Rietz impact mill is also a type of hammer mill and prepares a reasonably well reduced and uniform coffee particle fraction. The Entolator is an impact type of mill that effects breakup of the bean particles by giving them high velocity and smashing them against a metal wall. The Entolator also yields reasonably good coffee particle uniformity in its fractures.

*Courtesy National Coffee Association—U.S.A.*

FIG. 8.28. COMMERCIAL DISK TYPE COFFEE GRINDER

Most coffee roasting plants use the Gump type of mill. Since it gives satisfactory performance, when grinders are chosen for soluble coffee plant operation (where coarser grinds are used), it is natural to keep the Gump grinder in use. Also, the personnel in the roasting plant know its operation and maintenance.

It is worthy of note that coffee ground for espresso machines is done on disk type mills as shown in Fig. 8.28. Also, grinding is usually done only in small amounts just prior to espresso machine extraction. The grind often is extremely fine, bordering on powder. This is an essential part of the preparation of espresso coffee since the pulverized coffee gives a high yield of aromatics, solubles and oils, and hence, a heavy, rich flavored and textured cup.

## Chaff Normalizer

When the chaff collector and grinder-mixer system was altered in the 1930's, the chaff with granulated coffee leaving the reduction rollers was mixed with a high speed mixing screw located below the roller assembly. Figure 8.29 shows this Gump No. 888 granulizer. The mixing of the coffee granules with the chaff causes the chaff to break up and to be less conspicuous against the background of ground coffee. As stated before, this whipping action and incorporation of the chaff is in fact a deleterious action upon coffee flavor and appearance. It is simply an

*Courtesy B. F. Gump Company*

FIG. 8.29. GRANULIZER WITH NORMALIZER

advantage to the roaster to save the 1 percent chaff weight. Usually the same roaster also waters his roast beans with a few percent water which accelerates staling as well as denies the consumer full coffee value.

In the top trio of breaker rolls, the large roll is spring-loaded so that if an incompressible foreign material like a stone or metal nut passes through, the rolls will not bind, jam, or be damaged. One of each pair of regrind rolls is also spring-loaded for the same reason. A tightening nut on each spring allows more or less tension to adjust the grinding. Too little tension will not effect the fullest cutting action, whereas too much tension may cause crushing of the beans without much cutting action.

## Maintaining Grinder Rolls

Normally before, and then periodically after the grinder is placed into operation, it is necessary to check the spacing between the rolls for uniformity longitudinally. Rolls out of line may cause uneven grinding. In the extreme case, large particles pass and most of the grinding will take place at the largest gap; in another case touching rolls may be scored and seriously damaged. Such uneven wear of the rolls may mean their early replacement. Normally one to three million pounds of roast coffee may be ground before rolls need to be recut or replaced. However, this will depend on the care given the grinder, its setting, maintenance, protection from stones and foreign metal items, fineness of grind, darkness of roast, and so on. It is customary to have a permanent magnet above the grinder feeder duct so that any steel objects will be picked up before they reach the grinder. It is also the function of the stoner to protect the grinder rolls. To run a grinder with dull teeth is false economy because the grinder uses more power to achieve a certain grind; grinds are not as uniform due to poor cutting and shearing so that more roll pressures and smaller gaps have to be used. Once the roller gets to be sway-backed or bowed due to uneven wear, it is past time for replacement. It is also bad practice to start the grinder with whole roast beans in the cracker rolls since they act as a resistance to the grinder as it attempts to come to operating speed. The speed of the rollers exercises important impact and shearing action that is missed on start up of the grinder under load. Thus, when shutting down the grinder, any whole roast beans in the feed duct should be run out. Starting the grinder under bean load also causes an unnecessary strain on the motor and wear on the drive linkages.

## Roll Cuts and Speeds

Figure 8.27, which shows the section of the Gump grinder rolls, indicates the progressive steps in grinding that the beans and cracked particles sustain in their passage through the grinder. In this case, the rolls discussed are from a grinder in a soluble coffee plant. The longitudinal LePage cuts per inch for the three passes are: 4½, 8, and 10 whereas the ring or circumferential cuts on both rolls are 10 per in. The top longitudinal cut roll travels at a peripheral speed 50 percent higher than its paired ring cut roll but somewhat lower in the peripheral speed than the second pass longitudinal cut breaker roll. The first and only pair of reduction rolls travel about four times as fast peripherally as the breaker rolls, thus effecting a markedly greater shearing and cutting action. The roller spacings reduce with each pass and diminish from about 80 mils on the first pass, 60 mils on the second pass to about 20 mils on the third pass. Of course, the spacings are adjusted in the field to suit the operation and the screen analysis results desired. Normally neither the speed nor the sizes of rolls is altered. Thus, higher roller speeds and smaller passages for the particles are used as the size reduction progresses.

## Granulation Variations

When the Gump 88 granulizer is used at 2,500 lb per hr for preparing soluble coffee grinds, and spacings are reset to attempt to achieve the same size reduction at 4,000 lb per hr, several percent more fines (through 20 mesh) result. Higher

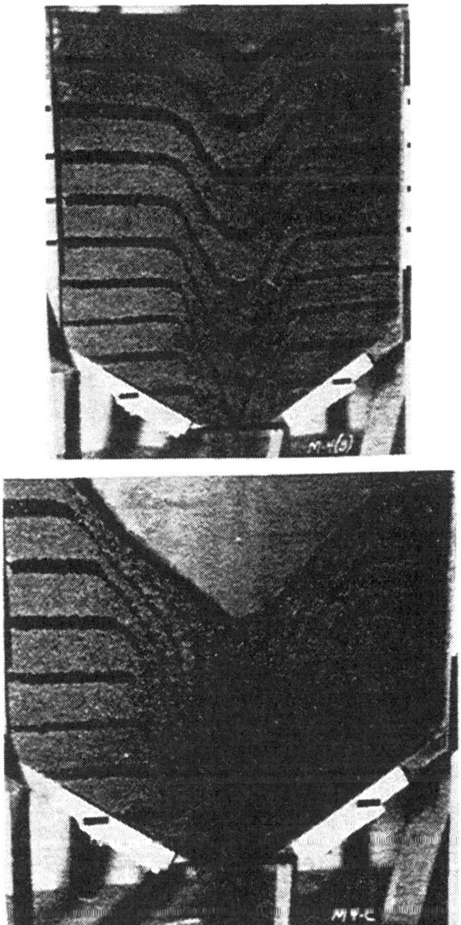

*Courtesy Howe Richardson Scale Company*

FIG. 8.30. TOP AND BOTTOM. GRANULES FLOWING FROM BIN

rates of granulizing result in more fines. Where this is critical, the grinder should not be run at its full capacity. More uniform particle size will be attained at lesser granulizing rates. Varying the throughput through the grinder will cause variable particle size reductions. This also occurs when starting a grinder when the rolls are at room temperature; later the rolls become warm and the grind changes noticeably. Here again this is important only in cases where high grind uniformity is critical.

From a soluble coffee processing viewpoint, the Fitzpatrick cutting mill does produce, for a given particle reduction, about a 10 percent higher bulk density. This gives a more compact coffee fill of the percolator column and contributes to more efficient extraction with higher extract concentrations. There is no doubt that stage reductions with screening will achieve more particle uniformity than any other type of grinding, but the question is, "What is the goal?" For household brewing where drip or vacuum methods are used, uniformity of grind with coffee bed porosity is desirable because the object is to separate the extract from the grounds. In the soluble coffee plant percolator, uniformity is not so necessary in coarse grinds, but higher bulk densities without fines are. For soluble coffee operations cut beans are best. Crushing weakens the bean structure and may cause earlier softening of the coffee particles in the percolator hydrolysis stages. This contributes to excessive hydraulic pressure drop. Cutting also keeps bean cells intact, whereas crushing pushes oils to the particle surface causing smearing and faster deterioration before packaging or extraction. A loss in effective tooth sharpness becomes evident with the reduced capacity of the grinder at a given cracker roll space setting. Dull teeth have less bite; hence, less beans are pulled into the grinder and more scarred (crushed) beans and fines result. Thus, records that are kept on grinder spacing adjustments and grinding results reveal trends of tooth dulling or roll misalignment. An alternate grinding machine, set and ready for production, is desirable security in a large grinding operation. In any case a set of replacement rolls should be on hand in case of gross damage to the roll teeth or as the time for roll replacement approaches.

## Grind Standards

Table 8.3 lists the three commercial grinds used in the United States; European roasters normally prepare only a coarse or fine grind, and sometimes only one all-purpose grind.

TABLE 8.3. U.S. DEPARTMENT OF COMMERCE RECOMMENDED COFFEE GRINDS[1]

| | Amount of Coffee Retained on | | | Tolerances on Amount of Coffee Passing Through Control Sieve, 28- Mesh | |
|---|---|---|---|---|---|
| Grind Designation | 10 and 14 Mesh Sieves, % | 20 and 28 Mesh Sieves, % | Amount of Coffee Passing Through Control Sieve, 28-Mesh, % | Not Less Than, % | Not More Than, % |
| Regular | 33 | 55 | 12 | 9 | 15 |
| Drip | 7 | 73 | 20 | 16 | 24 |
| Fine | 0 | 70 | 30 | 25 | 40 |

[1]U.S. Department of Commerce R 231-48.

Soluble coffee grinds vary with time and place. Some use a cracked bean, others a very coarse grind, while a few use a grind only slightly coarser than a commercial regular grind. However, by using two screen sizes, grinds used in soluble coffee plants may be largely categorized as 25 to 50 percent on 8 mesh (U.S. Std.) and less than 5 percent through 20 mesh. The important point about grind in soluble coffee manufacture is to be able to place such grinds into the percolator columns without particle segregation in conveying and storage.

For the vending machine trade, a very fine grind is used to get higher yields of solubles from 1 cup brewers.

In Europe, a third set of rollers is added to the Gump grinder to produce the very fine European grind at industrial rates of 2–3 tons per hour.

The roast bean properties have a great deal to do with grinding results. Beans cannot be ground directly after roasting as they are too soft and would be crushed, flattened, and scarred. When the beans are cool, hard, and brittle, they may be ground. Beans are physically softer when they have several percent moisture; they are most brittle when air cooled without moisture addition. Light roast coffees are tenacious, pliable, and tough; they do not break down as easily as hard, brittle, dark roasted beans. Table 8.4 shows how roast influences the results of granulizing. The differences obtained are sometimes marked, and wider roller gap settings for dark roasts will give grinds similar to narrower roller settings in ligher roasts. Dark roasts, however, will always yield more "fines" than lighter roasts. Even the natural origin of the coffee makes a difference in granulizing results. New crop coffees give less fines than old crop coffees for the same degree of roast. Robustas differ in the grinding characteristics and the particle size distributions from mild coffees as they differ from Brazilian coffees. Knowing these differences exist helps in making grinding adjustments that will yield the desired results. The nature of the roast coffee bean structure is that it fractures along natural concoidal surfaces when it passes the breaker rolls. Moist, pliable beans are flattened and scarred but not reduced in size when passed through rolls. Sometimes water quenching of roast beans to minimize fines only moistens the surface; then with closer roll spacings the inside of the beans are crushed to fines.

TABLE 8.4. HOW ROAST AFFECTS GRINDING[1]

| Mesh No. | Setting Cinna-mon | No. 7 Italian | Setting Cinna-mon | No. 6 Italian | Setting Cinna-mon | No. 5 Italian | Setting Cinna-mon | No. 4 Italian |
|---|---|---|---|---|---|---|---|---|
| 10 | 14.7 | 13.6 | 5.8 | 3.0 | 0.5 | 0.1 | 0.0 | 0.0 |
| 14 | 43.5 | 37.2 | 42.7 | 31.5 | 27.9 | 18.3 | 15.4 | 6.47 |
| 20 | 21.8 | 23.2 | 30.3 | 33.4 | 43.7 | 41.4 | 48.9 | 45.27 |
| 28 | 9.9 | 12.4 | 10.8 | 16.0 | 15.3 | 20.9 | 20.6 | 26.32 |
| PAN | 10.0 | 13.0 | 10.2 | 15.4 | 11.9 | 18.9 | 14.8 | 21.67 |
| Total | 99.9% | 99.4% | 99.8% | 99.3% | 99.3% | 99.6% | 99.7% | 99.7% |

[1]These figures are for a particular type of coffee. Differences in type and grade of coffee will affect degree of roast and its effect on grinding.

## $CO_2$ Release When Grinding

Grinding is accompanied by a very large release of carbon dioxide and volatile aromas. Standing near a grinder for long periods may be hazardous at times due to the high $CO_2$ content of the air. Concentrations of up to 5 percent have been developed in such locations, especially where ventilation is poor or nonexistent. Even at 3 percent $CO_2$, a person on prolonged exposure will become silly and act oddly. The coffee aroma volatiles contribute additionally to this physiological condition. It is hazardous for an employee to enter a storage bin of ground coffee located immediately after it has been ground. The $CO_2$ gas in air may be higher than 5 percent. A person may be overcome by the $CO_2$ even if oxygen is in the same atmosphere at only a few percent less than the normal 21 percent concentration. The physiological overcoming effect of $CO_2$ is so subtle that the person experiencing it does not realize the danger. For every hundred pounds of roast coffee ground, at least ½ to 1 percent of $CO_2$ is released which is equivalent to 4 cu ft $CO_2$ NPT.[2] As a rule of thumb, one volume of pure $CO_2$ gas is released per bulk volume of the roast coffee. A $CO_2$ gas volume equal to the initial volume will continue to come out of the ground coffee for a day or two depending on storage temperatures and ventilation. By this time, ground coffee will have undergone noticeable staling, but this is not a normal storage condition. Ground coffee is usually extracted or packed within 8 hr of grinding.

## Summary

Ground coffee readily absorbs atmospheric moisture and is a good desiccant. In evaluating the nature of roast coffee during the grinding process, it is desirable to consider the bean properties, namely: moisture content, roast, hardness, pliability, strength, resilience, fiber, brittleness, particle size, and flavor development. Grinding as done in the coffee industry is a relatively satisfactory operation except for the normalizer (whipper), which creates fines and darkens the ground coffee by bringing oil to the surface and incorporating bitter tasting finer chaff. Care should be taken in ventilating $CO_2$ gas from grinding areas.

# CONVEYING AND STORAGE OF GROUND COFFEE

After the coffee is ground, there is a considerable tendency for the grind mixture to separate into coarse particles, fines, and chaff during conveying, and filling and discharge of bins. In normal handling of the ground coffee in screw conveyors and elevators, particle segregation is not usually a problem. Classification of particles, however, does occur if the ground coffee is allowed to fall freely into a storage bin. Chaff and fines drift off into the air and accumulate at the top periphery of the bin. If the bin is emptied completely, chaff and fines will lodge in a single place in the

[2] Normal temperature and pressure.

*Courtesy Jabez Burns & Sons, Inc.*

FIG. 8.31. GROUND COFFEE "TRUE FLOW" BINS OVER PACKAGING ROOM

percolator. In commercial percolation, this condition contributes to excessive hydraulic pressure drop and channeling of flow.

Of particular design significance is the bin of R & G coffee from which the percolator column is filled. The aforementioned segregation of fines and chaff is accented in bins that are not round and that are large (several times the volume of a percolator column). A bin design that minimizes fines accumulation is round and holds only about 25 percent more R & G coffee than the percolator column loading. In addition, the inside of this bin preferably should have a cascade system of slides, so that the ground coffee does not fall freely but slides down in a zig-zag path and thereby minimizes particle segregation. In any case, there is always an accumulation of coarse particles at the periphery wall of the bin or percolator column because the coarse particles roll better than the fines. A further technique for obtaining a uniform and compact fill of R & G coffee in the percolator column is to fill as quickly as possible. This can be accomplished by prior evacuation of the percolator column to more than 24 in. of mercury vacuum and then to provide as large a top opening valve as practical. For example, with the 6 in. top Cameron valve, the filling time for a 600, 1,200, and 1,800 lb percolator column is approximately 30, 60, and 90 sec. The vacuum drawn during the R & G filling that sucks in the ground coffee may be taken through an internal perforated pipe (bayonet) at the bottom or the top of the percolator. Such a vacuum fill usually adds 10 to 15 percent to the bulk density in the column compared to a free fall fill. Such higher column loadings help give higher extract concentrations and better extract filtering action. It is desirable to maintain a minimum time between grinding and use of the ground coffee in the percolator column so as to allow the least flavor loss from the ground coffee. Staling of freshly ground coffee is noticeable within a few hours after grinding. Serious stale coffee flavor does not occur for a day or two depending on initial moisture and exposure. A good system of ground coffee handling usually incorporates only about a 2-hr retention of ground coffee.

Fine grinds have a higher bulk density than coarse as, for example, a grind used in an instant plant having 20 percent above eight mesh and a bulk density of 19 lb per cu ft. When ground finer to nothing on eight mesh, this gives a bulk density of 23 lb per cu ft. The vibrated density of free fall coffee at 19 lb per cu ft may be increased to 21 lb per cu ft by the finer grinding. In conjunction with faster R & G coffee loading, a more uniform grind will also give a higher column loading density.

Immediately after grinding, the release of carbon dioxide gas, coffee aroma and flavor volatiles from the ruptured and exposed coffee cells, forms an intense aroma and a temporary protective $CO_2$ atmosphere. For example, airveying ground coffee would expose the coffee to faster oxidation due to removal of these protective gases. Finer grinds give faster wetting and extraction in the percolator and also act as a better filter bed for holding back hydrolysis tars and colloids. Prewetting of the ground coffee is more efficient than wetting the coffee bean, but results in coffee flavor deterioration in the presence of air.

The bins holding R & G coffee are called "live bins." They have vibrators to activate the movement of the R & G coffees.

## ADDITIONAL READING

Sivetz, M. 1977a. Coffee: Origins and Uses. M. Sivetz, Corvallis, Oregon.
Sivetz, M. 1977b. New developments in coffee roasting. Tea and Coffee *16* No. 2, 32–64.

# 9

# Packaging Roasted Ground Coffee

## Freshness of Roasted Beans

Many coffee roasting firms forget that the main objective of packaging is to retain the fresh aroma and flavor after roasting and grinding the beans, until the R&G coffee is converted to beverage. Some claim that freshness is not lost in vacuum or even plastic pouched packages at any significant rate. However, this will be shown to be false in the following discussion of staling factors. Recently, the matter of freshness has become more important than ever before. Lower grade beans, especially less aromatic Robusta beans have increased to over 50 percent in the commercial blends, so there are not enough aromatics or desired flavor factors to sustain the aroma loss and oxidation and chemical changes that have always existed in such packaging methods. It is not easy to retain freshly roasted bean aromatic aldehydes in the R&G coffee nor to effectively delay their chemical changes and oxidation by air, which is why the cost factor overrides the quality aspect in packaging. Now, the public is generally insensitive to genuine roasted coffee fresh flavor, and has acquired a taste for the noticeably stale and smokey flavor of vacuum canned coffees.

One must realize, when dealing with roasted coffee, that the delicious aromas and flavors (which are very volatile organic chemicals that react with themselves and oxygen), are volatilizing continually from the moment roasting is completed. The most pleasing ones are so volatile, that when pure, they evaporate at room temperature; others are gases at room temperature and pressure. Accompanied by carbon dioxide, these aromatics vaporize from ground coffee and the oily roasted beans in a week or two. These aromatics are replaced by less volatile, less desirable aromas in several days. Free ventilation of air about the R&G coffee accelerates removal of volatiles and enhances oxidation rates. So a can of R&G vacuum packed coffee becomes fairly stale-tasting a few days after the can is opened to air. These chemicals are unstable—continually changing. The rates of volatilization, reaction and oxidation can only be retarded by sealing the R&G coffee in a vessel, removing as much oxygen as possible and by reducing the temperature as much as possible.

Freezing coffee beans and R&G coffee is not often practiced, nor its usefulness appreciated. But freezing to −10 F or −20 C, as in the household freezer, is a very effective way for extending the freshness of coffee aromatics for several reasons: 1) water reactivity is immobilized; 2) volatility of aromatics is reduced by about 4-fold; 3) rates of oxidation are reduced about 50 percent for each 10° C reduction, hence about 15-fold; and 4) the vegetable-like coffee oil is congealed, thereby reducing movement of volatiles dissolved therein from convective rates to diffusional rates. Altogether, roast beans keep for months and R&G for many weeks, which is at least a 30-fold freshness factor for beans and over a 15-fold freshness factor for R&G. Storage must be in a sealed container; coffee aromatics permeate a polyethylene bag and so does oxygen.

At the moment roasting is completed, the coffee bean has its maximum concentrations of the enjoyable, desirable aroma and taste chemical constituents. Some of these are aldehydes:- aceteldehyde, proprionaldehyde, butyr-aldehyde and valer-aldehyde. Acetaldehyde boils at 21 C, room temperature. Proprionaldehyde boils at 49 C. These important aroma and taste constituents rapidly evaporate from the cells of the roast coffee bean and are readily oxidized by air when it enters the bean cell structure. The oxidized aldehydes become corresponding acids with higher boiling temperatures and different, often disagreeable, odors and tastes. Therefore, there is aroma loss and change in aroma composition. Both occur simultaneously but at varying rates. These rates depend on the degree of bean roast, bean storage temperature, moisture content, percentage of oxygen in packages, and other factors.

**Carbon Dioxide Gas.**—Fortunately, whole roast coffee beans have some self flavor-preserving properties. The 2 percent carbon dioxide gas within the cell structure is slowly desorbed. The generated $CO_2$ atmosphere keeps oxygen (of the air) away for several days. If roasted coffee beans are kept warm (e.g. 90 F) and moist, so that water is absorbed, these two factors will each accelerate the liberation of the carbon dioxide protection and cause faster aroma loss and faster staling.

**Whole Roast Coffee Beans.**—At 70 F in dry air flavors will stay relatively fresh for 1 week. But when the roast coffee bean is broken into smaller pieces, exposing 10 to 20 times as much cell area (relative to the bean) to the atmosphere (with 21 percent oxygen) and the internal cells to air, aromatic constitutents are rapidly volatilized and oxidized within hours. The coffee is stale in a few days.

**Ground Roast Coffee.**—Ground roast coffee is normally stale-tasting within a few days after it is exposed to air. Staleness and freshness are relative taste terms. These terms depend on the experience, sensitivity and standards of the taster or coffee drinker. The taste standards and sensitivity of a coffee bean buyer for a large roasting firm are quite different from those of the average person. The layman unconsciously expects staleness as part of the coffee taste. Due to the lack of taste training and orientation, the consuming public has low standards of expectations.

**Desirable Coffee Aroma**.—There are dozens of volatile chemical constituents that contribute to what one recognizes as coffee aroma. Indeed, over a hundred aromatic constituents have been isolated and identified from coffee. Some aromatics occur in parts per billion in the R&G coffee. Some aromatics are potent and are readily recognized: as for example, di-methyl sulfide, or hydrogen sulfide. Others occur at 10 to 20 parts per million in R&G coffee. These are less potent in aroma detection, e.g. acetone, methyl ethyl ketone, and methy acetate. There are in coffee a great number of distinct volatile chemicals. Threshold concentration of aroma or taste recognition can vary many fold from person to person. Hence, we have the complexities of the composition of coffees: their varying aromatic constituents and the differences in sensitivity of detection by individuals.

**Consumer Influences on Freshness**.—Even when the consumer does purchase a reasonably fresh aromatic roast coffee, he often buys 3 lbs. when he can only use 1 pound in one week. The unused two pounds of R&G coffee are exposed to air, and undergo aromatic loss and oxidation while the first pound R&G is being used. Placing the two pounds into the refrigerator does not retard the staling process. Placing R&G coffee in a sealed container in a freezer at −10 F does help to retard the aroma loss rate and oxidation rate. In any case, the U.S. consumer is overbuying R&G coffee to save a few cents per pound. The net result is a stream of beverages noticeably less than fresh. The permeability of the R&G coffee storage container to gas is important. The diffusion of aromas out of the roasted beans will be retarded if the coffee aromas cannot leave the container. Also, if only 10 percent oxygen is in the storage container due to slow liberation of carbon dioxide gas from the coffee grounds, the rate of oxidation relative to air with 21 percent oxygen will be reduced. Roast coffee storage on a 90 to 100 F shelf and the presence of more moisture in the summer accelerates the loss of freshness.

The shift in 1963 from the key-opening can to a plastic cap reclosure was a useful step in retaining freshness of the R&G coffee in the opened can. The plastic capped can has largely eliminated the need for transferring the roasted coffee grounds out of the original can. An air-tight plastic closure stops the free diffusion of air in and aromas out as had occurred in the key-opening can. Also, by keeping the coffee grounds intact in the original can, the lower portion of R&G coffee is better protected from aeration as carbon dioxide is liberated.

**Commercial Oxygen Levels (in Package)**.—For highly effective flavor protection, less than 1 percent oxygen in the containers is desired. U.S. Army studies have shown that "zero oxygen" levels are ideal for effective preservation, but these as yet are not commercially used for coffee.

**Aroma Constituents.**—Few people today taste beverage from freshly roasted coffee beans. Every hour whole roast coffee beans are left in air, aroma losses occur. It can take several hours from the time the freshly roast coffee beans are commercially processed to grounds to when they are sealed in an evacuated can. On occasion, some firms leave roast coffee beans or grounds overnight before packaging them. Every minute ground coffee granules are exposed to air, they are oxidizing and liberating $CO_2$ and aromatics. Therefore, ground roast coffee beans enter the evacuated can at something less than their top flavor level. Evacuating the can also draws aromatics away from the coffee. After vacuum packaging, 1 percent oxygen is still in the sealed can. This residual oxygen, after 2 to 3 months storage, may react with the delicate aromatic aldehyde constituents to reduce their original concentrations several fold and introduced oxidation products (acids). It takes 5cc oxygen to react with 400 ppm aldehyde volatiles in one lb of R&G coffee. Aroma losses assume more importance in U.S. markets now using ⅓ to ½ non-aromatic Robusta type coffees. It is common today to open a commercial can pack of R&G coffee and get very little coffee aroma. The vacuum packed tinned steel can of R&G coffee beans, however, is still usually fresher and/or more aromatic than the bagged roast coffee beans or grounds. Hence, bag types of packages are progressively vanishing in the U.S.A.

## PACKAGING COFFEE

Excluding green coffee packed in jute bags, there are three types of coffee that are packaged: roast, instant, and liquid (extract or brew). For these, three general classes of materials are used: flexible (plastic, paper, or foil), metal, and glass. Flexible materials are usually formed on the filling machine at the coffee processor's plant. Metal cans and glass jars are made by outside suppliers.

In the case of roasted coffee, whether ground or unground, for short shelf storage periods (up to two weeks), lined paper bags may be used. If the time interval between roasting and consumer purchase is longer, vacuum packed cans or cans containing inert gas substantially free from oxygen are necessary. This additional protection, however, almost doubles packaging costs and is unnecessary when delivery may be made before staling occurs. Indefinite protection is not needed when the hold time is short and storage conditions are known with great certainty.

Unfortunately, the 1-lb roast coffee package does not maintain its freshness after the package has been opened. There has been little effort in the industry to promote the use of smaller protected packages of roast coffees. Partly, this is because the smaller protective package costs more and partly for the reason that the consumer has not been sufficiently discriminating in discerning flavor changes ensuing after the package has been opened. Although packaging firms claim that staling does not occur during the first three weeks, the truth is that staling occurs within a few days in ground roast coffee. After exposure to air, coffee does not, however, become foul and objectionable for 2 or 3 weeks depending on its storage conditions. In the home, the only way to store roast and ground coffee without serious deterioration for a week or more is to keep it frozen at 0 to 10 F (−18 to −12 C) in a sealed container. Refrigeration at about 40 F (4 C) also slows down deterioration.

The machinery at the coffee roasting plant fills and seals the containers which are then encased in cartons. There are different types of filling machines for bean, ground, and instant coffees as well as for extract. Depending on the protective atmosphere surrounding the coffee, this involves purging and sealing techniques in the machine after the filling. Most plants test their sealed packages for leaks.

The coffee product must be protected at all times from the effects of high temperature, moisture, and oxygen. Storage time must also be minimized.

Green coffee is reasonably stable when it contains 10 to 12 percent moisture and needs only to be kept in fiber bags porous to air movement. The air, however, should be cool, e.g., below 70 F (21 C) and with 40 to 60 percent relative humidity. As long as the green coffee moisture content remains within this range and the green coffee is not exposed to contaminating odors, its preservation is largely assured. Bags for green coffee are usually made of jute from India. Sisal, sisal hemp, and other fibers have been used in Central America from time to time, although jute is the preferred fabric for packaging.

This chapter will cover packaging of coffee: its importance in selling the product, the changes that occur in coffee without the protection offered by packaging, as well as packaging materials and the manufacturing of many types of packages. Also, special packages that have less commercial application will be reviewed. These include vented bags, evacuated bags, large pouches and special cans for vending machine brewing.

Only superficial reference is made here to auxiliary packaging equipment such as labelers, can and jar casers, carton sealers, coders, stencilers, palletizers, and strappers which are used after the unit package is sealed and which involve packaging processes common to many industries.

Most retailed coffee is unitized in some standard weight, but imported instant coffee is usually bulked for retail packaging in the import country.

It is not the purpose of this book to go into great detail about the broad and varied field of packaging. However, enough data will be developed to demonstrate an order and pattern in the coffee packaging field. It is very hard to generalize because packaging patterns are changing. For example, today in Europe most retail coffee is sold as whole roast bean to be ground before use in the home much as it was in the United States 40 or 50 years ago. Even in the United States today, firms like A & P and other chain stores sell whole roast beans in bags to be ground at the point of purchase and sometimes in the home. However, in-store grinding of coffee is rapidly disappearing because of delays, shopper inconvenience, and other disadvantages of the operation. In the United States most ground coffee is sold in vacuum pack 1-lb cans, with lesser amounts sold in inert gas-pressurized cans and jars. During World War II, when steel and tin were in short civilian supply, vacuum packed glass jars were used almost exclusively.

Roast coffee for restaurants is put up in various grinds, roasts, and weights to accommodate the preference of the owner and the type of brewing equipment he uses. In Europe there is considerably more catering to the customer's taste in special blends, roasts, and grinds than there is in the United States, which is why the small custom roaster and packer continue in business in Europe. In the United States one restaurant coffee merchant usually sells only a few blends and roasts, sometimes only one, to his entire restaurant clientele.

The packaging of coffee products may become quite involved and often it is best to seek advice on specific procedures from the manufacturers as well as from the firms supplying materials and equipment in the field. In labeling, printing, and packaging, not only must firm advertisement be considered but local customs and laws and export arrangements must be investigated. Many firms have shipped products to distant locations only to find that consumer customs were different or that local laws prohibited the sale of the package as it was prepared. Various factors such as machines and weather conditions (heat, cold, humidity) must be considered in designing packaging.

In addition to the aspects of coffee packaging already covered, there are such matters to be considered as pallets, strapping machines and materials, stencils, vibratory conveyors and packers, code markers on labels, roller conveyors, pallet stackers of cartons, depalletizers of cans or jars, case openers and sealers, handling cartons by lift truck without pallets, disposable pallets, counters, and so forth. Nearly all equipment is designed to minimize labor costs. Heat sealers for plastics are of varied design. Some use heat conduction while others use high frequency heating waves. Glue and adhesives are best explained by the manufacturers. (See Figures 9.1 to 9.21)

*Courtesy FMC Corporation*

FIG. 9.1. BAG MAKER, FILLER, AND SEALER (SIG-FMC)

*Courtesy F. R. Hesser*

FIG. 9.2. BAG MAKER, FILLER, AND SEALER; ALSO CARTON MAKER

## Coffee Roasting Industry

The principal functions in the industry are performed by wholesale roasters and packers, including institutional or specialty firms, wholesale grocers, chain grocers, and delivery route dealers. The retail coffee distribution units include chain stores, independent retail grocers, department stores, general stores and retail coffee stores. Some retailers, particularly the chains, feature their own brands and do their own roasting and packaging. Other retailers have their brands roasted and packed for them by trade roasters in addition to selling national and local brands. There are two types of wholesale coffee packers, national and regional, with the latter in the majority. Some of the more important coffee roasting centers are New York, Northern New Jersey, New Orleans, Houston, Chicago, and San Francisco.

## Changes in Coffee Flavor with Aging

When coffee ages, the aroma degenerates from flat to old to sharply rancid; then, in an advanced stage, a cocoa odor appears. Parallel to these changes in aroma, the brew becomes flat at first, e.g., the bloom of fresh coffee disappears; then a bitter flavor develops, accompanied by an old and rancid composition, at which point the limit of salability has been passed. Coffee which smells like cocoa, always tastes bitter or rancid on brewing. For unprotected coffee, the border of salability is reached between 7 and 14 days, depending on storage conditions and the sensitivity of the palate. After the loss of aroma (flat taste), the rise of an unpleasant pyridine taste and aroma appears much quicker with ground coffee than with whole beans.

Another occurrence which is related to the storage of coffee is release of $CO_2$ and volatile aromas (more than 40 different compounds), and adsorption of $O_2$ and moisture. The liberation of $CO_2$ and aromas occurs during the roasting. A part of the $CO_2$ is freed when the cells are crushed by grinding e.g., $CO_2$ is given off quite rapidly during the first 24 hr after grinding, and then more slowly. Not much is known about the process of $O_2$ adsorption.

**Factors Affecting Coffee Flavor Change.**—The most important factors affecting roast coffee shelf-life are oxygen and moisture.

For instant coffee, moisture is a far greater enemy of flavor stability than oxygen. When instant coffee (without aroma) is packed at a low moisture content, flavor stability is equal whether packed in air, nitrogen, carbondioxide, or vacuum. Accelerated tests involving storage at high temperatures and under normal ambient conditions for periods up to two years, demonstrate that a nominal lowering of the flavor level occurs during the aging of instant coffee (when packaged and maintained in a dry atmosphere) regardless of whether it is a gas or vacuum pack. No rancidity or unpleasant off-odors develop under these test conditions even in atmospheric packs.

Moisture is critical in causing staling in regular roasted coffee. When roast coffee (unpackaged) is brought into contact with air at predetermined relative humidities, an "old" taste is observed after 3–4 days when the R.H. is 100 percent; an old taste is not detected until 7–8 days at 50 percent R.H., and at 0 percent R.H., not until 3 to 4 weeks. When the moisture content of roast coffee is raised only about 1 percent (from 1.4 to 2.6), a definite stale odor is detected in 14 to 20 days. Moist nitrogen and dry oxygen are not as harmful as moist oxygen. Even a relative humidity of 25 percent, considered dry by ordinary standards, will cause the moisture content in roasted coffee to reach 5 percent, unless it is protected in some way. A moisture-proof package is definitely needed. It does not suffice merely to restrict this to moisture-proof packaging materials. The package seals must also be moisture-proof. The entire package must pass less than 2 gm of water per square meter per day at 65 percent R.H., unless the coffee is to be sold quite soon after roasting.

It also has been established that 14 cc of oxygen (70 cc air) will cause staling in 1 lb of roast coffee. Therefore, for a four-week shelf-life, the packaging material must permit no more than 0.4 ml oxygen/100 sq in/day to penetrate the package. It has been demonstrated that ground coffee will stale at a much faster rate than whole coffee beans due to the much higher surface area.

When one opens a can of coffee, the strength of the aroma is quite pronounced. This fragrance dissipates fairly rapidly because much of the aroma escapes into the atmosphere and essentially is lost. Aroma is important from a merchandising standpoint. However, the loss of a pronounced aroma from ground coffee still leaves acceptable aroma and taste in the coffee brew. Aroma, then, must be protected by the packaging material so that when the consumer opens the package or during subsequent use, he will associate the full aroma with fresh, high quality coffee.

A further important point is that 1 lb of roast and ground coffee in a bag has an interstitial volume that may hold considerably more oxygen as air than it takes to stale the coffee. Taking the figure cited of 14 ml oxygen or 70 ml air as being a correct order of magnitude for serious staling of coffee flavor, the interstitial air in the bag as packed is about 300 to 500 ml, or about four to seven times as much as is needed for serious staling.

The variables that enter into the process of coffee staling in a bag are many more than those related to the bag material and sealing of the bag. For example, blend, roast, and grind are very important factors. A less aromatic coffee will have less flavor to lose or change. A dark roast coffee has less $CO_2$ to lose and is more porous to gaseous diffusion, and hence is more susceptible to staling. It is well recognized that the greater surface of the roast coffee that is exposed to air allows the release of $CO_2$ gas more rapidly; hence, whole beans do not stale as quickly as coarse or regular grind, and regular grind does not stale as fast as fine grind.

An equally and possibly more important factor is the amount of air in the bag at the time the roast and ground coffee is sealed therein. Ground the coffee stored in bins continually gives off $CO_2$ gas which substantially displaces the air about it. The grounds movement from the bin into the bag filler is such that much of the surrounding $CO_2$ will remain about the grounds. Hence instead of 21 percent oxygen around the coffee grounds, there may be only 10 percent oxygen or less. The amount of oxygen sealed into the bag will therefore strongly influence the subsequent rate of staling.

Plastic films are not perfect and pin holes allow air movement into the bag. The temperature of storage of the coffee may markedly influence rates of staling. The difference between 85 and 68 F (30 and 20 C), a differential of 17 F or 10 C, may double the rate of oxidation at the higher temperature. If the roast coffee had a water quench or was allowed to absorb moisture on a humid day before sealing the bag, this might markedly accelerate the staling. The chemical reaction rates and equilibria at any time within the bag will be governed by every one of these factors.

## Package Requirements

A suitable package for roast coffee must meet the following requirements: (1) low moisture-vapor transmission rate; (2) excellent oxygen barrier characteristics; (3) greaseproofness; (4) impermeability to aromas and odors; (5) slight permeability to carbon dioxide; (6) durability (should withstand handling in distribution and at retail); (7) ability to perform well on coffee package forming and filling machinery; and (8) low cost.

The importance of moisture and oxygen barrier properties and aroma protection of the package must be stressed. Greaseproofness is also essential. Ordinary papers may have a detrimental effect on coffee due to the oil "sweating" quality of coffee. Paper, which has a large internal surface area, tends to absorb coffee oil and allow a great amount of contact with oxygen. Cellophane, aluminum foils, and greaseproof papers, tend to eliminate this problem much better than sulfite papers.

**$CO_2$ Permeability**.—Permeability of the package to carbon dioxide released from roasted coffee may be critical in flexible coffee packages, especially when coffee has been packaged immediately after roasting and grinding. It is necessary for a certain amount of the $CO_2$ to be released from the package in order to prevent ballooning and possible rupture of the walls. The American Can Corporation found that the sealed areas in bags and pouches constructed of the better flexible gas-barrier materials, such as aluminum foil laminates and Saran-coated cellophane, permit enough $CO_2$ to escape to eliminate this potential difficulty. The $CO_2$ gas escapes through the fine folds present in seal areas at the sides and bottom of the packages.

There is no flexible packaging material known that will freely permit the transmission of $CO_2$ yet remain a barrier to oxygen. Of the atmospheric gases, in general, carbon dioxide is quite diffusible, i.e., it passes through films in relatively large volume per unit time; oxygen is less diffusible than $CO_2$. Coffee aromas and some organic vapors are also quite diffusible.

## Status of Packaging Materials for Coffee

The vacuum can is the most widely used container for packaging roasted ground coffee by United States coffee roasters. About 85 percent of the coffee sold in retail stores is packaged in vacuum cans; the remainder is sold in bags. Institutional sales are equal to about one-third of total retail sales; bags and flexible pouches are commonly used. Instant coffee is sold almost entirely in glass, although foil-laminated pouches are also used in the institutional and vending machine markets. Flexible vacuum packages for roasted coffee are a new development in the United States, although they have been widely used by European coffee roasters.

## Advantages of the Vacuum Can

From a protection standpoint, the vacuum can is probably the best container currently available for maintaining the shelf-life of roasted coffee; it is durable and handles well on filling and packaging machines. The chief disadvantages of the can are its cost, its weight, and its shape. The last two factors, weight and shape, contribute to increased costs of fiberboard shipping cartons, warehousing, and transportation; also cans waste valuable shelf space in retail supermarkets. The U.S. Quartermaster Food and Container Institute indicates that flexible packages of the same capacity at the present state of their development weigh only about 50 percent as much as cans. Cylindrical cans, when fitted into oblong boxes, leave an estimated 21 percent of voids. The potential savings in initial package cost plus savings in shipping containers, transportation, storage, and a retail shelf-space, make it worthwhile for coffee roasters and retailers to consider flexible vacuum packages of rectangular shape for coffee.

## Glass Jars

Used almost entirely for instant coffee in the United States, glass jars do a very satisfactory job of maintaining the shelf-life of soluble coffee. The main disadvantage of glass is its fragility. Glass jars also have the same disadvantages as cans with respect to cost, weight, and cylindrical shape. Glass containers are heavier and, therefore, cost even more to transport than cans. Glass jars are resealable and reusable.

## Flexible Packaging Materials

The most common flexible package structure for roasted coffee is the bag. Coffee bags are usually printed and fabricated by converters of paper, film, and foil. Some of the coffee roasters have their own bag-making and filling equipment, using both plain and printed roll stock, which they purchase from paper and film suppliers. The bags are usually constructed with an outer bag of printed white sulfite paper and an inner-liner of Cellophane, glassine, or Pliofilm. Cellophane and acetate may also be used as outer bag material. Flexible pouches containing 2 to 3 oz of ground coffee (8 to 12 cups) and larger are used mostly for the institutional coffee market. These pouches are constructed from laminates of printed and plain films, or may have a printed sulfite paper outer layer with an inner liner of polyethylene.

**Flexible Vacuum Packages.**—Light-weight flexible packages, consisting of an extruded or laminated film have been introduced for vacuum-packaging coffee. Vacuum packages and the equipment for forming, filling, and evacuating them have been developed and used primarily in Europe. Shelf-life of up to three months has been reported by European coffee packagers using flexible vacuum packages. A few United States coffee packagers acquired this equipment but it was discontinued due to poor consumer acceptance vs. cans.

**Properties of Flexible Packaging Materials.**—Bags are the predominant form of flexible packages used by United States coffee packagers. Coffee bags are usually constructed with a 47 to 50 lb white sulfite paper outer bag. The outer bag may be surface laminated with Cellophane or acetate for additional protection and attractiveness. Coffee bags usually have an inner-liner constructed of 29 lb glassine (coated with nitrocellulose or polyethylene), 80- or 120-gauge Pliofilm (rubber hydrochloride), or 195-gauge Cellophane (nitrocellulose or Saran-coated). High density polyethylene and polypropylene have also been considered for use as liner materials.

As shown in Table 9.1 Saran-coated Cellophane is equal to Pliofilm as a moisture barrier and is much superior to glassine. As an oxygen barrier, Cellophane is superior to glassine and much superior to Pliofilm. Since 14 ml of oxygen cause staling in 1 lb of roast coffee, glassine and uncoated rubber hydrochloride have little to offer as an oxygen-barrier material to protect the freshness and extend the shelf-life of non-air packed roast coffee.

TABLE 9.1. MOISTURE AND OXYGEN BARRIER PROPERTIES OF FLEXIBILE FILMS[1]

| Material | Gauge, thickness or weight | MVTR[2] | Gas permeability[3] $O_2$ | $CO_2$ | $N_2$ |
|---|---|---|---|---|---|
| 1. Cellophane | | | | | |
| (a)Saran-coated | 195 | 0.008–0.03 | 0.5 | 3 | 0.5 |
| (b)nitrocellulose-coated | 195 | 0.008–0.03 | 0.5 | 3 | 0.5 |
| (c)plain | 195 | 1.4–2.7 | 1.0 | 3 | — |
| 2. Glassine | | | | | |
| (a)nitrocellulose-coated | 22/26 | 0.6–1.0 | 4.8 | — | — |
| (b)polyethylene-coated | 25/33 | 2.5–3.5 | — | — | — |
| (c)plain | 25 | High | 20 | 42 | — |
| 3. Greaseproof paper | 30 | High | 160 | — | — |
| 4. Rubber hydrochloride(N2) | | | | | |
| (a) | 80 | 0.007–0.23 | 19 | — | — |
| (b) | 120 | 0.007–0.23 | 7.–10. | — | — |
| 5. Polyethylene | 1 mil | 1.2–1.4 | 1,900 | — | — |
| (high density) | 1 mil | 0.3–0.4 | 450 | — | — |
| 6. Polypropylene | 1 mil | 0.06–0.97 | [4] | — | — |
| 7. Saran | 1 mil | 0.01–0.03 | 2.4 | 12 | — |
| 8. Mylar | 1 mil | 0.05–0.15 | 0.5 | 0.5 | 0.5 |

[1]Plastics World, Modern Packaging Encyclopedia, Know Your Packaging Materials (A.M.A), Avisco, and Various Suppliers.
[2]Where available—gm per 24 hr per 100 sq in per mil @ 95 F 90% relative humidity. Data for a few of the materials were reported at slightly different temperatures and relative humidities. MVTR means moisture vapor transfer rate.
[3]Cc per 100 sq in per 24 hr per mil.
[4]Polypropylene performed about the same as high density polyethylene in some Avisco studies.

**Resistance to Grease and Oils**.—Of the three most commonly used liner materials, Cellophane is impermeable, and Pliofilm and lacquered glassine have excellent resistance to coffee oils. Sulfite bag papers and pouch papers are poor barriers to coffee oils without liners of some type.

**Pouching Laminates**.—In the late 1960's there was the commercial application of plastic laminates like polyesters (Mylar) and polypropylene (EVA ethylene vinyl acetate) with thermal sealing polyethylene on the inner pouch surfaces for R&G coffee pouches. With Saran films, oxygen permeability was reduced without the use of aluminum foil, and flex-cracking was greatly diminished. It was a technological packaging breakthrough; however, even until 1978, its commercial feasibility was clouded because the R&G coffee distributed by most major U.S. producers was stale. Why? The reasons were several, and in most cases have not yet been corrected. What was and is feasible in the laboratory, was not executed in

| Flexible Materials | Resistance to Grease and Oils[1] |
|---|---|
| (1) RS-Cellophane | Impermeable |
| (2) MS-Cellophane | Impermeable |
| (3) Plain Cellophane | Impermeable |
| (4) Glassine (lacquered) | Excellent |
| (5) Pliofilm | Excellent |
| (6) Acetate (cellulose) | Excellent |
| (7) Polyethylene (high density) | Excellent |
| (8) Polypropylene | Excellent |
| (9) Mylar | Excellent |
| (10) Glassine (plain) | Very good |
| (11) Greaseproof paper | Very good |
| (12) Pouch paper | Poor |
| (13) Waxed paper | Poor |
| (14) Polyethylene (low density) | Poor |

[1]Modern Packaging Encyclopedia

the field. Most of the fault lay with the roaster, because he continued the following practices:

1. Using light roast coarsely ground coffee, susceptible to staling.
2. Laying on too much water to increase yield and to degas. This act contributes heavily to accelerated staling.
3. Holding R&G coffee in bins to degas, to prevent ballooning of pouches. Holding also loses aroma and flavor, and causes oxidation.
4. Inadequate inert gas flushing, leaving 4 to 5 percent oxygen in pouches.
5. Poor integrity in pouching, e.g. seal leaks, rough handling, etc.
6. Random collating in cartons causing flex cracking and leaks. Curwood reports that test show that 40 percent leakers can occur this way during handling and transport.
7. Assuming that he has a pouch with integrity, when he does not. This contributes to months of elapsed time between roasting and use.
8. Use of lower grade and Robusta coffees that lack aroma and reveal staling much faster than better quality mild coffees.

The result of the above circumstances has contributed to some roasters going to less costly films, and depending on faster delivery and use for freshness. The net result has been a massive confusion in performance, a massive production of stale unsatisfactory tasting coffee, as documented in two major consulting surveys made for the most prominent hotel and restaurant chain in the USA in 1976. A further manifestation of these circumstances was the dissatisfaction of OCS (office coffee service) operators and their customers, which has encouraged some to do their own roasting.

**Pouching Machines.**—In the U.S.A. there are a half dozen prominent manufacturers, e.g. Hayssen Div., of Bemis in Sheboygan, Wisc.; Package Machinery Co. of East Longmeadow, Mass.; Triangle Package Machinery Co. of Chicago, Ill.; General Packaging Equipment Co. of Houston, Texas; Wright Machinery Co. of Durham, N.C. and others. From Europe we have Rovema for

Pneumatic Scale of Quincy, Mass. Hesser of Stuttgart and Hoeller near Cologne, Germany; SIG, Swiss Industrial Co. of Switzerland, Hamac-Hansella, Bosch group in Germany, and others e.g. Sweden.

In the U.S.A., every coffee bean roasting plant has several pouching machines to fill 1.5 to 2.5 oz R&G coffee per pouch at rates of 60 per minute on single tube and 120 pouches per minute to twin tube machines. The principles of form-fill-seal are similar for most machinery fabricators. A roll of plastic film 8.75 to 9.25 inches wide and about 12" diameter before use, is unwound over tensioning bars and is then formed into a tube over a chrome hardened stainless steel form. See Fig. 9.11. For example, a 4" wide pouch, will have a ⅜" wide seal area, requiring a 8.75" web.

The adjoining illustrations show the jaw area of the machine, and 4 drawings showing the jaw and pouch relative positions during filling, sealing and dropping.

A single tube machine in 1978 costs from $20,000 to $30,000 with auger or volumetric filling of the R&G coffee, twin tube machines cost $35,000 to $45,000. In many cases weigh fillers are used and this adds $15,000 to $20,000 to the above costs. An additional $2,000 cost is for digital instead of cam actuation, a '76 development which increases performance accuracy and reliability.

Before 1978 most pouchers had volumetric feeders because drip grind R&G coffee is rather free flowing, however, it also dusts and causes fines to be sealed in the pouch end seals. But when darker roast finer grind R&G coffees are processed, an auger, e.g. Mateer, can very acurately dispense 1.6 oz. with little to no dusting.

Most machines are fitted with inert gas (nitrogen) flush to reduce oxygen in pouch.

**Ballooning Pouches.**—If most of the carbon dioxide gas is not removed from the R&G coffee before pouching, the pouch may subsequently balloon and possibly rupture. Of course, a ballooned pouch shows that it has integrity, and, if inert gas packed, shows that it is not a leaker (flat pouch). In and since 1968 KENCO Co. in London has offered its patents U.K. 1,024,214 and U.S. 3,333,963 for licensing as a means of degassing without equivalent aroma/flavor losses. In view of what is about to be disclosed, it would appear that the expenses of the KENCO system could be easily obviated, and are being easily avoided in 500 gram and kilo packs of finely ground R&G coffee in Scandinavia, esp. in Malmo and Copenhagen. Here, the good quality blends are dark roasted (17% weight loss), and finely ground, finer than U.S. Fine, and within 1 hour pass from grinder to gas flushed pouch without any subsequent ballooning problems. The reason for this is that fine grind has a more broken cell structure, as does darker roasts, and so the $CO_2$ gas is liberated first and mostly within 1 hour, before being sealed in the pouch. A pertinent part of this process is that the finely ground R&G coffee is poor flowing, and a "live" bin is required to assure movement of this fine ground coffee to the auger filler. A few OCS operators in the USA have installed their own systems. It will be noted on the other hand that some of the major U.S. roasters in their 2 oz or 16 oz R&G coffee pouches have no ballooning, and also have no freshness. This is unfortunate since these firms represent the major share of the hotel and restaurant business in the USA, and are using good quality laminates.

**Metalized Pouches.**—In 1976 and 1977 there was a strong introduction of metalized polyesters, e.g. St. Regis' Alure as illustrated in accompanying Fig. 9.6 and Table 9.1. Metalizing gives a luxurious appearance, while costing less than aluminum foil.

## COFFEE-BAG PACKAGING MATERIAL

At one time bagging whole bean or roast and ground coffee was a relatively simple matter: the coffee was placed in a glassine-lined paper bag. Glassine is a very fine fibered, highly calendered paper stock that does not allow grease or moisture to penetrate readily as compared with ordinary papers. Moisture-proofing is attained by a lacquer or wax. Glassine prevents the transmission of odors; hence, the coffee is less likely to pick up off-odors and to lose its coffee aroma. Creasing cracks glassine, hence the bag liners are designed to avoid too many creases. Although plastic films have been used for a long time, their widespread application to commercial packaging occurred after 1950. Since that time many types of plastic film have been applied to commercial packaging of roast coffee.

Most damaging to roast and ground coffee aroma and flavor in storage are heat, moisture, and then the oxygen of the air. With the advent of better, more varied, and more economical techniques for film processing, suitable properties were sought for the coffee bag material. Where neither of two plastic films was suitable, when laminated together they might be economical and able to deliver the service required. The high cost of film properties might be reduced by using thinner laminates. For example, Saran might be used for gas impermeability; polyethylene for heat sealing properties and cellulose film or aluminum as the carrier material. Thus evolved a three-layer film. The manufacture of multilayered films is developing rapidly and new types are being continually evolved.

The most interesting development since 1960 has been the Hesser machine. The bags are made, filled, evacuated, heat sealed, and packed in cartons all on one machine. It is the evacuation feature which is new. The film chosen for the vacuum service must be strong enough to withstand the atmospheric compression about the granules as well as any slight swelling of the bag. This may occur due to $CO_2$ gas from the coffee liberated after packaging. The precaution is taken to hold the ground coffee for a time in order to liberate the major amount of $CO_2$ gas from the coffee so that the bag film is always snug about the ground coffee until the seal is broken in the home. In practice, the evacuated bag affords full protection for the ground coffee for at least six weeks. Such protection is far beyond that provided by any other flexible packaging of ground coffee in air. Ground coffee in air pack is usually stale within a few days to two weeks.

A gas-tight coffee carton for $CO_2$ gas fill was developed in 1960 by Esselte-For-packaging A.B., Norrkoping, Sweden. The material used was a paper, aluminum foil, and polyethylene laminate. In countries that find steel cans too expensive, such laminates have been successful.

## Coffee Bag Packaging

An Italian firm has adopted a packaging system that has proven successful without the use of costly tin or glass containers. The "Fres-co System" of Goglio Luigi Milano S.p.a. has been used in a number of locations throughout the world, and the response to this vacuum packaging system has been quite positive.

To attack the problems of roasting and packaging coffee, one must keep certain essential factors in mind. Of foremost concern is the preservation of the product; coffee must be maintained in the best condition as regards fragrance and freshness; moreover, for ground coffee, while it is undoubtedly simpler to use, it is much more perishable. However, one must be concerned with not losing the aromas, produced and released during roasting, through evaporation.

To avoid these two important problems, packers usually follow two simple methods. The first consists in protecting coffee from the air which causes a taste cut-off: this can be achieved by creating a vacuum in the container. The second consists in adopting an air-tight bag which will not allow outside-inside gases to be exchanged. (See Figs. 9.1 to 9.5)

Tin and glass containers have been able to tackle the problem very well, but the expense involved in utilizing containers of such make is very high. Paper or polyethylene bags are cheaper, but they do not solve the problem either: because the volatile substances of roast coffee need room to expand, the distortion of the container or even its breakage is often probable.

Until recently, in order to pack coffee in air-tight containers, it was necessary, before packaging, to keep it out of contact with the air for the period of evaporation, this being a very costly operation, while in addition, causing significant loss of flavor.

The solution which Goglio Luigi Milano has found utilizes a one-way degassing valve which allows the coffee to be roasted, ground, and vacuum-packed at one time without considering any successive degassing problem.

This plastic degassing valve is made of three parts: a plate, a rubber disc, and a cap; it is applied to bags by a heatsealing process. This valve permits the gas to come out at a fixed pressure, thus avoiding the air which normally enters the bags.

The application of this valve became useful when a flexible, resisting, air-tight container was created. This container is composed of a three-layer roll whose outer, polypropylene one, is mechanically hit proof; the middle layer is made of aluminum foil designed for impermeability; and the inner layer, the polyethylene

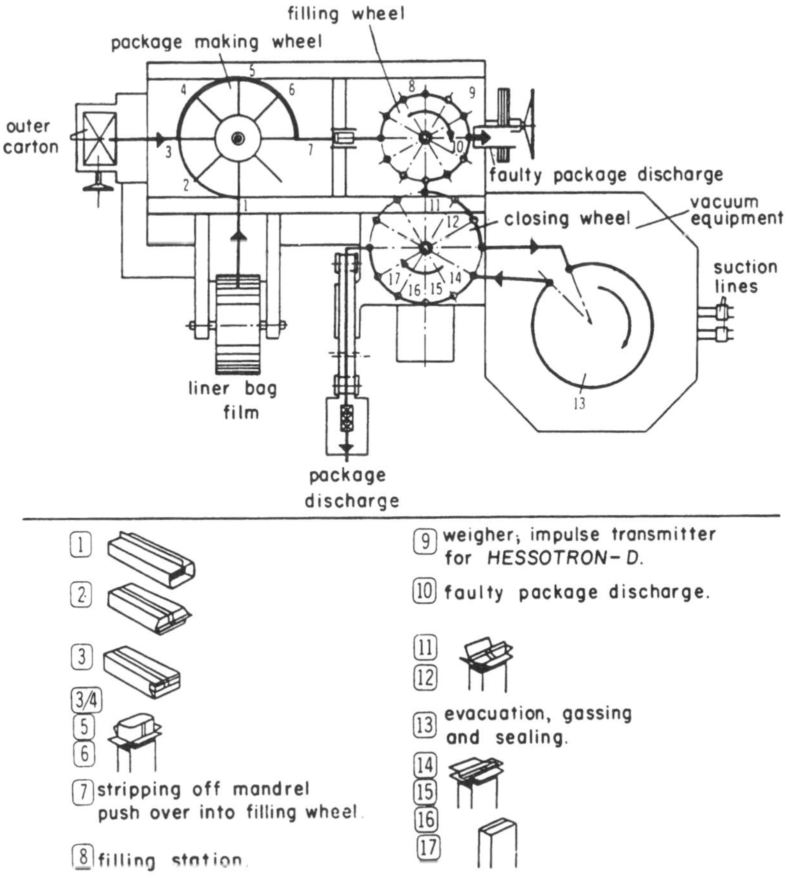


*Courtesy F. R. Hesser*

FIG. 9.3. FLOWSHEET FOR HESSER BAG AND CARTON MACHINE

one, allows an easy sealing and because of its neutral characteristics does not alter the contents of the bag.

Additionally, a special machine has been patented and designed in order to seal valves to the laminate which is coming off a reel. This machine can be installed on the production line with a bag forming machine and a filling and vacuum one. Bags designed for this operation have square bottoms which enable them to easily stand wherever they are used, including ultimately shelf display in the retail store.

Packaging is completed when the several filled bags are placed in cardboard cartons, thus avoiding the inflating process and allowing also an easy storage set.

The valves and bags together have at least the following possibilities: It is possible to roast, grind, and vacuum pack coffee in one single operation without intermediate degassing stops, thus employing investments otherwise lost in storage. Better quality and preservation of the finished product is reached in a single

*Courtesy F. R. Hesser*

FIG. 9.4. EVACUATED BAG OF GROUND COFFEE

*Courtesy B. F. Gump Company*

FIG. 9.5. BAG OPENER AND FILLER

packaging procedure, while transport and storage spaces are able to be utilized more economically.

Coffee has a better potential shelf-life, and will remain as fresh as long as a normally fresh coffee would, once the container has been opened. The cost in utilizing such a packaging system is significantly lower when compared to the cost of equivalent metal or glass container systems.

The "Fres-co System" is available with a wide range of machines, to be suited to individual customer requirements. There are fully automatic lines forming the bags starting from rolls, with stations for weighing the product and filling the bags, keeping them in shape, vacuum-packing or gas-flushing them, and placing the bags into multiple or single cartons, or into heat-shrinkable trays.

There are also smaller, non-automatic machines available for stores, laboratories, or small-medium production roasters. Both of these packaging lines, the automatic and non-automatic, have been designed with only one aim, claims the company's representative, "for manufacturing strength, easy maintenance, and the highest reliability in order to allow the best results even with a non-expertly skilled staff."

## Over 1½ Billion Pouches for Coffee

How many pouches are used to package coffee for *public feeders* in the USA?

More than 1,500,000,000 in 1964 alone!

That's what informed estimates indicate. Authorities having direct, continuing national contact with public feeding market say some 1/3rd of total poundage for these outlets goes to point of use in pouches.

Moreover, these pouches, which are mainly for bottle brewers, range from 2½ oz to 3½ oz.

Taking volume into account, the average pouch would be 2¾ oz, say these authorities.

Some 276,000,000 lbs of coffee went into consumption outside the home in pouches last year.

This adds up to 4,416,000,000 ozs, requiring some 1,606,000,000 pouches.

## Pressvac Container

It's square in shape, lightweight, takes up 25 percent less space and most of all, costs less. What is it? None other than the latest introduction by Cirkel Kaffe in Sweden—a paperboard, aluminum foil and plastic container made of Pressvac. Manufactured for Kooperativa Forbundet by Esseltpac, the new container is said to be as tight as conventional tins. Esseltpac supplies the container in four prefabricated parts: 1) A container body made of paperboard, with an inner lining of aluminum foil, 2) Plastic top frame, 3) Inner cover of aluminum foil and 4) Plastic top cover—which eventually will be sold separately.

So far the Pressvac operation has received the praise of employees on the line who find it much quieter than tins.

In supermarkets where the product has been tested, it has found to take up less space, stack neater in displays and looks more attractive.

The container has an aluminum foil pull-off inner lid under high vacuum and an extra plastic lid top so the consumer can preserve the fresh aroma and taste after opening.

The films that are least permeable to gases are Saran, Mylar, Cryovac and some laminates of Cellophane, Aluminum and Saran have very low permeability to gases. These properties are important where oxygen and moisture are to be kept out and carbon dioxide is to be retained. Some films are selective in that oxygen may enter more easily than moisture, e.g., polyethylene. There are many examples of such differing gas permeabilities. These data are not generally available and the film laminate manufacturers must be consulted. Modern Packaging magazine and annual encyclopedias offer recent information on laminates and suppliers. Usually water vapor transmission through various film thicknesses is given in grams per 24 hr per 100 sq in., for example:

| MVTR[1] | Laminates |
|---|---|
| 1.2 | 1 mil polyethylene |
| 0.2 | 4 mil polyethylene (or 2 ply glassine) |
| 0.14 | Paper backed 0.35 mil aluminum |
| 0.03 | 1 mil aluminum foil and 1 mil polyethylene laminate (also Saran and aluminum) |

[1]MVTR, moisture vapor transmission rate.

A 1 lb coffee bag may have an area of about 100 sq in. and in 100 days in 1 mil polyethylene the coffee would pick up at 100 F (38 C) and 95 percent R.H. (an extreme moisture situation), 1.2 gm moisture or about one-quarter of 1 percent water.

Other laminates are encountered with jar inner liners and labels. The composition of the label laminate may cause curling. For example, lithograph paints will not adhere to aluminum so a Cellophane or lacquer coating is applied first. The laminate paper side (for strength) absorbs moisture and expands. The result is curling of labels. On high speed labeling machinery, this causes real difficulties and impractical operations. In fact, the choice of any film or laminate must be consistent with the machine use. This is called machineability or workability in the machine. Obviously an excellent laminate cannot be used if it is not workable in the machine at high speeds. Cellophane, by its nature, has a high affinity for water and solvents. Just as we have seen the transition of the milk carton from waxed paper to polyethylene coated paper, we can expect new developments in coffee bag materials.

## Coffee Filling and Sealing Machines

After the bags are formed and are opened for filling, the proper weight of coffee must be measured. The techniques or mechanics of doing this vary but most commonly used are counterweighted systems where the bulk weight runs out in a few seconds and the final weight comes out in a dribble. This method is common to most weight discharge machines. Machines are available that operate manually or automatically.

The Hesser machine handles 60 1-lb bags per minute. The SIG (Swiss Industrial Company) machine handles 70 1-lb bags per minute with inner liner and outer paper carton. Figure 9.1 shows the SIG machine. Figures 9.2 to 9.4 show the Hesser evacuated bag flow sheet machine. Another model SIG machine handles 55 pre-fabricated bags per minute. SIG machines offer an assortment of bag closures, as do some other machine manufacturers.

The B.F. Gump Company Bar-Nun Automatic Bag feeder-opener-weigher (Fig. 9.5) with one attendant runs at 60 bags per minute followed by a heat seal unit. Another series of B.F. Gump Company models that are semi-automatic handle 28 1-lb bags per minute and weigh to a fraction of an ounce. Thus, the whole bag fabrication, weighing, and sealing may be processed by one machine or the steps may be individualized by manually feeding the bags and then placing weighed bags into a sealer. The machinery investment is related to the volume of the packaging operation. Geo. H. Fry Company makes a bag tucker to work between filler and bag sealer. The Benco Bag Closing Machine, manufactured by the Shellmar-Betner Package Division of Continental Can Company Incorporated, closes bags automatically either by steel band or glue at a rate of 32 per minute.

After bags of coffee have been sealed, they must be unitized for shipment to retailers. This may be done by placing the 1-lb bags in a larger bag or envelope or by case loading the bags into a carton. The Union-Bag Camp Paper Corporation makes an automatic coffee bag baling machine. It collates bags from two lines, picks up a baler bag from the hopper, opens the bag, moves collated bags inside, closes the bag, and seals it with paste, heat or tape. The machine may be adapted to various size bags; it runs at 6 bales per minute (12 1-lb bags per bale).

The Standard-Knapp Division of Emhart Manufacturing Company builds an automatic case loader for 1-lb bags which fills cartons 3 by 4 in one layer, or in two layers making 12 or 24 per carton. Bags are sorted into loading patterns and then pushed into the case (or bale). The cartons may then be sealed on a Standard-Knapp case gluer, closer and sealer (32 cases per minute) before being sent on a roller conveyor to pallet storage before shipment out of the plant.

Where higher bag filling speeds are desired from a machine, the bag filler units are set up in multiples. For example, the Pneumatic Scale Corporation manufactures a bag maker, net bean coffee weigher, and metal tie mechanism. A bag liner is formed from a roll of liner material, and the carton shell is formed from pre-printed blanks. The speed is 70 bags per minute from a 4 head weighing system.

Checkweighing scales for cartons, bags, jars or cans with reject mechanism are supplied by Thayer Scale Corporation. The Minnesota Mining and Manufacturing Company makes a flat surface applicator for printed tapes containing special notices and discounts. A solenoid operated head dispenses exact lengths of pressure sensitive tapes at 75 packages per minute. Permacel Company and Crown Zellerback Corporation also make pressure sensitive printed tapes. If tapes are used to seal cartons, machinery is built for this service by General Corrugated Machinery Company (Palisades Park, N.J.) and other firms.

# PACKAGING COFFEE IN CANS

## Can Manufacture

The metal can as a container for preserving food was invented in 1810 in England. It was based on Nicolas Appert's idea of preserving food in bottles after immersion in boiling water. The first commercial canneries were in operation in 1812 in England and in 1819 in New York City. Until the Civil War, cans were made by hand in the United States at the rate of about six per hour. Can manufacture at this time rose from 5 to 30 million cans per year. By 1870 machines were producing cans at 60 per hour. By 1900 the can production rate in the United States rose to 6,000 per hour and in the early 1960's it was 30,000 cans per hour. The Continental Can Company was founded in 1904 with one can-making machine. It was about this time that Hills Brothers in San Francisco produced their first vacuum pack of roast coffee for Alaska.

## Coffee Can Production

According to the Can Manufacturers Institute, Incorporated the amount of steel used for metal cans for roast coffee was 172,000 short tons in 1950 and about 230,000 short tons in 1960, which is equivalent to about 1,333,000,000 1-lb coffee cans. Of course R&G coffee is also packed in $\frac{1}{2}$-lb, 2-lb, 3-lb, and 12-lb cans. The largest volume sold is in 1-lb cans. The American Can Company has offered a can to hold 2-oz fresh R&G coffee for preparing 6 cups of coffee at a time. Fundamentally, this is a good idea for circumventing the staling of R&G coffee stored in an open 1-lb can, and it is suitable for sales in units of 8 per pound or for introductory offers.

## Can Size

The 1-lb coffee can is standardized in size and weight in the United States. It is $5\frac{1}{8}$ in. in outside diameter and $3\frac{4}{8}$ or $3\frac{5}{8}$ in. high with a volume of about 66 cu.

in. or about 1 liter. It is usually contoured at the periphery and dished in the ends. This gives the cylinder walls greater structural strength against buckling under outside atmospheric pressure and varying degrees of vacuum or pressure from the inside. Even in a vacuum sealed can, coffee may liberate enough $CO_2$ to cause a positive internal gas pressure of a few psig, and, with inert gas, 15 psig. The bulk specific gravity is about 0.45 or 28 lb per cu ft. Most coffee cans have some slack space. A full can of R&G coffee is achieved through a fast roast which causes larger swelling of the beans to give lower density. The pound of R&G coffee must be tamped into the the can to make it fit into a standard sized can. Sometimes when low bulk density occurs from new-crop coffees that are roasted rapidly, the standard amount willl not fit into the can without tamping. If the roaster does not have means for tamping, a real packaging problem results. The can is standardized in volume to take the bulkiest roast coffee. Vibrators are usually located under the cans during filling to settle the coffee particles to their greatest compaction. Natural fall bulk densities of ground coffee would be only about 0.3 specific gravity or 18 lb per cu ft. The volumes of the 2- (5⅛ in. diam by 6⅝ in. high outside) and 4-lb cans are almost proportionately larger. The 4-lb can is susceptible to collapse from atmospheric pressure with full vacuum within the can.

## Can Weight and Steel Gauge

There are about 6,000 1-lb cans per short ton (or 3 cans per pound) of steel. Each can weighs about 150 gm (5.3 oz) and has about 90 sq in. of surface. This is equivalent to about a 10.9 or 12.5 mil steel thickness corresponding to about a United States Standard 31 or 30 gauge. Actually, tin plate is specified in terms of the weight of a base box unit, which is equivalent to 112 sheets measuring 14 × 20 in. or 112 × 280 = 31,360 sq in. This area varies weight from 55 to 135 lb for plates in common use. Table 9.2 relates the weight of the base box to sheet thickness.

## Tin Plate

Most tin plated cans have less than 1 lb of tin plate coating per base box, and for coffee ¼ lb of tin is common. This corresponds to a tin thickness of 15 millionths of an inch. Tin coating may vary in thickness on each side of the sheet, and hence of the can. Some recently developed cans use no tin coating at all. During World War II black (non-tinned) plate cans were used until glass jars were substituted.

Before World War II, most tin plating was done by dipping sheets into molten tin baths, but after the war tinning at much lesser thickness and with greater adherence was done continuously through electrolytic baths. After the steel is tin-coated it is further protected by a lacquer coating. The continuous steel sheet passes through the lacquer baths; this is followed by heating and drying. The grade AAAAA has the greatest thickness and luster of tin plate; grade A, the least. The

TABLE 9.2. STANDARD WEIGHTS AND GAUGES OF TIN PLATED STEEL

| Trade Term, Lb | Nearest U.S. Standard Gauge | Thickness in Ten Thousandths of an Inch | Pounds per Sq Ft | Weight of "Base Box" (14 in. × 20 in.) Pounds, 112 sheets = 31,360 sq in. |
|---|---|---|---|---|
| 55 taggers | 38 | 62 | 0.253 | 55 lightest plate |
| 60 | 37 | 66 | 0.276 | 60 |
| 65 | 36 | 70 | 0.298 | 65 |
| 70 | 35 | 78 | 0.321 | 70 |
| 75 | 34 | 86 | 0.344 | 75 |
| 80 | 33 | 94 | 0.367 | 80 |
| 90 | 31 | 109 | 0.413 | 90 |
| 95 | 31 | 109 | 0.436 | 95 |
| 1 CL | 30.5 | 117 | 0.459 | 100 |
| 1C | 30 | 125 | 0.491 | 107 |
| 118 | 29 | 141 | 0.542 | 118 |
| 1 XL | 29 | 141 | 0.588 | 128 |
| 1 X | 28 | 156 | 0.620 | 135 |
| DC | 28 | 156 | 0.638 | 139 |
| 2 XL | 27 | 172 | 0.680 | 148 |

original steel is cold rolled to the desired thickness to achieve uniform ductility, elongation, and temper. It is also more resistant to rusting from moisture. After rolling, edges are cut to desired width and the steel sheet is stored in rolls like a long tape. The cold rolled sheet has been annealed at 2,300 F (1,260 C) continuously. No manual handling is involved up to this point.

Next, the tinned, lacquered roll is cut into accurately dimensioned 20 by 28 in. sheets on large automatic shear machines. These sheets are piled on platforms, inspected visually, and then automatically weighed. Rejects are set aside. The sheets are accurately squared off before the lithographic process.

## Can Lithography

The coffee cans usually have a design printed directly on metal. An original sketch from the can company's art department, after customer approval, is photographed in multiples for photographic transfer to the master printing plate negative. A separate printing plate is made for each color. These master press plates are used on high speed presses for printing the coffee can art work on the proper sized tin plate. The tin plate first receives its background color. The sheet from the painting roll slides between two moving wire frames on an endless rotary device that carries the plates through an oven. The discharged sheets are repeatedly painted and dried until all colors are accurately superimposed, and finally are varnished and baked to protect the painted design underneath. The sheets are then ready for cutting into the component parts of the can.

## Forming Can

The lithographed plates are automatically picked up from a stack and are fed into a punching or slitting machine which cuts them into body blanks of the right size. The blanks are fed into a body maker and side seamer (Fig. 9.6) which edges and notches them so that opposite sides will lock together when the blank is formed into a cylinder over the body maker horn. Soldering flux is applied and the side seam is hammered or "bumped" tight; the seam is then soldered. A flanger then flares out both ends of the cylindrical body. This prepares the can cylinder for end seaming; one end is seamed by the can manufacturer; the other end by the coffee roaster after the can is full. The can manufacturer seams the bottom of the coffee can with key attached. The coffee roaster seals the top of the coffee can after it is filled with coffee in the upright position. This has the slitted strip which will be removed by the consumer when the key curls the steel from around the top of the can. In 1963, a keyless can with a polyethylene cover was introduced.

The ends of the can are punched, dished to contour, and then passed through a machine that curls the outer edges in preparation for the double seam closing step. Fig. 9.7 shows a collar maker for coffee cans. Each can with the end is tested for leaks by placement on a rotary wheel which has a rubber gasket sealing the open face of the can. A preliminary pressure has been applied to the inside of the can. If that pressure changes, the can is automatically rejected on discharge from the tester. See Fig. 9.8.

*Courtesy American Can Company*

FIG. 9.6. BODY MAKER AND SIDE SEAMER (SOLDER) FOR VACUUM COFFEE CAN

*Courtesy American Can Company*

FIG. 9.7. COLLAR FORMING MACHINE FOR VACUUM COFFEE CAN

*Courtesy American Can Company*

FIG. 9.8. ROTARY AIR LEAK TESTER FOR VACUUM COFFEE CAN

## Can Transport

Since the coffee cans are formed by the can maker, but must be used by the coffee packer, the transport of cans is often awkward. For example, the open ended cans must be palleted in bags or cartons, stacked manually in trucks or railroad cars, and transported to the roast coffee packer. When they arrive, the open end cans must be conveyed to the filling machine. In a recent United States West Coast plant, a depalletizing machine was installed to sweep off layers of the 1 lb cans from cardboard sheets between the can layers. Figures 9.9 and 9.10 show this process. The most satisfactory solution of the problem of handling empty coffee cans has been devised by General Foods in Hoboken, N. J. Here the can company plant is adjacent to the coffee roasting plant. Cans are conveyed over without palletized handling. Empty cans are usually elevated by metal fingered conveyors, and then roll by means of gravity to the coffee filling machines via slatted contoured ways.

## Can Filling and Sealing

Some of the weigh fillers mentioned under bag filling may be used for filling cans with a small modification. Figure 9.11 shows such a coffee can filling machine. To attain a coffee filling speed of 120 1-lb cans per minute, eight B. F. Gump Company "Bar-Nun" weigh fillers fill 8 cans per each 4-sec interval. Filling systems may be designed to 200 to 300 1-lb coffee cans per minute. With multiple filling, a timing unit synchronizes movement of cans to and from net

*Courtesy J. A. Folger and Company*

FIG. 9.9. FORK TRUCK MOVING PALLETIZED EMPTY COFFEE CANS

*Courtesy J. A. Folger and Company*

FIG. 9.10. DEPALLETIZING EMPTY COFFEE CANS

weighers. Some sort of vibration is needed during the coffee can filling step to settle the ground coffee volume below the top level of the can lip. Figure 9.12 shows the can weight fill of coffee being checked manually.

## Can Evacuation and Closing

The Continental and American Can Companies supply coffee can closing equipment for vacuum or inert gas pressure pack. Can covers are fed automatically from a stack and are loosely clinched on top of the cans before the entrance valve of the evacuation can closing machine. When the can is within the machine, about 29 in. mercury vacuum is drawn; then the rotary clinches make the double metal seam, and the sealed can passes into the exiting compartment. The entrance and exit compartments are air locked to prevent any unnecessary air flow. The evacuation and can closing compartment operates only under a small fluctuating air influx. A 28.5 in. vacuum leaves 1 percent oxygen in the can, and this is about the maximum possible if deterioration is not to be apparent the first month.

Figure 9.13 shows the five steps in vacuum sealing a R&G coffee can: (1) the can is filled with R&G coffee and a loosely set cover is partially crimped before it approaches the machine; (2) the can passes into the entrance valve of the evacua-

Courtesy B. F. Gump Company

FIG. 9.11. COFFEE CAN FILLING MACHINE

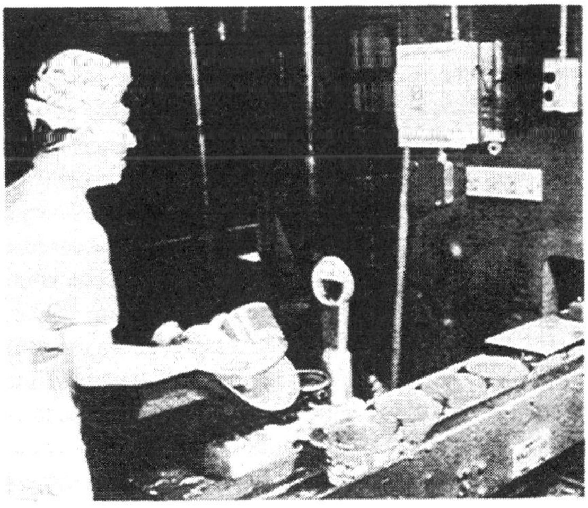

Courtesy Tea & Coffee Industries Magazine

FIG. 9.12. CHECK WEIGHING COFFEE FILLED INTO CAN

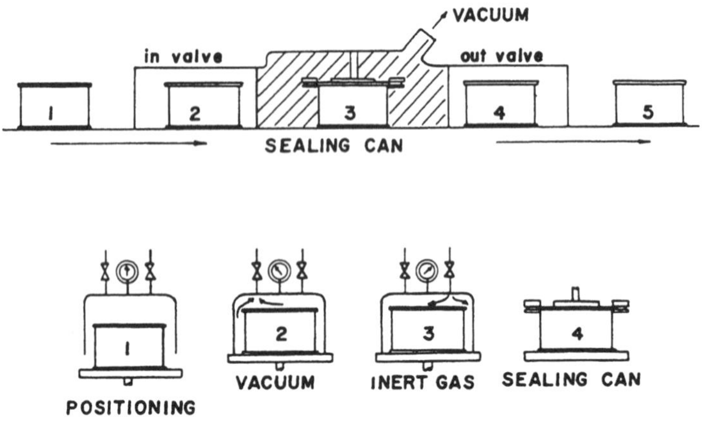


FIG. 9.13. STEPS IN EVACUATING AND INERT GAS PACKING OF CAN

tion machine; (3) full vacuum is drawn and the cover is clinched to the can cylinder wall by rotating wheels; (4) the sealed can enters the exit valve; and (5) the sealed evacuated can leaves the area.

The four steps in obtaining a sealed inert gas packed coffee can are: (1) the can is filled with R&G coffee and a loosely set cover partially crimped is set on a platform which seals against the upper dome; (2) vacuum draws air out of the sealed chamber and can; (3) vacuum valve is sealed and inert gas is let into the chamber and can; (4) the can is released from the chamber and the top cover is clinched to the cylinder can wall immediately in air surroundings. Figure 9.14 shows a vacuum capper for jars.

The can is on a rising platform with a dome sealing the can and platform; thereafter the dome space is evacuated, the vacuum is cut off, and inert gas is added at atmospheric pressure. The cover is then double seamed after the can comes into the air. Practically no air, hence no oxygen, enters the can. The vacuum gasser has the advantage of producing lower oxygen contents with less vacuum pump capacity than with the vacuum systems. The inert gas packed can does not have the structural compressive stress on it that the evacuated can has. These machines operate at over 120 cans per minute. Figure 9.15 shows a can evacuating and seaming machine. Figure 9.16 shows the same machine in a production plant. Figure 9.17 shows the inside of the vacuum pump commonly used with these can evacuating machines. Figure 9.18 shows a bank of these vacuum pumps in an operating plant. Figure 9.19 shows the semi-automatic vacuum closing machine for the 5-gal can, which size is used for military and institutional packing of R&G coffee. Figure 9.20 shows in-line testing of the top rim of vacuum sealed cans in a plant built in 1957. Figure 9.21 shows an underwater manual test for air leakage out of a faultily sealed pressure can. The sealed cans are routed to case filling

Courtesy Owens-Illinois Glass Company

FIG. 9.14. GLASS JAR EVACUATION AND CAPPING MACHINE

machines where they are usually loaded twenty-four to a carton. The carton flaps are glued (or taped down with an applicator) or flap sealed. The cartons are palleted for storage until they trucked away from the plant.

## ROAST COFFEE IN GLASS JARS

Few firms pack their R&G coffee in glass jars. During World War II, Owens-Illinois made air evacuating machines (see Fig. 9.14) sealed with gum gasketed screw caps. Glass under vacuum is, of course, dangerous, but was a necessity during a war emergency metal shortage. Allowing sunlight to enter through the glass is undesirable. Ultraviolet energy accelerates oxidation of aromatic coffee

*Courtesy American Can Company*

FIG. 9.15. CAN EVACUATION AND SEALING MACHINE

*Courtesy J. A. Folger and Company*

FIG. 9.16. CAN EVACUATION AND SEALING MACHINE IN OPERATING PLANT

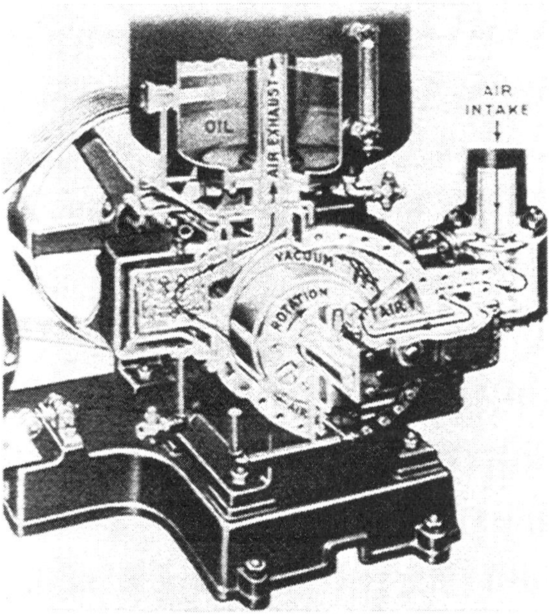


*Courtesy Beach-Russ Company*

FIG. 9.17. SECTION OF VACUUM PUMP FOR CAN SEALING MACHINE

*Courtesy J. A. Folger and Company*

FIG. 9.18. BANK OF VACUUM PUMPS IN PRODUCTION PLANT

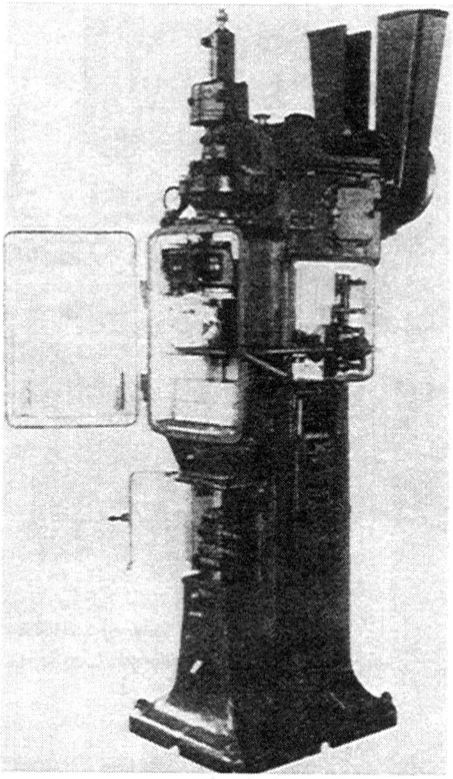

*Courtesy American Can Company*

FIG. 9.19. SEMI-AUTOMATIC VACUUM CAN CLOSING MACHINE

constituents. Glass jars may be filled with coffee with the standard type of weight fillers; but vacuum sealing of the cap requires a rotary clutching action in an air evacuated chamber. An inert gas fill of a glass jar with loose fitting cap, followed by prompt securing of the cap in air, is just as feasible as the same procedure with the inert gas-filled can. R&G coffee is rarely packed in jars when metal cans are available.

## ADDITIONAL COMMENTS ABOUT PACKAGING COFFEE

### Bagged Green Coffee Beans

Green coffee beans are reasonably stable for one year's storage if they are kept at about 70 F (room temperature) and at 40 to 60 percent relative humidity. The

*Courtesy J. A. Folger and Company*

FIG. 9.20. IN LINE LEAK TEST OF FILLED, EVACUATED, AND SEALED CAN OF COFFEE

bean's nominal moisture of 12 percent should not vary more than ± 1 percent. The burlap, hemp or sisal bags are porous enough to allow air movement through the bag to equilibrate bean moistures, air compositions and temperatures. Moisture condensation on the bag or on the beans can cause localized mold and fermentation of beans. In cases of high moisture, as in the mountainous areas of some coffee growing countries, moisture pickup by the coffee beans can cause the beans to swell sufficiently to rupture the bags and to spill beans on to the moldy earth. Consequently, severe bean damage may result. In shipping bags of beans, care must be taken not to contaminate the green coffee beans with foreign odors, tars, oils, chemicals, etc. Green coffee beans can deteriorate in taste, and aroma quality, as well as appearance in weeks in hot, humid seaports and in 6 to 8 months in drier, cooler areas.

*Courtesy American Can Company*

FIG. 9.21. MANUAL UNDERWATER LEAK TEST OF FILLED, EVACUATED, AND SEALED CAN OF COFFEE

# Part IV

# Instant Coffee Technology

# 10

# Percolation: Theory and Practice

## HISTORY

Roasted coffee beans were not used by the U.S. Army until 1832, when President Jackson ordered coffee to be substituted for rum. Before 1861 coffee was not considered an essential food item, by military or civilians. The soldiers that learned to keep themselves awake with poor-tasting coffee during the Civil War became regular coffee users after the war.

The soldier's coffee ration was often and effectively used for barter. Indeed, instant coffee, because of its stimulation and ease in use in such times, was a very negotiable commodity.

In England before 1900 liquid coffee extracts or "essences" which were made by batch extraction of solubles from the roast and ground coffee, followed by addition of sugar and vacuum evaporation, were in common use. In the coffee growing countries of Colombia and Mexico, strong liquid coffee extracts (without sugar or evaporation), were and are still commonly used.

However, it was the dried coffee solubles that had flavor stability during storage which were the first instants to be commercialized. In 1865, L. D. Gale received a U.S. Patent for wetting R&G coffee and squeezing out a concentrated coffee solubles extract; the extract was mixed with sugar and resulted in solid cakes. Gale Borden, commercializer of evaporated milk, also developed an instant coffee.

Although dry soluble coffees were made commercially in the U.S.A. about 1900, smaller, less commercial quantities of soluble coffees were made 10 to 20 years earlier. A Japanese chemist, Dr. S. Kato, in Chicago 1903, received a U.S. Patent for one such process and commercialized the sale of soluble coffee powder. This powder was first sold at the 1901 Pan American Exposition.

Shortly later, in 1906, G. Washington observed soluble coffee powder deposits on the lip of a silver serving pot while he was living in Guatemala. Water evaporates readily in the sunny 5,000 foot above sea level rarified air atmosphere. The soluble coffee powder deposits from the beverage made of high grade Guatemalan coffee, deposited slowly by low temperature evaporation of the water, tasted very good on reconstitution. This was the start and motive for a commercial manufacture and business. G. Washington set up this first processing plant at Bush Terminal, Brooklyn, but in 1927 he moved the manufacturing to Morris Plains, New Jersey, where it remained until about 1960.

## World War I

It was World War I that gave soluble coffee manufacture and use its impetus. The U.S. War Department called instant coffee one of the most important articles of subsistence. During World War I the U.S. Army bought all available instant coffee; production was increased 30-fold from 1400 to 43,000 lbs instant per day. This occurred primarily at the G. Washington processing plant. After World War I, G. Washington soluble coffee maintained commercial civilian sales. Then the Baker Importing Co. of Minneapolis, Minnesota, which sold "Barrington Hall" brand, also wholesaled instant coffee in aluminum tubes of 1 or 2 cup portions. These instant coffees were relatively expensive, especially during the depression years of the 1930's.

In those days G. Washington used prime quality coffee beans and low extraction temperatures, and took low solubles yields (of only about 20% from R&G coffee). For this reason, the instant product was "tacky" when exposed to moisture. It was dark in color, but tasted fairly good.

It was the late 1930's when Nestle commercialized instant coffee. Nestle took higher solubles yields than G. Washington at higher extraction temperatures. Both firms used a battery of percolator columns. Nescafe mixed its extracted coffee solubles with corn sugars, on a 50/50 by weight solids basis, before spray drying. The addition of these simple sugars had the major benefit of producing a free flowing, light colored powder. This powder did not readily absorb moisture from the air to become tacky.

The term "instant" was an advertising slogan developed in the U.S.A. shortly before World War II but not fully adopted until 1951. "Soluble" coffee was the terminology used prior to that time and soluble is still the more common term in the coffee industry.

## World War II

Nestle produced 25 million pounds of soluble coffee product for the U.S. Army during World War II. No instant coffee was sold on the civilian market from July, 1944 to June, 1945. During the war period new soluble coffee producers like General Foods, Standard Brands, and Bordens were asked by the U.S. Army to produce additional soluble coffee. However, no processing "know-how" was transferred from Nescafe or G. Washington to the newcomers. Some of the newcomers made terrible tasting instant coffee during World War II. Nestle's pre-war soluble coffee product (50/50) sales were estimated to be 6 million pounds. Nescafe's production was initially at Sunbury, Ohio. The second Nescafe plant was installed at Granite City, Illinois.

## After World War II

After World War II Nescafe built a large processing plant in Freehold, New Jersey. Nescafe sued various companies claiming that they used its percolation patents. However, a 1949 suit was found to be without merit. Immediately thereafter competitors installed batteries of percolators and processed therefrom commercially.

Percolators are a battery of 5 to 7 vessels with an elaborate piping manifold; they range in diameter from 12 in. to 30 in. and in heights from 6 to 20 feet. Pressurized hot water enters the first column, and concentrated coffee extract (25-20 percent solubles) leaves from the last column. Water flows through a series of columns. This gives tons of coffee grounds (100 foot length for example) about a 4 hour hot water rinse.

When one company started percolation in 1950, they found that they could take a 36 percent or more solubles yield from R&G coffee, and make a free flowing, spray dried powder without adding carbohydrates. This became the first free flowing, 100% pure instant coffee powder sold in the U.S.

In this chapter the technology of percolation will be explored. This will be followed in Chapter 11 by discussions of spray drying; in Chapter 12, aromatizing; and in Chapter 13, freeze drying.

# PERCOLATION: BASIC PRINCIPLES

A theoretical basis is needed to integrate rationally the various complications which in the past have been handled by trial and error. In defining the percolation process and its dependent and independent variables, it is to be noted that (1) an equilibrium must take place between wetted, saturated particles and free extract, (2) additional solubles come from solubilization by hydrolysis of normally insoluble coffee carbohydrates by chemical reaction, dependent only on time and temperature. To accomplish this, water must enter the coffee particle before extract can form and solubles diffuse out. The dry roast coffee particles selectively absorb water into their fibrous structure acting as a desiccant, thereby increasing the concentration of the passing extract.

Percolation consists of three distinctly different processes. Figure 10.1 shows a solubles concentration profile obtained from sampling extraction from a single percolator column over the whole residence time of the coffee grounds.

The process is explained as follows:

(1) Wetting. Gas is evolved from the coffee particles and interstitial voids as

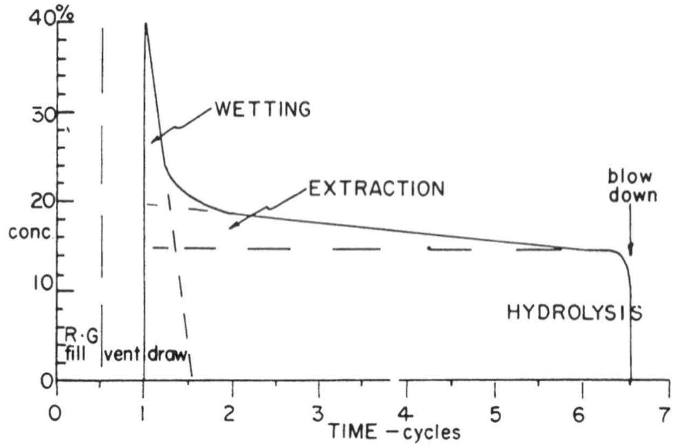


FIG. 10.1. PERCOLATION CONCENTRATION PROFILE—THREE-STEP PROCESS

they are wetted with hot extract. Desiccation by the coffee of the rising extract raises its solubles concentration. Selective water and solubles absorption by the coffee particles prepares the particles for solubles extraction. Similar phenomena may be demonstrated with sucrose solutions. Taller coffee beds and finer grinds accent the increased concentrations attained at the beginning of solution drawoff. Faster flows also give higher initial solubles concentrations, at the expense of attaining less than the full selective water absorption at this point. Finer coffee particles retain slightly more selectively absorbed water than do coarser particles.

(2) Extraction of coffee solubles occurs rapidly after the particles are wetted with selective water absorption and absorbed solubles. The solubles content of the coffee particle is higher at this time than ever before or after. A ternary diagram for solubles removal from a solid in a solvent medium has been used for decades to calculate batch stage equilibria. Figure 10.2 shows such a diagram. The factors controlling extraction and the use of the ternary diagram in such stage calculations are referred to at the end of this chapter.

(3) Hydrolysis in coffee grounds is mostly the breakdown of large molecules of water insoluble carbohydrates into smaller molecules which are water soluble. These are mostly reducing sugars, but there are also larger sugar molecules. Proteins also undergo hydrolysis under these conditions.

Hydrolysis is a basic process in sucrose inversion, starch, and dextrin solubilization to corn syrup. During World War II, wood was hydrolyzed with acid in a countercurrent battery of extractors in Germany to produce reducing sugars which were in turn fermented to alcohol.

Since percolation is a sequencing of three different processes, there is no classical mathematical treatment for percolation. However, calculations based on

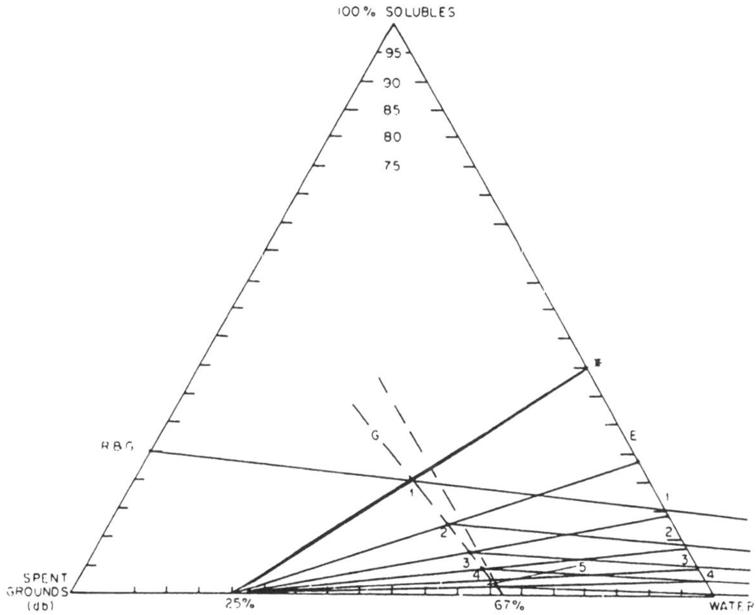


FIG. 10.2. TERNARY EQUILIBRIUM SOLUBLES DIAGRAM FOR ROAST COFFEE IN WATER

the three-step process may be projected for illustrating the importance of each process step and each operating variable in a material and heat balance over the percolator system. This will be done after more background is presented in regard to the independent and dependent operating variables.

## INDEPENDENT AND DEPENDENT PROCESSING VARIABLES OF PERCOLATION

The independent percolation variables are those influences which can be imposed on the percolation system within a broad range, namely:

1. **Geometry of percolation equipment.**
   This means the number of columns, size of columns, heights and diameters.
2. Green coffee blend, roast, grind, and water quality (materials in process).
3. Processing techniques: (a) Mode of R&G fill into column, (b) rate of percolation (cycle times), (c) ratio of water to R&G coffee used (material balance), and (d) mode of heat input (amounts and place).

With these elements under control, certain results are desired in the issuing extracts and spent grounds. These product results are the dependent variables since they depend on the controllable ones.

The dependent variables or results are:

1. Taste quality of coffee extract.
2. Yield of solubles from R&G coffee.
3. Concentration of solubles in the drawn off extract.
4. Temperature profile across percolator system (from material and heat balance).
5. Resistance to flow in the R&G coffee bed.

It is important that all these variables be understood in order to know which ones may be changed at will and which may be changed only indirectly. The processing results are dependent on a proper choice of independently set conditions. The following factors remain to be recognized:

1. The practical limits of the variables.
2. The direction of influence of each independent variable.
3. The degree of change in the independent variable needed to effect a desired degree of change in the resulting variable.
4. The degree of equivalence of different independent variables in achieving a given degree of change in a resulting variable.

The fact that there are three processes—wetting, extraction, and hydrolysis—occurring in the percolators simply means the influence of each imposed variable must be regarded in respect to its influence on each of the three process steps. Then the optimizing of each imposed variable so as to achieve the best results from the overlapping processes may be visualized, tested, and confirmed in practice. The value of this explanation is that it lends harmony to a multitude of percolation processing factors that otherwise have a loose and often incomprehensible relationship. With an understanding of these relationships it is possible to design and operate the percolator system with confidence, efficiency, and most of all to produce a far better natural coffee flavored extract. The percolation operating limits are not only competitive with other processor's product flavor and price, but are absolutely limited by the natural laws which govern the physical and chemical properties of coffee and the process. A knowledge of the properties of the substances in process is essential to a study of the theory and practice of processing.

Table 10.1 summarizes the cross influences of independent variables on wetting, extraction, and hydrolysis. Table 10.2 summarizes the cross influences of the dependent variables from each percolation step of wetting, extraction, and hydrolysis. Table 10.3 relates the independent to dependent percolation variables. Each dash is represented by a paragraph in the text that details the relationships. Due to the numerous overlapping influences, certain phenomena are discussed and repeated several times in order to have the essential data complete in each category.

The efficiency of percolation is reflected in the final results of taste, solubles yield and concentration, and physical properties.

TABLE 10.1. THE THREE PERCOLATION PROCESSING STEPS VS INDEPENDENT PROCESS VARIABLES

| Independent Variable | Wetting | Extraction | Hydrolysis |
|---|---|---|---|
| Equipment | | | |
| Geometry | — | — | — |
| Materials | | | |
| Blend | — | — | — |
| Roast | — | — | — |
| Grind | — | — | — |
| Water | — | — | — |
| Process | | | |
| R&G Fill | — | — | — |
| R&G preheat or prewet | — | — | — |
| Rate percolation | — | — | — |
| Ratio: water/R&G | — | — | — |
| Mode heat application | — | — | — |

TABLE 10.2. THE THREE PERCOLATION PROCESSING STEPS VS RESULTS

| Results | Wetting | Extraction | Hydrolysis |
|---|---|---|---|
| Temperature profile | — | — | — |
| Concentration profile | — | — | — |
| Yield of solubles | — | — | — |
| Uniformity of cycles (frequency of high pressure drops) | — | — | — |
| Taste | — | — | — |

TABLE 10.3. INDEPENDENT PERCOLATION VARIABLES VS RESULTS

| | Independent Percolation Variables | | | | | | | | | | |
|---|---|---|---|---|---|---|---|---|---|---|---|
| | Equipment, Coffee Materials | | | | | Processing Techniques | | | | | |
| Results | Geometry | Blend | Roast | Grind | Water | R&G Fill | Coffee Heat | Wet | Rate Perc. | Ratio: Water/ R&G | Mode Heat Applic. |
| Temperature profile | — | — | — | — | — | — | — | — | — | — | — |
| Concentration profile | — | — | — | — | — | — | — | — | — | — | — |
| Yield solubles | — | — | — | — | — | — | — | — | — | — | — |
| Uniformity cycles due to freq. high press. drop. | — | — | — | — | — | — | — | — | — | — | — |
| Taste (tars, clarity, etc.) | — | — | — | — | — | — | — | — | — | — | — |

## Percolation Independent Variables

**Geometry of Percolators.**—*Number of Columns.*—A one-column percolation process is a batch process. A two-column system is little better. Solubles depletion is 100 percent for one column and about 75 percent for two columns, but very little flexibility is available for obtaining high extract concentrations and fast process time. The amount of solubles removed in the extract drawn off relative to the total solubles in the system must be a small fraction so that the process is not unbalanced. In a continuous roast coffee feed and extract drawoff, the incremental solubles depletion is small. Solubles depletion in practice will vary according to plant design and how the plant is operated. High solubles depletion on drawoff is desirable, because this is equivalent to a short coffee flavor residence time, but it must be consistent with high extract concentrations. Normally, in a multiple column system, solubles depletion on drawoff may be only one-third of the fresh column solubles, and one-half of the next to freshest column solubles. The last three percolator columns have practically only hydrolyzed solubles. The total solubles present in the system (solubles inventory) may be only two-thirds of what would be present if all columns had their original coffee fill. It may be more or less. Solubles inventory is governed by concentration profile which, in turn, is influenced by the number of percolator columns. There should be enough columns and enough solubles inventory, so that extract drawoff does not cause extreme variances in temperature and concentration profiles. The solubles drawn off from a system with a large number of columns is small; hence deviation from operating equilibriums is small. It would be too costly and awkward to build and operate too many columns. The length of the percolator system must be enough to effect wetting, extraction, and hydrolysis under a given productive rate; to that extent, overall length governs column length.

As one column is out of service one-half to one cycle, the length of the system is the sum of all columns minus one. The percolator column height chosen is based on ease of installation as well as ease in filling, emptying, and operating.

The column height should be at least five times the column diameter so that entrance and exit dilution effects are less important. Column height should not be as much as 20 ft, because this would exclude the use of finer grinds. Figure 10.3 shows the end-of-column dilution effects. A squatter column is filled more quickly with ground roast coffee than a tall column.

*Total Column Length.*—The choice of independent variables must further the attainment of the desired dependent variables of extract concentration and solubles yield. Fixing equipment design thereafter governs processing results. For example, the need for 25 percent solubles concentration and 44 percent solubles yield, with or without intercolumn heating of extract, and other specified conditions, imposes a minimum length for the percolation system based on experience.

A 20 ft overall length will not work, but a 60 or 80 ft length will. The physical properties of roast and ground coffee also influence the rate of solubles and heat exchange and, consequently, wetting, extraction, and hydrolysis result.

The overall length may be shortened with lower extract concentrations, lower solubles yields, finer grinds, longer process time, intercolumn heating of coffee extract, reduced channeling, and higher feed water temperatures and profiles, etc.

In a percolator system where heat enters only with feed water, and 20 to 30 min cycles are used at the 25 or 30 percent solubles concentrations and 44 percent solubles yields, the first 30 ft of a 60-ft system essentially effect particle wetting and extraction. The grounds are brought up to hydrolysis temperatures in about the next 15 ft and hydrolysis occurs in the last 8 ft. The use of intercolumn heaters may effectively reduce the twelve feet needed for natural heat exchange. Heat loss from small columns often necessitates the use of intercolumn extract heaters. For the same solubles productivity from two percolator systems, the longer flow path effects a better rinsing action, similar to reduced process time from a given system. Therefore concentration profile is steeper and solubles inventories are less. The length of columns and system chosen must not be so long for the conditions to be used that excessive pressure losses occur.

The number of columns required for commercial use must be at least four; normally more columns than seven are not justified.

*Column Diameter*.—This should be less than one-fifth the height to minimize end-of-column extract dilution effects. Naturally, ground coffee that falls into a column suffers segregation of coarse and fine particles which may be minimized by faster loading of the coffee under air suction. Coarser particles more loosely packed against the wall of the column offer less resistance to hot extract flow; hence, channeling occurs especially in the fresh coffee column. Figure 10.3 illustrates this point. The flow of extract through the center of the bed is retarded due to lower temperatures, higher concentrations, and less permeability. Uniformly-sized particles and finer grinds each reduce channeling. Channeling may cause 1 to 4 percent loss in extract concentration, e.g., 22–25 instead of 26 percent.

Less wall support for the coffee bed in large diameter columns causes more frequent occurrence of pressure loss; small diameter columns (less than 1 ft) give looser bed densities and hence, are more permeable to flow.

Larger diameter percolator columns lose less heat than smaller diameter columns relative to their heat content, because relative heat loss is inversely proportional to diameter. These heat losses are small but noticeable in altering temperature profiles, and consequently, percolation performance. A heat loss of 200 BTU per hr per sq ft is about 25 percent of the heat throughput on a 1-ft diameter column, 12½ percent on a 2-ft diameter column, and 8 percent on a 3-ft diameter column.

Blowing out spent coffee grounds from small diameter percolator columns may be more difficult, because the ground coffee forms rings (doughnuts) which are harder to dislodge and break up on the smaller than on larger diameter columns.

One larger diameter percolation system demands about one-third to one-half less capital investment and is more economical to operate than two smaller systems of the same total capacity.

**Process Materials**.—*Blends of Coffee*.—The controlling factors in choosing green coffees for preparing instant coffees differ from the choosing of green coffees for use in roast coffees for the following reasons:

1. Only a small fraction of the natural coffee flavor in brewed coffee appears in the cup of instant coffee.
2. Solubilization of normally insoluble carbohydrates and proteins and the higher temperatures, higher extract concentrations, and longer process times alter the coffee flavor and composition to give a different coffee beverage.
3. Drying percolated coffee extract drives off most of the volatile natural coffee aromas and flavors, ordinarily enjoyed in brewed coffee, but leaves noticeable aromas and flavors of hydrolysate products. Freeze dried instants more closely approximate the blends used.
4. Hydrolysis forms more and different acids than the natural acids lost in processing.

Thus, the green coffee blends for instant coffees are not as yet as critical as green coffee blends for roast coffee brewing. However, blend differences are noticeable in flavor among different instant coffees produced by the same process, and the instant coffees carry characterizing notes of flavor from the coffee of origin. Instant coffees, prepared from high and low quality green coffee blends, taste far less different than corresponding cups brewed from corresponding roast coffees. Hence, many instant coffees are prepared and sold from relatively low priced green coffees. An advantage of the soluble coffee processing is that it is able to use lower grade green coffees and it frequently makes an acceptably flavored instant coffee. Better processing methods that carry through more natural coffee flavor are more suitable for processing the better grades of green coffee.

Since Brazil has manufactured much soluble coffee, which has been exported to the U.S. and Europe, it has become a practice to "neutralize" the Brazilian product's taste and relegate it to a "filler" role for consuming country blends.

Where there is significant natural coffee flavor carry-through to the instant coffee, the resulting instant coffees are preferred in the same order as brewed coffees from roast beans, that is, high, then low grown mild Arabicas, Brazilian, and, finally, Robusta coffees. By processing a single type of green coffee to instant coffee, the fullest appreciation of the contribution of each type of green coffee is obtained.

When brewing coffee in the United States, the standard of 50 cups of brew per pound of roast coffee is recommended by the Coffee Brewing Institute. But instant coffee yields 100 cups from the same pound of roast coffee. The higher coffee flavor in the instant coffee cup must come from the process as well as from heavy bodied coffees.

Since different types of green coffees roast and percolate differently, they also give disproportionate results when processed as a blend, and these disproportionate results influence extract quality, atomizing properties, powder color, solubility and fluidity, size of particles, powder bulk densities, and cup cleanliness.

*Robusta.* This is a botanical variety of coffee bean commonly cultivated in Africa. Robusta coffees vary in flavor but all are less acceptable than Brazilian and mild coffees in the world market. Hence, they are available at a lower price. Robusta coffees have the advantage of easily yielding higher solubles yields. An associated influence on percolation is higher extract concentrations and lower processing temperatures. Because they are naturally more dense, Robusta coffees give higher roast coffee column loadings. Some undesirable processing results are: more tars, softening of the grounds with greater difficulties in obtaining complete spent grounds blowdown, and more frequent excessive pressure drops. Tarry rings on instant coffee cups are often due to Robusta coffees especially at higher solubles yields. Ambriz Robusta coffees from Portuguese West Africa-Angola may contribute a noticeable, persistent foam in the cup of instant coffee. Robustas have 2 percent caffeine compared to 1 percent for other types of coffee. About 15 percent Angola Robustas can be substituted for Brazilian or low quality mild coffees in a blend without causing a very noticeable taste change. However, the roast has to be darkened to retain cup flavor, since Robustas are more difficult to roast than Brazilian coffees. The percolation temperatures may have to be increased a few degrees to maintain temperature profile with the slightly higher column loadings. In another case, a Robusta addition may give higher extract concentrations and yields at the same temperature profile. Robusta coffees are less acid than Brazilian and mild coffees and hence, influence taste accordingly. Due to their higher carbohydrate yields, Robustas tend to foam in the extract and cup as well as cause spray drying to light densities. When Robustas are blended with physically harder mild beans, percolation yields disproportionately higher solubles.

*Brazilian Coffee.* These beans behave similarly to Robustas in physical strength, but do not give up as much solubles or tars. Soft Brazilian beans, especially at dark roasts, contribute to excessive pressure drops. They usually roast darker in a blend with mild coffee beans and Robustas. Brazilian roast coffees form more fines on grinding and this contributes to slight cup rings and fine sediment. Brazilian instant coffees are preferred over Robustas, but are inferior to mild instant coffees.

*Arabica Mild Coffee.* With this coffee variety, the types grown at the highest altitudes are usually better flavored and command the best prices. The highest grown beans are physically hard and take higher temperatures or more time for hydrolysis than milds from low altitudes. The bean strength is related to pressure drop. Low grown mild coffees behave more like Brazilian coffees in the percolator. Mild coffees have more body and flavor which carry through into the instant coffee cup. Instant coffees prepared from mild coffees are easily distinguishable and preferred over Brazilian coffees. High grown mild coffees like Medillin from Colombia or Coatepec from Mexico are preferred to the lower altitude mild coffee from Central America, for example.

*Coffee Quality Differences*.—Blend changes and percolation influences may usually be noted when 15 to 20 percent of the coffee blend has been significantly changed. This may be evidenced in higher or lower extract concentrations, a different solubles yield, more frequent excessive pressure drops or foaming.

The substitution of about a 20 percent new-crop green coffee of the same general type as the old-crop being used may cause excessive pressure drops or reblows. A similar effect will be experienced when substituting darker roast coffees, imperfect coffees, softer bean coffees, aged coffees, pre-treated or pre-extracted coffees (decaffeinated), and so on.

*Roast*.—The limits of roasting are that the bean flavor development must make a palatable cup; the roast development must not be too light (underdeveloped) nor burnt (overdeveloped). For uniform flavor and degree of roast the resulting ground coffee must be continually examined for color and cup flavor both as R&G coffee and instant powder.

Light roasts yield denser beans and, therefore, higher percolator column loadings. Dark roasts yield more swollen beans, hence lower percolator column loadings and lower extract concentrations. Dark roast beans have had more cells damaged and destroyed; hence, the residue structure is weaker and more likely to soften under hydrolysis. Darker roasts also produce more fines which may contribute to progressive blocking off of flow paths with higher pressure drops. Fines mostly form in grinding, but also occur in conveying. Darker roasts have a higher content of fatty acids. The broken cell structure allows oils and fatty acids to come to the particle surface which, in turn, results in higher fatty acid content of extracts. Fatty acids, like soapy films, reduce surface tension during spray drying, causing finer particles and heavier densities of bulk powder. The carry over of oil and fatty acids also helps carry over coffee flavor volatiles dissolved therein.

Freshly roasted coffee has no free moisture and, in fact, is hygroscopic when exposed to air. When coffee is kept in an air-tight jar, moisture cannot continue to enter as it is absorbed, as occurs in paper bags or an unsealed vacuum can. Absorption of moisture by roasted coffee left in a warm humid atmosphere may reach about 5 percent in a few days. The absorption of moisture is associated with a liberation of the coffee flavor volatiles and $CO_2$ gas. The loss of flavor volatiles with the entrance of oxygen results in a staling that makes a very unpalatable cup of coffee. Water quenching of beans leaving the roaster (so that several percent moisture is retained within the bean) has the same effect as slow moisture absorption, except that staling is more rapid and gives an undesirable pyridine-like flavor to the R&G coffee entering the percolators. Quenching reduces fines on grinding, but it also causes flavor loss when used at the 5 to 7 percent level.

The swollen broken cell structure of darker roast beans, which have less $CO_2$ content, allows for faster wetting and solubles extraction. This is due to the reduced gassing and greater permeability of the extract into the cells.

Dark roasts often upgrade gamey flavored natural and other off-flavored coffees by driving out the major offensively flavored volatiles. The dark roast technique is not effective with Rioy medicinal flavored coffees. Their flavor may be detected in the cup at 5 percent or less in clean tasting blends. Dark roasts drive out the volatile acids and are less acid tasting than lighter roasts of identical blends. For example, a brewed cup from R&G coffee may vary from 5.1 pH for medium roast to 5.3 pH for dark roast; Robustas are generally in a higher pH range—about pH 5.4 for light roasts and pH 5.7 for dark roasts.

Darker roasts often do not yield as much solubles by 1 or 2 percent as lighter roasts. For example, a fully developed roast may give a 44 percent yield of solubles from R&G coffee, whereas the dark roast may yield only 40 percent solubles. Lighter roasts give higher solubles yields at lower temperature profiles, and result in a cleaner cup of instant coffee with less pressure drop in percolation. Degree of roast may cause a 5 to 10 percent difference in R&G coffee column loadings. For example, a bulk density change of from 18 to 20 lb per cu ft may occur in making a lighter roast which is equivalent to a column loading increase of from 1,800 to 2,000 lb R&G coffee.

The percolation of dark roasts with greater interstitial void results in higher temperature profiles due to the greater amounts of hot extract that move through the coffee beds. This also gives extracts of lower concentration.

Darker roast coffee flavors carry through the soluble coffee process better than lighter roast flavors. A light roast gives an instant or brewed cup that is weak flavored and nutty.

*Grind.*—The degree of fineness to which a roast coffee bean is reduced may be discussed in terms of fine grind, coarse grind, and uniformity of particle size, insofar as they influence percolation. Then there is the mode of grinding and/or screening, which may influence grind uniformity as well as the disposition of fines.

*Fine Grind.* Finer coffee grinds due to the large reduction in the absolute size of the particle structure are intrinsically weaker and less resistant to compression under percolation flow conditions. Fines are also less resilient after compressive forces have been released. Hence, fines are more easily compacted, and therefore cause progressively higher pressure differentials and ultimately shut off flow for all practical purposes.

The relation between cell and coffee particle size is 20 microns (mu) for the cell and 800 mu for a 20 mesh particle. A mm is 1,000 mu. Thus, there are about 30 to 40 cells per edge on a 20 mesh R&G coffee particle.

The bed resistance to flow is small in short beds. For example, in an espresso machine, R&G coffee that is almost pulverized may be used (80 mesh is 34 mu or about 6 cells wide), but the depth of the coffee bed is only about ¼ to ½ in. The extraction time is a few seconds. The solubles yield compared to instant coffees is relatively low since no hydrolysis occurs. The greater surface per unit weight for finer particles causes more flow resistance. The finer grinds give higher bulk densities. Since particles are more closely compacted there is less flow area. Finer particles swell more than larger particles. Smaller particles absorb more bound water, hence effect greater increases in extract concentrations. All these factors contribute to greater flow pressure drops and coffee bed compression.

The wetting and extraction of fine particles are rapid: (1) little $CO_2$ gas is left (espresso brewed coffees are ground immediately before use); (2) there is a large surface area; (3) there is only a short distance to wet into the center of the particle; and (4) there is the same short distance for solubles to diffuse out of the coffee particle. The colloidal and large molecules that are not able to diffuse through the cell structure of larger particles can be swept off the surface of finely ground coffee particles. Fine particles increase solubles yields without as rigorous a hydrolysis, and solubilization is almost immediate. Hence, temperature of the water is a less important but still an effective process variable. Similarly, the concentration of the contacting extract is less important in effecting solubles diffusion out of finer grinds. In percolation, finer grinds move the solubles concentration profile toward the fresh coffee end, and lower temperature profiles are required to achieve the same solubles yield. Consequently, less heat damage to the coffee flavor results since shorter process times are required. The resulting extract coffee flavor is more closely related to the brewed cup from the same R&G coffee than from coarser grinds. The finer particle bed with less flow area filters out hydrolysis tars.

Due to the greater surface exposure of finer grind coffee, a greater release of oils and fatty acids result (at any given roast), which reduces extract surface tension (as with darker roasts), carries through more coffee flavor, and yields more of the denser and finer spray dried powders. This technique is often advantageous in controlling powder densities; for example, coarse coffee grinds produce bulky powder densities. Fatty acids from finer grinds thus help reduce foam in the cup which occurs for various reasons.

Finer grinds with higher column loadings of the coffee cause higher extract concentrations at the fore part of the percolation system with reduced temperatures and increased viscosities.

*Coarse Grind.* Whole coffee beans may be extracted, but the resulting flavor is not as normal as with ground coffee. Cracked beans, for example, 65 percent on 8 mesh, expose enough internal surfaces and shorten diffusion distances so that extraction without pressure drop difficulties is effected due to the greater structural strength of the larger R&G coffee particles. However, such coarse grinds allow fine particles to migrate and these may sometimes cause plugging of flow passages by the migration of fines and their lodgment at a point, thus reducing the effective flow path. Fine grinds lock the finer particles in their local environment and hence, fines do not drift to plug up normal flow paths. Halving the average particle size halves the penetration distance and doubles the surface, with the end result that extraction time is at least halved. Since percolation is a three-part consecutive process, finer particles hasten wetting rates but do not hasten hydrolysis.

For a given processing condition where increasing flow rates and therefore, shorter process times, cause a higher frequency of excessive pressure drops, coarsening the grind may eliminate the excessive pressure drops. Control screen analyses on a routine basis, as well as operator attention to grind analyses changes, help to maintain control over percolation conditions and to maintain uniform operational cycles. For example, unauthorized or misguided changes in the grinder roll spacings may cause an abrupt reduction in the coarse fraction with consequent increase in column loading weights and greater resistance to extract flow.

Grinds used in large percolator columns that are more than 2 ft in diameter and 12 ft high usually have more than 30 percent of the ground coffee on an eight mesh screen. Smaller diameter and shorter columns may use progressively finer grinds down to a few percent on an eight mesh screen, at about 30-min cycle times.

Grinds with less than 10 percent on eight mesh show in their percolator concentration profiles a more gradual transition from extraction to hydrolysis than coarser grinds. The extraction ends earlier and hydrolysis starts earlier with finer grinds. Fine grinds, at times, cause a dirty cup ring. Less hydrolysis flavors come through with percolation of finer grinds. The transition from extraction to hydrolysis is the section in the concentration profile that shows no concentration rise.

*Solubles Yields.*—Coarse grinds yield less oil, carbohydrates, and proteins in hot water extraction, and less solubles at 212 F (100 C). Pulverized coffee with 212 F water extraction will yield over 30 percent solubles, whereas coffee particles larger than 1 mm (20 mesh) will produce less than 23 percent and as little as 15 percent solubles. Finely ground coffee of 0.1 mm particle size has 2 or 3 cells per edge, and Soxhlet extraction at 212 F overnight will yield over 30 percent solubles. Soxhlet water extraction of defatted, finely ground coffee shows increased solubles yields of from 30 to 33 percent solubles. The fat prevents portions of the carbohydrates from going into water solution.

*Uniformity*.—The uniformity of particle sizes coming from the grinder will influence channelling of extract flow in the fresh R&G percolator column. The less uniform grinds tend to segregate more during slow filling of columns under gravity fall. Faster rates of grinding produce more fines and hence, a less uniform grind. Extracts from finer grinds taste somewhat bitter, while those from coarse or whole bean grinds taste rather flat.

Darker roasts yield more fines, which can cause high resistance to flow. These fines may appear in the extract and the final powder. Extract tankage capacity for adequate hold time will allow these fines to settle out.

Screening of the fine particles is done occasionally in times of excessive pressure drops, but normally is not necessary. The problem then arises as to what to do with the fines that are separated out. Usually these may be extracted separately or pelletized and returned to the R&G coffee entering the percolator. Neither procedure is satisfactory, and such schemes are usually temporary. These fines usually carry chaff with them in the separation and their flavor is inferior compared to the granular coffee from which they came.

*Pelletizing*.—High coffee bulk densities may be attained with fine grinds, such as 100 percent through 8 mesh (2.0 mm) and 85 percent through 30 mesh (0.6 mm). The less than 30 mesh particles will be pelletized and mixed into the plus 30 mesh fraction. Pelletized fines will break apart after they become wet and swell. But due to the fineness of the rest of the grind on 30 mesh matrix, the released fines cannot migrate far.

The increased density of the on 30 mesh is about 20 percent higher than a grind of 30 percent plus 8 mesh; that is, 22 instead of 18 lb per cu ft. With about one-sixth pelletized fines by weight, the overall coffee bulk density is about 26 lb per cu ft. The increased coffee loading per percolator column and the greater particle surface area produce a few percent increase in drawn off extract concentration and solubles yield.

Sprout-Waldron and other firms make pellet mills. The pelletizing forces may be varied according to the chamber fill of coffee. That is, more coffee will be compressed more. A water or instant coffee binder may be desirable. Absolute pellet densities may be 60 lb per cu ft. Pellet size depends on mold size, e.g., ¼-in. diameter × ⅝ in.

*Water*.—Acid and demineralized water accelerate hydrolysis and soluble yields. Alkaline waters retard percolation rates and hence, the extraction method in drip preparation of brewed coffee. But the water to coffee ratio is about fifteen to one. Alkaline waters at this high ratio to coffee influence pH noticeably, and produce a blander tasting cup and a darker cup color. However, in commercial percolation, water to coffee ratio is less than 4 to 1, and the spent grounds coffee column has a pH of about 3.4. Since higher acidities make for more rapid hydrolysis, this small neutralization may slow down the hydrolysis rate slightly.

More acid conditions allow hydrolysis to occur at an acceptable rate at lower temperatures. This is desirable since fewer side chemical reactions and products will occur. Some soluble coffee plants use neutral, (7.0 pH) demineralized water, and others add a little organic or mineral acid to the feed water. Acids may be neutralized readily by ammonia.

The main precaution necessary for the entering water is to rid it of scale-forming calcium and magnesium salts. This is usually done by softening and substituting sodium cations to protect the steam heater from losing heat transmission capacity.

Less than one-fifth of the entering water leaves with the extract. The water has passed through more than 60 ft of ground coffee, which is a strong adsorber itself. The quality of the extract leaving the percolators is relatively unaffected by the quality of the water entering. During World War II boiler water with tens of thousands of parts per million of solubles was used directly as feed water to the percolators, but the issuing extract was not noticeably affected.

**R&G Coffee Fill of Percolator Column**.—Particle segregation occurs in coffee grounds movement, storage, and unloading. There are numerous factors that affect the proper loading of a coffee column in preparation for extraction. Some are percolator wall temperature, rate of coffee fill into the percolator column, size and shape of ground coffee bins, uniformity of grinds, valve size on percolator, etc.

*Particle Segregation*.—Coarse particles in a free fall tend to roll to the edge of a bin. Fines and chaff settle later in the corners of square bins and at the edges of round bins. This phenomenon occurs in all the bins in which the ground coffee is handled, as well as in the percolator column under gravity fall of ground coffee. Faster filling of a bin or percolator column traps particles as they fall, so segregation is minimized. Fast filling of the column is attained by sucking the ground coffee in with air and removing the air from inside the column. The fast fill has the advantage of allowing higher coffee loadings in the column by about 10 percent over gravity fill. Higher coffee column loadings mean higher productivity from the percolator system and a few percent higher extract concentrations. Denser column loadings reduce channeling of extract flow; this is more important in larger diameter columns. Uniformly fine grinds also cause less particle segregation and create higher density of fills.

*Cooling Percolator Wall*.—If fresh coffee grounds are allowed to rest against the wall of a column which has just been blow down, their temperature is 330 F (166 C) or higher. Fresh coffee grounds stale and deteriorate from heat in minutes under these conditions. Hence, it is important to keep both the wall temperature and the residence time of the coffee to a minimum before wetting with extract. A tap water flush on the column wall will bring the temperature down below boiling.

*Slack Fill*.—This is not important in downflow, but in upflow the fines may be flushed out of the bed and become plugged at the top outlet. Fines are highly compressible. A slack coffee bed also results in extract mixing and channeling which reduces extraction efficiency and gives lower extract concentrations and lower solubles yields. An incomplete fresh coffee fill may also occur in the percolator column if spent grounds from the prior load are not completely discharged. Such a situation leads to disproportionate yields, and excessive heat treatment and hydrolysis of the residual grounds, which may make reblow hard. The values of the snug high density vacuum fill of dry ground coffee are that the coffee bed is rigid, a swelling of 7 percent after wetting further reduces void, and extract flow is through a uniformly supported bed and flow path. Prewetting of ground coffee before filling the column will reduce frequency of excessive pressure drop, but this type of slackness results in stale flavored coffee, looser beds with channeling and lower extract concentrations. A very fine grind has very little slack space and may be successfully percolated to give high extract concentrations.

*Bin Size and Discharge*.—Ground coffee leaving a bin to enter a percolator column draws coffee from the center core, not from uniform layers. If the bin size is such that all or most of the grounds run out, the tailings, which would then reside at the top of the percolator column, are high in chaff and fines. With extract upflow these fines would be compressed enough to seal off flow. By making the round ground coffee bin used over the percolator column about one-third larger in volume than the percolator, about 20 percent of the ground coffee including all the fines will be left in the bin after the percolator is filled. Any fines or chaff accumulation will be mixed in with the next column fill.

**R&G Coffee Pretreatment**.—Two treatments that have some beneficial effects on the ground coffee are preheating with steam and prewetting with water.

*Preheat*.—Heating the ground coffee in the fresh percolator column may be done in a few minutes with low pressure (20 psig) saturated steam. Heating grounds to near boiling temperatures raises drawn off extract temperatures and concentrations. Warmed grounds are wetted more easily with extract, have less gases, and are already swollen, forming a more uniform particulate bed for flow of extract. They also drive off coffee aroma and flavor volatiles for easy recovery. Faster solubles extraction from a preheated column of coffee helps keep solubles inventories down.

*Prewetting*.—Although prewetting may be done outside the percolator column, the practice is bad because it causes staling of ground coffee in less than an hour, accompanied by a heavy, undesirable flavor and a loss in natural coffee volatiles. Particle prewetting in the column allows more fresh coffee solubles to be released before hydrolysis products and tars push through. Prewet coffee yields greater coffee flavor, higher cup pH, and less hydrolysis acid taste. Column prewetting reduces extract concentrations 1 or 2 percent, but cup flavor and cup cleanliness are noticeably improved. Prewetting liberates more gases from the wetted coffee, and a frontal flow of lower extract concentration assists gas liberation in the rest of the coffee granules. This treatment also reduces channeling and contributes to lower frequencies of excessive pressure drop. Prewetting, however, does reduce temperatures in the fresh coffee column. The cooling effects may be best offset by intercolumn extract heating. With rich coffee flavor attained in the fore part of extract drawoff, prewetting techniques may be united with vacuum film evaporation of the back part of the extract drawoff so as to send a more concentrated extract to the spray drier. Prewetting allows more fatty acids and oils to be released, resulting in more flavorful powders.

When beaker quantities of roast and ground coffee, as used in commercial coffee percolators, are added to boiling water, the coffee particles become wet in 5 min with continued boiling, but wetting the particles with boiling extract of 15 to 20 percent concentration takes over an hour. The same situation exists within the percolator column.

In another test, extract of 15 to 20 percent solubles concentration is added to coffee particles containing ten and to another batch containing 20 percent moisture. When the particles and extracts were mixed and maintained at boiling temperatures, practically all the highly prewetted coffee particles had sunk in 10 min and their gas was completely evolved. Most of the lesser prewet coffee articles floated and gave off gas for over an hour.

**Water to Coffee Grounds Ratio**.—The ratio of water to coffee is mostly governed by the concentration of solubles in the withdrawn extract. It is also influenced by the density of coffee column loading, or the void space. Any low extract concentration may be drawn, but the numerous operating variables in percolation must be carefully adjusted to obtain the highest extract concentrations. Water flow at lower concentrations of extract is an independent variable; at highest extract concentrations, water flow to coffee flow ratio is a dependent variable. The more water and extract that pass over a coffee particle, the better the rinsing action. Temperature history, however, may be more important. If the concentration of the extract drawn off were not important, a high water throughout relative to R&G coffee throughput would broaden the limits of percolator operation. It would allow more heat and flushing action to be carried to the fore part of the system where R&G coffee with delicate flavor chemicals may be more easily wetted, extracted, and hydrolyzed. This method produces, at lower solubles concentrations, a better flavored coffee extract.

But low solubles extract concentrations may not be dried without evaporating large amounts of water per unit of solubles with the associated large loss in coffee flavor volatiles. Hence, the amount of associated water with the solubles must be kept to a minimum. Ordinarily, so much volatile coffee flavor is lost in spray drying (or other water evaporation and concentrating methods) at solubles concentrations below 25 percent that solubles concentrations over 25 percent are sought. Thirty-five percent is noticeably better than 30 percent in coffee flavor retention in the powder.

This, then, sets the limit to the amount of water that may be used in relation to R&G coffee, because water entering the percolator system leaves only with the extract drawn off or in the discharged spent R&G coffee.

Thus, the concentration of coffee solubles in the extract drawn off must be attained by thermal, grind size, or other methods from within the percolator system. Only one-fifth of the water entering the percolator system leaves with the extract drawn off; the other four-fifths leave with the discharged spent grounds. With such a water and heat flow pattern a considerable portion of the heat may not be properly distributed through the rest of the coffee columns; therefore, intercolumn heating is necessary to effect desired temperature variations. The ratio of feed water to coffee by weight is from 3/1 to 4/1.

**Modes of Heat Application**.—Temperatures across the percolator system (profiles) are the most important processing variable for wetting grounds, extracting solubles, and hydrolyzing solubles. Temperature is an intensive measurement that only reflects heat input and its distribution among water, solubles, and coffee grounds flowing in opposite directions. The distribution of heat is governed by: (1) rate of water and coffee movement and cycle time, (2) heat losses and additions, (3) entering and leaving heat content of each flow stream, and (4) heats of wetting, solution, and hydrolysis. It is difficult to generalize about temperature profiles, but Fig. 10.5 shows the temperature and concentration profiles that may result from feed water heating as compared to intercolumn extract heating. With feed water heating, the curvature of the temperature profile will be influenced by cycle time, system heat losses, and hence, size of columns. Factors such as steam preheat of coffee grounds, prewet, and column wall flush coolings add or remove heat from the percolator system.

*Feed Water Heat Input*.—Water has a heat capacity 2.5 times that of coffee particle or solubles. The water in the final extract usually has a ratio of 2 or 3 to 1 (water to solubles). The heat contribution of the feed water, therefore, in passing through the percolator system is an important heat input factor. Successful percolation at commercial yields and concentrations may be carried out with the heat content of the feed water being the only heat supply, at cycles of less than an hour, and with adequate system insulation. With heat supplied only through the feed water, control of temperature profile is limited. Depending on heat losses, column sizes, and cycle times, feed water temperatures of from 320 to 350 F (160 to 177 C) are used. Where high extract solubles concentrations are sought, less heat comes to the fore part of the percolation system which contributes to higher solubles inventories.

*Intercolumn Heating of Extract*.—This method introduces heat into the central part of the percolator system and allows more flexibility in control of temperature profile. Intercolumn heating of extract requires lower feed water temperatures which causes less rapid softening of coffee grounds. Fortunately, the net heat input to effect marked percolation processing changes is small. The heating is best done in the early stages where the extract is most dilute, least viscous, least flavor-holding and has the highest heat capacity, i.e., near the spent coffee grounds end of the system. Intercolumn extract heating allows similar performance to be obtained from large and small columns, and from fast and slow cycles. A 3 F (2 C) rise in intercolumn heater extract temperature at the feed water end of the system may noticeably increase solubles yield, and drawn off extract concentrations on the order of 1 or 2 percent. The influence of heat input closer to the fresh coffee column is more marked on drawn off extract concentrations. However, heating 30 percent extract concentrations may cause a caramelized flavor as well as an increased rate of heater tube scale deposition. If there are no weekly shutdowns of the plant, it may become important to substitute a clean intercolumn heater for a scaled one. Steam heating temperatures must be kept as low as possible. Accelerated reduction in rate of heat transfer often results in raising steam heating temperatures to continue effective heat transfer. Such a condition invariably leads to a badly fouled and ineffective heat exchanger. High velocities of dilute extract flow at the tube wall give the best heat transfer with minimal fouling rates. High velocities are attained by using suitably sized tubes in a multipass heat exchanger. Tars coalesced by heating also deposit on the inner walls of the piping and valves, all of which cause much expense to clean.

## PERCOLATION PROCESS STEPS

A detailed and illustrated discussion of the wetting of R & G coffee with extract, extracting the solubles and then hydrolyzing the normally insoluble coffee substances will be presented in that order. Thereafter, the factors influencing overall percolation rate will be discussed.

### Wetting Roast Coffee Particles

Extraction may not begin until coffee particle wetting is complete. Contacting a dry coffee particle with 20 percent coffee extract will not allow the extract to be readily absorbed into the particle. One reason for this is that the particle is made up of 30 to 40 mu cells filled with $CO_2$ and coffee flavor volatiles. Also, carbon dioxide gas is not very soluble in hot coffee extract. Air surrounds each particle and acts as a barrier to extract contact whether air is absorbed on the surface or forms a foam with the extract. The surfaces of a whole roast bean, due to a finer cell structure and somewhat different waxy composition, do not absorb extract as readily as the broken surfaces with exposed cell structure. Also, as hot extract rises through a bed of coffee particles, it gives its heat up to the particles. The cooled extract is then more viscous and wetting does not occur. Coffee particles of the size used in commercial percolation do not wet with extract at temperatures below 170 F at a reasonable rate for commercial processing. Channeling of extract flows causes bypassing of particles, occluding of gases, and reduction of rates of wetting—all of which produce lower extract concentrations. Freshly roasted and ground coffee gives off more gases and foam, especially when dry.

Coffee particles are wetted through their fibrous cell walls. Free extract will eventually occupy the cell cavities. Pressing drained coffee grounds saturated with extract squeezes extract out of the interstices of the particles and of the cell cavities. Release of pressure while the particles are in contact with free extract results in the free extract being reabsorbed into the cell cavities.

**Bound Water**.—Roast coffee, like most natural granular and roasted or dried matter, will hold a significant percentage of water when in equilibrium with air moisture. For example, grains commonly hold 10 to 15 percent water under ambient conditions; others hold more or less. Roast materials such as coffee, chicory, wheat, and sugars have lost all this bound water, and due to their nature they usually are able to hold higher moistures, perhaps 25 or 30 percent. In other words, roast coffee is a desiccant; it is very dry. Mixing the roast coffee particles with coffee extract results in a higher coffee extract concentration because free water is removed from the extract.

In the percolator column wetting process, this results in higher extract solubles concentrations on drawoff. The more complete the equilibration between the particles and the extract, the more effective is the desiccating. Equilibration is hastened by the use of finer particles and higher temperatures.

**Free Water**.—The water that does not selectively become bound to the coffee cell wall, and thereby become immobile, is free water. This is the water associated with the extract and it is free to move into and out of coffee particles.

**Equilibrium Wetting**.—This does not normally occur in practice; in fact, the coffee particles are often not fully wetted with extract until the second percolator column. Equilibrium wetting of coffee particles is much more than selective bound water absorption; it is the equilibration of the entering coffee solubles in the extract with the solubles in the particle. Table 10.4 itemizes the conditions and results quantitatively in equilibrium wetting.

TABLE 10.4. EQUILIBRIUM WETTING OF COFFEE PARTICLES WITH COFFEE EXTRACT

*First equilibrium stage (at base of fresh roast coffee column)*

Dry fresh coffee grounds are mixed with three times their weight of 20 percent coffee extract solubles to form almost half the weight of 28 percent solubles extract as wetted but drained coffee grounds.

| | | | | | | |
|---|---|---|---|---|---|---|
| | | | | 25 bound water | | |
| 75 dry grounds | | 240 water | | 125 free water | | 90 water |
| | + | | → | 75 dry grounds | + | |
| 25 solubles | | 60 solubles | | 50 solubles | | 35 solubles |
| 100 dry coffee | | 300 (20% extract) | | 275 wet coffee | | 125 (28% extract) |

Basis of calculation:
1. The coffee particle bed has a density of 100 gm per 300 $cm^3$
2. The interstitial void is 150 $cm^3$
3. The coffee is dry
4. After the particles swell in a confined volume, the interstitial void is 100 $cm^3$
5. Dry spent coffee grounds hold twice their weight of water
6. Bound water is one-third of dry spent grounds weight

Note that the concentration of the solubles in the free water in the particles is about 28 percent. The solubles content of the wetted grounds are now double that of the fresh coffee. The weight of more concentrated free extract is now only 0.4 of that at the start, and will rapidly diminish in the following equilibrium stages.

**Non-Equilibrium Wetting**.—Extract is able to rise faster in the fresh coffee column than bound water can be completely absorbed and faster than extract can permeate each particle. The result is a steep solubles concentration profile during extract drawoff, for example, from 40 to 20 percent solubles. A shallow or flat solubles concentration profile may be due to high solubles inventories and/or channeling. Table 10.5 shows a three step non-equilibrium wetting of particles with the resulting progressive increase in concentrations.

TABLE 10.5. NON-EQUILIBRIUM WETTING OF COFFEE PARTICLES WITH COFFEE EXTRACT

| | | | | | | | |
|---|---|---|---|---|---|---|---|
| Step 1. | 75 dry grounds | | 240 water | | 20 bound water | | 220 water |
| | | + | | → | 75 dry ground | + | |
| | 25 solubles | | 60 solubles | | 25 solubles | | 60 solubles |
| | 100 dry coffee | | 300 (20% extract) | | 120 wet coffee | | 280 (21.5% extract) |

The extract concentration has changed from 20 to 21.4 percent solubles.

Step 2. The next extract contact with 100 weight parts dry roast particles gives 60/260 or 23.1 percent solubles.

Step 3. The next contact step gives 60/240 or 25.0 percent solubles.

Step 4. And so forth.

Basis: Similar to Table 49 except that only the bound water effect is used.

Note that as less water is associated with the extract solubles, each increase in extract concentration is higher than the one before.

**Maximum Extract Concentration**.—Prewetting roast coffee for any purpose reduces the desiccating effect as well as the maximum possible extract concentrations attainable. Invariably the first extract that leaves the percolator column is higher in concentration than any other extract. When a low extract concentration appears first, it is a certain indication of channeling of extract flow. The slope of the extract concentration during drawoff reflects the performance of wetting inside the column. A relatively uniform extract concentration during the half cycle of drawoff indicates a high solubles inventory within the percolator system.

Maintaining extract drawoff temperatures close to boiling water temperature helps effect better wetting and solubles equilibrium. Intercolumn extract heaters at the fore point of the percolator system and an extract flow that is not too fast help achieve this purpose. The leading edge of the extract flow will always undergo cooling and become more viscous, and then mix with extracts that follow. Coffee extracts at 50 percent solubles concentration are sufficiently viscous so that their syrupy nature reduces the effectiveness of particle wetting. Hence coffee extract concentration higher than 50 percent is not normally produced. Although steam preheat raises the coffee particle temperature to effect better wetting, such heating usually adds about 5 or more percent moisture to the coffee. This moisture addition reduces the desiccating potential of the roast coffee.

**Wetting Comments.**—The heat of wetting is sufficiently exothermic that when room temperature particles and water are contacted for extraction, the extract and coffee bed become noticeably warm to the touch. There are some indications that when coffee solubles enter into the coffee particle, the large molecules and colloids do not diffuse in at the same rate nor to the same degree; hence, extract outside the coffee particles is richer in these types of water solubles.

## Extraction of Coffee Solubles

Solubles extraction or diffusion does not start until the coffee particle is satisfied with bound water, saturated with free extract, and free of gases. Then the influencing factors of grind size and uniformity, column loading, temperatures, and associated concentrations act to determine the rates of extraction, the completeness of extraction, and temperature and concentration profiles. Ordinarily the ternary diagram of water, coffee solubles and coffee grounds particle may be effectively used for presenting graphically the stage extractions of solubles. The ternary diagram and its use are presented here. The practical application to percolator extraction is small but the mechanics of its use are instructive in regard to pointing up influencing variables. It is difficult, if not impossible, to discuss one variable because in any situation, more than one variable is involved. The reader must, therefore, expect discussion of several variables under one topic.

**Concentration Profiles.**—To visualize where solubles extraction occurs in percolation and how it varies one must examine the concentration profiles. Figure 10.1, a schematic diagram, shows extraction tailing out by a gradual fall in

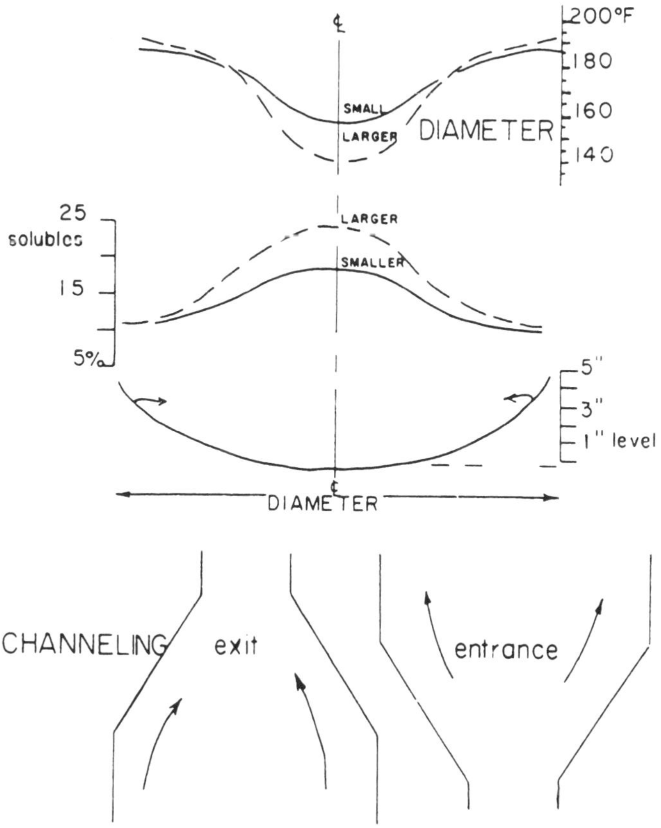


FIG. 10.3. CHANNELING OF EXTRACT FLOW IN PERCOLATOR

solubles concentration after wetting and up to hydrolysis. Figure 10.4 shows a more typical concentration profile where there is no solubles increase from cycles two through four. This means extraction has been relatively complete since cycle two. The high concentration profile in Fig. 10.5, with intercolumn heating of extract, is due to building up a solubles inventory for gaining higher extract concentrations on drawoff, and in fact represents a retardation effect in the solubles extraction rate. This is a good example of the concentration of the flowing extract being the controlling factor in the rate of solubles extraction from coffee particles. Figure 10.6 shows the effect of faster cycles in higher temperature profiles, and hence, improved extraction and wetting rates. Slower cycles increase solubles inventories. The increased solubles concentrations and inventory with finer grinds are due to increased column loadings and lower temperatures at the

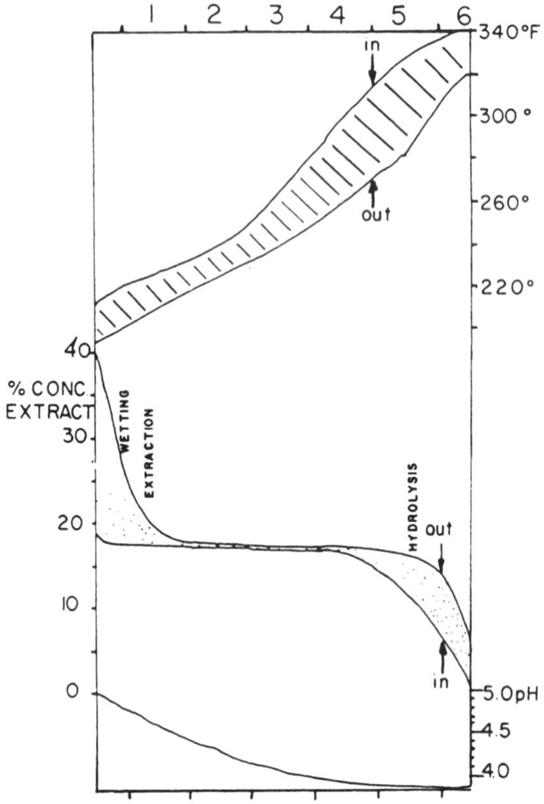


FIG. 10.4. PERCOLATOR COLUMN PROFILES: TEMPERATURE, CONCENTRATION, AND pH

front end where wetting and extraction are retarded. Also, the plateau separating wetting from hydrolysis disappears due to the tailing back of solubles. The fine grind concentration profile may be pushed forward by faster processing; hence, there are higher rates of water throughout.

*Factors Influencing Concentration Profiles*.—From the foregoing figures it may be seen that concentration profiles are the result of three concentration influences: (1) concentration of extracting media, (2) solubles concentration in the coffee particle, and (3) the difference in the above concentrations. The concentration of the extracting medium will depend on yields and effluent concentrations taken. The higher these are, the less driving force there is for solubles to move from particle to free extract.

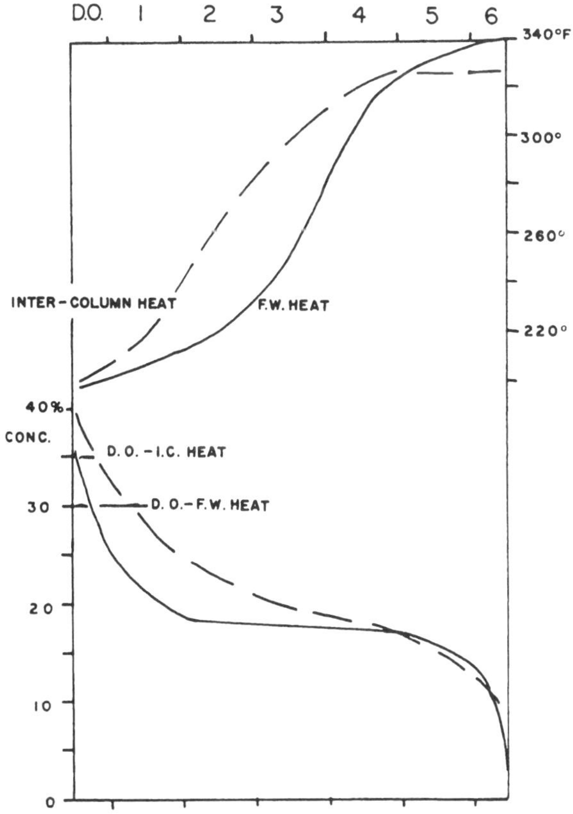


FIG. 10.5. COMPARISON OF PERCOLATOR PROFILES: INTERCOLUMN VS FEED WATER HEATING

Higher temperatures have a marked influence in reducing extract viscosity; hence, temperature is a controlling factor in the higher concentration extraction regions. The absolute temperature also increases diffusion rates of solubles and water, but this is a less important factor than viscosity. The major driving force for solubles movement is the difference in concentration inside and outside the coffee particle. A prewet coffee will not attain as high a solubles concentration inside the particle as a dry particle fully wetted with the same extract. A loosely packed column and channeling will reduce the sectional efficiency of extraction by subsequent extract dilutions. The driving force for solubles diffusion out of the particle diminishes as its solubles concentration relative to the passing stream is reduced.

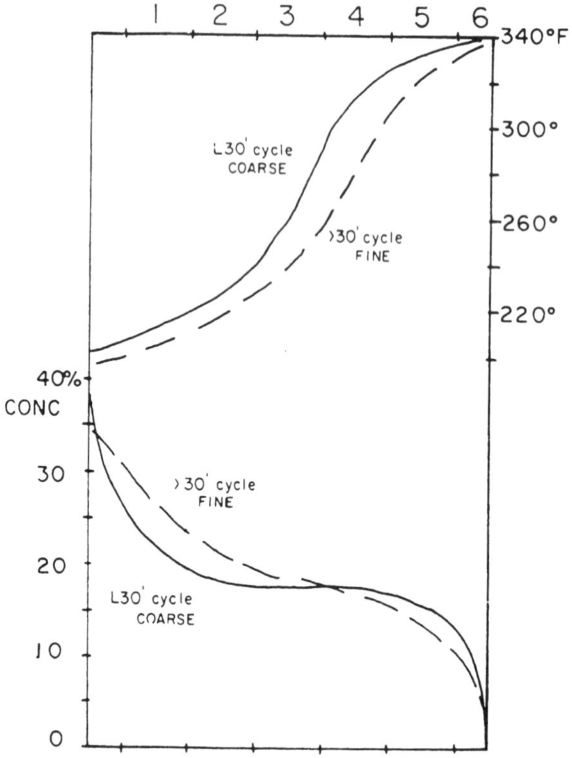


FIG. 10.6. COMPARISON OF PERCOLATOR PROFILES: COARSE VS FINE GRINDS; FAST VS SLOW PROCESS TIME

**Coffee Particle Size**.—The fact that extraction time is short on the percolation profiles is related to the fact that coffee grinds, even as used in commercial percolators, are sufficiently fine to effect solubles removal faster than other controlling factors such as wetting and high solubles inventories. From Table 8.2, Chapter 8, it may be seen that percolated coffee has over 30 $cm^2$ per gm of surface area, and the greatest diffusion distance for the solubles is less than 1 mm. This constitutes a labyrinth of less than 30 cells, and probably about 15 cells. At almost two square meters of particle surface per pound of roast coffee, the rate of solubles movement outward is controlled by the free extract solubles concentration.

**Rate of Solubles Extraction**.—The temperature and concentration profiles already show that extraction, even with 15 to 20 percent hydrolysis solubles, takes only one cycle for the most part, and this may be 30 to 40 min. Figure 10.7 shows a schematic profile of the solubles extraction rate with boiling water on one per-

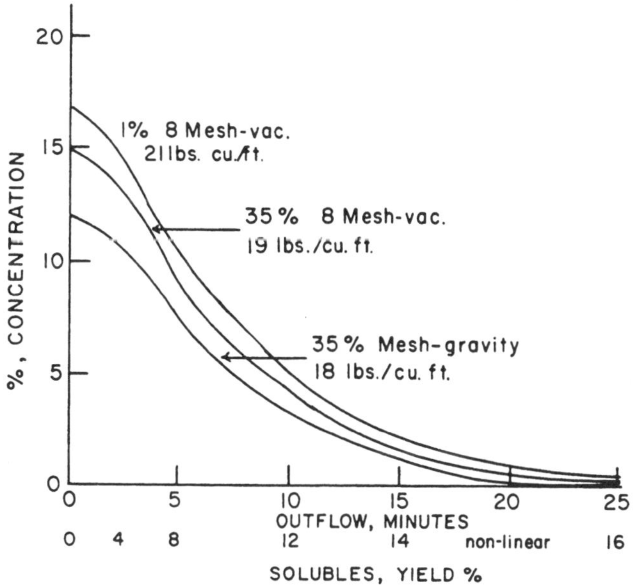


FIG. 10.7. MODE OF ROAST AND GROUND COFFEE FILL VS RATE OF SOLUBLES EFFLUENT

colator column of ground and roast coffee. Although somewhat better and faster extractions are obtained with finer grinds, and with vacuum loaded finer grinds than with gravity loaded columns, the general course of the extraction takes less than 20 min. More than half the solubles are extracted in 5 min, one quarter of the solubles in the next 10 min, and so forth. Although the above solubles yield is only about 15 percent, finer particles, about 20 mesh or 0.1 mm, give up 20 to 25 percent solubles in about 15 min with boiling water. The rate of solubilization from these particles of such surface area is so great that temperature is not controlling. Solubles yield does not vary much even if 150 F (66 C) water is used. Coarse particles of 2 mm size (8 mesh) are influenced by water temperature. Boiling water will accomplish a good solubles extraction in half an hour, whereas 120 F (49 C) water may take three-quarters of an hour. Cracked beans take about an hour to get the same solubles in boiling water, while the same yield may take 2 hr in 120 F (49 C) water. Hence, fine grinds are less sensitive to temperature influence. All solubles do not diffuse out of the coffee particles at the same rate, and this may be called a chromatographic separation of different types of chemical substances.

**Chromatographic Solubles Extraction.**—Rates of diffusion (distance traveled per unit time per unit area) have been measured for individual chemical substances in different media, and under different circumstances. Diffusion is a molecular phenomenon. Small compact molecules diffuse faster than large branched ones. Diffusion is also influenced by the composition of the solvent or extract. The first issuing extract from a coffee column tastes bitter and in no way represents the coffee flavor. Subsequent fractions taste differently. But not until all fractions are put together, and then tasted, do they represent a natural and wholesome coffee flavor. Small molecules that may contribute bitterness and are readily extracted are caffeine, trigonelline, chlorogenic acid, and free salts. Simple sugars, although not polar, are also quickly extracted with water. Larger molecules, like hydrolysis fragments, polymerized sugars and caramelized carbohydrates and proteins, which are water soluble are the last to diffuse out. Colloidal material does not appear in percolator extract concentrations over about 15 percent solubles because the electrolytic density (conductivity) attains a level to make the "brew colloids" of oily phosphoproteins irreversibly insoluble. This action is not sharp at 15 percent solubles, but it occurs progressively until hardly any colloids exist at this concentration. Colloid behavior is, no doubt, influenced to a lesser extent by the pH of the extract, the botanical type of the coffee, its blend, roast, and other factors.

Ordinarily, caffeine should be thoroughly extracted from the coffee particles, but it may appear in the spent column extract samples at ½ to 1 percent (about 3 to 4 percent in soluble coffee) of solubles content. Such a situation may occur due to high solubles inventories, allowing caffeine to exist at that concentration in the last column, or due to some portion of coffee particles that were extracted late due to earlier channeling. Corresponding degrees of exhaustion may be measured for other individual constituents, and their profiles may be related to the overall solubles profile variations.

**Ternary Diagram**.—The mixing of wetted coffee grounds with the countercurrent batch flow of extract may be calculated to show the stages required to attain specified extract concentrations and solubles exhaustions of grounds, or the calculations may be done graphically on a ternary diagram, as shown in Fig. 10.2. The batch mixing of wet grounds and extract gives only intermediate compositions but tells nothing about how fast such a process will occur. The diagram does not take hydrolyzed solubles into consideration, although it does account for concentration increases in extract due to removal of bound water by dry coffee grounds. Since there are three substances, water, coffee solubles, and coffee grounds (without solubles) under consideration, their relationship may be developed on a ternary diagram. The pure substances are located, one at each peak. The lines between peaks represent compositions of the end substances. For example, the base line represents the dry spent coffee grounds free of solubles with varying percentages of water. Two key points in the base line are readily identified: (1) Spent grounds with one-third bound water absorption are located at 25 percent water; and (2) spent grounds with both bound and free water (but not drainable) contain two parts of water to one part of grounds or 67 percent water. All extract compositions lie on the line between pure water and pure solubles. Dry roast coffee lies on the grounds-solubles line at about 20 to 25 percent solubles. All extract to coffee grounds equilibria will lie on lines radiating out from the grounds equilibria point at 25 percent moisture on the base line. These radiating lines are called tie lines. By experiment, an equilibrium line representing grounds, water and solubles mixtures can be determined. This equilibrium line lies very close to one-third dry spent grounds at all compositions, meaning that the other two-thirds composition is water and solubles.

*Equilibrium Line.*—This line may be determined from a number of tests in which fresh roast coffee with known solubles is added to extract with known solubles. The mixture is heated without water vapor loss until the free extract concentration is at a maximum. The weight composition of the mixture lies on the line joining the extract concentration and the focal point of the tie lines. The linear distance from extract composition to mixture composition is proportional to the weight of 25 percent moist spent grounds, and the weight of extract is proportional to the rest of the linear distance. The reason the equilibrium line does not stay at 67 percent water is that at higher solubles concentrations the solubles compete for the available water.

The wetting of dry roast coffee with extract (point 1) results in a mixed composition on the equilibrium line. By drawing a line from the 25 percent bound water spent grounds base point through this mixture, the resulting extract concentration of about 40 percent is obtained.

Due to the natural variation in coffees, roasts, grinds, etc., the experimental determination of the equilibrium line cannot be exact, but will vary several percent from the indicated locations. The location of the equilibrium line is also influenced by the amount of non-diffusibles in solution, such as are dissolved from the coffee particle surfaces.

*Equilibrium Calculations Vs Practice.*—Assuming that hydrolyzed coffee solubles enter the extraction equilibrium, and that three-fourths of the hydrolyzed solubles participate in solubles equilibria, the ternary diagram may be used to calculate mass and heat balances graphically, and reasonably similar solubles concentrations and temperature profiles can be evolved using practical operating rates. The batch calculations assume a constancy of weights moving countercurrently, but this is not so in practice. The extraction example, however, illustrates what is happening as well as the important factors that control a mass and heat balance. This understanding can then be translated into operating and design decisions advantageous to plant operation.

*Graphic Calculation*.—The ternary diagram terminal conditions are (1) dry coffee and water entering and (2) extract and spent grounds leaving. A minimum solubles level in the blown down or last stage extract is also set. In the diagram, the distance between the extract concentration point and equilibrium grounds composition cuts the tie line distances inversely to the weight ratios. It is assumed for simplicity that there is a constant weight ratio of extract to wet grounds, which is the case in the percolator.

Figure 10.2 shows dry coffee grounds containing 25 percent solubles mixing with a 15 percent solubles extract. This results in 20 percent solubles, but if the quantity of contacting extract is small, the highest attainable solubles concentration is 40 percent.

By taking the mid-point of the tie line (extract wetted particle to 15 percent solubles extract) and projecting through the mid-point a line from the focal point of tie lines to extract, a 23 percent solubles extract results.

An operating focal point is located far to the right (off the page) and slightly below the horizontal base line at the intercept of lines G to $E_1$ and $G_5$ to water apex. Now $G_2$ is connected to this far right operating point to locate $E_2$. By equally splitting line $G_2$ to $E_2$ and connecting the midpoint with the focal grounds point at 25 percent water and extending to the extract to give $G_3$ to $E_1$, the next solubles equilibrium stage is prepared. The following stages are developed in the same way until $G_5$, the chosen solubles depletion discharge of spent grounds and liquor, is attained. Point $E_1$ must coincide from lines $E_1$ to $G_3$ and $G_1$ to $E_1$ by trial-and-error solution.

## Hydrolysis—3rd Step in Percolation

There are many aspects to the subject of coffee grounds hydrolysis and these will be covered under the headings which follow.

**History Of Hydrolysis**.—The commercial solubilization of starches with acid has been used for over 100 years and is today a very large industry. Corn and other grain starches after acid hydrolysis give weakly sweet syrups. The higher the dextrose equivalent of the syrup, the more simple the sugars that have been formed, and the less viscous is the syrup. Moderate acid hydrolysis of starches results in large molecular units which make the resulting syrups viscous and almost tasteless. Starches, naturally water insoluble, are complex molecules of simple sugars, called hexoses. Cellulose substances are even less soluble than starches, but are also complex fibrous structures made up of hexoses. Celluloses are classified according to their degree of solubility in various media. Some of the classes of celluloses are: pentose; xylose; alpha-, hemi-, and holo-celluloses; and cello-biose, -triose and -tetrose. Acid hydrolysis of the starches and celluloses results in fewer by-products such as levulonic and other acids and acetone and other carbonyls. For every simple sugar molecule formed, one molecule of water joins every broken cellulose molecule; the water augments the solubles yield by about 10 percent for every such simple sugar formed.

Not until World War II did hydrolysis of spent coffee grounds assume importance or find application. Prior to that, solubles yields from roast coffee were less than 25 percent and were primarily solubles leached out with boiling water. Progressively higher solubles yields from roast coffee were taken, rising to 35, then 40 and even 45 percent for Robusta coffees. The taking of higher solubles yields is directly attributable to reducing the contributing cost of the coffee raw material and has been accentuated by the extremely competitive pricing in the United States. The increased solubles yields from hydrolysis were necessary in the period 1948 to 1950 when carbohydrate additions to soluble coffees were abandoned in the United States and 100 percent coffee products were marketed. These hydrolysis solubles and volatiles characterize the taste and aroma of resulting coffee extracts and powders in an unappealing way. The common labeling of *100 percent pure coffee* on instant coffee jars is somewhat of a misnomer.

**Blend and Roast Influences on Hydrolysis**.—The nature of the starch to cellulose composition in different green coffees will influence the rate and products of hydrolysis. Physically hard high grown mild coffee beans require a longer time, higher temperatures or more acid condition to effect a given solubles yield than a softer mild bean from lower altitudes, or than Brazilian or Robusta coffees. Beans that are not uniformly graded (quaker, blacks, etc.) will hydrolyze differently. Aged, fermented, decaffeinated, pre-treated coffees are all relatively softer and hydrolyze more readily, but do not necessarily produce any higher solubles yields although they frequently give up by-products such as tars. Hence, mixtures of soft and hard beans will, under the same hydrolysis conditions, undergo different degrees of physical softening and give up disproportionate amounts of solubles.

**Water Hydrolysis**.—The Morgenthaler U.S. patent 2,324,526 of July 20, 1943, cites temperatures of 320 to 347 F (160 to 175 C) for hydrolysis solubilization of spent coffee grounds to give 27 percent solubles. Actually, much higher solubles yields are possible at this temperature, but the patent does not state this directly. The patent states that a total of 50 percent solubles are obtained (23 percent additional) by acid hydrolysis.

**Acid Hydrolysis**.—Clough and Benner have two patents on the acid hydrolysis of water-extracted coffee grounds. These are U.S. patent 2,573,406 (1951) and U.S. patent 2,687,355 (1954) and are assigned to the National Research Corporation of Cambridge, Mass. This process was only part of an overall soluble coffee operation using Kennedy extractors and vacuum belt drying. The production of solubles by acid hydrolysis after the extraction process has the advantage of obtaining higher solubles yields without pressure drop problems as in percolators, and allows purification of the hydrolyzed solubles to form a tasteless substance.

The coffee grounds that have had about 25 percent of their solubles removed are placed in a mixing tank with enough phosphoric acid to make the pH 1.5 to 2.0. This is set forth in the patent by weight as 40 parts dry spent grounds, 200 parts water and 2.4 parts phosphoric acid. Hydrolysis continues for 1 hr with agitation at 212 F (100 C). Then the grounds are centrifuged free of liquor, which is neutralized with calcium oxide while agitated to a pH between 5.5 and 7.0; a foul flavor occurs at a pH above 7.0. Due to the poor solubility of calcium oxide only three-quarters of its theoretical required amount is added at first; the final neutralization is done with limestone. The insoluble calcium phosphate that results is rotary vacuum filtered at 150 F (66 C), yielding a clear amber filtrate. This 2.5 percent solubles solution is passed through an absorption bed such as silica gel, carbon, or fullers earth to remove foreign tastes. The solution is then concentrated in a vacuum or atmospheric evaporator to about 30 percent solubles. The primary coffee extract is mixed with these hydrolyzed solubles in a ratio of about 3 or 4 to 1.

Actually, most mineral acids cause cellulose hydrolysis. Sulfuric acid is often used because it is cheap and can be handled hot in more alloy vessels than can hydrochloric acid. However, both sulfuric and hydrochloric acids are difficult to precipitate or remove. Phosphoric acid has the advantage of forming calcium phosphate, which is very insoluble, and the precipitate may be developed into particles that readily filter. The resulting hydrolyzed solubles are said to be more palatable than those prepared by elevation of temperature on already atmospherically extracted coffee grounds and water in a pressure vessel. The acid hydrolysis apparently causes fewer side reactions to form the less pleasant-tasting products.

The pressurized water hydrolysis occurring in the spent end of the percolation system at pHs of about 3.5 allows a self-purification of the hydrolyzed solubles during their passage through 50 to 100 ft of R&G coffee. The advantage of the water hydrolysis is that auxiliary acids, tankage, labor, and handling are not required.

The solubles yields cited in the phosphoric acid patents are no doubt minimal, since overall solubles yields from R&G coffee with phosphoric acid treatment for longer times and higher temperatures approach the Morganthaler results of 50 percent of the R&G coffee being solubilized. Of this, about 25 percent is obtained with temperatures less than 250 F (121 C); the rest of the solubles are obtained at temperatures up to 350 F (177 C) (water hydrolysis). Commercial solubles yields from coffee in percolator systems may reach more than 40 percent before the grounds soften. About 5 percent solubles may be discharged with the spent coffee grounds column. Variations of the patented phosphoric acid hydrolysis are (1) higher acid concentrations, (2) higher temperatures in a pressurized vessel, or (3) both; all of these conditions accelerate the rate of hydrolysis of coffee solubles.

**Laboratory Hydrolysis Measurements.**—Taking coffee grounds that have had about 20 to 25 percent of their solubles extracted with boiling water, and heating them further in a household pressure cooker at 20 psig or 260 F (127 C) will yield only a few percent more solubles in a few hours. In fact, such a test in a sealed pipe rotating about its axis in an oven at 280 F (138 C) will show very little additional solubles yield. Several such tests may be made at different temperatures and times to determine solubles yields and rates.

**Percolator Hydrolysis Measurements.**—In the plant, the method of measuring hydrolysis solubles yields is from temperature and concentration profiles as in Fig. 10.5. If the temperatures in and out from one percolator column are within 5 or 10 deg, the solubles concentration increase from the profile multiplied by water flow rate gives the rate of solubles yield at that mean temperature. Such results are easy to confirm since percolation conditions are usually uniform for days or weeks, and even the influence of different coffees may be inferred. Taking a 1,000-lb column of roast coffee through which passes 12 gpm water or 100 lb water flow per minute, an average 5 percent increase in solubles concentration is achieved with a 330 F (166 C) inlet and 340 F (171 C) outlet temperature in 20 min. This is a 13 percent solubles yield at a rate of about 0.0065 parts of solubles per part of original coffee grounds per minute. The total percolation hydrolysis yield can be allocated in this manner. The rates of solubilization may also be calculated this way for different temperatures.

**Hydrolysis Rate.**—Since the rate of solubilization of the grounds under hydrolysis is a zero order chemical reaction, rate is dependent only on temperature. This may be plotted according to the Arrhenius equation:

$$S = (\text{empirical constant}) \times e^{-(E/RT)}$$

where
*S* is the rate of solubilization
*e* is the logarithmic constant, 2.73
*R* is the gas constant
*T* is the absolute temperature (Kelvin)
*E* is the activation energy required for cellulose hydrolysis

The log of the rate of solubilization is solely dependent on the absolute temperature. All other terms are constant. The addition of acid increases the rate of hydrolysis. This is what is done in the phosphoric acid application. Since hydrolysis is a chemical reaction only dependent on temperature, it means that particle size, agitation, and water to R&G coffee ratio have little or no bearing on hydrolysis rates. Even though there is protein undergoing hydrolysis at the same time as the cellulose, hydrolysis of proteins is also a zero order reaction and would be influenced and controlled by temperature. The solubilized protein to carbohydrate weight ratio is about 1 to 3.

Figure 10.8 is a plot of the Arrhenius equation, with semi-log coordinates of percentage of original R&G coffee solubilized per minute (rate) vs $1/T$, the reciprocal of absolute temperature over the effective range of hydrolysis temperature, 290 to 360 F (143 to 182 C). Hydrolysis rates can be verified under numerous and varied percolation conditions. They are well suited for determination in percolator systems that are insulated and have intercolumn heaters to control temperatures. Note in Fig. 10.8 that the parallel rate lines above the water line show a 25 percent increase in hydrolysis rates for dilute acids and double the rate for concentrated acids. Although lines are drawn to show rates, it must be understood that coffee varieties, grades, roasts, and treatment (e.g., decaffeination), may determine higher or lower values. But the hydrolysis rate lines give the general order of magnitudes. Thus, the disadvantage of alkaline waters used to neutralize the acids in the hydrolysis column is apparent, as is the advantage of using acid to reduce hydrolysis temperatures, time, or both. From the nature of the Arrhenius plot, only one experimental point is needed to fix a rate line. However, two points that fall on the same rate line are confirmatory.

The hydrolysis rate plot is instructive. For example, assuming a certain process time and temperature profile, the fraction of original R&G coffee that is solubilized by hydrolysis may be calculated:

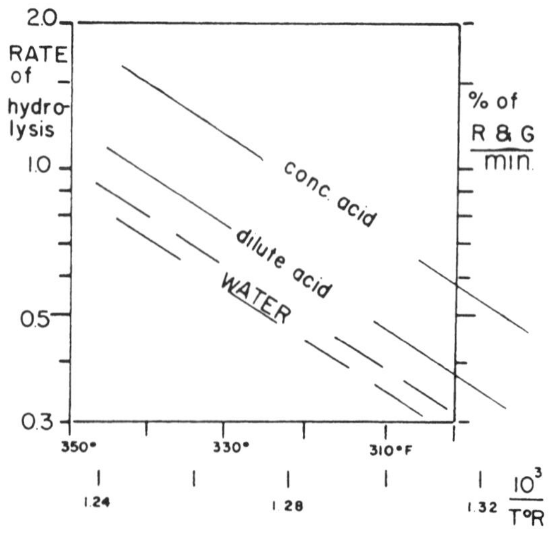


FIG. 10.8. RATE OF ROAST COFFEE HYDROLYSIS VS RECIPROCAL OF ABSOLUTE TEMPERATURE

| Step no. | Time, min | | Rate, fraction per min | | Solubles yield, lb per lb original R&G coffee | Temperature F | Temperature C |
|---|---|---|---|---|---|---|---|
| 1 | 5 | × | 0.007 | = | 0.035 | 340 | 171 |
| 2 | 10 | × | 0.0045 | = | 0.045 | 320 | 160 |
| 3 | 15 | × | 0.002 | = | 0.110 | 290 | 143 |
| | | | | | 0.110 | | |

Since 25 percent solubles have already been taken, the added 11 percent makes a total solubles yield of 35 percent solubles from the original R&G coffee.

**Hydrolysis Degradation**.—The holo- and hemi-cellulose fractions of the coffee grounds are hydrolyzed to solubles, but only about one-third of this is reducing sugars. The celluloses are dissolved, leaving a soft granule of original size. Reducing sugars held at temperatures over 360 F (182 C) can form tars; hence, there is a reduction in net solubles yield. A general rule in organic chemistry is that most all organic compounds undergo rapid thermal degradation and change at temperatures approaching 400 F (204 C), e.g., roasting. Since the oil content in the spent grounds becomes higher as hydrolysis proceeds, oiliness is characteristic of highly hydrolyzed spent coffee grounds. Fingering such grounds reveals a weak, plastic, granular, structure-free carbon. Continued hydrolysis causes progressive softening of the coffee grounds. At 360 F (182 C) and higher, coffee grounds, with 20 to 25 percent solubles already extracted, undergo severe degradation in ¼ to ½ hr, as do the solubles formed. The result is a soft, carbonized mass of foul smelling coffee grounds residue with insoluble tars. About 20 percent solubles are normally hydrolyzed in percolator systems, but 25 percent solubilization is feasible from slurries, especially if acids are used and solubles are promptly removed from the thermally degrading environment. Besides reducing sugars and water soluble carbohydrates, hydrolysates contain degraded and reacted proteins, amino acids, caramelized substances, furfural, carbonyls, and other volatiles. The content and composition of these substances will depend on raw materials and hydrolysis conditions.Although the hydrolysis solubilization is considered in gross, its side reactions are complex. Factors in percolation that impede water flow increase residence time of hydrolyzed solubles at the hydrolysis temperatures with greater opportunity for acids, tars, and side reaction products to form. Such contributing factors are high solubles (low water) in the extract drawn off, prewet grounds and longer process times. On the other hand, too fast percolation can flush through these objectionable tasting products to extract drawoff. This is one reason why it is more difficult to obtain high solubles drawoffs with high hydrolysis yields.

**Hydrolysis Taste**.—Hydrolysis solubles and volatiles in instant coffee represent one of the major differences in taste between brewed and instant coffee. The non-volatile acids formed in hydrolysis compensate for the acetic acid homologues lost in spray drying. This bitter-acid flavor allows instant coffees at 80 cups per original pound of roast coffee, and the acidity is useful in alkaline water areas. Coarse grinds, channeling, and higher hydrolysis solubles yields cause a more prominent hydrolysis taste and lower cup pH, e.g., 4.85 instead of 4.95. Lower hydrolysis temperatures for longer times (lower forward flow rate) show less acidity in the extract drawn off. Whether grounds soften more with longer time, lower temperature or shorter time, higher temperature hydrolysis is probably not a controlling factor.

**Rate of Percolation**.—This is dependent on the rates of wetting, extraction, and hydrolysis as well as on the objectives sought in drawn off extract concentration and solubles yield. Wetting the coffee particle and heating to hydrolysis temperatures take the longest times.

Percolation process rates decades ago took 10 to 20 hr; temperatures of extraction were low and no hydrolysis solubles were recovered. These percolation rates were mostly governed by the time required to attain solubles equilibrium between the R&G particles and extract. During World War II the merits of higher feed water temperatures and intercolumn heating were recognized and applied. Increased soluble coffee production was needed from existing facilities and attention was focused on faster processing rates. The net result was that temperatures and, consequently, hydrolysis solubles yields were raised. Process times were cut in half, but total R&G coffee residence time in the columns was still over 4 hr.

The worst obstacle to faster percolation rates was the frequent occurrence of excessive pressure drops which caused erratic process flow rates and erratic process times. Now that the causes for excessive pressure drop through the coffee bed are better understood, faster percolation is possible—that is, cycles of less than an hour down to 20 min on six-column systems. It takes only a few minutes of percolation time for extract to wet the R&G coffee if it has been prewetted. If it has not been prewetted or preheated, wetting the dry coffee depends a great deal on extract temperatures, extract concentrations, and grind size. At the least, wetting will take 30 min or one cycle. In other words, natural wetting of particles is very incomplete in the extract draw off column. Extraction of solubles takes about one cycle or ½ hr. Hydrolysis rates depend on temperature but normally the process takes about one cycle or ½ hr. Extraction and hydrolysis may even occur in half that time for each.

Without judicious application of intercolumn heaters, as much as an hour or more will be taken to effect natural countercurrent heating of the R&G coffee to hydrolysis temperatures of over 300 F (149 C) by the heat and mass transfer occurring. Thus, the percolation rates are limited by the natural properties and the use of equipment, materials, and processing techniques.

An important objective, then, is to obtain the most and best coffee flavor in the issuing extract and subsequent instant coffee powder.

Faster processing has the advantages not only of reducing R&G coffee residence times, but also of reducing coffee solubles and flavor inventories in the percolator system, a complementary effect. More and better coffee flavor is obtained from the percolation process with relatively less heat loss as less heat needs to be applied to the percolation system and this, in turn, reduces heat damage to coffee solubles and flavors.

Rates of solubles and water absorption into the grounds, rates of solubles diffusion out of the grounds, and the chemical rate of hydrolysis solubilization sequentially govern the overall percolation rate. In viewing fast cycles, there are practical limits as to how fast percolator columns may be filled and emptied, how fast R&G coffee pre-treatments may be done and the time required to change valve positions accurately. For these reasons one extra percolator column is needed. If R&G coffee bed preparation is inadequate, faster flows accentuate channeling. Faster processing usually calls for coarser grinds and may require the sacrifice of a few percent (say 38 instead of 40 percent) in solubles concentration in drawn off extract. Smaller columns (still commercial size) under proper temperature profiles, grinds, and cycles allow the widest latitude in percolator performance and product quality.

The ability to achieve specified percolation results of extract concentration, yield, and product quality at a faster rate is a direct measure of percolation efficiency.

## Percolation Dependent Variables

So far, the independent variables and the three processes occurring in percolation have been discussed. The *various results desired* (Dependent variables) govern the limits and types of independent variable conditions used. Since it is possible to achieve a given end several ways e.g., substitution of one independent variable for another, the independent variables may be adjusted differently to achieve similar results. With roastings for soluble coffee production at 18 percent of all green coffee used in the United States, the maintenance of high standards of taste quality for instant coffee is important. For this reason, the taste factor will be the last of the dependent variables to be discussed. The best taste is not necessary to a profitable business. The public supports multitudes of poor products sold at a fraction less than the best available.

**Temperature Profile.**—This is the reflection of heat distribution, materials distribution, countercurrent flow rates, and thermal properties of the water, grounds, and solubles in the percolator system. It includes heat losses, heat inputs, heats of solution, wetting, and hydrolysis. The temperature profile explains what happens in the percolator thermally. Although temperature profile cannot be calculated exactly, it may be calculated with sufficient accuracy to be meaningful in showing the contributions of each factor in the thermal process. Certain data are needed to carry out such calculations and, where data is not available, reasonable assumptions need to be made.

**Coffee Properties**.—Based on known values for other carbohydrates, a specific heat of 0.4 for solubles and R&G coffee and 1.0 for water may be used. Properties of mixtures of water and grounds and water and solubles will have specific heats proportional to the weight contributions of their components. The heats of wetting and solution may be neglected in these calculations since they are relatively small. Dissolving instant powder in water or R&G coffee in water in a thermos bottle will show that these are about 5 to 10 BTU per lb for powder or for R&G coffee. Heats of dilution can be derived from heats of solution, but these, too, are negligible for these calculations. Heat of hydrolysis of cellulose and starchy substances, which is a breaking of linkages and then reaction with water, is a reversal of the photosynthesis process; hence, it it can be inferred that energy is liberated. Since it appears that this is not a large heat change, it is assumed that it does not contribute significantly to the overall heat balance. Wetting is assumed to be only partial in the fresh R&G coffee column. The ternary extraction diagram may be used to determine batch equilibrium stages of solubles and grounds movement countercurrently. Specific volume of the extract changes as concentration of solubles changes. The sucrose specific gravities and temperature gradients for specific gravity may be used. Only a portion (about one-third) of the hydrolyzed and atmospheric solubles at 212 F (100 C) goes back into the R&G coffee during the particle wetting process. The heat loss per column depends on (1) cycle time, (2) temperature profile, (3) the size of the percolator vessel, and (4) the insulation on vessel and piping, the importance of which is often underestimated.

The bulk density of the R&G coffee is about 20 lb per cu ft and about 50 percent of the bed volume is interstitial void which varies with grind and column filling technique. Since R&G particles swell about 7 percent in linear edge dimension, after wetting, the interstitial void will be reduced to about 40 percent. The dry roast particle absolute specific gravities are about 0.63. The dry spent grounds hold twice their weight of water making 67 percent moisture or about 60 percent water and solubles together. Preferential water absorption is about one-third the dry weight of discharged coffee grounds.

**Assumptions in Balance**.—Other boundary conditions about the percolator system may be taken as:

1. No moisture in the entering R&G coffee, an atmospheric solubles yield of 25 percent, and a hydrolysis solubles yield of 18 percent more to give a total solubles yield of 43 percent;
2. A 1,000 lb R&G coffee column content on a ½-hr cycle with a 1,750 lb

extract drawoff at 25 percent solubles concentration. This is a water-to-coffee ratio close to five to one;
3. Room temperature, 80 F (27 C), may be used as the enthalpy (heat content in BTU per lb) base temperature, i.e., may be considered as zero;
4. Feed water temperature may be taken as 350 F (180 C). Intercolumn heating may or may not be applied. Other process conditions may be varied as desired. Variations in terminal conditions should be used to gain appreciation of the influence of changes in variables.

**Percolator Material Balance**.—With the foregoing data and assumptions, a material countercurrent balance may be drawn up with the aid of stage calculations from the ternary diagram. First, the overall material balance is prepared, then the internal stage balances, and then the thermal heat balance. In percolator practice, the material and heat balances interact to give the overall and internal concentration and temperature profiles. Internal profiles derived cannot be exact because rates of solubles exchange are not known exactly, but reasonable rates of wetting, extraction, and hydrolysis based on experience may be used. This analysis does not interfere with getting an appreciation of the influence of key variables operating within the percolator columns. Figure 10.9 shows, respectively, the type of graphic presentation that will result from the stage to stage calculations. Figure 10.9 is instructive because it shows the progressive buildup of autoclaved and then atmospheric solubles at 212 F (100 C) in the water flow, while the coffee grounds moving in the opposite directions (with associated water) are shown being progressively depleted of atmospheric solubles and then hydrolyzed solubles. The buildup of solubles in the coffee grounds during wetting is the true starting point for extraction of solubles.

**Percolator Heat Balance**.—The heat content of both the extract and the grounds is markedly less in the fresher coffee columns. With intercolumn extract heating of the two columns before the most spent column, hydrolysis solubles are formed over three columns at lower uniform temperatures. In essence, these block diagrams of Fig. 10.9 represent schematically the measured concentration and temperature profiles. Here the important influence of heat losses and intercolumn heat input may be seen relative to the heat content at each stage. The limited influence on the temperature profile of supplying heat only in the feed water is seen. Also the influence on flush water cooling of the fresh coffee column wall may be seen. The temperature profile is evolved by setting up all equations of heat balance across each stage. These are all related because the extract leaving one stage enters the next. There is a heat content or enthalpy balance about each stage; and this set of simultaneous equations may be solved by trial-and-error to give the proper temperatures at each stage.

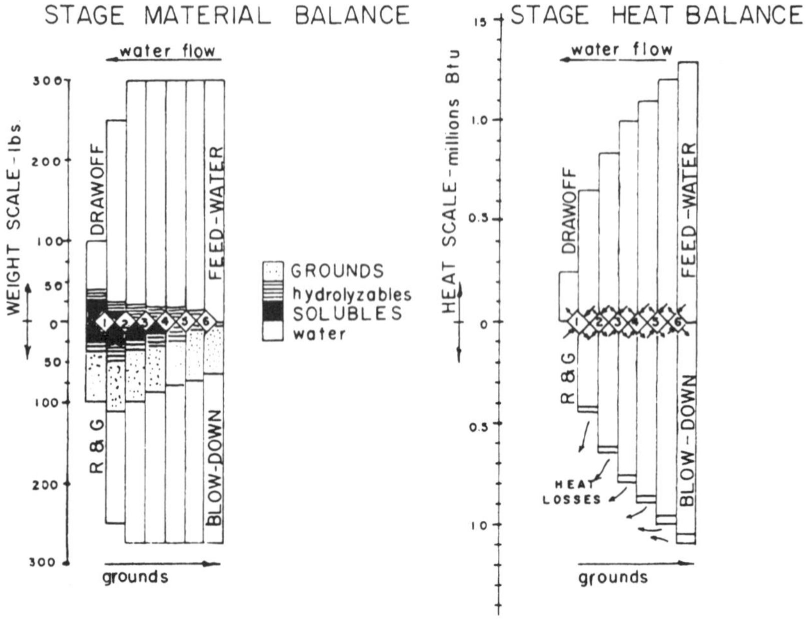


FIG. 10.9. GRAPHIC PRESENTATION OF PERCOLATOR MATERIAL AND HEAT BALANCES

**Heat Losses and Gains**.—Heat losses occur through any direct metal contacts of piping or columns to structural steel which act as good radiators. Thousands of BTU per hr may be lost by such seemingly unimportant metal connections. Water flush cooling of the column after blow-down can remove tens of thousands of BTU. This quantity enters significantly into the temperature and heat profiles.

A comparison of the effect of change in drawoff extract concentration on the heat content of the drawoffs (between 20 and 30 percent) relative to the heat entering by the feed water alone, shows a marked loss in heat input. The advantage of using intercolumn extract heaters under such circumstances is obvious. Faster cycles with more heat throughput per unit time, and essentially the same rate of heat loss from the metal system, leave an advantageous heat retention in the percolation system. This elevates the temperature profile over that obtained from feed water alone. Heat transfer rates between grounds and water are instantaneous for all practical purposes.

## Mass and Heat Balance Example

Although inter-stage calculations are not included, the following figures give the overall heat and material balance about the percolators.

## EXAMPLE: Overall Percolator Material and Heat Balance

Basis:

1. 100 lb of R&G coffee per cycle.
2. Ratio: weight drawn off extract/weight R&G coffee taken as 1.0.
3. Total yield solubles: 35 percent of R&G coffee. (45% usually taken)
4. Concentration of solubles in drawoff: 35 percent. (20% usually taken)
5. Entering heat content of R&G coffee at 80 F taken as zero.
6. Overall water/R&G coffee ratio, 3/1.

| | |
|---|---|
| *Input feed water heat at 345 F (174 C)* | |
| 300 lb × 1.0 sp ht × 265 F = | 79,500 BTU |
| *Output drawn off extract at 200 F (93 C):* 100 lb | |
| 35 lb solubles × 0.4 sp wt × 120 F = | 1,700 BTU |
| 65 lb water × 1.0 sp ht × 120 F = | 7,800 BTU |
| Total heat content | 9,500 BTU |
| *Output blowdown spent grounds at 342 F (172 C) average:* 300 lb | |
| 235 lb water × 1.0 sp ht × 262 F = | 62,600 BTU |
| 60 lb grounds × 0.4 sp ht × 262 F = | 6,300 BTU |
| 5 lb solubles × 0.4 sp ht × 262 F = | 500 BTU |
| Total heat content | 69,400 BTU |

Heat loss = input heat − output heat = 79,500 − 78,900 = 600 BTU/100 lb/cycle

Water flush cooling the metal walls of a fresh percolator column after grounds discharge may remove more heat than that cited for insulated equipment. Note that only about one-fifth of the entering water leaves with extract drawoff, and only about one-eighth of the heat entering in the feed water leaves with drawn off extract. In a 5 column system on the line, a loss of 3,000 BTU per 100 lb is about one-third of the net heat moving through the system, hence it should be compensated for by intercolumn heaters or heat losses must be kept to a minimum. At lower drawn off extract concentrations, more feed water heat moves through the percolator system, and temperature profiles are accordingly higher. Faster cycles have less heat loss and give higher temperature profiles. Heat losses from the percolator system can be much higher than noted depending on the size and geometry of columns, amount of insulation on vessel and piping, processing time, temperature profiles used, etc. Variation of terminal conditions will quickly show their effects on the system material and heat balance, and hence, on the temperature profile.

**Temperature Influences**.—Temperature is the most important variable controlling rates of wetting, extraction, and hydrolysis because temperatures control extract viscosity. Temperatures above 170 F (77 C) are needed to effect rapid wetting of the R&G particles with extracts of about 20 percent solubles. Then, the viscosity of extracts passing over the R&G coffee particles after wetting controls rates of solubles diffusion out of the cellular particle structure. Temperature is the single factor that governs the rate of hydrolysis. An increase of several degrees Fahrenheit in hydrolysis temperature may increase hydrolysis yield 10 percent (say from 25 + 18 to 25 + 20). Increasing hydrolysis time from 30 to 36 min will respectively give 10 percent more solubles yields. Percolator operators seldom sufficiently appreciate the fact that more rapid softening of the grounds at elevated temperatures will cause severe flow resistance until flows become so low that R&G coffee is lost through premature blowdowns, or cycle times are made too long. The troublesome effect is that non-uniform quality extract is produced characterized by a heat-treated and hydrolysis taste.

Extract temperature control through intercolumn heaters is an effective way to attain extract concentrations of over 35 percent. Warm extracts effect rapid disgorging of gases and wetting of the R&G coffee particles. Preheat and prewet techniques also may be used successfully in a limited way. Auxiliary heating methods are useful when dealing with conditions such as high R&G coffee column loadings, and small percolation systems. One reason small percolation systems (having columns holding less than 100 lb of R&G coffee) do not give representative extracts from "pilot" runs is that temperature profiles are so different from commercial systems that solubles concentrations and yields are low. Unless supplementary heat is applied, higher concentration extract drawoffs have lower temperatures.

**Flow Resistance**.—Uniform sequential percolation may be carried out only if there is no fluctuating resistance to feed water flow. Such variable flow resistances develop from a diminution of flow area within the percolation path and are more usual in the spent R&G coffee undergoing hydrolysis. The solubilization of the cellulose type of substances naturally weakens the cell framework of each particle. The R&G particles become softer and more compressible. As they compress together under a hydraulic pressure differential, the void flow path for passing extract is restricted. Such restrictions cause further increases in hydraulic pressure differential, because it takes more pressure differential to maintain constant flow, constant cycle time, and constant process time. If uniformity of process time is not maintained, hydrolysis solubles yields rise and grounds particles become softer. Many soluble coffee processors take high hydrolysis solubles yields, and they frequently run into long cycle times as well as occasionally plugged columns. Such circumstances invariably upset the thermal and solubles equilibrium in the percolator system. Solubles are not uniform in concentration nor taste during such periods. In the worst instance, the plugged column may have to be cooled and emptied manually with concomitant losses in product and quality.

**Compressibility of the Coffee Grounds**.—This may be demonstrated by taking (1) grounds whose solubles have been extracted at atmospheric pressure, (2) then hydrolyzing the same grounds to various solubles yield levels, as well as (3) blowdown grounds from normal operations. Placing these grounds in a laboratory Carver hydraulic press the following data may be gathered:

(1) *R&G particle bed height vs applied pressure (psig)*. The originally well compacted column of grounds will be reduced to 40 to 50 percent of its original height or volume at pressures of over 500 psig, with only slightly greater compac-

tion with pressures at 1,000 or 2,000 psig. The grounds that are only atmospherically extracted to 20 to 25 percent solubles yield, show greater resistance to compression by almost twice compared to that of commercially discharged spent grounds with solubles yields of 37 percent or more from R&G coffee. Results will vary with roast, grade, and variety of green coffee, and grind. There are indications that for about the same hydrolysis solubles yields, those coffee grounds that have been hydrolyzed at higher temperatures are softer. Pretreated coffees, such as decaffeinated coffees, show higher compressibilities than directly roasted green coffees. Compressibilities of the softest grounds are such that the original height can be reduced to about one-fifth before further compression is negligible at 1,000 or 2,000 psig. Leveling out bed density occurs at over 500 psig and is about 1.3 gm per $cm^3$ on a dry basis. This is almost twice the absolute density of the original R&G coffee, indicating that little, if any, interstitial space is present.

(2) *R&G coffee bed volume change and expelled extract collected vs applied pressure*. High hydrolysis (45 percent) of R&G coffee shows that the volume of water squeezed out is almost the same as bed volume compression. This occurs at pressures as low as 100 psig and well below 200 psig for most commercially spent grounds. This does not occur, however, in atmospherically extracted spent grounds until pressures exceed 500 psig.

(3) *Change in bed porosity vs applied compressive pressure (psig)*. A prime objective in percolation is to maintain continual uniform flow through 60 to 150 ft of percolator bed under a limited pressure supply. This is usually done without experiencing excessive pressure drop across any portion of the R&G coffee bed which would result in bed compaction. The degree of bed porosity (or flow area) is a function of differential pressure across the coffee bed. For example, doubling the velocity of flow (due to cutting the flow area in half) increases pressure drop four times (velocity squared is proportional to the pressure differential). The resistance to flow through various packed beds is discussed in most chemical engineering texts, filtration theories and practice. Initial bed porosities of coffee grounds vacuum-filled into a percolator column, before swelling, is 45 to 55 percent interstitial space. Compressibility of 75 psig on atmospherically extracted R&G coffee may reduce flow area about 10 percent, but on hydrolyzed softened grounds it may reduce flow area to less than 50 percent its original value. Actually, pressure drop rises to hundreds of psig when the interstitial void is reduced to below 25 percent of the bed area. This pheonomenon may occur deep in the coffee bed, e.g., about the exit port. Fines and chaff accumulating at one point have only a fraction of the compressive resistance of coarse ground coffee. Hence, plugging may occur with low differential pressures and much lower extract flows.

**Particulate Bed Flow**.—Flow areas with even 20 percent open area across the R&G coffee bed are ample to accommodate the throughput flow at small pressure differentials, if the area available has an equivalent pipe diameter. However, when this flow area is broken up into the minute flow paths, bordered by the soft R&G particles, the spaces may only be a millimeter wide and resists flow.

**Reverse Flow Relief.**—Ordinarily, high differential pressure drops are due to localized flow restrictions, and these can usually be eliminated by reverse flowing a column for a few minutes. This establishes new flow paths and helps to restore the normal upward flow. In more severe cases, reverse flow may have to be maintained until the column is discharged, which may have to be done earlier than usual.

**Fines and Fine Grinds.**—It is unusual to have excessive (over 50 psig) pressure drops in a fresh R&G coffee column. This may, however, occur if fines migrate with viscous extracts to plug off the normal flow area. Reverse flowing is awkward so far front in the percolator system, and may lead to serious fall off of flow. Such high pressure drops in themselves show that selective flow paths have been established, which also means less efficient extraction of the coffee bed. When excessive pressure drops occur near the fresh coffee end of the system, they are seldom completely eliminated.

Fine grinds in themselves are not always the cause of excessive pressure drops. Fines do contribute to such a condition when combined with large diameter columns, dense loading, faster flow rates, taller columns, high concentration extract drawoffs, high hydrolysis yields, or any combination of these. Fine grinds may be percolated successfully under the proper circumstances.

Grounds that are prewet before loading leave more void space for flow. Too much void space allows fine particles to migrate and plug localized areas. Slack-filled columns allow the finer particles to segregate. This causes inefficient extraction and plugs the outlet point with soft compressible fines and chaff.

**Column Differential Pressures.**—The foregoing discussion of behavior in the compressibility of the coffee grounds warns that differential pressures should be kept less than 50 psig on any column. Invariably when over 50 psig percolator differential pressures occur before the hydrolysis stages vigilance is required and reverse flow may be needed intermittently or continually to see that the column leaves the system without causing excessively low flow rates. The incidence of one such column often may cause others to soften under prolonged holdup.

When 50 to 75 psig pressure loss already exists across one column, pressure losses naturally and progressively grow on that column to seal off even more flow area. For this reason, a maximum feed water pressure supply of, for example, 200 psig limits the rate and degree of harm that may occur.

As a matter of interest, when a dry bed of R&D coffee is pressed with several hundred psig, wetting of the R&G particles immediately relieves the pressure and softens the rigid particles as they reorient themselves. A counterpart of this observation is that only a fraction of 1 psig is required to retain a loosely filled R&G coffee bed (8 ft high) when wetted from below. The wetting reorients the particles to fill in the void and produces a small outward pressure on the vessel. Without such a restraining pressure at the top, the coffee bed will float.

**Column Size and Shape.**—In regard to percolator diameters and pressure drop at 20 to 30 min cycles, a 3-ft diameter column will in general contribute to more pressure loss than a 2 ft diameter column. The same is true in comparing the 2-ft with the 1-ft diameter column.

A percolator column 20 ft tall will cause much more frequent pressure losses than a 15-ft column, and a 10-ft column causes hardly any pressure loss. This observation, of course, depends on grind, yields, etc., but may be used as a "rule of thumb." The swelling of wetted grounds at the bottom of the taller column does not compress the grounds at the top because they are dry and rigid. But coarser grinds need to be used with taller columns. If taller columns contribute to a longer system, the longer flow path will contribute to a larger overall pressure drop. Also, for a given process time the linear flow velocities will have to be greater.

**Stoppage of Percolation.**—There is a balance between process time, hydrolysis yield, and softening of grounds. If a shutdown in percolator operations occurs, e.g., a power outage, the hydrolysis continues. Such a spent coffee column must be discharged at its normal time, otherwise the coffee granules will become too soft. The next forward column at a lower hydrolysis temperature may not have to be discharged, depending on its temperature and hold time.

An excessive pressure drop is always caused when a percolator operator, during throttling of extract to the fresh R&G column, shuts the throttling valve; in other words, when he stops the flow. Then the full water pressure is progressively applied as a hydraulic pressure across each column. This situation is similar to the Carver press compression of grounds. The end result is a generally diminished flow area and a high pressure drop when normal flow is re-established. Flow surges with an open throttle valve may also cause such bed compression. Steam will form in the hottest columns if hydraulic pressures fall below saturated steam pressures corresponding to existing water temperatures. In this case, the system must be placed back into a hydraulic condition carefully so that flow surges do not contribute to bed compression and a reduction in flow area under abnormally large pressure differentials.

*One Column to Battery Stoppage.*—It requires only one column affected by a restricted flow area and absorbing a disproportionate part of the available pressure drop to place the whole battery of columns into potentially the same condition. Release of a differential pressure by stopping flow does not solve the problem because the particles are not resilient enough to form new flow paths; only reverse flow can do that.

*Incomplete Blowdown*.—If the coffee is not completely discharged from a percolator column and fresh R&G coffee is added, the residue coffee will be so consolidated that little or no flow passes through it. Flow will be channeled about it at higher than average velocities. Such incomplete blows are often ring shaped, and the open central area for flow through the fresh R&G coffee may only be a fraction of the normal flow area. This flow pattern contributes to subsequent high pressure drops as well as to the likelihood that the coffee will be incompletely blown again. A short coffee column fill will result in lower extract concentrations being withdrawn. Successful spent grounds reblows may be achieved by feeding water at 330 F (166 C) or higher into the unsuccessfully blown coffee column (while it is isolated) until the hot water covers the residue grounds and the equilibrium vapor pressure is at least 100 psig. This second blow is usually successful. In some cases, several steam reblows may be needed. In extreme cases soaking in cold water and taking the coffee column off the line for a complete process cycle with additional hot water reblows may be needed to remove deposits.

Nothing can be substituted for a visual inspection of the interior of the percolator column to give assurance that the blow and reblow are complete. After the column is filled with ground coffee, a preliminary technique for alleviating potential flow area restrictions about the discharge bayonet is to flush back several gallons of water or extract about the top bayonet before starting upflow from the base bayonet.

**Concentration Profile**.—The concentration of solubles in the extract flowing forward is, at every point in the system, a reflection of wetting, extraction, and hydrolysis performance. It not only reflects temperature influences, but also grind, blend, roast, R&G column loading, water to coffee ratio, rate of countercurrent processing, column and system design, and solubles yield being taken. Concentration profile is, in many cases, a counterpart of temperature profile since they are related to each other in material and heat balances. For example, when concentrations are high in the fresh R&G coffee column, temperatures are low. Similarly when high concentration extract drawoffs are taken, the initial and average drawoff temperatures will be low. Reference to Figs. 10.4, 10.5, and 10.6 illustrates how faster processing moves the concentration profile forward, how finer grinds with higher column loadings increase concentration profile, and how especially intercolumn heating helps to raise solubles concentrations at the fore part of the percolator system.

**Extract Drawoff Influences**.—The one variable that to a large extent controls the concentration of solubles in the drawoff is the weight of extract drawn off in relation to the weight (dry) of R&G coffee used per column. Large extract drawoff ratios yield lower extract concentrations because more water moves through the system. The temperature profiles are high, solubles inventories are low, and taste of the extract is very good.

Since higher extract concentrations are needed for better flavor retention during drying, this aspect of percolation calling for less water with the solubles in the extract drawn off has been substituted for by evaporation of extracts. Higher solubles concentration may be achieved in various ways: finer grinds, higher R&G coffee column loadings, a lighter roast, higher solubles yields, blends with more Robusta, etc.

Depending on the existing conditions in the percolator process, reducing the weight of extract drawn off sometimes allows the concentration to rise at the same yield. When reducing the ratio of extract drawoff to R&G coffee has little or no effect in raising the extract concentration, and yield falls off, then yield may be restored by using higher temperatures within the system. Reducing water flow coming forward also reduces solubles coming forward.

**Solubles Inventories**.—Solubles accumulate at the fore part of the percolation system, and drawn off solubles concentrations are achieved by higher solubles inventories. The rate of solubles extraction is reduced, but the driving force of solubles movement (between R&G and extract) is augmented by higher temperatures and reduced viscosities which increase solubles diffusion rates. Higher extract temperatures help particle wetting to give higher initial particle solubles.

Sometimes, the amount of extract drawoff may be increased without loss in

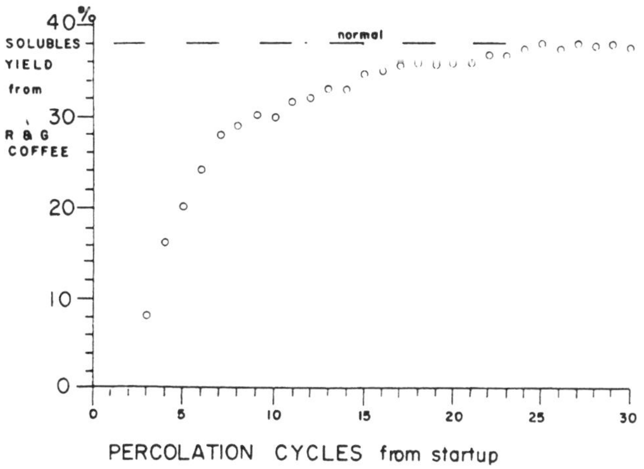


FIG. 10.10. PERCOLATION STARTUP SOLUBLES YIELD PROFILE

solubles concentration, indicating that the increased heat flow forward and new equilibrium conditions are sufficient to give a somewhat higher solubles yield. Depending on concentration profile, extract solubles drawoff may represent one-fifth to three-quarters of the solubles inventory.

**Solubles Equilibrium Factors.**—The equilibrium that exists among grounds, solubles and the water varies with changes in drawoff ratio, temperatures, grinds, etc., but it takes considerable time to reach the new equilibrium. This may take ten or more cycles, depending on the degrees of change in the variables.

The maximum drawn off extract concentration is a little over 50 percent and is governed largely by extract viscosity which depends on temperature in the particle wetting process.

Drawoff extract concentrations vary from cycle to cycle but only about ±1 percent in a well-operated percolation system. These differences are due to variations in grind, roast, fill, blend, channeling, temperature profiles, column individuality, etc.

Under a given set of percolation conditions, the drawoff weight ratio is directly influenced by the water to R&G coffee ratio being used. The rate of attaining solubles equilibrium is faster with large water to coffee ratios.

**Measuring Concentration Profiles.**—Concentration profiles may be measured experimentally either by taking extract samples at one point in the percolator system over the whole process time, or by taking extract samples entering every column simultaneously. The former technique is preferable. The area under the concentration profile is a quantitative reflection of the pounds of solubles at each point in the system and is, therefore, a solubles inventory. The coffee flavor inventory is mostly in the fresh and next-to-fresh R&G coffee column.

**Solubles Yield.**—Overall solubles yields are governed by pricing competition. Today, overall solubles yields on commercial products are about 40 to 50 percent from R&G coffee. Higher solubles yields are not feasible in the percolator without causing compression of grounds, reducing flow rates and hence, process rates, and otherwise jeopardizing productivity and quality. Even solubles yields at the top of this range contribute tars and objectionable flavors, and are usually taken from Robusta blends. Pulverized coffee may produce solubles yields of well over 30 percent with 212 F (100 C) water, but percolation is impractical.

Solubles yields over about 25 percent of R&G coffee are attained by hydrolysis solubilization of the cellulose matter in the cell structure of the particle. Hydrolysis is a chemical reaction dependent on temperature. It is illustrated in the concentration profile as the rapid rise of solubles from the feed water end which levels out until solubles extraction begins.

**Flavor and Aroma**.—Although the independent and dependent variables of percolation have been covered, the key factors of flavor and aroma of the drawn off extract remain. In spite of often unacceptable taste qualities of the instant coffee powder, the percolator extract has a high natural coffee flavor level. The extract, although somewhat bitter acid makes a cup similar in coffee flavor to that brewed and filtered from the same R&G coffee. In fact, a cup of coffee brew that is filtered through paper, which has no colloids, is quite similar to the cup of coffee from percolated extract. The strong natural coffee aroma and flavor are characteristic of the coffees from which it is prepared. Robusta extracts at cup concentration taste similar to brewed Robusta coffee without colloids. High grown mild coffee extracts at cup concentration taste like brew (without colloids) from high grown mild coffees. There is no confusing extract and brew, nor is there any confusion in identifying the blend, roast, and grind contributions of each.

When a reasonably good blend of coffee is used, aroma and flavor overtones develop in the percolation process. High solubles yields due to hydrolysis produce a bitter-acidy cup. Heat treatment of coffee flavors and solubles to attain high extract concentrations produces weak, caramelized, and less characteristically flavored products compared to the R&G coffee used. On the other hand, some naturally bad off-flavors of green coffees will be noticeably toned down in the percolation process. Rioy flavors, however, are very prominent in coffee extract flavor. The higher hydrolysis yields produce a weaker cup flavor strength. Long processing, such as six or more hours of percolation time, yields weak coffee flavor in the extracts. Heavier coffee flavors (winey) from relatively clean naturals are desirable. Darker roast coffee flavors carry through the percolation process more faithfully. The best flavored percolator extracts are at the lowest concentration profiles.

## Miscellaneous Aspects of Percolation

**Substitution of Independent Variables**.—More than one independent variable may be used to modify a dependent variable. Therefore, one independent variable may be substituted for another, depending on the latitude for variation and the results desired. For example, flow impediments causing erratic cycle times may be alleviated by coarsening the grind, slowing the percolation rate, or reducing hydrolysis temperatures.

Intercolumn extract heaters are not needed in a percolator design that uses fast percolation cycles of 20-30 min and well insulated vessels and piping. Higher extract solubles concentrations may be attained by intercolumn extract heating or higher column loadings of R&G coffee.

High extract drawoff temperature such as 210 F (99 C) may be reduced by water flush-cooling the column walls that has just been blown down and prior to its being refilled with R&G coffee. The temperature may also be lowered by reducing feed water temperature and/or intercolumn heating.

If fresh R&G coffee column temperatures are too low, they may be increased with a longer drawoff, steam preheat of R&G coffee, faster cycles, more intercolumn extract heat, and/or higher feed water temperature.

**Startup and Shutdown**.—Figure 10.10 illustrates the solubles and yields with progressive cycles from startup. There is no drawoff or solubles yield for the first two cycles; the yield is only about half normal on the third, two-thirds normal on the fourth, three-fourths on the fifth, and not nearly normal until about the tenth cycle. The reasons for this are that it takes that many cycles to establish thermal and solubles equilibrium. No drawoff is taken on the first two columns so that solubles may buildup in the percolator system and extract concentrations are over 20 percent initially, and over 25 percent after the fifth cycle. Since the initial extracts have little or no hydrolysis solubles and higher than average oil content, the first ten cycles would spray dry and taste differently than later cycles. If, for example, the column of drawoff is about 100 gal for each cycle, then the first five drawoffs may go into one 1,000-gal tank and the next ten drawoffs into the other 1,000 gal tank. Drawoffs 16 through 20, which are practically normal, may be mixed with the first five drawoffs. The mixed extracts are then more nearly normal in solubles concentration and yield; hence, they spray dry and taste more normal. The net effect of such a startup is that the hydrolysis yields from about 3 or 4 columns are lost. On shutdown, the equivalent of three full extract drawoffs may be taken from the last fresh R&G column depending on the solubles inventory of the system.

On shutdown, the final drawoff must not be allowed to get so hot that it foams and spumes on entering the plate cooler. This may be prevented by reducing feed water temperature slightly and refusing any solubles concentration below 15 percent. These startup and shutdown losses as well as the loss in equilibrium during such transition periods, make it impractical to start and stop daily. Percolation must be carried out 24 hr a day, preferably for weeks at a time. When production warrants only a five day week, there are net solubles losses which represent about 1 to 2 percent yield.

**Productivity**.—Theoretically, and often in well-operated plants, percolation goes on like clockwork. If a ½-hr cycle is being used, there will be 48 cycles per day, every day. In some plants the number of cycles per day may be 40 one day, 46 the next, 42 the next, etc. These variations are associated with variable concentrations, yields, and flavors. Aside from the reasons cited already, other factors often influence variances in productivity. For example, the spray drier may be overloaded, and percolation production must be held back. There may be an electrical outage for minutes or hours as during electrical storms. Mechanical failures of pumps, flanges, pipes, instruments, valves, heaters, etc., may shut down the system temporarily. Such interruptions are usually expensive in idled labor as well as in lost quality and production. For these reasons, it is usually more economical in plant design to use the most reliable equipment, techniques and labor. Unattended changes in grind, roast, blend, and blowdowns may also cause variable flow areas and time cycles. Steam failure, reblows, loss of vacuum, or scaling of heat exchanger tubes may cause processing to stop. Preventive maintenance during scheduled shutdowns to correct recognized percolator equipment shortcomings avoids the more costly shutdowns during production periods.

**Useless Techniques.**—Often, variances in percolation techniques are used experimentally or offered as superior methods. Since some of these methods have been tested and have little or no value, it is pertinent to report on them to avoid future duplication.

Rupture of the coffee cell structure is a popular means of aspiring to attain greater solubles yields and rates of percolation. Fast roasting to obtain the greatest swelling, hence, rupture, of beans as well as freezing and thawing of wetted roast or green coffee grounds produces no significant change in the coffee extraction characteristics. Steam heating wetted green and roast beans under pressure and then releasing them to atmospheric pressure to effect puffing are useless techniques. Use of enzymes to solubilize cellulose portions of green or roast coffee is impractical. Roasting spent grounds does not yield more solubles, nor do supersonic vibrations applied to R&G coffee and water mixtures. Defatting green or roasted coffee shows no increase in available solubles nor any practical increase in extraction rate.

**Transfer and Storage of Extract.**—The extract leaving the fresh R&G coffee column must be cooled to 40 F (4 C) promptly to preserve the chemically unstable flavors and prevent enzymatic activity and bacterial growth in the sugary medium. Holding extracts over 80 F (27 C) causes flavor deterioration in a fraction of an hour. Such temperatures are ideal for bacterial growth and fermentation reactions. Higher extract solubles concentrations (less free water) and higher acidities noticeably retard bacterial action. All dead end pipe locations should be avoided since these invite bacterial growth and enzymatic action. Sometimes ultra-violet lamps are used to sterilize extract surfaces in the tanks, but short residence time and cold temperatures are more effective methods for retaining extract freshness. Extract lines and tanks should be scoured and cleaned with hot water (and caustic if necessary) as frequently as is practical to minimize opportunities for biological changes.

## PERCOLATION EQUIPMENT AND PROCESSING

Percolation has emerged as the industry standard practice because in one process it offers (1) high extract concentrations, and (2) the complete solubles yield competitively required, with the least time, labor, and equipment. But, more importantly, the issuing extract is not oxidized, does not require vacuum concentration, and thus has the best natural coffee flavor carry-through.

The only elaborations on the percolation process that have evolved since G. Washington's time have been the use of higher feed water temperatures and faster processing time to achieve solubles hydrolysis as well as extraction, within the same battery of percolator columns. The extract which becomes progressively concentrated as it passes through several columns also achieves a self purification by (1) precipitating the brew colloids, and (2) filtering out oils, suspended and particulate matter so that not even centrifuging or clarification of the issuing extract is required.

Figures 10.11 and 10.12 illustrate the equipment used.

A battery of percolator coffee columns is similar to batteries of columns used for sugar beet extraction. There are usually 5 to 8 columns; the height may vary from 6 to 20 ft and the diameter may vary from 1 to 4 ft. Hot water at about 325 F (163 C) enters the column with the coffee grounds that are most spent and progressively moves upward through each column of less spent coffee grounds. The flow is in series through each column. The extract, moving forward continuously, fills the interstitial voids between the grounds in the freshest coffee column, simultaneously wetting the grounds and displacing gases preferably upward and outward. The coffee extract leaves the freshest column of roast and ground coffee as a syrup, having a concentration of coffee solubles ranging from 25 to 35 per cent. The hot feed water flows continuously into the column with the most spent coffee grounds. The extract from the freshest roast and ground coffee column is withdrawn intermittently. The time for extract drawoff alternates in time periods with the time for extract filling of the fresh dry roast and ground coffee column. The time for extract to fill the freshest coffee column (until drawoff), is about the same as the time required to draw off the extract. For example, 15 min to fill, 15 min to draw off.

"Cycle Time" is the period between any operating step in one column and the same step in the next freshest column, e.g., extract is fed to a fresh roast and ground coffee column and then to the next fresh R&G coffee column; cycle time is this interval.

"Process Time" begins when extract enters a particular column of fresh R&G coffee and ends when that column of coffee is discharged and has been taken off-stream. During this process time, coffee grounds have been wetted, extracted, and solubles have been hydrolyzed. For a seven-column system, this would be 3 hr. One column is off-stream for discharge, inspection, cleaning, and refill. The design of the percolation system, column size, shape, and number are based on: solubles yield, solubles concentration, solubles production rate desired, and operating experience.

## Example of Geometric Design of Percolator Vessel System

In a typical case, specifications require the percolators to give 700 lb of solubles per hour. If the cycle time is 30 min, then each percolator column must yield 350 lb of solubles per column. If specifications require a 35 percent solubles yield from the R&G coffee, then each column must hold 1,000 lb of R&G coffee. Since the nominal bulk density of ground coffee may be taken as 20 lb/cu ft, the volume of each percolator column must be 50 cu ft. Next the column volume must be proportioned in height and diameter.

The shorter the column height, the less difficulty there will be in excessive pressure losses at constant flow but too short a column height will result in too short an overall R&G coffee bed distance in the column battery; hence, extract concentrations will be too low. At this point, it must be stated that seven columns are chosen for the design, since experience has shown that four or less columns give too low extract concentrations at fast cycles and at moderate column heights. Although six columns are adequate, seven allow one column to be off-stream.

With fine grinds, the column may be smaller in diameter because of faster extraction. This also gives greater wall support for the coffee bed; the column should also be shorter to minimize further excessive flow pressure losses due to bed compression. Although grind may be adjusted during operations, we will assume a grind that is about 20 percent on eight mesh (United States Standard) and less than 5 percent through 20 mesh. Since tall columns allow greater compression of the R&G coffee bed, due to swelling of particles on wetting, the tallest columns are not as desirable. Thus a medium height of 10 ft may be chosen; this results in a 2.35-ft internal diameter for a 50 cu ft volume when the top and bottom cones are included.

If specifications require a 35 percent solubles concentration in the extract drawoff, this will be achieved through control of the temperature profile across the percolator column battery either by insulation, cycle time, and feed water temperatures, or by intercolumn heating of extract. In this case, the drawoff weight of the extract may be 1,000 lb with 350 lb solubles per column.

## Objectives Governing Layout of Percolator Vessels and Piping

Having established the number of columns, their height and diameter, we must determine how to orient the columns and manifolding. Some designs, especially in Europe, arrange the columns in a square or circle so that R&G coffee may be run into each column from a central overhead duct with rotating discharge duct. This facilitates central loading of grounds but is not good for monitoring of the column temperatures or valve positions, which the operator must check continually. Such rotary layouts also make for awkward installation of piping and often do not achieve the close arrangement of columns necessary to minimize heat loss in intercolumn piping. In addition, blow-discharge of the spent grounds from the columns is difficult. Experience with numerous systems has indicated that the alignment of all percolator columns in a straight line, and as close as possible, is the most workable arrangement because a symmetrical line-up minimizes operator confusion. It allows a single blow-line on the center line of the columns which is most direct and compact. It permits valving on the front operating face while supply lines are either at the front or rear of the columns. Furthermore all manifolding is symmetrical and square, facilitating pipe installation as well as the tracing of lines by trainee operators. This design also allows any R&G coffee bin to be moved over the center line of columns during filling and inspection of columns, and enables the operator to see all valve positions, dial thermometers, and pressure gauges quickly. Whether the columns are operated from the top, bottom, or by remote control is a matter of choice that barely affects the percolation operation under normal circumstances. The important factor is to have clear full view and communication between the R&G coffee loader and the percolator operator; it is better yet to have the same man do both jobs and to have quick access to the extract drawoff weight tank and storage tanks.

## Percolator Column Vessel Specifications

The vessel should be designed according to the ASME Code of unfired pressure vessels; when fabricated it should be ASME inspected and stamped. This assures a degree of workmanship and materials which without extra cost offer security to operating personnel and compliance with insurance company standards for minimal rates.

# 11

# Spray Drying and Agglomeration of Instant Coffee

## INTRODUCTION

In the mid 1960's virtually all instant coffees were sold as a powder; solubles yields from R&G coffee were in the low 40's per cent. In the early 1970's, due to increased solubles yields to the high 40's, agglomeration of instant coffees took over. The increased solubles yields were accompanied by lower extract concentrations which formed finer powders with poorer solubilities. An awkward interesting by-product manufacturing problem occurred in Brazil, a principal world supplier of instant coffees in the late 1960's and early 1970's. So many fines were being produced in spray drying that often ⅓ of the spray dried powder had to be removed as "fines" (below 40 mesh) and be redissolved for redrying. Naturally, such recycling caused marked downgrading of the final instant coffees as well as causing quality control problems for the manufacturer and the buyer.

An additional problem was that with increased solubles yields, more "tars" were formed. These were largely proteinaceous in nature and virtually insoluble in water. Tars caused carbonization during spray drying and left a black sediment in the cup and on the cup walls. Further, the tars occluded coffee solubles and this constituted significant solubles production losses, e.g., as high as 10% of solubles production in some cases. The tars ordinarily accumulated in the bottom of the extract settling tanks to a depth of several inches and had to be manually shoveled out of the tanks as a total loss. In this period, it became standard processing procedure to use continuous desludging centrifuges to remove colloidal tars, and this assured cleanliness, even though there was still a loss of a few percent solubles. Also by 1970 most of the Brazilian instant coffee processors had already installed either APV flash evaporators or De Laval Centritherm film evaporators, (vacuum operated), to increase extract concentrations from about 18-20% to 30-33%, thereby reducing the evaporative load on the spray drier, and in most cases increasing the nominal production capacity of the spray driers 35 to 75%.

The physical changes that occurred in spray drier construction for instant coffee manufacture as with milk drying, was that a spray box or short drier was no longer feasible. To make large particles, the driers had to be over 100 feet high, and 4 meters to 6 meters in diameter. These correspond to instant coffee production rates of 400 and 1200 lbs per hour from 30% solubles extracts.

In addition, long drying runs, without frequent cleanings or washdowns, that might last 1 week or several weeks were very much dependent on using extracts near 30% solubles which were less dusty than 20% solubles extracts, as well as good inlet air distribution so that fines did not stick to the hot ceiling or walls and caramelize or even carbonize after some hours or days. Such carmelized-carbonized deposits when falling into production powder cause black insoluble contaminates, making the product not marketable. Good inlet air distribution systems were first developed by Swenson Evaporator Co. of Harvey, Ill. in work done with Folger in Houston. In the 1970's, NIRO Atomizer of Copenhagen, Denmark had advanced these systems, characterized by an inverted cone air entrance at the top of the drier, with higher air velocities keeping walls clean. Many details of spray drier design may be found in the book, "SPRAY DRYING" by K. Masters, a Niro employee, published by Leonard Hill of London in 1972.

Another important design feature of the taller, larger instant coffee drier is its bustle for removing exiting air, while allowing particles to fall downward into the cone. When combined with the inlet of cool dry air into the bottom cone, issuing particles are cooled.

A properly designed and operated spray drier will collect less than 10% fines outside of its bustle area in the cyclones. A characteristic of some driers is to allow a trail of fines to be liberated to the environment near the drier tower, and this can constitute sometimes serious pollution as well as material loss.

Another aspect of the tar removal and evaporation was the preparation of the coffee extracts for freeze drying. Virtually no tars could be tolerated in freeze concentration of extracts, otherwise solubles losses became high. It is noteworthy to mention that the tars not only came from hydrolysis, but also from the polymerized aldehydes deposited on the roasted beans in the roasting process with recycled gases. For freeze drying, solubles concentrations of at least 33 per cent were sought, and some manufacturers went as high as 40 per cent solubles accompanied by gassing the extract before freezing to obtain the corrrect final granule density. Therefore the tar removal and evaporative procedures ran in parallel for spray and freeze drying.

Another odd variation occurred initially when extracts were evaporated; a number of large processors evaporated all the extract drawn from the percolators. Naturally, this resulted in almost all of the aromatic constituents being stripped from the extract, so that subsequent spray or freeze drying resulted in very bland products.

Shortly thereafter, most instant coffee processors increased their extract tankage so that the first half of the solubles drawn off were stored at their naturally higher solubles concentration (about 30%), and the 2nd half of the draw off (about 17% solubles) only was evaporated. Later the original higher solubles portion was mixed proportionately back with the evaporator concentrated portion, resulting in much higher aromatics contents, which were significantly retained after spray or freeze drying from say 33 per cent solubles mixtures.

It is important to note that all these steps, taken to increase solubles yield and reduce tars, resulted in additional plant investment and operating costs; the higher solubles yields were obtained progressively but with off-setting costs.

Therefore, prior to discussing spray drying, *extract treatment* will be covered, as it would be encountered in a process plant. *Agglomeration,* which occurs after spray drying, will be discussed at the end of the chapter.

## Coffee Extract Treatment

**Higher Solubles Yields**.—Since about 1969, dilute extract effluents have been taken in order to increase solubles yields, and then the dilute product is evaporated from 18 percent solubles to 30 percent solubles for spray drying or 33 percent for freeze drying.

Since 1952 processors have taken progressively higher solubles yields from R&G coffee by hydrolysis in extraction. This has usually been done by passing more water through the extractors which yields lower solubles concentrations in the drawoffs. The lower solubles concentrations of extract are concentrated by water removal before spray or freeze drying, thus increasing solubles yields.

About 1960 the ICA (International Coffee Agreement) established a 1 lb of solubles yield of instant coffee from 3 lbs of green coffee beans as a basis of export quota control. This corresponds to 39 percent solubles yield from roast coffee beans. Few instant coffee processors today can compete cost wise on this basis.

In 1972 a new solubles yield relationship was agreed to between the customs officials of the European Common Market countries at 1 lb solubles per 2.5 lbs green coffee beans. The large processors like General Foods and Nestle concurred.

Increasing solubles yields have resulted in the production of more acid tasting instant coffees. The solubles have characteristic furfural aromas and tastes. The coffee flavor is diluted with carbohydrates.

To compensate for these taste deficiencies, it was logical to benefit from the thermal fusion-agglomeration process. Agglomeration usually not only diminished the acid/furfural taste, but subsequent addition of coffee oil improved the beverage taste and aroma.

The taste downgrading, when increased solubles yields were taken, were somewhat offset by: 1) drying more concentrated extracts, which lose less aromatics (prepared by freeze concentration or evaporation), and 2) by fortifying the more concentrated extracts with collected aromatic coffee volatiles.

The cost of the coffee bean (raw material) is the most expensive part of the instant coffee product. Therefore, it was more economical to do more elaborate instant processing with flavor recycling to maintain an acceptable instant coffee taste, while taking solubles yields as high as possible.

These patterns of instant coffee processing resulted in the general use of vacuum evaporators for distilling off water, bad hydrolysis flavors, and excess acidity. The resulting higher concentrations of solubles in the extract leaving the evaporator produced increased (spray or freeze drier) production capacity and produced a harder, less fragile granule. Since freeze concentration of extracts resulted in an ice by-product (with several percent associated coffee solubles), water evaporators were also important for recovering the solubles from the melted ice.

The use of low grade green coffees in the growing countries for making instant coffee further accentuates the solubles yield as the sacrifice to the consumer in caffeine deficiencies and coffee purity. Low caffeine levels have been common in some production from Brazil using low grade, defective beans.

**Centrifuging Away Tars**.—Progressively higher solubles extraction yields cause insoluble tars in the withdrawn coffee extracts. To achieve cleanliness of the instant coffee beverage in the cup, it was essential to use automatic de-sludging centrifuges like the DeLaval BPRX types or the Westfalia SAM-5036 types. (See Fig. 11.1–11.5).

FIG. 11.1 EXTRACT PLATE COOLER-DELAVAL

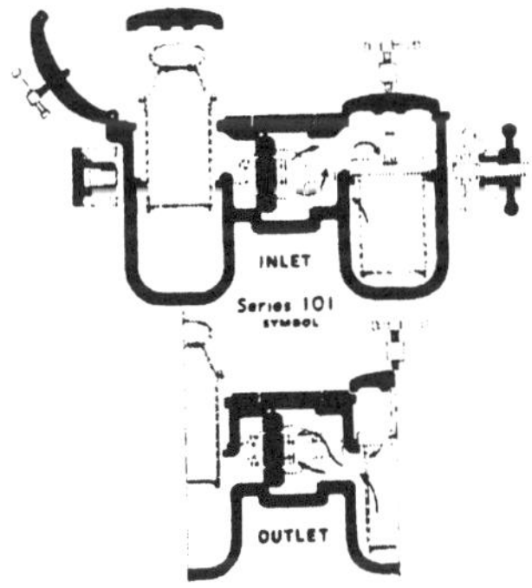


FIG. 11.2. DUAL BASKET STRAINER

FIG. 11.3 AUTOMATIC DESLUDGING CENTRIFUGE-DELAVAL

FIG. 11.4 EXTRACT SCALE AND STORAGE TANKS—CIA CACIQUE—BRAZIL

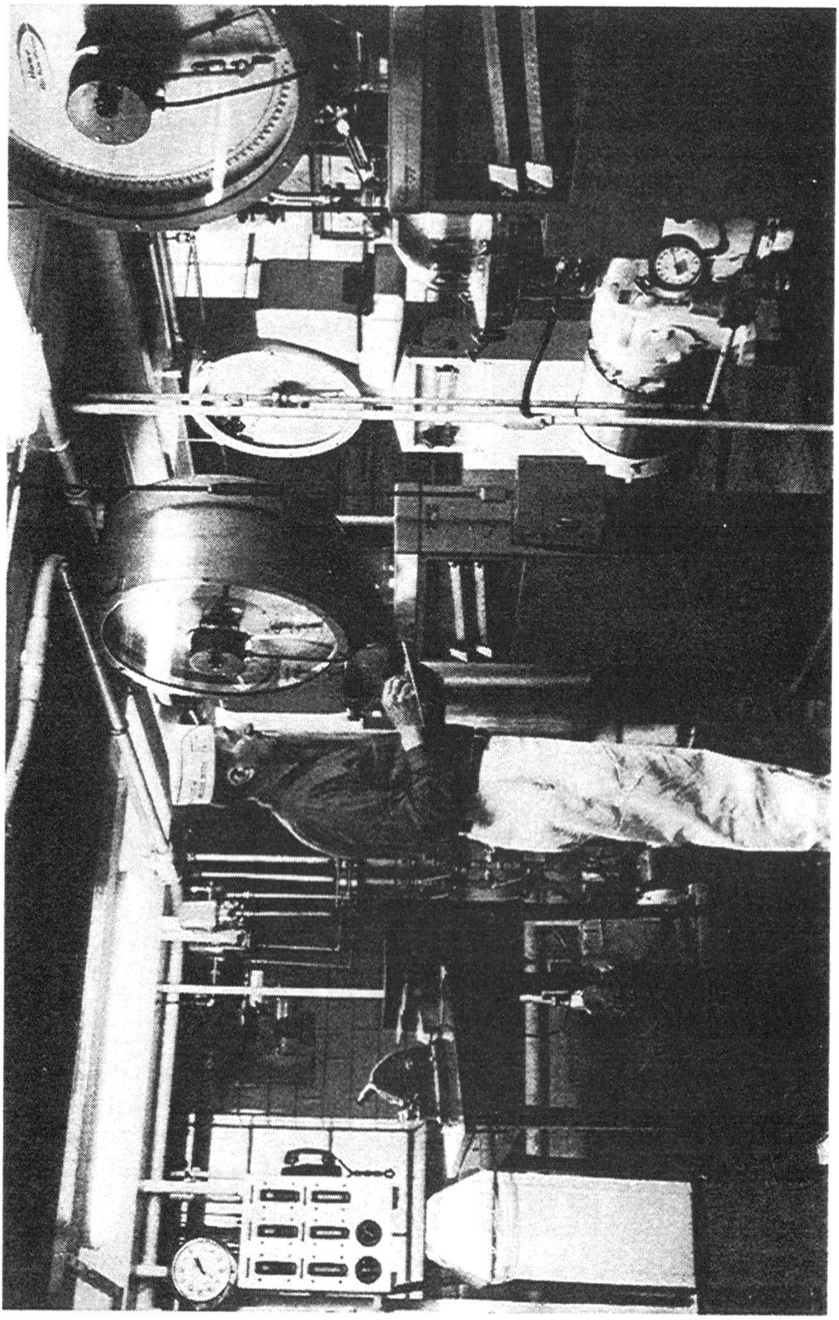

FIG. 11.5 EXTRACT TANK WEIGHING/REFRACTOMETER AREA

Settling tars from the extracts in tanks reduces tar load on the centrifuge. The tars centrifuged away carry noticeable coffee oils, coffee flavor, and aroma for the price of beverage clarity. Settling these tars in cold storage tanks for 4 to 6 hrs removes most of the tars when good high grown coffee beans are used at reasonable yields. But settling is inadequate at high solubles yields from Robusta beans and Brazil beans; hence, centrifuging is required.

Historically, it is important to know that the automatic de-sludging centrifuge for coffee applications was not commercially realized until the early 1960's. For example, Geo. Harrison in Paterson, New Jersey, processed mostly robusta coffees to instants, was one of the first firms to use such a centrifuge (Westfalia) in 1958. However, until about 1963 no centrifuge supplier guaranteed performance. The de-sludging centrifuge had been used for decades on shipboard for recovering oil from bilge water, but clarifying coffee extract was another matter. Hence, the successful use of the automatic de-sludging centrifuge was evolved by DeLaval and Westfalia as demand increased with higher yields and more Robustas.

The removal of this black tar is a normal and essential step in processing the extract whether or not freeze concentration is used. The instant coffee extract and powder cleanliness are measured on a 1.25" diameter filter paper disc against the "milk standard". Freezing, concentrating and thawing can often coalesce additional tars out of the coffee extract.

**Increasing Solubles Concentration of Coffee Extracts.**—This is usually and was initially done by distilling water away from the percolator coffee extract in a wiped wall vacuum evaporator. Later freezing out ice crystals, as in a "slush making" machine, was followed by a centrifuging step to separate the small ice water crystals from the now more concentrated solubles in the product extract. Solubles concentrations in extracts from the percolators were at 20 to 25 percent and were increased to 33 to 40 percent. Evaporation of water is normally associated with a considerable loss of acid and aromatic flavors.

There is also a substantial loss and change in flavor and aroma of the concentrate by the more expensive procedure of freeze concentration when centrifuging is used. In 1973 Grenco in Hergonbosch, Holland, offered for sale a freeze concentration system that did not use a centrifuge so did not expose the delicate coffee concentrate to oxidation. This process should yield an improved tasting extract, but technical difficulties delayed commercial applications until 1976 when an improved unit was installed in Hamburg. It is doubtful if freeze concentration is worthwhile; it is not competitive with evaporation.

The **film evaporator** usually works under a 90 percent vacuum (2 to 3" Hg pressure absolute) and has a hot wall rotating wiper inside of a 2 to 3 foot diameter tube. The outer wall is heated with steam. The film evaporator may be constructed for vertical, inclined or horizontal use. Distilate passes through a water cooled condenser and then through a wet mechanical vacuum pump. The coffee extract may be in contact with a 240 F wall for less than a minute or more, depending on the methods used.

In the **APV** plate type heat exchanger the dilute coffee extract is heated as it flows over stainless steel plates. Then the super-heated extract is allowed to flash evaporate into a separation chamber, releases and separates the evaporated water to a condenser. The concentrated coffee extract is then cyclone collected. The extract may be reheated and reflashed as many times as is necessary to remove the desired amount of water. (See Fig. 11.6.)

The **Alfa Laval Centritherm** evaporator design allows only a thin film of entering percolator coffee extract to spread over a set of steam heated, high speed rotating cones in a vacuum chamber. Extract residence time is only 0.1 second. The centrifugal type evaporator is much more expensive, but is used by DEK,GF HAG in Germany and Iquacu, Vigor, Brasilia and Dominium in Brazil.

In any case, no matter how gentle the evaporation, the process of water distillation also steam distills coffee aromas. Where low grade, bad tasting coffee beans have been used and/or where high hydrolysis yields and volatile taste by-products have been made, evaporation is not such a flavor degrading step as when only good quality coffee beans and reasonable solubles yields have been taken. (See Fig. 11.7.)

The coffee processing industry found that larger and taller spray drier vessels made larger spray dried particles (averaging 0.4 mm and ranging up to 1 mm). These larger particles retained more natural coffee flavor and aroma than coffee extracts dried in small milk type driers. One company called their product "flavor buds." The larger particles had better solubility, better flavor and less dust.

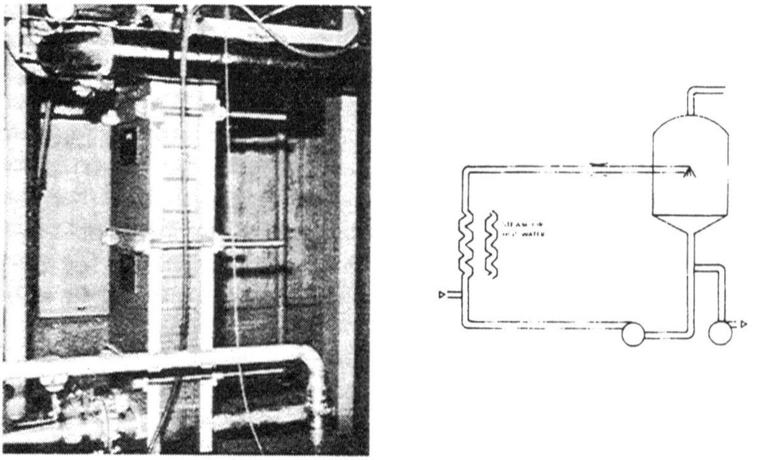

FIG. 11.6 APV PARAFLUSH EVAPORATOR SYSTEM

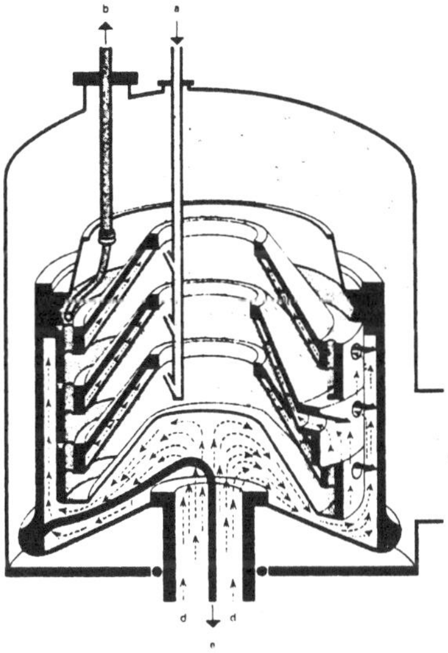

FIG. 11.7 DELAVAL CENTRITHERM EVAPORATOR

## SPRAY DRYING PROCESS

It is now of the interest to learn how the equipment for spray drying makes a dry, hollow bead product of about 3 percent moisture. To explain this, the first consideration is a heat-mass balance in the exchange of heat from the air to the evaporated extract. This fixes the inlet and outlet air temperatures and air moisture levels.

Figure 11.8 shows a heat-mass balance flow diagram for a spray drier handling 7200 cfm NTP (580 lb air per min). The air enters at 480 F (249 C) and leaves at 240 F (116 C). The coffee extract is 33 percent solubles at 40 F (4 C) entering at 1,800 lb per hr. The powder leaves at 600 lb per hr with 3 percent moisture at 90 F (32 C). The 40 F (4 C) extract temperature is used as the thermal base temperature.

The heat content of the entering air at 400 F (244 C) above base temperature is 59,000 BTU per min; extract input heat is zero.

The heat content of the dry exit air at 240 F (116 C) is 26,600 BTU per min. The heat content of the 20 lb per min water vapor is 4,000 BTU per min sensible and 20,000 BTU per min latent heat. The heat content of the 10 lb per min exit powder is 200 BTU per min. With a heat loss of 8,200 BTU per min, the entering and exit heat contents are 59,000 BTU per min or 3,540,000 BTU per hr.

The only difference between the entering and exit mass flows is that the water entering with the extract leaves mostly with the air. The incoming air is assumed to

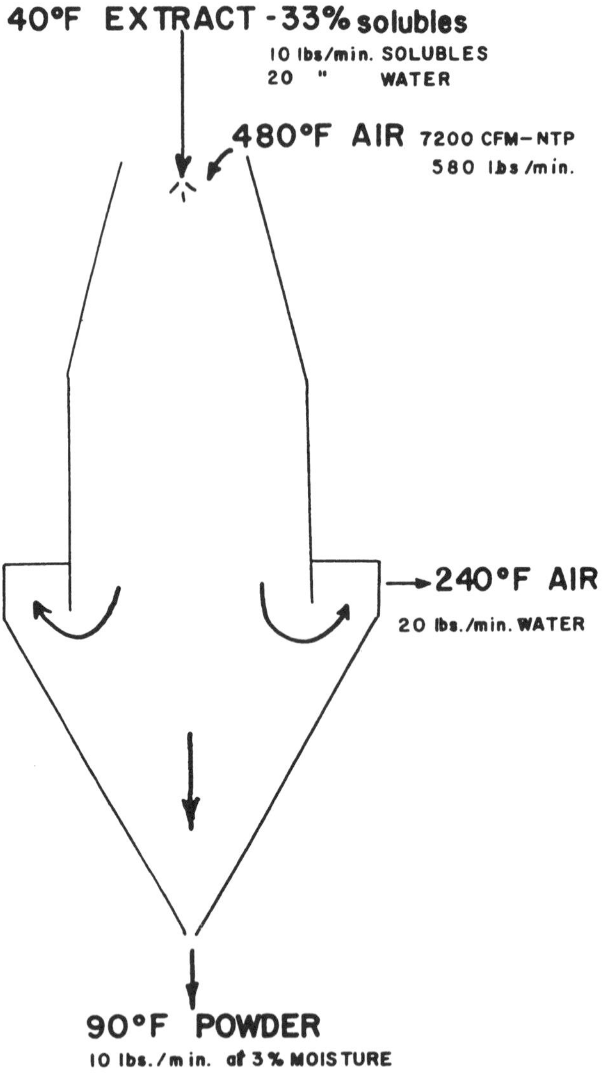


FIG. 11.8. HEAT AND MASS BALANCE ABOUT CO-CURRENT SPRAY DRIER

be dry for simplicity. Actually, the moisture content depends on plant intake air conditions; the incoming air at 85 F (29 C) and 40 percent R.H. holds up to 0.021 lb water per lb air.

## Drier Profiles

The heat-mass balance is simple, but the profiles of air and particle temperatures, moistures, and vapor pressures are more informative. Figure 11.9 shows

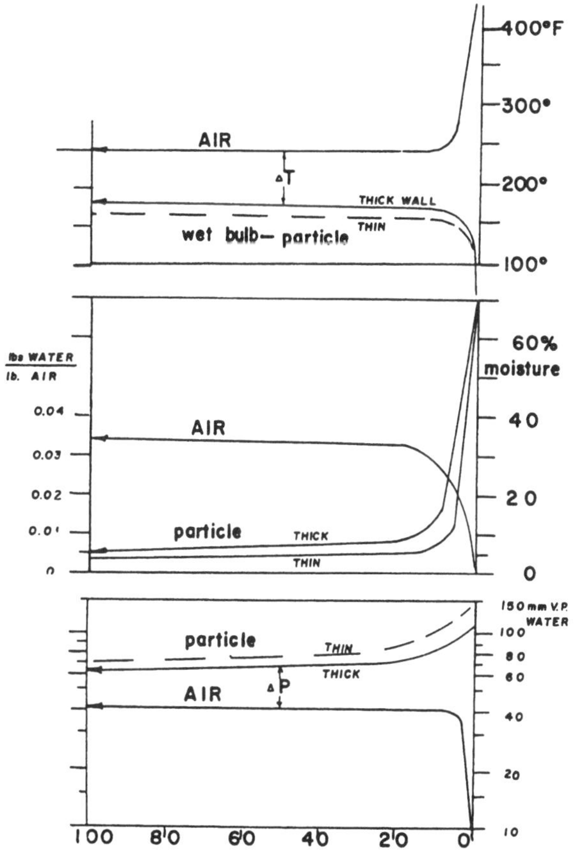


FIG. 11.9. CO-CURRENT SPRAY DRIER PROFILES; TEMPERATURE, MOISTURE, AND VAPOR PRESSURE

these three profiles plotted against drier height, from atomizer to collection bustle. Actually, some drying is effected below the bustle, but it is of little importance. The profiles are schematic in so far as the absolute wall thickness of the particle and the exact shape of the curves are concerned.

The temperature profile controls the rate of drying once the atomized droplets are formed in the turbulent, hot air. Since heat transfer is almost instantaneous, the droplets attain only the wet bulb temperature. In other words, the water vapor freely moves out of the droplet as heat moves in. The evaporative equilibrium temperature is the wet bulb temperature, in this case 92 F (33 C). See Fig. 11.10 for the high temperature psychrometric chart. The low or normal psychrometric chart, Fig. 11.11, is used for determining moistures of intake air and for other air-conditioning problems.

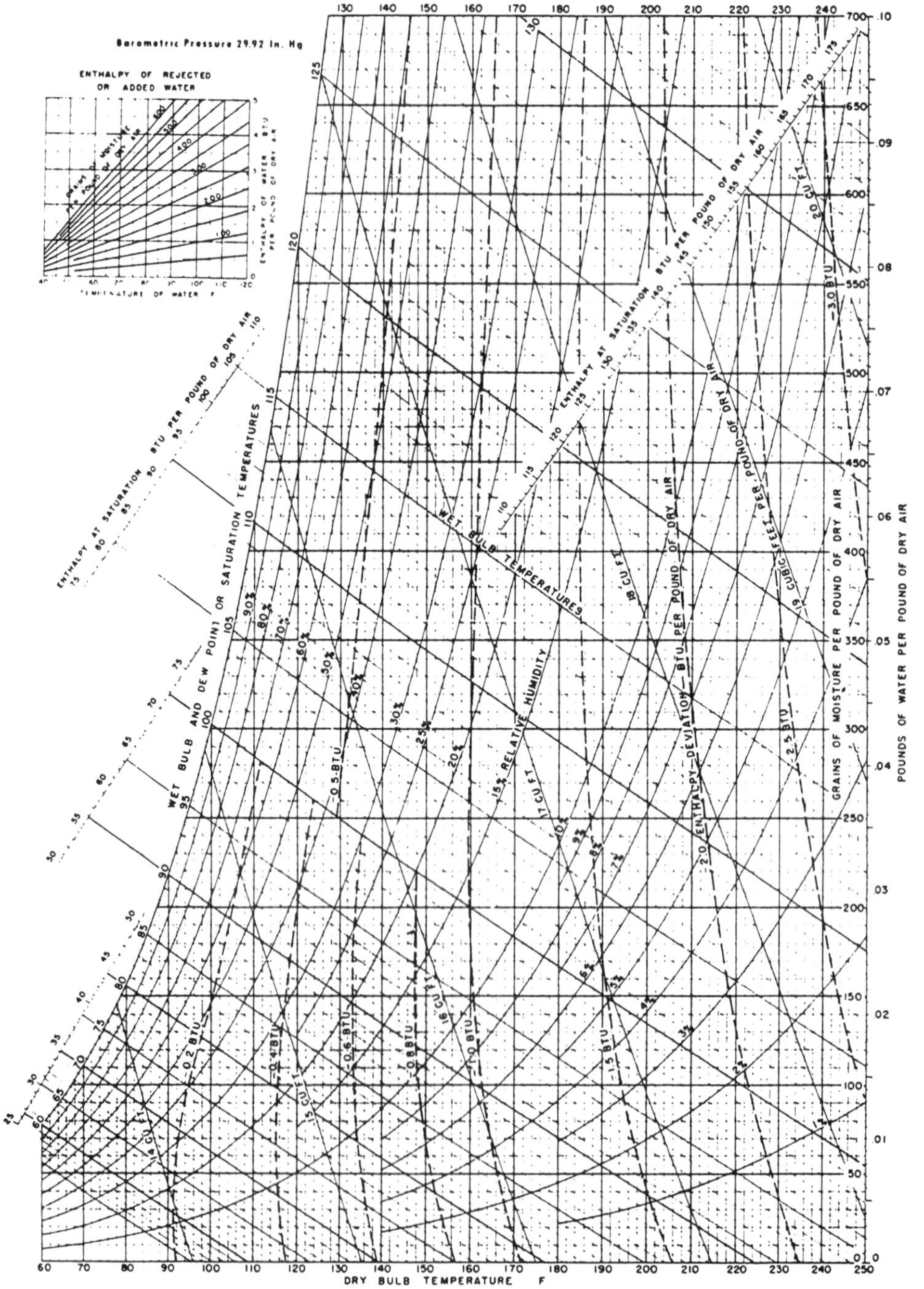


Fig. 11.10. Psychrometric Chart: High Temperature

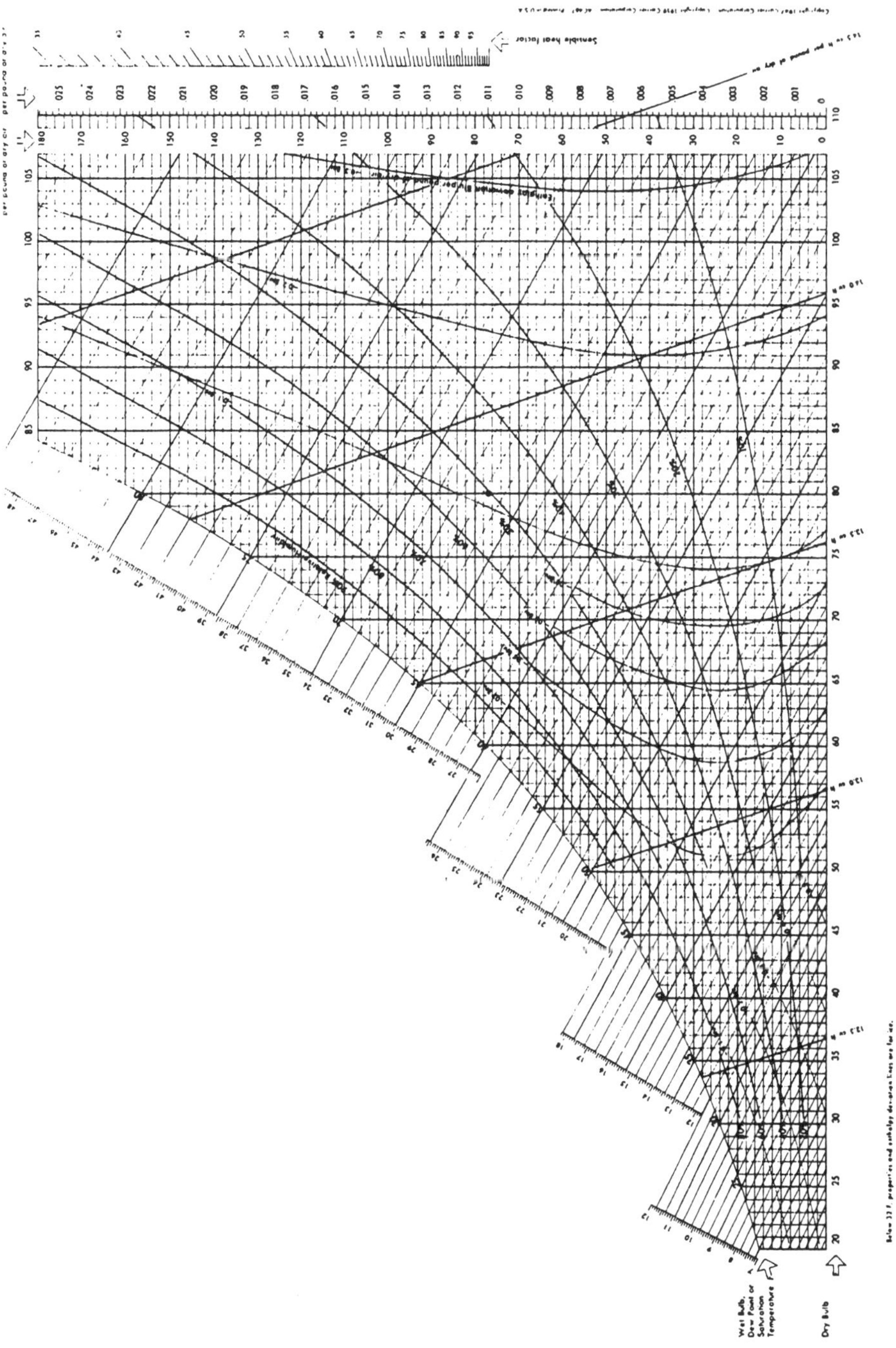


*Both Psychrometric Charts Courtesy of Carrier Corporation*

FIG. 11.11. PSYCHROMETRIC CHART: NORMAL TEMPERATURE

The vapor pressure from tiny droplets of micron size is higher than from a flat liquid surface. This fact is based on capillary measurements of vapor pressure of curved surfaces and from Rayleigh's derivations which are found in most physical chemistry texts. These higher vapor pressures become important on particle sizes smaller than a few microns. There is reason to suspect that, in spray drying, when the equivalent of 10 sq ft or more of liquid surface is exposed per gram of extract, vapor pressures somewhat higher than those from flat surfaces may occur. Atomized droplets are mostly larger than one micron. It is recognized that physical properties gradually change as substances are subdivided to this extent. In fact, this theory may explain the successful results attained in flavor and biochemical stability achieved by spray drying.

**Blossoming**.—As long as the droplet is liquid and has internal turbulence, and there is external air turbulence, a high rate of water vaporization will be maintained. When enough water evaporates and plasticity occurs, the rate of moisture release is retarded. The formation of hollow spheres shows that, even though the rate of water vapor release outwardly has been retarded, the inward release of water vapor continues, as evidenced by the hollow spheres produced.

**Particle Temperature**.—At this point, the hollow, spherical wall is about 1 to 10 mu thick. The residual rate of moisture movement outward is likely to be fast due to the short distance of travel. The water vapor pressures will be fixed by the particle temperature. This temperature will lie between the air temperature and the wet bulb temperature. Table 11.1 lists the temperature-vapor pressure relations in this spray drying region. It is reasonable to assume that the heat transfer to the particle will be at a fast rate. The temperature of the particle will be at a point that is a controlling factor in the particle heating rate; that is, the temperature differential between air and particle. Since the area for heat transfer changes little, and the air film about the particle governing heat transfer changes little, the temperature differential is controlling.

The temperature of the particle, on the other hand, will also be governed by the rate of evaporation of water. For a given particle, this is substantially controlled by the water vapor pressure differential between the particle and the air.

**Rate of Water Evaporation from Particle**.—Drying of the surface reduces the mol fraction of water but, in any case, the principles of heating the particle and evaporating water from the particle to attain some equilibrium particle temperature apply.

The following equation relates the circumstances:

| | | | |
|---|---|---|---|
| (1) | Factor governing heating rate of particle | = | factors governing water evaporation rate from particle |

(2) (air temp. dry bulb − temp. particle) = (constant) × (mol fraction water) × (particle vapor pressure at $T_p$ − air vapor pressure)

(3) $(T_a - T_p) = K \times F_w \times (\text{v.p.}_p - \text{v.p.}_{\text{air}})$

The vapor pressure of water rises a great deal for small increases in temperature as shown in Fig. 11.12. Thus, the above equilibrium equation may easily be brought into balance by raising the particle temperature.

TABLE 11.1. TEMPERATURE VS VAPOR PRESSURE OF WATER

| C | F | mm Hg | State of Instant Coffee |
|---|---|---|---|
| 5 | 41 | 6.5 | |
| 10 | 50 | 9.2 | |
| 15 | 59 | 12.8 | |
| 20 | 78 | 17.5 | |
| 30 | 86 | 31.8 | |
| 40 | 104 | 55.3 | |
| 50 | 122 | 92.5 | |
| 55 | 131 | 118 | |
| 60 | 140 | 149 | |
| 65 | 149 | 188 | solid |
| 70 | 158 | 234 | plastic |
| 75 | 167 | 289 | |
| 80 | 176 | 355 | |
| 90 | 194 | 526 | |
| 100 | 212 | 760 | |
| 125 | 257 | 1,740 | |
| 150 | 302 | 3,570 | |

*Drying Equilibria*.—Higher particle temperatures reduce the rate of heating of the particle and increase the rate of evaporation from the particle. Tables 11.2 and 11.3 show the water vapor pressure relations for particles at various moistures and temperatures vs the water vapor pressure of the air. The fraction of particle area for water vapor permeability is taken as the water mol fraction at any given solubles content during the drying process. Surface drying effects are not taken into account, although case hardening may reduce water vapor permeability. In order to maintain a higher water vapor pressure in the particle than in the surrounding air, the particle temperature must rise. Water vapor permeability area is lost progressively as the particle approaches 3 percent moisture at the dryer discharge. However, water remains a high mol fraction.

In the example given in Tables 11.2 and 11.3, the particle temperature attains

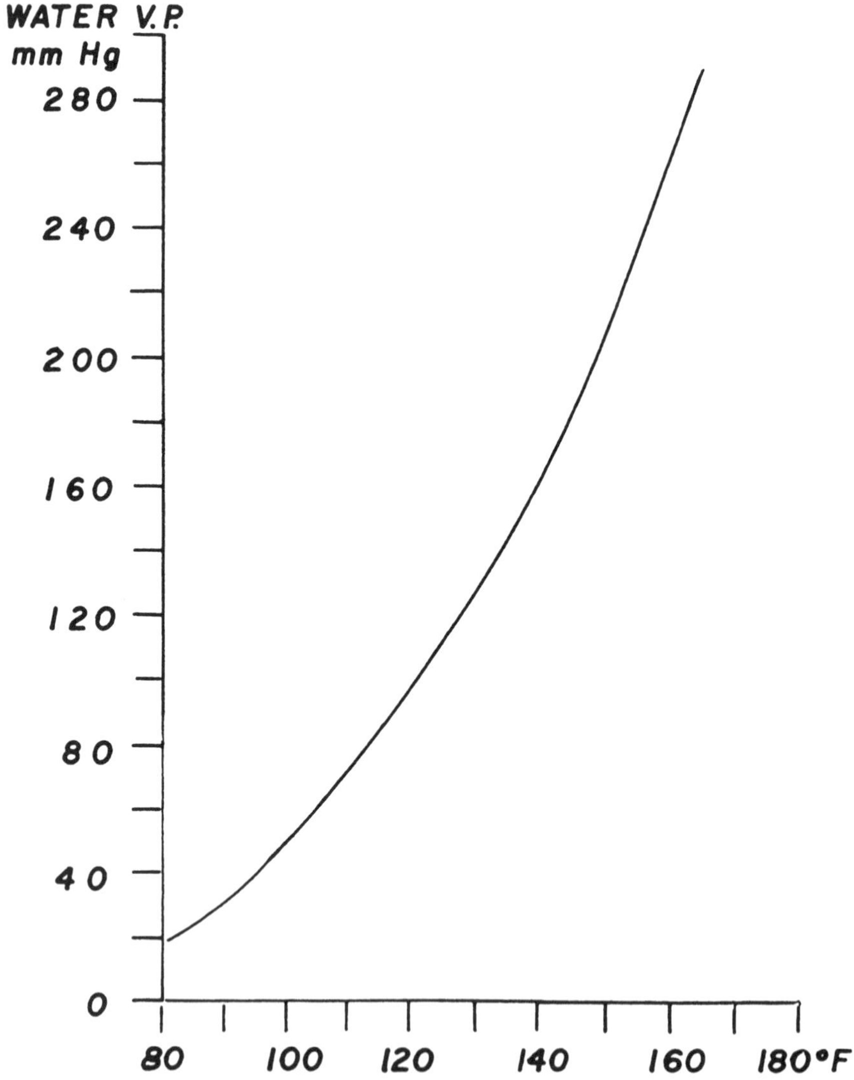


FIG. 11.12. WATER VAPOR PRESSURE VS TEMPERATURE

160 F (71 C) which is logical for two reasons. (1) Particles (depending on moisture and coffee solubles) start to fuse in a melting tube at about 145 F. The physical change is that of a cool, rigid, hollow bead becoming plastic, and then fusing with another particle or itself. The time of fusion is significant; it may take minutes depending on moisture, solubles, temperature, particle size, and wall thickness. (2) A falling particle (in the spray drier) which is almost dry has an air film about it. If it collides with another particle in its fall, the particles are unlikely

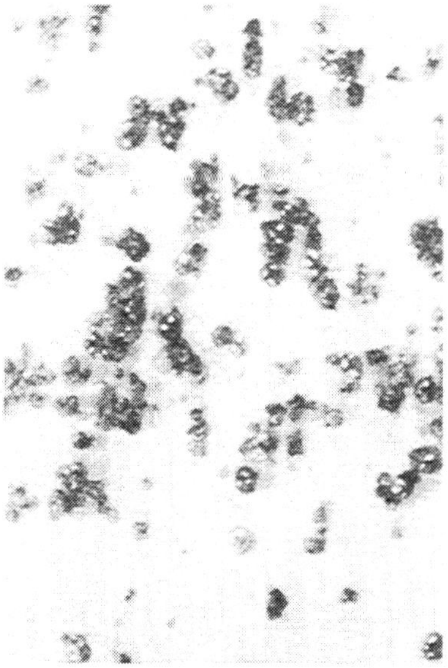

*Courtesy of W. R. Marshall, Jr., Univ. of Wisconsin*

FIG. 11.13. PHOTOMICROGRAPH OF SPRAY-DRIED INSTANT COFFEE PARTICLES

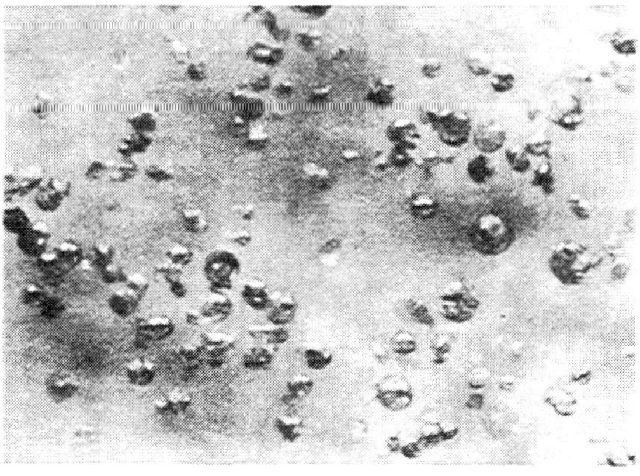

*Courtesy of Delavan Manufacturing Company*

FIG. 11.14. PHOTOMICROGRAPH OF SPRAY-DRIED INSTANT COFFEE PARTICLES

TABLE 11.2. CO-CURRENT SPRAY DRIER

*Water vapor pressure[1] of particles and air[1] vs particle moisture, temperature, and wall thickness*

| Weight Fraction Solubles | Weight Ratio W/S[2] | Mol Fraction Water | mm v.p.[1] Water at F(C) Thin mm | Thin F | Thin C | Thick wall mm | Thick wall F | Thick wall C | Water in Air lb Water/lb Air[3] | Water in Air Mol Fraction[4] | Water in Air mm v.p. Water | Air dry bulb temp. F | Air dry bulb temp. C |
|---|---|---|---|---|---|---|---|---|---|---|---|---|---|
| 0.33 | 2/1 | 0.96 | 144 | 140 | 60 | 188 | 149 | 65 | None | None | None | — | — |
| 0.50 | 1/1 | 0.91 | 135 | 140 | 60 | 211 | 155 | 68 | 0.018 | 0.029 | 22 | 240 | 116 |
| 0.67 | 1/2 | 0.83 | 134 | 143 | 62 | 234 | 158 | 70 | 0.026 | 0.042 | 32 | 240 | 116 |
| 0.90 | 1/9 | 0.53 | 100 | 149 | 65 | 289 | 167 | 75 | 0.033 | 0.053 | 40 | 240 | 116 |
| 0.95 | 1/19 | 0.35 | 83 | 158 | 70 | 355 | 176 | 80 | 0.034 | 0.054 | 41 | 240 | 116 |
| 0.97 | 1/32 | 0.24 | 60 | 160 | 71 | — | — | — | 0.035 | 0.056 | 42 | 240 | 116 |

TABLE 11.3. COUNTERCURRENT SPRAY DRIER

| Weight Fraction Solubles | Weight Ratio, W/S[2] | Mol Fraction Water | Particle v.p. mm | Particle v.p. F | Particle v.p. C | Fraction Heat Load Used | Water in Air lb Water/lb Air[3] | Water in Air Mol Fraction[4] | Water in Air mm v.p. Water | Air dry bulb temp. F | Air dry bulb temp. C |
|---|---|---|---|---|---|---|---|---|---|---|---|
| 0.33 | 2/1 | 0.96 | 10 | 55 | 13 | None | 0.0033 | 0.0053 | 4.0 | 80 | 21 |
| 0.50 | 1/1 | 0.91 | 11 | 60 | 16 | 0.50 | 0.0016 | 0.0016 | 2.0 | 177 | 81 |
| 0.67 | 1/2 | 0.83 | 12 | 70 | 21 | 0.75 | 0.0008 | 0.0012 | 1.0 | 225 | 107 |
| 0.90 | 1/9 | 0.53 | 11 | 80 | 27 | 0.88 | 0.0004 | 0.0006 | 0.2 | 250 | 121 |
| 0.95 | 1/19 | 0.35 | 13 | 90 | 32 | 0.97 | 0.0001 | 0.0002 | 0.1 | 267 | 131 |
| 0.97 | 1/32 | 0.24 | 12 | 98 | 37 | 1.00 | None | None | None | 273 | 134 |

[1] Vapor pressure in mm Hg.
[2] Water to solubles ratio by weight.
[3] Water to air ratio by weight.
[4] Assuming air molecular weight is 29.

to fuse in such a contact which may take only a fraction of a second. Further, spray dried instant coffee particles (see Figs. 11.13 and 11.14) examined under a stereomicroscope appear as hollow, glassy spheres that resemble balloons that are not quite full of air. The surfaces are undulating, collapsed, or blown out. Because the temperatures of the hollow beads are high enough to be changing the beads a plastic state, formation of these undulations, holes, ellipsoids, etc. while the particles are falling is reasonable. These particle shapes "freeze" after passing through the cooler air of the drier cone outlet zone.

**Co-current Drier Profile.**—In the co-current spray drier profiles in Fig. 11.9, it may be seen that particle temperatures at the end of the drying period are higher than corresponding wet bulb temperatures. Two particle temperature profiles are shown—one for a thin-walled particle (lower temperature) and the other for a thicker-walled particle. The curves are relative. From diffusion theory, doubling the diffusion distance doubles the diffusion time (i.e., halves the diffusion rate). In these profiles, for the sake of simplicity, particles of different wall thickness are subjected to the same air temperature profile. In this case, the thicker-walled particle will discharge at a higher particle temperature (176 vs 160 F, 80 vs 71 C) and at a higher moisture (5 vs 3 percent). The thicker walled particle attains a higher particle temperature sooner and remains at a higher temperature throughout the drying period; yet it leaves at a higher moisture level. If particle wall thickness is greater, higher outlet air temperatures and longer drying times are required to attain a final moisture near 3 percent. This, however, is not harmful to the final powder flavor since the time-temperature differences are in seconds and only 10 to 20 F higher.

In Fig. 11.9, the three sets of curves illustrate the temperature of air and thick or thin walled particles while passing through the drier, the buildup of air moisture and reduction in extract moisture, and the water vapor pressure in air and particles. The differential temperatures and vapor pressures of water between particles and air reflect in the driving forces in drying.

These curves simplify what is occurring, but they are definitely illustrative of the drying factors involved. The increased particle temperature increases water vapor pressure, fluidity, and molecular movement within the particle wall, and hence, increases the effectiveness of water diffusion. The facts are that considerably thicker-walled, larger, hollow particles, with excellent coffee flavor retention (far better than is commercially attained at present) and with about 3 to 4 percent terminal moistures may be attained. Thus, the limits of the spray drying process for better quality soluble coffee may be extended. This point has been somewhat obscured and neglected since the addition of coffee oils to spray dried powder was introduced. No addition of coffee oil, however, will produce the same quality of coffee flavor as achieved by the natural retention of coffee flavor in the spray drying process.

*Particle and Air-Water Vapor Pressure.*—Figure 11.9 shows the profiles of corresponding water vapor pressures in the air to thin- and thick-walled particles. If the air enters with 0.022 lb of water per lb of air, then it would have 0.056 lb of water per lb of air upon leaving. The vapor pressure of water in the air, being proportional to molar volumes, would be $0.056 \times 29/18 \times 760$ mm $= 68$ mm or considerably more than is shown using the assumption of dry air entering. Thus, the importance of the moisture in the entering air in limiting the rate of water evaporation from the drying particles may be appreciated. For example, air with 68 mm Hg water vapor pressure cannot dry the particles on the profiles shown. Higher outlet air temperatures would be required to increase the water vapor pressure from the particles to effect an evaporative driving force (differential water vapor pressure).

Figures 11.15 and 11.16 show respectively a LEWA extract piston pump at ground level of the drier, with three extract lines at top and three nozzles inside a 6 meter drier. Fig. 11.17 shows the outside of a commercial installation.

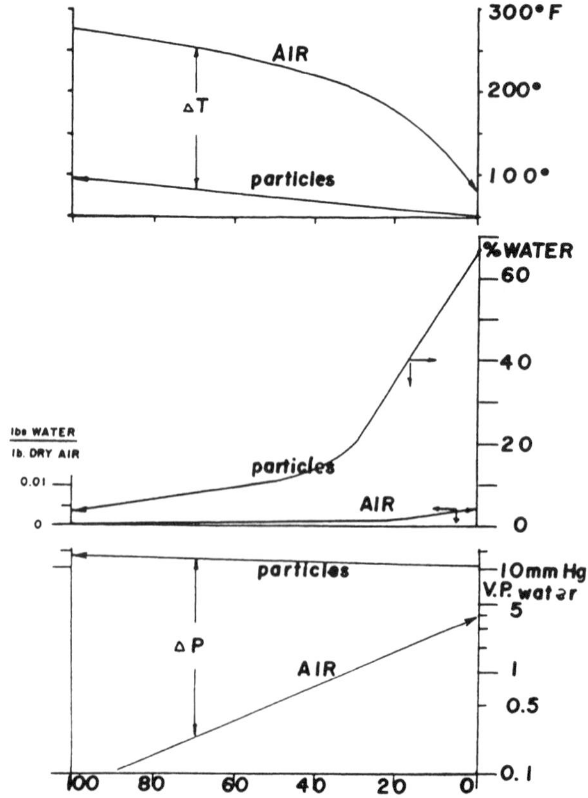


FIG. 11.15. COUNTERCURRENT SPRAY DRIER PROFILES: TEMPERATURE, MOISTURE, AND VAPOR PRESSURE

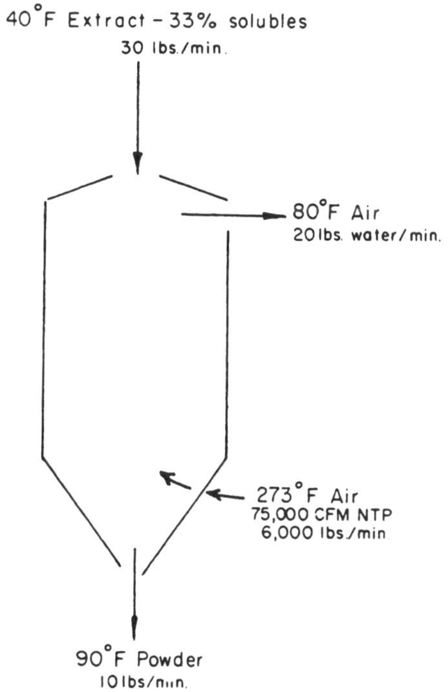


FIG. 11.16. COUNTERCURRENT SPRAY DRIER FLOWSHEET; HEAT AND MATERIAL BALANCE

FIG. 11.17. COUNTERCURRENT SPRAY DRIER

*Temperature and Volatiles Retention.*—Low temperature drying, whether by spray drying or by a modified or true freeze drying, does not offer any better volatiles retention. In fact, there is evidence that freeze drying loses more volatiles than spray drying. This may be attributed to the much longer time that freeze drying takes (24 hr) compared to spray drying (seconds). The point is that some drying variables are more important to flavor retention than others; low temperature in itself is not a guarantee of obtaining a better flavored product.

*Particle Buoyancy.*—Buoyancy or air drag on particles is an important consideration in countercurrent driers. Some particles will be levitated upward. Air drag forces show what size and density of particles will be levitated out with the air. In co-current spray driers air drag pulls downward on the particle falling under gravity. In a countercurrent spray drier levitation may lengthen the time of suspension of particles, thereby increasing their drying time.

*Stokes' Law.*—The equation for calculating the terminal or constant falling velocity of a small, spherical particle in air is as follows:

$$\text{Force} = 6\pi r V \eta$$

The force, $F$, is the frictional force of the air through which the particle is falling, and when this is counterbalanced by the force of gravity acting on the particle, i.e., its weight, the velocity becomes constant.

$F$ = dynes when
$r$ = radius of particle in cm,
$V$ = terminal velocity in cm per sec and
$\eta$ = coefficient of viscosity of air in poises
One poise = one dyne-sec per $cm^2$

The weight of the particle in grams is its volume in $cm^3$ times its density, $d_1$, in gm per $cm^3$. The weight in gm is the acceleration of gravity, $g$, times its weight in dynes. An average value for $g$ is 981 cm per $sec^2$.

We have

$$F \text{ (dynes)} = 6\pi r V \eta \text{ and } F \text{ (dynes)} = \frac{4}{3}\pi r^3 g(d_1 - d_2)$$

or

$$6\pi r V \eta = \frac{4}{3} g r^3 (d_1 - d_2)$$

where $d_2$ is the density of the air.
Solving this equation for $V$
we have

$$V = \frac{2gr^2(d_1 - d_2)}{9\eta}$$

But particle density is surface area multiplied by wall thickness multiplied by density of wall divided by particle volume.

$$d_1 = \frac{4\pi r^2 tp}{4/3\pi r^3}$$

where

$p$ = wall density = 1.6 gm per cm$^3$ used for carbohydrates
$t$ = wall thickness in cm
r = radius of particle in cm

Therefore

$$d = 4.8 \frac{t}{r} \text{ gm per cm}^3$$

For normal spray drier practice, $t/r$ will range between 0.1 and 0.004; hence, $d$ will range roughly from 0.5 to 0.02 gm per cm$^3$. Air density, $d_2$, at 273 F (134 C) is 0.0009 gm per cm$^3$; hence, in relation to particle density, $d_1$, it may be neglected, and by substitution we have

$$V = \frac{2 \times 4.8\, tr}{9\eta} = \frac{1.07\, tr}{\eta}$$

In other words, the terminal falling particle velocity in air becomes solely a function of particle density and size. As a result of doubling the particle size with the same particle density, wall thickness is doubled. This quadruples terminal falling velocity; the air drag is doubled, but the weight of the particle is 8 times more at constant $t/r$. Countercurrent air flow forms little or no hollow spheres; hence, the particles are denser and smaller and are not easily levitated. Agglomerated particles are larger but give more drag.

The viscosity of air, $\eta$, is
183 $\times$ 10$^{-6}$ poises at room temperature

and

219 $\times$ 10$^{-6}$ poises at 212 F (100 C).

Note that gas viscosities rise while liquid viscosities fall with rising temperatures.

Figure 11.18 is a plot of the last equation on parameters of 1, 5, and 10 mu wall thicknesses. With a countercurrent spray drier, air velocity is 60 fpm. This about 0.7 mph, which is hardly a breeze. If a horizontal air velocity line is drawn in Fig.

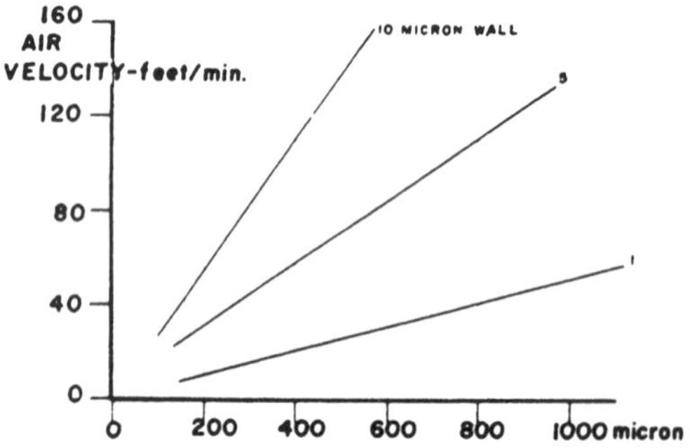


FIG. 11.18. PARTICLE DIAMETER VS TERMINAL VELOCITY

11.11, the particles above and below the line fall are levitated. These data are also useful for a co-current drier by showing how the air-drag influences the free fall velocity. The ranges of particle size and wall thickness for most commercial instant coffees are included in the plot. The velocities of these particles would be modified by the indicated air velocities whether up or down, as the case might be.

Figure 11.19 is a similar plot, but here terminal particle velocity is plotted against particle diameter on parameters of *t/r*, wall thickness/radius.

## Spray Drying Operating Variables

Now that spray drying equipment and temperature-moisture profiles have been discussed, it is appropriate to list the independent and dependent processing variables:

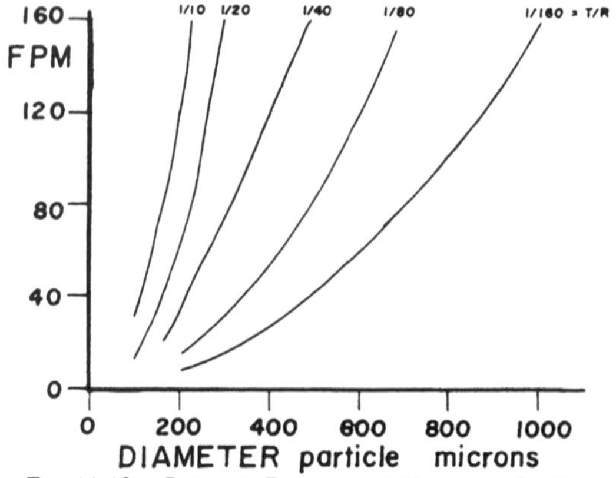


FIG. 11.19. PARTICLE DIAMETER VS TERMINAL FALLING VELOCITY

*INDEPENDENT*

1. ratio of mass flow: air/extract
2. mode of heat application
3. size and shape of spray drier and auxiliaries
4. air flow rate and mode of introduction
5. extract flow rate and mode of atomization
6. moisture of inlet air
7. air inlet and outlet temperatures
8. alteration of extract properties

*DEPENDENT*

1. moisture profile
2. temperature profile
3. powder properties:
   a. flavor
   b. bulk density
   c. moisture
   d. solubility
   e. color
   f. fluidity
   g. clear cup—no turbidity
      —no floating substances
      —no sediment
      —no rings on cup rim
      —no foam
   h. appearance (geometry of particle)

Once the spray drier equipment is designed for a certain processing capacity, its performance may be changed only in limited ways. For example, inlet and outlet air temperatures may be varied somewhat, rate and mode of atomization can also be altered; the extract may be gassed to get lighter powder densities (bulkier product) or fatty acids from dark roast coffee may cause heavier powder densities. Higher extract concentrations give increased drier productivity.

**Powder Properties.**—The leaving powder must have a moisture close to 3 percent. If it is higher, particles may fuse in storage at room temperature, as well as deteriorate in flavor with time. The bulk density of instant coffee before packaging in jars must be about 14 lb per cu ft or about 22 gm per 100 cm$^3$. If it is too bulky, it will not fit into the jar unless the hollow beady particles are broken to increase density. Light density powder may also be blended with heavier density powder. Instant coffee powder must be soluble within no more than 10 sec after boiling water is added to it, but this is less important when agglomerated.

An advantage of beady powder with thick walls is that it undergoes less particle breakage during transfer and packaging. Broken particles are not only physically undesirable but have less flavor. This may be demonstrated, as we have said before, by tasting a cup of instant coffee before and after particle crushing. Similar taste results are found by screening out the coarse and fine fractions; the coarse fraction has superior and more natural coffee flavor.

Color is not a critical property of the powder, but most people perfer a dark, reddish powder. Many instant coffee powders are a light color, pale tan, due mostly to high hydrolysis yields and fines. Powder fluidity is a relative property, but is important as it causes the powder to flow properly in a vacuum-type Pneumatic Scale Company jar filler. If powder does not flow properly, jar fills will be incomplete, which will disrupt a high speed packaging line. High speed jar filling machinery requires that the powder has both good fluidity and proper bulk density and little dust. Agglomerates are usually weighed.

Although the cleanliness and clarity of the cup of instant coffee are due largely to the properties of the extract being spray dried, spray drying contributes to some of the cup shortcomings such as foam, sediment, and floating specks. Agglomerated and burnt particles in the powder are abnormal and undesirable. When present, these off-standard powder properties must be corrected during processing. The ability achieve and maintain all the powder properties listed is not easy and usually preference must be given to the two more commercially important physical properties, bulk density and moisture, and then to solubility and fluidity. Cup cleanliness is, of course, essential too. In this choice, unfortunately, there is a loss in the flavor retention in the powder. For example, breakage of particles is often used as a means of increasing powder bulk density, but particle breakage deteriorates instant coffee flavor. Lightening bulk density is often done by use of $CO_2$ gas in the extract or by heating extract to produce more beady particles of more uniform size. Both methods result in a larger surface per unit weight of powder with associated natural coffee flavor losses. Few soluble coffee plants maintain all desired powder properties consistently. Off-standard powder must be dry to be blended back into production powder to dilute the effect of such variances. The properties of the extract have the greatest influence over spray drying results.

**Controlling Variable Extract Properties in Spray Drying**.—There have been numerous instances when a spray drier operator has not been able to keep powder properties within standards. Properties may differ at different times. The drier operator does not know what coffee blend, roast, grind, or percolation conditions have produced the extract. Hence, he does not know the variation in surface tension and other physical and chemical extract properties that may occur.

*Plant Supervisor*.—In such circumstances it is the responsibility of the shift foreman or plant processing superintendent to interject his knowledge as to what changes have occurred in processing the extract that would change extract properties and hence, powder properties. The foreman must know how to bring powder properties back within standard. If the superintendent does not know the cause of the circumstances, it is difficult to suggest a remedy. Such situations may result in "out-of-standard" powder for 8 hr or longer, depending on the rigidity of product standards. It is the aspect of soluble coffee processing that makes the testing of product quality a dynamic procedure; it may be called process control. It takes an experienced supervisor to know how to evaluate product properties and processing, and then to take the facts and make the proper adjustments that will result in a powder within standards.

*Solubles Concentration*.—The spray drier operator does have some meaningful knowledge about the extract properties such as solubles concentration and extract temperature. Figure 11.13 shows the relationship between extract solubles concentration and the bulk density of the resulting powder under relatively similar atomizing conditions and the same extract temperature.

Bulk powders densities increase rapidly at extract solubles concentrations over 32 percent. At extract solubles concentrations over 35 percent, maintaining bulk densities at about 14 lb per cu ft is no problem. This is an ideal bulk density to maintain. Invariably, soluble coffee processing will form extracts that cause too light a powder bulk density; the higher solubles concentration will remedy such a deficiency. Higher powder bulk densities occur because the extract is more viscous, which results in a thicker-walled particle. It also means that higher outlet air temperatures are needed to effect a 3 percent moisture in the discharge powder.

*Extract Viscosity*.—The increase of extract viscosity with rising solubles concentrations is not great until over 30 percent is attained; the viscosity is much higher at solubles concentrations over 40 percent. Viscosity increases are more accentuated at temperatures below 70 F (21 C) than near 200 F (93 C). The viscosities of coffee extracts of about 35 percent solubles at 50 to 70 F (10 to 21 C) may be simulated in 30 percent extract by cooling to below 40 F (4 C). In the range of 35 to 40 percent solubles concentration, a 3 percent increase in solubles concentration will have about the same effect in increasing viscosity as a 20 F (11 C) chill. In extracts of 35 to 40 percent concentration at temperatures below 50 F (10 C), a 10 F (6 C) chill will produce a viscosity rise equivalent to that produced by several percent increase in solubles concentration. Viscosity increase may be attained by cooling the extract or by higher solubles concentrations.

The absolute viscosities of coffee extracts are important properties. Viscosities may be readily measured by timing extract flows at various temperatures and concentrations through a given length of small standardized bore glass tube, or through an Ostwald viscosimeter. The coffee extract viscosities are made relative to a liquid for which the viscosities are known, for example, sucrose solutions. The viscosities of sucrose above 20 percent solubles are much lower than for coffee extracts. The reason for this is that a large number of the coffee extract solubles have large molecules, whereas sucrose is only a disaccharide. The larger molecules cause more friction (higher viscosity) in flowing over each other and, because of their caramel nature, they hold some water chemically or physically. At 10 or even 20 percent solubles in extract, the viscosities are not much different from water. But at over 30 percent solubles, when the amount of free water that helps molecular mobility is reduced to the point where molecular flow friction is increased, viscosities rise rapidly. The influence of temperature on coffee extract viscosity is not very marked until below 50 or 40 F (10 or 4 C) and above 30 and 40 percent solubles.

The viscosity-temperature-solubles concentration relationship will vary with different instant coffees because of the blend, roast, grind, mode of percolation, etc. One coffee may be more caramelized than another. Caramelization is the polymerization to complex saccharide structures with large molecular weights. The viscosities of coffee extracts in different plant processes, therefore, will vary within a small range for a given temperature and concentration. Also, since they are natural substances undergoing variable treatments to assorted molecular structures their viscosities cannot be as constant as those of pure chemical substances.

In the coffee solubles viscosity data, viscosity may be represented by two equations in terms of temperature and concentration. The extract concentration viscosities between 0 and 10 percent solubles behave similarly to water. The viscosities of concentrations between 10 and 30 percent behave as designated by one equation, and the extract viscosities between 40 and 60 percent behave as designated by another equation. Parameters of extract concentration radiate out from one point. Dilute coffee extracts behave like water; concentrated coffee extracts probably behave more like large molecule carbohydrates. In the region between 30 and 40 percent solubles, where much higher powder bulk densities are produced in spray drying. Coffee solubles and sucrose viscosities are similar to about 20 percent solubles, but the viscosities differ markedly at higher solubles concentrations.

*Extract Density*.—The solubles concentration of the extract alters density so that the volume contributions of water are 1 gm per ml and solubles are 1.6 gm per ml. In other words, a 32 percent solubles extract is 0.68 ml water and 0.20 ml solubles; this 0.88 ml coffee extract has a specific gravity of 1.135. This relationship holds quite well up through 60 percent coffee soluble concentration. Brix density, which is a direct percentage for sucrose solubles in water, yields about a 15 percent higher Brix percentage than actual coffee solubles. For example, 36.0 Brix is 30.0 percent soluble coffee solids, or 40.0 Brix is 34.0 percent coffee solubles. The spray drier operator, therefore, may quickly measure the solubles concentration of the extract. The operator is not able to change the solubles concentration, but he can appraise the spray drying conditions as for sucrose being difficult, normal, or hopeful for correcting bulk density. High concentration solubles extract and cold viscous extracts produce thick-walled hollow spheres, a more beady powder, a richer, darker color, and much better water solubility and fluidity. The extract density enters into the atomization phenomena, but is not of much significance apart from the fact that high powder densities result from high extract concentrations; high viscosity is the effective extract property influencing the character of the final powder.

*Surface Tension*.—The surface tension of coffee extract has a great deal to do with how the extract atomizes. It may vary considerably and is usually directly due to the amount of fatty acids and oils in the extract. The surface tension of a liquid is like a surface film that pulls the material together like a skin. The film's ability to pull its molecules together is dramatically demonstrated by mercury which has a surface-to-air tension of 470 dynes per cm. Water has 75 dynes per cm at 32 F

(0 C) in air but only 60 dynes per cm at 212 F (100 C). Liquefied helium (0.1 dyne per cm), nitrogen (8 dynes per cm) or carbon dioxide (1 dyne per cm) have low surface tensions. For example, a liquid nitrogen film will climb the side of a Pyrex beaker and flow over. Matter naturally assumes its lowest energy form. Hence, substances that contribute to reduced surface tension in a solution concentrate at the surface. The homologous fatty acids all reduce the surface tension of water. The higher molecular weight fatty acids are more effective in reducing the surface tension for given molar concentrations. Soap, which is the sodium or potassium salt of fatty acids, is a cleanser because in reducing surface tension it "wets" the grease that holds dirt particles to one's hand or to oily surfaces. Water alone is unable to do this. The amount of soap dissolved in the water to effect this wetting action is very small, e.g., of the order of magnitude of parts per million. For example, fatty acids in coffee extract at 10 to 100 ppm reduce surface tension from about 50 dynes to 25 dynes per cm.

Within the range of spray drying temperatures, temperature has little influence on changes in surface tension. There is a 10 percent reduction of surface tension in a 50 F (28 C) rise in temperature. Solubles concentration has little to do with the surface tension of the coffee extract except indirectly. That is, the amount of fatty acids in the extract depends on how the extract was prepared.

The surface tension, hence the wetability, has much to do with the palatability of beer and the quality of its foamy head (surface boundary energy).

The reduced surface tension due to fatty coffee acids is used to advantage in U.S. pat. 2,929,716, by Barch and Reich (1960). Not only may the fatty acid content of the coffee extract be enhanced by darker roasts, finer grinds, and by omitting the centrifuging of the drawn off extract, but additions of fatty (oleic) acids at about a 100 ppm level may effect marked increases in powder bulk density. This is achieved because the atomized conic film is thinner (like a soap bubble), breaks up into more fine droplets but forms thicker-walled hollow spheres. This gives a darker powder with poorer fluidity but better coffee flavor retention in the spray drying process. The fatty acids also reduce persistent foam in the cup. The cup of instant coffee has a smoother taste texture and better overall flavor. Reduced surface tension of the droplet skin makes smaller sized droplets. Surface tension is a direct influence in jet or drop breakup. Reduced droplet film strength under a wave disturbance effects easier breakup of the projected film. Reduced surface tension may also contribute to more satellite droplets which may become fines or agglomerates and contribute to poorer powder fluidity. The liquid conic film breaks up as the surface tension (skin retention force) is overcome. Theoretical equations developed for drop size distribution show that the mean droplet diameter is reduced in proportion to the square root of the surface tension.

*Coffee Fatty Acids*.—Fatty acids, such as the unsaturated types of oleic acid, oxidize and yellow when exposed to sunlight, especially when they are in water solution in warm storage. Fatty acid aqueous solutions should be prepared only as they are about to be used. Solution of fatty acids in water is effected by stoichiometric neutralization of a sodium hydroxide solution with the acid at concentrations of about 141 gm oleic acid (mw 282) to 20 gm NaOH in one liter distilled water. This concentration of fatty acid may be diluted with distilled water to $^1/_5$ or $^1/_{10}$ this concentration for metering into coffee extract. Usually a liter of concentrate diluted to a gallon of metering solution is convenient.

Fatty acids with oils that naturally occur in coffee extracts may hinder drying of atomized droplets by coating droplet surfaces. If there is as much as 3 percent fatty acid-oil mixture, powder bulk density may be as high as 30 lb per cu ft, and it becomes difficult if not impossible to dry the extract completely. The oil coats each droplet and reduces permeability to water vapor removal. The extract must be centrifuged before atomizing for successful spray drying. Percolation, by its nature, filters the extract, and the higher solubles concentrations of extract do not allow the colloidal oils to pass out with the extract. The electrical conductivity of the coffee extract becomes so high at concentrations over 15 percent solubles that the oily colloids do not stay in solution. Instant coffee in the cup is clear, not turbid like brewed coffee, except for aromatized instant coffees with oil add-back in the powder.

*Tars and Fatty Acids*.—Hydrolysis tars that are carried out of the percolators in the extract are colloidal in size, coagulate on standing, and subsequently settle out. If standing time is too long or if extracts are centrifuged, practically all the oils and fatty acids (as well as flavor) are removed with the tars. This clarified extract yields powder with a light bulk density. It is simpler and better to allow the coffee extract to settle out the largest tar particles and then to spray dry the extract while there are still some fatty acids and coffee flavor remaining. This procedure also keeps powder bulk density in the range desired for packaging. When espresso or Italian dark-roast coffees are percolated, the amount of fatty acids and oils carried with the extract produces high powder bulk densities and also contributes to poor powder fluidity.

*Dark Roast*.—Oleic acid at 100 ppm of coffee solubles is effective in increasing powder bulk densities 1 to 2 lb per cu ft. Oleic acid taste is not detected until 200 or 300 ppm. In blending 10 to 15 percent Italian dark roast coffee with the normal soluble coffee plant roast and blend, percolated extract will have enough fatty acids and oil to effect satisfactory control of powder bulk density. This will also darken powder color, improve cup flavor, and reduce foaming as the cup is prepared.

This circumstance occurs naturally when low grade or imperfect types of coffees are used. This is because the lack of bean uniformity causes some beans to burn or over-roast, thus contributing fatty acids and oils. Darker roasts appear to give higher extract viscosity, which might be due to a higher degree of caramelization of the soluble coffee carbohydrates. Opposed to this, light bean roasts tend to give light powder bulk densities, light powder colors and very poor coffee flavor retention in the powder. High powder bulk densities are directly due to high extract viscosities (high solubles concentrations or low temperatures of extracts) and dark roasts. Low bulk densities of powder are directly due to low extract concentrations, high solubles yields from R&G coffee, light bean roasts and high gas content of extract (usually $CO_2$ is naturally present in R&G coffee). High extract viscosities are a controlling factor in the resulting powder properties at concentrations of solubles of over 30 percent, whereas at lesser concentrations the surface tension (fatty acid content) is more of a controlling factor on average particle size, color and flavor of powder. Coffee extracts have lecithin-type phospholipids, which also reduce surface tension of the atomized droplets. Addition of expelled coffee oil (500 ppm) has only a slight influence on surface tension because there is much more water insoluble oil than water soluble fatty acids.

In working with the coffee oils from darker roasts, it is noted that their flavor contribution is due to the fatty acid content. The higher fatty acid content holds more fine tars in suspension; dark roast coffees show more floating specks and sediment. Extract must settle for about 2 hr before spray drying, otherwise tars will carry over into the product powder.

An oil slick or tiny droplets of oil are sometimes noted in cups of instant coffee. These are usually associated with dark roasts or portions of the coffee blend that are dark roasted. In themselves, some oil droplets do no harm; the oil helps carry coffee flavor through the spray drying process.

*Gas in Extract*.—$CO_2$ gas dissolved in the extract lightens powder bulk densities by a few pounds per cubic foot. The effect of $CO_2$ gas in the extract may be noted when freshly roasted coffee is extracted and sent promptly to the spray drier. This usually causes foaming, light bulk densities of powder and light colored powder. De-gassing of coffee extract is not normally done, but is certainly worthy of inclusion in the extract pre-treatment before spray drying. Foaming of the coffee extract must be circumvented. This may be done in moderate vacuums in a film evaporator or in a soft beverage carbonator machine. De-gassing would have to be done without loss of coffee flavor volatiles. Resorting to $CO_2$ gassing of the extract to lighten powder bulk density in such a case results in spheres blossoming into fragments, which not only give the powder a "fuzzy" appearance but also produce poor fluidity. Starvation of extract flow to the piston pump pressurizing the atomizer will cause foam and erratic atomization, as well as fines, "fuzzy" powder, and "gold" specks.

*Startup and Shutdown Extract*.—The effects of the fatty acid and hydrolysis solubles are best noted in startup and shutdown extracts and in the powders. Percolator startup extracts have little or no hydrolysis solubles, are lower in solubles concentration, and higher in fatty acids; they spray dry to a dark, dense powder having good flavor. Percolator shutdown extracts have more than average hydrolysis solubles, are about average in solubles concentration, but are low in fatty acids; they spray-dry to a powder that is light in color, light in bulk density, and poor in coffee flavor, which produces much foam in the poured cup often with a persistent foamy ring. Therefore, it is desirable to dilute startup and shutdown extracts with normalized production extract to diminish these end differences in extract properties.

An example of how startup extract properties change is the case in which 17–19 lb per cu ft powder densities resulted with about 5 percent plus 40 mesh and 5 percent through 80 mesh. Particles were fine and agglomerated with poor fluidity. Raising and then lowering atomizing pressure and increasing air inlet temperatures, from 475 to 540 F (246 to 282 C) had no appreciable influence on powder properties. When the next sequential batch of extract was spray dried (concentration 27 percent instead of 23 percent), at closer to normal solubles yields, the powder densities abruptly leveled out at 13 lb per cu ft without any other change in spray drying conditions. The particles became beady with 20 percent through 40 mesh and 3 percent through 80 mesh; agglomerates practically disappeared; and fluidity was good. Hydrolysis carbohydrates control the distensible character of the extract to produce larger particles and lower powder bulk densities; the fatty acid content of this extract is naturally less. There is a region where the carbohydrate content controls fatty acid content; when carbohydrates are altered by darker roasts, fatty acid content is controlling. For example, 100 ppm fatty acids at the high carbohydrate level of light roasts have little influence, while at a darker roast level they are very effective in increasing powder bulk density. High hydrolysis solubles content of extract is usually associated with poor powder solubility and residual foaming.

*Direct Atomizing of Extract*.—Why is percolator extract not spray dried immediately upon removal from the percolator system? When this is done the extract is sent to the spray drier before tars and fine R&G particles have been settled out, and its hot nature and low viscosity yield a foamy powder with many specks, cup rings, and a high loss in natural coffee flavor.

If coffee extract sent to the spray drier is not clean and free of suspended matter, the resulting cup will have sediment, cup wall film and/or cup ring. Further, there may be floating carbonized specks which are most obvious when cream or milk is added.

*Extract and Powder Densities*.—Figure 11.20 illustrates schematically the change in powder bulk density with particle surface to weight ratio. It shows that high powder bulk densities may be obtained with the production of a non-uniform assortment of small sized particles such as are produced by high pressure atomiza-

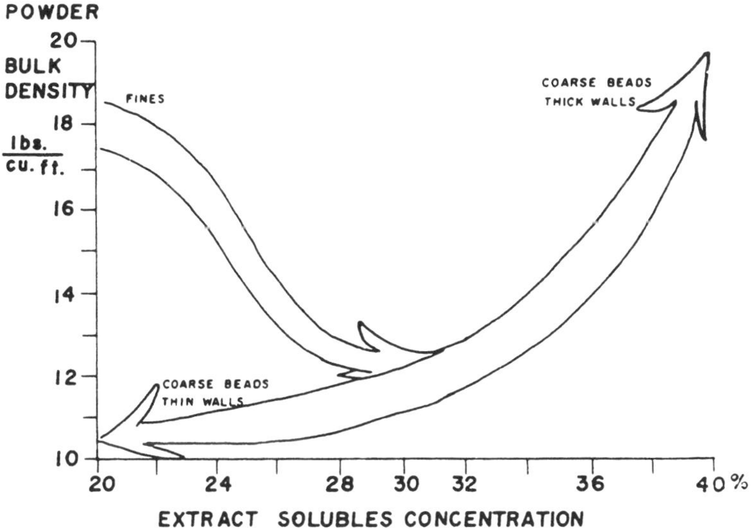


FIG. 11.20. POWDER BULK DENSITY VS EXTRACT SOLUBLES CONCENTRATION

tion of low viscosity extracts warm extracts or low solubles concentration extracts). Such densities may also result from low atomizing pressures on more viscous extracts, expecially with high carbohydrate levels (from light roasts) which give light bulk densities at about 27 percent solubles concentration. With extracts of over 35 percent solubles concentration, the powder densities increase at still lower atomizing pressures. The graph does not include all the extract variables involved but shows the contributing extract variables influencing powder bulk density variations. The particle surface to weight ratio is a key to degree of natural coffee flavor retention. For example, a trend to lighter powder bulk densities may be traced back to a possible change in blend, roast, extract concentration, solubles yield, fineness of grind, extract temperature, extract holding time, etc. The individual extract drawoffs, collected from the percolators, are mixed when the decanted extract has been transferred to the spray drier supply tank; this prepares a uniform extract concentration and temperature. Otherwise, there is marked variation in spray dried powder properties due to stratified extract layers. Extract layers stratify even when they are transferred from one tank to another.

Figure 11.20 not only shows the striking increase in bulk density that is obtained with extract concentrations over 30 percent, but the marked change in physical properties of the spray dried particles. The coarse, thick-walled beads may also be formed by higher extract viscosities due to chilling of the extract, reduced surface tension of extract as with dark roasts and fine grinds that release more fatty acids, as well as with lower solubles yields containing less carbohydrates. These coarse beads carry 5 to 10 times as much natural coffee flavor volatiles through the spray drying as low concentration extracts. The coarse beads have about one square meter surface per gram of powder while the thin walled spheres and fines have about 25 square meters surface per gram of powder. The percentage of coarse beads with thin walls as obtained from low concentration extracts can also be increased by high carbohydrate yields, light roasts, and reduced extract viscosities obtained by heating extracts. Fines cause higher bulk densities due to their greater bulk compaction, but they exhibit poor solubility, foaming, low flavor level, and poor powder flow.

**Volatile Coffee Flavor Retention (From Extract to Spray Dried Powder)**.—The major problem of improved instant coffee flavor is the large coffee flavor loss in transforming extract to powder as it is currently processed in spray driers. Unfortunately, leading soluble coffee companies have chosen to fortify the spray dried powder with expelled coffee oil rather than to work toward natural coffee flavor retention in the resulting powder at the spray drier.

An examination of the variables influencing losses of volatiles will point out the appropriate steps for minimizing coffee losses of volatiles in current spray drying practices as well as to improve flavor retention in the future. It is remarkable that any low boiling volatiles that constitute the coffee flavor are retained during spray drying, but the fact is that they are. Volatiles retention may be increased several-fold in various ways.

The most important factor in the change from extract solubles with their associated water to dry solubles with only 3 percent moisture in powder form is the fraction of original to final water that must be evaporated. From the general laws of vapor pressure of mixtures of volatiles, the more water that is associated with a volatile, the higher will be the associated coffee volatile losses. This also depends on the volatility relative to water of the specific coffee volatile. Figure 11.21 graphically depicts the vapor pressures over the pertinent temperature range of most of the currently identified and important coffee aroma and flavor volatiles.

*Volatiles Retention Mechanisms*.—Relative volatiles are largely a function of vapor pressures and molar composition in the evaporating water. Relative vapor pressures of these substances may vary at very low concentrations in a coffee extract medium for at least the following seven reasons. (1) The coffee solubles, for example, increase solubility of aldehydes occurring in coffee. There is real affinity for the solubles and this class of volatile compounds. This attraction would reduce the relative vapor pressure of aldehyde to water, compared to what is found in a pure aldehyde water system. (2) Other coffee organics mixed with the volatiles will influence the volatility behavior of each; hence, the true vapor pressure curve should be different from that of the pure volatile. (3) Some volatiles form constant

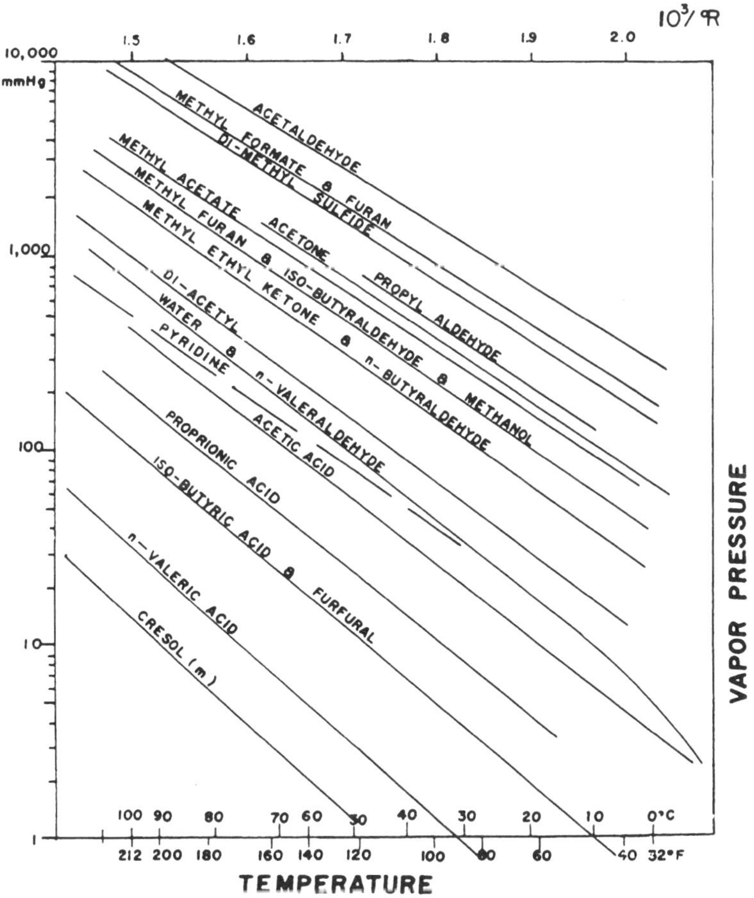


FIG. 11.21. VAPOR PRESSURE OF COFFEE VOLATILES VS TEMPERATURE

boiling mixtures, azeotropes; this might be another factor influencing distillation; (4) Some volatiles dissolved in coffee solubles behave like water, physically-chemically attracted; this is analogous to free and bound water. (5) Also coffee solubles exert adsorption and absorption powers. For example, some phospholipids are slighly water soluble, yet they hold organic carbonyls, fatty acids and petroleum ether solubles, better than water. Oils are immiscible with water or coffee extract, yet oils have a very high solubility distribution (coefficient) for aliphatics and long carbon chain carbonyls. For example, the oils may hold 1,000 times the concentration of some coffee flavor volatiles that the contacting soluble coffee extract holds. If the final instant coffee powder has 0.1 percent oil (petroleum ether solubles), the one part of oil per 1,000 parts of powder will hold as much of that coffee flavor volatile as all the powder. Petroleum ether residues in percolated coffee extracts, spray dried to powders, run to about 0.5 percent while still attaining packable bulk densities, but 0.2 to 0.3 percent is more common. Therefore, if the volatiles distribution concentration is only about 300, the fraction of 1 percent oil in the solubles holds as much coffee volatiles as the powder. (6) Invariably, the extract, unless it has been centrigued, has fine particles of coffee which also contain oil; these are carriers of volatiles. Adding a few percent of finely ground roast coffee to instant coffee powder is sufficient to leave a noticeably improved aroma and flavor impression. (7) Distillation in a fraction of a second is hardly enough time for full equilibrium to occur.

*Taste Sensitivity.*—It takes only 1/50 of the original volatile flavor content of the roast coffee (which, for the most part, is carried over into the extract in good percolation practice) to impart a suggestion of natural coffee flavor in instant coffee. Nature has been generous with coffee flavor and has also endowed man with great sensitivity in the perception of minute amounts of volatile chemical substances. This order of magnitude of perception in flavor level change may be better appreciated from taste tests on residue concentrations in spray dried powders.

*Calculated Volatiles Retention.*—Table 11.4 gives the results of a calculation. It is assumed that each coffee volatile listed is distributed according to the vapor pressures of the pure components relative to water at 212 F (100 C) and to their mol fraction in solution. Converting weight relations (ppm) of volatiles to molar compositions reduces the molar concentration of the coffee volatile relative to water. This is because all of the coffee volatiles have molecular weights much higher than water. (See Table 11.5.)

The vapor pressure of the coffee volatiles relative to water is smaller at 212 F (100 C) than at 90 F (32 C). The relative vapor pressures and molecular weights are shown in Table 11.4. The column headed *adjusted (c)* differs from volatility as used in distillation practice due to allowance made for volatiles retention by coffee solubles. The ratio of pure volatile to water vapor pressures divided by the ratio of molecular weights of volatiles to water is called *(c)* in the *simple distillation equation:*

$$\left(\frac{W}{W^0}\right)^{c-1} = \frac{x}{x^0}$$

where

$W_0$ = the original weight of solution being evaporated
$W$ = the final weight of solution being evaporated
$x_0$ = the original molar composition of volatile component
$x$ = the final molar composition of volatile component

TABLE 11.4. PROPERTIES OF SELECTED COFFEE VOLATILES

| Coffee volatile | B. pt D | B. pt C | Mol. wt. | Mw/18 | Rel. vp ppm in solubles | Vol/$H_2O$[1] 90 F (32 C) | Vol/$H_2O$[1] 212 F (100 C) | c[2] 212 F (100 C) | Adjusted c-1 |
|---|---|---|---|---|---|---|---|---|---|
| Acetaldehyde | 70 | 21 | 44 | 2.4 | 90 | 36 | 18 | 7.5 | 6.5 |
| Methyl formate | 90 | 32 | 60 | 3.3 | 12 | 21 | 12 | 3.3 | 2.3 |
| Furan | 90 | 32 | 68 | 3.8 | 7 | 21 | 12 | 3.3 | 2.3 |
| Dimethyl sulfide | 100 | 38 | 62 | 3.5 | 3 | 18 | 10 | 3.0 | 2.0 |
| Propylaldehyde | 120 | 49 | 58 | 3.2 | 30 | 10 | 6 | 2.0 | 1.0 |
| Acetone | 133 | 56 | 58 | 3.2 | 60 | 9 | 5.5 | 1.8 | 0.8 |
| Methyl acetate | 136 | 57 | 74 | 4.1 | 25 | 8 | 4.6 | 1.1 | 0.1 |
| Methyl furan | 146 | 63 | 82 | 4.5 | 10 | 5.3 | 3.5 | 0.8 | −0.2 |
| Isobutyraldehyde | 144 | 62 | 72 | 4.0 | 30 | 5.3 | 3.5 | 0.8 | −0.2 |
| Methanol | 149 | 65 | 32 | 1.8 | — | — | — | — | — |
| Methylethylketone | 176 | 80 | 72 | 4.0 | 15 | 3.5 | 2.5 | 0.6 | −0.4 |
| N-butyraldehyde | 169 | 76 | 72 | 4.0 | 15 | 3.5 | 2.5 | 0.6 | −0.4 |
| Diacetyl | 190 | 88 | 86 | 4.8 | 200 | 1.5 | 1.3 | 0.25 | −0.75 |
| Water | 212 | 100 | 18 | 1.0 | high | 1.0 | 1.0 | 1.0 | 0.0 |
| N-valeraldehyde | 217 | 103 | 86 | 4.8 | | 1.0 | 1.0 | 1.0 | 0.0 |
| Pyridine | 216 | 102 | 79 | 4.4 | 300 | 1.0 | 0.7 | 1.0 | 0.0 |
| Acetic acid | 244 | 118 | 60 | 3.3 | 3000 | 0.64 | 0.56 | 0.15 | −0.85 |
| Propionic acid | 286 | 141 | 74 | 4.1 | 300 | 0.24 | 0.26 | 0.06 | −0.94 |
| Isobutyric acid | 309 | 154 | 88 | 4.9 | 200 | 0.11 | 0.13 | 0.026 | −0.974 |
| N-buytyric acid | 327 | 164 | 88 | 4.9 | 200 | 0.11 | 0.13 | 0.026 | −0.974 |
| Furfuraldehyde | 324 | 162 | 96 | 5.3 | ?[3] | . . . | . . . | | |
| N-valeric acid | 369 | 187 | 102 | 5.7 | 100 | 0.03 | 0.04 | 0.007 | −0.993 |
| M-cresol | 394 | 201 | 108 | 6.0 | 100 | 0.01 | 0.02 | 0.003 | −0.997 |

[1]Relative vapor pressure, volatile/water.
[2]$c$ = mol percent volatile in vapor/mol percent volatile in liquid adjusted downward relative to water.
[3]Furfural yield varies with degree of hydrolysis solubles formed.

Table 11.5, volatiles having almost the same temperature-vapor pressure curves are grouped together because they are likely to behave similarly on volatilization.

TABLE 11.5. CALCULATED FRACTION ($x/x_0$) OF COFFEE FLAVOR VOLATILES CONCENTRATION RETAINED IN INSTANT COFFEE POWDER RESIDUE FROM 33 PERCENT SOLUBLES COFFEE EXTRACT DURING SPRAY DRYING

| | | | | |
|---|---|---|---|---|
| Final moisture | 13% | 6.3% | 3% | |
| Fraction water retained | 1/12.5 | 1/25 | 1/70 | |
| Log of water fraction | −1.1 | −1.4 | −1.8 | |
| | | | | c-1 |
| Acetaldehyde | $7 \times 10^{-8}$ | $1 \times 10^{-9}$ | $2 \times 10^{-12}$ | 6.5 |
| Methyl formate | $3 \times 10^{-3}$ | $6 \times 10^{-4}$ | $8 \times 10^{-5}$ | 2.3 |
| Furan | | | | |
| Dimethyl sulfide | $6 \times 10^{-3}$ | $2 \times 10^{-3}$ | $2 \times 10^{-4}$ | 2.0 |
| Propyladehyde | 0.08 | 0.04 | 0.02 | 1.0 |
| Acetone | 0.13 | 0.08 | 0.44 | 0.8 |
| Methyl acetage | 0.8 | 0.7 | 0.5 | 0.1 |
| Methyl furan | 1.6 | 2.0 | 2.5 | −0.2 |
| Isobutyraldehyde | | | | |
| Methanol | | | | |
| Methylethyl ketone | 3 | 4 | 5 | −0.4 |
| Butyraldehyde | | | | |
| Diacetyl | 7 | 10 | 23 | −0.76 |
| Water | 1 | 1 | 1 | 0.0 |
| N-valeraldehyde | | | | |
| Pyridine | 1 | 1 | 1 | 0.0 |
| Acetic acid | 2 | 3 | 4 | −0.67 |
| Propionic acid | 10 | 16 | 34 | −0.85 |
| Isobutyric acid | 12 | 23 | 57 | −0.974 |
| Furfuraldehyde | | | | |
| N-valeric acid | 12.5 | 25 | 70 | −0.993 |
| M-cresol | 12.5 | 25 | 70 | −0.997 |

The term $c$ would normally be taken from an experimental slope of mole percent volatile in vapor/mole percent volatile in liquid. In the absence of such data, it may be assumed that the ideal laws (Raoult) hold and the ratio of pure vapor pressures may be used until better or confirmatory data are obtained. For example, for acetic acid, $c = 0.75$, but from comparative vapor pressures at 90 F (32 C), $c = 0.64$, and at 212 F (100 C), $c = 0.56$. These are not far off and, in fact, are somewhat more nearly correct for dilute acetic acid solutions in water. The division of the vapor pressure ratios by the molecular weight ratios is based on the assumption that the solubles in coffee extract will have a depressing effect on the vapor pressure of the organic volatiles. Since $c$ is in the exponent, its effect on the amount of residual volatile concentration with different $W/W_0$ degrees of water evaporation will be important. The reduced value of $c$ due to division by molecular weight ratio is in better agreement with observed results. The equation used is based on differential amounts of water solution evaporating under a changing equilibrium molar concentration, and represents an integrated solution to the evaporation problem; that is, a reasonable projection of the evaporative drying of the spray dried droplet of coffee extract.

The term $c$ is influenced by terminal compositions of the evaporating mixture and, in this case, by the solubles retentive properties. When $c$ is more than 1, the coffee component is more volatile than water and its concentration is reduced with evaporation. When $c$ is less than 1, the coffee component is less volatile than water, and the residue becomes relatively richer (compared to residue water) in concentration of that volatile coffee flavor component. For example, when $c = 3$, and $W/W_0 = 1/10$, then $(W/W_0)^2 = x/x_0$ or 1/100.

A sample problem is illustrated in Table 11.5 as follows:

*Problem:* Find the fraction of coffee flavor volatiles left after instantaneously spray drying a soluble coffee extract with 67 percent water to droplets having 13.8 percent (16/166), 7.4 percent (8/108), and 2.9 percent (3/103) final moistures. It is assumed that water diffusion drying will continue to about 3 percent from these higher moistures. There should be few volatiles lost in the secondary drying since water has the smallest molecule by far and will diffuse faster than the volatiles. This corresponds to the constant drying rate period. The original coffee volatiles contents in the extract are given in Table 11.4 with the ratios of volatiles to water vapor pressure.

*Fraction of water mixture retained*

$$\text{Case I} = W/W_0 = 16/200 = 0.080 \text{ or } 1/12.5$$
$$\text{Case II} = W/W_0 = 8/200 = 0.040 \text{ or } 1/25$$
$$\text{Case III} = W/W_0 = 3/200 = 0.015 \text{ or } 1/70$$

Where the values of $x/x_0$ are greater than one, as in the lower part of the table, the concentrations of the coffee volatiles are increased in relation to water. The absolute amount of each coffee volatile in mg retained from each kg of solubles in 33 percent solubles extract is the original volatile content in mg/kg (ppm), from Table 11.4, times the listed fraction ($x/x_0$) retained for each degree of concentration from Table 11.5 multipled by the fraction of water retained. For example, for diacetyl at the various moisture levels, we have

$$200 \times 7 \times 1/12.5 = 112 \text{ mg/kg}$$
$$200 \times 10 \times 1/25 = 80 \text{ mg/kg}$$
$$200 \times 23 \times 1/70 = 66 \text{ mg/kg}$$

These values are shown approximately in the order of magnitude of volatiles retention column in Table 11.6.

*Boiling Point Rise*.—The boiling temperature of coffee extract is very close to that of water at atmospheric pressure. The boiling point rise is small because the molecular weights of most of the coffee solubles relative to water are high. The major extract solubles have the following molecular weights:

| Percent of coffee solubles | Molecular weight |
|---|---|
| 2 Trigonelline | 137 |
| 8 Simple sugars | 171 |
| 4 Caffeine | 194 |
| 14 Chlorogenic acid | 354 |
| 15 Ash | 1,000—associated with other molecules |
| 15 Protein | 1,000 or more |
| 25 Hydrolyzed sugar | thousands |
| 17 Others | thousands |
| 100 | 1,000 average, estimated |

Then mol fraction solubles is 0.33/1,000 or 0.00033, and mol fraction water is 0.67/18 or 0.03700.

*Dimethyl Sulfide Retention*.—Table 11.5 shows the fraction of volatiles retained in the solution. For the state problem Case I shows, under the most optimistic conditions, that $0.08 \times 6 \times 10^{-3}$ or 0.5/1,000 ths of the original dimethyl sulfide (only noticeably present in high grown mild coffees) is left. Assuming 10 ppm DMS in the original coffee extract solution, this leaves only about 5 ppb DMS or barely a taste threshold concentration (6 ppb DMS) in the

TABLE 11.6. COFFEE FLAVOR VOLATILES

*Natural Abundance in R&G Coffee[1]—Magnitude of Retention in Powder, Taste Thresholds*

| Coffee flavor volatile | Occurrence in R&G coffee, ppm | Order of magnitude factor of volatiles retention | Observed volatile retention, powder ppm | | Taste threshold, ppm: Coffee powder | Taste threshold, ppm: Coffee[2] beverage |
|---|---|---|---|---|---|---|
| Acetaldehyde | 50 to 100 | Negligible | . . . | . . . | 20 | 0.20 |
| Methyl formate | 10 to 20 | 1/1000 | 0.2 | 0.02 | 3 | 0.03 |
| Furan | 5 to 15 | 1/1000 | 0.4 | 0.02 | 6 | 0.06 |
| Dimethyl sulfide (DMS) | 3 to 6 | 1/1000 | . . . | 0.006 | 0.6 | 0.006 |
| Propylaldehyde | 10 to 20 | 1/200 | 4 | 0.1 | 5 | 0.05 |
| Acetone | 40 to 80 | 1/100 | 55 | 0.8 | 30 | 0.3 |
| Methyl acetate | 5 to 10 | 1/30 | . . . | 0.3 | 2 | 0.02 |
| Methyl furan | 5 to 20 | 1/10 | 16 | 1 | 5 | 0.05 |
| Isobutyraldehyde | 10 to 30 | 1/10 | 14 | 1 | 6 | 0.06 |
| Methanol | . . . | 1/10 | . . . | . . . | 100 | 1 |
| Methylethyl ketone | 10 to 20 | 1/5 | 11 | 4 | 10 | 0.1 |
| N-butyraldehyde | 5 to 10 | 1/5 | 0.4 | 0.4 | 1 | 0.01 |
| Diacetyl | 15 to 30 | 1/2 | 24 | 15 | 4 | 0.04 |
| Acetyl propionyl | 10 to 20 | 1/2 | 11 | 10 | 10 | 0.1 |
| N-valeraldehyde | 10 to 20 | 1/2 | . . . | 10 | 10 | 0.1 |
| Isovaleraldehyde | 2 to 5 | 1/2 | 2 | 2.4 | 5 | 0.05 |
| Pyridine | 200 to 500 | 1/2 | . . . | 100 | 10 | 0.1 |
| Furfuraldehyde | 100 to 400 | 1/2 | . . . | 50 | 10 | 0.1 |
| Methyl mercaptan | 2 to 5 | 1/1000 | 2 | . . . | 0.2 | 0.002 |
| Hydrogen sulfide | 2 to 10 | 1/1000 | 1 | . . . | 0.2 | 0.002 |

[1]There are about 200 to 400 ppm volatiles (not counting acids) in R&G coffee.
[2]Cup concentration taken as 1/100 powder.

powder and 1/80 of this concentration in the cup. The dimethyl sulfide in the natural coffee extracts is above taste threshold. The use of high grown mild coffees for instant coffee blends is not yet justified by sufficient flavor retention in the instant coffee powder. The dimethyl sulfide aroma and taste are easily recognizable in the coffee extract.

Additions of 100 ppm dimethyl sulfide to coffee extract (Case I) prior to spray drying, according to the calculations, yield about 50 ppb DMS in the powder or perhaps one ppb in the cup and should not be detectable by taste.

Actually, DMS is prominent in the powder aroma. This means that DMS volatiles retention is in fact better than shown in Case I by at least a magnitude of ten. This volatile retention result would have to be attributed to the seven causes previously cited. There must be great attraction between the DMS at low concentrations and the non-volatile coffee solubles. The accuracy of the other calculated volatile retentions may also be compared to observed experience. In every case the level of volatile retention is very much influenced by the fraction of water evaporated.

*Fraction of Water Retention*.—In raising the solubles concentration from 25 ($3W/1S$) to 50 percent ($1W/1S$) the associated water is reduced to one-third. It is most important that the fraction of water remaining in the powder increases relative to the amount of water in the extract at the start of drying. The water reduction ratio depends on initial and final moistures of the extract and powder respectively. In Fig. 11.22 this water reduction ratio is plotted vs solubles concentration on parameters of final percent moisture. The area for normal spray drying is below 4 percent moisture and above 35 percent solubles. Depending on which part of this area is being worked, the water reduction ratio may vary from 150 to 50, a three-fold factor. Taking moistures down to 1 percent increases the water reduction factor. The higher the water reduction factor, the higher the volatiles reduction factor. The parameters of over 4 percent moisture are shown because viscous extracts will flash dry to these higher moisture levels and then continue drying by diffusion of water through thicker hollow bead walls. With 50 percent solubles in the extracts and about a 10 percent residual moisture when diffusion drying starts, the initial water reduction factor is only about 5. Thus, the water reduction factors work in two ways. The apparent way is the reduction of extract moisture to final moisture, but the more retentive way for volatiles is with larger, thicker walled particles which retain volatiles by a factor several times higher.

*Observed Volatiles Retention*.—Table 11.5 shows the calculated coffee volatile reduction factors in relation to water reduction factors. Table 11.6 lists the same volatiles as to their natural abundance in roast coffee, their observed retention factor in spray drying instant coffee, and the corresponding amounts of these volatiles that are actually found in the powder. For comparison, the amounts

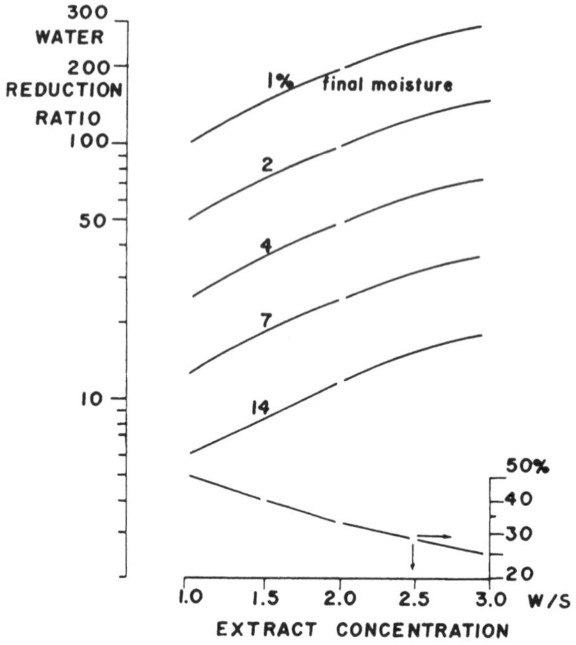


FIG. 11.22. WATER REDUCTION RATIOS VS EXTRACT CONCENTRATIONS ON PARAMETERS OF POWDER MOISTURE

of these volatiles found in instant coffee are listed alongside. To bring further meaning to these flavor levels, the order of magnitude of taste threshold for each volatile in a cup of coffee is listed. Orders of magnitude are given for three reasons: (1) the exact values are not known in some cases; (2) in general, it is hard to obtain unanimous agreement as to taste threshold when only one concentration is used; and (3) taste threshold will vary with the brew quality (roast, blend, grind, etc.). The order of magnitude of volatiles concentration for flavor appreciation serves this discussion quite satisfactorily. Corresponding to the cup concentration is (the column) flavor volatiles concentration in the powder, which is simply taken as 100-fold higher. Instant coffee is consumed from the cup at concentrations varying from about 1.2 to 1.5 percent solubles. In order for any one coffee flavor volatile to be clearly perceived in the cup, the residual amount of this volatile constituent must be at threshold concentration or higher. Since taste threshold figures are not exact, it is fair to say that if the threshold concentration and the volatiles residue concentration are within a factor of 2 or 3, there must be some flavor contribution.

An examination of Table 11.6 volatile concentrations, retained by calculation and observation (taste thresholds), shows remarkable agreement. Hydrogen sulfide and methyl mercaptan are simply shown for completeness. They are formed from reaction products with water and dimethyl sulfide,and do not, in the true sense, occur as natural constituents in the roast coffee bean. They are noticeable in the brew. The longer a pot of coffee is held, the more prominent are these components. Two volatiles have been added pyridine and furtural. The former is naturally present in roast coffee, while furfural is a product of hydrolysis of carbohydrates.

*Hydrolysis Volatiles.*—An examination of Table 11.6 shows that certain volatile concentrations in instant coffee are considerably higher than research has indicated might be anticipated from the fraction of volatiles expected to be retained after spray drying. These are the volatiles formed on hydrolysis of carbohydrates, namely acetaldehyde (70–100 ppm), furan (2–5 ppm), propionaldehyde (3–4 ppm), acetone (40–60 ppm), methylethyl ketone (10–20 ppm), and furfural 200–500 ppm). Except for the less volatile furfural and methylethyl ketone, hardly any of the other volatiles would remain in the instant coffee powder since their volatile retention in conversion of extract to powder would be hundredths to thousandths as much. For the orders of retention of the volatiles in the instant coffee powder determined from research, the original solubles before spray drying must have had over 10,000 ppm acetaldehyde, over 1,000 ppm acetone, furan and furfural, and hundreds ppm methylethyl ketone and propionaldeyde.

*Instant Coffee Taste.*—A taste acquaintance with these hydrolysis coffee flavor volatiles, in such different proportions from roast coffee, combined with the almost complete loss of methyl formate, dimethyl sulfide, and comparable volatile constituents in roast coffee, reveals the major flavor-contributing sources of instant coffee, see Table 11.6. About 14 ppm isobutyraldehyde is in the instant coffee powder. This is an important and recognizable part of good coffee flavor, but this level of retentivenss is questionable. The only source of isobutyraldehyde in cellulose hydrolysis is acetaldehyde. The partial loss of diacetyl, acetyl-propionyl, isovaleraldehyde, and normal and isobutyraldehydes still leaves them at perceptible levels of taste; their reduction factors from extract to instant coffee powder are less than ten. For diacetyl and acetyl-propionyl there may be as much as one-half volatile retention.

With the relatively high residual or small volatiles reduction factor for pyridine and furfural, these volatile flavors are the chief components of the aroma and cup flavor of instant coffee. Instant coffee powder (without add-back of expelled roast coffee oil) smells sweet-to-sickening and greatly resembles the aromas of pyridine and furfural with backgrounds of aldehyde sweetness, perhaps best related to acetone-type sweetness. All the aldehydes are sweetish in aroma and taste. Whether the major taste contribution is from acetone or methylethyl ketone, the result would be similar. Furfural, with its hay-like odor, is characteristic of instant coffees. Adding furfural to brewed coffee makes it taste more like instant coffee.

Pyridine has a very sickening odor and represents the residual odor of both instant and stale coffees. Staleness is noticed in part by a dominance of pyridine odor after the more sensational odors of dimethyl sulfide and mixed aldehydes have oxidized and volatilized away.

The residual concentration of volatiles in instant coffee are low, according to the volatiles reduction factors for methyl acetate and furan; yet their concentrations are within or above the taste threshold range. All the lesser volatiles are carried through to the instant coffee from the roast coffee in relatively significant fractions. Hence, they contribute their aromas and flavors to the instant coffee accordingly. The net result is a different aroma and flavor for instant coffee than for brewed coffees.

*Taste Threshold*.—The volatile components listed have not yet been considered from the viewpoint of their contributions to a specific intensity of coffee flavor. Even at parts per billion flavor volatiles retention in the powder, some potently odorous volatiles are very important in contributing to the taste impression of natural coffee flavor. For example, dimethyl sulfide, which has a threshold of about 10 ppb, has a very potent taste at 1 ppm or 1,000 ppb, whereas acetone (at a taste threshold near one part per million in a cup of coffee) is not a very important coffee flavor contributor even at several ppm in the cup. Thus, the focus of attention on taste is brought to the more important natural coffee flavor contributors. Taste and aroma thresholds are not the only consideration. There are also flavor tone, strength, quality, and concentration. Further, there is some balance or range of compositions of volatile flavor components that defines coffee aroma and taste. If these vary too greatly in disproportionate volatiles losses in spray drying extract to powder, then the similarity to coffee flavor will diminish. Four of the important flavor contributors to natural roast coffee flavor occur in the following proportions and taste thresholds in roast coffee:

| Coffee flavor volatile | Brewed coffee, 5 gm R&G per 100 ml water | | Concentration factor over threshold | Instant Coffee, 1.5 gm per 100 ml water, ppm volatile in instant |
|---|---|---|---|---|
| | ppm in R&G | Taste Threshold | | |
| (1) Dimethyl sulfide | 4 | 0.10 | 40 | Nil |
| (2) Methyl formate | 12 | 0.50 | 24 | 2 |
| (3) Isobutyraldehyde | 20 | 0.50 | 40 | 14 |
| (4) Diacetyl | 40 | 1.00 | 40 | 20 |

Not only are the volatile proportions in the instant coffee much reduced and out of balance, but, in brewed coffee, the natural volatiles abundance is 20- to 40-fold higher than the taste threshold.

Taste threshold concentrations are influenced by the substances with which they are associated. For example, acetaldehyde has little aroma or flavor. Yet when acetaldehyde is part of a mixture with lesser volatile substances, it accents the presence of the latter. So aroma and taste thresholds must be considered in the context of their natural use environment, namely, a cup of brewed or instant coffee. The aroma and taste threshold for some volatiles is difficult to establish from a cup of coffee. For example, to evaluate the doubling of acetone content among carbonyls demands the nasal detection of a particular carbonyl in a similar background. Further, a conditioned taster looking for a particular nasal sensation associated with a particular compound will detect it at several-fold lower concentrations in the cup than the less educated sniffer or taster.

The foregoing discussion is only an introduction to coffee flavor volatile retention in spray drying. More accurate and more useful data may be obtained with the gas ionization chromatographic detectors when analyzing for specific coffee flavor volatiles before and after spray drying, in conjunction with aroma and taste tests. This will provide exact volatile retention factors and guide constructive changes in the process of spray drying as well as lead the way toward chemically defining a more pleasing instant coffee flavor.

*Natural and Hydrolysis Volatiles Retention*.—With the objective as well as the ability to retain natural coffee flavor volatiles through spray drying, the unfortunate contribution of the hydrolysis volatile flavors, which are not desirable, results. Nevertheless, if more volatiles are retained, the natural coffee flavor volatiles dominate the overall volatiles flavor impression in spite of the background flavors of the hydrolysis volatiles. In the development of a better flavored instant coffee, hydrolysis volatiles removal or diminution is sought.

*Means of Volatiles Retention*.—By better recovery of percolator vent gas volatiles and their reincorporation into the coffee extract, a several-fold increase in gross coffee flavor intensity may be attained in the extract. Increasing natural water solubles and volatiles with reduced hydrolysates also shows noticeably greater volatile flavor retention in the spray dried powder. These means for retaining volatile flavors from extract to powder are often subsequently offset because the powder is not immediately packed in a protective atmosphere free of oxygen. Excellent flavored instant coffee powder, with many times more natural coffee flavor than is now commercially available, may be prepared and marketed today with little additional plant investment or processing costs. Spray drying can be modified to dry large, thick-walled particles, possibly at lower air temperatures and with drier air. Flavor improvement lies simply in operating in the more flavor retentive areas. If necessary, extracts fortified with those volatile components that have a low retentive character may alter the flavor of instant coffee so that it becomes not necessarily like brewed coffee, but better. With control and design of coffee flavor, an improved coffee flavored beverage will result. Better retention of coffee flavor in the drying step encourages the use of better quality coffees.

Solubles concentration of extracts may be increased appreciably to produce a stronger flavored instant coffee powder. Evaporative concentration of extracts (evaporation of water and volatiles takes place slowly) prior to spray drying drives off the volatile coffee flavors. Spray drying freeze concentrated extracts (up to 55 percent solubles) causes increased volatiles flavor retention noticeable to untrained tasters. Extract solubles concentrations of 5 percent increments, e.g., 30 to 35 percent, show noticeable improvements in flavor retention.

If a 40 percent solubles extract is divided into three batches of which two are diluted to 30 percent and 20 percent, and the three batches are spray-dried separately, startling differences are shown in flavor retention. The batch of highest concentration shows the best retention. Another means of carrying the volatile coffee flavors through is to carry as much fatty acid and oil as possible with the extract through drying. Under such circumstances, the final spray dried powder bulk density may be high (30 lb per cu ft), but the volatiles flavor retention may be five times the norm. Such dense, flavor-rich powders can be used alone or be blended with less dense, weaker flavored powders to produce acceptable densities and a higher overall coffee flavor.

**Process Controls in Spray Drying**.—Extract properties are the most important factors in controlling the properties of the resulting powder. Since the extract properties are interwoven with processing temperatures, air flows, and atomizing conditions, it is appropriate to discuss each.

*Green Coffee Blend*.—This has no direct influence on drying, other than as it might influence roast and percolation. For example, imperfect coffees yield some burnt beans with a release of fatty acids and oil that reduce extract surface tension. This release produces higher powder bulk densities, oilier powder, better volatiles flavor retention, and poorer powder flow. The powder color is usually darker and less foam is in the cup. The particle wall is relatively thicker. A very dark roast will usually yield a lower extract drawoff concentration, but powder densities of 15 lb per cu ft are easily maintained, perhaps even requiring gassing. Robustas yield high carbohydrates, and low powder densities.

*Gassing Extract*.—Gassing the extract may be done by simply injecting carbon dioxide gas into the coffee extract line on a volume ratio (gas to extract) of about 1 to 10, depending on the results. In a more elaborate extract gassing setup, a soft beverage type of carbonator can be used. Here the extract cascades over a series of plates while being maintained at a thermostatically set outlet temperature in a pressurized atmosphere of $CO_2$ gas. The $CO_2$ pressure, extract temperature, and solubles concentration govern the volume of $CO_2$ gas dissolved. Excessive gassing of the extract causes fragmented particles with poor powder fluidity and marked coffee flavor loss. Gassing the extract may cause a reduction in bulk density of up to several pounds per cubic foot, but normally gassing is not recommended due to the shortcomings cited. The higher the surface to weight ratio that results in the powder, regardless of how it is attained, the lighter is the powder color, the less acid is the cup, and the less aroma and flavor are held by the powder.

*Dark Powder Colors*.—These are often due to beady, thick-walled particles with a few fines, and are formed usually from higher solubles extract concentrations (35 percent) or from extracts rich in fatty acids. Lower hydrolysis yields of solubles produce darker powders. The powder color is not a direct indication of roast. A light colored powder, due to fines or thin particle walls, may result from a dark roast coffee.

*Inlet Air Temperatures*.—These may sometimes influence the degree of blossoming of atomized particles, depending on the surface tension of the droplet. For example, an extract low in fatty acids (such as might be obtained with high hydrolysis solubles yields) will blossom noticeably to cause reduced powder bulk densities with 20 to 40 deg F (11 to 22 C) higher inlet air temperatures. But if fatty acids are controlling, increased inlet air temperature will cause little or no powder bulk density change. Inlet air temperatures over 600 F (316 C) may cause charring of some particles. In general, reduced inlet air temperatures do not cause a marked change in flavor retention and properties of the spray dried particles until inlet air temperatures are reduced to about 400 F (204 C) or less. Such a reduction in inlet air temperature reduces spray drier evaporative capacity to half. The reduced temperature allows less blossoming of droplets and the resulting thicker-walled particles contribute to better flavor retention.

*Air Flow*.—As the inlet and outlet air temperatures are ajdusted, adjustment in air flow may be necessary. Reduced air flows may not pull down the heavy droplets soon enough, and they strike the wall. On the other hand, too much air flow will carry over more fines (depending on bustle design) to the collectors resulting in their breakage and deterioration. For such reasons it is good practice to inspect the drier walls for powder buildup and the powder "split" from time to time. "Split" is the weight distribution of powder collected in the drier vs. primary cyclone.

*Outlet Air Temperatures*.—These have to be maintained high enough to effect the drying of the particles (to 3 percent moisture) leaving the spray drier. This temperature is usually 225 to 250 F (107 to 121 C). A popular belief is that setting the outlet air temperature will directly control the outlet powder moisture and hence, that the inlet air temperature will be controlled from the set outlet temperature control. This is true only when all spray drying conditions are the same. However, hardly 24 hr pass before extract quality changes, and atomizing conditions may have to be changed. The net result may be a thicker-walled droplet which, at the set outlet air temperature, leaves the drier about 1 percent higher in moisture. The spray drier operator who is repeatedly measuring powder moisture, powder density, and cup tests must recognize the particle property changes occurring and change the outlet temperatures accordingly.

*Powder Cooling*.—The powder leaving the spray drier without a cold air dam in the cone or air cooling in the outlet conveyor may exceed 140 F (60 C) which is close to its fusion temperature. If the powder is collected in bulk Tote Bins, for example, the powder temperature does not fall. The powder is self-insulating. Powder temperature may rise from 125 F in the bin to 135 F (52 to 57 C) before it cools to room temperature. This liberation of heat, as noted from the temperature rise, is due to oxidation reactions at that temperature ($CO_2$ is evolved). The instant coffee flavor deterioration is noticeable under such conditions.

*Increased Powder Density*.—Table 11.7 shows several means for increasing the bulk density of instant coffee powder; the reverse steps would be taken to decrease its bulk density. A clear transition point when more uniform particle sizes are obtained cannot be defined, except from measurement of bulk density, particle size distributions (+40 mesh/−80 mesh) and microscopic particle examination under drying conditions (pressure of atomization, whizzer-nozzle combinations, extract quality, etc.). Uusually a higher tangential velocity will cause more uniform particle distribution; powder density may go up or down.

With high extract atomizing pressures, a non-uniform and fine particle distribution results which is usually dense. Lesser atomizing pressures produce more beady, uniform-sized particles of lesser density and darker color. The least atomizing pressure produces large, beady or agglomerated particles with thick walls; the powder will be dark colored with good solubility and more dense than with median atomizing pressure. The high pressure atomizing may be considered a misting while the low pressure gives splashing.

**Water Evaporating Rate**.—The water evaporating rate is only one factor related to attaining a given particle size. Up to a point one may sacrifice water evaporative capacity to obtain coarser particles at a lesser rate of production. Most driers today base their evaporative capacity on 28 percent solubles in the extract. If, for example, a given evaporative rate is guaranteed at this concentration, reducing water content of the extract to half (45 percent solubles extract) will not double the drying capacity with the same particle properties. In fact, depending on particle size distribution actually obtained, the drying rate will be about the same as with 28 percent solubles extract. The benefits will be mainly improved flavor

TABLE 11.7. SPRAY DRIER OPERATING CHANGES FOR INCREASED POWDER BULK DENSITY

| Operating change | Principle influencing higher bulk density | Means required for change |
|---|---|---|
| 1. Increase solubles concentration of extract. | Increased weight per particle volume | Higher solubles concentration in drawn off percolator extract. |
| 2. Reduce inlet air temperature. | Reduced particle blooming. | Reduced set inlet air temperature with possibly higher air flow or reduced rate of atomization. |
| 3. Increase extract temperature. | Reduced vicosity causes finer atomization. | Heat extract before atomizer. |
| 4. De-gas extract. | Removes dissolved gases which blossom particles. | Evacuated chamber with film spread of extract. |
| 5. Increase pressure drop of extract across atomizer. | Reduces finer, less uniformly atomized particles. | Smaller atomizing orifices. |
| 6. Decrease pressure drop of extract across atomizer. | Produces coarser particles with thicker walls at reduced drying rate. | Reduce extract pump speed to obtain reduced feed rate. |
| 7. Raise and lower pressure of atomization and observe effect on particle uniformity ratio (+40/−80) and density. | Wide range, less uniform particle sizes cause higher density. | Increase or decrease pumping rate with same atomizer. |

retention in the powder and better physical properties. There will be some increase in fuel costs. Air velocities may have to be increased to keep particles away from the walls.

Spray drier throughput of solubles and water may be increased by means of finer atomization; finer atomized droplets dry more quickly. This is at a sacrifice of flavor retention and there is increased fragility of the thin-walled particles.

*Powder Sterility*.—Coffee extracts are usually not sterile. Bacteria, yeast, and mold may grow in coffee extract, although it not a very good medium for such growth. Higher solubles (less free water), lower pH, and lower storage temperatures retard these growths. After spray drying, the powder is essentially sterile regardless of the count of bacteria in the extract.

*Corn Syrup Startup of a Spray Drier*.—This requires a special attention and procedures. For a new soluble coffee plant with a new spray drier and operating personnel, it is good practice and good economy in coffee powder use to start the training and spray drier operation with corn syrup solubles. These resemble coffee solubles. Coffee solubles are about ⅓ polysaccharides and about 8 percent reducing sugars. The corn syrup may be obtained from firms such as American Maize Company, Staley & Company, Corn Products Refining Company, etc. The American Maize product of 30 DE (Dextrose Equivalent) or about 30 percent reducing sugars is called Lodex and is quite suitable for this application. It comes in steel drums (650 lb/50 gal) at 80 percent solubles (43 Be), which make it easy to dilute to the desired solubles concentration. To obtain a more viscous corn syrup to simulate cold coffee extract, some percentage of Frodex powder (100 lb bags) of DE 15 or 24 may be dissolved into the diluted Lodex solution.

*Routine Normal Spray Drier Startup*.—Weekly or longer periods of operation require the following startup steps: all manholes must be closed; the spray drier, auxiliaries, and powder conveyor must be cleaned and dried; instrument air supply must be checked; and all electrical switches to all operating equipment must be closed. Fuel gas pressures and fuel oil supply must be inspected; all fuel lines must be opened. All automatic controls must be placed on manual position for startup. Fans, airdrier and rotary valves must be started. The filters must be clean at startup. Air flows must be adjusted to normal positions and the burner must be started. Air pressure control must be set manually to desired inlet air temperature and brought up to 500 F (260 C). The piston pump must be prepared for pressurizing water for atomization when outlet air temperature approaches 300 F (149 C). Pistons need water lubrication and the fluid drive oil needs water cooling. With inlet air at about 475 F (246 C), it must be switched to automatic inlet air temperature control. The inlet air signal must be manually controlled by manual control of the outlet air temperature control instrument. Care and attention at this time are necessary to avoid causing the burner to cut back or open up too abruptly; in other words, the drier must be warmed gradually, with stability and control. Atomizing of water may be started when outlet air reaches 300 F (149 C); manual outlet air temperature control is maintained until overall stability in temperatures is attained. Then the outlet air temperature is placed on automatic control. The atomizing (pump) pressure is placed on automatic control initially. The inlet air temperature controller should have a cutoff at 600 F (316 C). The change to extract from water may be made with only a small increase in atomizing pressure. The rotary valve sleeves are now connected for discharge to the oscillating powder conveyor, as is rework powder. The cone vibrator air supply is turned on with the timing sequence motor. When extract supply is tapped, the booster centrifugal pump is started. The steam and cooling water for the air drier are supplied. The time lapse until extract begins drying is 1 to 1.5 hr.

*Normal Shutdown of the Spray Drier*.—This is effected by first changing from extract to water and reducing atomizing pressure. The extract feed centrifugal pump and then the water supply are shut off when all the extract has flushed through the atomizer. Then the atomizer pump and air heaters are shut off. All fuel lines are closed. When exit air temperatures fall below 200 F (93 C), air fans are stopped. This takes about 20 min. All the powder must run out. Then, manholes are opened for visual inspection of the interior of the drier for any spattering patterns of powder buildup. Any hung powder is recovered by rubbing it off the wall with long sticks. Then the drier is washed down with hot water. The residual solubles are not worth saving. Instruments are reset on manual control. Washdown liquors are routed by means of a base adapter from the bottom drier cone to drainage. All powered equipment is shut down, except the rotary valves, while collectors are inspected and washed. Water pressure must project hot hose spray about 20 ft to impact on the opposite walls.

*Abnormal Shutdown of the Spray Drier Operation.*—This may be caused by mechanical difficulties with the equipment or by spattered walls. Wall deposits of soluble coffee powder will fall off after charring and will contribute to lumps, black specks, and agglomerates. Carbonized powder will float on the surface of a cup of instant coffee. Gold specks in the cup are due to foaming of the atomized extract. Prolonged operation under such circumstances only becomes worse. The solution is to shut down, wash and dry the tower, and start over under conditions that do not cause wall deposits. Too wide a spray angle with too little air flow or too big droplets at too high a rate of extract feed causes wall deposits. Washing down a drier and placing it back in operation takes about 6 hr and is to be avoided. Hence, enough tank storage capacity for extract must exist between the drier and the percolators to take up such accumulation of extract production during drier washdown.

*Lumps of Powder.*—These sometimes may be broken down by hand and reworked dry after screening. In cases where so much carbonized material is present that dry dilution of powder is impractical, the lumps and carbon contaminated powder must be redissolved in the rework tank, and then be centrifuged and chilled for rework as solution. This may be done at some dilution not noticeably different in final taste from that of normal production extract. Plastic, syrupy solubles that might accumulate in the drier under poor operating conditions would be reworked similarly. Sometimes lumps need to be thrown away.

*Data During Spray Drying.*—This is of three types: (1) log of unusual conditions handwritten by the supervisor; (2) instrument recorded equipment operating data; (3) powder and extract physical tests. Table 11.8 shows a type of powder inspection sheet and Table 11.9 shows the operations sheet. In addition, records of the time of tank fillings and emptying of extract, condition of tars and sediment in emptied tanks, summaries of each drying run time, weight of powder dried, and amount of rework formed must be kept. The status of Tote Bin packable and rework inventories must be maintained in conjunction with shift accountability of solubles entering and leaving.

## Agglomeration of Spray Dried Instant Coffee

**Product Introduction.**—In the 1966 to 1969 period General Foods and Nestle expanded their freeze dried instant coffee sales with promotions. Some

TABLE 11.8. SPRAY DRIER CONTROL LABORATORY
*Properties of instant coffee powder*
*Permission for Emptying Tote Bin for Packaging.*

| Run. | Shift | Date | Tote bin no. | Ap-pear-ance | Color | Mois-ture | Den-sity | Mesh +40 −80 | Solu-bility | Taste | Specks and oil | Curdle | Sedi-ment | Flow | Packable Yes No | Operator's Name |
|---|---|---|---|---|---|---|---|---|---|---|---|---|---|---|---|---|
| | | | | | | | | | | | | | | | | |
| | | | | | | | | | | | | | | | | |
| | | | | | | | | | | | | | | | | |
| | | | | | | | | | | | | | | | | |
| | | | | | | | | | | | | | | | | |
| | | | | | | | | | | | | | | | | |
| | | | | | | | | | | | | | | | | |
| | | | | | | | | | | | | | | | | |
| | | | | | | | | | | | | | | | | |
| | | | | | | | | | | | | | | | | |
| | | | | | | | | | | | | | | | | |

TABLE 11.9. SPRAY DRIER OPERATING DATA

Date ______ Shift ____ Operator________ Run No. ____ Diameter Whizzer Orifice____
Temperature, ° F Diameter Vertical Orifice____
Pressure, inches water

Hour

1. Extract tank No.
2. Tote bin No.
3. Rpm piston pump (gph)
4. $T_1$ air leaving preheater
5. $T_2$ air leaving main heater
6. $T_3$ air after atomizer
7. $T_4$ air leaving tower, east
8. $T_5$ air leaving tower, west
9. $T_6$ air leaving cyclone
10. $T_7$ air leaving dust collector
11. $T_8$ air in drier cone
12. $T_9$ air entering drier cone (cool)
13. $T_{10}/T_{11}$ ambient air: dry bulb/wet bulb
14. Percent relative humidity-air
15. Barometric pressure: mm mercury
16. Pressure fuel oil pump-psig-flow
17. Pressure gas fuel, psig (points ½)-flow
18. Pressure piston pump-extract-psig- at nozzle
19. $P_1$ air leaving preheater
20. $P_2$ air entering tower
21. $P_3$ air after atomizer
22. $P_4$ air leaving tower (bustle)
23. $P_5$ air after cyclone
24. $P_6$ air after Dustex
25. $P_7$ cool dry air at cone
26. $P_8$ combustion air supply
27. $P_9$ combustion chamber
28. $P_{10}$ combustion gas supply
29. Powder rework, lb per hr
30. Extract soluble concentration, percent, F (C) temp.
31. Powder bulk density, gm per cu cm
32. Powder moisture, percent
33. Time of startup ________ Time of shutdown. ________
34. Other comments ________

consumers resisted buying the freeze dried instant coffee because of its higher unit price. In addition, some spray dried instant coffees took 20 to 40 seconds to dissolve in boiled water, and they often left a ring of foam at the coffee beverage level surface in the cup. With increased solubles yields and more Robusta beans in the blends, these cup foaming properties took on a worse aspect relative to freeze dried coffees. The freeze dried instants dissolved much more easily and with little to no foam residues on the cup.

In early 1968 Nescafe introduced an agglomerated (spray dried) granular product. That is, the 0.1 mm spray dried particles were coalesced into 3 mm clusters. The purpose of this property change was to improve granule solubility over the fine (milk drier) spray dried hollow spheres (0.1 to 0.3 mm diameter) and to minimize foam and ring in the cup.

At the same time the instant coffee clusters were darker (called richer) in color than the finer spray dried particles. There were few fines when the consumer was spooning out the powder at home.

**Purposes of Agglomeration**.—The relative importance of each of the several purposes depends on the processor's outlook:

| | | |
|---|---|---|
| COMPETITION | 1. | To meet retail offerings. |
| APPEARANCE | 2. | To look like R&G coffee, even freeze dried coffee. |
| COLOR | 3. | Consistent with (2). |
| SOLUBLITY | 4. | To minimize foaming. |
| DUST | 5. | To eliminate. |
| FRAGILITY | 6. | To have a physically strong particle—will not break. |
| DENSITY | 7. | Comparable to spray dried instants (0.25 gm/cc packed). Actually most agglomerates are about 20 percent denser and require a smaller volumed jar. |
| TASTE | 8. | The beverage taste from agglomerate must not be noticeably inferior to its originating spray dried powder, or at least not inferior to aromatized. |
| COST | 9. | Initial equipment cost and operation costs must be low. |
| LOSSES | 10. | Coffee losses in process must be low (less than 2%). |
| STABILITY | 11. | Final moisture must be near 3 percent. |

**Microphotos** (agglomerated and freeze dried).—Examination of agglomerated particles under a microscope will reveal how the agglomeration process has affected the original spray-dry powder. Usually, the spheres are broken so that no air is entrapped. This eliminates foaming when the agglomerated instant coffee is dissolved. (See Fig. 11.23.)

Examination of the texture of freeze-dried particles can reveal gassing, a process in which nitrogen is used to "foam up" highly concentrated (40 percent solubles) extracts to achieve final densities comparable to spray-dried products. Bulk and granule color variations and evidence of ice crystal formations (striations or cavities) are related to different rates and methods and freezing coffee extracts.

**Principles in Agglomeration**.—Principles in agglomeration were originally based on work done with milk and sugary products, which are much easier to agglomerate by simple turbulent impact of warm moist particles than instant coffee is. By agglomerating broken particles of spray dried coffee, contact surface was increased with high moisture (to 10 percent) and high spray drier temperatures, followed by vibratory conveying finishing drying and cooling. The result—a fairly fragile granule or like R&G coffee, and a high equipment investment. Folger's

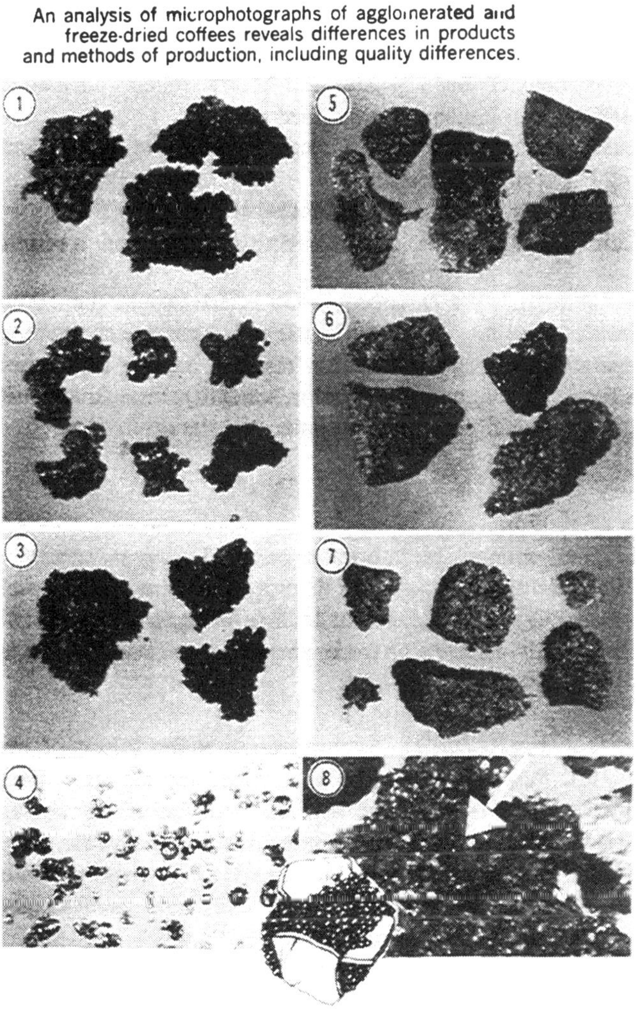

An analysis of microphotographs of agglomerated and
freeze-dried coffees reveals differences in products
and methods of production, including quality differences.
1
2
3
4
5
6
7
8


FIG. 11.23. MACRO PHOTOS OF AGGLOMERATED AND FREEZE DRIED GRANULES

flaking a spray dried coffee, with steam glazing the flakes to strengthen them followed by vibratory conveying and cooling, was a similarly involved and expensive process yielding a rather caramelized tasting end product.

*Basic Principles Used In Instant Coffee Agglomeration*.—Fusion is used in all the agglomeration processes even though the machinery and sequence of steps may vary. Also influencing variables are utilized to different degrees:

1. Addition of moisture (steam, water or extract) to spray dried hollow sphere surfaces or broken spheres.
2. Elevate temperature of moist particles. Drying occurs at the same time.
3. Increased contact area, by breaking hollow spheres in hammer mill or forming fewer fragmented spray dried particles.
4. Increased pressure of contact. Compact particles to cause adhesion, e.g., flaking between smooth rolls.
5. Time of contact (or impact) to effect bond strength.

Combinations of the above 5 factors will vary agglomeration results.

*Interchangeability of Variables*.—To effect fusion, the 5 variables are interchangeable to a point. For example, use of less particle moisture will require higher temperatures for fusion, and may result in adverse tastes. In any given process the properties of appearance, color, solubility, taste, and density must be controlled, be reproducible, and be optimized for the results desired.

*Degrees of Fusion*.—

Stage 1. Fragile—spheres intact, point to point adhesion at surface.
2. Moderately fragile—spheres deformed, but shapes discernible.
3. Moderately hard—spheres all deformed, original shapes not discernible.
4. Quite hard—Brittle with force, amorphous to glassy surface, appearance like lava rock-pitted cavities.

**Process** *(see Fig. 11.24)*.—Agglomeration is the fusion of many small, instant particles. Fusion is caused by heat, time, and moisture. Fusion can be achieved with varying methods and will attain varying results and agglomerate properties. Any heat treatment of the spray dried instant powder will degrade its taste and aroma. However, flavor degradation due to fusion was relatively minor compared to the flavor degraded by cheaper coffee blends and higher solubles yields. Adding volatile aromas and/or expelled coffee oil to the agglomerate usually ameliorated any flavor degradation due to fusion.

General Food's Maxwell House spray dried instant coffee powder is pulverized in a hammer mill, which breaks up the tiny hollow spheres. The broken spheres are then steam wetted, heated, and coalesced. The moist clusters are then dried in a warm air conventional spray drier using 450 F inlet air. This is followed by secondary air heating and air cooling of the clusters on an outlet oscillating conveyor.

Fusion agglomeration is also carried out on a stainless steel belt passing through a series of heated and humidified compartments. In this case, the fused particles come off the belt as a sheet. The sheet is granulated to specifications and the product looks very much like R&G coffee.

TATING DISC AGGLOMERATOR:

. Feed hopper.
. Powder feeder.
. Rotating disc.
. Coarse sieve.
. Vibro fluidizer.
. Sieve.
. Cyclone.
. Fan.
. Pressure conveying system for fines.
. Pressure conveying power feeding system.
. Liquid feed pump.
. Blower.

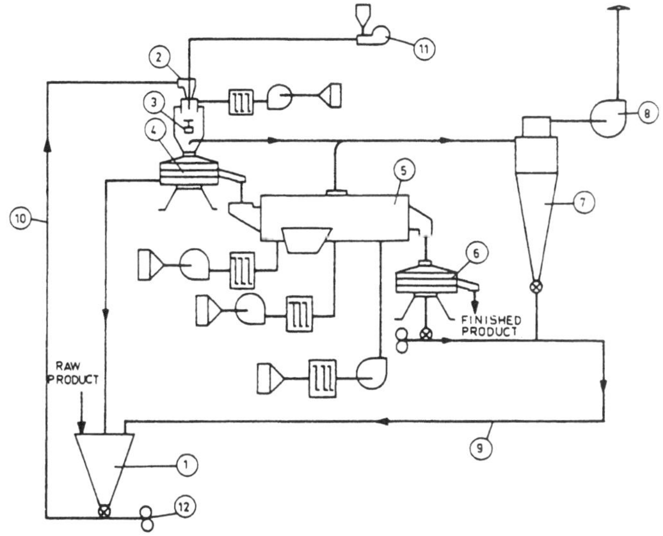


FIG. 11.24 AGGLOMERATION PROCESS

**Properties and Pricing of Agglomerate**.—The result was a chunky instant coffee product with faster solubility in hot water, less foam in the cup, a darker granule color, and no fines. These agglomerates were offered to the public in the face of growing freeze dried instant coffee sales.

By 1972 freeze dried coffee sales in the U.S.A. totaled about 60 million pounds, almost 30 percent of all instant coffee sold. By 1977 freeze dried instant coffee sales were about 45 percent of all instant sales.

By 1972 about ¾ of the instant coffees sold in the U.S.A. were agglomerated, aromatized (with expelled coffee oil), and packaged with inert gas in jars. The preparation of instant agglomerates in Europe was also beginning to dominate the retailed jars by 1973. Some agglomerates, like Yuban, were priced on retail shelves at almost freeze dried coffee prices.

## ADVANCES IN AGGLOMERATION OF INSTANT COFFEE

In the field of agglomeration, Niro Atomizer has been successfully agglomerating instant soluble milk products for a long time, and over the past few years has successfully marketed a fully developed industrial size coffee agglomeration unit (Fig. 11.25).

As agglomeration of sugary products like milk, chocolate milks, etc. is much easier than with instant coffee, Niro Atomizer developed a new technique to handle the hygroscopic and thermoplastic properties of instant coffee.

FIG. 11.25 AGGLOMERATION OF INSTANT COFFEE

The Niro Atomizer coffee agglomerating unit features a fast rotating disc (Fig. 11.26) on to which agglomerate falls immediately after being moistened. The flow and distribution of powder feed plus water and steam is accurately controlled for optimum moistening.

In addition, the quantity and temperature of warm air flowing through the agglomeration chamber creates an environment conducive to the agglomeration process.

Powder on its way from the feeder to the rotating disc is superficially moistened both by liquid from a spray nozzle and by steam.

Immediately afterwards the powder impinges upon the central area of the rotating disc. Particles collide on the surface of the disc, and the agglomerates thus formed are ejected from the disc edge by centrifugal forces. The agglomerates pass to a vibro fluidizer (vibrated fluidbed), which acts as a combined dryer/cooler.

The principle of the rotating disc has proved very successful for agglomerating instant coffee, and several plants have in the past few years been supplied to commercial concerns in South Africa, Philippines, Belgium, Germany, Switzerland, and Spain.

The rotating disc technique can also be used for products that are even more hygroscopic and thermoplastic than coffee, and which were previously considered impossible to agglomerate, e.g. coffee substitute such as chicory mixed with coffee.

Sivetz visited three of these installations including the most recent, which features Niro Atomizer's most up-to-date equipment.

The processing conditions fulfill most, if not all, of the objectives a processor seeks in agglomeration of instant coffees: namely capital investment is relatively low, and product losses are small since the plant can be started or stopped virtually "at will" without frequent cleaning.

The powder system is hermetically tight during shut-down. One man can operate the system which produces 150 kg/hour of product or more.

Powder recycling is minimal and thermal treatment of the powder is virtually not detectable by taste.

What is even more satisfying is that particle size can be effectively controlled to produce a uniform attractive reddish brown particle looking much like freeze-dried or even ground coffee granules.

Niro Atomizer has received a U.S. patent No. 3.966.975 which describes the process in greater detail.

With the latest Niro Atomizer technology, it is now possible to produce instant coffee agglomerates of excellent quality. The availability of such process equipment can contribute to new marketing ramifications both for consumers and for coffee growers alike. It will be interesting to see how such new effective technology will influence world coffee trade.

FIG. 11.26 ROTATING DISC IN AGGLOMERATOR

# BIBLIOGRAPHY

## Spray Drying


ALLCOCK, H. W. 1959. How to make coffee-er coffee. Instrumentation *12*, No. 6, 9-11.

ANON. 1956. Origin of spray drying. Am. Perfumer Essent. Oil Rev. *67*, No. 3, 62.

BATEY, R. W. 1960. Drying techniques to improve instant coffee flavor and aroma. World Coffee and Tea *1*, No. 2, 39–41.

BUCKHAM, J. A. 1953. Spray drying study. (Ph.D. Thesis) Dept. Chem. Eng. U. Wash. Seattle.

BUCKHAM, J. A., and MOULTON, R. W. 1955. Factors affecting gas recirculation and particle expansion in spray drying. Chem. Eng. Prog. *51*, No. 3, 126–133.

BULLOCK, K., and LIGHTBOWN, J. W. 1943. Spray drying of Pharmaceutical products. Quart. J. Pharm. and Pharmacol. *16*, 213–226.

CLELLAND, W. M. 1954. Spray drying. Ind. Heating *21*, 544–551. Also Battelle Tech. Rev. *3*, 9240.

CROSBY, E. J., and MARSHALL, W. R. JR. 1958. Effect of drying conditions on properties of spray dried particles. Chem. Eng. Prog. *54*, No. 7, 56–63.

DOMBROWSKI, N., and FRASER, R. P. 1954. Disintegration of liquid sheets. Phil. Trans. Roy Soc. London. Sept. Cambridge Univ. Press.

DOUMAS, M., and LASTER, R. 1953. Liquid film properties for centrifugal spray nozzles. Chem. Eng. Prog. *49*, 518–526.

FOGLER, B. B., and KLEINSCHMIDT, R. V. 1938. Spray drying. Ind. Eng. Chem. *30*, 1372–1384.

LAPPLE, C. E. 1954. Elements of dust and mist collection. Chem. Eng. Prog. *50*, 283–287.

LAPPLE, C. E., and SHEPHERD, C. B. 1940. Calculation of particle trajectories. Ind. Eng. Chem. *32*, 605–617.

LASTER, R. 1953. Factors in selection and design of spray drier. Food Technol. *7*, 264–267.

LEE, S. 1961. Spray drying vs vacuum drying. Tea and Coffee Trade J. *121*, No. 3, 24, 32–36.

LEE, S. 1959. Heat transfer in spray drying coffee. Tea and Coffee Trade J. *116*, No. 3, 28, 58–60.

MARSHALL, W. R. JR. 1955. Heat and mass transfer in spray drying. Trans. Am. Soc. Mech. Engrs. *77*, 1377–1385.

MARSHALL, W. R. JR. 1954. Atomization and spray drying. Monograph. Chem. Eng. Prog. Ser. *50*, No. 2.

MARSHALL, W. R. JR, and SELTZER, E. 1950. Principles of spray drying. Chem. Eng. Prog. *46*, Part I. Fundamentals of spray drier operation 501–508; Part II. Elements of spray drier design. 575–584.

MASTERS, K. 1972. Spray Drying. Leonard Hill, London.

METCALFE, L. 1956. Progress in spray drying. Coffee and Tea Inds. *79*, No. 12, 28–29.

MEYER, F. W. 1956. Some problems involved in spray drying. Tea and Coffee Trade J. *111*, No. 2, 37, 42.

MEYER, F. W. 1955. Soluble coffee plant construction. Tea and Coffee Trade J. *109*, No. 4, 70, 76.

PARKER, M. E., HARVEY, E. H., and STATELER, E. S. 1954. Spray Drying. Vols. I, II, and III. Reinhold Publishing Corporation, New York.

PERRY, J. H. 1976. Chemical Engineers' Handbook. McGraw-Hill Book Company, New York.

RHOADES, J. W. 1958. Sampling method for analysis of coffee volatiles by gas chromatography. Food Research *23*, 254–261.

RHOADES, J. W. 1960. Volatile constituents of coffee. J. Agr. Food Chem. *8*, 136–141.

ROBINSON, C. S., and GILLILAND, E. R. 1939. Elements of Fractional Distillation. McGraw-Hill Book Company, New York.

SELTZER, E., and SETTELMEYER, J. T. Advances in Food Research. Vol. II. 399–520. Academic Press. New York.

SIVETZ¿ M. 1976. Advances in the agglomeration of instant coffee. Tea and coffee Trade 5. Aug. 1976. p. 25.

SIVETZ¿ M. 1977. Coffee: Origin and Uses. Sivetz Coffee Enterprises, Inc. Corvallis, Oregon.

SMITH, D. A., and DOWNING, A. R. 1956. Trends in building spray driers. Tea and Coffee Trade J. *111*, No. 2, 32.

TATE, R. D., and WILCOX, R. L. 1962. Spray drying nozzles for soluble coffee production. World Coffee and Tea *2*, No. 11, 29, 31.

TURNER, G. M. 1950. Spray droplet size frequency distributions for hollow cone spray. (Ph.D. Thesis, Chem. Eng. Dept.) Univ. Washington, Seattle.

TURNER, G. M., and MOULTON, R. W. 1953. Drop size distributions from spray nozzles. Chem. Eng. Prog. *49*, 185–190.

WALLMAN, H., and BLYTH, H. A. 1951. Product control in Bowen-type spray drier. Ind. Eng. Chem. *43*, 1480–1486.

WOODCOCK, A. H., and TEISSIER, H. 1943. Laboratory spray drier. Can. J. Research 21, A. No. 9, 75–78.

## Other Types of Drying

ANON. 1958. Fundamental aspects of the dehydration of foodstuffs. J. Soc. Chem. Ind. London No. 26, 821–826.

ANON. 1955–1960. Accelerated freeze drying (AFD). Ministry Agr. Fish. & Food. Aberdeen, Scotland; York House, London.

ANON 1947 High vacuum process for soluble coffee. Food Inds. *19*, No. 2, 198.

BONNEL, J. M. 1957. Continuous vacuum dehydration of citrus juices. J. Florida Eng. Soc. *11*, No. 4.

FIXARI, F., CONLEY, W., and BARD, G. 1959. Continuous high vacuum drying techniques. Food Technol. *13*, 217–220.

FLORSDORF, 1943. Freeze drying. F. J. Stokes C., Philadelphia, Pennsylvania.

LAWLER, F. K. 1962. Foam-mat drying goes to work. Food Eng. *34*, No. 2, 68–69.

MORGAN, A. I. 1959. Technique for improving instants. Food Eng. *31*, No. 9, 86–87.

NAIR, J. H. 1962. German freeze dry conference. Food Eng. *34*, No. 4, 15, 44–45.

STASHUN, S. I., and TALBURT, W. F. 1953. Puffed powder from juice. Food Eng. *25*, No. 3. 59–60.

# 12

# Aromatizing Soluble Coffee

Since aroma is an essential component of soluble coffee, this chapter will be devoted to an analysis of the properties of coffee aroma and flavor, as well as the various methods described in patents and published articles for their recovery, retention, and storage.

Aromatizing instant (soluble) coffee means preserving or restoring the characteristic aroma of roast coffee in the preparation of the water soluble powder. Ever since coffee processing machinery has been in operation, inventors have endeavored to isolate, retain, or restore the coffee aroma. The patent literature is, therefore, full of objectives and claims in this direction. Yet, the elusive and fugitive character of the aroma has defied capture and definition to this very day. The aroma and flavor are such transient entities that progress in their isolation, analysis, and chemistry has been slow.

To aromatize instant coffee, the chemical and physical nature of the volatile aroma must be known. Knowledge of the odor potency of each volatile constituent and the odor impression of each combination of volatile constituents are needed. The problem is further complicated by the unbalance in coffee aroma, readily brought about by oxidation, volatilization, diffusion, and chemical reactions.

In the past decade, significant progress has been made in isolating and fixing coffee aroma. Future progress is expected at an accelerated pace. After these are reviewed, an explanation of the part played by the colloidal matter in brewed coffee flavor and the reasons for change of flavor in instant coffee extract will be presented. With the idea that retention of the coffee volatiles is preferable to their restoration in the final instant coffee powder, freeze concentration of extracts and distillates will be covered. Since coffee oil, extracted or expelled from roast coffee has come into prominent commercial use, the methods for its recovery and reapplication will also be discussed. In addition, because coffee aroma is composed of volatile chemical substances, the techniques for their recovery will be investigated together with the sources of the aroma which include grinder gas, vacuum distillation from roast and ground (R&G) coffee, dry vacuum aroma

(DVA), atmospheric distillation from R&G coffee, and distillation from coffee extracts.

Finally, incorporation of the recovered coffee essence into the coffee extract or into other media such as gelatin, gums, molten carbohydrates, and capsules will be explained.

In summary, aroma and flavor can be added to instant coffee in numerous ways. Present commercial methods of aromatizing coffee will no doubt change as more dependable and acceptable techniques of flavor retention are developed. The methods used at any time depend not only on technical feasability but also on successful advertising and the costs of the operation. Every aromatization technique involves special equipment and analytical methods. It is not inconceivable that instant coffee fortified with flavor will evolve once suitable fortifiers have been determined. Analogous developments concerning aromas and flavors in dairy products, fruits, tobacco, bakery products, vegetables, smoke, and other industries influence the coffee flavor field. Scientific research in coffee aroma and the gap between commercial taste identities of green coffees and trained tasting associated with the chemical constituents of coffee aroma and flavor becomes smaller, a basic definition of coffee flavor that is acceptable and valuable will be forthcoming. Because of the fundamental nature of flavors in chemistry as well as in human responses, these problems offer an exciting and possibly lucrative field of endeavor.

## A REVIEW OF PATENTS

Until 1957 soluble coffees were processed by percolation and spray drying, often with much loss in flavor resulting from only fair to poor practices at each process stage. There had been no significant instant coffee flavor improvement until about 1952 when the larger hollow bead particles with better coffee flavor and physical properties was marketed as 100 percent coffee. Instant coffee still lacked the delectable aroma of roast coffee. The phenomenal growth rate of instant coffee sales in the early 1950's was not maintained and, as sales competition increased, the industry recognized that coffee aroma had to be introduced into the instant coffee jar in order to stimulate consumption.

### Expelled and Distilled Coffee Oil

The industry's largest research resources were directed to that end. Cole's U.S. patent 2,542,119, assigned to General Foods Corporation, was applied for in 1948 and issued in 1951. It dealt with separation of coffee aroma volatiles by film distillation from coffee oil. Canadian patent 572,026, issued in 1959 to Clinton and Pitchon and assigned to General Foods Corporation, dealt with "aromatizing the head space of a jar of soluble coffee"; this patent was applied for in the United

States in 1954. This was followed by U.S. patents 2,875,063 and 2,947,634 (five-fold coffee oil concentration) to Feldman *et al.* for General Foods Corporation in 1959. U.S. patent 2,931,728, issued in 1960 to Franck and Guggenheim and assigned to General Foods Corporation, refers to extraction of oil from roast coffee.

In 1957 Maxwell House commercially prepared and marketed an expelled-oil-coated instant coffee which had aroma in the jar and was packed in inert gas. Thus, the needed coffee aroma was added back to the conventionally prepared powder. The aroma volatiles were, however, quickly lost, and the instant coffee flavor staled within a day or two. Much of the coffee aroma in the jar never appeared in the cup except for an oily, smoother texture in the cup taste. Other soluble coffee firms since 1957 have made similar aroma additions by means of coffee oil add-back to powder. A 1950 U.S. patent by two Swiss workers, Schaeppi and Mosimann, expelled coffee oil and sprayed it back on instant coffee powder. In 1960, a U.S. patent by Barch and Reich of Standard Brands (Chase & Sanborn) proposed fatty acid addition to reduce foam formed when mixing instant coffee powder in the cup; fatty acids can also hold coffee aromas. Gilmont's 1949 application for U.S. patent 2,563,233, issued in 1951, uses solvent extraction of coffee oil. Solvent is stripped in a molecular still to fractionate the most aromatic coffee oil fraction which is sprayed back over the instant coffee powder. Kellogg has five U.S. patents, dated 1926, 1942, 1944, 1945, and 1951, which deal with aspects of expressed coffee oil add-back to dry coffee solubles. It has often been observed that a novel idea is likely to be put into practice about 20 years after the original principles are defined in a patent or otherwise published. This frequently holds for instant coffee developments.

## Petroleum Solvents

Trigg *et al.* had a 1919 U.S. patent, 1,292,458, that used petroleum jelly or mineral oil to capture coffee aromas. By solvent transfer these aromas are placed on the instant coffee powder. There were two more Trigg patents in 1921. One refers to lactose as a coffee aroma adsorber. The other uses petroleum ether extraction of aqueous coffee condensate to remove coffee aromas and add them back to the instant coffee powder. Use of petroleum ether as a transfer medium for coffee aromas is not greatly different from using coffee oil for the same purpose.

In 1938, Sylvan's Swedish patent covered the extraction of R&G coffee with petroleum ether. The flavorful portion was then transferred to alcohol extract of ether and then to instant coffee in a process not unlike Trigg's 1919 patent using enfleurage into petroleum jelly.

Coffee oil add-back to powder was neither a new approach nor the only way to transfer coffee aroma onto the instant coffee powder and into the jar. Since this was the method chosen by a leading producer of instant coffee, it will be covered first. Thereafter, other and possibly better aromatizing methods will be discussed.

## Powder Aroma Retention

The procedure already outlined in Chapter 11 for retaining greater coffee aroma and flavor volatiles through spray drying powder particles with a lower ratio of surface to weight is a direct method with no complicating side processes. More complicated processes entail greater plant investment and higher operating costs, and make oxidation, deterioration, and loss of coffee volatiles more likely.

## Roast Coffee Contact

Two aromatizing U.S. patents, Heyman (1935) 2,022,467 and Meyer and Haas (1931) 1,836,931, use R&G coffee to contact soluble coffee powder and thus transfer coffee oil and aroma to the soluble coffee powder. In 1961 the Chock Full O'Nuts instant coffee marketed in the New York City area added about 1 percent fine ground coffee to its instant product. This was packed in a can under vacuum to protect the natural coffee aroma and flavor from oxidation.

## Spray Drier Techniques

Blench has two interesting patents. The earlier one (1956) covers aromatizing instant coffee powder by countercurrently cascading it against rising aroma-laden gases. In the 1959 patent one adds petroleum ether to coffee extract before spray drying it. Blench claims marked coffee flavor retention that effects better volatiles retention through spray drying by this latter process.

## Polar Solvent Transfer

Only a step away from the method of using direct contact of R&G coffee with soluble coffee powder is to transfer aroma by means of solvents. This is done by polar solvents such as liquid carbon dioxide or sulfur dioxide as explained in Brandt's (1944) U.S. patent 2,345,378, or by ethylene oxide reviewed in Brandt's (1942) U.S. patent 2,286,334. Solvent extracted coffee oil is also dispersed over soluble coffee powder by vibration. Etaix of Paris in U.S. patent 1,251,359 of 1917, transfers aroma from R&G coffee to soluble coffee powder by cold nitrogen gas recirculation. Supposedly, moisture is needed to effect transfer. This does not transfer coffee oil as do the methods outlined in Brandt's patents. Methylene chloride and Freon are also polar solvents for aroma and oil transfer. In 1930, I. G. Farben Industrie took out a French patent for liquid ammonia extraction of R&G coffee (especially for alkaloids); evaporation of the ammonia left a water soluble coffee with good aroma and flavor. The similarity between this method and the later Brandt (9144) patent using liquefied polar solvents should be noted.

## Alcohol Solvents

Another class of instant coffee aromatization patents is based on R&G coffee extraction with alcohols or aldehydes. These patents are of interest because they show that petroleum solvents that remove coffee oil do not remove all the coffee flavor from the R&G coffee. Further, since the coffee volatiles identified to date contain many aldehydes, a carbonyl solvent probably has merit. Alcohol extractions of R&G coffee are covered in U.S. patents by Musher (1950); Cohen (1950); and Anhaltzer (1921). Musher had some earlier U.S. patents in 1939, 1940, and 1942. In 1948 Medial Laboratories acquired a British patent using a three-step extraction of R&G coffee with three solvents (petroleum ether, alcohol, and water) which is similar to Cohen's 1950 U.S. patent. Medial Laboratory has a 1949 Swiss patent using alcohol extraction of R&G coffee, followed by chilling the solution to remove fats and then adding to coffee powder with alcohol evaporation. The Schott Swiss patent 277,290 (1951) uses alcohol extraction of R&G coffee and is similar to the Medical Laboratory 1949 patent and Cohen's 1950 patent.

## Distillation

The patents discussed so far have more or less referred to coffee oil transfer, since this is the technique widely used today. There are, however, many patents for adding aroma and flavor to instant coffee through recovery of coffee volatiles by steam distillation with heat or under vacuum. These should be considered even though their commercial application is not yet as wide as coffee oil add-back. Coffee aroma gases and especially the steam distillates have been a major source of coffee flavor recovery. As pointed out in the percolation section (Chapter 10), the return of vent gas condensate is a notable contribution to coffee flavor in the instant coffee powder.

Patents assigned to Nutting and Darling Hills Bros. relate to steamed R&G volatiles recovery. In 1956, Hills Bros. was one of the first companies to process and sell an aromatized instant coffee that was so markedly different from others that, in the following years, its flavor was altered to conform to its competitors. Other steam distillation patents are owned by Wendt (1939) and Heyman (1947). The Bonotto patent (1959) steam distilled R&G coffee by collecting and concentrating the aromatics on silica gel. These are then driven off by steam and reincorporated into instant coffee. This patent is assigned to McCormick and Company. Bacot's (1950) U.S. patent also related to pre-steam distillation of aromatics before water extraction of R&G coffee. These are only a few of the patents that use steam distillation of coffee volatiles. They are listed here to indicate continuing interest in this method of coffee flavor recovery. Such persistence over the years is usually an indication of merit.

Lemonnier (1954) was granted a patent for dry vacuum distillation (DVA) of coffee volatiles to be added back to the powder. Johnston (1942) has a similar patent for an atmospheric recovery system for grinder gases. Two 1949 Swiss patents deal with recovery of R&G coffee distillates and reincorporation of them into the extract. A French patent by Laguilharre (1950) concentrates volatiles from the first part of the distillate from coffee extract. Thereafter, the extract is dried with concentrated volatiles restored. In 1950, a Swiss patent similar to the Lemonnier dry vacuum aroma (DVA) recovery was taken out by Schott, followed by two more Swiss coffee patents in 1951.

Some of the percolation patents previously mentioned also aromatize. For example, Barotte (1890) and Whitaker and Metzgar (1915) used steam distillation prior to extraction of R&G coffee and reincorporated the volatiles into the extract.

## Molten Carbohydrates

Another process incorporates the flavor into a molten carbohydrate which is immediately cooled to form a protective powder. Patents by Turkot (1957 and 1959), Eskew (1960), Dimick (1959), Makower (1959), and Epstein (1948) relate to this process which has not, however, been successful with coffee aroma or flavor even though some of these patents refer to coffee applications. Similar patents for coffee have been taken out by Chase and Lee (1958), Lorand (1932), and G. Washington (1924). A 1944 British patent covers mixing R&G steam distillate and absorbing non-condensables in corn syrup. Roaster gases are similarly collected and all are mixed into coffee extract which is vacuum dried in thin layers. The residue is 50 percent each of coffee solubles and carbohydrates.

A British patent by Nyrop in 1950 introduced up to 2 percent hexose sugars with phosphoric acid, claiming that the spray dried product had better coffee flavor retention. Girardet has a 1951 British patent similar to Nyrop's, pointing to the addition of carbohydrates and gelatin to help retain flavor achieved through spray drying. Girardet also has a 1950 Swiss patent on coffee flavors. In 1948, Nestlés Afico, S.A., obtained a Swiss patent for adding a quantity of polyhydric alcohol to coffee extract equal to the amounts of solubles obtained in spray drying to produce a low hygroscopic powder with aroma retention. A 1956 British patent 744,757 fixes extracted coffee flavor in a carbohydrate melt. Since soluble coffee use has not grown in Europe (except for England) as quickly as in the United States, fewer patents and less research work have come from Europe.

# COFFEE FLAVOR CHEMISTRY

European soluble coffee patents go back to 1925 with Staudinger, followed in 1926 and 1928 by a Swiss firm, International Nahrungs und Genusmittel. All three are British patents. With Staudinger's analyses of coffee aroma in the 1920's and the patents for synthetic coffee by Swiss firms at about the same time, knowledge of coffee chemistry was then more advanced in Europe than in the United States. Since World War II, however, the United States has taken the lead as a result of research sponsored by the Coffee Brewing Institute, a branch of the National Coffee Association in New York City, supplemented by private research in the field of flavor and aroma of instant coffees.

In the food industry most flavor developments result from trial-and-error tests rather than scientific research directed toward gaining basic knowledge. The application of gas chromatography, mass spectroscopy, and other technical analyses has made investments in coffee chemistry studies more basic and sound. After knowledge is accumulated about coffee chemistry and flavors, it is often years before such information is commercially used. The study of coffee flavor is not without its parallels in other flavored products such as smoked meat and fruit flavors. *Freeze concentration* and *locked-in flavor* preparations are discussed later in this chapter.

## Means of Aromatizing

A very satisfactory aromatized instant coffee can be prepared in many ways. Tastes may differ yet each type retains much of the pleasurable aroma sought from the original R&G coffee. Aromatization involves two basic steps (1) recovery of the flavor portion in some concentrated form, and (2) its reincorporation into or onto the instant coffee powder. Placing fugacious and unstable volatiles like coffee aroma *on* the powder is of questionable value.

## Difference Between Roast and Instant Coffee

Hydrolysis products deviate in flavor from regular coffee; the higher the hydrolysis yields of solubles, the greater the deviation in flavor. A second point is that aroma is low in instant coffees. The lower it is, the less it tastes like brewed coffee.

The third point is all-important. Even if hydrolysis yields of solubles are low and as much aroma as necessary has been added, instant coffee still will not taste like brewed coffee because it has no brew colloids. While the colloidal portion of a cup of brewed coffee contributes 5 to 10 percent oil to the solubles portion, instant coffees have only a few tenths percent petroleum ether solubles at most before spray drying and less than 1 percent after coffee oil add-back.

## Coffee Brew Colloids

Colloids provide a tactile sensation of smoothness to the cup and also increase opacity which can be measured by the amount of transmitted light. Since these colloids contribute so much to appearance, taste, and flavor of brewed coffee, it is desirable to understand what part brew colloids play in coffee flavor before attempting to define the term "aromatizing." Aromatizing is often understood to mean making instant coffee resemble regular brewed coffee. To explain brew colloids is best done by describing their properties. It will then be possible to understand why brew colloids do not exist in instant coffee, to evaluate how their absence affects it, and to determine whether it is worthwhile to attempt to place them in instant coffee. Most of the physical properties of brew colloids can be ascertained by relatively simple brewing and bench top tests.

**Filtration of Brew Colloids**.—Passage of a coffee brew through a paper filter removes practically all colloidal particles. The amount removed will depend on the porosity of the filter paper and the method of preparation. True colloids, less than about 1 mu, will pass through the paper; hence, it appears that the brew colloid particles are larger than 1 mu and may be considered as pseudo-colloids. In other words, their colloidal properties are real only in so far as the life of the brewed cup of coffee is concerned. Further evidence that brew colloid particles are not true colloids is that some rise to the top as a fatty layer and some fall to the bottom as sediment. This separation is noticeable as soon as brewed coffee cools to room temperature.

Coffee passing through filter paper has no "brew flavor." In fact, the flavor of the filtered brew bears a striking resemblance to the percolator coffee extract diluted to cup strength. This resemblance at times can be used as an index of the kind of extract cup that will be obtained from some coffees. Care must be taken that the filter paper used has not absorbed foreign odors (especially when paper is stored in a coffee aroma area) and that it is rinsed with boiling distilled water to remove any residual foreign flavors before use. Removal of the brew colloid leaves a harsh, typically acid, instant coffee flavor; the colloids give a smooth taste texture to the cup. Gravity funnel filtration is more effective than Buchner pressure filtration in removing the brew colloids.

Rinsing the colloid collected with hot distilled water washes away considerable coffee flavor; hence, the particles are ad- or absorbers of coffee flavor constituents. Transferring the filtered colloids from the paper to a cup of instant coffee imparts an appreciable brew flavor and turbidity. The instant coffee flavor is then weaker and the taste texture is smoother. Colloid ad- or absorption also produces a less acid cup (0.05 pH) as noted by taste and pH measurements. This type of phenomenon is consistent with the general properties of colloidal substances.

Filtration removes most, but not all, of the colloids. The addition of $^1/_{10}$ to $^1/_{20}$ of a cup of brewed coffee or the filtered out colloids from this quantity of brewed coffee to a cup of instant coffee is enough to make a perceptibly smoother and less acid cup.

**Yield of Brew Colloids**.—If brew colloids can be transferred to instant coffee, it is of interest to know how much is perceptible in taste and how much is made available under different modes of brew preparation. For example, some obvious preparation factors are roast, grind, and the ratio of water to R&G coffee.

Darker roasts and finer grinds release more colloids, as is evidenced by greater turbidity of the cup. The addition of a brew solution equivalent to $^1/_2$ to 1 gm of R&G coffee will impart perceptible brew flavor to a cup of instant coffee. But since each cup of instant coffee originates from about 6 gm R&G coffee, this is 8 to 16 percent of the R&G coffee, which is a significant fraction. Brew colloids will result in water to R&G coffee ratios from normal brew strength (about 20 water to 1 R&G or 1.1 percent solubles) to much stronger ratios (about 5 to 1 ratio or 7 percent solubles). The greater-than-proportional strength in the second case results because the R&G coffee absorbs water selectively, i.e., in preference to the solubles. Hence, the higher the coffee to water ratio, the more accentuated is the solubles concentration, but the less the solubles yield.

The water to R&G coffee ratio, however, does not seem to influence the yield of brew colloids. This might follow from the fact that the colloidal particles come from the surface of the R&G coffee, and the roast and particularly the grind are more important factors. The water to R&G coffee ratio does influence the solubles content of the resulting extract; hence, it influences the pH and the electrolytic density (electrical conductivity) of the solution. Colloidal phenomena are sensitive to pH and electrolytic density and, therefore, will appear different under such changed conditions. For example, the 5 to 1 water to R&G coffee ratio may result in a pH of about 4.90, whereas the 20 to 1 household brew ratio may show a pH of 5.0.

**Nature of Brew Colloids.**—It is a common observation that the hydrolyzed solubles cause turbidity. About one-fourth of the solubles hydrolyzed are from proteins. Addition of alkali will dissolve these colloids which are in acid solution suspension at about pH 4.0. The brew colloids are oily in nature, but they behave like proteinaceous oils which show increased turbidity at pH values below 5.0 Proteins adsorb acids, a process which certainly occurs with coffee brew colloids. Different proteins have minimum solubilities or isoelectric points at different pH's, mostly between pH 4.7 and 5.2. Some proteins can act as acids (carboxylic groups) or bases (amino groups), depending on the solution in which they exist. These are described as amphoteric. Some proteins are soluble only in alkali, such as corn zein. It is beyond the scope of this text to discuss protein chemistry, but it is apparent that proteins play an important part in brew colloid stability and in hydrolysis solubles from which tars are formed.

Freezing usually causes colloid particles to coalesce and they cannot be restored to their original condition on thawing. This is partly true for coffee brew colloids; however, it depends on the degree of colloid coalescence, rate of freezing, and, hence, the size of ice crystals. For example, quick frozen brew colloids that are freeze dried will reconstitute practically to their original colloidal condition.

The colloidal nature of coffee brew is not limited to brewing with boiling water; room temperature extractions of R&G coffee yield good brew flavor and cup turbidity.

Repeated boiling of brewed coffee breaks down the colloidal stability, with oil coalescing to droplets and rising to the surface. Brew colloids that collect as cream on the top of a cooled, quiet brew show on evaporation a separation of oil droplets and a black, tarry protein residue that is insoluble in water.

**Centrifuging Brew Colloids**.—Since the brew colloids are not true colloids because they separate out on standing, the process of separation can be accelerated by centrifuging. For example, by taking 10 ml of a strong Silex brew (40 gm fine grind roast coffee with 400 ml water) and placing it in a laboratory centrifuge, depending on blend and roast, about 1 percent of tan colored buttery oil and a few percent of sediment will result. The middle portion will be a clear solution tasting similar to instant coffee but without hydrolysis flavors. The oil-emulsion colloid portion has excellent and strong coffee flavor. The sediment has strong brew flavor when added to a cup of instant coffee. The potency of coffee flavor of the "cream" and the sediment portions is such that either one at a 5 to 10 percent level from original brewed cup imparts a characteristically noticeable brew flavor to a cup of instant coffee. The sediment may carry some R&G fines (fiber) and a slight amount of carbohydrates and oil (petroleum ether solubles), but is otherwise protein.

Proteins are divided into simple proteins which yield only amino acids on hydrolysis, or conjugated proteins in which some other component is linked to the protein. The non-protein portion is called a prosthetic group. Casein from milk is classed as a phosphoprotein and contains phosphoric acid as the prosthetic group; the protein of coffee colloids appears also to be a phosphoprotein.

The composition of the floating, buttery or creamy portion will depend on the degree of coalescence of the colloidal oil-protein. The greater the coffee oil content, the greater the colloid coalescence will be. For example, petroleum ether extraction of the "cream" can yield from 15 percent down to a fraction of 1 percent of oil. The portion insoluble in petroleum ether is a black tar containing protein and phosphorus.

## Lecithin

Phospholipids (substituted fats containing phosphoric acid and proteins such as lecithin, cephalin, and sphingomyelin) are found in seeds of plants. Lecithin is obtained commercially from soybeans and is a mixture of the diglycerides of stearic, palmitic, and oleic acids linked to the choline ester of phosphoric acid. It contains about 4 percent phosphorus. Commercial lecithin has about 2.2 percent phosphorus according to the Merck Index. Its importance is that it is insoluble but swells in water, forming a colloidal suspension. It is soluble in fatty acids but insoluble in vegetable oils. These properties make lecithin a very effective surfactant and emulsifier. Lecithin is used for these purposes in the food industry.

The brew colloid particles appear to be in the range of 1 to 10 mu, varying with the degree of coalescence. The fact that the brew colloids are not entirely stable means that a major fraction of the colloids coalesce to particles larger than 1 mu which then may rise to form a creamy emulsion film. Heating this emulsion on a steam bath causes emulsion breakup into a clear, almost tasteless yellow oil and black tarry deposit. The oil has a flavor that is bitter and similar to that of fatty acids in vegetable oil. The black-brown deposit is probably the protective colloid (phospho-protein) layer that surrounds the minute oil particles. The emulsion has a physical texture like dairy cream and a rich coffee flavor with a note of unsweetened chocolate.

## Test for Phospholipids

Lecithin colloid with coffee oil provides instant coffee with qualities of smoothness and body. Without this ingredient, its flavor is flat. Brew colloids are tested positively for phospholipids as follows: (1) acetone precipitates colloids; (2) methyl acetate precipitates colloids (result: upper layer clear green; lower layer clear brown); (3) cadmium chloride precipitates brew colloids using ethanol; and (4) ethyl and pretroleum ether emulsify in colloids.

The brew colloid retains its natural characteristics only if it is not concentrated. If colloids coagulate, their tactile taste and basic brew flavor properties are lost. There is separation of free vegetable oil and proteinaceous, tarry sediment. Concentration of brew colloids may be effected in several ways: (1) percolation to solubles concentrations over 15 percent and especially 20 percent; (2) evaporation of water by any means, e.g., still or spray drying. As already pointed out, concentration of the colloidal emulsion either by evaporation or centrifuging simply coalesces the colloidal micron-sized particles until larger and larger bodies result. This simultaneously reduces their potential for creating a smooth, tactile sensation. An emulsion exists until water solubles are perhaps 15 to 25 percent, but by then the emulsion particles coalesce irreversibly.

## Freeze Dried Emulsion

If the brew colloid solution were freeze dried as a means of reintroducing some noticeable brew flavor into spray dried instant coffee, there might be some merit in concentrating the brew colloids to a certain degree to minimize evaporation costs of freeze drying. By adding only about 5 percent of freeze dried, brew flavored instant coffee to 95 percent spray dried instant coffee, the resulting product will be more like brewed coffee. Colloidal proteins will precipitate in the rising acidity with increased solubles concentration and electrolytic density. Brewed coffee has characteristically a pH of 5.0 and contains about 1 percent solubles. Hence, the electrolytic density or conductivity of the coffee solution is relatively poor. Since the protective oil-protein colloid is sensitive to breaking down in more acid environments, the pH range is an important consideration in colloid stability. Percolator extracts with high solubles concentrations coagulate brew colloids.

Turbidity is no assurance of brew flavor, but without turbidity there is neither brew taste nor flavor. Also, brew colloids that are spray dried cannot survive the solubles concentration without losing stability; hence, spray dried powders have no brew character. But since freeze drying, with prior quick freezing, prevents gross coalescence of the brew colloid, subsequent ice sublimation from interstitial voids may produce soluble coffee powders with brew taste character. Some brew colloidal taste character can be produced through spray drying by using high pressure atomization as for cream, whole milk, egg whites, egg yolk, whole eggs, coffee oil emulsion in gum, and other food products.

Atomization at 3,000 to 5,000 psig of brew tasting coffee extract (several percent solubles) can bring some brew flavor into the powder. But because of the fine atomization, practically all the coffee flavor volatiles are lost. This indicates a need to balance blended powders to achieve a desired coffee flavor and taste texture. The colloids in brewed coffee are delicate, so homogenization of brewed coffee may coagulate the colloids.

## Simulation of Colloidal Oil

This can be done by homogenizing at about 5,000 psig, for instance, one part of coffee oil with one part of a 25 percent solution of gum arabic and atomizing at high pressures (3,000 psig). Micron-sized oil droplets coated with gum arabic result. Lecithin does not have to be used but is desirable. In the cup of instant coffee containing ½ to 1 percent of homogenized oil, a smooth tasting cup with brew flavor is achieved.

Coffee oil does not homogenize with coffee extract and cannot be spray dried even at high atomization pressures without having oil separation.

The high pressure atomization of cream is used to prepare dry cream powders. The protein protective layer around the oil must be conserved for restoration of tactile emulsion sensations on reconstitution. The oil droplets produce a powder with a sugar-protein coating.

## Review of Brew Colloids

In summary, brew colloids are micron-sized oil droplets (in a protective protein-phosphate medium) that are unstable in acid media, high electrolytic media, and in concentrated solutions. These droplets break down the emulsion to liberate free oil and protein-phosphate tars. The colloids are buffers, adsorbing acids to yield a less acid and harsh tasting cup of coffee with a smooth tactile sensation. The oils help to carry coffee flavor and aroma. The colloids are unstable at elevated temperatures and concentrated solutions such as occur in spray drying. High pressure atomization spray drying brings some brew colloids through but hardly any volatile flavors. Freeze drying brings many brew colloids into a powder form. Brew flavor colloidal effect is simulated by emulsifying coffee oil in gum arabic and freeze or spray drying at high pressures.

## FREEZE CONCENTRATION OF COFFEE SOLUBLES

The less water in the extract that is sent to the spray drier, the less that has to be evaporated. Also, the greater the percentage of water that is retained in the powder, the greater is the retention of volatile flavors, Furthermore, because the higher extract concentrations are more viscous, they yield thicker walled particles with less surface per unit weight, and their water evaporation occurs by diffusion more than by flash evaporation. Thus, the theoretical and practical evidence is all in favor of freeze concentration.

No one freeze concentrates coffee extract commercially. Strongly supporting this statement is the fact that extract solubles concentrations of over 30 percent are readily obtained by normal percolation. With intercolumn heating of extract 40 percent solubles can readily be obtained. However, these higher concentrations of solubles are obtained at the price of coffee flavor loss caused by high percolation solubles inventories or percolator residence time.

The best tasting percolator extracts with minimum hydrolysis flavors are made with the highest water flow throughputs at the lesser extract concentrations of about 20 percent solubles. These concentrations are too low to carry much volatile coffee flavor through spray drying. Therefore, freeze concentration of the lower extract concentrations improves instant coffee powder quality.

### Cost

There are some limiting technical features to freeze concentration, but investment is not one of them. The cost of freeze concentration equipment is approximately the same as the cost of expelling oil, spraying it back on the powder, and finally, packaging the product in inert gas. Freeze concentration of orange juice has been the technique used to obtain a far superior flavored orange juice concentrate. One difficulty is that there is no "off-the-shelf" equipment to buy, and an installation would require some minimal development work. There are numerous patents and articles on freeze concentration for fruit juices, vinegar, and other foods, and there are several plants freeze concentrating these items commercially.

## Viscosity, the Limiting Factor

The physical factor limiting freeze concentration is the increased viscosity of coffee extracts at solubles concentrations of over 50 percent at the freezing temperature, about 14 F (–10 C), which is required to separate out ice crystals. Figure 12.1 shows a schematic phase diagram for ice and coffee solubles. It can be seen, for example, that at 23 F (–5 C), the equilibrium mixture is about 6 percent coffee solubles in the solid phase and 40 percent solubles in the liquid phase. This means that the ice phase must go to another stage of freezing out ice for higher concentrations of solubles. When the rejected ice has only a few percent solubles, it would be practical to vacuum evaporate the dilute solution to recover these solubles. Figure 12.2 illustrates such a batch countercurrent freeze concentration process. The process of freeze concentration should be protected with inert gas to avoid oxidation of coffee flavors. The freeze concentration process has the advantage of coagulating the tarry coffee substances between the ice crystals. The resulting extract has a markedly reduced hydrolysis flavor and taste.

## Ice Crystal Growth

Sharp separation of ice and more concentrated coffee extract is obtained with large ice crystals which entrain little extract in centrifuge separations. The growth of ice crystals takes time. Large crystals efficiently separate ice from the concentrated coffee solution. Moderate-sized ice crystals can be grown from refrigerated metal surfaces, but as they grow the rate of heat transfer diminishes.

Freeze concentration requires a turbulent flow of extract around the freezing surfaces so that as ice crystals grow, the adjacent concentrated extract layers are

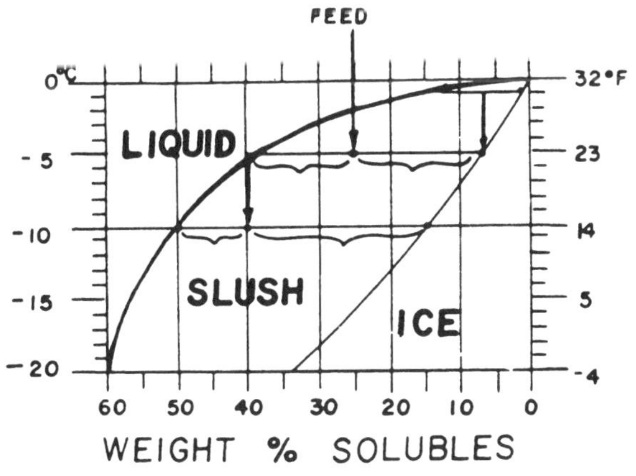


FIG. 12.1 PHASE DIAGRAM FOR COFFEE EXTRACT

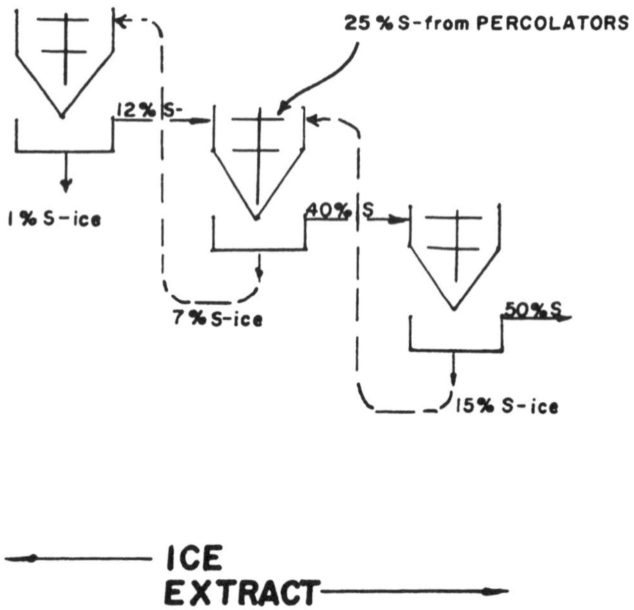


FIG. 12.2. COUNTERCURRENT MULTI-STAGE FREEZE CONCENTRATION

removed. Quick freezing of the extract is not suitable for ice crystal separation because only small ice crystals form. Small crystals grow in a rotary film freezer (such as is used for ice cream, and final concentrated orange juce). Research on saline water purification shows that the growth of ice crystals in a chilled suspended medium is feasible.

## Batch Freezing of Ice

Preparation of commercial 100-lb blocks of ice includes using a center air bubbler to mix the concentrating solubles in the water. Coffee extract also can be frozen in the same way with brine at about 18 F (−8 C) circulating outside the can. However, after ice crystals 2 in. thick have formed, the rate of ice growth is very slow. It is better to have an immersion freezing plate with extract circulating around it, and to remove the plate with about 1 in. of ice crystals. The ice crystal growth rate is high up to that thickness. The cooling rate depends on the coolant temperature. The concentration of coffee solubles influences extract viscosity and the rate of heat transfer. The ice crystals also occlude extract. Thus, the growth of free ice crystals in a turbulent fluid extract solution which is cooled from a side stream of the extract is more attractive. A pure, 1 in. thick ice crystal can be grown in about 24 hr from a chilled wall at 18 F (−18 C).

**Example I**.—A 25 percent solubles concentration extract (3 lb water per pound solubles) can be concentrated to 55 percent solubles concentration extract (0.8 lb water per pound solubles). This concentration removes about four times the final water content associated with the solubles. The water removed from solubles, when these extracts are dried to 3 percent moisture, are respectively 100/1 and 26/1. Hence, retention of volatile flavors is improved at least four times. The thicker particle wall factor may actually increase volatile flavor retention in spray drying by as much as 1000 percent. Particles spray dried from such high extract solubles concentrations are dark, shiny, and beady, have good fluidity, and excellent water solubility without much foaming. The ice crystals on chilled plates can be thawed off the plate and placed into a basket centrifuge for recovery of coffee solubles. The ice phase can be discarded or the ice can be further recrystallized, depending on the solubles content of the ice. The phase diagram, Fig. 12.1, shows a 25 percent extract feed concentration with a tie-line at 40 and 6 percent solubles; the phase weight distribution is, respectively, 55 to 45 and shows 85 percent of the solubles in concentrated extract phase.

*Preparation of Phase Diagram*.—By taking several coffee extract solubles concentrations and cooling them, the rate of temperature decrease alters slope once ice has formed. That paint is the liquid-slush equilibrium temperatures point. Further cooling causes a decreasing slope again. When the latent heat of ice has been absorbed, the solution is solid; this is the solid-slush equilibrium temperature point. This is somewhat more difficult to determine exactly. The temperature transition readings become more difficult to determine with higher extract concentrations. Since mechanical separation of ice crystals and extract is the commercial objective, phase data over this range are easily obtained.

**Example II**.—Continuing with the separation of phases at 14 F (−10 C), the 35 weight parts of 50 percent solubles extract and 10 parts of 15 percent solubles separate to 95 percent solubles in the concentrate and 5 percent solubles in the ice phase. Ice crystals should be at least 2 mm in size for effective centrifuge separation of extract and ice. Ice crystals 1 mm long can be grown in about an hour.

## Volatile Flavor Protection

Some deterioration of extract flavor occurs in freeze concentration. It is best to set aside the volatiles condensed from the percolator vent and to return them to the freeze concentrated extract just before spray drying. This minimizes the deterioration of the delicate volatiles. Freeze concentration of brew colloids in extract causes coalescence of the colloid, oil release, and loss of brew taste in the resulting concentrated extract and hence, in the spray dried powder. Significant volatile flavor deterioration is noted in freeze concentrated extract in the form of a heavy coffee flavor similar to that obtained by pre-wetting R&G coffee and leaving it in air. The spray dried powder from high concentration extracts usually has a strong natural coffee aroma and flavor in the cup.

## Centrifuging

The separation of extract from ice can be carried out in a few minutes in a basket centrifuge operating at a few thousand revolutions per minute. The process is similar to the separation of sugar crystals from molasses.

## Rotary Chilled Freezing Tube

This type of unit will produce 3,000 lb per hr of an extract slush containing about 50 percent of ice crystals by chilling a 33 percent solubles extract from room temperature to 14 F (−10 C) with a refrigeration consumption of 240,000 BTU per hr or a rating of 20 tons per day. With a 165 BTU per hr per sq ft per deg F heat transfer rate, about a 30 F (17 C) mean temperature difference (refrigerant at −14 F or −26 C) and about 48 sq ft of freezing tube surface will be required. The small ice crystals formed are useful only if they can be grown to larger crystals.

## Freeze Concentrating Volatiles Distillate

Since the freeze concentration of coffee extracts is limited by the increased viscosity of the ice crystal extract mixture at lower temperatures, an alternative procedure for concentrating coffee flavor volatiles may be preferable. The coffee extract can be concentrated in a vacuum film evaporator, and the distillate can be frozen immediately. The distillate containing about 0.6 percent of acetic acid (depending on initial extract concentration and fraction distilled) and lesser amounts of homologous carboxylic acids, aldehydes, and other volatiles, can be freeze concentrated. Figure 12.3 shows the phase diagram for acetic acid. This can be used as a guide when freezing coffee distillates. For example, a separation of equal parts by weight of ice crystals and concentrated acetic acid, starting with 0.6 percent acetic acid, will yield about 1.2 percent acetic acid with hardly any acetic acid in the ice crystal phase. A second stage separation of equal weight ice crystal to acid solution separation will again almost double the concentration of the acid. The ice crystal phase at these low acid concentrations has hardly any acid. After three such freeze concentration stages, seven-eighths of the original distillate water will have been removed, (see Fig. 12.3).

## Evaporation of Extract

Coffee extract can be concentrated by vacuum evaporation from 25 percent solubles (3 lb water to 1 lb solubles) to 50 percent solubles (1 lb water to 1 lb

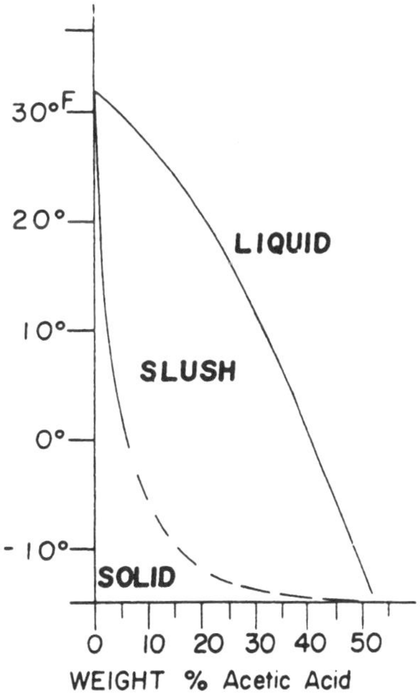


FIG. 12.3. PHASE DIAGRAM FOR ACETIC ACID

solubles). Then, the 2 lb of distilled water are freeze concentrated to ¼ lb and returned to the extract. The water now to be evaporated in spray drying is about one-third of its original content in the 25 percent solubles extract.

Freeze concentration of the evaporated distillate is efficient because the distillate has water-like properties. Therefore, freezing, crystal growth, and phase separations are easy. However, such handling of the flavor volatiles often is accompanied by oxidation, deterioration, and loss. Further, the aroma of less desirable hydrolysis volatiles (furfural, furan, acetone, etc.) becomes more prominent.

## Elimination of Hydrolysis Volatiles

This is desirable before freeze concentrated distillate is returned to the evaporator concentrated extract. The concentration of the aldehyde volatiles in the evaporator distillate may be undesirable because in the distillate the volatiles can react among themselves. Such reactions create a yellow to reddish and then black colloid which eventually precipitates. By excluding air and keeping temperatures below 32 F (0 C) however, these reactions can be reduced. Evaporation of the flavor volatiles produces in the extract residue a caramelized flavor; restoration to the extract of the stripped volatiles largely restores the original flavorful cup (without caramel flavor).

### Other Distillates

The freeze concentration of coffee extract distillates can also be carried out with percolator vent gas distillates, R&G coffee steam distillates, and dry vacuum aroma distillates. All have considerable water associated with the volatiles.

## COFFEE OIL

### Solvent Extraction

Coffee oil can be extracted from R&G coffee with numerous solvents and with various results. The properties of the extracted oil vary because each solvent is selective in what it extracts. The selectivity of the solvent will depend on its nature. For example, the solvent may be a pure hydrocarbon such as butane or hexane; or it may be a polar chlorinated hydrocarbon, e.g., chloroform, Freons, carbon tetrachloride, etc. The solvent may also be polar, such as carbon dioxide, sulfur dioxide, or ammonia. What is extracted from the roast coffee will naturally depend on the affinity of the solvent for constituents in the roast coffee. Therefore, these solvents will remove coffee oil to varying degrees and with varying amounts of aromatics and waxes. The aroma and taste character of the resulting oil after evaporating the solvent will also vary, as will its aromatizing effect on the instant coffee powder. Furthermore, the removal of the solvent from the oil may not be complete in some cases. Any residual solvent in the coffee oil will affect the subsequent taste and aroma of the aromatized instant coffee. Solvent extraction is influenced by blend, roast, and grind. Time of solvent contact, temperature, and ratio of solvent to R&G coffee are also factors influencing the quality and quantity of coffee oil removed from the roast coffee. Solvent extraction of the R&G coffee should be done within a few hours after roasting. To avoid volatiles loss no water quench should be used.

**Polar Solvents**.—The compounds $SO_2$, $CO_2$, and $NH_3$ are polar (ionized) solvents soluble in water. They are gases at atmospheric pressure and room temperature. When these gases are compressed and cooled, however, they become useful as liquid solvents. Carbon dioxide in the gas phase is a solvent for oils. As $CO_2$ gas pressures are increased, the solvent effect is greater. It is at a maximum for liquid $CO_2$. Since $NH_3$ and $SO_2$ are likely to leave a pungent residue after coffee oil extraction and are more irritating to work with, they are less useful as solvents for coffee oil. Sulfur dioxide however, does have the advantage of being a reducing agent and thus it extracts coffee oil without an oxidizing atmosphere. Purity of the solvent can be checked by evaporating it from a neutral sugar powder, e.g., dextrose and instant coffee. One solvent may not yield a wholesome coffee oil-aroma residue after evaporation. Then a mixture of (low boiling) solvents can be used to obtain more wholesome-tasting coffee oil and aroma extraction. Other polar solvents not previously mentioned are: nitrous oxide, bp −130 F (−90 C); methane, bp −258 F (−161 C); ethane, bp −128 F (−89 C); propane, bp −44 F (−42 C); methyl chloride, bp −75 F (24 C); and methylamine, bp −45 F (7 C).

*Liquid $CO_2$.*—Liquid $CO_2$ is very volatile at room temperature.[2] It is ideal for use as an aroma transfer medium. Carbon dioxide is non-flammable, non-toxic in foods, and relatively inert chemically; it excludes oxygen in its use, and is relatively cheap. Other solvents approach the performance of $CO_2$, but they are flammable. Figure 12.4 illustrates a system used by some firms for aromatizing their instant coffee. The solvent can be $CO_2$, $SO_2$, ammonia, low boiling chlorinated hydrocarbons (Freons), or even low boiling hydrocarbons (pentane). When the solvent extracted oil carries aroma and flavor properties that differentiate high grown premium quality coffees from others, the transfer process is sensitive. Effective transfer of aroma and flavor can be achieved only with pure solvents. The extraction system must be clean and the coffee oil transfer must be complete for a balanced coffee flavor.

**Coffee Oil Yield.**—Top quality coffee should be used as the source of coffee oil. Roast coffee has about 10 to 12 percent of coffee oil and aromatic coffee oil yields of over 5 percent are readily attainable. If only ½ percent coffee oil is sprayed on the instant coffee powder, then only $^1/_{10}$ the oil times $^1/_{2.5}$ (solubles yield) or $^1/_{25}$ of the roast coffee percolated will have oil extraced. Recovery of a

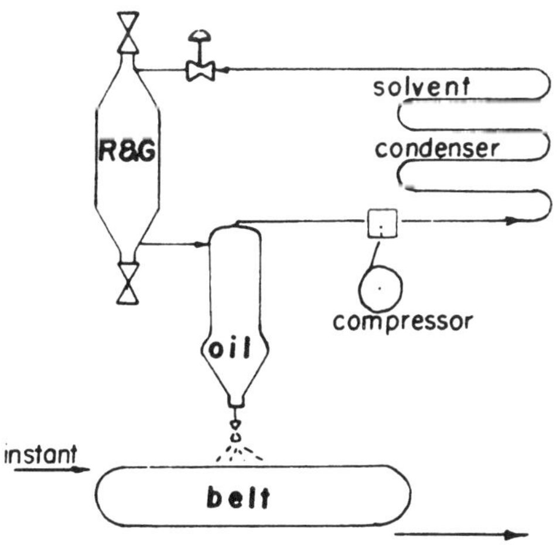


FIG. 12.4. CLOSED CYCLE AROMATIZING OF INSTANT COFFEE FROM SOLVENT EXTRACTED ROAST COFFEE

[2]The vapor pressure of $CO_2$ at 77 F (25 C) is 938 psia (48, 523 mm Hg).

costly solvent in a closed cycle system is essential to make this type of coffee oil extraction process economical. Otherwise solvent costs would be too high. A recycle system of solvent compression and condensation also assures that the aromatics of the coffee are held in a closed system free of air. Once coffee volatiles are in the oil, they are relatively secure and stable.

**Solvent-Extracted R&G Coffee**.—This coffee must be stripped of its solvent residue before water extraction. Thus, low boiling solvents like $CO_2$ are advantageous. $CO_2$ leaves neither residual odor nor taste. If there is a solvent residue, then the R&G coffee may be heated under vacuum or may be steamed to remove noticeable traces of the solvent. Normally such a procedure is feasible but undesirable.

Solvent-extracted R&G coffee with its oils removed is noticeably lighter in color. Brewed, solvent extracted R&G coffee is lower in coffee flavor and aroma; some coffee flavor remains, but coffee aroma is mostly depleted. The solvent-extracted R&G coffee makes a much darker brew.

Solvent-extraction of R&G coffee does not degrade the ground coffee particles as is the case in oil expelling. Solvent extractions of roast coffee make subsequent water extraction easier. If $CO_2$ is used as the solvent, the coffee oil after solvent evaporation is left in an inert atmosphere.

**Coffee Oil Stability**.—The coffee oil with aroma, whether extracted or expelled, is quite stable chemically. It shows no rapid tendency to stale or deteriorate in a sealed brown bottle or tin can. Keeping the oil refrigerated helps extend its period of freshness.

Once the coffee oil is spread on the instant coffee powder, it stales noticeably in air at 80 F (27 C) during the first day and to an objectionable level within the first week. In air at less than 40 F (4 C), the coffee oil on instant coffee powder remains fresh for at least a month.

**Cleanliness of Extraction System**.—Acetone-ethanol mixtures effectively clean coffee oil residues. Equipment used for handling solvents and coffee oil must be kept scrupulously clean. A little residual stale oil in the system may contaminate coffee oil subsequently prepared. Pipe sealing compound and lubricating oils in new piping systems can contaminate coffee oil. Flanged connections are more sanitary than screwed pipe threads.

**Oil Add-Back**.—Coffee oil add-back, at more than 0.6 percent of instant coffee powder, releases oil droplets. These appear as oil slicks on the cup and create a heavy oil taste. Coffee oil additions make a cup of instant coffee taste smoother and more like natural coffee in taste and aroma. The coffee aroma in the jar is especially good and desirable. Consumer testing of the fresh coffee oil sprayed onto instant coffee powder will usually show a significant perference over the untreated instant coffee powder. This is due to the initial coffee aroma from the jar and the small enhancement in flavor in the cup. The addition of coffee oil can sometimes mean the difference between consumption or rejection of instant coffee. With oil add-back, undesirable taste and odor caused by hydrolysis are diminished. The R&G coffee used to prepare the coffee oil carries its aroma and taste qualities to the cup of instant coffee. The aroma is usually more noticeable at some distance from the cup when the powder is dissolved in boiling water.

**Hydrocarbon Solvents.**—The solvents discussed have been very volatile. They mostly require pressurized systems at room temperature to maintain the solvent in liquid condition. Some solvents, however, do not need pressurized systems. These solvents can be useful for extraction of coffee oil, and yet can be successfully evaporated from the coffee oil and the roast coffee. The hydrocarbons pentane, bp 97 F (36 C), and hexane, bp 136 F (58 C), extract oil from R&G coffee efficiently in a Soxhlet extractor. The solvent evaporates from the coffee oil under mild stripping conditions, yet leaves most of the coffee aroma in the oil. Carbon disulfide, bp 115 F (46 C), is an excellent coffee oil solvent but is highly flammable. Chlorinated solvents such as ethylene dichloride, bp 183 F (84 C); trichlorethylene, bp 189 F (87 C); carbon tetrachloride, bp 171 F (77 C); and chloroform, bp 142 F (61 C), are good extractors of coffee oil but may leave an undesirable residual odor and taste.

**Columnar Extraction System.**—The chemistry and physical properties of coffee oil will be covered later but it is pertinent to describe here some of the equipment and operating variables in solvent extraction of coffee oil from freshly roasted and ground coffee. Soxhlet extraction of coffee from R&G coffee is satisfactory for small samples. But columnar solvent extraction of larger amounts of R&G coffee is more efficient and more rapid and produces high coffee oil yields. For example, a bank of extraction columns, as used for percolation, works well with hexane solvent. The size of the columns depends on the rate of processing desired. Hexane at 80 F (27 C) will extract from a single R&G coffee column about half the coffee oil in 15 min. With multiple column extraction, the full coffee oil yield can be obtained in a 15-min cycle time. Figure 12.5 shows a generalized profile of hexane-coffee oil drawoff plotted vs time or relative bed volume. Figure 12.6 shows the relation between specific gravity of the hexane-coffee oil solutions as taken by hydrometer or balance vs the weight percent of coffee oil at room temperature. Similar coffee oil concentration—specific gravity curves can be prepared for any solvent to be used.

The generalized curves (Figs. 12.5 and 12.6) are for a R&G coffee column several inches in diameter and several feet high. Note that about 80 percent of the oil yield is in the wetting part of the curve; the overall coffee oil yield in this case is

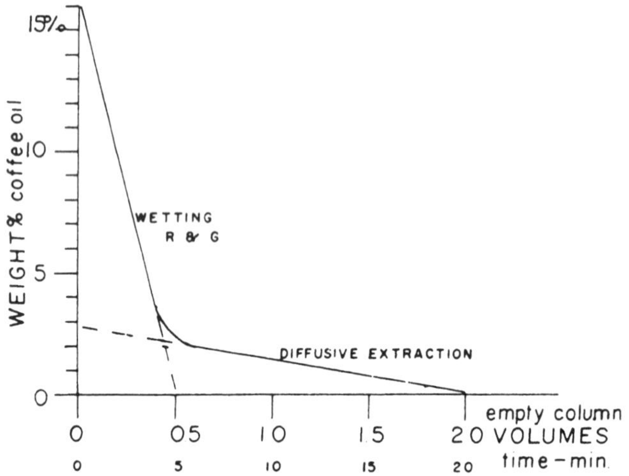


FIG. 12.5. COFFEE OIL CONCENTRATION PROFILE FROM COLUMNAR EXTRACTION OF ROAST COFFEE

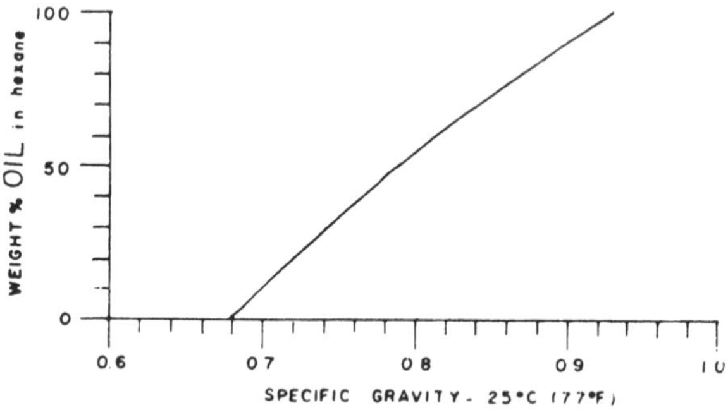


FIG. 12.6. COFFEE OIL CONCENTRATION IN HEXANE VS SPECIFIC GRAVITY

only about 10 percent. Hexane occupies about two-thirds of the empty column volume, but only about one-third of the empty column volume of hexane drains off freely. Note the similarity in hexane concentration profile to percolation water solubles concentration profile. The initial steep slope is the ground wetting with solvent while the oil concentrates in the free solvent. The later shallow slope is the slow diffusion of coffee oil out of the particle into the passing solvent.

Perforated tube entrances and exits with hole sizes relative to the grind size being processed may be used in these percolation columns. The solvent filled oil is drained from the spent coffee ground column. The R&G coffee is then steamed free of solvent. The immiscible layers of water and solvent in the distillate are separated and the solvent is reused.

**Independent Extraction Variables.**—(1) *Temperature:* This is an important variable since it governs vapor pressure and volatility of the solvent and thus, flammability and stripping. Although hexane can be stripped from coffee oil, pentane and especially butane are easier to strip. The temperature of extraction governs the amount of waxes removed with the solvent.

(2) *Particle size:* Finer grinds release their oil more readily. Higher coffee oil yields are obtained faster, but more color is extracted.

(3) *Moisture:* The R&G coffee must be low enough in moisture not to interfere with wetting of the particles by the hydrocarbon. Moistures between 2 and 10 percent are suitable. About 5 percent is optimum.

(4) *Ratio of solvent to R&G coffee:* Extraction is so rapid (as shown in Fig. 12.5) that large amounts of solvent are needed only in exhaustive oil removal.

(5) *Construction materials:* For protecting the coffee oil, especially the fatty acids, all materials such as vessels, pumps, and piping should be free of iron and copper. Stainless steel or glass are best for this service.

**Dependent Extraction Variables.**—These variables include yield and quality of coffee oil, rate of extraction, and concentration of the oil in the effluent solvent. The degree of roast governs the amount of fatty acids in the oil. Coffee oils after roasting have about 5 percent fatty acids; after percolation hydrolysis they have 12 to 15 percent fatty acids. The coffee oil content of fresh dark roast coffee is 10 to 12 percent, but after percolation it is 15 to 20 percent. The fresh R&G coffee oil has fresh aromatic coffee volatiles, but the coffee oil from R&G coffee after percolation has none. If higher-than-room temperature is used for solvent extractions, waxes and fatty acids settle out from the coffee oil.

## Expelled Coffee Oil

Since 1957, major coffee processors have expelled coffee oil from freshly roasted coffee beans and then sprayed it on instant coffee powder which is packed in a jar with $CO_2$ atmosphere containing only a few percent of oxygen. Figures 12.7 and 12.8 show a photo and a sectional view of the V.D. Anderson oil duo-expeller. The unit on top of the expeller is a Sharpless Super-D-canter centrifuge for immediate clarification of the expelled oil. Similar types of expellers can also be used to squeeze water out of percolated spent grounds reducing moistures to about 55 percent.

**Expeller Operation.**—The entering R&G coffee may be steam heated, 150 psig–375 F (190 C), by the entering screw jacket or moistened to help soften the coffee. A motorized choke valve on the discharge end of the expeller controls the

*Courtesy The V. D. Anderson Company*

FIG. 12.7. COFFEE OIL EXPELLER

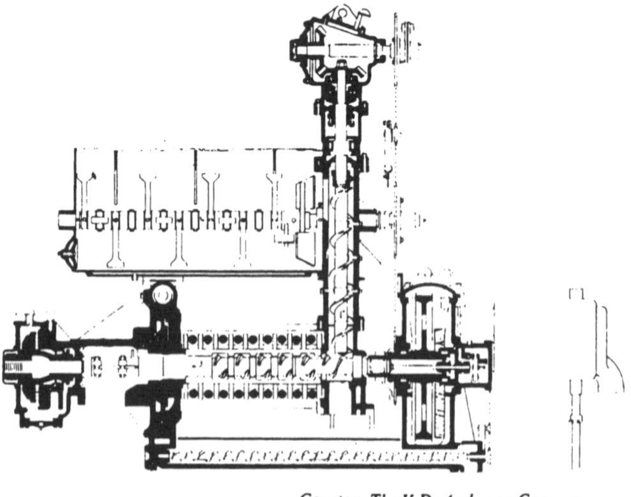

*Courtesy The V.D. Anderson Company*

FIG. 12.8. COFFEE OIL EXPELLER

back pressure (50 tons) on the extruded expelled cake. An ammeter control on the main driving motor (about 50 hp), if overloaded, reduces the vertical rate of coffee feed. The expelled cake can be extruded into strings and cut into pellets for reintroduction into the percolators with the normal R&G coffee. Because of the great amount of frictional heat developed during the expelling process, the central shaft of the expeller is water cooled. A series of short replaceable blades rotating at 20 rpm compresses the coffee as it is pushed and pressed forward. The coffee oil comes out of peripheral slots that are only a few thousandths of an inch wide. The coffee oil is hot, about 180 F (82 C), when it leaves these slots and is then immediately cooled. Centrifuging away "foots" makes the aromatic coffee oil more stable. The coffee oil can be stored at room temperature when it is clear. A molecular still has been used according to the Cole patent (1951) to strip volatiles and to concentrate them in a lesser amount of coffee oil. The coffee oil carries a sulfurous portion of the coffee aroma as well as phenols which stain the hands yellow and have a barbecue odor.

**Oil Yield.**—Roast coffee is a very hard material as compared to palm kernels. The duo-expeller will reduce palm kernel oil from 50 to 15 percent in the vertical squeeze and from 15 to 5 percent oil in the horizontal squeeze. Roast coffee contains about 12 percent coffee oil. At a rate of roast bean coffee feed of 500 lb per hr and a residual 5 percent oil left in the expeller cake, about 35 lb per hr of oil can be produced. If a 7 percent coffee oil yield is obtained from 7 percent of the percolated roast coffees, the overall coffee oily yield is ½ percent of the original R&G coffee. This is about 1.3 percent of the instant coffee. About 3 percent moisture in the roast coffee also helps to soften the granular structure.

**Expeller Construction Materials.**—All expeller oil contact surfaces should be stainless steel or nickel. Copper or bare steel will react with fatty acids and dissolve, resulting in subsequent deterioration of the coffee oil aroma. Some expellers are also run with an inert gas blanket.

**Oil on Powder.**—The coffee oil can be sprayed onto the instant coffee powder as shown in Fig. 12.9. The powder takes on a darker, "wet snow" or "sandy" appearance with poor fluidity. When "plated' correctly, several tenths of 1 percent coffee oil on the instant coffee powder imparts a desirable taste texture and cup smoothness. This coffee oil retains aroma volatiles and reduces foaming in the cup. Some dark roast (espresso) instant coffees have about 1 percent coffee oil, as normally percolated and spray dried. Similar coffee oil contents result from percolation of fine grind coffee. The dark roast oils on instant coffee powder stale more slowly than oils from lighter roast coffees.

**Nature of Oil.**—In general the chemical, physical, and yield properties of the coffee will vary with the type of original coffee. Robusta coffee oil emulsifies easily with extract while good mild coffee oils hardly emulsify. Combining lime with the fatty acid emulsifiers can break an emulsion. The coffee extract pH rises and color darkens, producing a disagreeable taste.

*Courtesy The Johnson-March Corporation*

FIG. 12.9A. COFFEE OIL SPRAY EQUIPMENT FOR INSTANT COFFEE POWDER

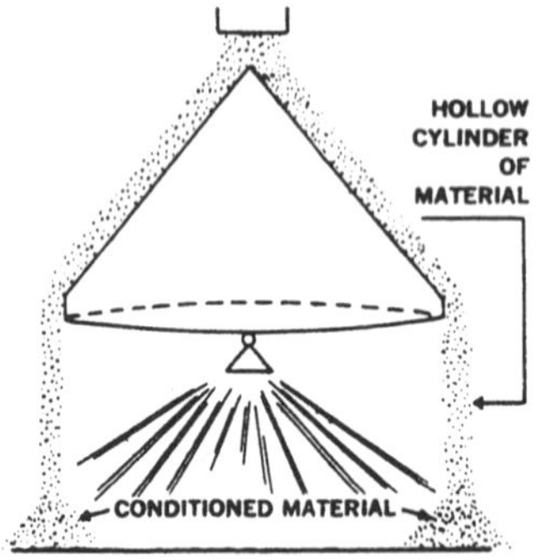


FIG. 12.9B. COFFEE OIL SPRAY EQUIPMENT FOR INSTANT COFFEE POWDER

# COFFEE VOLATILES RECOVERY

## Aroma Sources

Most coffee patents are concerned with recovery of the volatiles which produce the aroma so enjoyable in coffee preparation. There are also the pleasant, naturally liberated coffee aromas from coffee roasting and grinding, and from the freshly opened (vacuum or inert gas packed) tin can. Each of these aroma sources has been investigated with varying results. Coffee aroma can also be induced (1) by placing a vacuum over the fresh R&G coffee; (2) by wetting and/or heating the roast coffee; and (3) by distilling coffee extract. An inert gas stream, steam, or vacuum may carry the liberated volatiles to condensers.

## Coffee Aroma Essence

Without knowledge of the coffee volatiles' chemical and physical properties, it is scientifically impossible to evaluate their performance in the phases of liberation, collection, stability aromatization, and pleasure.

The first work of real merit was done in about 1925 by Staudinger, a chemist, who isolated the coffee aroma essence in a series of cold traps. He then identified many of the individual chemical compounds. Considering the nature of the analytical equipment and the methods available at that time, the results were good and many aroma constituents were identified. These have been duplicated, confirmed, and improved upon since then by other workers with better equipment and methods. The British patent 246,454 (1926) assigned to International Nahrungs und Genussmittel Aktiengesellschaft of Schaffhausen, Switzerland, is informative. It describes a series of coffee aroma condensers: ice water, dry ice in solvent, and liquid nitrogen, at respectively 32 F (0 C), −112 F (−80 C), and −292 F (−180 C). Similar aroma recoveries have been made by other workers. The coffee aroma has been identified as a clear, yellow-green liquid that floats on water at 0 C.

**Essence Composition**.—Gas chromatography for qualitative and quantitative volatiles identification, gas chromatographic infra-red and mass spectroscopy provide reliable information about coffee aroma. An examination of the coffee aroma-flavor volatiles shows that there are about 400 ppm of these substances in roast coffee. They are discrete chemicals, mostly aldehydes and ketones, but also sulfides, furans, esters, and alcohols. The aldehydes are predominant in quantity, but the sulfides are often predominant in taste and aroma. Many of the components boil at room temperature, which accounts for the difficulty in their isolation. Also, these components are more volatile than water and are only partly recovered with water in an ice water cooled condenser.

## Substances Less Volatile than Coffee Essence

These substances boil in the temperature range of water or higher. They are primarily the homologous acids: formic, acetic, propionic, butyric, and valeric. Acetic acid is dominant at about 0.35 percent in roast coffee and about 0.5 percent in the water portion of percolated extract. Percolator vent condensate might have up to 1.4 percent acetic acid before the R&G coffee becomes wet with water.

Under vacuum and/or heating of extracts or R&G coffee, some higher boiling esters, aldehydes, and alcohols as well as furfurals may be distilled.

### Dry Vacuum Aroma

Dry vacuum aroma (DVA) corresponds to the coffee volatiles driven off with enough water so that they are soluble in the water condensate or aqueous phase. These are differentiated from coffee essence which is more volatile and potent with limited solubility in water. Such distillates are usually about 99 percent water. The major portion of the organics are homologues of acetic acid with only about tens to hundreds of parts per million of coffee essence volatiles.

This aqueous condensate of volatiles is very unstable. It is clear upon collection, but turns clear yellow within an hour and develops a sickening, tobacco-like odor. It turns colloidal reddish in a few hours and becomes colloidal black overnight at room temperature. The black colloid after coagulation will settle out, and appears to be a phenolformaldehyde or furan precipitate. The precipitate is a polymer, because in more concentrated solutions (such as ethyl ether extracts with ethyl ether evaporated) containing about 25 percent water, a solid black gel develops in storage at room temperature (less rapidly at lower temperatures). These lesser coffee volatiles are not as readily lost in the spray drying of instant coffee.

## TECHNIQUES OF AROMA AND FLAVOR RECOVERY

Numerous patents describe methods for the recovery of coffee volatiles. They refer to ice water or chilled brine, dry ice in solvent, and liquid nitrogen condensers. Most of these aroma collections were more qualitative than quantitative, because much of the earlier work was done by chemists. Their laboratory equipment was largely for qualitative results, not for complete recovery of volatiles. The difference between collecting coffee aroma volatiles qualitatively and quantitatively lies largely in design of the condensers, cooling methods, construction materials, gas throughput rates, and temperatures of coffee essences collections. Even low effluent gas temperatures from a condenser do not assure complete condensation. There may be entrainment of water condensate droplets with effluent gases. One may cool with liquid nitrogen at −292 F (−180 C) and have little heat transfer surface in the condenser; but when the condenser surfaces collect ice in the process of collection, the heat transfer rate decreases. The surface in contact with the condensing essence then may be far above liquid nitrogen temperatures.

### Volatility of Coffee Essence Aroma

The most volatile constituent, hydrogen sulfide, boils at −76 F (−60 C). Methyl mercaptan, the next most volatile constituent, boils at 45 F (7 C). The rest of the volatiles boil at higher temperatures. Since $H_2S$ is not an important part of good, fresh coffee aroma, it need not be collected. But if one chooses to collect it, the effluent gas temperature from the last condenser must be less than −76 F (−60 C). In practice, with effluent gas temperatures of −50 F (−45 C), the effluent condenser gas has no odor other than the pungency of carbon dioxide. This gas effluent temperature establishes the performance required from the last condenser.

## Solid $CO_2$ Coolant

With carbon dioxide coolant at −109 F (−78 C), liquid nitrogen coolant is not needed in an adequately designed and sized heat exchanger for essence recovery.

Dry ice can be used to cool many liquid media to −108 F (−78 C) or to some higher temperatures if eutectic ices form. Examples of liquid media are isopropyl alcohol, Cellosolve, and the Freon-11 or Genetron-11 chlorinated hydrocarbons.

Evaporation of $CO_2$ gas circulates the coolant to effect a uniform temperature. Pump circulation of the coolant is more positive and may have to be used in some designs. Care must be taken to seal the cooled liquid media from atmospheric water condensation. This condensation would cause ice and foul media, pumps, and strainers. Usually a desiccant in the liquid medium suffices to absorb small amounts of moisture. The advantages of the Freon-11 type medium are that it is not flammable and ice floats to its surface. The $CO_2$ has good solubility in Freon-11. However, $CO_2$ volatilizes and separates from the coolant when warm. Such dry ice coolant systems are suitable for temporary testing.

## Refrigerant Coolant Systems

For permanent coolant installations, it is more economical to have a two- or three-stage refrigeration system. This may operate, for example, at −60 F (−51 C) or −100 F (−73 C) (Freon-22). A closed cycle coolant system loses no refrigerant and is always under operating control. At these temperatures, no $CO_2$ gas from aroma sources will freeze out as dry ice snow in the trap as will occur with a liquid nitrogen coolant. Assuming a $CO_2$ flow of 8.8 lb/min, a three-stage Freon-22 system would have the following operating characteristics:

| | Aroma Temperature, $CO_2$ Gas | | | | Refrigerant Temperatures at Suction | | Refrigeration Cooling Load | |
|---|---|---|---|---|---|---|---|---|
| | In | | Out | | | | | |
| Stage | F | C | F | C | F | C | Tons | Motor hp |
| 1 | 200 | 93 | 35 | 2 | 30 | −1 | 1.0 | 1.5 |
| 2 | 35 | 2 | 0 | −18 | −25 | −32 | 0.5 | 1.0 |
| 3 | 0 | −18 | −50 | −46 | −80 | −62 | 0.4 | 15.0 |

Note the high horsepower requirement at the lower temperature compressor stage. Actually, a two-stage system will work if the final heat exchanger is made large enough. A desirable feature in the coolant system would be a refrigerant gas at 100 F (38 C) to defrost ice in the cooling tubes. All condensate lines, valves, fittings, heat exchangers, etc. should be of stainless steel construction to withstand the thermal stresses at these low temperatures. All low temperature surfaces should have 8 in. of polystyrene foam insulation.

**Aroma Condensing Sequence**.—Ice water or brine can be used instead of the 30 F (−1 C) refrigerant in the shell of the first gas condenser. Condensate from the 35 F (2 C) effluent gas can be separated by a cyclone receiver. The aqueous condensate should be frozen and/or added back to the extract just before drying. The primary condenser gas outlet temperature is 35 F (2 C) so that considerable water removal will occur without ice forming on the cooling surfaces. The second stage series condenser may cool the exit gas to 0 F (−18 C), which removes almost all the water vapor. It may even be desirable to use a larger heat exchanger and a colder gas outlet temperature such as −20 F (− 29 C) to assure that the least possible amount of water vapor carries over into the coldest condenser collecting the coffee aroma essence. The less water associated with the coffee essence aldehydes, the easier it is to isolate the essence.

The boiling temperature at atmospheric pressure of the coffee aroma volatiles must be noted. Most of the aroma volatiles will be recovered in the intermediate condenser cooled to −25 F (−32 C). The most cooled condenser is a scavenger to assure (by the odor of its effluent gases) that no odorous coffee substances escape. With this in mind, the aromatic coffee essence is condensed at −20 to −30 F (−29 to −34 C) effluent gas temperatures from the second condenser. No third condenser is needed. Effluent gas temperatures from each trap can be recorded to monitor aroma collection performance. If ice builds up, showing rising effluent gas temperatures, then either the trap can be thawed and its contents recovered, or an alternate clean trap can be placed in line. The gas line should have a flowmeter. A gas flow recorder is desirable until condensation effectiveness is established. Essence condensation can also be effected at higher condensation temperatures and pressures, but this is not usually practical.

This type of aroma trapping system may also be used for grinder gas, percolator vent gases, and steamed gases from R&G coffee.

**Essence Stability**.—The collected coffee essences can be stored safely at temperatures below 20 F (−7 C) as long as there is no free water to provoke the coffee essences' chemical deterioration.

**Small Condenser Heat Load**.—The first gas condenser has the major heat load of condensing water. This can be estimated as saturated moisture in the entering $CO_2$ gas at the entering temperature. The heat load for condensing coffee essence in later condensers is very small, and most of it may be due to inadequate insulation. The cooling loads and the heat exchangers are small in size; hence, the overall coffee essence recovery equipment needs only a small investment. It does require special engineering, however, which adds to the equipment cost.

**Quality and Potency of Condensates.**—It is convenient to use a microsyringe to measure out hundredths of a milliliter into a cup of instant coffee or larger brew batch to gain insight into the aroma, potency and taste properties of the recovered volatiles. The person recovering the coffee essence must not do the tasting for two reasons: (1) the condenser operator or unloader is saturated with essence (including hands and clothes) so that he becomes insensitive to cup evaluations, and (2) the person collecting essence may be so involved in the research effort that he has become biased. An independent yet objective group of tasters should evaluate the aroma and flavor merits of recovered coffee essences.

**Recovery of Roaster Gases.**—These gases are, as a rule, impractical to recover because their properties represent more of what is occurring chemically in the roasting process than the final volatiles composition of the roasted coffee bean.

**Recovery of Aromatics from Coffee Oil.**—Recovery of essence by molecular or any other type of distillation or separation from coffee oil is not practical because the coffee aroma constituents are tenaciously held and are well protected in their natural habitat. Concentrating the aromatics from coffee oil is of dubious value since not all the coffee flavor volatiles are in the oil. They do not pass from coffee oil to water to improve cup flavor.

**Coffee Aroma Recovery from Grinder Gas.**—In some cases, the roast coffee bean grinder is placed in an inert gas atmosphere to eliminate air oxidation as well as moisture. The grinder gases are rich in coffee aroma and $CO_2$ gas, but the portion of coffee volatiles liberated is only a small fraction of that available in the rest of the coffee bean. For example, a regular grind may release only about one-third of its $CO_2$ content and perhaps only one-tenth of is volatiles content. after all, $CO_2$ is by far the most abundant volatile roasting product. The $CO_2$ content of the roasted bean varies from 1 to 1.5 percent. The aromatic coffee essence is only about 0.05 percent. The $CO_2$ sublimes at $-108$ F ($-78$ C); dimethyl sulfide, the most odorous coffee volatile, boils at 100 F (38 C); and the volatile acetaldehyde boils at 68 F (20 C). During essence collection, if the moisture of the air is not excluded or if the coffee beans are wetted with water, the amount of associated moisture collected can be one-hundred-fold greater than the coffee essence. For example, there is only a fraction of 1 percent coffee aroma essence in the $CO_2$ gas. With moisture present, the coffee essence is dissolved in the much larger amount of condensed water.

## Coffee Essence Stability

If air enters the coffee aroma system, the naturally fruity, pleasant, sweet, and fresh coffee aroma constituents will develop a tobacco-like, stale coffee odor and a strong, brew-like taste. The condensate will develop objectionable odors and become sufurous (like $H_2S$). These condensates (if not too diluted with ice crystals or frozen $CO_2$ crystals in liquid nitrogen traps) are usually light to dark yellow. The lighter essence colors that are clear, deep yellow to yellow-green are purer and fresher. The yellow color confirms the presence of sulfur compounds; the green color is related to acetyl-carbonyl and diacetyl. The pungency and fruitiness of the aroma is caused by acetaldehyde; fruitiness is caused also by other aldehydes. The coffee essences will deteriorate in minutes if they are not kept below 32 F (0 C), keeping the associated water frozen. If coffee essence and little water are collected, the essential oil may separate out as a floating phase over the ice. This separation can sometimes be completed by centrifuging. Coffee essences and their contacting aqueous solutions may be evaluated in flavor, aroma, and potency by dilution in instant coffee. Water condensates saturated with $CO_2$ are acidic at 4.0 pH. The collection of grinder gas coffee aroma essence is informative but is not in the main source of coffee aroma.

**Carbon Dioxide**.—With 0.5 to 1.5 weight percent of $CO_2$ in bulk roast coffee beans, the volume of $CO_2$ gas at NTP[1] is about 2 to 3 times the bulk volume of the roast coffee. For example, 100 cu ft roast coffee at 1.1 percent $CO_2$ has about 200 cu ft of $CO_2$. This situation has been described as a roast coffee bean having 100 psig or about six atmospheres of $CO_2$ pressure. This would be true if the $CO_2$ gas were in the voids in the beans; but the roast bean is openly porous, as can easily be seen under the microscope. Therefore, the $CO_2$ gas must be ad- or absorbed on the bean surfaces or within the the coffee oil or both. The charcoal adsorption capacity of $CO_2$ is about 50 volumes $CO_2$ gas per volume of charcoal. Depending on the degree of roast, more or less charring occurs and not necessarily on the outside of the bean. With a 16 percent roast weight loss, 11 percent may result from "free water" liberation, 2 percent from bound water and 1.5 percent from $CO_2$ from carbohydrates ($C_6H_{10}O_5$). There is a small part of carbon that forms during pyrolysis of organic substances. If there is 0.3 percent of carbonaceous matter in the bean, 100 gm (300 ml) of bean could adsorb about 50 ml $CO_2$ gas. This partly explains the $CO_2$ adsorption. Carbon dioxide is soluble in vegetable oils, amines, and carbohydrates, all of which are present in the roast bean. Cellulose adsorbs gases. The roast coffee readily combines with water, hydrocarbons, aldehydes, etc, which is evidenced by the solvent desiccation effect on the first leaving a column of dry roast coffee when extract water is admitted.

[1]Normal temperature and pressure.

## Coffee Volatiles Adsorption

Further, almost every patent dealing with the removal of aroma and flavor volatiles from roast coffee uses (1) higher coffee temperatures, and (2) lower atmospheric pressures around the coffee. Adsorption isotherms for charcoal, silica gel, and other adsorbents show reduced capacity of retention with increasing temperatures and decreasing pressure. Charcoal is an excellent adsorber for acetone, holding half its weight in acetone when saturated. Charcoal is also a good adsorber for aldehydes, esters, acetic acid and its homologues, also methyl, and ethyl alcohol. Furthermore, commercial adsorbers like charcoal and silica gel are reactivated by steam to release their adsorbed volatiles. Charcoal is a very good adsorbent of the coffee essences; it immobilizes the adsorbed aroma, so that they neither stale nor escape.

Adsorption liberates heat; the more chemical in nature are the affinities, the more physical adsorption becomes a chemical reaction. The affinity roast coffee has for water is greater than its affinity for other substances, like aldehydes, $CO_2$, ketones, and sulfides. Therefore, added water may displace them.

The addition of water to roast coffee (directly or from the atmosphere) releases volatile aromatics and $CO_2$, because of the selective preference of the fiber structure for water. The wetting of the roast bean with water destroys the adsorptive power of the dry cell surfaces for aldehydes and other coffee essence constituents.

## Factors in Liberating Volatiles

The variables controlling the amount and rate of coffee volatiles liberated are blend, roast, water content, grind, temperature of grounds, and system pressure. If the water, acetic acid, etc., were free, they would readily distill in minutes. The fact that the volatiles distill over only partially in hours means that the water, acids, and essences are held on the coffee grounds by physical, and perhaps chemical, forces. The use of a vacuum necessitates a liquid air coolant for condensing and is a disadvantage of the vacuum system. Heat transfer coefficients also are poorer under vacuum and require larger heat exchangers. The use of liquid nitrogen for such an essence recovery system is too expensive. The Lemonnier method for recovery of "dry" vacuum aroma is relatively inefficient (because of the method and low essence yields from R&G coffee) as compared to steaming R&G coffee.

## DVA Method

Another way to collect grinder gas coffee aroma is to use vacuum and heating of roast coffee. This draws off the coffee volatiles as described by Lemonnier (1954), who used a medium roast with about two or more percent of moisture. The patent lists neither acidity evolved, percent organic substances, nor the aromatic potency of the distillates. Lemonnier emphasizes the rapid rate of coffee essence deterioration. He does not explain, however, how he can spread this coffee essence on the instant coffee powder without it becoming oxidized. The product specified in the patent is based on collecting the essence-water condensate at yields of 1 lb per 200 lb batch of R&G coffee, so is largely water. If 200 ppm true coffee essence were in the Lemonnier traps it would be 20 gm essence and 454 gm water. Figure 12.10 shows one type of cold trap for coffee essence.

Aroma laden coffee gases, whether from grinders or percolator vents, after being cooled to about 35 F (2 C), will release the volatile coffee essence by passing them through dry ice traps at −108 F (−78 C) so that the effluent $CO_2$ gas is below −40 F (−40 C). The sketch shown (Fig. 12.10) illustrates one such system in which a 1-in. pipe forms a labyrinth for gas passage inside a 2- or 3-in. tube. Quick-coupling connectors help to disassemble the lower thimble where the essence and water ice crystals are formed. A Freon-dry ice bath effectively keeps the tubes at bath temperature; a thermocouple indicates exit gas temperature. All surfaces must be of smooth, sanitary stainless steel to facilitate cleaning the equipment thoroughly between runs. For handling about 1,000 cu ft of percolator vent gases (after an ice water condenser) in 10 min, about 15 linear ft of flow path are required so that coffee aroma is not detected at the $CO_2$ exhaust point. About 6 to 8 in. of styrofoam insulation are required about the bath shell for proper insulation. Under the removal circumstances described, it is doubtful if more than 5 gm of true coffee essence are collected in 500 gm of water. This is about 1 percent coffee essence solution (no free essence phase) which would be highly unstable in the water phase at room temperature. To overcome the disadvantages of a batch essence collection, the system should be continuous. Continuous movement of R&G coffee through a heated screw and trough jacket followed by a cooling screw and jacket will suffice. A heat transfer coefficient of about 2 BTU per hr per sq per ft per deg F can be expected. Lemonnier's patent describes conditions of the distillation for acetic acid and its homologues with small amounts of phenol-type compounds.

**Completeness of Volatiles Recovery.**—The coffee volatiles liberated in any kind of coffee aroma recovery process must be carried over to the instant coffee without substantial loss or change so as to retain a well-rounded coffee flavor and aroma.

Wetting ground coffee liberates most of the coffee volatiles promptly. When coffee temperatures rise to 300 F (149 C), oil flotation and darkening of the grounds will result. Under such circumstances sublimation of caffeine and some oils occurs. Caffeine sublimes at 351 F (177 C) at atmospheric pressure, and at considerably lower temperatures under vacuum.

As water is driven off from the R&G coffee and its moisture decreases, the rate of liberation of flavor acids and aromatic volatiles diminishes. Water is the key to

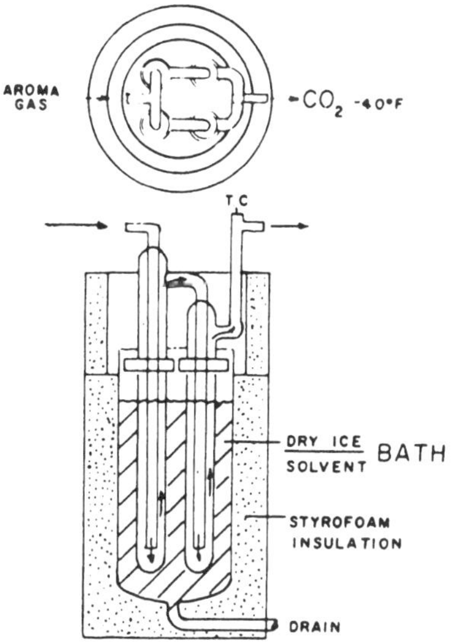


FIG. 12.10. DRY ICE COOLED COFFEE ESSENCE RECEIVER

freeing coffee volatiles. The concentration of acetic acid distilled over from the roast coffee based on starting concentrations of 0.30 percent of acetic acid in the coffee and 4.0 percent moisture is about 7 percent of acetic acid. Table 12.1 shows acetic acid vapor-liquid equilibrium data. Mild coffees yield more volatile aromas and acids than Brazil or Robusta Coffees. Acid yields under this type of collection system are only about three-fourths of the 0.30 percent acetic acid in the roast coffee. These aqueous acid distillates with coffee aroma essence are unstable. They form gray, cloudy colloids and turn to reddish-black precipitates in hours at room temperature with an accompanying flavor deterioration.

**DVA Water Removal.**—Water, time, temperature, and oxygen (air) contribute to such chemical changes. Hence, it is of interest to see the effect of water removal. Solvent extraction concentrates the organic essence portion of the dry vacuum distillate. Ethyl ether or other solvents immiscible with water can be used to transfer about three-fourths of the acidity and coffee flavor to the solvent phase. This can be done by countercurrent extraction followed by solvent stripping. The resulting solvent concentrate may have 25 percent water; the original DVA distillate has 95 percent water. Storing at 20 F (−7 C) this "lesser volatile portion" (as differentiated from essence portion), which is about half acetic acid

TABLE 12.1. ACETIC ACID VAPOR—LIQUID EQUILIBRIA DATA

| BP | | Percent Acetic Acid | |
|---|---|---|---|
| F | C | Liquid | Vapor |
| . . . | . . . | 1 | 0.6 |
| . . . | . . . | 2 | 1.2 |
| . . . | . . . | 3 | 1.7 |
| . . . | . . . | 4 | 2.3 |
| . . . | . . . | 6 | 3.6 |
| . . . | . . . | 8 | 4.8 |
| 212.4 | 100.2 | 10 | 6.2 |
| 212.7 | 100.4 | 20 | 13 |
| 213.1 | 100.6 | 30 | 19 |
| 213.8 | 101.0 | 40 | 26 |
| 214.9 | 101.6 | 50 | 34 |
| 216.0 | 102.2 | 60 | 44 |
| 217.9 | 103.3 | 70 | 54 |
| 221.0 | 105.0 | 80 | 68 |
| 227.7 | 108.7 | 90 | 82 |

(equivalent), shows it is relatively stable for many months. A polymerized resin settles out of solutions held at higher temperatures. The dry vacuum condensate has some of the coffee essence (or more volatile portions). If an effective preliminary ice water cooled heat exchanger is used to condense out the bulk of the water before allowing the aromatic gases to pass on to the liquid nitrogen traps, coffee essence with little water is frozen out as ice, as reported by Lemonnier. It is not uncommon to obtain sweet, pleasant sulfurous odors from the dry ice collected in liquid nitrogen traps, and to have these odors change almost instantly to foul cabbage-like odors. R&G coffee after DVA processing still has some coffee flavor and aroma.

**Cupping DVA.**—All DVA distillates enhance the flavor of a cup of instant coffee at a proportioned add-back level. Figure 12.11 shows a micro-syringe for such work. For example, one drop of distillate per cup of instant coffee containing 2 gm of powder corresponds to a collection of 1 liter of DVA distillate per 250 lb roast coffee. At 80 cups of instant coffee per pound of roast coffee, this distillate will fortify 20,000 cups of instant coffee. The instant coffee flavor is largely removed. Such fortified instant coffee powders have richer, smoother, and fuller coffee aroma and flavor. It is difficult to add an aqueous DVA back into dry instant coffee powder; hence, the ether-extracted and ether-stripped DVA is used for this purpose.

**DVA Distillates.**—These carry a strong smoked-ham odor that is attributed to phenol-type coffee compounds. This is verified by adding 1 percent ferric chloride to the distillate which forms a black solution. Other tests for phenol are also

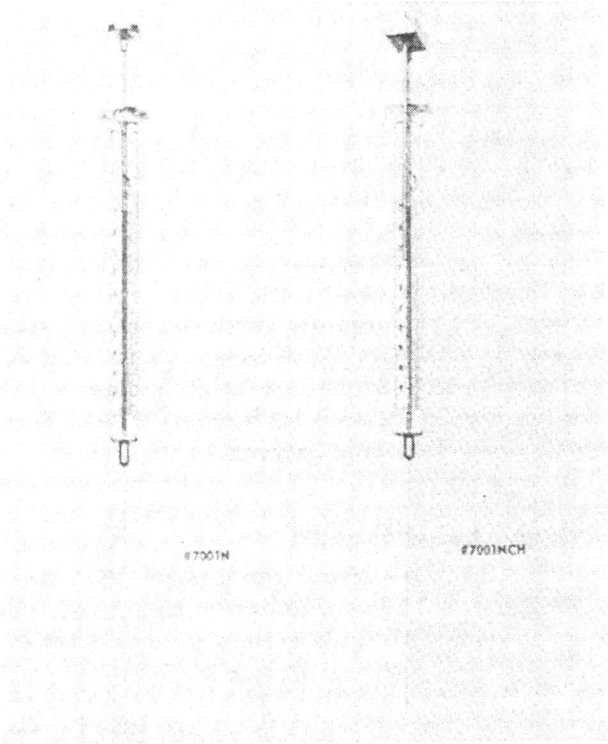


*Courtesy Hamilton Company, Inc.*

FIG. 12.11. MICRO-SYRINGE

positive: (1) the addition of hydrochloric acid to a furfural-containing solution forms a resinous precipitate; (2) aniline turns red in presence of acetic acid and furfural. The red resin that forms in DVA is not extractable with ethyl ether—indicating a polymer. Diacetyl, bp 189 F (87 C), can be isolated by fractional distillation from the DVA ether extract.

**Resins in the DVA Ether Extract.**—When resins form, water solubility decreases and the solution darkens to a deep reddish color. Vacuum evaporation of the DVA ether extract containing resin will separate out a pale yellow distillate. The resin residue remains behind. This distillate has a strong, pungent, acetic acid odor and is soluble in water. DVA ether extract is soluble in 95 percent ethanol, acetone, and glacial acetic acid, which would make an attractive, stable solution of coffee flavor. These solvents tie up the water, making for greater stability of coffee volatiles. Pure acetic acid will lend a smooth flavor to instant coffee.

**DVA Properties**.—The DVA ether extract represents about 0.2 percent of the R&G coffee. The more coffee essence in the DVA ether extract, the less stable it is in storage.

Other tests on DVA ether extract show a white needle crystal deposit from yellowed chloroform extract. The needles are soluble in caustic soda and bicarbonate and might indicate a polyhydroxy benzene. Hydrochloric, sulfuric, and phosphoric acids all cause resin formation in the DVA distillate. The resin is soluble in caustic soda and contains about 0.1 to 0.3 percent sulfur, possibly occluded; the resin could also be a furan ($C_4H_4O$) polymer. About three-fourths of the dry DVA is acetic acid equivalent to about five percent propionic acid; it also contains furfural and its derivatives. Dry DVA is very pungent. Redistillation of DVA ether extract down to 1 mm hg in about six fractions at 302 F (105 C) eliminates water fractions, leaves others that freeze at 0 F (−18 C), and float out droplets of a pale yellow-green coffee essence.

## Steam Distillation of Roast Coffee

Steam distillation is used as a general term for such coffee aroma displacing methods as: (1) wetting roast coffee particles with moist gas; (2) steaming roast coffee at varying pressures; and (3) adding hot water to roast coffee. The effects are all the same, and coffee volatiles released can be condensed through a series of cold traps.

The wetting of R&G coffee in the percolator with extract at 220 F (104 C) causes some steam to lead the extract flow. The steam wets the R&G coffee and drives off coffee aroma and flavor volatiles before the particles are covered by extract. In the case where steam is used to preheat the R&G coffee in the percolator column, the production of coffee aroma and flavor volatiles occurs faster and in greater quantity. The vented gases pass through an ice water condenser and leave at about 35 F (2 C) to remove as much moisture as possible. This richly flavored condensate is restored to the drawn off extract.

**Vapor Composition**.—Each pound of moisture-saturated air at 35 F (2 C) carries 0.004 lb of water. Assuming about the same for liberated $CO_2$ gas, every 1000 lb roast coffee yields 90 to 160 cu ft of $CO_2$. This is 10 to 18 lb $CO_2$, 0.06 lb water and 0.4 lb coffee essence, theoretically.

In practice, this ratio of essence to water is never realized; it is more nearly 10 to 20 times as much water as essence in the dry ice or liquid nitrogen-cooled condenser. The reasons for this are: (1) the effluent gases entering the dry ice or liquid nitrogen-cooled condensers are at much higher temperatures than 35 F (2 C); (2) the displacement of coffee aroma essence is not complete; and (3) there is water vapor dilution from air. However, when efficient volatiles collections are made, the coffee essence to water ratio can be reduced to about one. Batch essence recoveries are less practical than continuous recoveries. Percolator vent gases, whether produced by steam or the rise of hot extract, are the best source of coffee essence for collection.

**Properties of Volatile Condensates.**—The first ice-water-cooled condenser trappings are usually water, clear when fresh to slightly colloidal gray when old. The dry ice or liquid nitrogen-cooled traps, if they do not have too much water carryover, collect yellow ice with potently aromatic coffee essence and perhaps droplets of water and an insoluble dark yellow liquid. The odors emanating during the handling of these traps range from fruity to cabbage-like to tobacco-like. The tobacco odor is associated with conditions that have allowed the coffee aroma to become oxidized. Coffee essence recoveries of the yellow-to-green oil that floats on water are about 50 ppm of the roast coffee, with possibly another 50 to 100 ppm dissolved in the aqueous phase. The yellow-green coffee essence will mostly evaporate in a few minutes at room temperature and will turn dark reddish with a barbecue-tobacco-like odor. Once exposed to air, the essence is no longer protected from deterioration by its cooling. Nitrogen gas protection is helpful at above freezing temperatures, but best coffee essence stability is obtained at temperatures below 20 F (−7 C). This may be the melting point of the eutectic of coffee essence with water. At this temperature coffee essence is ususally a beautiful pale green.

Aqueous distillates from R&G coffee have about 2 percent acetic acid and 2.8 pH. This is because it takes about 10 lb steam condensate to heat 100 lb R&G coffee to 212 F (100 C). This is 0.3 lb acetic acid in 10 lb of water, or about 3 percent acetic acid. Table 12.1 shows the liquid-vapor phase concentrations for acetic acid; distillation separations of water and acetic acid are not efficient.

Titration of aqueous coffee distillates is practially identical to 3 percent acetic acid when pH is plotted vs milliliter 0.1 *N* NaOH. Removal of the distillate acid with anion resin causes an amine (possibly tri-methyl amine) odor in the alkaline effluent. It has a definite coffee flavor when it is added to instant coffee. The alkali equivalent is about 10 ppm of the R&G coffee.

**Polymers.**—The percolator vent gas condensates form resinous red colloids that become black and later settle out. These are polymers of phenols and aldehydes or furan formed in the acetic acid. There is an assortment of aldehydes and phenols in the aqueous coffee distillates. Polymerization rates are faster at higher acid concentrations and higher temperatures. Phenols are unstable and redden on exposure to air and light. Depending on the mode of percolator venting, distillates may have only about 0.25 percent acetic acid (pH 4.0). The flavor factor in the aqueous coffee distillates is primarily phenols and acetic acid, respectively smoky and pungent. Where large fractions of water are frozen out as ice in the coffee essence cold traps, the essence concentrates to a slush. The richer coffee essence portion which melts at about 20 F (−7 C) can be centrifuged away.

**Steaming R&G Coffee.**—Steam distillation of R&G coffee at 15 to 20 psig is carried on only until the coffee has been heated to the steam temperature corresponding to the pressure used. If the distillation is carried beyond this point, the flavor and aroma of the distillate are lowered in quality as are the non-volatile solubles flavor portion left in the coffee. Prolonged steam distillation of R&G coffee causes a bitter flavor that is associated with a greater acidity; the pH of distillates falls from about 4.0 nearly down to 3.0. The evolution of R&G coffee aroma should occur in less than 5 min at 212 F (100 C) at atmospheric pressure, much as it occurs when household coffee is prepared.

**Oxidation.**—Oxidation of an aqueous percolator vented distillate, which is relatively free of aroma essence aldehydes, causes little to no flavor change, even with formation of black precipitate. But oxidation of the coffee essence results in an unpleasant aroma and flavor in the coffee cup. The volatile aqueous and essence (aroma and flavor) constituents differ, but both improve flavor in instant coffee. The former provides more flavor and smoothness, and the latter provides more aroma. Percolator vent distillates that discolor from yellow-to-red-to-black are accompanied by a loss in fresh coffee odor. They develop a tobacco odor and unpleasant cup taste when added to instant coffee. During these precipitation changes pH does not change, which indicates that acetic acid is not directly involved. Prolonged steaming of R&G coffee causes the distillate to develop an odor similar to a warmed-over pot of coffee. The brew pH from that R&G coffee may fall from 5.0 to 4.7.

**Acid and Essence Volatiles.**—By not allowing the percolator vent distillate gases to reach 212 F (100 C), only about one-twentieth of the total volatile acids (0.35 percent in the R&G coffee) is distilled. By holding the vent gas temperatures several minutes at 212 F (100 C), about three-quarters of the volatile acids in the R&G coffee can be distilled. Thus, the coffee essence volatiles are distilled mostly below 212 F (100 C), and the volatile acids and probably the phenols are distilled above 212 F (100 C) with prolonged steaming. The harsh taste of coffee extracts from R&G coffee that has been steamed for several minutes at 212 F (100 C) can be smoothed by adding back to the extract the removed aqueous and essence coffee distillates. Neutralizing the acid distillate with NaOH, distilling off water to concentrate the acid salt, and finally reacidifying with HCl or cation resin concentrates the coffee acid portion before it is added back to the extract prior to spray drying. Such evaporative concentration of the extract solubles before add-back of acid and essence reduces volatilization losses during spray drying.

## FLAVOR FORTIFIED COFFEE EXTRACT

Restoration of the aqueous and essence distillate trappings to the coffee extract before spray drying is essential to obtain the best balanced coffee flavor and aroma from the powder. Since the coffee essence constituents readily deteriorate, however, the resulting powder with essence residue must be packed with inert gas immediately after spray drying to preserve its enhanced coffee aroma and flavor. Promptness in protective packaging is essential because once oxidation starts, it seems to be an autocatalytic chain process. Oxidation will continue after vacuum and inert gas packaging.

## Distillation of Coffee Extracts

The procedures of the Milleville (1948 and 1950) patents, which have been useful in fruit juice flavor concentration, flash distill 10 percent of the feed juice under vacuum. This is followed by column distillation with reflux. This method suggests concentrating the volatile coffee essences in coffee extract in the lighter boiling product. Acetic acid, which has a higher boiling point, would not be in the distillate residue. A major difference between commercial coffee extracts and fruit juices is that percolator extracts have components formed by hydrolysis. These are objectionable in aroma and flavor, so their recovery is undesirable. Flash and still distillation of coffee volatile components, using coffee extracts without hydrolysis solubles, may have some disadvantages. Coffee essence constituents are much more reactive than fruit juice. To protect the coffee essence aldehydes, distillation should be under a vacuum which corresponds to room temperature boiling points, and preferably lower. Furthermore, the coffee essence constituents in water solution and with the least amount of air will still oxidize and polymerize to form precipitates. The use of vacuum requires liquid nitrogen-cooled recovery traps, and the distillation system must be free of air. This is not easy to achieve.

**Azeotropes**.—The separation of the coffee essence constituents by distillation is complicated because of azeotropes. For example, methyl furan and water boil at 137 F (58 C), and 78 percent methyl furan and 22 percent methanol boil at 125 F (52 C). There are also ternary azeotropes, such as methyl furan-acetone-water, bp 132 F (56 C). It would take considerable testing to determine all the azeotropes that may form from the coffee essence constituents. They may represent only about 100 ppm (0.01 percent in the distillate) even with a high reflux ratio. To recover significant amounts of distilled coffee essence, one must feed the evaporative system continuously. Altogether, the number of problems and techniques to overcome are sufficient to discourage this distillation approach.

**Oxidation**.—Any air in the distilling system will oxidize the essence and water mixture. Then the still system must be thoroughly cleaned with caustic soda to free it from its foul, nasal-irritating, pyridine-like odors. Because furan polymerizes, it will eventually foul the packed distillation column. Some distillation odor notes are similar to fufural or dried fruits (apricots and/or prunes). Rubbery odors are not uncommon when working with coffee volatiles.

The most volatile fractions carry over into the vacuum system cold traps. Sufficient concentration of coffee essence is accomplished to condense yellow distillate with perhaps green to orange colorations and a strong coffee aroma. The flash extract distillate residence time in the still must be very short to separate essence as well as to minimize time for deterioration of the essence in water solution.

The essence in water is still quite dilute and very unstable when it is taken from the top of the fractionating column. Hence, essence separation is difficult to achieve. Since the flash distillation of the coffee extract releases only a small fraction of its essence, the overall coffee essence yields are small. These are also not representative of the original coffee aroma and flavor.

**Partial Extract Evaporation.**—Removing some water from coffee extract does have some commercially applicable features. For example, by taking the second half of the percolator extract drawoff, which is at a lower concentration of solubles than the first half, the extract can be concentrated by a vacuum evaporator from about 20 to 50 percent solubles. This technique removes most of the hydrolysis volatiles. The more concentrated coffee extract can then be combined with the richer, natural-flavored first half of the extract drawoff for spray drying. Using this procedure, less natural coffee volatiles are lost in spray drying. A further improvement in the flavor of instant coffee powder may be obtained by rendering the flavor acids of part of the coffee distillate non-volatile by neutralizing them with NaOH, removing most of the water by distillation, and then acidifying the concentrated residue by cation resin treatment. The released concentrated acids along with condensed volatiles from the distillation and/or other sources may then be added back to the regular coffee extract before spray drying.

**Caramel Taste.**—As volatiles are progressively removed during vacuum evaporation, coffee extract flavor improves. For example, it is often found that the foul off-flavors of low-grade coffee as well as the hydrolysis volatiles are removed. Then a caramel or burnt taste develops increasingly with the progressive removal of water. If distillate is returned at this point, the caramel-burnt taste is greatly reduced or even eliminated.

**Vacuum Evaporators.**—Figure 12.12 illustrates a type of laboratory vacuum still that can be used for such preliminary investigations. Figure 12.13 illustrates a commercial-sized vacuum rotary wiped wall film evaporator.

**Coffee Aroma Essence and Distillates.**—The removal of volatile, natural off-flavors and undesirable hydrolysate flavors which come off in the first stages of vacuum distillation could be carried out commercially to eliminate the caramel-burnt flavors if a practical way could be found to avoid damaging the

*Courtesy Buchler Instruments, Inc.*

FIG. 12.12. LABORATORY VACUUM ROTARY FLASK FILM EVAPORATOR

*Courtesy Rodney Hunt Machine Company*

FIG. 12.13. INDUSTRIAL VACUUM ROTARY WIPED FILM EVAPORATOR

desirable acidic and other flavor components. Then coffee essence could be added to the concentrated extract, which could subsequently be spray dried with a high degree of coffee aroma and flavor retention. Figure 12.14 shows a schematic diagram of a film evaporator.

Ethyl ether extract of aqueous distillate, after stripping off the ether under vacuum, contains about 15 percent water; it consists of an oily and an aqueous layer. The two layers contain about 40 or more volatile organic chemicals, mostly aldehydes, carbonyls, sulfides, esters, alcohols, and unsaturated hydrocarbons such as isoprene. The aqueous layer contains acetic acid and its homologues, phenols, furans, furfurals, and others. The layers cannot be quickly and sharply separated. The oily layer precipitates a furan polymer while the aqueous phase precipitates a furfural-phenol-aldehyde polymer. Such resinification removes coffee volatiles. When dimethyl sulfide oxidizes to sulfoxide, bp 374 F (190 C), one of the most important coffee aroma and flavor volatiles is lost. This reaction is related to what is commonly known as staling. The instability requires careful processing techniques to recover the volatiles without change. Distillation of aqueous media does not appear feasible.

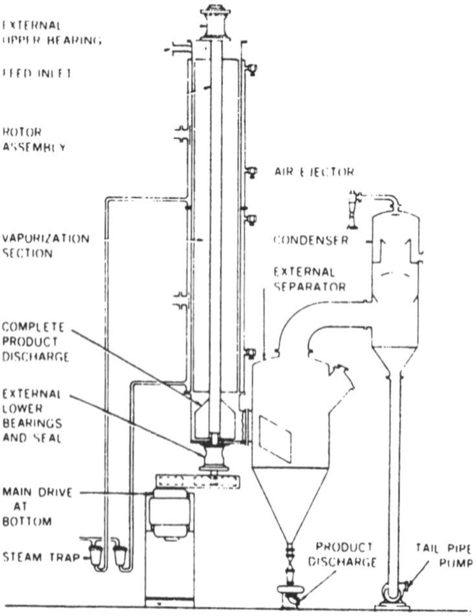


*Courtesy Blaw-Knox Company*

FIG. 12.14. INDUSTRIAL VACUUM ROTARY WIPED FILM EVAPORATOR

**Process Rules.**—The coffee essence volatiles are low boiling compounds, and if a wholesome coffee aroma and flavor is to be retained, *all* natural coffee volatiles must be recovered. They must be recovered without being chemically changed. To do this, the volatiles must be released naturally (brewing or steam distillation of R&G coffee) and be frozen immediately. Freezing removes water and lowers temperatures, thereby lowering rates of chemical reactions. The coffee volatiles must receive minimum exposure to oxygen, air, oxidizing conditions, and daylight. There must be no iron, copper, or other metallic or foreign contamination. For stability, the coffee essence should be removed from the bulk ice and be maintained below 20 F (−7 C). Coffee aroma volatiles are best recovered from continuous process systems which also allow larger amounts of roast coffee to be stripped, and shorter process periods.

The coffee aroma and flavor volatiles recovered must be promptly fixed so that they do not deteriorate or disappear before the coffee reaches the consumer.

## LOCKED-IN FLAVORS

The sealing of volatile flavors in a solid gum structure is usually referred to as "locking-in" the flavor. However, the sealing of flavors as used here also includes gelatin encapsulation and fixation in solid carbohydrate sugars.

### History

The technique of spray dry sealing volatile flavor essences is relatively new. It was commercially applied shortly after 1950, but did not gain full importance until about 1955. Earlier work dealt with emulsifying fruit flavors in gelatin as typified by an Olsen and Seltzer (1945) patent assigned to General Foods Corporation. It discloses how gelatin flavors were prepared for gelatin desserts like JELL-O. A Griffin (1951) patent injected flavor oils into sorbitol, a molten carbohydrate, which on cooling to room temperature, solidified into a brittle film. Griffin used spray chilling of droplets. Other inventors used gums instead of sorbitol and spray drying instead of spray chilling. Figures 12.15 and 12.16 show an homogenizer.

### Phases

The key factor in these flavor fixations is the immiscibility of the oil and aqueous gum solution. The aqueous gum solution is the continuous phase, and the droplets are the discontinuous phase. If the flavor constituents are soluble in the aqueous-gum phase, they are evaporated in spray drying. By adjusting the concentration of the aqueous gum solution and its ratio to the essence or oil so that there is considerably more than enough flavor essence to saturate the aqueous phase, an immiscible essence phase is obtained. In spray drying, the oily droplets

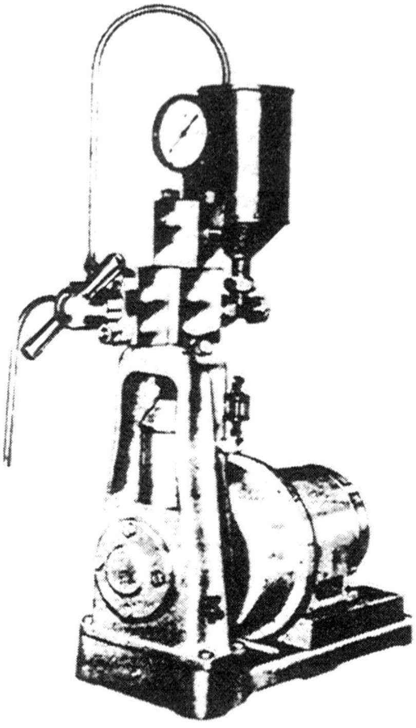

*Courtesy Manton-Gaulin Mfg. Company, Inc.*

FIG. 12.15. HOMOGENIZER

are left coated with the dry gum. Spray drying preserves most of the volatile essences. The essence is in cells within the dry gum lattice.

## Stability of Locked-in Coffee Flavor

Coffee essence is a very unstable mixture of extremely volatile organic compounds. Yet it is retained relatively faithfully within this gum cell structure for years in air at room temperature. Liquid coffee essence can be kept stable only at temperatures below 0 F (−18 C) for any length of time. The removal of air, water, and acids are key factors in stabilizing the coffee essence. The impermeability of the gum cell wall prevents volatility losses.

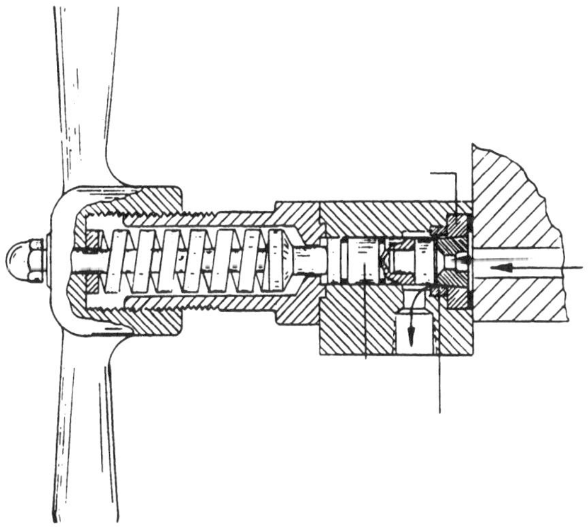

FIG. 12.16. HOMOGENIZER VALVE DETAIL

**Gum Emulsion**.—The modes of emulsification and spray drying are not critical. For example, one part of coffee essence can be emulsified in a gum arabic solution consisting of four parts gum and eight parts water. A homogenizer can reduce the essence particles to micron size. Figures 12.15 and 12.16 show a homogenizer and a detail of the valve. The emulsion can then be spray dried in air at 300 F (149 C) inlet and 200 F (93 C) outlet to remove the water to about 1 3 percent moisture residual. The same technique can be used with coffee oil. A dye (du Pont oil blue in acetone) can be added to the emulsion to examine the size of homogenized droplets under a microscope. The dried powder will not release the volatile essence until the protective gum is dissolved away in water. The fixed coffee essence powder can be added to instant coffee powder at a fraction of the original amount of essence found in nature. For example, 100 ppm of the coffee essence (50 ppm powder) is sufficient to produce an appreciable aroma, flavor, and smoother cup of instant coffee. No antioxidants are required.

## U.S.D.A. Candy-Making Method

The U.S. Department of Agriculture system, patented by Turkot (1959) and Eskew *et al.* (1960), offers a continuous way to inject volatile fruit flavor essences into molten sugar, followed by cooling within seconds. Figure 12.17 shows a flowsheet of the method used. Coffee oil, or coffee oil fortified with essence and emulsified with invert sugar, is quite stable. It gives off natural coffee aroma and imparts a natural coffee flavor after months of storage at room temperature in air.

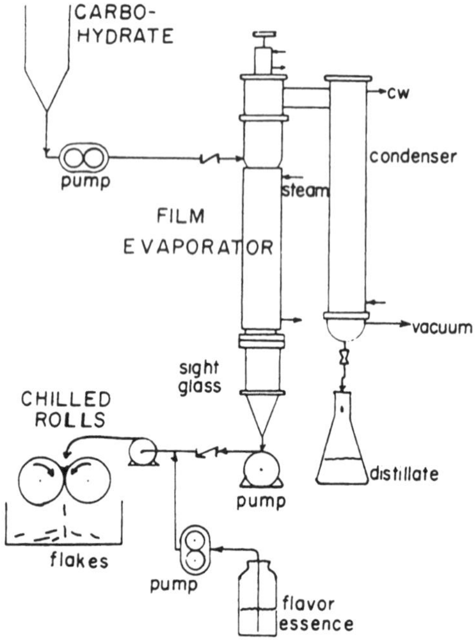


FIG. 12.17. FLOWSHEET U.S.D.A. FLAVOR ESSENCE INJECTION INTO MOLTEN SUGAR

Candies flavored with such a creamy mixture have a fine natural coffee flavor that cannot be duplicated in any other way. Coffee flavor as such is not a commercially available substance. Hence, coffee candies are often artificially flavored. This coffee flavor stability in a sugar-oil emulsion probably stems from the fact that no free water is available for deteriorative reactions.

## Capsules

Since about 1957 the National Cash Register Company of Dayton, Ohio has commercially prepared encapsulated chemical liquids. When the gelatin capsules are ruptured, the chemicals react with other chemicals from ruptured capsules to form a color. This substantially represents their capsule-coated papers that make impressed copies without use of carbon paper. The upper side of the lower paper is coated with one encapsulated reagent, and the under side of the upper paper is coated with the other encapsulated reagent. Typing or writing ruptures the capsules.

Many other substances have been encapsulated and the spray dried locked-in flavor technique is the means used to attain such encapsulations. The Southwest

Research Institute (1959) at San Antonio, Texas encapsulated gasoline and other volatile substances by preparing discrete droplets in aqueous gum solution, and then drying them until only a gum film covered the droplets. These systems can successfully encapsulate coffee essence, since the techniques are only variations of a gum seal about the many tiny volatile essence cells.

## Summary

With the numerous methods of adding-back or retaining natural coffee flavor, a broad and acceptable coffee flavor appeal can be presented to the consumer. Figure 12.18 shows the inroads that instant coffees have made in the roast coffee market. It is expected that flavor-fortified and flavor-modified coffee beverages will continue their growth.

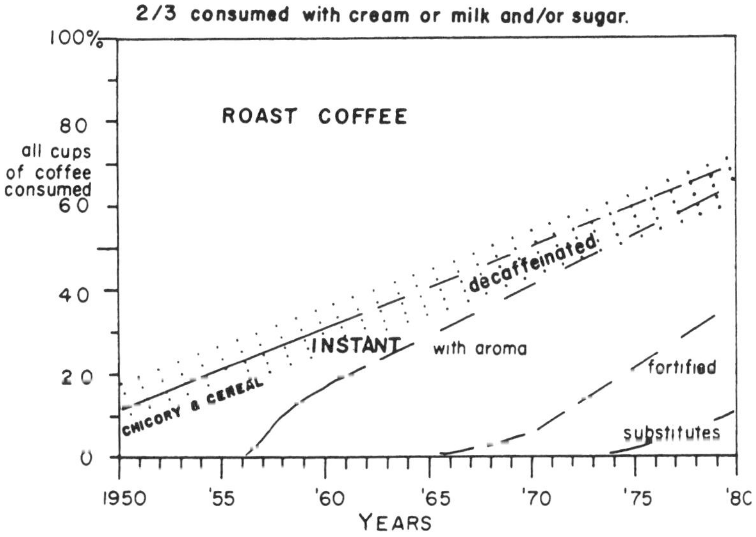


FIG. 12.18. CURRENT AND PROJECTED USE OF COFFEE BEVERAGES

# 13

# Freeze Dried Coffee Production

## FREEZE-DRYING OVERVIEW

Industrial freeze-drying has already become an important part of modern food technology. The importance of this new food conservation technique has influenced the plant manufacturing industry to develop process equipment aimed at economic production.

Freeze-drying is a special drying process for conserving products which contain water or solvents. One must differentiate between two process stages: freezing of the goods and removal of the ice or solvent crystals. A characteristic of freeze-drying is that the liquid is gently separated from the solid material. The frozen liquid is removed without thawing it—that is, sublimation. This means that the phase transition which normally takes place in drying, "liquid-gaseous", is replaced by the transitions "liquid-solid" (= freezing) and "solid-gaseous" (= sublimation). Separation does not occur during sublimation but during the freezing stage.

Freeze-drying can be used for most products, regardless of whether water or solvent is to be removed. In practically all of its industrial applications, i.e. food conservation, the extraction of water is the primary consideration. Freeze-dried, water-free products can be stored for a practically unlimited time if the packaging is impermeable to water vapor and gases. This significant technical and economic advantage makes freeze-drying the ideal method of conservation for many products. Compared with deep-frozen storage, for example, freeze-drying eliminates the high energy costs for transportation and storage (the refrigeration chain) and the risks connected with power failure or the costs of measures to avoid such risks.

The freezing process influences the quality of the finished product due to direct consequences of freezing rate and ice—or solvent crystal size to later freeze-drying rates.

Various freezing methods are used for products in pieces, liquid or pasteous:

fluidized bed, freezing drum, freezing belt, freezing under vacuum, and freezing trays. (See Fig. 13.1)

More and more of these freezing methods which lead to so-called "ice-granules" are used.

This is because of: (1) market requirements; (2) small particle size provides optimum conditions for economic freeze-drying; (3) grinding of already dried product is avoided. Freezing of products in pieces, liquid or pasteous, depending on the method used, producers uniform and optimum products for sublimation.

The freeze-drying process is carried out in drying chambers or tunnels. Plants for discontinuous or continuous operation may be designed according to product and capacity requirements. The freeze-drying operation naturally depends on the economical aspects. Considering all market requirements manufacturers of freeze-dried products have to equilibrate parameters such as (1) prices of raw material, (2) production capacities desired, and (3) production costs. Especially by also using continuously operating freeze-drying installations, high production capacities may be achieved. The production costs consist of the following items: energy costs, labor costs, depreciation of investments over a period of 10 years, and capital services, based on an interest burden of 10 percent p.a.

The costs which arise in connection with the operating personnel are also an economic aspect. These costs may be reduced by using continuous plants.

energy costs, labor costs, depreciation of investments over a period of 10 years, and capital services, based on an interest burden of 10 percent p.a.

The costs which arise in connection with the operating personnel are also an economic aspect. These costs may be reduced by using continuous plants.

## Freeze and Other Types of Drying

In its true sense, freeze drying is the freezing of an aqueous food or beverage and the sublimation of the ice away from the non-volatile solubles. Unfortunately, such drying rates are low and form about 1 mm of ice per hr. Shelf drying, e.g., using 12 mm thickness for an area of 100 sq ft is 4 cu ft or about 300 lb coffee extract of which 180 lb is water (40 percent solubles), would produce 120 lb of dry solubles each 12 hr. This is a very low rate of drying and is much more expensive (2 vs 0.2/lb water) than spray drying. If, for example, 10 percent of soluble coffee production were freeze-dried, it would be practical and the final product would have some brew colloid flavor not possible to obtain by spray drying large, hollow beads.

The National Research Corporation of Cambridge, Mass., worked a great deal to develop a faster and continuous process for vacuum drying during World War II. Currently, the Chain Belt Company (Carrier Division of Girdler Corporation) manufactures a continuous drying unit that operates under vacuum. Drying is carried out on a steel belt and represents something between freeze and vacuum

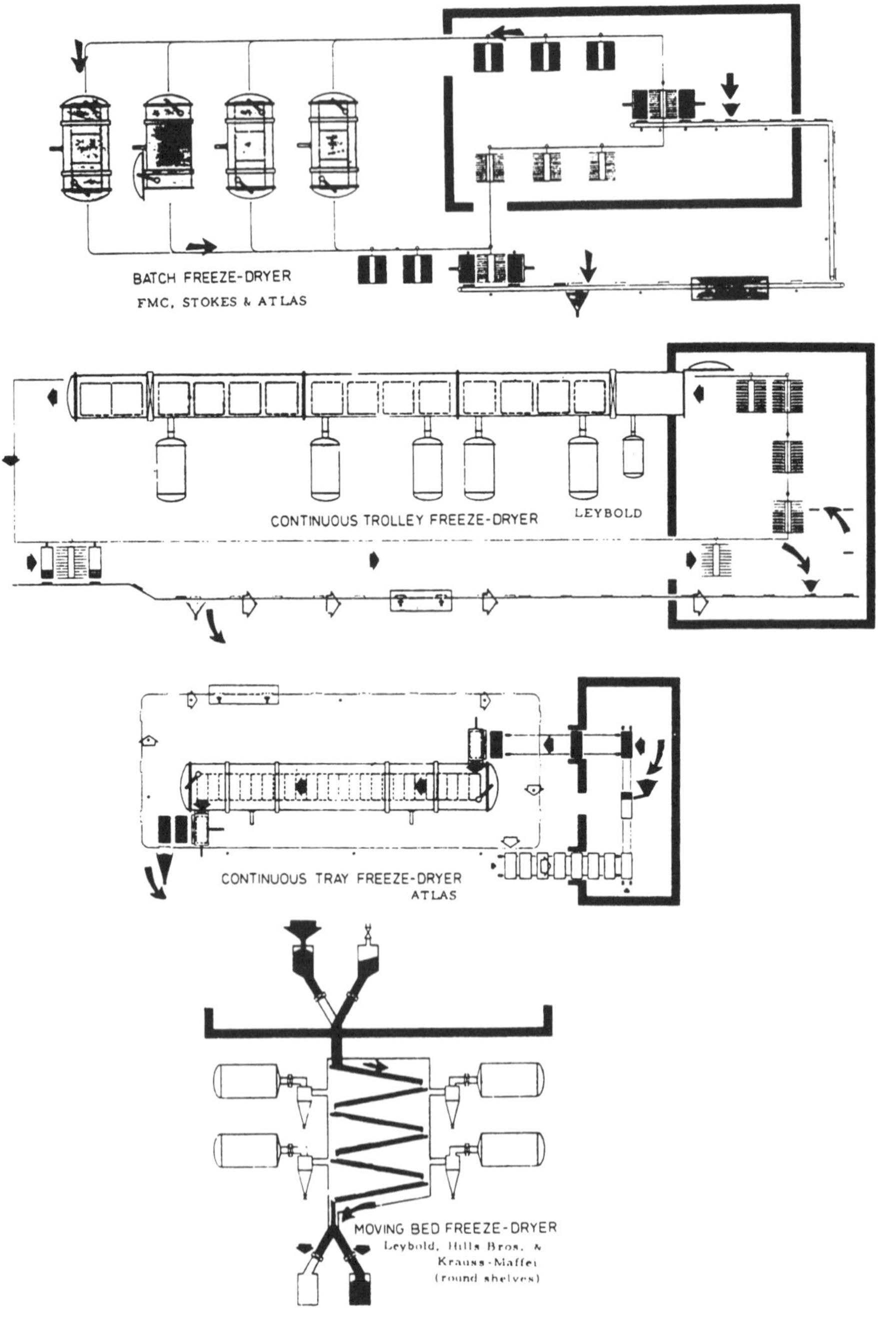


FIG. 13.1. FREEZE DRYING SYSTEMS

drum drying to speed up drying rates over true freeze drying. Figures 11.16, 11.17 and 11.18 illustrate the batch tray and belt vacuum driers. Vacua of less than 5 mm, which is just about the vapor pressure of water at the freezing temperature, are maintained in the vessel. See Table 11.1 and Fig. 11.14.

Vacuum drying is carried out in several stages. The first stage is on a drum heated above 140 F (60 C), where over 80 percent of the drying occurs. The second stage is on the belt under radiant heaters. Then the belt is chilled against another drum and the film of dried instant coffee flakes is scraped off. The current Butternut instant coffee is made in this manner by Penndale.

The advantage of flaked coffee from this process is that it has excellent solubility even in cold water. This high solubility is because flakes have less surface per unit weight; spheres occlude air. Further, the moisture has been withdrawn from within the flake at slower rates than in spray drying; hence, the flake has a porous structure for water absorption. Flake processing may bring through any brew colloids. However, even under ideal freeze drying conditions, the volatiles of coffee aroma and flavor are not retained any better, if as well, as in spray drying. In freeze and belt vacuum drying, water content of extracts should be less than 60 percent. This reduces water evaporation load and drying costs.

**Laboratory Freeze Drying**.—These experiments are simple to carry out. A thin film of coffee extract is frozen on the inner surfaces of a one-liter flask. By connecting the flask to a vacuum pump that maintains less than 5 mm pressure (with dry ice trap between the pump and flask to trap evaporated moisture), the extract will be dry in less than 16 hr. The flakes may then be broken away from the flash surface for evaluation. Invariably the brew colloids, but not much of the coffee volatiles, will be preserved. The taste of the resulting instant coffee will be rather flat but it will have a colloidal texture. The less water evaporated to attain dryness, the more volatile aroma and flavor constituents are retained within the powder. Freeze-dried flakes mixed at a 10 to 15 percent level with spray dried powder produce a brew-like coffee flavor, turbidity in the cup, and a smoother tasting cup. Hydrolysis flavors are diminished and pH in the cup is raised in proportion to the amount of freeze-dried powder mixed with spray-dried powder. A pH rise of from 4.90 to 4.95 or higher at a 10 percent mix level is feasible. Even though freeze-dried samples lack volatiles, they otherwise carry over a fair amount of the flavor characteristics of the coffee blend used.

With the retention of brew colloids in freeze-dried flake soluble coffee, the petroleum ether solubles are much higher than in spray-dried instant powders which have less than 1 percent. For example, freeze-dried powders have as much as 4 percent petroleum ether solubles. The associated phosphorus and protein due to the phospholipids are also higher than for spray-dried instant coffees.

In general, it can be said that in the preparation of freeze-dried instant coffee powders, blend, roast, colloid content, and water content are important properties that carry over to the dried flakes.

**Foam Drying**.—Variations of vacuum of film drying have been experimented with on puffed powder and on foam drying at atmospheric pressure. Both works are from U.S. Department of Agriculture laboratories. The foam method prepares a thin film that will dry quickly in ambient air. Although the patent gives the drying of a coffee extract as an example, there has been as yet no commercial application of this method.

# FREEZE DRYING PROCESS STEPS

## Freezing Coffee Extracts

The rate of freezing extract establishes the color of the dried product. The size of ice crystals and pattern of growth control color.

There are four basic freezing methods in use:

SANDVIK Stainless Steel Belt. This is a moving stainless steel belt with brine or glycol spray cooling the underside of the belt. There may be a cold air blast from above the belt, or simply insulation. These freezing belts can be operated at ambient room temperature.

FMC Teflon/Rubber Belt. This is a moving Teflon coated rubber belt operated in a −45 F cold room. Air blasts onto the belt.

Tray System. Teflon coated aluminum trays are manually filled with coffee extract. Filled trays are on aluminum racks. The coffee extract is frozen with an air blast of −45 F to control color.

ATLAS Ice Slicer Machine. Coffee extract is frozen on the surface of a vertical, slowly rotating stainless steel drum. The inside of the drum is refrigerated. A knife continuously removes the frozen slab.

Each of the four freezing systems has its advantages and disadvantages. Freezing extract at an ambient room temperature is very advantageous, especially for operator comfort.

**SANDVIK Stainless Steel Belt Freezing System**.—This is the most expensive system. This method is used in the largest processing installations, e.g. General Foods, Nestle and the Colombian Federation. Belts are 4 to 6 ft wide and the pulley drums are 80 to 100 ft apart. The stainless steel belt is usually 0.040 in. thick and is tensioned with springs. The under belt coolant spray chambers are usually 12 ft long.

If 3 cooling spray chambers are used, each chamber can be operated at a different cooling temperature. Because 60 percent ethylene glycol-water mixtures are very viscous at −55 F, glycol coolant is recirculated with a helical (Roper) pump. Coffee extract slush can also be applied to the belt to reduce its fluidity. The side edges of the belt have Silicone bands to keep the liquid extract or slush from running off. Extract slab thicknesses are about ½ in. thick, depending on the belt speed, cooling area, and coolant temperatures.

Coffee extract starts to freeze at 27 F, and is completely solid at −14 F. However, the frozen extract slab is not brittle enough until −35 F, and preferably −45 F. The type and concentration of extract solubles determine the exact temperature. The belt speed is easily adjusted by means of a variable speed drive. Freezing time is from 20 to 30 minutes. The frozen extract slab, after it separates from the stainless steel belt at the discharge drum end, is broken into pieces 2 in. by 3 in. A Jacobsen finger breaker may be used. A single freezing belt can produce frozen extract slab at 3,000 lbs/hour. The face of the stainless steel belt is highly polished, so that the slab easily separates from it.

**FMC Teflon/Rubber Belt**.—FMC has provided two sizes of Teflon coated rubber belts. These are 30 or 60 ft long with a 36 in. effective freezing width. Belts are operated in a −45 F cold room, with air blasts onto the belt. Productivity can be 300 and 600 lb slabs per hr, respectively. If mechanical problems such as torn belts, warped belts, or non-uniform layering occur, production rates fall off.

**Tray System**.—For the small to medium sized freeze drying plant, freezing the coffee extract on racked trays requires the least investment and maintenance. Aluminum trays racked on carts are lightweight, easy to clean and require no painting. Racks can be moved on tracks or rollers (skates). A standard tray rack can accommodate on 2 in. of space 36-16 gauge thick aluminum trays (26 in. by 18 in. by 1.125 in.) with a ¾ in. thick layer of coffee extract. Racks and trays must be very level for uniform layers of coffee extract to be produced for freezing as well as for spillage avoidance.

In a −45 F air blast, a ¾ in. thick layer of 30 percent coffee extract in a rack will freeze into a hard slab of correct color in one hour. When a 31 percent solubles concentration extract is used, no gassing for density adjustment is required in the U.S.A., but gassing of slush is needed at this concentration in Europe.

Manual labor is needed mostly to unrack the trays, invert them, and tap out the frozen slab into a breaker. One weak feature of this system is the eventual damage of the aluminum trays from repeated manual handling and slab disgorging. Filling the trays with a metered volume of extract, such as with a Bursa pump, is accurate and rapid. Racked trays of liquid extract cannot be moved until frozen; otherwise extract will invariably spill, causing slop and waste. Tray freezing allows good ice crystal development and better aromatics recouperation into final product.

**ATLAS Ice-Slicer Machine**.—ATLAS in Copenhagen, Denmark uses a slice ice machine for freezing coffee extract, but without controlling color. Coffee extract is sprayed and frozen continuously on a vertical refrigerated cylinder; ice chips are removed with a sharp drum blade. The frozen ice chips fall into a granulator below in a −45 F room. Freezing capacities are 500 to 1,000 lbs chips per hour. Since the drum freezing time is very short, desired color is actually determined later by a freeze-thaw sequence. "Melt-back" technique and fast freezing cause the extract to lose coffee aroma. Foaming slush is used to control color and density.

## Granulation of Frozen Coffee Extract

This is an important step in preparing the finished granule so that it looks like roasted and ground coffee beans. Four types of granulators are in commercial use.

**Hammer-Mill.**—This is best illustrated by the Fitzpatrick mill in the U.S.A. and the BAS granulator in Europe. Stainless steel arms, about 6 in. long, with knifechamber on striking edge, are rotated at 1200 rpm to break up the frozen chunks of coffee extract. At the lower semi-circular periphery, to the edge of the blade movement, is a perforated semi-cylinder. Perforations may be ⅝ in. diameter or may be square openings. Most of the granules that pass through are less than 6 mesh (0.132 in.) in granule size. On this size machine, 600 lbs broken slab granules can be produced per hour.

However, 35 percent of these granules are finer than 20 mesh (0.033 in.) and need to be recycled. Recycling fines usually means remelting and refreezing. One company uses a compaction press and then regrinds the granules. This ⅓ recycle flow is higher than is desirable. Fines complicate processing at −45 F due to frequent blinding of the classification screens and conveyors. The recycled fines undergo oxidation and deterioration in flavor, as well as contribute to extra work and losses. Fines and dusting leads to material losses. Jamming, heating of the hammer-mill or coffee slab to above −35 F, softens the frozen coffee slabs so that they become tacky (not brittle). The soft coffee slabs block the hammer mill, requiring stoppage and cleaning.

The hammer-mill is a sanitary machine, and is easy to clean. However, the time and labor involved in removing a mill from a −45 F room to thaw it for cleaning and washing, recooling it, and reinstalling it are problems one wants to avoid. The shaft bearings require a special grease lubricant for use at −45 F. Normal lubricants are too viscous to be used at these temperatures. Similarly, motor bearings must have special greases.

A Fitzpatrick hammer-mill of the 6 in. radial size would normally take a 5 HP, 1750 rpm motor driving a pulley to reduce the speed of the blades to 1200 rpm. However, in the −45 F environment a 7.5 HP motor is needed. Sometimes mill resistance to rotation is built up, especially on starting, when frozen coffee granules adhere between moving and stationary parts. To lower the temperature of a hammer mill from ambient to −45 F temperature can take an hour. For these reasons, two hammer mills are usually installed. One mill is always working, if the other mill requires repair or cleaning. The rate of frozen slabs introduced to the hammer-mill is controlled by a screw feeder. Otherwise the hammer-mill would be overloaded, causing slabs to over-heat, soften and jam. The feed chute to the hammer-mill must be a labyrinth so that pieces cannot slip out to strike persons or be wasted on the floor.

**Oscillating Bar Against Screen.**—This reduction machine breaks down medicinal particles. It can be used on frozen extract chunks in a −45 F room. The granulator produces less than 10 percent (minus 20 mesh) fines. In fact, no fines need to be removed at all. It does this by using a low speed (134 oscillations per minute) on about a 6 in. radius bar cage. The moving metal bar cage does wear out the tensioned stainless steel screen (6 mesh) so this screen change becomes a significant operating expense. In some cases the costs of screen replacements are tolerable and offset because no fines are recycled. The smaller the chunks of frozen coffee extract fed, (⅝ in. to ¾ in. size), the less wear and tear there is on the 6 mesh screen.

A typical model can produce (−6 mesh) granules at a rate of 300 lbs/hr with a ¾ in. chunk feed. A larger model can produce 1,500 lbs/hr. The same particle reduction machine can be used at room temperatures to crush the freeze dried product also. When used in a −45 F room, the crankcase oil box has a several kilowatt electric heater therein to keep the oil fluid. Warm oil also minimizes motor work load, due to otherwise excessive oil viscosities at these low temperatures.

**Rotary Slicer or Cutter.**—A slicer originally designed for potato chip slicing has been adapted by one company to cutting frozen (soft) coffee extract with the least production of fines. A typical slicer has a capacity of 5,000 lbs/hr on soft frozen coffee extract.

**Roller Mill Lepage Cut.**—Used with roast coffee beans, this has also been used for granulating frozen coffee extract at −45 F. It has serrated perimeters on the rollers. The rollers rotate at 300 to 500 rpm through progressively reduced roller spacings, which achieves gradual (staged) cutting and crushing of the larger granules. Nevertheless, some 30 percent fines (−20 mesh) are produced and require recycling.

## Screening Fines

Screening of fines at −45 F is usually done on vibratory screens. Over 6 mesh particles are recycled for regranulation. Under 30 mesh particles are recycled for melting. The 30 to 40 mesh screens are subject to "blinding". So vibration, bouncing balls and frequent inspection are needed to assure adequate fines separation. These same classifying screens can be used for particle size separations of the dried granules.

## REFRIGERATION

The need to freeze, granulate, screen and load trays in a −45 F room and building requires effective designs for these low temperatures. These are lower temperatures than ice cream blast-freezing rooms.

### Design and Construction

This is best carried out by firms specialized in this field. Elimination of atmospheric moisture from rooms and their insulated walls is a key feature of these designs. Using insulations like fiber glass (most economical initially), foam glass, Styrofoam and Urethane calls for specific design considerations. Insulation design of sub-concrete floor areas is important to avoid freezing subsoil moisture and causing cracking and heaving of concrete floors. Doors and windows also require special design construction, insulation, and heating elements to keep open door contact points from freezing together.

Drainage and ventilation during use and during thaws are additional design features that require consideration in this overall special construction. At these low operating temperatures, it is usually better to defrost the refrigerant cooling coils in isolated compartments.

### Refrigerants Used—Ammonia, Freon 12 or 22

In these systems ammonia, Freon 12, or Freon 22 have been used. Due to physical properties, Freons require more compressor work input than does ammonia. Since liquid Freon is dense, oil floats on it. Oil sinks in liquid ammonia systems, however, requiring different types of lubricant recouperation systems.

To attain −60 F refrigerant temperatures, two stages of mechanical compression are used. The lowest refrigeration temperatures are attained under vacuum conditions. The refrigerant gas is quite voluminous.

### Low Pressure Stage Blowers

Therefore, blower type or turbine type compressors are used to recompress the gaseous refrigerant from these voluminous low pressure conditions to less voluminous near atmospheric pressure conditions. One example of this type of gas compressor is the Fuller rotary blower. Turbine or screw compressors can also be used.

## High Pressure Stage Compressors

Now the near atmospheric pressure ammonia gas is turbine or screw compressed to near 150 psig. The compressed gas is cooled by cooling tower water to near ambient temperatures. Compressed and liquified refrigerant is stored in large steel tanks; this represents the refrigerant surge supply for the systems.

Between the low and high pressure refrigerant stages, there is usually an intermediate self-cooling refrigeration system and liquid refrigerant storage reservoir.

## Low Stage Refrigerant Reservoir

A system of compressor and liquid refrigerant level controls, actuates or shuts down the compressors, while maintaining the −60 F coldest liquid refrigerant reserve. There is also a 0 F intermediate liquid refrigerant reserve.

Piston mechanical compressors are usually avoided on refrigeration systems of over 500 tons, because they require high maintenance. In over 500 T.R. (tons of refrigeration) installations, screw or turbine compressors are usually used.

## Work Per Ton of Refrigeration (T.R.) Electrical Needs

For −60 F coolant, 4 HP per T.R. are used, whereas at −45 F, 3 HP per T.R. are used. For a 500 T.R. plant, 2,000 HP is required for the refrigeration system alone. When refrigeration loads are equally distributed on two low pressure and two high pressure turbine or screw compressors, each compressor may be driven by a 500 HP motor.

In batch freeze drying, a 1,000 sq ft tray chamber may require 80 T.R. at the beginning of the drying cycle to condense the large sublimed water vapor load. With four such chambers 250 T.R. are adequate, since each drying chamber is at a different stage in the drying process.

The following graphs of gas volume vs absolute pressure illustrate the amount of work required to pump air or water vapor at less than 0.500 mm (or 500 microns).

Another important operating consideration during freeze drying is to keep the vacuum low enough so that the coffee extract will not melt, as sometimes occurs. The following table and graph show the absolute pressures of ice vs its sublimation or boiling temperature.

# INDUSTRIALIZATION

Although the art of freeze drying was known and practiced in pilot plants in the 1950's, it was not until General Foods and Nestle pioneered commercial freeze dried coffee that meaningful progress was made. Vacuum pumps, chamber loadings, condenser design, refrigeration systems, instrumentation, and freeze concentration methods were all improved.

By 1970, the 10 min continuous freeze drier with improved vacuum drying efficiencies had been put into use. Large firms were first to make large capital investments in machinery, preliminary development, and marketing. Hills Bros. made the technological investment but apparently shied away from the "marketing battle".

Most of the smaller instant coffee processors continued to compete by using lower priced but taste-acceptable Brazilian spray dried instant coffees. This price competition was, and is, very effective in keeping the consumer.

One of the essential features in freeze drying coffee extracts is the need to keep the water frozen while subliming water vapor from ice at the highest possible rates (driven by platen temperatures). The continuous oscillating conveyor vacuum driers required uniform sized frozen extract granules, hence required better granulation methods or more fines recycling.

Their high rates of drying carried many fines off the moving conveyor. This "carry over" limits production rates and can cause significant coffee losses. Filters in front of the vacuum pumps became a necessity.

The freeze drying process alone does not govern quality. The instant end quality may not be due to the drying equipment, since the original coffee bean qualities, solubles yields, and extraction and concentrating methods are also important influences.

"Trays-on-cart" batch drying yields a non-uniform product. The continuous tray drying system is an improvement, but it is not as fully uniform as the truly continuous granule drying system. True continuity of drying requires particle uniformity and with all its attendant accomplishments.

Excessive final drying temperatures like 130 F have been used in efforts to reduce drying times. Potentially good products have been ruined by this type of zeal to run a 6 hr cycle instead of 8 hr cycle.

Excessively high temperatures noticeably downgrade coffee taste, even when compared to 100 F or even 110 F finishing temperatures. Many companies push their production rates, only to downgrade their final instant; they depend on oil aromatization afterwards to coverup for such deficiencies.

Smaller granules dry faster, e.g., 6 mesh granules dry in 20 percent less time than 5 mesh granules. Uniform granules dry faster than non-uniform granules.

Final dried coffee product under vacuum is filled with nitrogen gas. Granules should be kept under nitrogen gas until they are packaged in nitrogen gas in the final retail jar. Jar packaging should be done as quickly as possible after granules are removed from the dryer to retain aromas and flavors. However, many firms ship and package freeze dried coffee in air, which downgrades its taste quality. Coffee oil additions to the freeze dried granules must be made under nitrogen gas only just before retail packaging.

Moistures of freeze dried granules are 1 to 2 percent, somewhat lower than spray dried instants which are 3 to 3.5 percent water. Lower powder moistures imply that volatiles and aroma content are less.

## Productivity

A four chamber (1,000 sq ft each) FMC freeze dry plant should produce over 1.25 million Kg dry coffee product per year. The Colombian 8 chamber Stokes plant produced 2.5 million Kg freeze dried coffee per year in 1973. Figure 13.6A shows the Stokes batch drying chambers.

The General Food's installations at Houston, Texas and Hoboken, New Jersey and the Nestle installations at Sunbury, Ohio and Freehold, New Jersey each probably produce in excess of 20 million lbs freeze dried coffee per year, since U.S. consumption by 1973 was 33 percent freeze dried coffee or 67 million lbs/yr.

In batch freeze drying chambers in 1974 capital investment ranged from $200–$400 per sq ft of tray-platen surface (which can yield about 3 lbs dry coffee product per 24 hours).

Refrigeration plant costs can be slightly higher than freeze dry plant costs.

**Tray Loadings of Frozen Granules.**—These loadings are between 2 and 3 lbs per sq ft. The granules may be loaded onto the trays manually or, preferably, automatically. A 1,000 sq ft drying chamber can produce on 6 to 8 hr cycles at least a ton of dried coffee per 24 hours. To operate economically, the plant must run 24 hours per day as many days per year as possible.

**Flavor Losses.**—From a volatile flavor retention aspect, freeze drying is a gentle process. However, much coffee volatiles retention is related to aroma affinity to coffee solubles.

Many volatile coffee aromas and flavors are removed during vacuum freeze drying. Some volatile aroma restoration is made with condensed percolator volatiles. These volatiles are collected separately and are added after freeze concentration. Instead or in addition, coffee oil additions are made to the freeze dried granules. Nestle pioneered liquid $CO_2$ extraction of coffee aromatics followed by dry ice sublimation.

## Drying and Refrigeration Plants

Although FMC, Stokes, Leybold and Atlas can offer a complete package drying installation, it is best to separate the freeze drying costs from the refrigeration costs to keep the quotations competitive. Since these are large investments, it is important to review in detail the equipment and performance specifications of each bidder, as well as his experience in the field via currently operating plants.

Continuous drying processes are only advantageous if they materially reduce equipment investment and operating labor costs. This has not yet been clearly demonstrated. The refrigeration load requirements for batch or continuous freeze drying call for the same ice sublimation heat removals.

Truly continuous freeze drying produces a uniform flow of dried coffee product. This should be a more uniform product than the quality variations experienced with batch tray freeze drying.

## FREEZE DRYING COFFEE EXTRACTS

MAXIM—A Case History—General Foods Corp. initiated commercial freeze drying in 1964 in the U.S.A. In 1965, General Foods installed 10 FMC batch (1,000 sq ft each) drying chambers at the Maxwell House coffee plant in Hoboken, New Jersey. In 1967–1968 large scale marketing and taste tests for MAXIM were carried on for two years while major investments and equipment installations were made in Houston, Texas, Canada and Germany.

At first, frozen slabs of coffee extract on trays were freeze dried in the FMC carts. Shifting to granulated frozen coffee extracts on trays doubled drying production rates. The 1968 Pfluger patent describes their next improvement. Trays were replaced with vertically heated panels for the granules, which doubled cart loading and productivity per chamber.

Increased condenser areas were obtained with aluminum finned condenser tubes. Vertical panels of coffee granules made it easier to load frozen granules and unload dried granules.

## Pricing and Quality of Freeze Dried Coffees

The relatively more expensive price of the freeze dried coffees caused some consumer resistance to buying this new product initially. The fact that less freeze dried coffee was needed per cup of beverage was not adequately communicated to the public and, in fact, worked against the advantages of the freeze dried coffee.

On a cup for cup basis, freeze dried instant coffees were only about 10 percent more expensive to the consumer than spray dried coffees. Freeze dried coffees entered the market in a period (1971) during which less expensive agglomerates were introduced and when Brazil instant coffees were entering the U.S.A. (at a rate of 25 to 40 million lbs per year). Use of lower priced spray dried Brazil instants and their modifications (agglomerated and aromatized) offered good tasting instant coffees. These Brazils were priced well below the U.S. freeze dried instant coffees, and no doubt hindered growth of freeze dried instant coffee sales.

The higher costs to produce freeze dried instant coffees were later offset by increased solubles yields and increased use of cheaper Robusta bean varieties.

This resulted in a marked downgrading of taste in freeze dried coffee qualities, diminishing and at times eliminating their improvement over agglomerates.

By 1973 the freeze dried coffee market had grown to almost 35 percent of all instant coffee use, or about 75 million lbs per year. By 1977, in the U.S.A. freeze dried coffee was 45 percent of all instant sales—almost 100 million lbs/yr.

## Density of Granules

Percolator extracts were freeze concentrated to retain aroma and flavor in the finished product as well as to increase productivity of the vacuum drying chambers. Hence, extract concentrations of 40 percent solubles were often used.

Maxim 40 percent extracts were freeze dried "as-is". Taster's Choice 40 percent extracts were gassed to yield final granule densities comparable to spray dried instant coffees. The increased granule density compounded the increase in flavor intensity of MAXIM freeze dried coffee. At first, many consumers were preparing their freeze dried coffee beverages too strong.

## Laboratory Tests

A great many physical, chemical and taste tests are made before, during and after instant coffee manufacture. Instruments include light refractometer for solubles content of extracts; light reflectance meter for color of R&G and instant coffees; screening equipment for particle size analyses; measurer of oxygen in gas space of jar; and pH meter with expanded scale for measuring beverage acidity to within 0.01 pH units.

## RELEVANT TERMS

*Fines in powder:* These are usually indicated at less than 10 percent fines through an 80 mesh screen.

*Solubility—speed:* This used to be an important criteria, but in itself it is not so important since agglomeration has become so widespread.

*Foam in beverage cup:* This too, like speed of solubility, has become a secondary criteria because very rapid powder solution is obtained without foam after agglomeration.

*Agglomerability:* Some instant powder buyers place significance on this property. It is usually associated with the level of sugars or simple carbohydrates. In other words, some instant powders stick together better than others, which is also a function of the type of agglomeration process.

*Acidity and pH beverage:* Many buyers like acid beverages with pH's in the range of 4.80. Surprisingly, however, many major instant buyers are rather oblivious of the importance of cup pH and allow wide pH ranges like 4.90 to 5.30.

The instant processor can adjust the product pH usually in a narrow range, and a specification like pH 4.95 to 5.05 is not unreasonable, production wise.

*Antifoam:* Most U.S. instant buyers do not want antifoam silicones added to the coffee; however, some few do and many do not care. Levels of use are about 60 ppm.

*Flowability:* Powders should flow well for packaging directly or for feeding to agglomeration systems.

*Color of powder:* Once a powder color range has been specified, the buyer expects all his purchased instant coffee to stay within these specifications. This is more important to the buyer packaging powder directly than to the one agglomerating. Color is important when different blends and sources of powder are being mixed.

*Uniformity:* This quality is probably the most difficult one to produce consistently day after day. There are at least a dozen powder specifications: physical, chemical and taste. If any one specification is off, the powder can be rejected by the buyer. Sometimes the buyer rejects the coffee because he has "made a better buy" elsewhere, or has made a contract and discovered that spot prices are lower than his contract prices. Other times there are genuine faults in the powder such as physical dirt, bad taste, and non-uniformity.

*Appearance of powder:* This is usually associated with a dark, beady, free flowing property but is otherwise difficult to define.

*Beverage color:* Reddish color is preferred. Any other shade is objectionable.

*Insolubles:* Floating specks or oil on the cup of beverage, especially noted with milk, is not acceptable. Sediment in the bottom of the cup is not acceptable. A foamy, specky ring around the cup is not satisfactory and normally, turbidity is not acceptable.

*Taste:* As was already explained, some buyers prefer a neutral, non–aromatic, even caramelized taste for blending purposes rather than much sign of coffee flavor.

*Cleanliness:* Cleanliness is very important. Cartons that contain rubber bands, pieces of paper, buttons, tools, etc. worry the buyer.

*Chemical properties:* Most buyers do not include these, except for occasionally caffeine. Moistures must be less than 3.0 percent, and usually 2.5 percent is specified. There should be practically no heavy metals like lead. Bacterial analyses are seldom specified.

*Physical property:* Bulk density is the most important and stringent property, usually at 200 to 220 grams/liter vibrated to facilitate packaging.

## FREEZE CONCENTRATION

Freeze concentration has been used in conjunction with freeze drying for two reasons only. First, it was said that higher solubles concentrations in coffee extracts would retain more natural flavors and aromas (volatiles) in freeze drying, and second, that there would be less water to sublime while vacuum drying.

The performance of freeze concentration did however, match its expectations; thus, better economy and better quality would result. Solubles requiring recovery in the reject ice and flavor stability was never achieved. Also, the cost of the investment and its operation seemed unwarranted when less than 40 percent solubles resulted.

At least 29 percent solubles concentration in the extract is needed to be freeze dried to obtain proper bulk densities in the final dried granules. Freeze concentration of vinegar, fruit juices, sea water and beer are used commercially. The application of this technique to coffee extracts in 1964 was a logical extension. In Europe lower bulk densities of granules are used with somewhat larger jars, hence extracts of even 22 percent solubles are freeze dried directly in one Swiss plant.

General Foods and Nestle both have concentrated the solubles in their coffee extracts to almost 40 percent solubles concentration. General Foods initially (1968) freeze dried this coffee extract "as is". Nestle (Taster's Choice) gassed the 40 percent solubles extract to obtain a granule "in-jar" density comparable to spray dried coffee. Later, General Foods also gassed the 40 percent solubles extract, which is done in a slushing tube. General Food's first MAXIM, therefore, was a denser granular product and was packaged in a smaller sized jar than Nestle's product. This was an unfortunate decision since the consumer felt he was getting less (volume); he also used too much of the granules per cup, making the beverage too strong tasting.

In 1964, General Foods used the Struther-Wells ice crystallization tube with internal oscillator scraper. But by 1967 General Foods was using the Votator ice crystallizer tubes (also used for ice cream) commercially in their Houston, Texas plant. Ice crystals of about $^{1}/_{5}$ mm in size are produced. The ice crystals are separated from the concentrated coffee solution by means of a continuous or batch centrifuge. A single ice crystallization step can increase solubles concentrations from 25 to 30 percent. Three ice crystallization and separation steps are required to concentrate 24 percent solubles to 38 percent solubles. At higher solubles concentrations of lower temperatures, the coffee extract becomes very viscous. This high viscosity reduces separation efficiency by allowing more solubles to pass out with the ice crystal phase. Close temperature control and adequate mixing of ice and solution are important variables in the freeze concentration process. The type coffee bean, degree of roast, solubles yield, etc. will affect the freezing properties somewhat.

The temperatures at which ice crystal formation starts and the corresponding solubles concentrations are: 30 F–20 percent S; 25 F–35 percent S; and 20 F –42 percent S. In jacketed chill tubes, brine or propylene glycol is recirculated in a 3 stage crystallization at thermostatically controlled temperatures: 28 F, 24 F, and 20 F. This leaves only 2 to 3 F temperature differential between coolant and slush.

Table 13.1 relates extract specific gravity and light refraction (°Brix) to percentage coffee solubles. The percentage of solubles is usually about 16 percent less than the °Brix.

## INSTANT COFFEE SPECIFICATIONS

Although each processor makes instant coffee to suit his own finances, quality, and public image, the great amounts of instant coffee that have become available from Brazil since 1966 have resulted in some standardization of coffee specifications. For example, Brazil exports to the U.S.A. are mostly according to General Food's standards, whereas U.K.'s shipments from Brazil are more to Lyon's standards—not too different from G.F.-U.S. standards. However, instant coffee specifications do vary according to the importing country.

Instant coffee specifications have varied in different periods. For example, in

TABLE 13.1. COFFEE EXTRACT CONVERSION TABLE—ABBREVIATED

| °Brix | Percent Solubles | Specific Gravity, 60 F (16 C) | Weight Ratio, Water/ Solubles | Extract Weight, per gal | Weight Powder, lb per gal | Gal Water | Ml Extract for 2 gm Solubles |
|---|---|---|---|---|---|---|---|
| 2 | 1.8 | 1.009 | 54.5 | — | — | — | — |
| 4 | 3.4 | 1.016 | 28.3 | — | — | — | — |
| 6 | 5.0 | 1.024 | 19.0 | 8.53 | 0.43 | 0.97 | 39.0 |
| 8 | 6.5 | 1.030 | 14.4 | 8.58 | 0.56 | 0.96 | 30.0 |
| 10 | 8.3 | 1.037 | 11.0 | 8.62 | 0.72 | 0.95 | 23.2 |
| 12 | 9.6 | 1.043 | 9.4 | 8.68 | 0.83 | 0.94 | 20.0 |
| 14 | 11.4 | 1.050 | 7.8 | 8.75 | 1.00 | 0.93 | 16.7 |
| 16 | 13.0 | 1.056 | 6.7 | 8.79 | 1.14 | 0.92 | 14.6 |
| 18 | 14.8 | 1.064 | 5.8 | 8.86 | 1.31 | 0.91 | 12.7 |
| 20 | 16.5 | 1.071 | 5.1 | 8.93 | 1.47 | 0.90 | 11.3 |
| 22 | 18.2 | 1.079 | 4.5 | 8.99 | 1.64 | 0.88 | 10.2 |
| 24 | 20.0 | 1.087 | 4.0 | 9.05 | 1.81 | 0.87 | 9.2 |
| 26 | 21.5 | 1.093 | 3.7 | 9.11 | 1.96 | 0.86 | 8.5 |
| 28 | 23.2 | 1.101 | 3.3 | 9.16 | 2.12 | 0.85 | 7.8 |
| 30 | 25.0 | 1.106 | 3.0 | 9.21 | 2.30 | 0.83 | 7.2 |
| 32 | 26.6 | 1.115 | 2.8 | 9.26 | 2.47 | 0.82 | 6.7 |
| 34 | 28.5 | 1.123 | 2.5 | 9.35 | 2.67 | 0.80 | 6.3 |
| 36 | 30.2 | 1.130 | 2.3 | 9.41 | 2.84 | 0.79 | 5.9 |
| 38 | 32.0 | 1.139 | 2.1 | 9.48 | 3.04 | 0.77 | 5.5 |
| 40 | 33.8 | 1.148 | 1.96 | 9.55 | 3.23 | 0.76 | 5.2 |
| 42 | 35.5 | 1.155 | 1.32 | 9.62 | 3.42 | 0.75 | 4.9 |
| 44 | 37.2 | 1.164 | 1.69 | 9.70 | 3.61 | 0.73 | 4.6 |
| 46 | 39.0 | 1.173 | 1.56 | 9.78 | 3.82 | 0.72 | 4.4 |
| 48 | 41.0 | 1.183 | 1.44 | 9.86 | 4.04 | 0.70 | 4.1 |
| 50 | 42.8 | 1.190 | 1.34 | 9.92 | 4.25 | 0.68 | 3.9 |
| 52 | 44.6 | 1.198 | 1.24 | 9.97 | 4.45 | 0.66 | 3.7 |
| 54 | 46.5 | 1.207 | 1.15 | 10.1 | 4.70 | 0.65 | 3.6 |
| 56 | 48.3 | 1.216 | 1.07 | 10.1 | 4.88 | 0.63 | 3.4 |
| 58 | 50.0 | 1.224 | 1.00 | 10.2 | 5.10 | 0.61 | 3.3 |
| 60 | 52.0 | 1.232 | .92 | 10.3 | 5.36 | 0.59 | 3.1 |

1966 the Brazilian instant coffee was of better quality, lower yield and relatively lower price. From 1966 to 1968 100,000 to 250,000 pound lots of instant coffee (even more on a continual basis) with hardly any specifications were ordered. Then in the next period, buyer specifications tightened up as more Brazilian instant coffee plants started competing with each other for sales. When agglomerated instant coffees started to take over the U.S. retail market in 1971, instant coffee specifications, although still existing, were not strictly enforced because many physical properties became unimportant once imported instants were destined to be pulverized for agglomeration.

An interesting development in this regard was the establishment of a rather flat, noticeably caramelized, rather aroma-less instant coffee that became the hallmark of the 1970's. This taste developed because the large Brazilian producers vacuum evaporated their dilute percolator extracts to obtain higher yields, thus creating lower selling prices at higher profits and a neutral or near neutral product. Large buyers like General Foods, Folger, Hills Bros., and Tenco often preferred a neutral tasting Brazilian instant coffee. Because they were buying from several Brazilian producers, a neutral common denominator made it easier for their production people to blend any Brazilian product. Also, retention of the low grade natural coffee flavors from beans used by many of the Brazilian instant coffee producers was undesirable from a marketing viewpoint. These circumstances created the situation that the buyers wanted a uniform flat product over any that might have coffee flavor and aroma.

On the other hand, Brazilians shipped to the U.S.A. some very nonuniform and off-standard lots of instant coffee from time to time to see what the trade would bear. Dominium, under changing management and technical personnel, suffered the greatest losses when U.S. buyers refused almost a million pounds of instant coffee that arrived during 1971-1972. Small amounts of off-standard powder were bought by some firms from time to time when their needs exceeded their quality desires, but Dominium sent too much powder that was too much off standard, both in physical properties and in taste. Much of this powder was returned to Brazil for rework at a considerable expense, resulting in financial losses and decreased production.

A method of determining the *Purity Index* of green coffee beans used in instant coffee manufacture has been devised. In early 1973 Brazilian instant coffees ranged from 60 to 90 percent purity index. The most obvious deficiency in the Brazilian instant coffee has been the consistently low caffeine contents as compared to wholesome green beans. Neither the Brazilians, the buyers abroad, nor even the U.S. FDA has shown much interest in this purity.

In general, instant coffee buyers look for clean powder; that is, free from sediment, floating oil, specks, and turbidity. In addition, some buyers (e.g. Hills Bros. in the U.S.A. and Lyons in the UK (where milk is used 50/50) look for a very red powder and beverage color. Redness has been an attribute pursued since instant coffee was commercially successful in the early 1950's. Redness can be achieved by using good quality, clean coffee in an iron-free processing system. Carbonization and iron contamination cause a gray color when the coffee is used with milk. Robustas tend to yield a brown, dark brown to black beverage color, whereas Arabicas yield a reddish color. Potassium instead of sodium neutralization of extract acidity also produces a more reddish color. Beverages with pH's closer to 4.90 are more reddish than those at higher pH's. Lesser solubles yields produce redder colors; e.g. from the same coffee, a 34 percent yield from green beans is redder than a 38 percent yield from green beans. Carbonization of the coffee at any stage of the processing will diminish reddish color. Higher extract concentrations of solubles help to yield a redder coloration of powder.

Although some instant buyers want powder densities higher than most or colors darker than most, these requests are infrequent and considered abnormal. The Brazilian producer prefers to process one type of bean at a given roast color and yield because continual changes in product specifications add to the work load and costs.

## PACKAGING INSTANT COFFEE

### Glass Jars

Glass jars have been the standard package for instant coffee in the United States since the rapid sales growth of instant coffee after World War II. In fact, the geometry and shape of the glass jars were amazingly uniform until about 1961, when Maxwell House introduced a contoured jar. This change in jar shape was quickly appropriated by many other soluble coffee firms. The shape of the jar was used as a merchandising tool with carafes and carafe shapes appearing from time to time.

**Making Glass Jars**.—Glass is a fused combination mixture of silica, quick lime, and soda ash with glass rework or cullet. It is molten at 2700 F (1482 C) and is fed to a rotary mold forming machine which in several steps during each rotation, can make about 20 jars per minute.

The approved jars are packed in their final printed cartons that will also be used to transport them out of the instant coffee packaging plant. Usually jars are loaded mouth down in the carton so that they will be upright when dumped on a table in the soluble coffee plant.

**Jar Sizes**.—Jar sizes were determined mainly before 1950 when carbohydrates were added to soluble coffee; they have remained mostly unchanged since that time when the practice of adding carbohydrates was abandoned. The internal volume of the jar holding 6 oz of instant coffee, or 170 gm of powder, for example, is 600 to 610 ml. The glass weighs about 230 gm for one glass supplier, 275 gm for another, and perhaps something else for a third. Jar weights are usually within a few grams of each other. This large difference in weight for an almost identically shaped jar means that one jar supplier makes a stronger glass or uses less glass weight to obtain a satisfactory jar. The shape of the jar has bearing on the weight of glass that must be used.

## PACKAGING INSTANT COFFEES

### Vacuum Volumetric Fill (Powder With and Without Oil)

In the early 1950's spray dried instant coffee particles were fine and dusty. The most popular jar-filling machine was the Penumatic Scale Co. or Albro vacuum filling machine with a single stationary head or many rotating heads. Instant coffee powder was sucked into the jar. The net weight of powder fill was manually monitored. If more or less weight was desired, more or less vacuum was applied. See Figs. 13.2 and 13.3.

**Oil Aromatized Powder.**—In 1956–1957 General Foods introduced aromatized instant coffee powder coated with coffee oil; this gave the powder poor flowability. The filling machine reservoir for powder had to be nitrogen gas purged, as had the aromatized oily powder fed to the filling machine. Further, the entering jars had to be nitrogen (or $CO_2$) gas flushed to remove air. Pneumatic Scale Co., working closely with General Foods, modified their suctioning powder filler to achieve inert gas jar packaging with 2 percent (or less) oxygen content. This oxygen level could preserve coffee oil aroma freshness for a few months. Due to the poor flowability of the oiled powder, larger powder entrance orifices (about ½″) were provided. This flowability was also counteracted by reducing the powder temperature, e.g. to 65 F, in the packaging room so that the oil was a solid fat and could not hinder powder flow. Other firms used volumetric powder fillers without vacuum draw to fill their jars.

### Weight Fill (Agglomerates and Freeze Dried Granules)

In 1970–1971, with broad introduction of agglomerates and freeze dried granular instant coffee (which was more fragile, and in which fines were more objectionable), new types of granule filling machines were needed. Pneumatic Scale Co. introduced a pneumatically balanced weigh head for net weigh filling each jar. This was a much more expensive type filler than the suction type. Some expense was offset by making an 8 or 12 head rotary filler with only one weigh head (which

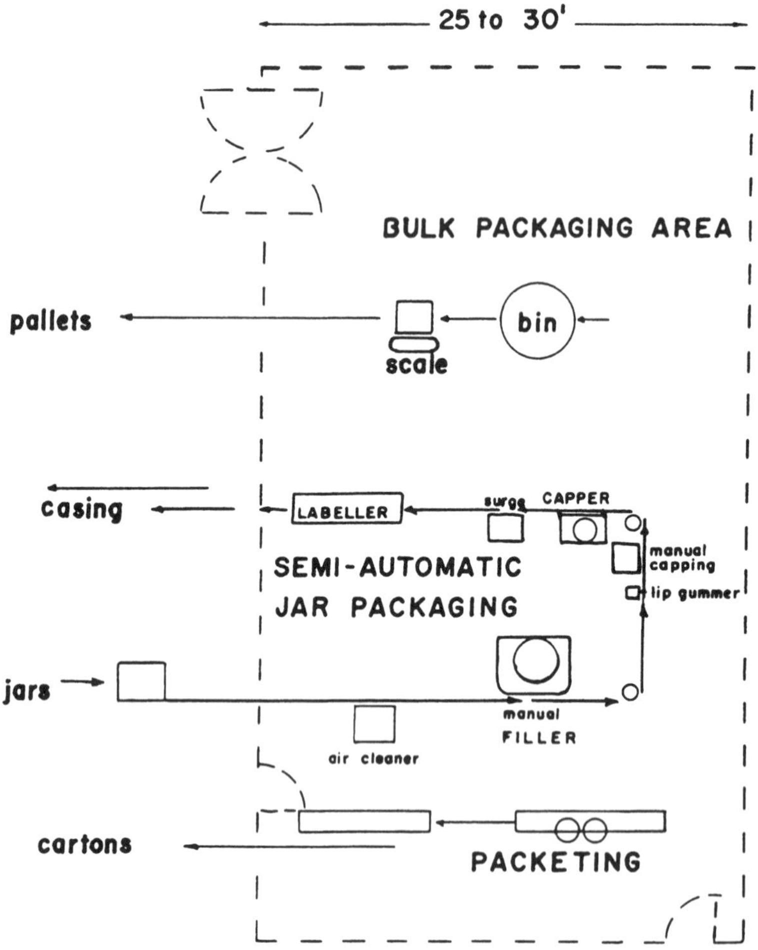


FIG. 13.2. PLAN VIEW OF SEMI-AUTOMATIC JAR PACKAGING LINE

was recording). These rates of filling production were 180 to 300 jars per minute.

It is illegal to sell merchandise that is underweight. It is the manufacturer's responsibility to minimize his "over-fills". The recording weigh-filler gave the filling machine operator immediate net weigh fill knowledge and control. The savings from "over-weights" could pay for the weigh filling machine in a year, or less time.

A later volumetric filler cost about half as much as the Pneumatic filler machine and delivered close to 1 percent oxygen residual in the jar. However, it depended on manual checking of net weighings for fill control, so manual monitoring or a

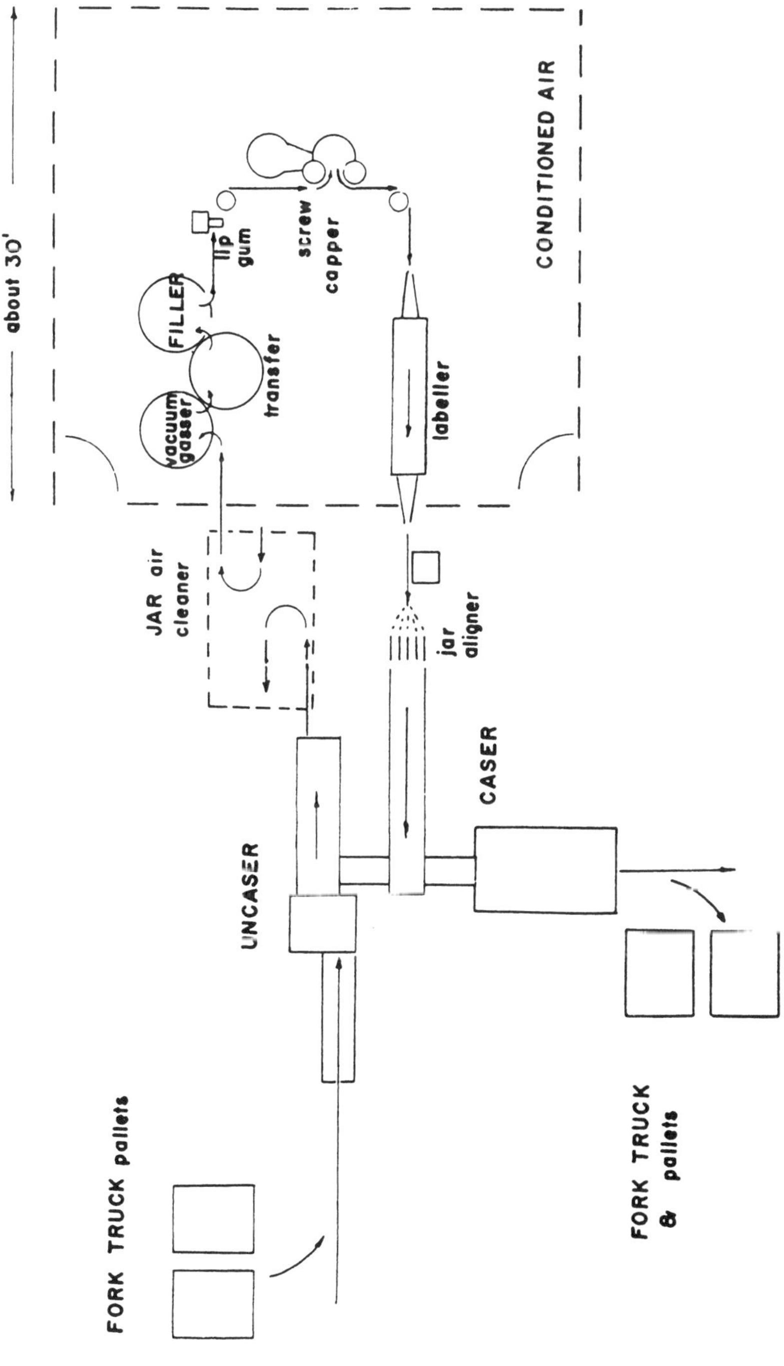


FIG. 13.3. PLAN VIEW OF FULLY AUTOMATIC JAR PACKAGING LINE

sophisticated net weigher machine was needed. Glass jars are not very uniform in weight, especially when different lots are used out of sequence.

Systematic recorded weighings on the Pneumatic machine ($8.00/lb freeze dried instant coffees, 1977 prices) reduced "give-away" costs. Even with agglomerated and aromatized instant coffees weighing the net jar fills was a quickly rewarding operation.

An additional feature offered by Pneumatic by 1971 was the synchronized jar filling line. That is, the jar air cleaner, jar coffee filler, jar capper, jar labeller and jar caser steps were all synchronized so that no surges of jars would occur between steps. Such surges had caused glass breakage, powder waste, glass waste and coffee contamination. Synchronized lines require only one operator. Crises situations and cleanups are avoided. (Figs. 13.4 to 13.20)

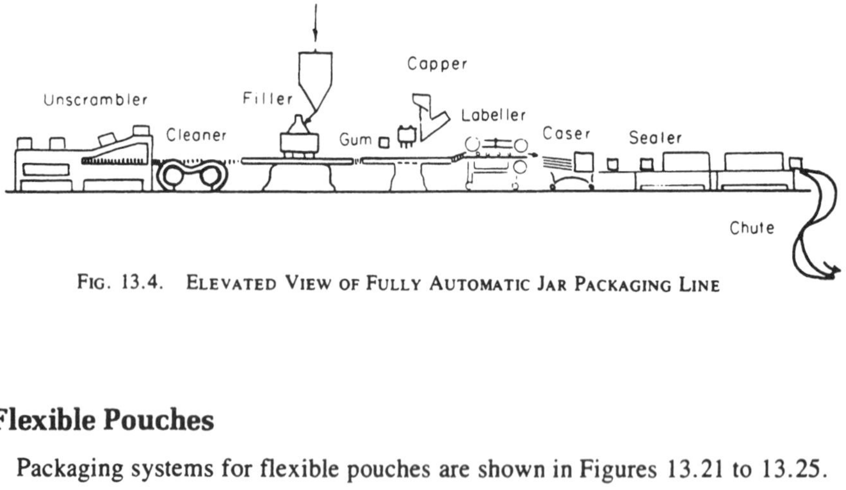


FIG. 13.4. ELEVATED VIEW OF FULLY AUTOMATIC JAR PACKAGING LINE

## Flexible Pouches

Packaging systems for flexible pouches are shown in Figures 13.21 to 13.25.

# U.S.A. INSTANT COFFEE USE, BLENDS AND YIELDS

In the U.S.A. since the mid 1950's, instant coffee use has been about 4 million bags (60 Kg) green coffee beans per year. The total U.S. imports are about 21 million bags of green coffee per year.

High solubles yields (40 to 50 percent from green beans) are being obtained today. The level of instant coffee use in the cup is 2.0 to 2.25 grams per 5 fl oz beverage. At least ⅓ of all cups of coffee consumed are from instants. This level of sales and use of instants has been largely because of convenience to the consumer. The acceptable quality attained is relative to the quality of brewable beverages prepared. In the early 1950's good quality coffee beans were being used by the major processors for preparing instant coffees. Price competition in sales has

*Courtesy Pneumatic Scale Corporation Ltd.*

FIG. 13.5. SINGLE HEAD SEMI-AUTOMATIC JAR FILLING MACHINE

progressively reduced blend qualities used and has caused increased solubles yields to be taken.

Recovering volatile aromas, drying liquid coffee extracts from higher solubles concentrations, and adding back expelled coffee oils (from roast beans) at 0.3 percent wt onto the final powders, spray and freeze dried, have been attempts to compensate for flavor losses due to severe extraction conditions, lower grade beans, and the diluting effects of higher solubles yields. But the processing technology developed industrially has been used primarily to reduce processing costs, and secondarily to improve product quality.

Robusta coffee beans represent ⅓ of the European and ⅓ of the U.S. imports (about 7 million bags to each). Since the imported coffee beans are the major processing cost, cheaper beans like Robustas from Africa now constitute a large

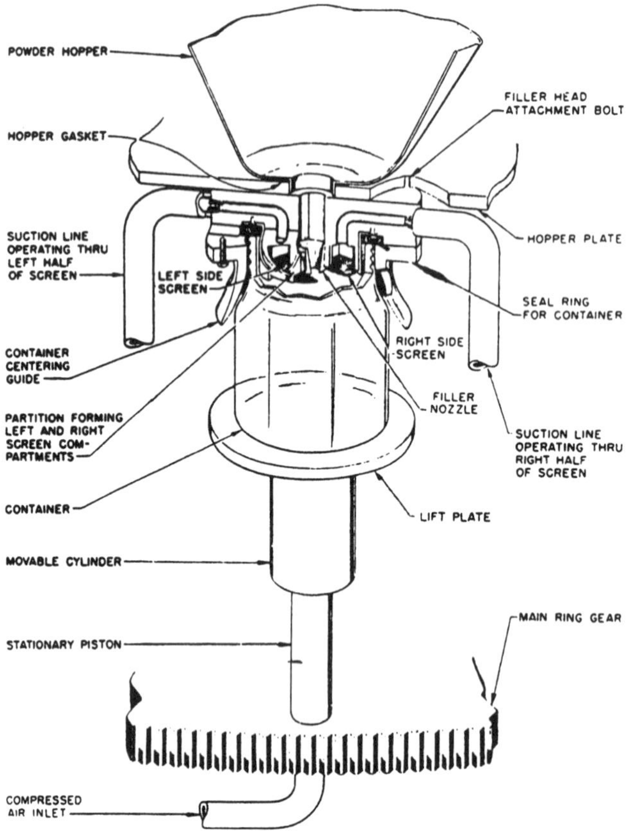


FIG. 13.6. DETAIL VIEW OF VACUUM POWDER FILLER HEAD MECHANISM

portion (over 50 percent) of commercial instant coffee blends in the U.S.A. and Europe. Some instant blends are 100 percent Robusta. Some lower priced supermarket private-label brands are 100 percent Robusta, and much Robusta is also used in R&G coffee vacuum can blends. The flavor retention benefits of the freeze drying process and the freeze concentration processes have been mitigated by the increased use of Robusta beans and higher solubles yields.

The demand for Robusta beans has approached its limit. Robusta bean prices have never been very far below other bean prices and especially Brazilian bean prices. This is a high price level for Robusta beans—an encouragement to African producers to grow more Robusta coffees. World demand for coffee is on the increase even though per capita use in the U.S.A. has fallen in the past decade (due ostensibly to noticeably poorer coffee qualities).

*Courtesy Coffee & Tea Industries Magazine*

FIG. 13.7. RECORDER FOR CHECKING JAR NET WEIGHT

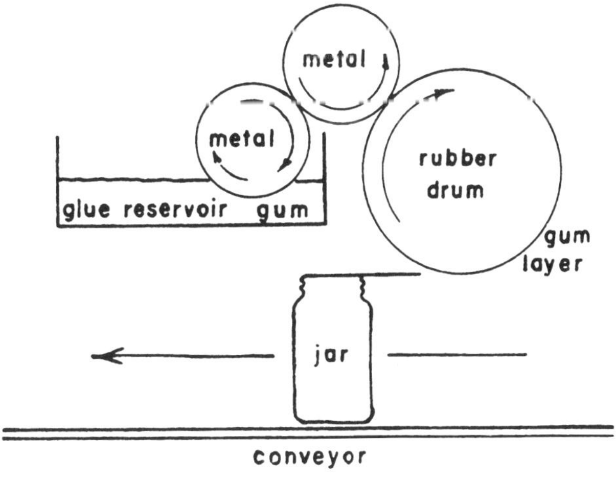


FIG. 13.8. JAR LIP GUM APPLICATOR

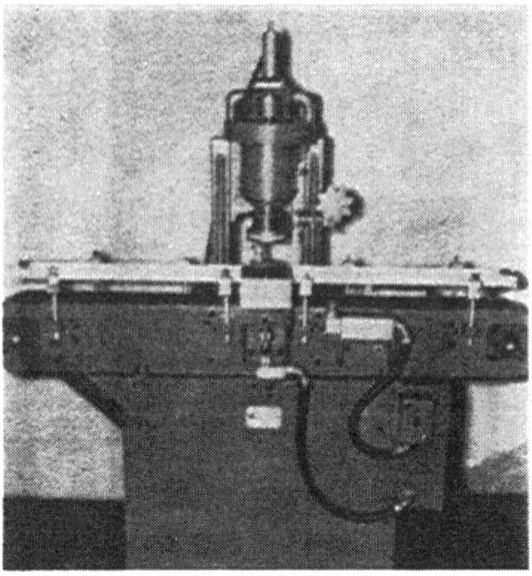

*Courtesy Pneumatic Scale Corporation, Ltd.*

FIG. 13.9. JAR CAPPER FOR SEMI-AUTOMATIC JAR PACKAGING LINE

*Courtesy Pneumatic Scale Corporation, Ltd.*

FIG. 13.10. CONTINUOUS AIR BLAST JAR CLEANER

*Courtesy Pneumatic Scale Corporation, Ltd.*

FIG. 13.11. VACUUM GASSER FOR INERT GAS JAR FILL

*Courtesy Pneumatic Scale Corporation, Ltd.*

FIG. 13.12. ROTARY HEAD POWDER FILLING MACHINE

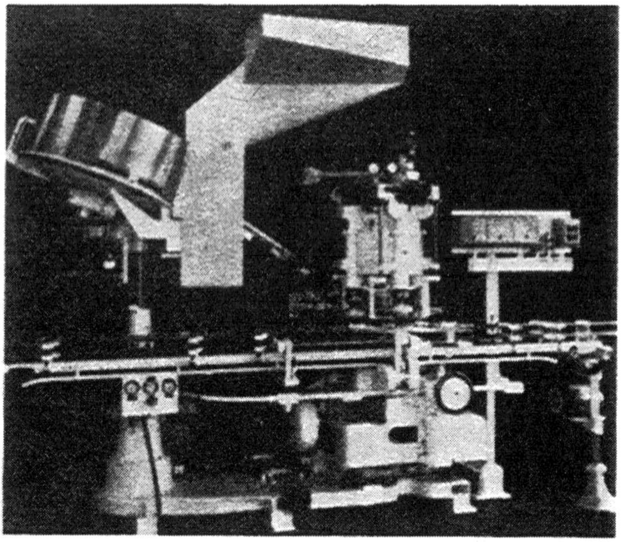

*Courtesy Pneumatic Scale Corporation, Ltd.*

FIG. 13.13. JAR CAPPER FOR FULLY AUTOMATIC LINE

*Courtesy of Standard Knapp*

FIG. 13.14. JAR LABELING MACHINE

*Courtesy of Griffin-Rutgers, Inc.*

FIG. 13.15. LABEL EDGE CODE MARKER

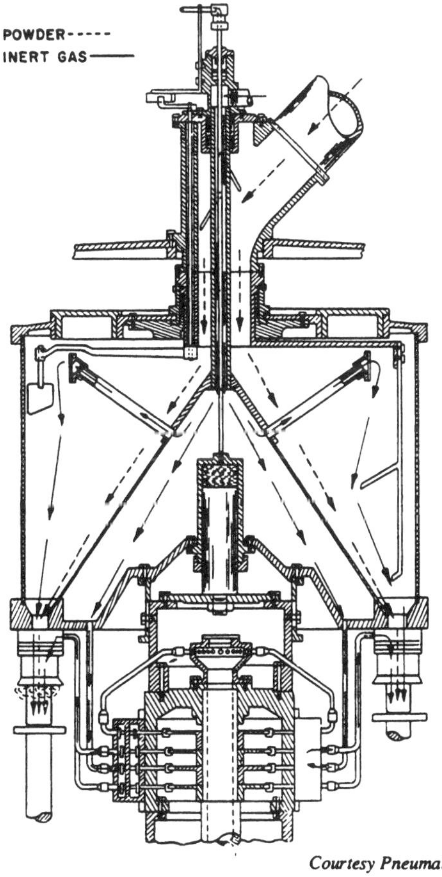


*Courtesy Pneumatic Scale Corporation Ltd.*

FIG. 13.16. SECTION OF INERT GAS ROTARY HEAD POWDER FILLING MACHINE

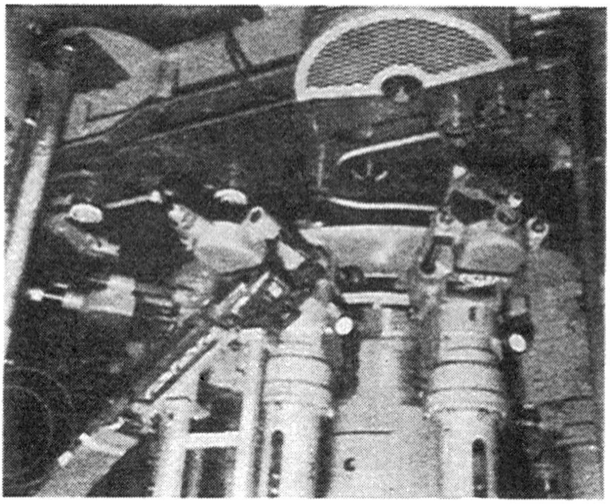

*Courtesy of Continental Can Co.*

FIG. 13.17. VACUUM CAPPER FOR INSTANT COFFEE GLASS JAR

*Courtesy of Alfa Laval*

FIG. 13.18. ALFA LAVAL CT-6 OR CT-9 CENTRIFUGAL VACUUM EVAPORATOR-CONTINUOUS.
(a) feed; (b) concentrate; (c) vapor to condensor, (d) steam and (e) condensate.

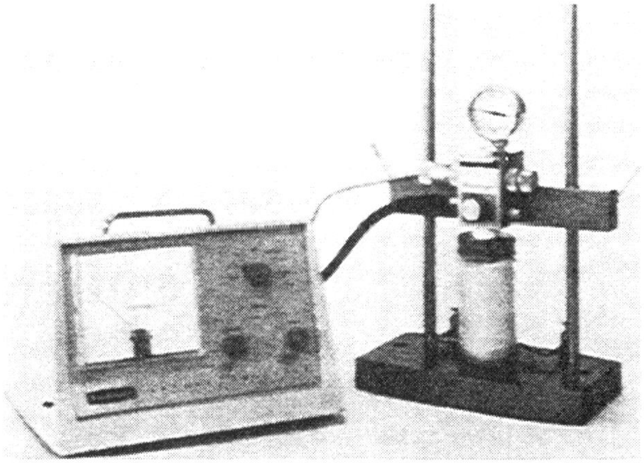

*Courtesy Beckman Instruments, Inc.*

FIG. 13.19. MEASURING OXYGEN IN GAS OF INSTANT COFFEE JAR

*Courtesy Syntron Company*

FIG. 13.20. VIBRATING TABLE FOR BULK INSTANT COFFEE IN CARTON

TABLE 13.2. COFFEE ROASTINGS FOR SOLUBLES MANUFACTURE FROM GREEN BEANS IMPORTED (INTO U.S.A.); SOLUBLE COFFEE IMPORTED (INTO U.S.A.); AND FREEZE DRIED COFFEE PRODUCED (IN U.S.A.) 1961 to 1970

| Solubles Year | Annual Roastings millions TOTAL | Annual Roastings millions SOLUBLES | Roast equivalent 60 Kg bags millions lbs. | Imported Solubles c/ as solubles b/ | Imported Solubles c/ millions lbs. | (% of | Annual Total total use) million lbs. | Produced Freeze Dried Solubles (% of all solubles) | Produced Freeze Dried Solubles (millions lbs.) |
|---|---|---|---|---|---|---|---|---|---|
| 1960 | 21.9 | — | — | — | 4.6 | = | — | — | nil |
| 1961 | 22.0 | (18.2%) 4.0 | 193 | +(1.7%) | 3.4 | | 195 | — | nil |
| 1962 | 22.7 | — | — | — | 4.0 | | — | — | nil |
| 1963 | 22.8 | (17.3%) 4.0 | 193 | +(3%) | 6.3 | = | 199 | — | nil |
| 1964 a/ | 22.4 | (16.9%) 3.8 | 182 | +(3%) | 5.3 | = | 188 | — | 1 |
| 1965 | 21.7 | (17.5%) 3.7 | 177 | +(2%) | 2.8 | = | 180 | (1%) | 2 |
| 1966 | 21.3 | (16.3%) 3.5 | 158 | +(6%) | 10.5 | = | 179 | (3%) | 6 |
| 1967 | 21.3 | (15.0%) 3.2 | 152 | +(15%) | 27.4 | = | 180 | (7%) | 12 |
| 1968 | 21.2 | (15.9%) 3.4 | 162 | +(12%) | 22.5 | = | 186 | (14%) | 24 |
| 1969 | 20.8 | (16.9%) 3.5 | 168 | +(20%) | 40.0 | = | 208 | (25%) | 48 |
| 1970 | 20.0 | (16.5%) 3.3 | 158 | +(18%) | 35.7 | = | 194 | (30%) | 58 |
| 1971 | 19.6 | (17.4%) 3.4 | 163 | +(18%) | 36.5 | = | 200 | (33%) | 66 |
| 1972 | 20.8 | (16.3%) 3.4 | 163 | | 56.3 | | 200 | estm. (35%) | 70 |
| 1973 | 21.8 | (14.7%) 3.2 | 152 | | 68.0 | | 200 | estm. (37%) | 74 |
| 1974 | 19.2 | (15.7%) 3.0 | 138 | | 72.2 | | 200 | estm. (39%) | 78 |
| 1975 | 20.3 | (14.8%) 3.0 | 138 | | 48.8 | | 186 | estm. (42%) | 80 |
| 1976 | 19.8 | (15.2%) 3.0 | 135 | | 80.0 | | 200+ | estm. (45%) | 90 |
| 1977 | | | | | | | | estm. (47%) | 94 |

a/Maxim freeze dried coffee introduced in U.S.A.
b/Assume 2.75 lbs solubles per lb of green (0.363 yield from green).
c/Exported solubles not considered here.

TABLE 13.3. EXPORTS OF SOLUBLE COFFEE FROM BRAZIL (IN BAGS OF 60 KILOS, GREEN EQUIVALENT)

| Destination | 1975 | 1974 | 1973 | 1972 | 1971 | 1970 | 1969 | 1968 |
|---|---|---|---|---|---|---|---|---|
| United States | 908.180 | 966.400 | 888.376 | 914.738 | 583.659 | 431.831 | 653.646 | 433.063 |
| United Kingdom | 560.236 | 540.735 | 716.186 | 489.868 | 435.545 | 320.250 | 116.576 | 71.868 |
| West Germany | 57.166 | 87.644 | 101.214 | 74.528 | 34.504 | 51.612 | 57.207 | 33.996 |
| Netherlands | 15.367 | 64.123 | 85.823 | 63.823 | 21.172 | 31.050 | 13.300 | 5.607 |
| Italy | 1.856 | 48.515 | 14.101 | 2.249 | 12.153 | 12.017 | 7.394 | 5.375 |
| Canada | 16.142 | 37.553 | 65.531 | 63.468 | 22.821 | 127.878 | 34.466 | 1.632 |
| Japan | 39.370 | 32.092 | 10.107 | 9.490 | 1.981 | 5.201 | 5.450 | 6.634 |
| France | 12.180 | 21.071 | 10.522 | 10.376 | 15.109 | 9.604 | 10.288 | 7.672 |
| Soviet Union | — | 7.500 | — | 22.500 | 5.778 | 3.165 | — | 2.996 |
| Switzerland | 17.693 | 7.294 | 7.495 | 7.514 | 4.278 | 4.250 | 750 | 700 |
| Bulgaria | 5.000 | 7.250 | 5.000 | 2.500 | 1.250 | 2.500 | 1.750 | 1.000 |
| East Germany | 12.605 | 5.000 | 6.350 | 10.002 | 10.022 | 22.600 | 12.250 | 3.250 |
| Austria | 45.191 | 3.482 | 34.014 | 6.980 | — | 50 | 50 | — |
| Norway | 4.219 | 3.164 | 2.619 | 452 | — | — | — | — |
| Hungary | 6.078 | 2.883 | 889 | — | — | — | — | |
| Belgium-Luxemburg | 270 | 1.834 | 470 | 7.516 | 3.079 | 66 | — | 60 |
| Spain | 853 | 1.511 | 3.706 | 5.183 | 4.862 | 1.183 | 1.444 | — |
| Greece | 11.099 | 1.150 | 2.600 | 2.424 | 947 | 1.622 | 1.725 | 677 |
| Lebanon | 500 | 1.014 | 1.572 | 1.526 | — | — | — | — |
| Czechoslovakia | 1.758 | 540 | 1.456 | 324 | — | — | 750 | 750 |
| Poland | — | 403 | 403 | 192 | — | 13.500 | 2.000 | — |
| Cyprus | 250 | 350 | 325 | 388 | 126 | — | 50 | 71 |
| Paraguay | — | 292 | — | — | — | — | — | — |
| Formosa | 98 | 158 | 263 | 223 | 55 | 594 | 100 | 400 |
| Australia | 2.934 | 91 | 298 | 101 | 124 | 36 | 11 | — |
| Kuwait | — | 72 | — | — | — | — | — | — |
| Denmark | 85 | 27 | 50 | — | 2 | — | — | 6 |
| Hong Kong | 32 | 7 | — | 9 | — | — | — | — |
| Chile | — | — | 1.080 | 15.306 | 35 | — | — | — |
| Yugoslavia | — | — | 390 | — | — | — | — | — |
| | 1.723.798 | 1.842.155 | 1.960.966 | 1.711.954 | 1.161.159 | 1.041.127 | 922.917 | 576.915 |

### Coffee Sales: Shares of Instant Coffee Market

In the U.S.A., General Food's brands of Maxwell House, Yuban and Sanka represent 53 percent of all U.S. instants sales, while Nestle's brands of Taster's Choice and Decaf represent 25 percent of all instant sales. The TenCo Division of The Coca-Cola Co. is the major private label instant coffee producer. Instant coffee sales in the U.S.A. in 1977 amounted to 1,500 million dollars. Seventy-eight percent of the instant coffee sales were dominated by 2 firms, General Foods and Nestle.

## CHEMICAL COMPOSITION OF INSTANT COFFEES

Instant coffees, due to the great fraction of solubilized celluloses and the introduction of hydrolysis by-product acids and aromatics, is very different in composition from brewed coffee solubles compositions. Hence, they differ in taste.

Brewed coffee solubles are about 33 percent large molecular carbohydrates, 8 percent reducing sugars, 33 percent non-volatile acids, mostly chlorogenic, 17 percent minerals (ash), 5 to 10 percent caffeine, 4 percent trigonelline and 1 to 3 percent oil (with phospho-lecithin). Caramel or melamine are large complex molecules.

Instant coffee powder solubles are 61 percent carbohydrates (including 6 percent reducing sugar), 17 percent non-volatile acids (mostly chlorogenic), 10 percent minerals (ash), 2 to 3 percent trigonelline, 3 to 6 percent caffeine (Arabica to Robusta blends), 0.1 to 3 percent oil (with phospho-lecithin), and 3 percent moisture.

## SPENT COFFEE GROUNDS DISPOSAL AS FUEL

Since the 1973 OPEC price increases on petroleum products ranging to 500 percent through 1977, the inefficient and wasteful disposal of spent coffee grounds from instant coffee plants has become a financial burden for many plants. Some plants are recovering only a portion of the fuel value of the spent grounds due to inefficient dewatering, drying and combustion systems in boiler applications.

The following table compares 3 different sized instant coffee plants and the fuel costs from recovered grounds:

### Preparing Spent Coffee Grounds for Burning

Industrially, most grounds burning systems have been poorly designed, built and operated. The reasons for this are that such systems are made up of several

TABLE 13.4.

| PROCESSING VARIABLES | | Three Locations | |
|---|---|---|---|
| | Nicaragua | Brazil '67 | Brazil '77 |
| 1. BAGS (60 Kg) green coffee/year | 60,000 | 180,000 | 360,000 |
| 2. BAGS (60 Kg) green coffee per month | 5,000 | 15,000 | 30,000 |
| 3. BAGS (60 Kg) green coffee per day | 167 | 500 | 1,000 |
| 4. Kg green coffee per day | 10,000 | 30,000 | 60,000 |
| 5. Kg INSTANT COFFEE per day | 4,400 | 13,200 | 26,400 |
| 6. Kg dry spent coffee grounds per day | 5,600 | 16,800 | 33,600 |
| 7. Kg dry spent coffee grounds per hour | 233 | 700 | 1,400 |
| 8. HEATING VALUE (millions BTU hr) | 3.6 | 10.8 | 21.6 |
| basis: 7,000 BTU/dry lb or 385 KCal/Kg | | | |
| millions (KCal per hr) | 0.9 | 2.7 | 5.4 |
| 9. EQUIVALENT STEAM PRODUCTION at | | | |
| (60% thermal efficiency) Kg/hr | 1,633 | 4,500 | 9,000 |
| lb/hr | 3,600 | 10,000 | 20,000 |
| 10. BOILER HORSEPOWER | 100 | 300 | 600 |
| (34.5 lbs stm/hr=1 HP) | | | |
| 11. EQUIVALENT FUEL OIL REQUIREMENT | | | |
| gal per hr | 20 | 60 | 120 |
| liters per hr | 75 | 225 | 450 |
| 12. EQUIVALENT FUEL OIL COSTS: at $1.00/gal (26¢ per liter) | | | |
| per hour $20 | | $ 60 | $ 120 |
| per day 480 | 1,500 | 3,000 | |
| per month 15,000 | | 45,000 | 90,000 |
| (360 days) per year | 180,000 | 540,000 | one million |

*Courtesy United States Rubber Company*

FIG. 13.21. COLLAPSIBLE RUBBER CONTAINER FOR BULK POWDER SHIPMENTS

process steps, usually involving large and expensive components, and few firms have invested enough thought and money to do a proper job. A great number of installations that require fuel oil to burn up the grounds, because the feed grounds are too wet have resulted. These types of systems invariably use Dutch ovens to dry the grounds or under feed stokers to build up a pile of wet grounds within the furnace chamber adjacent to boiler tubes. In fact, Folgers' use of fuel oil to incinerate its waste grounds actually cost money and recovered no heat. TENCO and its affiliate in the 1960's used under feed stokers which were rather inefficient when spent grounds with 50 percent or more water were used. Even General Foods, with its massive wealth in R&D, wasted fuel by feeding wet grounds to Dutch ovens.

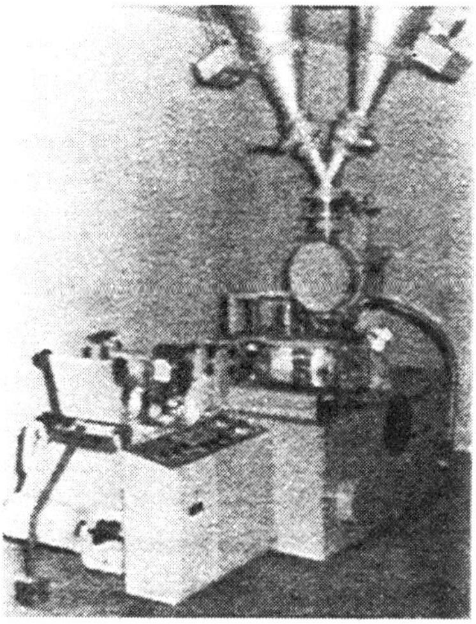

*Courtesy Industrie-Werke Karlsruhe*

FIG. 13.22. POUCH FORMER, FILLER, AND SEALER

*Courtesy Hayssen Manufacturing Company*

FIG. 13.23. POUCH FORMER, FILLER, AND SEALER

*Courtesy Bartelt Engineering Company Inc.*

FIG. 13.24. POUCH FORMER, FILLER IN INERT GAS, AND SEALER

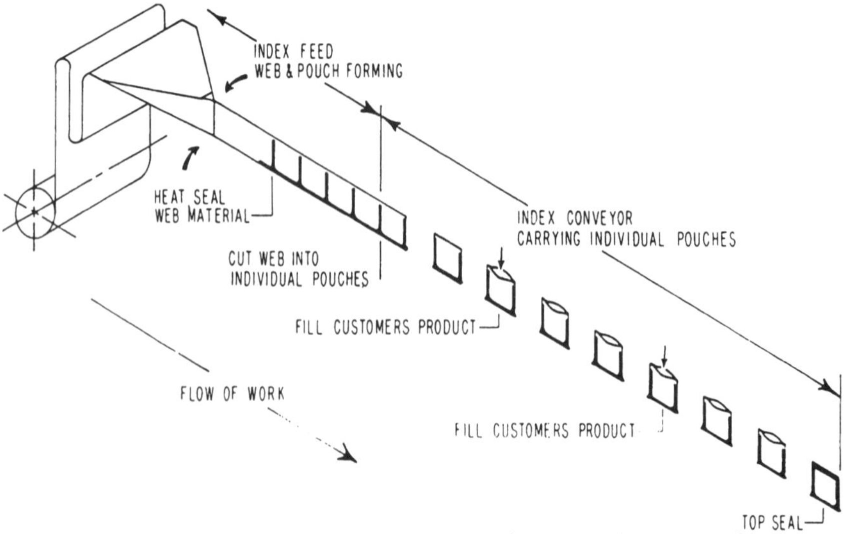


*Courtesy Bartelt Engineering Company Inc.*

FIG. 13.25. FLOWSHEET OF PACKET FORMATION FROM ROLL STOCK

Another shocking aspect was that well established engineering firms made massive blunders, because they had had no experience in the burning of spent coffee grounds. Some of the errors were:

1. Drying wet (75 percent moisture) grounds when presses were much cheaper for reducing water content.
2. Using hot air rotary driers that dried unpressed grounds to only about 50 percent moisture.
3. Using fire tube boilers that quickly had their tubes scaled up, and then lost their capacity to deliver adequate amounts of steam unless they were shut down and cleaned.
4. Using water tube boilers and underestimating the increased ventilation load due to moisture released from the cellulose in wet coffee grounds.
5. Not including dewatering screens to remove loose water, which otherwise passed thru.
6. Not using surge bins before presses, thereby negating press performance.
7. Invariably not drying the spent grounds enough, thereby losing much heat value.

The last point is the key to efficient fuel use. The grounds must be dry enough. The following table itemizes the relationship between the 80 percent moist blown down spent grounds and the final dry product to be used as fuel. Drying to below 30 percent cannot be done safely when high heating temperatures are used; otherwise fire occurs.

TABLE 13.5. SPENT GROUNDS LOCATION

| | #W per # dry grounds | % water in grounds | # Water removed #dry grounds | % Water removed | Cum. |
|---|---|---|---|---|---|
| 1. Blown and drained | 4.0 | 80 | | | |
| | | | 1.0 | 25 | 25 |
| 2. After dewatering screen | 3.0 | 75 | | | |
| | | | 2.0 | 50 | 75 |
| 3. After pressing | 1.0 | 50 | | | |
| | | | 0.5 | 12.5 | 87.5 |
| 4. Dried grounds | 0.50 | 33 | | | |
| | 0.33 | 25 | 0.67 | (17) | 92 |
| | 0.25 | 20 | 0.80 | (20) | 95 |
| | 0.15 | 12.5 | 0.85 | (22) | 97 |

**Types of Driers Used.**—The drier requires a significant investment, and there are several types that can be and are used.

1. FURNACE: Dutch Oven, stoker onto grill, pile of grounds (worst and most inefficient method).

2. ROTARY DRUM with tumbling granules:
   a) HOT AIR (less efficient, and can cause fires).
   b) STEAM TUBE (more expensive and massive).
3. TRAY DRIER with Rotary Plows (for mixing and exposing granules).
   a) HOT AIR (Wyssmont type).
   b) STEAM JACKETED SHELVES
4. FLASHDRIER: Raymond type requires pulverization of granules and recycling.
   It poses fire and explosion hazards when granules are dry (oil content).
5. FLUID BED DRIER: Possibly the least expensive drier, it produces a uniformly dried product at low temperatures and low product moistures without hazard of fire. Fines are carried off at higher moistures, but are only less than 15 percent of throughput.

CAUTION: There are are two situations that can cause serious problems:

1. Fire is accompanied by a smoke that does not allow access to the fire source, hence steam quenching from installed nozzles is a necessity in most driers.
2. Underestimating the importance of dried spent grounds feed to fuel economy.

# Part V

# Coffee and Its Influence on Consumers

# 14

# Physical and Chemical Aspects of Coffee

## PHYSICAL PROPERTIES OF COFFEE

A knowledge of the physical as well as chemical properties of coffee at all stages of processing is necessary (1) to design process equipment; (2) to make material balances for coffee solubles and insolubles; (3) to make heat balances for heat and temperature distributions in the process; and (4) to control and develop processing conditions and products for desired coffee flavor and uniformity.

Quality control tests are closely associated with coffee properties. Because coffee is a natural substance, its properties are variable, and it is frequently necessary to state ranges of properties rather than exact figures for constituent concentrations.

The manner of presentation in this chapter is to review a property, for example, density of green coffee through roast coffee beans and grind to powder and spent grounds. Extract densities are covered under coffee extract.

This chapter covers coffee densities, light reflectance or color, particle size distribution, granular fluidity, thermal properties, swelling of grounds and water adsorption, solubles availability and extraction, moistures, instant coffee powder solubilities and, finally, extract properties. In addition, there is the factor of cleanliness of the cup of instant coffee as it is affected by sediment, floating specks, dirty cup rings, and oily droplets or films.

### Densities

**Cherry.**—The coffee cherry, when it is picked red, has a bulk density of about 50 lb per cu ft; after pulping and fermentation but while the green beans are still wet (50 percent moisture), they retain a bulk density of about 50 lb per cu ft. When the green coffee bean in pergamino or hull is dried to about 13 percent moisture, its bulk density is about 25 lb per cu ft. The dry, hulled, polished green coffee bean has a bulk density of about 44 lb per cu ft.

A recapitulation of the bean weight yield relations from cherry are: 550 lb of fresh cherry yield 225 lb of wet pergamino, which yield 120 lb of dry pergamino; the 120 lb of dry pergamino yield 100 lb of dry hulled and polished green coffee beans (wet processing). In dry processing, 550 lb of fresh cherry yield 200 lb of dry cherry and then 100 lb of dry hulled and polished green coffee beans.

**Green Beans.**—The absolute dry green bean specific gravities vary from 1.30 for high grown mild coffees to 1.25 for lower grown mild coffees and 1.20 for Brazilian and Robusta coffees.

**Roast Beans.**—Absolute specific gravities will depend on the bean type and the rate and degree of roast, but they are about 60 percent of the original green bean specific gravities for palatable light roasts, and 50 percent for medium to dark roasts. Robusta coffees do not swell in volume quite as much as mild coffee of comparable roast colors. The absolute bean specific gravities of Portuguese West African Robustas are more similar to high grown mild coffees than to Brazilian or low grown mild coffees. For example, the denser beans are about 0.1 specific gravity unit more dense before and after the same type of roasting. Decaffeinating green coffee beans before roasting them causes them to swell less, resulting in greater bulk densities. For example, instead of 18 to 22 lb per cu ft, decaffeinated coffee may be 22 to 26 lb per cu ft over a comparable roast color range. These swelling differences may be related to differences in chemical composition and physical structure of these beans. Darker roasts, in general, yield lighter bulk densities for a given coffee or blend. For example, a 23 lb per cu ft bulk density for a given blend at the lightest roast that is palatable may yield 18 lb per cu ft at a dark palatable roast.

**Ground Roast Coffee.**—Grinding invariably raises the bulk density of roast coffee. A more uniform particle size and a more uniform particle size distribution in the bed or column also yields higher bulk densities. Vibration, vacuum loading, and other techniques for settling the particles among themselves to minimize void increase the bulk density of ground roast coffee. For example, vacuum loading R&G coffee into a percolator column increases column loadings by about 10 percent, as compared to a natural fall fill. Lower bulk densities occur in smaller diameter vessels (e.g., less than 1 ft) than in larger diameter vessels. This is due to greater wall support for the granular coffee bed in the smaller diameter column. Vibrated coffee beds may increase bulk density by 15 percent. Finer grinds form denser beds. For example, a coarse grind (25 percent on 8 mesh) may yield 19 lb per cu ft, but a home-use, regular grind yields 23 lb per cu ft, and a fine grind yields 24 or 25 lb per cu ft. Too fine a grind may become less dense due to its fluffy nature, i.e., entrained air. Moisture content of the roast coffee will affect its bulk density. For example, R&G coffee with 6 percent moisture is more dense than R&G coffee with no moisture. Localization of chaff can cause very low bulk densities and occurs when fines and chaff accumulate in bin corners. Uniform particle distribution (across the bed diameter of a 2 or 3 ft diameter percolator column) increases loading densities 5 to 10 percent. Finer grinds may increase loading densities 10 to 15 percent.

Fine grind coffees yield bulk densities of 25 to 30 lb per cu ft; after vibratory settling, the bulk densities increase to 27 to 33 lb per cu ft.

Bulk densities of whole beans can be measured by natural fill or vibratory weight fill of a cubic foot box of beans leveled off at the top or in a large steel beaker (see Fig. 14.1). Usually natural fall fills are taken because that is the way beans or R&G coffee are collected in bins or columns.

In Fig. 14.1 R&G coffee falls freely from funnel into beaker and overflows. The bottom of the funnel is about 4 to 6 in. above the top of the beaker. The beaker must be made of rigid stainless steel. The spatula is for leveling the freely fallen pile of grounds on top of the beaker.

**Instant Coffee Powder.**—This bulk density is an important property that is continually monitored and controlled in the spray drying of extract. It can be measured in various ways. The volume of powder sample must be sufficient to give an accurate result. For example, 100 gm of powder are weighed and put into a 1,000 ml accurately indexed Pyrex graduate. The graduate is then placed on a vibratory table or tamping mechanism (see Fig. 14.2) to yield the settled powder volume in 15 sec. If this settled powder volume is 500 ml, then the bulk density is 0.20 gm per ml or 12.5 lb per cu ft. Powder bulk densities of 14 to 15 lb per cu ft are used with present packaging practices. Low extract concentrations, low oil contents, and high carbohydrates yield powder bulk densities of 10 or 11 lb per cu ft. In the case of extracts heavily laden with oils and fatty acids, bulk densities of 27 lb per cu ft may occur. Packaged instant coffee powders (which have undergone much particle breakage) are about 18 lb per cu ft in the jar. Powder bulk densities are invariably the result of the coffee extract properties. The appearance of the particles (and their screen analysis) can be seen under a stereo-microscope (see Fig. 14.3). High powder bulk densities characterize the powder with poor fluidity because of fines. Low bulk densities with discrete beady particles usually have good fluidity. Agglomerated particles give lower powder bulk densities and poor fluidity. Figure 14.4 shows one means of measuring powder fluidity.

Powder falls freely from a funnel onto an accurately leveled and dimensioned disc. The less fluid the powder, the more it accumulates on the disc. Powder density variations must be taken into account. The bottom of the funnel and the accumulating disc must be maintained at a set distance. The accumulated powder is tipped into a collection cup for weighing to obtain the fluidity index.

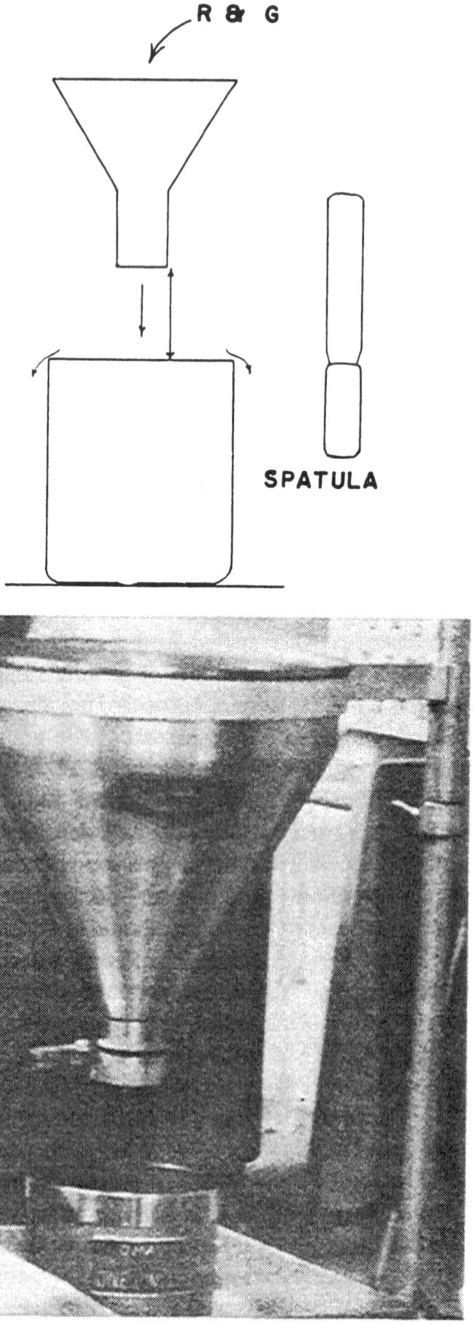


*Courtesy Coffee and Tea Industries Magazine*

FIG. 14.1. APPARATUS FOR DENSITY MEASUREMENT OF GROUND COFFEE

*Courtesy Syntron Company*

FIG. 14.2. VIBRATING TABLE FOR POWDER BULK DENSITY MEASUREMENTS

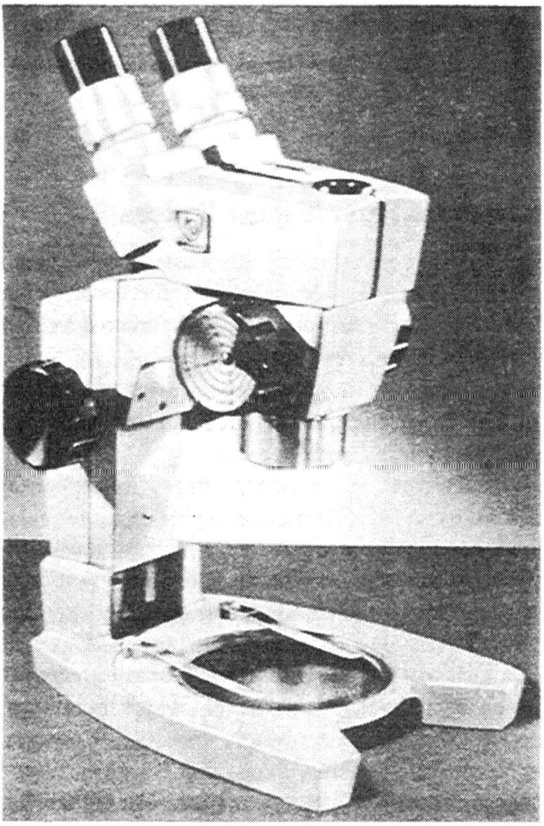

*Courtesy American Optical Company*

FIG. 14.3. STEREO-MICROSCOPE

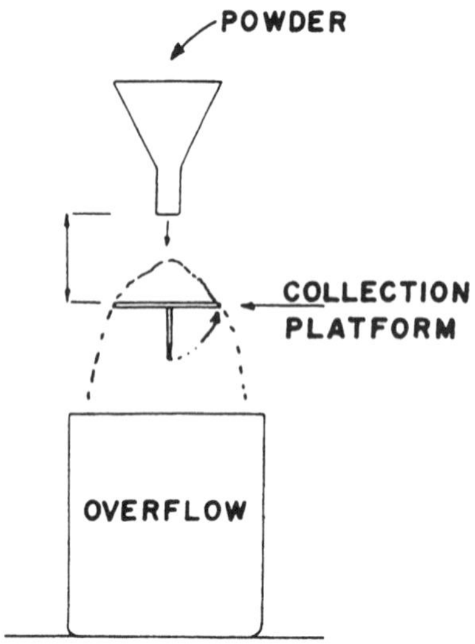


FIG. 14.4. APPARATUS FOR FLUIDITY MEASUREMENT OF GRANULAR MATERIAL

## Color: Roast, Ground, and Instant Coffees

Ever since green coffee beans have been roasted, the degree of coffee flavor development in the roaster has been judged by the color of the roasted beans. Roasted bean color is a fair criterion of coffee flavor development, especially when one is working with a given green coffee.

Since the end of World War II, there has been some use and limited acceptance of reflectance of light from ground, screened, and smoothed surface roast coffee particles as measured by a photoelectric cell.

In brief, the light intensity cell transforms the reflected light into an electrical voltage and is at least as sensitive as the eye; it gives a numerical reflectance value as a point of reference useful for day-to-day, month-to-month control. The eye cannot register a numerical reference point. The one limitation to an economical light intensity measurement is the color spectrum being examined. Roast coffee beans have different colors, depending on blend, age, moisture, and other factors. Therefore, to obtain a comparable light intensity measurement for varying colors, the light source must be a compromise representing the majority of the coffee roast colors encountered. This is not accurately possible. However, the Photovolt Corporation of New York produces a tri-stimulus glass filter which gives an acceptable light spectrum for such practical work. The light reflectance measurements are useful and accurate as long as reference coffee sample measurements are made frequently enough.

A major problem in applying light reflectance measurements lies in the preparation of the ground coffee samples. The outside surfaces of roast beans are not a good index of the degree of bean roast. They can be dark on the outside but much lighter on the inside, as well as the reverse. For color comparison by eye, the bean sample must be ground and spread on a tray next to the control sample, preferably in daylight. Since some fluorescent lighting does not give a natural light spectrum, comparisons of grounds by daylight are invariably better than comparisons by electric lights at night.

For the light reflectance meter it does not matter whether the surroundings are day or night. Figure 14.5 shows the instrument and the reflectance probe. Usually a porcelainized plate of suitable reflectance is used as a standard.

For duplication of reflectance results, the ground coffee should be screened to obtain a uniform particle size; for example, through 30 mesh, on 40 mesh. Furthermore, the selected grounds fraction should be gently purged with air to remove any chaff, which is light colored and will cause a high reflectance reading. Then the screened fraction, which is free of chaff, is compressed to about 1,000 psig on a Carver press (Fig. 14.6) to form a disc 2 in. in diameter and about ¼ in. thick. Compression of the fine particles orients them and forms a smooth reflecting surface. Compression must not be so great that oil is released to the surface of the particles, for this will cause a low reflectance reading. The reflectance probe is placed directly on the smooth coffee grounds surface for the reading. The probe lens should have no fines or dust to interfere with the passage of reflected light. Care should be taken to use fresh reference coffee samples, because oil comes to the surface of roast beans that have been stored for several days. Roast coffee reflectance standards are altered to accommodate the best flavor results as well as changes in blend and moisture.

*Courtesy Photovolt Corporation*

FIG. 14.5. LIGHT REFLECTANCE PROBE AND INSTRUMENT

*Courtesy Fred S. Carver, Inc.*

FIG. 14.6. CARVER HYDRAULIC PRESS

The preparation and measurement of reflectance color by this means takes some effort and time. But in any medium to large roast coffee operation where uniformity is important, this method is useful. For example, in instant coffee processing a small darkening in the roast can increase fatty acid content of extract and change atomizing performance.

Roast colors will vary from any single roaster, but color variations are somewhat greater with batch than with continuous roasters. The objective is to prevent the range of roast color variance from being so great that taste differences are noted. A deviation of one shade from standard reflectance is not too serious, but a deviation of two shades definitely reveals flavor changes.

The equipment used to measure roast coffee light reflectance are: (1) Model 610

Photovolt reflection meter with amber tri-stimulus filter and calibrated standard enamel reflector; (2) a Carver laboratory press, with a 2-in. diameter by 1-in. high cylinder of steel; (3) several dozen steel blocks ⅜ in. deep with 2¹/₁₆ in. circle, ¼ in. deep; (4) W. S. Tyler Ro-Tap or equivalent with 30 and 40 mesh screens (see Fig. 14.7); and (5) laboratory grinder and chaff air blower with a spatula for leveling R&G coffee.

The same reflectance probe and instrument can be used to measure the reflectance of spray dried instant coffee powder. Since such powders are dusty, however, the powder is placed in a clean glass beaker with flat bottom and then placed on the upright probe. Care must be taken that no fine dust adheres disproportionately to the bottom glass surface, because this causes more light reflectance.

Powder light reflectance, as a numerical index of color, is measured to determine uniformity of production as well as to compare reflectance in color with competitors' products. Here again, reflectance is measured by the tri-stimulus light source, and does not indicate differences in coffee color spectrum. For example, some instant coffee powders are reddish while others are tan; such light spectrum differences are not differentiated on this instrument. Redness may reveal itself only as a darker color with less light reflectance. The oil-surface instant

*Courtesy W.S. Tyler Company*

FIG. 14.7. SCREENING ROTARY TAPPING MACHINE

coffee powders are much darker than those to which oil is not added. Powders, before oil addition, that are light in color usually have many fines. Instant coffee powder color or light reflectance is *not* an indication of roast color but only of the physical particle size of the spray dried powder. For example, a dark roast instant coffee may have a very light powder color because the particles are fine. On the other hand, a light roast instant coffee may form dark, beady particles.

The crushing of beady particles results in more light reflectance and consequently, lighter colored powders. Screening of particles shows the fines to be light colored and the coarse, beady particles to be dark. The public tends to prefer darker powders, possibly because they resemble the familiar appearance of roast coffee. Agglomerates of fine particles appear light. An oily fine powder may be darker than a non-oily, large beaded powder. High carbohydrate content, as from a 100 percent Robusta instant coffee at a 40 percent solubles yield from R&G coffee, causes lighter particles and more light reflectance than similar sized particles from lesser yields and other blends. Light powder densities of 12 lb per cu ft are concentrated to 17 lb per cu ft, which results in very light colored, dusty powders. A 3-color line spectrum recorder, which is useful for confirming color contours of color lithographed labels, caps, and so on, has been placed on the market. Figure 14.8 shows the Photovolt unit being used for the same purpose.

## Particle Size Distribution

**Green Coffee Bean Sizes and Shapes.**—Bean sizes and shapes vary, as was discussed earlier. In Brazil, coffees are classified by screens. No. 20 is a very large bean; No. 19 is an extra large bean; No. 18 is a large bean; No. 17 is a bold bean; No. 16 is a good bean; No. 15 is a medium bean; and Nos. 13 and 14 are small

*Courtesy Coffee and Tea Industries Magazine*

FIG. 14.8. LIGHT REFLECTANCE FROM COLORED LABEL

beans. Other coffee growing countries have their own bean size standards. Mild coffee beans are, in general, larger averaging than Brazilian beans. Bean lengths are 7 to 11 mm; widths are 6 to 7 mm; boldness is 3 to 5 mm, most beans 4 mm. Beans weigh 0.1 to 0.2 gm each, averaging about 0.15 gm.

Roasting coffee causes about a 16 percent weight loss and a 50 to 80 percent increase in bean volume. The resulting bean dimensions may be determined from this volume increase.

**Ground Coffee.**—Figurc 14.9 shows cumulative screen analyses of grinds, ranging from cracked beans to those used in commercial percolation as well as in the home. Figure 14.10 shows the weighing of screened fractions of roast coffee. Knowledge of ground coffee screen analyses is important in the processing of soluble coffee to guard against excessive pressure drops in the water flow through the percolator bed.

Figure 14.7 shows the RO-Tap machine used for screen analyses. Three screen fractions, such as on 8 through 20 mesh, are sufficient to know the quality of the

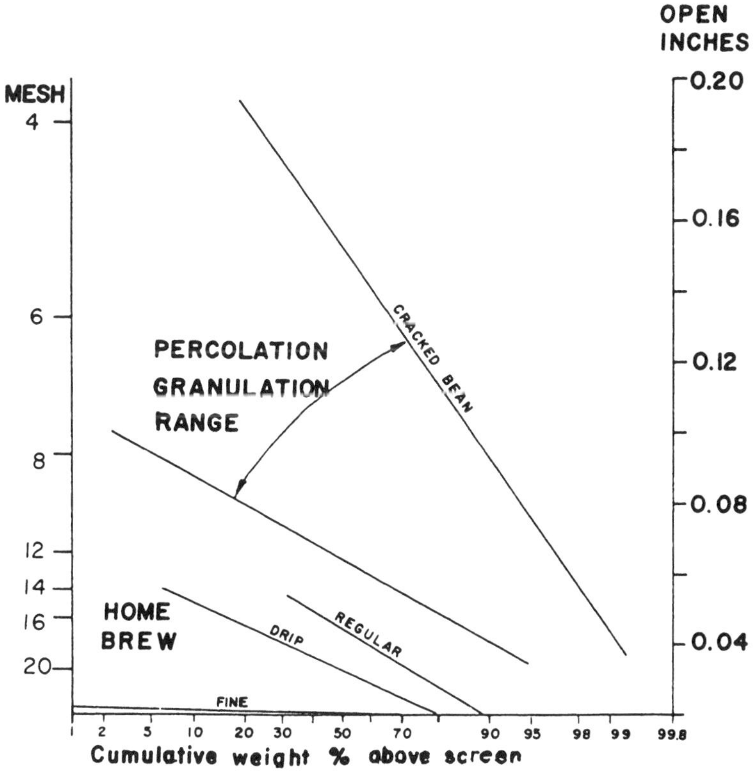


FIG. 14.9. CUMULATIVE SCREEN ANALYSES OF DIFFERENT COFFEE GRINDS

*Courtesy National Coffee Association–U.S.A.*

FIG 14.10. WEIGHING SCREENED FRACTIONS OF ROAST COFFEE

grind. If, for example, 5 or more percent of a grind used for plant percolation falls through 20 mesh some flow difficulties may occur, depending on process conditions. Usually a 3-min rotary tapping of a 50 gm ground coffee sample yields the desired weight fractions. Grinder samples should be run every few hours.

**Instant Coffee.**—A similar screen analysis can be made by using finer screens for instant coffee powder; for example, 40 and 80 mesh. Beady spray dried powders have 10 to 35 percent on 40 mesh and up to 10 percent through 80 mesh. After powder is vacuum tamped into jars this same powder analysis is only about 1 to 10 percent on 40 mesh and as much as 30 percent through 80 mesh.

Screen analyses of powders while they are spray dried provide powder bulk densities, microscopic examination of particles, spray drying results, and possible influences of extract properties. Agglomerated particles yield high weight fractions on 40 mesh, but the quality of the particles and fluidity must be examined.

The 40 mesh opening corresponds to about 350 mu and the 80 mesh opening, to about 180 mu.

An example of a beady spray dried powder is about 5 percent of 600 mu, 10 percent of 500 mu, 20 percent of 400 mu, 25 percent of 300 mu, 25 percent of 200 mu, 10 percent of 100 mu and 5 percent of 50 mu and less.

An intrinsic property of fine particles is their large surface per unit weight; 1 gm of powder averaging 200 mu in diameter has a total surface of about 1 sq meter. As particles are made finer, their physical and chemical nature changes markedly when the ratio of surface to apparent volume exceeds 10,000. Soluble coffee particles are in this order of magnitude. It is simple to demonstrate that (1) on 40 mesh, (2) through 40 mesh on 80 mesh, and (3) through 80 mesh, instant coffee powders each taste different.

## Instant Coffee Powder Solubility

Until 100 percent coffee in the form of spray dried, beady particles was placed on the market in 1951, the extracted coffee solubles were called "soluble coffee product." They were termed "soluble" because they were water soluble—slowly, not instantly—and "coffee product" because they had about 50 percent carbohydrates. However, when the 100 percent beady powder appeared, it was named instant coffee because it dissolved in boiling water in a few seconds, while the earlier powders did not. Instant coffees from spray driers dissolve quickly at temperatures down to 150 F (66 C). Some instant coffees, dissolve fairly quickly at 120 F (49 C), but none except the foam or pseudofreeze dried flakes dissolves instantly in 60 F (16 C) water. Normally, if an instant coffee dissolves in less than 10 sec upon addition of boiling water, it is considered fast enough. But sometimes instant coffees do not dissolve in 30 sec or a minute, or even longer. To dissolve the powder, one should place powder in the cup and pour boiling water over it. The powder should not be placed in the boiling water because more lumps may form.

If the water has not been boiled, it will contain dissolved air. This gives the resulting instant coffee a flat taste and also makes rapid and complete solubility more difficult. The undissolved particles absorb moisture and trap air bubbles, which prevent some powder from being wetted. These clumps of powder float about the cup without dissolving, and often leave a residual foam. High percentages of fines also hold air so that they do not wet and dissolve. Poor solubility is often caused, in part, by high carbohydrates (high hydrolysis yields) and low fatty acids. The latter help to break up the foam.

Agglomerated fine particles dissolve quickly because they are fused together. The thicker walled, larger beady particles also dissolve well since they have less surface per unit weight for air adsorption. Shutdown powders[1], which have high carbohydrate contents, usually have poor water solubility. Mixing powders of good and poor solubilities does not improve the poorer solubility fraction very much.

Powders resulting from extracts of concentrations over 30 percent usually have better solubilities than those from extracts containing less than 30 percent solubles, due to the thicker walls and larger sizes of particles formed. Powders that have had considerable particle breakage and increase in bulk density during packaging usually have poorer solubilities. Reworking powders of poor solubilities usually is at less than 20 percent and often at less than 10 percent levels. That is, the normal product will tolerate only these levels of reworked substandard powder without noticeable damage to quality.

## Water Solubles From Green and Roast Coffee

The fraction of coffee soubles from green and roast coffees that will dissolve in boiling water varies somewhat with type of coffee and degree of roast. Even different lots of the same variety of green coffee or lots of different moistures yield somewhat different amounts of solubles. Green coffees yield about 26 percent solubles and Robustas yield about 2 percent more. Water extraction of green coffee is slow; grinding of the green beans hastens the operation.

**Grinding Roast Coffee Finely.**—The fineness of roast coffee grind influences the solubles yield due to the increased release of oil-proteinaceous colloids from exposed surfaces. However, in darker roasts, coffee solubles decrease 2 or 3 percent from about 26 percent in the palatable range of roasts. Decaffeinated coffees, due to solubles losses during decaffeination, yield 2 to 3 percent less solubles than before decaffeination. Before attaining very dark roasts, solubles yields rise about 1 percent before they fall. This might be partly caused by resolubilization of celluloses, carbohydrates, and denatured proteins. Darker roasts have less volatile and titratable acidity. This can be discerned from odor, pH measurements, and taste.

Breakdown of roast coffee bean cells speeds up solubles extraction and increases solubles yields. For example, boiling water can remove 35 percent solubles from pulverized roast coffee beans. The increased solubility is due to the breakdown of the coffee bean cell structure, allowing free solution without the necessity for diffusion of molecules and colloids out of the cells.

**Molecular Size of Solubles.**—The coffee bean cells are about 30 mu in diameter. The boiling point temperature rise and freezing temperature depression for coffee extract solubles are quite small. This means that most of the solubles have rather high molecular weights. Concentrated, cold extracts are viscous. Some hydrolyzed solubles have large molecular weights, also, which can be measured by diffusion experiments through calibrated permeable membranes, such as Visking plastic dialysis sheets.

When considering water solubility data, one must also consider the grind and modes of extraction. Some water soluble molecules are within the particle cells, but are too large to diffuse out rapidly, if at all. These large molecules can be dissolved only by opening the cell walls. The greater the number of cell walls opened, the greater the solubles yield without hydrolysis. The finer the grind, the greater is the percentage of cells ruptured.

Because of the buildup of non-diffusible water soluble molecules (too large to diffuse readily into the coffee cell) in the extract moving from water feed, there are more large molecules in the extract solubles outside the roast coffee particles than within. This phenomenon can be measured by mixing a dextrose (30 D.E.) solution and coffee grounds. The dextrose portion is diffusible; the partially hydrolyzed, large starch molecules are not as diffusible. Dextrose has a molecular weight of 198 and is 10 A long, while caffeine has a molecular weight of 194 and is 15 A long. The 30 mu cell is 30 by $10^4$ A, but permeability through the cell wall, i.e., the size of the cell openings, may be only a fraction of this or a few hundred Angstroms. The caramelized molecules and hydrolyzed molecules are hundreds of Angstroms in size.

Temperature and time of extraction are not important variables with pulverized roast coffee. Temperature and time of diffusion of solubles are important variables with coarser particles.

**Ternary Equilibrium Diagram for Water Solubles and Insolubles.**— Solubles equilibrium between coffee grounds and extract is a physical phenomenon. It has been previously discussed, but will be amplified in this section to show that perhaps 50 percent of the coffee solubles will not diffuse back into the grounds. This results in a higher solubles concentration outside the coffee grounds. Solubles equilibrium occurs with the wet (not dry) coffee grounds (see Fig. 14.11). The equilibrium line varies slightly with the amount of non-diffusibles solubles in the free extract and with the grind used. Solubilization of large molecules, such as those of corn syrup and coffee extract, forms a true solution. Solubles yield is increased by several percent by the presence of colloids. Before molecular solubles (large or small) or colloids can move away from the ground coffee particles, the roast cell walls must have absorbed about one-third their roast weight in water.

**Wetting.**—Selective water absorption concentrates the adjacent extract: (1) by selective water absorption; (2) by the inability of non-diffusible solubles to enter the cells; and (3) by the surface contribution of non-diffusible solubles. Selective water absorption by coffee grounds is about four times greater from a sucrose solution than from a salt solution. Salt ions hold water more strongly. Selective water absorption by the coffee grounds in the percolator accounts for about a 3 percent concentration rise which would otherwise not occur. The wetting and swelling give off a little heat and cause a 7 percent increase in roast coffee particle

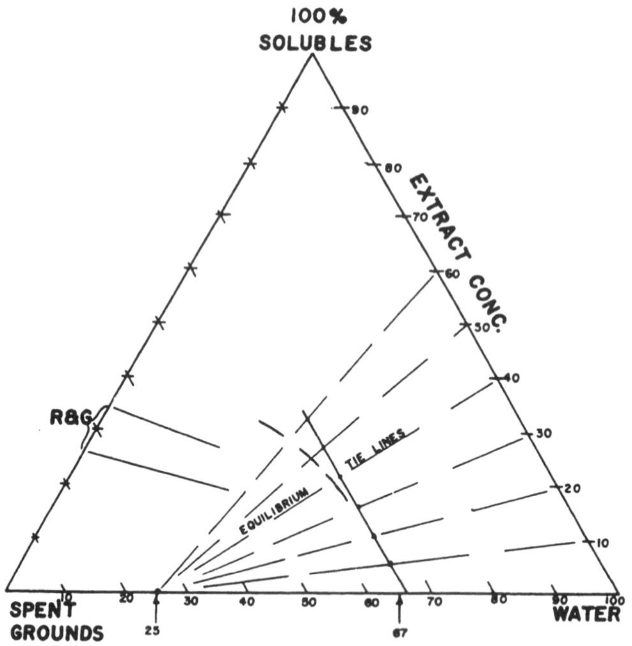


FIG. 14.11. TERNARY EQUILIBRIUM DIAGRAM FOR WATER AND COFFEE

size. This can be seen under a microscope with a calibrated slide. In a coffee bed, this particle swelling fills the interstitial void.

**Hardness of Wet Coffee Particles.**—Compressibility of a bed of grounds is related to the fineness of grind and the degree of breakdown of the cellular wall structure. The bed of wet coffee grounds is more compressible when it consists of finer grinds, darker roasts, and higher hydolysis yields of soluble components. Easier compressibility results in reduced interstitial voids for extract flow with subsequent cumulative sealing off of flow passages. Such compressive forces of granular beds become effective with fine grinds at 50 psig differentials.

## Moistures: Green, Roast, and Instant Coffees

Green coffee contains about 13 percent moisture when it leaves the coffee growing country and usually a few percent less when it is processed in the consuming country. The equilibrium moistures of green, roast, and instant coffees at various relative humidities are listed in summary form in Table 14.1. The rates of moisture absorption and liberation vary with temperature, air circulation, and other factors. Soluble coffee absorbs moisture up to equilibrium in hours to days, whereas roast coffee requires days to weeks. Green coffee equilibrates more slowly. The rates of moisture absorption are usually much faster than the rates of desiccation. With greater water-holding capacity in air of warmer temperatures, faster rates of moisture absorption occur. Green coffee that has been stored for months to years in a lower temperature and lower relative humidity than those in equilibrium with twelve percent moisture will have its moisture reduced to perhaps 8 percent. New crop coffees have the highest moisture consistent with safe bagging and exporting. Past crop coffees are bleached or dark and have much less than 10 percent moisture if storage conditions have been dry.

**Measuring Moisture**.—Obtaining green coffee moistures from whole beans is a slow process. Therefore, it is usually desirable to grind the coffee before making oven moistures or toluene distillations. There are several instruments that are used for measuring moisture of grains. These work usually by measuring dielectric strength. They should be calibrated against toluene or oven moistures periodically. Once calibrated, they give quick and reasonably reliable results. Some of these instruments are Seedburo or Steinilite, Moisture Register or Tagliabue, Division, Weston Instruments. The Universal measures the conductivity of a pellet compressed in a standard manner, and the Alladin II by Klock, Incorporated, uses a direct weighing and the infrared heating of a small ground sample to obtain results in a few minutes. As it is not uncommon even with toluene and oven moistures to obtain somewhat variable moistures, moisture allowances and the use of sample references are always good practices.

Roast coffee moistures are also run by toluene reflux (1 to 2 hr) or oven (24 hr). With heater standardization, coffee grounds moisture can be measured on an Ohaus electric coiled balance (see Fig. 14.12) or Genco Infrared lamp balance in 20 to 30 min.

Roasted coffee normally has hardly any moisture unless it has been water quenched. Some vacuum packed coffees have up to 3 percent moisture, which causes them to stale faster once the can is opened. Some soluble coffee processing operations use even higher roast coffee bean moistures to keep fines low during grinding, as well as to pre-moisten the particle fibers for better wetting during percolation. The 20 to 30 min heat balance moisture methods are accurate enough for such moistures. This method can also be used on partially dried spent coffee grounds, but may need the full 30 min heating on the balance. The heat balances can usually be adjusted so that the rate of heating can be altered to suit the sample being dried. For example, soluble coffee would have a gentle rate of heating, while moist spent grounds would have a higher rate. Drying rate curves can be run to set the best rate of heating without carbonizing or liberating bound moisture. Soluble coffee powders usually have less than 4 percent moisture. Spent coffee grounds may have 70 percent moisture initially and as low as 15 percent after drying. Soluble coffee powder moistures can also be measured by benzene or toluene

*Courtesy Ohaus Scale Corporation*
FIG. 14.12. MOISTURE BALANCE

reflux in cases of verification (see Association Official Agricultural Chemists, A.O.A.C.).

Spray dried powders should have their moistures recorded once an hour even if drier operating conditions appear normal. At comfortable temperatures the relative humidity in the room where moistures are measured must be less than 40 percent. Otherwise, the powder during sampling and weighing absorbs enough moisture to yield high results as, for example, 4.0 percent, when moisture is really 3.0 percent.

Depending on moisture and composition, instant coffee powders can fuse at temperatures above 145 F (63 C) in a few hours. Fusion may occur in less than an hour at 160 F (71 C). Temperatures of 130 F (54 C) may cause instant powder fusion in less than 24 hr. Powder fusion in a glass jar causes the fused mass to shrink and pull away from the wall. It is possible to use 2 percent calcium gluconate or "Micro-cel" (calcium silicate) for moisture absorption in instant coffee powders.

## Thermal Properties of Coffee

Design of any equipment that heats or cools coffee requires a reasonable knowledge of the associated thermal changes. Many of these have already been covered under roasting and spray drying, but the thermal properties will be summarized here.

**Heat of Wetting.**—Coffee moisture is mostly free, but since heat is liberated in wetting dry green, dry roast, and dry solubles, this indicates that some percentage of the coffee holds the water in strong physical or chemical combination (bound water). About 2 to 5 cal per gm (4 to 9 BTU per lb) for heat of wetting of cellulose has been reported.

**Heat Capacity**.—For green or roast coffee, heat capacity can be taken as about 0.4. Coffee compositions and (specific heats) are: $^{1}\text{fi}_{10}{}^{3}$ water (1.0); $^{1}/_{10}$ oil (0.45); 0.4 cellulose (0.32); and 0.3 carbohydrate (0.3). Extract specific heat can be approximately calculated by using (0.4) for solubles and (1.0) for water at different concentrations.

**Heat of Pyrolysis**.—The heat liberated in roasting pyrolysis can be approximated from the amount of $CO_2$ liberated (for example, 4.0 percent, or about 100 cal per gm green coffee or 180 BTU per lb green coffee). The exothermic heat depends on the degree of roast. Darker roasts may give up 300 BTU per lb. Roasting coffee is mainly pyrolysis, not oxidation.

**Heat of Combustion**.—Burning of spent coffee grounds is oxidation. The heat liberated per unit weight can be estimated from the constituents of the grounds: 15 to 20 percent oil and the rest, mostly cellulose. This gives about 9,000 to 10,000 BTU per lb dry grounds, based on 18,000 BTU per lb for oil and 8,000 BTU per lb for cellulose.

**Heats of Solution**.—Carbohydrates liberate about 3 to 5 cal per gm (5 to 9 BTU per lb) of solubles. It is about the same for coffee solubles, depending on the degree of dilution. Since the initial water added to carbohydrates is held more securely, there are higher heats of solution evolved from higher concentration extracts (e.g., 30 to 40 percent) than from lower concentration extracts (e.g., 0 to 10 percent). This means dilution would absorb heat slightly or cool the solution.

## Cleanliness of Cup of Instant Coffee

Instant coffee in the cup is expected to be clear with no floating nor sedimentary particles, no oil slicks, and no cup rings. Also, the beverage should not be turbid nor foamy. To meet all these requirements is sometimes difficult and some allowances have to be made.

**Floating Oil Slicks**.—These are undesirable, but with many of the coffee oil "aromatized" instants, tiny oil droplets are characteristic of the beverage cup. Dark roast coffees and fine grinds sometimes allow enough oil into the extract to cause small slicks. This oil carries considerable amounts of coffee flavor through spray drying.

**Floating Specks**.—These are usually black and are most often due to powder buildup on the drier walls which has been burnt or carbonized. Direct-fired carbonized fuel may also cause floating specks. The black carbon specks, and sometimes coffee powder agglomerates are objectionable foreign matter. They are especially noticeable in a cup of coffee with cream or milk. These specks usually do not alter coffee flavor and often are very minute. Inspection of spray dried powder for such floating specks by adding milk to the cup is a routine test.

**Foam**.—This is often due to high hydrolysis solubles yields and low oil content. Robusta coffees are often associated with a foamy cup on pouring. Foam that disappears in 5 to 10 sec is not objectionable, but is undesirable when it lasts for several minutes. Foam that persists on cups of espresso coffee is sometimes sought; otherwise, it is undesirable. Even in an espresso cup, foam is undesirable in taste and appearance when Robustas are in the blend. A foamy extract spray dries to yield golden specks in the powder.

**Turbidity**.—This is seldom encountered in soluble coffee unless it is contributed by (1) the oil content of dark roasts; (2) oil add-back in aromatized coffee; or (3) freeze dried brew colloids.

**Cup Ring**.—This is frequently related to foam. It is a surface tension phenomenon that seems to draw insoluble substances to the cup wall. Removal of extract tars by centrifuging also removes oils (and flavor), results in excessive foam, and yields light bulk powder densities.

**Foam and Cup Ring**.—Cup rings are not infrequently associated with foamy rings. The tars tend to cling to the cup wall, presenting an unsightly appearance. The cup rings are often very difficult to wash out. Tars can be minimized by (1) taking lesser hydrolysis yields; (2) eliminating Robustas from the blend; and (3) settling out the tars. Centrifuging to remove the tars usually removes most of the coffee oil but tars settled in extract can minimize the cup ring yet not remove so much cup flavor. Settled tars can represent several percentage of extract volume. Some tars are oily and occlude air, so they float on the extract. Complete tar and oil removal causes poor powder solubilities and foams. Foam color can be tan to black. The tan foams are usually associated with Robusta blends. Black particulate tar rings may occur in any blend. Blends and light roasts that will contribute to instant coffee foaming can often be anticipated from brews prepared from the same roast coffee. A high ferric iron contamination also causes a black cup ring. Foam and cup rings are sometimes contributed by a cumulative series of circumstances. For example, separate factors such as high extract drawoff temperatures, channeling of extract, or too short a settling time for the extract before spray drying may cause tar rings on the cup.

**Sediment in the Cup**.—This is usually due to fine particles of roast coffee and is sometimes due to tars. A good procedure for measuring sediment content is to filter instant coffee through milk disk filter paper under aspirator suction. Insolubles are caught on the filter paper disk, and the number and size of particles can be compared to the milk standard disks 1, 2, and 3. Instant coffees should be like standard disk 1, seldom like 2, and never like 3. Figures 14.13 and 14.14 show the standard paper filter disks and the filter cup apparatus. Specks that float on a milk and coffee surface are distinctly different from sediment that settles to the bottom of the cup.

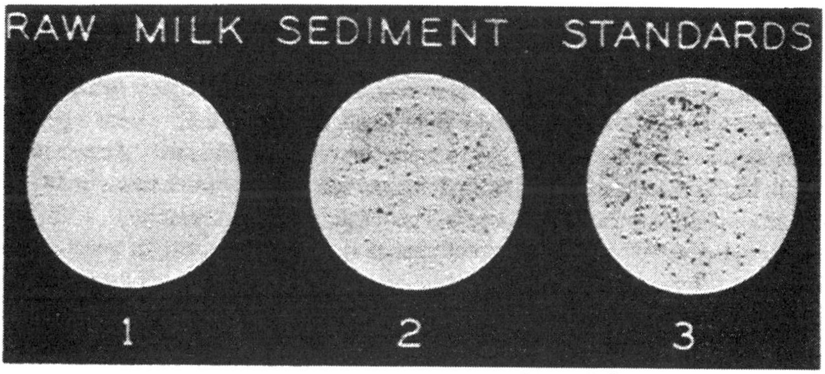


*Courtesy Sediment Testing Supply Company*

FIG. 14.13. SEDIMENT DISK FILTER STANDARDS

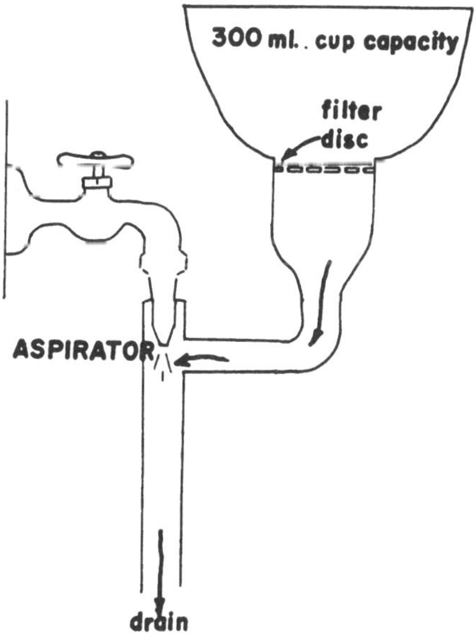


FIG. 14.14. APPARATUS FOR PREPARING VACUUM FILTER SEDIMENT DISK

## Light Transmission Through Brewed Coffee

This will be influenced by the degree of roast, blend, and concentration of solubles in the cup. Turbidity and pH will also be factors influencing the amount of light transmission. Dark roasts have less acidity and produce a much darker cup color. Robusta coffees, which are less acid, also yield a much darker cup color.

The visible light transmission spectrum for coffee solubles, regardless of the source of coffee, is similar. The color spectrum is similar to that for caramel and amber glass. Figure 14.15 illustrates such a curve in the visual spectrum from red (7,000 A) to violet (4,000 A). Since brown is the dominant color of coffee, most of the transmission of light wave lengths occurs in the red to yellow region with least light transmission in the violet to blue region. The roasted coffee with only 4 percent dry weight loss allows more light through than the coffee with 8 percent dry weight loss, which in turn allows more light through than the 12 percent dry weight loss. This corresponds to weight losses from "as is" green beans of 14, 18, and 22 percent, respectively. There is a pH change which is about 5.0, 5.15, and 5.30, respectively. The magnitude of change in percentage of light transmission with such differences in roast weight losses is small and varies over the spectrum. It is equivalent to about one-third to one-half of the effect of doubling solubles

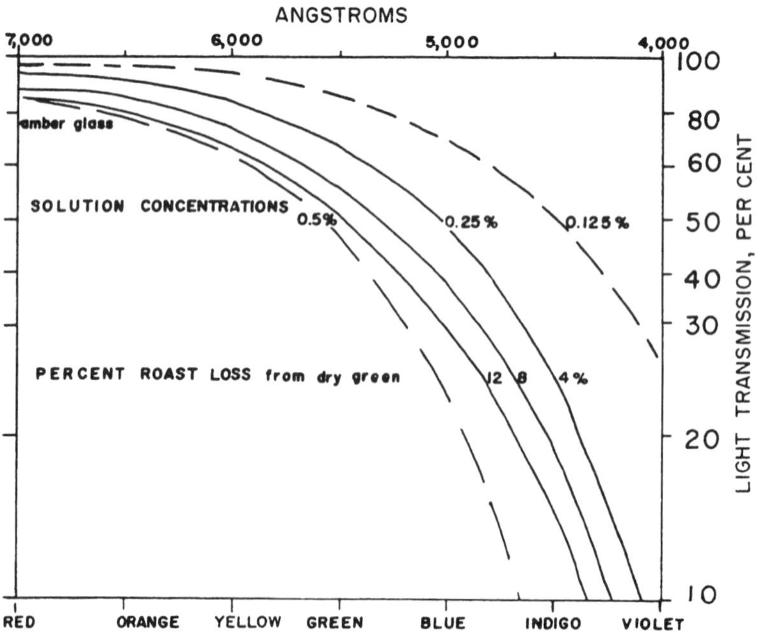


FIG. 14.15. VISIBLE LIGHT TRANSMISSION THROUGH INSTANT COFFEE SOLUTIONS

concentration from the median value (0.25 percent solubles) of 4 percent dry weight loss.

Deschreider's coffee extraction technique takes a little longer than measuring light reflectance of R&G coffee as described earlier, but is certainly a practical method. A standard coffee solution is used as reference in Deschreider's Stammer colorimeter, which is similar to a Klett or du Bosque comparator as shown in Fig. 14.16. This type of comparator colorimeter is used in the sugar refining and caramel manufacturing processes for color control purposes. In the case of a Beckman spectrophotometer, light transmissions can be made at any wave length from a rotating prism. In most instances, coffee solution light transmission measurements at 1 wavelength in the portion of highest light absorption area may be sufficient, e.g., blue at 510 mmu or 5,100 A or 430 mmu indigo-violet. Using 1 wavelength of light (filter) simplifies the measurement. An economical instrument that can be used for this purpose is shown in Fig. 14.17.

The use of glass color standards is not accurate unless the glass spectrum standard corresponds reasonably well with the spectrum of coffee. Such a glass standard can be prepared by the manufacturer at nominal cost. Glass color standards have been prepared for caramel, which has a slightly different spectrum from coffee, depending on its mode of preparation.

Most instant coffees at a concentration of ¼ percent at 430 mmu wavelength allow 18 to 22 percent light transmission; whereas, at 510 mmu at ¼ percent there

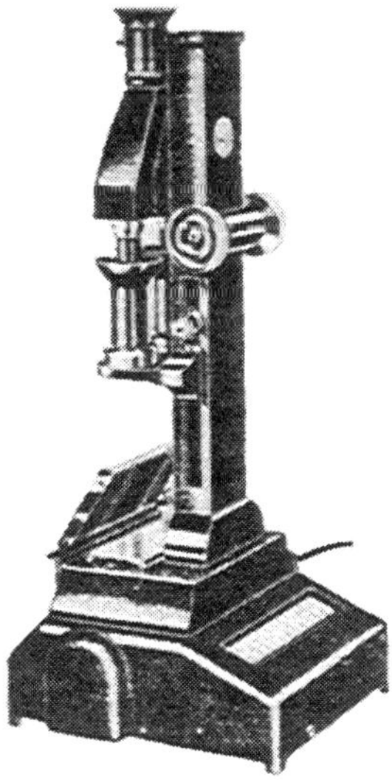

*Courtesy Klett Manufacturing Company*

FIG. 14.16. LIGHT TRANSMISSION COMPARATOR (DUAL TUBE)

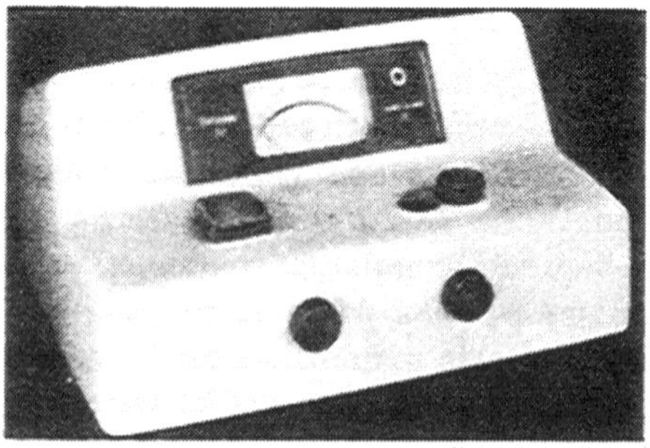

*Courtesy Bausch & Lomb, Inc.*

FIG. 14.17. MONOCHROMATIC LIGHT TRANSMISSION INSTRUMENT (340 TO 950 MMU)

is about 50 to 60 percent light transmission. Such light transmission measurements can be useful in comparing competitive brew cup color intensities. For example, for lower priced instant coffees (used in some vending machines and under private labels) a dark beverage color is desired with the least weight of instant coffee. A dark roast coffee can easily effect a solubles reduction of 25 percent in the cup, as far as matching beverage light transmission is concerned. Such competitive light transmissions of beverage are also influenced by pH; Robustas are less acid and darker in cup. High hydrolysis yields of soluble components increase light transmission by dilution of the light absorbing constituents.

Beer's law, in regard to transmitted light, states that halving the brew concentration doubles the percentage of light transmitted.

## Light Refraction

Light is refracted in passing through a sucrose solution or a coffee extract solution. For sucrose, degree Brix (density) is the same as percentage of sucrose in solution and can be read also from a refractometer. For coffee solubles in extract, degree Brix readings from a refractometer are about 15 percent higher. That is, 30° Brix is closer to 25.5 percent coffee solubles, or 60° Brix is closer to 51 percent coffee solubles. The degree Brix-percentage of solubles relationship does not vary much, for all practical purposes, with different blends, roasts, and yields. At any rate, no such exhaustive study has yet been reported. Accountability of coffee solubles in process is very much dependent on this relationship. There are also recording instruments that will record soluble solids from continuously monitored extract flow streams, but the high expense is seldom justified.

## Specific Gravity of Coffee Extract

This is a function of the solubles content and is about 1.5 to 2.5 percent higher than sucrose. This means that the water is held more tightly. For example, a 27 percent sucrose solution has a specific gravity of 1.100 at 60 F (16 C), but 1.115 for coffee solubles. A 53 percent sucrose solution has a specific gravity of 1.200 at 60 F (16 C), and 1.230 for coffee solubles.

Specific gravity can be readily calculated from cumulative composition. For example, a 32 percent coffee solubles solution, using a 1.60 specific gravity for pure coffee soluble (or carbohydrates), will yield 0.20 ml gravity equivalent for solubles. The 0.68 ml water with the same gravity equivalent gives a total of 0.88 ml or a specific gravity of 1.00/0.88 = 1.135. A 50 percent coffee solubles solution has a 0.31 ml equivalent volume for solubles and a 0.50 ml equivalent volume for water, making a total of 0.81 ml or a specific gravity of 1.00/0.81 = 1.230. Figure 14.18 gives specific gravities for sucrose and coffee solutions. Table 14.2 is an abbreviated coffee extract conversion table which can be interpolated. Other items in the table that are useful are weight of water per unit weight of solubles, milliliters of coffee extract needed for 2 gm of solubles when cupping, and weight of extract, solubles, and water per gallon.

Some plants use hydrometers for measuring specific gravity of coffee extracts. Accurate hydrometer readings can take ¼ to ½ hr due to occlusion of air in the

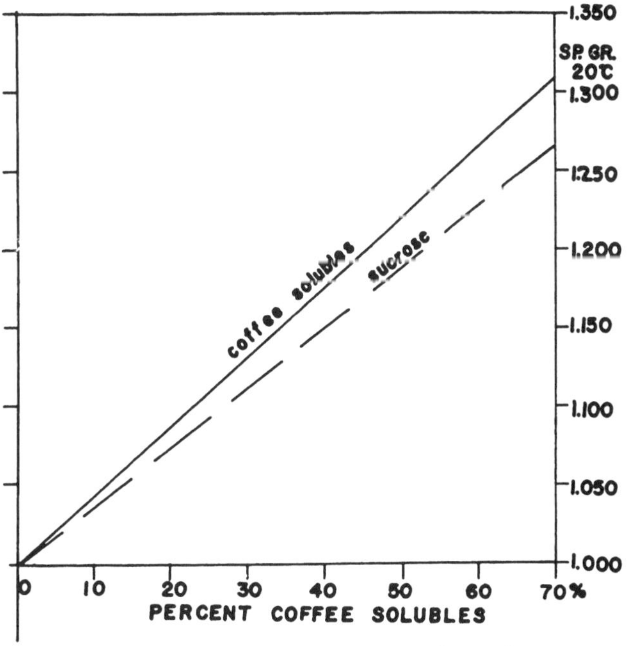


FIG. 14.18. SPECIFIC GRAVITY OF SUCROSE AND COFFEE SOLUTIONS

foam. The hydrometer rises continually as gases escape. Also, glass hydrometers break and are a continual expense, and there is an extract loss from samples which are not returned to production. Each 10 F (5.6 C) temperature change in the commerical range of percolation concentrations from 40 to 80 F (4 to 27 C) causes about a 0.002 specific gravity change. For example 35.0 percent coffee solubles at 50 F (10 C) is 1.160, or 1.156 at 70 F (21 C). Sometimes °Baumé or °Twaddell degrees are used instead of specific gravity, and these are related as follows:

| Percent Solubles | Degree Brix | Specific Gravity at 60 F (16 C) | Degree Baumé | Degree Twaddell |
|---|---|---|---|---|
| 0 | 0. | 1.00 | 0 | 0 |
| 11.4 | 14.0 | 1.05 | 7 | 10 |
| 23.2 | 28.0 | 1.10 | 13 | 20 |
| 34.0 | 40.0 | 1.15 | 19 | 30 |
| 44.5 | 52.0 | 1.20 | 24 | 40 |
| 56.5 | 64.2 | 1.25 | 29 | 50 |

## Viscosity of Coffee Extracts

The important property of viscosity of coffee extract can be measured easily at various temperatures by pipette or viscosimeter and a reference standard, such as sucrose solutions. Figure 14.19 shows a plot of viscosity (log scale) vs the reciprocal of absolute temperature (1/K). The radiating lines are for different concentrations of coffee solubles. The 0 to 10 percent viscosities are similar to water. The 20 to 30 percent viscosities are similar to sucrose solutions, while the viscosities for 40 percent and higher are like corn syrup. The 10 to 30 percent radial lines can be defined by one equation. The 40 to 60 percent lines can be defined by another equation. In the first case, the angle with the horizontal is thirteen degrees plus the concentration minus ten divided by two; in the second case, the angle is thirteen degrees plus the concentration divided by two. The X intercept, 0.0011, is somewhat of a focal point for the radiating lines, corresponding to a temperature of 450 F (232 C). Neither the reasons for the radial pattern nor the focal intercept temperature are significant to the author; sucrose and other solutions have similar patterns. Perhaps the meaning of this pattern will be clarified sometime. The fact that the two similar equations relating temperature and viscosity are separated between 30 and 40 percent solubles may be significant, because extract properties do change in this region. This has been observed in atomizing coffee extracts containing over 33 percent solubles to produce large, beady particles and better flavor retention.

Plotting the viscosities for 33 dextrose equivalent (D.E.) corn syrup shows that the corn syrup viscosities are somewhat less than viscosities of corresponding 40 and 60 percent coffee solubles solutions. This means that coffee solubles have a

TABLE 14.2. COFFEE EXTRACT CONVERSION TABLE—ABBREVIATED

| °Brix | Percent Solubles | Specific Gravity, 60 F (16 C) | Weight Ratio, Water/ Solubles | Extract Weight, per gal | Weight Powder, lb per gal | Gal Water | Ml Extract for 2 gm Solubles |
|---|---|---|---|---|---|---|---|
| 2 | 1.8 | 1.009 | 54.5 | . . . | . . . | . . . | . . . |
| 4 | 3.4 | 1.016 | 28.3 | . . . | . . . | . . . | . . . |
| 6 | 5.0 | 1.024 | 19.0 | 8.53 | 0.43 | 0.97 | 39.0 |
| 8 | 6.5 | 1.030 | 14.4 | 8.58 | 0.56 | 0.96 | 30.0 |
| 10 | 8.3 | 1.037 | 11.0 | 8.62 | 0.72 | 0.95 | 23.2 |
| 12 | 9.6 | 1.043 | 9.4 | 8.68 | 0.83 | 0.94 | 20.0 |
| 14 | 11.4 | 1.050 | 7.8 | 8.75 | 1.00 | 0.93 | 16.7 |
| 16 | 13.0 | 1.056 | 6.7 | 8.79 | 1.14 | 0.92 | 14.6 |
| 18 | 14.8 | 1.064 | 5.8 | 8.86 | 1.31 | 0.91 | 12.7 |
| 20 | 16.5 | 1.071 | 5.1 | 8.93 | 1.47 | 0.90 | 11.3 |
| 22 | 18.2 | 1.079 | 4.5 | 8.99 | 1.64 | 0.88 | 10.2 |
| 24 | 20.0 | 1.087 | 4.0 | 9.05 | 1.81 | 0.87 | 9.2 |
| 26 | 21.5 | 1.093 | 3.7 | 9.11 | 1.96 | 0.86 | 8.5 |
| 28 | 23.2 | 1.101 | 3.3 | 9.16 | 2.12 | 0.85 | 7.8 |
| 30 | 25.0 | 1.106 | 3.0 | 9.21 | 2.30 | 0.83 | 7.2 |
| 32 | 26.6 | 1.115 | 2.8 | 9.26 | 2.47 | 0.82 | 6.7 |
| 34 | 28.5 | 1.123 | 2.5 | 9.35 | 2.67 | 0.80 | 6.3 |
| 36 | 30.2 | 1.130 | 2.3 | 9.41 | 2.84 | 0.79 | 5.9 |
| 38 | 32.0 | 1.139 | 2.1 | 9.48 | 3.04 | 0.77 | 5.5 |
| 40 | 33.8 | 1.148 | 1.96 | 9.55 | 3.23 | 0.76 | 5.2 |
| 42 | 35.5 | 1.155 | 1.32 | 9.62 | 3.42 | 0.75 | 4.9 |
| 44 | 37.2 | 1.164 | 1.69 | 9.70 | 3.61 | 0.73 | 4.6 |
| 46 | 39.0 | 1.173 | 1.56 | 9.78 | 3.82 | 0.72 | 4.4 |
| 48 | 41.0 | 1.183 | 1.44 | 9.86 | 4.04 | 0.70 | 4.1 |
| 50 | 42.8 | 1.190 | 1.34 | 9.92 | 4.25 | 0.68 | 3.9 |
| 52 | 44.6 | 1.198 | 1.24 | 9.97 | 4.45 | 0.66 | 3.7 |
| 54 | 46.5 | 1.207 | 1.15 | 10.1 | 4.70 | 0.65 | 3.6 |
| 56 | 48.3 | 1.216 | 1.07 | 10.1 | 4.88 | 0.63 | 3.4 |
| 58 | 50.0 | 1.224 | 1.00 | 10.2 | 5.10 | 0.61 | 3.3 |
| 60 | 52.0 | 1.232 | .92 | 10.3 | 5.36 | 0.59 | 3.1 |
| 62 | 54.0 | 1.240 | .85 | 10.3 | 5.57 | 0.57 | 3.0 |
| 64 | 56.0 | 1.248 | .79 | 10.4 | 5.83 | 0.55 | 2.9 |
| 66 | 58.0 | 1.255 | .72 | 10.5 | 6.08 | 0.53 | 2.8 |
| 68 | 60.0 | 1.261 | .67 | 10.5 | 6.30 | 0.51 | 2.6 |
| 70 | 62.0 | 1.278 | .61 | 10.6 | 6.60 | 0.48 | 2.5 |

viscosity corresponding to less than 33 D.E. or perhaps 20 D.E., which is a 20 percent conversion of the starch to dextrose. The 20 D.E. corn syrup has larger molecules than the 33 D.E. corn syrup.

## Surface Tension

For water, surface tension is 75 dynes/cm; for coffee extract, it will vary primarily with fatty acid and oil content. Tension is usually reduced by surface-concentrating organic compounds to about 40 dynes/cm or less. Increased temperatures reduce extract surface tension, according to the Eotvos equation or its variant: $\sigma = \sigma_0 (1 - T/T_c)^n$ where $n$ is about 1.21, $\sigma_0$ depends on critical constants of the liquid, $T_c$ is the critical temperature of water, and $T$ is the temperature of

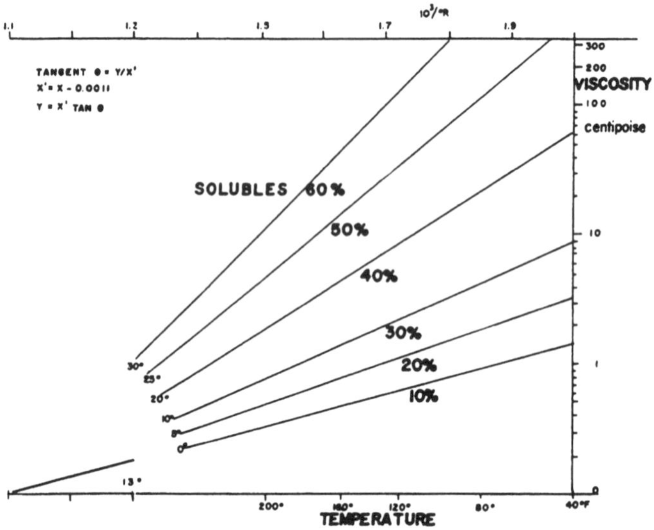


FIG. 14.19. VISCOSITY OF COFFEE EXTRACTS VS TEMPERATURE

interest; both $I_c$ and I are absolute temperatures. As *T* rises, surface tension decreases. Extract solubles concentration seems to have little influence on surface tension.

## Electrical Conductance of Coffee Extracts

There are indications that coffee solubles ionize. Specific electrical conductance is directly related to solubles concentration up to about 10 percent solubles; at this point the proportionality deviates. There is reduced ionization in more concentrated solutions, e.g., acetic acid. The fact that this deviation from linear relation occurs markedly at about 15 percent solubles could be related to the disappearance of brew colloids during percolation at this electrolytic density. Related to the ionization of extract constituents is the pH of extracts and brew cup strength.

## Hydrogen Ion Concentration (Acidity)

The pH of coffee extract is higher than the cup of instant coffee prepared from it. For example, a 35 percent solubles extract pH of 5.20 may correspond to a 1 percent solubles cup pH of 4.85. The sugar or caramel binding of water must reduce H ion dissociation. The pH of a cup from extract (e.g., 4.85) is more acid than the cup from spray dried powder (e.g., 4.95) because volatile acids are lost. The extract has about 0.9 percent of acetic acid relative to coffee solubles plus volatile acids formed in hydrolysis. The cup pH increase from 4.85 to 4.95 causes a noticeable taste difference. Cups of coffee from extract taste and are more acidic (with bite and bitterness) than those from powder. Table 14.3 shows the relation of pH to acidity and alkalinity. The pH of brewed coffees is about 5.1, depending on blend and roast (e.g., Robustas are higher at about pH 5.4). Instant coffees are about pH 4.9, but may be higher when made from Robusta blends. Holding brewed coffee hot in the presence of air causes acidity to rise as long chain compounds break down to form shorter chain acids.

## Curdling of Milk or Cream

Milk or cream added to a cup of hot coffee, may curdle if the cup pH is too low (acid). This sometimes happens to restaurant coffee that is held hot for several hours. Curdling can occur with cups of instant coffee that have a high hydrolysis solubles yield (acids). Such an acidic instant coffee can usually be detected before real harm is done by adding milk or cream to hot, water-diluted extract. Since the diluted extract cup of coffee is about 0.10. pH unit lower than powder, the dried powder will be safe if the extract does not cause curdling. Remedial steps in processing can be taken before curdling occurs in the dissolved powder. In a borderline case, curdling may not occur at 150 F (66 C), but will take place at temperatures of over 178 or 180 F (81 or 82 C). Milk and cream are also contributing factors. Drinkable yet slightly sour milk, as well as slightly acid cream, may cause curdling. It is of prime importance to avoid packaging and releasing to the consumer instant coffee powder that will curdle, because this can damage a firm's reputation.

The transition point at which milk or cream can curdle relates to a very small difference in pH. For example, the 100 ppm alkalinity of tap water may be sufficient to prevent curdling under some conditions. This is another good reason for using distilled water in testing instant coffee powders; it allows for a margin of safety. Natural waters are normally slightly alkaline.

An extract with pH 4.8 may have a 4.65 pH when diluted in distilled water. The same coffee extract diluted to cup strength in 100 ppm alkaline water may have a 4.75 pH. The instant coffee powder dissolved in distilled water may have 4.80 pH and in 100 ppm alkaline water, a 4.9 pH. Curdling may occur at 4.65 pH, but it will not occur at 4.75 pH. Processors of soluble coffee, using moderate solubles yields, considerable Robusta coffee in their blends, and a dark roast, seldom have a milk curdling problem. Instant coffees produced from one plant are usually very constant in the cup pH; for example, a cup may have a pH of 4.90 for many months, perhaps years, with a deviation of only ±0.05 pH.

It is not possible to determine by taste the point at which curdling will occur. Figure 14.20 shows a pH meter with electrodes.

More alkaline cups of coffee appear darker in color. For example, a cup of Robusta coffee may have a pH of 5.30 and will allow about 25 percent less light

TABLE 14.3 THE MEASURING AND MEASURMENT OF pH[1]

Water is seldom neutral. Usually it is alkaline and occasionally it is acid. The amount of alkali or acid present can be determined by analysis and is reported as total alkalinity or total acidity. In addition to the quantity measurement, however, there is also an intensity measurement of acidity and alkalinity; this is called the "pH value" of the solution.

A neutral condition exists when this pH value is 7.0; i.e., the solution is neither acid nor alkaline. When the pH value falls below 7.0, it indicates a greater *intensity* of acidity. When this value exceeds 7.0, it indicates a greater *intensity* of alkalinity. Thus, pH is a number which indicates the *intensity* of either acidity or alkalinity of a water solution.

*A–Showing the variation in pH when 100 ppm of different materials are added to pure water*

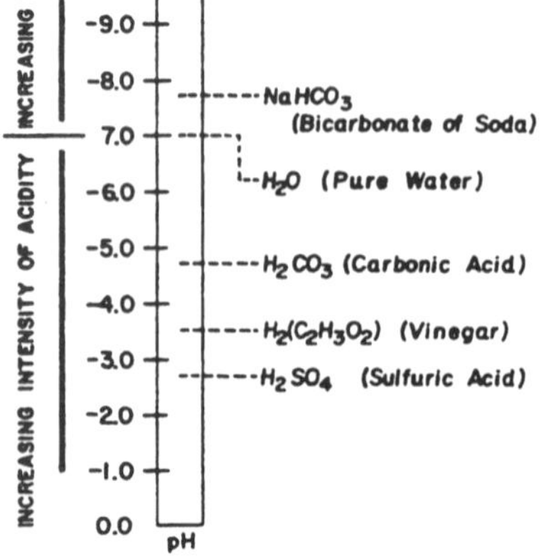


To illustrate that pH is not a quantity measure, Table A shows different pH values of common acids and alkalies—all at concentrations of 100 ppm. As an example, a solution containing 100 ppm of carbonic acid has a pH of 4.6, whereas a solution with the same amount of sulfuric acid has a pH of 2.8. This difference is due to the fact that sulfuric is a "stronger" acid than carbonic. Likewise it can be seen that a solution containing 100 ppm of bicarbonate of soda has a pH of 7.7, whereas the same quantity of caustic soda in solution has a pH of 11.4. This shows that caustic soda is a "stronger" alkali than sodium bicarbonate.

This scale adopted to show intensity of acidity of alkalinity (pH value) is somewhat misleading, for a change of 1.0 pH actually means that the intensity (either acid or alkaline) is multiplied by ten. This is more clearly shown by Table B, in which pH 7.0 is given a value of one.

*B—Intensity of Hydrogen Ion Concentration*

| | *pH Value* | *Values Showing Intensity* | |
|---|---|---|---|
| | 14 | 10,000,000 | |
| | 13 ↑ | 1,000,000 | |
| | 12 | ↑ 100,000 | |
| | 11 | 10,000 | |
| | 10 | 1,000 | |
| | 9 | 100 | |
| | 8 | 10 | |
| neutral | 7 | ——Alkalinity | Acidity |
| | 6 | | 10 |
| | 5 | | 100 |
| | 4 | | 1,000 |
| | 3 | | 10,000 ↓ |
| | 2 | | 100,000 |
| | 1 ↓ | | 1,000,000 |
| | 0 | | 10,000,000 |

[1]Courtesy of Allis-Chalmers.

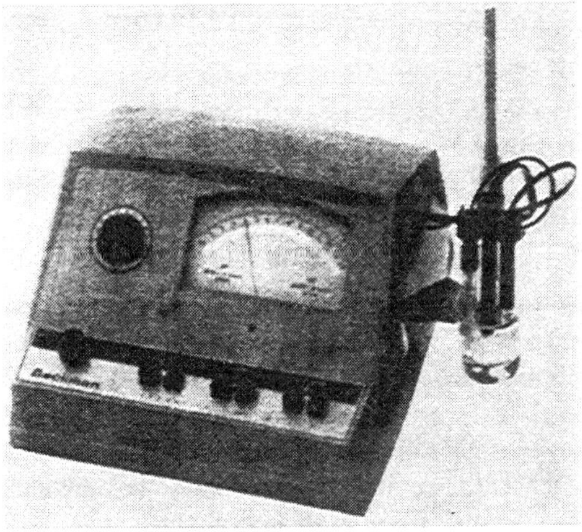

*Courtesy Beckman Instruments, Inc.*

FIG. 14.20. pH METER

transmission than will a cup of mild Brazilian coffee of the same solubles concentration at 5.00 pH. For the same reason, a cup of coffee brewed with alkaline water in the Midwest of the United States is not only flat-flavored, but has a pH well above 5.00—even with a mild, acidity coffee blend.

## CHEMICAL PROPERTIES OF COFFEE

The best sources of published information on coffee chemistry and composition are the bulletins of the Coffee Brewing Institute and reports of their research as listed in the bibliography. Articles since 1954 are available from coffee, food, and related industry journals, especially foreign journals from coffee-growing countries. In recent years, a great deal of analytical data about coffee has been reported from India. Germany has always been a highly interested and productive source of coffee investigation and publications. Brazil, with its large coffee surpluses, has done a great deal of work in attempting to use coffee beans for varied purposes. It was only after 1958 that truly significant quantitative analyses of coffee aroma volatiles were made. The processing of soluble coffee since World War II has revealed much about coffee; also, a great many new questions have arisen. Many chemists and chemical engineers were hired to operate, design, and develop soluble coffee process plants and processes. Since 1950, the chemistry of coffee has received more investigation and more constructive analytical work and processing experience than in all previous years of roasting and grinding coffee. Parallel progress has been made to determine the chemistry of other natural flavor substances.

## Variation and Chemical Complexity of Natural Products

One problem in evaluating reported chemical analyses (aside from the variance in composition of green coffees and degree of roast influences) is that chemical analyses for class compounds are obscured by the complexity of chemical combinations in nature. This is frequently why analyses of natural substances cannot be duplicated exactly.

**Carbohydrates.**—For example, carbohydrates vary from simple sugars to disaccharides, to trioses, etc. There are starches in the green coffee bean and dextrins (heat solubilized starch) in the roast coffee bean. Larger carbohydrate molecules are pentosans which yield mannose and galactose on hydrolysis. Pentosans are part of a larger group of polysaccharides called hemi-celluloses which are soluble in alkali. Then, there are the holo-celluloses—insoluble in alkali. Lignin, the binding agent for the celluloses, is an amorphous material of high molecular weight which is also soluble in caustic or bisulfite. The chemical formula for lignin is related to caffeic acid or coniferyl alcohol. Lignin is also related to vanillin as cinnamic acid is to benzaldehyde, and lignin can occur as a glucoside. The holo-cellulose is soluble in concentrated sulfuric acid and yields glucose. There are other types of celluloses designated as cellulose and alpha-cellulose. Cellulose is soluble in ammoniacal copper hydroxide. Absolute or definitive chemical analyses are difficult to make when there is such a complexity of compounds as well as variation in chemical compound size and molecular structure.

Furthermore, there are glucosides of protein; proteins are associated with oils or lipids. For example, a petroleum ether or hexane extraction of roast coffee removes fatty acids and oils, plus a significant percentage of phosphoproteins. Thus, there results some variation in the amount and composition of chemical class analyses. Results often depend on the procedure of analysis. Man has categorized each class of chemical compounds to find a useful work pattern. But natural phenomena and substances are not quite so sharply delineated. Chemical substances are interlocking and overlapping, thus confounding analyses and the analyst.

**Proteins.**—These are no less complicated than carbohydrates. Proteins are the source of much coffee flavor as derived in the pyrolysis volatiles from roasting. Proteins have a range of molecular size and properties resembling, in some ways, the carbohydrates. Some proteins are water soluble; they are the small molecules whose basic unit is the amino acid. Large molecules of protein are hydrolyzed with acids to make smaller water soluble molecules of amino acids. Some protein molecules are so large that they are insoluble in water and dilute acids.

Some of the identified proteins in coffee are methionine and cysteine. These contain sulfur which occurs in mercaptan groupings. Proteins are a source of nitrogen compounds in pyrolysis. Some of the pyrolysis products formed in roasting have nitrogen in the cyclic group structures, for example, proline and pyrrole. Tryptophane and indolen, which are not unlike pyridine in odor are also found. Roasting green coffee denatures most of the protein, render less water-soluble and amorphous. Some large protein molecules partly break down. The protein reactions are largely irreversible. These phenomena partly explain why water extract of green coffee gives higher solubles yields with room temperature water than with boiling water, e.g., 32 vs 27 percent solubles yield.

**Phosphates.**—The integrated relationship between the classes of chemical compounds in coffee can be seen from the fact that phosphorus, as phosphate, is associated both with lipids and glucosides. There are phospholipids, phosphoproteins, and phosphatides like fatty glycerides as lecithin.

**Mineral Constitutents.**—These play an important part in the structural growth of the plant and seed. They are part of the chemical structures of carbohydrates, lipids, and proteins.

Ash is not lost in roasting. Ash can be separated from the roast coffee in parts as water solubles and insolubles. It is well established that many coffee trees can be grown successfully in mineral solutions of phosphates, nitrates, ammonium or urea, sulfates, ferric, potassium, calcium, and magnesium ions. Trace elements are often required for plant health and good yield of cherry. Manganese, boron, zinc, copper and other trace elements are important.

## Analyses

**Green Coffee.**—Hence, the coffee bean is a complex chemical composition and as variable as nature and man's attendance to its cultivation. Although a critical examination of the published chemical analyses of coffee will not be made here, an approximate illustration of the chemical composition of green, roast, and soluble coffee is offered.

Table 14.4 gives the approximate chemical composition of green coffee. The table is arranged to show what is water soluble and what is not. Carbohydrates are listed in general categories from completely water soluble to practically water insoluble. Carbohydrates represent the largest portion of the green coffee bean—about 50 to 60 percent—whereas carbohydrates are only about 40 percent of the water solubles.

The coffee oil is insoluble in water as are most of the proteins. However, the portion of the proteins that dissolves is dependent on the fineness of the coffee particle during extraction, the temperature of the extracting water, and the exhaustiveness of the extraction. Although 26 percent solubles are shown from green coffee, over 30 percent can be attained by variations of the above procedures. In

TABLE 14.4. CHEMICAL COMPOSITION OF GREEN COFFEE
*Dry Basis—Approximate*

| Classes and Components | Water Solubility | Percentage of Green Coffee | | |
|---|---|---|---|---|
| | | Item | Total | Soluble |
| 1. Carbohydrates | | | 60 | |
| Reducing sugars | Soluble | 1.0 | | |
| Sucrose | Soluble | 7.0 | Pectins | Soluble |
| | | 2.0 | | |
| | | 10.0 | | 10 |
| Starch | Easily solubilized | 10.0 | | |
| Pentosans | Easily solubilized | 5.0 | | |
| | | 15.0 | | .. |
| Hemi-celluloses | Hydrolyzable | 15.0 | | .. |
| Holo-cellulose | Non-hydrolyzable fiber | 18.0 | | |
| Lignin | Non-hydrolyzable fiber | 2.0 | | |
| | | 20.0 | | .. |
| 2. Oils | Insoluble | | 13 | .. |
| 3. Protein (N × 6.25) | Depends on percent denatured | | 13 | 4 |
| 4. Ash as oxide | Depends on percent hydrolyzed | | 4 | 2 |
| 5. Non-volatile acids | | | | |
| Chlorogenic | Soluble | 7.0 | | |
| Oxalic | Soluble | 0.2 | | |
| Malic | Soluble | 0.3 | | |
| Citric | Soluble | 0.3 | | |
| Tartaric | Soluble | 0.4 | | |
| | | 8.2 | 8 | 8 |
| 6. Trigonelline | Soluble | | 1 | 1 |
| 7. Caffeine (Arabica 1.0%, Robusta 2.0%) | Soluble | | 1 | 1 |
| | | | 100 | 26 |

addition, the amounts of carbohydrates solubilized depend on the fineness of grind.

The fiber portion under carbohydrates is taken as 20 percent and consists of holo-cellulose and lignin; hemi-celluloses are considered hydrolyzable.

Discrete chemical substances like caffeine, trigonelline, and chlorogenic acid are the most accurately defined. Robustas have twice the caffeine content of Arabica mild, or Brazilian coffees. The oil fraction is fairly accurate but varies with coffees. Some of the colloidal proteins and phosphatides removed with the water soluble fraction and petroleum ether fraction. Usually there are more minerals in the water soluble portion than in the water insoluble portion, but more minerals are water soluble after roasting.

**Roast Coffee.**—A comparison of the analyses of the green and roast coffee compositions show what happens to the chemical classes (Table 14.5).

Some of the carbohydrates are destroyed in the roasting process. Almost all the sucrose disappears. Since the pyrolytic chemical reactions are complex, the water soluble portion of the carbohydrates is simply designated caramelized (browning type compounds). The water soluble and insoluble portions of caramelized carbohydrates are shown in a range related to fineness of grind and degree of hydrolysis. This shows a solubles yield of 27 percent from roast coffee of normal grinds and boiling water and 35 percent solubles yield from pulverized grinds.

**Composition of the Water Solubles of Coffee Powder.**—Table 14.6 lists the resulting chemical compositions. The composition of extracted and hydrolyzed spent coffee grounds are also listed. The composition will vary plus or minus about 15 percent with different blends, roasts, and extraction yields. The caffeine content of Robusta coffee is double that of the Arabicas. Chlorogenic acid is reported to vary in a few coffees. Soluble coffee composition can be evolved from the roast coffee composition. Over 90 percent of the soluble substances will

TABLE 14.5. CHEMICAL COMPOSITIONS SOLUBLE AND INSOLUBLE PORTIONS OF ROAST COFFEE (APPROXIMATE, DRY BASIS)

| | Percent | |
|---|---|---|
| | Solubles | Insolubles |
| 1. Carbohydrates (53%) | | |
| Reducing sugars | 1–2 | . . . |
| Caramelized sugars | 10–17 | 7–0 |
| Hemi cellulose (hydrolyzable) | 1 | 14 |
| Fiber (not hydrolyzable) | . . . | 22 |
| 2. Oils | . . . | 15 |
| 3. Proteins (*N* X 6.25); amino acids are soluble | 1–2 | 11 |
| 4. Ash (oxide) | 3 | 1 |
| 5. Acids, non-volatile | | |
| Chlorogenic | 4.5 | . . . |
| Caffeic | 0.5 | . . . |
| Quinic | 0.5 | . . . |
| Oxalic, Malic, Citric, Tartaric | 1.0 | . . . |
| Volatile acids | 0.35 | . . . |
| 6. Trigonelline | 1.0 | . . . |
| 7. Caffeine (Arabicas 1.0, Robustas 2.0%) | 1.2 | . . . |
| 8. Phenolics (estimated) | 2.0 | . . . |
| 9. Volatiles | | |
| Carbon dioxide | Trace | 2.0 |
| Essence of aroma and flavor | 0.04 | . . . |
| Total | 27 to 35 | 73 to 65 |

Note.—Volatiles may be classed chemically as acids, amines, sulfides, carbonyls (aldehydes and ketones), and others. Non-volatiles may be classed chemically as acids, carbohydrates, proteins, oils, phospholipids, minerals and others.

be removed from the R&G coffee. For example, if caffeine is 1.0 percent in green coffee and there is a 16 percent roast loss, then 1.2 percent caffeine will result in the roast coffee. Hardly any caffeine is lost in the roasting. Caffeine crystals sublimed into roaster stacks are accumulations from minute continual roaster losses. With a 38 percent solubles yield from R&G (dry) coffee, the caffeine content of the instant coffee (dry) is 1.2/0.38 or 3.2 percent. Similar calculations can be made for the nonvolatile acids, trigonelline and ash. Over 90 percent of the ash constitutents are water soluble; the spent grounds have only a few tenths of one percent ash, depending on efficiency of extraction. Some caffeine and other water solubles are lost with the spent grounds. This is an index of the inefficiency of extracting the truly water soluble portion of grounds.

**Coffee Ash.**—In green coffee, ash is about 60 percent $K_2O$, 15 percent $P_2O_5$, 15 percent CaO and MgO (in a ratio of about 1 to 2), and ½ percent $Na_2O$. The anions are about 5 percent $SO_3$, 1 percent $SiO_2$, and 1 percent chlorides. The ash contains about 1 percent iron and traces of other metals. Thus, sulfates and chlorides appear in soluble coffee at several tenths percent. Table 14.7 shows the estimated ash distribution on green and roast coffee, solubles, and spent grounds.

The function of minerals and their precise composition in the bean have been grossly overlooked since they have much influence on the course of the thermal pyrolysis during roasting. For example, research has shown that the presence of mineral salts in wood reduces pyrolysis temperature and markedly influences the composition and amount of volatile organic substances formed. Also, a marked difference in coffee flavors occurs with different fertilizer formulas.

TABLE 14.6. CHEMICAL COMPOSITION OF COFFEE SOLUBLES AND SPENT GROUNDS (INSOLUBLES) *(Approximate, Dry Basis)*

| Chemical Compound or Class | Percent Solubles | Percent Spent Grounds |
|---|---|---|
| 1. Carbohydrates (3 to 5 percent reducing sugars) | 35.0 | 65 |
| (browning complexes) | 15.0 | . . . |
| 2. Oils (and fatty acids) | 0.2 | 18 |
| 3. Proteins (amino acids and complexes) | 4.0 | 15 |
| 4. Ash (oxide) | 14.0 | Fraction of 1% |
| 5. Acids non-volatile | | |
| Chlorogenic | 13.0 | . . . |
| Caffeic | 1.4 | . . . |
| Quinic | 1.4 | . . . |
| Others | 3.0 | . . . |
| 6. Trigonelline | 3.5 | Few tenths per cent |
| 7. Caffeine | | |
| (Arabicas) | 3.5 | Few tenths per cent |
| (Robustas) | (7.0) | . . . |
| 8. Phenols (estimated) | 5.0 | Few tenths per cent |
| 9. Volatiles | | |
| Before drying—acids and essence | (1.1) | Nil |
| After drying | Nil | Nil |
| Total | 100 | 98+ |

TABLE 14.7. ESTIMATED COFFEE ASH DISTRIBUTION

| | Green Coffee | Roast Coffee | Soluble Powder | Dry Spent Grounds |
|---|---|---|---|---|
| Dry weight relations[1] | 1.176 | 1.000 | 0.380 | 0.620 |
| Percent ash content, dry basis | 4.00 | 4.71 | 10.00 | 1.47 |
| Weight ash per unit weight roast coffee, dry basis | 0.0471 | 0.0471 | 0.0380 | 0.0091 |

Percentage distribution of ash components

| Mineral Oxide | Percent of Total | | | | |
|---|---|---|---|---|---|
| | Green, Roast Ash | Solubles Ash | | Grounds Ash | |
| | % | % Totl. | % Sol. | % Tot. | % Grds. |
| $K_2O$ | 62.5 | 52.0 | 75.59 | 10.5 | 33.65 |
| $P_2O_5$ | 13.0 | 3.0 | 4.36 | 10.0 | 32.05 |
| CaO | 5.0 | 2.0 | 2.90 | 3.0 | 9.62 |
| MgO | 11.0 | 8.0 | 11.63 | 3.0 | 9.62 |
| $Fe_2O_3$ | 1.0 | 0.4 | 0.58 | 0.6 | 1.92 |
| $Na_2O$ | 0.5 | 0.4 | 0.58 | 0.1 | 0.32 |
| $SiO_2$ | 1.0 | ... | ... | 1.0 | 3.21 |
| $SO_3$ | 5.0 | 2.0 | 2.90 | 3.0 | 9.61 |
| Cl | 1.0 | 1.0 | 1.46 | ... | ... |
| | 100.0 | 68.8 | 100.0 | 31.2 | 100.0 |

[1]Assuming 15 percent weight loss on roasting.

**Coffee Volatiles Composition.**—The chemical nature of coffee aroma and flavor is a complex subject. There are numerous volatiles: some very volatile, some subject to rapid oxidation, and others subject to polymerization. Other volatiles are subject to resinification and precipitation. Recent work using gas chromatography and infrared absorption and/or mass spectroscopy have provided more accurate data about these coffee aroma volatiles.

The data are in reasonably good agreement. They show that about 50 percent of the volatiles are aldehydes, about 20 percent are ketones, about 8 percent are esters, about 7 percent are heterocyclic, about 2 percent are dimethyl sulfide, and lesser amounts are other organic and odorous sulfides. There are also small and fractional percentages of nitriles, alcohols, and low molecular weight saturated hydrocarbons and unsaturated ones, like isoprene. There are also furans, furfurals, acetic acid, and homologues. These are as volatile as water but not so volatile as to contribute much coffee aroma at 50 ft.

No one has yet marketed a synthetic coffee aroma. None of the commercial flavor houses in the world have ever prepared an acceptable sample to simulate coffee aroma. One reason is that flavor houses, in general, have been awed by coffee aroma as being too complex and chemically unstable to support investments to investigate it thoroughly. Further, flavor houses have, in general, very little practical working knowledge of coffee. Coffee is not one of their normal commercial products. The addition of individually identified chemical components to coffee extract or brews definitely enhances the cup flavor. Success in this field will be achieved in time through reasonable effort in systematic investigation of the aroma and flavor contributions of each chemical component. Data about combinations of chemical components in influencing aroma and flavor are needed. Eventually, acceptable coffee flavors, perhaps even better than natural, can be formulated. Such a study would probably also explain the cause of staling.

TABLE 14.8. ANALYSIS OF A COFFEE AROMA ESSENCE

| | Mol Wt | Per-cent | Bp | | Relative Flavor Importance[1] |
|---|---|---|---|---|---|
| | | | C | F | |
| Acetaldehyde | 44 | 19.9 | 21 | 70 | 1 |
| Acetone | 58 | 18.7 | 56 | 133 | 2 |
| Diacetyl | 86 | 7.5 | 88 | 190 | 1 |
| *n*-Valeraldehyde | 86 | 7.3 | 102 | 216 | 2 |
| 2-Methylbutyraldehyde | 86 | 6.8 | 91 | 196 | 2 |
| 3-Methylbutyraldehyde | 86 | 5.0 | 91 | 196 | 2 |
| Methylfuran | 82 | 4.7 | 63 | 145 | 2 |
| Propionaldhyde | 58 | 4.5 | 49 | 120 | 2 |
| Methylformate | 60 | 4.0 | 32 | 90 | 2 |
| Carbon dioxide | 44 | 3.8 | −78 | −108 | . . . |
| Furan | 68 | 3.2 | 32 | 90 | 1 |
| Isobutyraldehyde | 72 | 3.0 | 63 | 145 | 1 |
| Pentadiene (isoprene) | 68 | 3.0 | 30 | 86 | 2 |
| Methylethyl ketone | 72 | 2.3 | 80 | 176 | 2 |
| $C_4$–$C_7$ paraffins and olefins | — | 2.0 | 35 | 95 | 2 |
| Methyl acetate | 74 | 1.7 | 57 | 135 | 2 |
| Dimethyl sulfide | 62 | 1.0 | 38 | 100 | 1 |
| *n*-Butyraldehyde | 72 | 0.7 | 75 | 167 | 1 |
| Ethyl formate | 74 | 0.3 | 54 | 129 | 2 |
| Carbon disulfide | 76 | 0.2 | 46 | 115 | 2 |
| Methyl alcohol | 32 | 0.2 | 65 | 149 | 3 |
| Methyl mercaptan | 48 | 0.1 | 6 | 43 | 1 |
| | | 100.0 | | | |

[1] 1, large; 2, medium; and 3, small.

Some volatiles are more important than others in formulating the coffee aroma. On the other hand, some identified volatile components of coffee aroma are not odorously important at all. Known facts about coffee aroma come close to explaining staling. Given the chemical nature of coffee solubles and the coffee aroma portion, certain chemical reactions probably occur in staling. First, R&G coffee loses the low boiling volatiles of coffee aroma. The important dimethyl sulfide, acetaldehyde, propionaldehyde, and aromatic esters volatilize in a few minutes to a few hours. The boiling temperature of each volatile component identified is listed in Table 14.8. The R&G coffee absorbs water from the surrounding air.

**Staling.**—Moisture absorption (or water addition from quench) needs only to be 1 percent of R&G coffee on a tray to effect noticeable staling in less than an hour. After a vacuum coffee can is open for a day, the same stale flavor will be apparent. Staleness is a coffee taste term not yet defined by chemical composition change. The loss and alteration in volatiles composition is accounted for by simply holding the R&G coffee open to air 66 hr. Staling can be demonstrated by exposing ground coffee to air or to several percent moisture (by adding water to the coffee) for minutes to an hour before preparing a vacuum brew type extraction. Compared to freshly ground roast bean coffee, this coffee's taste difference is obvious.

At one time staling was associated with coffee oil rancidity, but it is now known that rancidity has little to do with staling. Staling is the loss and alteration in composition of the volatile coffee aroma constituents. Under packaging, ample evidence has shown that the 29-in. Hg vacuum in the can is essential to preserve the coffee aroma. A 27-in. Hg vacuum allows sufficient coffee aromatics to be oxidized and the stale taste to be detectable in the resulting brewed cup. Thus, the difference of a few milliliters of oxygen is enough to oxidize the aromatic aroma and flavor fraction of the 1 lb of roast and ground coffee in the can.

Aldehydes are known to be readily oxidizable, especially in sunlight (a can allows no light to enter). Dimethyl sulfide (DMS) is not only volatile, but also readily oxidizes to methyl sulfoxide; this loss of a potently aromatic component will downgrade most coffee aroma and flavor. DMS is an exhilarating part of the coffee aroma; the sulfoxide is not very volatile nor odorous. The latter has little taste character at ppm levels. The sulfoxide is also susceptible to further oxidation and chemical reactions.

Staling of roast bean flavor is retarded by the cellular protection given the chemical constituents by the AB- or adsorbed $CO_2$ atmosphere. However, as moisture penetrates the bean, $CO_2$ and coffee aroma volatiles are released and altered. Beans are moderately stale in a week, more stale in two weeks, and quite stale in three weeks. Darker roasts stale differently than lighter roasts. Coffee staling rates can be retarded noticeably by storing roast coffee beans or grounds in air-tight containers at cooler temperatures, $-10$ F ($-23$ C). Drip or vacuum preparation of coffees, which brings out coffee flavor better, reveals staling sooner than percolation, which distills off volatiles and leaves a caramelized brew. Extracts which have coffee volatiles stale rapidly. For example, extract held at 80 F (27 C) for 1 or 2 hr is noticeably inferior in flavor to extract held at 40 F (4 C). The greater the exposure of extracts to air and the lower their solubles concentrations, the faster the staling and coffee flavor deterioration. Frozen extracts at $-10$ F ($-23$ C) will not deteriorate for many months, or perhaps years.

A source of deterioration in coffee extracts and distillates is the chemical reactions that occur. Aldehydes in extract do not evaporate so readily as from the R&G coffee, but the aldehydes can be precipitated by phenols which are present. Aldehydes will condense or polymerize by themselves in an acidic medium.

Moisture added to roast coffee partially releases some aromatics through hydrolysis reactions.

The vacuum pack of R&G coffee accomplishes adequate protection of the coffee aroma and flavor. The subject of staling coffee has not been of much interest in recent years.

Antioxidants are used largely to protect fatty substances from oxidation. There is no indication that antioxidants retard the staling of coffee.

Ascorbic acid is used in military specifications for instant coffee, but this is not so much to protect the instant coffee from oxidation as to provide vitamin C to the military personnel.

Instant coffee prior to aromatization by means of coffee oil (and other methods) underwent little flavor change from the day of preparation. After the initial coffee volatiles were lost and some oxidation occurred, the instant coffee powders were relatively the same in taste for many months, possibly up to a year or two. Any flavor changes in instant coffee packed for six months to several years would be nominal if the coffee was properly packed in sealed, air-tight cans and stored at less than 70 F (21 C). But with coffee oil add-back and inert gas pack protection, staling of instant coffee is not uncommon before the product reaches the consumer. Staling occurs in a day or two after the jar is opened by the consumer, but it is not as objectionable as stale R&G coffee. Instant coffee, when not aromatized, is very stable in air storage-when instant coffees are aromatized (carrying noticeable coffee flavor and aroma), staling is easily noticed.

## Classification of Coffee Compounds

In discussing the chemistry of coffee, there is always the problem of how to organize the material for study and understanding. Substances are found in the green coffee and are converted in the process of roasting. Thereafter, some of these are water soluble and some are not. Some are gaseous and very volatile under normal conditions. Some compounds are chemically stable; others are not. Combinations of some coffee substances are reactive.

Classification has already been made into generally recognized groups of organic compounds found in foods. These are carbohydrates, proteins, oil, and ash. Acids are an important part of coffee and are classified as volatile and non-volatile. These were previously discussed. Phenols, as derived from chlorogenic acid, are classified because they are quite different in chemistry from other classes of compounds. Phenolic compounds are important chemically and taste-wise. Phenols also represent as much as 5 percent of the soluble coffee composition.

Trigonelline and caffeine each occur at about 1 percent in the green Arabica coffee beans. They are definite chemical compounds. Nicotinic acid, pyridine, pyrolle, methyl pyrolle, etc., are undoubtedly derivatives from trigonelline, caffeine, or their precursors. Furan, methyl furan, furfural, thiophene, and related compounds, as well as pyrolles, are pyrolysis products of the bean, e.g., furfural from pentosan hydrolysis.

The chemistry of these nitrogenous compounds must be differentiated from the protein nitrogen compounds and amino acids.

Similarly the sulfur-bearing compounds must be differentiated from their originating proteins.

Nitrogen compounds may be stable like caffeine, unstable like proteins, and volatile like amines. The sulfur compounds can be in solid (water soluble or insoluble) protein structures or in volatile sulfides and mercaptans.

The coffee oil fraction, least influenced by the other chemical changes in the coffee bean (because oil undergoes little change in roasting and extraction) can be examined almost as a physical and chemical composition that can be largely withdrawn from roast coffee or spent coffee grounds when desired. This subject will be developed on pp. 683 to 690.

**Chaff-Composition**.—Chaff is the bean cavity parchment (similar to the outer hull) liberated during roasting. Its chemical composition resembles the green coffee bean but it has a bitter and poor flavor. It contains about the same amount of caffeine as the bean.

**Cherry Composition**.—Research has shown that 40 percent of the ripe cherry is pulp, 20 percent is mucilage, and 40 percent is bean and parchment. The pulp contains 60 percent water, 28 percent organic matter, of which 1.6 percent is nitrogen (10 percent protein equivalent), and 1.3 percent ash. It is high in phosphate and potassium. The dry pulp contains 2 percent oil, about 33 percent fiber, 60 percent nitrogen-free extract, 10 percent protein, and 10 percent sugars. Work has been done in the coffee-growing countries to find use for the pulp, mucilage, and hulls removed from the bean. In general, other than returning the mucilage and pulp to the land for mulching and burning the hulls, no significant industry has developed from these rejected substances. Pulp can be used to prepare vinegar or for cattle feed, but neither application is used widely.

**Coffee Mucilage**.—This is about 85 percent water. The 15 percent solids are about 9 percent protein, 4 percent sugars, 1 percent pectic acid, and 0.6 percent ash. Mucilage is high in calcium and sulfur, with traces of manganese. The mucilage has a pH of 4.8 when the bean is ripe, 5.0 pH when it is green. The sugar content of mucilage solids is about 13 percent in green fruit, rising to 24 percent in ripe fruit. Corresponding solids are 14 and 18 percent. The gel structure of the mucilage is broken down by pectinases. Some sugar fermentation and protein decomposition also occur, which lead to the potent odors around green coffee bean fermentation areas.

Chlorophyll has a pyrrole chemical nature and magnesium within the structure. New crop coffee beans are often vividly green. The chlorophyll that the beans contain alters during the aging of coffees to yield derivatives of chlorophyll.

**Chemical Analyses of Coffee Constituents.**—These analyses are detailed in the Assoc. Official Agric. Chemists manual methods, except where improved methods are reported in other specific publications. This book does not cover such analyses. In each chapter many references to papers describing such analytical techniques are given. Many analytical methods recently applied, such as gas, paper, and column chromatography, supersede earlier analytical procedures. Furthermore, the use of the mass spectrometer, infrared spectrometer, and the emission spectrometer have made possible accurate analyses not heretofore possible. The use of gas and solution adsorption columns, techniques not known or seldom applied before World War II, has made possible isolation of individual coffee constituents.

## Coffee Oil

Green or roast coffee bean oil has a chemical composition similar to that of many edible vegetable oils, as shown in Table 14.9. It is liquid at room temperature and at 45 F (7 C), but fatty acid crystals will settle out slowly during storage. Roast coffee has more fatty acids than green coffee, and percolator hydrolyzed grounds have even more free fatty acids. Roast coffee oil may contain 5 percent free fatty acids.

**Literature on Coffee Oil.**—This is mostly European, especially in regard to chemical analyses. Much work in connection with using the coffee oil portion to recover some value from surplus coffees has been done in Brazil also. Such work was carried on in Brazil from 1937 to 1940 and in 1960.

Coffee oil has a relatively large portion of unsaponifiables, which are listed at about 7 to 12 percent. Unsaponifiable contents are a function of the type of solvent used. The unsaponifiable portion contains sterols and sterol derivatives useful in pharmaceutical preparation.

**Oil Market.**—Coffee grounds, even after soluble coffee extraction, do not have as much oil content as other oil-bearing seeds. For example, the following seeds have 50 percent or more oil: babassu, castor bean, copra, peanuts (shelled), sesame seed, and tung nuts. Those seeds having more than 20 percent oil are: corn (germ), flax, hemp, perilla, and rape. Cottonseed has 15 percent oil; kapok seed has 18 percent oil; and soybeans have 16 percent oil. Green coffee beans have 11 percent, and roasted coffee beans have 13 percent oil. Percolator extracted and hydrolyzed coffee has about 15 to 18 percent oil.

TABLE 14.9. CHEMICAL COMPOSITION OF FATS AND OILS (APPROXIMATE)

| Number Carton Atoms | Tri-glyceride | Number of Double Bonds | MP F | MP C | Cof-fee | But-ter | Cot-ton seed | Soy-bean | Corn | Coco-nut | Tal-low | Olive | Lin-seed | Tund |
|---|---|---|---|---|---|---|---|---|---|---|---|---|---|---|
| 8 | Caprylic | . . | 61 | 16[1] | . . . | . . . | . . | . . | . . | 8 | . . . | . . . | . . . | . . |
| 10 | Capric | . . | 88 | 31[1] | . . . | . . . | . . | . . | . . | 7 | . . . | . . . | . . . | . . |
| 12 | Lauric | . . | 109 | 43[1] | . . . | . . . | . . | . . | . . | 47 | . . . | . . . | . . . | . . |
| 14 | Myristic | . . | 129 | 54[1] | 3 | . . . | 1 | . . | . . | 18 | . . . | . . . | . . . | . . |
| 16 | Palmitic | . . | 145 | 63 | 28 | 28 | 21 | 7 | 7 | 9 | 35 | 15 | 4 | 2 |
| 18 | Stearic | . . | 156 | 69 | 10 | 25 | 22 | 4 | 3 | 2 | 40 | 10 | 4 | 3 |
| 18 | Oleic | 1 | 57 | 14 | 21 | 39 | 29 | 32 | 43 | 6 | 25 | 70 | 18 | 13 |
| 18 | Linoleic | 2 | 32 | 0[2] | 28 | . . . | 23 | 49 | 39 | 3 | . . . | 5 | 30 | . . |
| 18 | Linolenic | . . | 32 | 0[2] | . . . | . . . | . . | 2 | . . | . . . | . . . | . . . | . . . | (73)[3] |
| 20 | Arachidic | . . | . . . | . . | 3 | . . . | Tr. | 1 | 1 | . . . | . . . | . . . | . . | . . |
| | Other glycerides | . . | . . . | . . | . . . | 8 | . . | . . | . . | . . . | . . . | . . . | 44 | 9 |
| | Non-saponifiables | . . | . . . | . . | 7 | . . . | . . | . . | . . | . . . | . . . | . . . | . . . | . . |
| | Total | | | | 100 | 100 | 96 | 95 | 93 | 100 | 100 | 100 | 100 | 99 |

[1]Melting points of the free acids which are usually within 1° C of those of the corresponding triglycerides.
[2]Below freezing.
[3]Eleo-stearic acid, isomer of linolenic acid.

Sale of vegetable oils is very competitive. Good prices for vegetable oil for either edible purposes or fatty acids cannot be obtained. Vegetable oil for fatty acids are higher priced. The chemical composition of the coffee oil largely determines its commercial use as well as the demand for a particular oil composition. There is no demand for new edible oils or soap stock. Use of sterols has not encouraged any pharmaceutical firm to draw on the coffee oil supply.

**Solvent Extraction Plant.**—Coffee oil has not been available because every household discards its coffee grounds and, since 1955, most soluble coffee processors have been burning their spent grounds. For oil recovery by expelling or extraction, a sizable equipment investment is needed. The value of the product does not warrant building an oil recovery system unless there are at least 100 tons per day of dry feed coffee grounds, and preferably 200 tons per day. A factor in the cost of the plant is the amount of iron contacting the oil in process. Fatty acids attack bare iron, steel, brass, or copper. These metals have a degrading effect on oil flavor and color. Use of stainless steel is warranted in some of the oil handling equipment to avoid contamination. This brings a better price for the oil. Oils are frequently rated on properties such as clarity, odor, color, and fuming at elevated frying temperatures. One would have to investigate each property of coffee oil before investing in a coffee oil recovery plant.

Coffee oil must have uniform properties if it is to be recovered for sale. With the limited data available, it appears that differences in coffee oil composition for Arabic coffees are small. More data on coffee bean oil properties and composition are needed. Robustas are reported to have higher than average coffee oil content. More data, coffee varieties would have to be obtained to determine whether blend variations in a soluble coffee plant would cause significant enough variations in the recovered coffee oil to influence its end use. Depending on the use of oil, such variations in oil properties may be important. There are several soluble coffee locations (such as metropolitan New York City and Houston, Texas) that have sufficient spent coffee grounds to justify operating a 100 ton per day oil extraction plant for spent coffee grounds.

Another consideration that discourages the recovery of coffee oil is the fact that after percolation (hydrolysis) as much as 15 percent of the saponifiable portion is in the fatty acid form. This means that if an edible oil were to be processed by removing the fatty acids (as, for instance, an ammonium salt by centrifuging), it would result in entraining an almost equal weight of neutral coffee oil. The fatty acid portion would be sold for soap stock at a very low price. Thus, the total loss of oil would be about 30 percent: 15 percent as soap stock, 10 percent as unsaponifiables, and several percent as "winterizing" acid and wax crystals. The light color of the oil is an important factor in selling. Coffee oil, due to the high reddish coloration from the grounds, is not as light in color as other vegetable oils. This may mean a lower selling price. Pre-rinsing the coffee grounds with water results in a lighter colored oil. Subsequent bleaching out with Fuller's earth and/or charcoal or removal of color bodies from the oil may be required. This means additional operating costs and oil yield losses. Glycerine has been manufactured synthetically for years. The recovery of glycerine from oil splitting does not offer financial gain. Until there is some commercial recovery of coffee oil, it is unlikely that much of its chemistry and value to users can be realized.

**Commercial Oil Recovery Equipment.**—Expelling coffee oil is probably the cheapest way to remove the oil from the spent grounds. Expelling carries out with the oil a great many substances that are not wanted, e.g., color and "foots" that must be centrifuged away. Expeller oil yields leave about 5 percent coffee oil in the residual cake. Solvent extraction leaves only a few tenths of 1 percent coffee oil in the granular, stripped grounds. Coffee oil has been used in specialty soaps, characterizing them with good lather and detergent properties. It could possibly be used also for cosmetics. Coffee oil may sell best as a fatty acid or for edible use. Once coffee oil is available, other users will draw on its unique properties.

**Pilot Recovery of Coffee Oil.**—Preliminary development work can be carried out in bench top Soxhlet solvent extractors and pilot column extractors and pilot solvent strippers. When low boiling petroleum solvents are used, fire and explosion hazards are always present. The solvent can be stripped thoroughly from the coffee oil by heating the oil to at least 212 F (100 C) in a thin film at 29 in. Hg vacuum. Final solvent traces can be removed from the oil by placing the oil on a tray in a laboratory oven at 212 F (100 C) overnight.

Solvent extracted coffee oil is filtered to remove insolubles. Coffee oils having 35 to 75 percent light transmission in a 1 cm tube at 510 mm wavelength may result and may appear reddish to amber. Fine coffee grounds contribute more to coloring solvent extracted coffee oils. Light amber oils with 80 percent light transmission can be achieved by the grounds with water first, screening off finer particles, and by taking the full 17 percent oil yield. Lesser oil yields, such as 8 percent, carry a full red color. Caustic refining causes the greatest color lightening of coffee oil but causes a 30 percent soap stock loss. Refining losses, according to the Assoc. Official Agric. Chemists Manual, are 2.5 to 3 times the free fatty acid concentration. Hence, 14 percent fatty acids would mean at least 35 percent oil loss. Solvent extractions of coffee grounds with above room temperature hexane at, for example, 140 F (60 C) will cause cream colored crystals to deposit out from the oil overnight at 70 F (21 C); oil extractions at 80 F (27 C) yield many fewer crystals.

**Solvent Removal from Grounds.**—In commercial operations with rice bran and soybean, the solvent saturated fiber is heated in a screw jacket as the fiber tumbles forward until all solvent is removed. Moisture levels in entering oil laden coffee grounds should be less than 10 percent, or the solvent will have difficulty penetrating the coffee cell structure. The bulk density of dry spent coffee grounds is 30 to 40 lb/cu ft. Coffee oil content of spent coffee grounds varies from about 15 to 18 percent, depending on blend, roast, and solubles yields.

When handling combustible solvents, one should use inert gas for purging. Inert gases are used also to cover oils during storage to avoid oxidation.

Since granular fiber has a specific gravity of about 1.6, a unit volume of extractor will hold 0.6 weight unit of dry spent grounds and 0.6 volume unit of solvent, of which about one-third wets the coffee grounds.

Although oil extraction systems used are the Kennedy paddle type and Blaw-Knox rotary cell, the percolator column (as used for water extraction of coffee grounds) is well suited for solvent extraction of oil from the dry spent coffee grounds. It offers: (1) vacuum loading of granular coffee; (2) continuous operation of a battery of columns—similar to percolation; (3) a minimum solvent to coffee grounds (or oil) ratio; (4) a closed solvent system with no moving parts which is safer; moderate heating is possible; (5) steam purge, which can be used to remove solvent from coffee grounds after oil extraction; and (6) solvent stripped coffee grounds, which can be blown out of the percolator column at 10 to 15 percent moisture—ready for disposal or burning. The solvent percolation process for coffee oil is identical to water percolation.

Spent coffee grounds, as discharged from water percolation, still have some hydrolyzed sugars which will ferment. Hence, fly and insect infestation and foul odors may result from storing such wet grounds. Also, as the spent coffee grounds drain and dry partially, they are subject to spontaneous smouldering and possible combustion. For these reasons, special consideration and storage facilities must be provided if appreciable amounts of wet spent coffee grounds are to be stored.

Hexane extraction of freshly roasted and ground coffee yields a different oil than that from hexane extraction of percolated spent coffee grounds. The former oils are rich in coffee aroma, while the latter have virtually no desirable aroma. Freshly steamed R&G coffee after hexane extraction has no noticeable solvent residue. The condensed distillate of water and hexane, however, has a very strong coffee aroma. This strong aroma can be concentrated in a fraction of, for example, one-tenth or one-twentieth the hexane volume by adding coffee oil, followed by hexane evaporation. The coffee aromatics are held more strongly in the coffee oil than in the hydrocarbon hexane. Hexane can be stripped from coffee oil at 122 F (50 C) in a rotary film vacuum evaporator. Similar extractions for oil and coffee aromatics concentration from steam distillates can be made with solvents such as Freon-11, which is nonflammable. Indications are that the R&G coffee steam distillates with solvent are richer in coffee aromatics than the solely solvent extracted coffee oil.

**Commercial Solvent Extraction Process.**—The oil in solvent solution, which may contain 10 to 20 percent oil, can, after filtration, have the bulk of the hexane solvent removed under vacuum (or atmospheric pressure using 20 psig steam) so that the coffee oil still retains about 10 percent solvent. Various types of evaporators such as a film type or a flash type can be used. The coffee oil with 10 percent solvent is now sent to a bubble cap distillation column. Oil enters at 230 F (110 C). Tower trays are heated with 20 psig steam. The tower is at about 20-in. Hg vacuum, and steam is purged into the bottom of the column so that steam distillation of hexane occurs. All noticeable hexane solvent is driven from the oil.

The solvent stripped oil is stored in large tanks over a period of 3 to 6 days at about 40 F (4 C). This crystallizes out fatty acids and insolubles. This is called "winterizing"; insolubles are centrifuged or filtered out. The oil is then treated with ammonium hydroxide to neutralize the fatty acids which become the soap stock (creamy yellow precipitate), entraining about their own weight of neutral oil for centrifuge separation. The resulting clear oil is much lighter in color, and can be used for preparing margarine or other edible products. To use the oil at this point for deep frying or salad oil, one must deodorize it. This process is substantially a steaming of the hot oil under higher vacuum to remove volatiles and cooling before removing the oil from the deodorizer. Oxygen must be kept away from the hot oil to prevent off-flavor called "reversion." Bleaching of the oil with fuller's earth or a similar medium is then required to attain lighter colors.

General handling of coffee oil indicates that it is relatively stable chemically. Rancidity (oxidation of the unsaturated linkages) does not readily occur. However, oils that have been refined as described have also had most of their natural antioxidants removed. Such refined oils may be unstable and readily oxidized. For this reason, refined oils are stored under inert gas (nitrogen) and antioxidants may be added back to them.

As an index of the importance of materials of construction in such an oil processing plant, a comparison of oil quality achieved in glassware and stainless steel vs carbon steel can be made. Invariably the glass and stainless steel system makes a better quality and lighter colored oil. Iron contamination colors the oil a slight green.

A projected use of coffee oil is to convert it to fatty acids by fat splitting. This is quickly accomplished by high pressure steam in a fractionating column. This step would separate out unsaponifiables. The fatty acids can then be separated by crystallization of stearic acid from methanol and/or vacuum distillation. Fatty acid sale value is almost twice that for edible oils. Fatty acids can also be separated by high vacuum distillation. Molecular stills are used to recover Vitamin D from cod liver oil. Similar separations have been considered for coffee oil, but it is not rich enough in vitamins to make such a process commercially attractive.

**Oil Analyses**.—*Acid Number* is the percentage of free fatty acids. Oils in green coffees are said to have 1 to 2 percent; oils in roast coffees have 3 to 5 percent (depending on the degree of roast); and percolated coffees have up to 15 percent.

*Saponification Number* is the number of milligrams of potassium hydroxide required to saponify 1 gm of fat or oil. For coffee oil, it varies from 170 to 199.

*Iodine Number* is the number of grams of iodine which combine with 100 gm of oil or fat. For coffee oil 95 to 100 is the range reported.

*The Reichert-Meissl Number* is used mostly for butter. It is an index of volatile acids, which is usually very low for coffee oil. There are other oil test numbers reported, such as for esters, hydroxyls, Hehner, Polenske, and thiocyanogen.

The subject of oils, their properties and analyses, is a specialized field. The published literature on the chemistry of coffee oil remains sketchy. Coffee oil has powerful antioxidant properties, which may explain why the normally unstable coffee aromatics are quite stable in the oil.

In the references to studies on unsaponifiables are some references on coffee waxes. But, except for the confirmed observations that the surface of the bean has a high wax content, information is fragmentary. Although waxes have physical properties and solubilities somewhat similar to fats, they have different chemical compositions and undergo different chemical reactions. Waxes are usually a mixture of higher alcohols, with some esters. Waxes are not made soluble by boiling caustic solution and appear in the unsaponifiable fraction of oils.

Probably the largest commercial production of coffee oil was in Brazil in 1937. Oil was extracted from 2,000,000 lb of green coffee, which would be about equal to oil from 150,000 132-lb bags of green coffee.

There are less phospholipids extracted with the oil from roasted coffee than with oil from green coffee. Coffee oils are not oxidized during roasting.

Some have expressed the opinion that coffee oil is not suitable for consumption because the glycerides hydrolyze during deodorization; this is a matter that would need to be determined experimentally.

Hydrogenated coffee oil has a very oily nature, due to the unsaponifiables, which hinders the making of good flakes. With the current trend toward using unsaturated oils, hydrogenated properties are less important. Coffee oil is about 53 percent unsaturated and 39 percent saturated.

The physical properties of coffee oil are similar to those of other vegetable oils. Coffee oil has a specific gravity of 0.9440 to 0.9450, a refractive index of 1.468 to 1.469, and a viscosity variation with temperature slightly higher than soybean or linseed oil.

# 15

# Physiological Effects of Coffee and Caffeine[1]

*Michael Sivetz*

The subject matter in this chapter is treated in three distinct parts:

1. SENSES: Eyes, nose and throat, as applied to coffee odors and tastes, are subjects often least considered, and yet are the most fascinating and most important aspects to be considered. The body's senses assess the quality of green bean, roast bean, and the processed coffees. The senses frequently guide processing controls and help to establish physical quality controls. A key job for the green coffee bean buyer is the recognition of good and/or foul odors and tastes. These "feelings" have now been captured to some extent by analytical instruments. Now a chemist trained in the coffee field can identify discrete chemical coffee odors and tastes as they occur or as they are separated from coffee samples. The body sensors of smell and taste can and do give responses to control and guide actions. This is discussed in terms of human sensitivities to detecting flavors, and the identified chemical constituents by language (terminology).

2. STIMULATION: The other physiological aspect is how caffeine and roast chemical products affect the body functions over which the individual has no control, but simply reacts to. These aspects can be considered medical, pharmacological, or biological. In individual cases, people can be benefited or can be harmed by regular intakes of coffee.

Part I provides a background about human differences and tolerances. Why do people drink coffee at all (especially when it can taste bad)? How much caffeine is consumed with each cup? What are some of the toxic and excessive coffee use aspects?

Part II deals more specifically with the effect of caffeine and roasted coffee chemicals on specific organs of the body. What are the caffeine absorption rates and excretion rates? What are the short term influences on the mind, muscles, and blood circulation?

[1]Source: Sivetz, M. 1977. Coffee Origin and Uses. Coffee Publications, Corvallis, Oregon.

3. CAFFEINE: this stimulating chemical in coffee will be discussed in terms of its origin, chemical recovery, and processing. The history of its discovery, isolation, and production are discussed as well as caffeine's chemical properties. Some of the major selling commercial decaffeinated coffees will be illustrated.

The market for decaffeinated coffees for the aged has been growing both in a relative and in an absolute way; it represents about 25 percent of instant coffee sales in the U.S.A. and 15 percent of all coffee sales in Europe and the U.S.A. From a marketing point of view, the reactions people have to aromas and tastes, whether mental or physiologically influenced, in fact will govern what people buy, use and do.

There is no question that some people, in fact a great many, are influenced by coffee aroma and taste and physiological body effects, at some time in their lives. What many people do not realize is that a good deal of these irritating and unpleasant effects are actually due to maltreatment of coffees.

In the case of commercial coffees, it is not uncommon to ''not-like'' the taste. Commercial coffee has a roasting and grinding odor that is attractive, but the taste is usually not so. What has happened? There can be low grade and Robusta coffee beans, or unsanitary beans (1975 imports to the U.S.A. had 6 percent detentions) or improperly roasted beans. Further, the coffee can be stale, and often is. A stale coffee has a predominant pyridine odor, which is sickening. Brewing can even undo a good coffee, due to over-extraction, over-heating, incomplete and slow separation from the spent grounds, or chlorinated or brackish waters. The end result is not just the absence of a good taste, but the dominance of a very bad taste—sometimes even to the point of being undrinkable with dairy product additions.

## AROMA AND TASTE SENSES

### Influences on Sensitivity

Heredity, environment, and experience factors influence people's impressions of coffee aromas and tastes. People differ in their odor sensitivity. Some discrete odors and tastes are clearly objectionable. Usual odors do not draw reactions. Other odors are clearly likeable. The ability to detect different levels of aromas and tastes is often due to a person's age, health, body metabolism, odor conditioning, environment, sex, race, and training. Cultural customs, which are eductional factors, often strongly influence likes and dislikes of various aromas and tastes. In addition, likes and dislikes are influenced by circumstances such as price, peer attitudes, advertising, convenience and habit.

### Language of Aroma and Taste (Terminology)

In effectively communicating aroma and taste impressions, three factors are important:

1. A vocabulary pertinent to the item being tasted (see Table 15.1);
2. A group's common agreement on a taste or odor impression from the same coffee cup, e.g. several tasters sampling the same beverages around a common table;
3. A depth and breadth of coffee tasting experiences;

Further, the aroma/taste language for coffee (as well as other foods) has three aspects:

1. Trade or lay terminology by non-chemists, e.g. green coffee growers, importers, brokers, buyers, and connoisseurs;
2. Chemical terminology by chemists, chemical engineers, and food technologists;
3. Combinations of 1 and 2;

TABLE 15.1. TASTE AND AROMA TERMS
Used to describe the taste characteristics of green coffee beans.

| | |
|---|---|
| Acidity | A desirable flavor in high grown coffees—sharp and pleasing, but not biting. |
| Acrid | A burnt flavor—sharp, bitter, and perhaps irritating. |
| After-taste | A taste that remains in the mouth longer than usual. |
| Aged | Implies a controlled coffee storage to bring out a heavy body; not the same as "old" crop. |
| Aroma | Usually volatile, pleasant smelling substances with the characteristic odor of coffee. Chemically, they are aldehydes, ketones, esters, volatile acids, phenols, etc. |
| Astringent | A flavor that causes puckering and a bitter taste impression. |
| Bitter | When strong, an unpleasant, sharp taste; biting like quinine; some people get an acid taste impression while others notice a bitter taste from the same coffee. |
| Bland | Smooth and flavorless, such as alkaline water. |
| Body | A taste sensation or oral feeling of viscosity. Usually associated with heavy aged coffee flavor, but in no way an increase in true viscosity. |
| Burnt | Burnt carbohydrate, protein and oil, e.g., charcoal, meat and fatty acids. |
| Brackish | Distasteful, bitter, salty, that occurs in some waters. |
| Caramelized | Burnt-like flavor, like caramelized sugar. A desirable taste note if complemented with coffee flavor. Loss of coffee flavored volatiles enhances the caramelized flavor. |
| Dirty, Foreign | Undesirable "fuzzy" taste that dominates the coffee flavor background. |
| Earthiness | Undesirable taste or odor resembling the odor of freshly uncovered earth; usually due to molds. |
| Fermented | Chemical changes caused by yeast or enzymes on the green coffee sugars or proteins; like sugar fermenting aldehydes to alcohol or vinegar and proteins to amino acids. A pronounced fermented flavor is undesirable. |
| Grassy | A flavor often found in early pickings of new crop coffees and caused by immature beans. Suggestive of an intense, fresh greenness, such as newly mown hay or lush grass. |
| Green | Under-roasting beans, failing to develop the fullest coffee flavors. Somewhat pasty. A sourish flavor imparted by "green" beans, immature. Distinguish from grassiness. |
| Harsh | Unpleasantly sharp, rough, or irritating, e.g., Parana, Brazil coffees. |
| Hay | Odor and taste like dried grass. Hay flavor is common in dried foods. Furfural like. |
| "Instanty" | Characterized by furfurals and hydrolyzed cellular. Volatiles not normally found in brewed coffees at such high levels. |
| Musty | A taste akin to earthiness. Similar to a closed closet; moldy. |

| | |
|---|---|
| Natural | A "natural" coffee denotes an unwashed coffee bean, i.e., one prepared by drying the whole berry (dry process) rather than by fermenting the outer fruit (wet process) after pulping off skin. A "natural" bean does not have the development of coffee flavor and acidity characteristic of washed coffees of the same growth. Naturals often do not have the clarity and uniformity of cup flavor, but may have as much or more body. "Natural" most often refers to dry process unwashed coffees. Most Brazilian and Robusta coffees are naturals. |
| New Crop | A fresh light coffee flavor and aroma which enhances the normal characteristics of a coffee blend, particularly in flavor and acidity. Not to be confused with wildness or greenness, which are frequently present in new crop coffees. |
| Old Crop | A flavor and aroma in which the normal characteristics of a mature, greenish coffee are weakened or toned down—particularly less acidity and heavier flavor. This can also be a deterioration of these qualities into a woody or papery flavor with little or no body. |
| Pungent | A pricking, stinging, or piercing sensation; not necessarily unpleasant; e.g., pepper or snuff, fruity aldehydes. |
| Quakery | A "peanuty" flavor caused by undeveloped dead beans, which appear very light colored when roasted. |
| Rioy | An unpleasant, medicinal flavor. A property that cannot be hidden by blending. It is somewhat medicinal (iodine) with possibly woody or fermented overtones. |
| Robusta | Robustas have a bland flavor (lack acidity) and little aroma compared with mild coffee, desirable "true" coffee. At best, they can be neutral tasting, with a straw-like flavor. The majority have a rubbery flavor and aroma. The 2 percent caffeine content contributes to their distinctive flavor. Robustas are the lowest priced of all coffees. |
| Rubbery | An odor similar to heated rubber car tires on pavement. Robustas have this usually undesirable but characteristic odor. Instant coffee held at above 120 F (49 C) for days develops such an odor with sickening overtones. |
| Sour | Unpleasant flavor having a sharp, acid taste. Different from acidity. |
| Stale | A sweet but unpleasant flavor. Aroma of roasted coffee which reflects the oxidization of the pleasant volatile aldehydes and the loss of others. |
| Sweet | A pleasant, clean taste. |
| Wildness | Extreme flavors, sour or ferment, found in poorly prepared coffees, mostly "naturals". |
| Winey | Reminiscent of wine flavor and body, usually in high grown coffees. |
| Woody | A taste caused by deterioration of the coffee; akin to wood or paper. |

## Aromatic Constituents

Table 15.2 illustrates some of the chemical compounds found in aroma and flavor of roasted coffee beans. Well over 100 distinct chemical compounds that contribute to aroma and taste have been positively identified in roasted coffee beans. The mixtures are so complex and unstable that they defy synthesis to date.

## Thresholds of Aroma and Taste

Some aromatic chemicals found in coffee, are, of course, more important than other chemicals in influencing bodily responses. Some chemicals have extremely low threshold concentrations of recognition, while other chemicals do not (see Table 15.3). Hence, the concentration of a specific aromatic chemical is not directly related to its aroma or taste importance. For example, dimethyl sulfide is

TABLE 15.2 AROMATIC CHEMICAL CONSTITUENTS IN ROASTED COFFEE BEANS
(In parts per million)

| | Estimated averages | 1 Aroma | 2 Aroma |
|---|---|---|---|
| | ppm | Percent | Percent |
| *Aldehydes* | | | |
| Acetaldehyde | 80 | 17.9 | 19.9 |
| Propionaldehyde | 30 | 8.0 | 4.5 |
| Butyraldehyde | 4 | | 0.7 |
| Isobutyraldehyde | 15 | | 3.0 |
| 2-Methylbutyraldehyde | 30 | | 6.8 |
| Valeraldehyde | 30 | | 7.3 |
| Isovaleraldehyde | 20 | 18.2 | 5.0 |
| Acrolein | 2 | 0.6 | |
| Methyl ethyl acrolein | 5 | 1.4 | |
| *Ketones* | | | |
| Acetone | 80 | 0.5 | 18.7 |
| Methyl ethyl ketone | 15 | 14.2 | 2.3 |
| Diacetyl | 40 | 10.3 | 7.5 |
| Pentanedione | 1 | | |
| *Heterocycle Compounds* | | | |
| Furan | 10 | 2.5 | 3.2 |
| 2-Methyl furan | 20 | 5.1 | 4.7 |
| 2,5-Dimethyl furan | 1 | 0.3 | |
| Pyrrole | 2 | 0.5 | |
| *Sulfur Compounds* | | | |
| Hydrogen sulfide | trace | | |
| Carbon disulfide | 1 | 0.3 | 0.2 |
| Dimethyl sulfide | 4 | 0.6 | 1.0 |
| Methyl ethyl sulfide | 1 | 0.3 | |
| Dimethyl disulfide | 12 | 3.1 | |
| Methyl mercaptan | ½ | | 0.1 |
| Thiophene | ½ | | 0.1 |
| *Esters* | | | |
| Methyl formate | 20 | 4.9 | 4.0 |
| Methyl acetate | 15 | 5.7 | 1.7 |
| Ethyl formate | 2 | | 0.3 |
| *Nitriles* | | | |
| Acrylonitrile | 2 | 0.5 | |
| Allycyanide | 4 | 1.1 | |
| *Alcohols* | | | |
| Methanol | 4 | 0.9 | 0.2 |
| Ethanol | 1 | 0.3 | |
| *Hydrocarbons* | | | |
| Isoprene | 15 | | 3.0 |
| $C_4$-$C_7$ Paraffins | 10 | | 2.0 |
| | 400+ | 97.9 | 96.2 |

detectable by odor in parts per billion and occurs only at a few parts per million in the roast bean of high grown coffees.

Acetone can be 80 ppm in roast coffee beans, yet it is not a particularly strong odor. The physiological effects may be important only when acetone is taken into the body at significantly high concentration levels, e.g. by "glue snifters".

Most of the unstable and "flighty" or volatile aromas of roast coffee are the aldehydes. These are volatile, fruity, odorous chemicals. Aldehydes are readily

TABLE 15.3A. ODOR THRESHOLDS OF SELECTED CONSTITUENTS IN ROASTED COFFEE

| Chemical Compound | Odor Similarity | CONCENTRATION (Parts per Billion) (mg per 1,000 l of Air) |
|---|---|---|
| n-Butyric acid | Perspiration | 9.0 |
| Pyridine | Burnt | 1.0 |
| Hydrogen Sulfide | Rotten eggs | 0.2 |
| n-Butyl sulfide | Foul, sulfurous | 0.1 |
| Coumarin | Newly mown hay | 0.02 |
| Ethylmercaptan | Rotten cabbage | 0.0007 |

TABLE 15.3B. ODOR THRESHOLDS OF SOME VOLATILE CHEMICALS FOUND IN ROAST COFFEE

| Chemical Compound | Boiling Points C (F) | CONCENTRATION (Parts per Million,) (mg per l of Air) |
|---|---|---|
| Ethylacetate | 77 (171) | 0.70 |
| Ethylmercaptan | 37 ( 99) | 0.05 |
| Pyridine | 115 (239) | 0.03 |
| Valeric acid | 186 (367) | 0.03 |
| Butyric acid | 162 (324) | 0.010 |
| Propylmercaptan | 67 (153) | 0.006 |

TABLE 15.4. TASTE THRESHOLDS OF ORGANIC ACIDS FOUND IN ROASTED COFFEE

| Organic Acid | CONCENTRATION (Parts per Million,) (mg per l of Water) |
|---|---|
| Formic | 83 |
| Acetic | 168 |
| Butyric | 308 |
| Valeric | 378 |
| Oxalic | 252 |
| Tartaric | 480 |

oxidized and very reactive chemically. This explains why the pure coffee aroma and flavor are such fleeting constituents. When these volatiles react or disappear, staleness (the undertone constituents) takes over. Then the heavier layers of odors like pyridine and tobacco, which are distasteful features, appear.

## Associating Coffee Flavors Notes with Identified Chemical Constituents

To the chemist and flavor compounder, the ultimate sources of flavor and aroma are the constituent chemicals. Analytical techniques applied since 1960 have allowed separations of natural aromatic volatiles. This allows a nasal association of separated natural aromas with chemical entity. Simple compounding of a few volatile chemicals of these aromas can be blended to yield acceptable impressions of natural fruits, flowers, cheeses, and other foods. Better analytical methods, instruments, and experience have been developed to this end. It becomes a work of

TABLE 15.5. PERCENT CHEMICAL CONSTITUENTS BY TYPES OF AROMA IN ROASTED COFFEE BEANS (Based on 400 ppm total aromatic volatiles in the roast bean)

| | Aroma Types | Percent |
|---|---|---|
| *Aldehydes* | | RANGE |
| | 1. Acetaldehyde | 18.0-26.0 |
| | 2. Propionaldehyde | 4.0-8.0 |
| | 3. Butyraldehyde | 0.3- 0.7 |
| | 4. Isobutyraldehyde | 3.0- 7.0 |
| | 5. 2-Methylbutyraldehyde | 7.0 |
| | 6. Valeraldehyde | 7.0 |
| | 7. Isovaleraldehyde | 5.0-18.0 |
| | 8. Acrolein | 1.0 |
| | 9. Methyl, ethyl acrolein | 1.0 |
| *Ketones* | | |
| | 1. Acetone | 20.0 |
| | 2. Methylethyl ketone | 8.0-14.0 |
| | 3. Methyl vinyl ketone | 1.0 |
| | 4. Diacetyl | 6.0-10.0 |
| | 5. 2, 3-Pentanedione | 7.0 |
| *Heterocycle Compounds* | | |
| | 1. Furan | 3.0 |
| | 2. 2-Methyl furan | 3.0- 5.0 |
| | 3. 2, 5-Dimethyl furan | 0.3 |
| | 4. Pyrrole | 0.5 |
| | 5. N-Methyl pyrrole | Trace |
| | 6. Dimethyl pyrrole | Trace |
| *Sulfur Compounds* | | |
| | 1. Hydrogen sulfide | 1.0 |
| | 2. Carbon disulfide | 0.2 |
| | 3. Dimethyl sulfide | 1.0 |
| | 4. Methyl ethyl sulfide | 0.3 |
| | 5. Dimethyl disulfide | 3.0 |
| | 6. Methyl ethyl disulfide | Trace |
| | 7. Methyl mercaptan | 0.1- 1.0 |
| | 8. Thiophene | 0.1 |
| *Esters* | | |
| | 1. Methyl formate | 3.0- 5.0 |
| | 2. Methyl acetate | 2.0- 6.0 |
| | 3. Ethyl formate | 0.3 |
| *Nitriles* | | |
| | 1. Acrylonitrile | 0.5 |
| | 2. Allylcyanide | 1.0 |
| *Alcohols* | | |
| | 1. Methanol | 1.0 |
| | 2. Ethanol | 0.3 |
| *Hydrocarbons* | | |
| | 1. Isoprene | 3.0 |
| | 2. $C_4$-$C_7$ Paraffins | 2.0 |
| *Oxides* | | |
| | 1. Carbon dioxide | 4.0 |
| | | 100.0 |

Aroma compositions range depending on type of coffee, condition of coffee, degree of roast, mode of roast, etc.

art to blend these individual chemicals into a pleasing impression, as, for example, perfumes or fruit flavors for candy or beverage use. Flavor firms have shunned coffee aroma as too complex and too unstable a mixture to try to duplicate. No really good synthetic coffee flavor has been developed, although a few satisfactory flavor fortifiers have evolved in recent years. Table 15.2 shows the verified quantities of specific chemical constituents identified in coffee aromas based on the work of several investigators.

## Acidity in Coffee Flavor

**Content and Composition.**—Acidity in most foods and beverages is a key factor to its flavor properties and appeal. Table 15.6 shows the pH values of common foods and chemicals. Table 14.3 illustrated the meaning of pH and titratable acidity. In coffee flavor, acidity is important in apparent or free acidity, called pH, which is perceptible to the taste. Most freshly brewed coffee beverages have a pH of 5.0 to 5.1; pH is quite uniform as a result of buffering action of the weak organic acids.

Most commercial coffee buyers have little to no understanding of pH to taste relationships. Mild coffees grown at high altitudes have more natural acidity. Brazilian coffees have noticeably less acidity. Robusta coffees have the least acidity. Robusta beverages are pH 5.3 to 5.7; in alkaline waters Robustas may have a pH of 6.0, which is bland/flat.

Holding freshly brewed coffee beverage in an urn for hours forms acids. This reduces brew pH from 5.1 to below 4.8 and can cause curdling of milk or cream. Furthermore, some distinction between volatile and non-volatile acids in coffee is desirable. As a rule, organic acids are relatively chemically stable. Aldehydes can be oxidized to acids. Chlorogenic acid, when heated, breaks down to caffeic and quinic acids. This increases the acid content of the coffee beverage. Such acid increases and changes are also accompanied by detectable taste changes. Table 15.8 shows the volatile acids in roast coffee.

TABLE 15.6. pH VALUES OF COMMON FOODS AND CHEMICALS

| Foods | pH | Chemicals | pH |
|---|---|---|---|
| Limes | 1.9 | 0.1 N HCl | 1.0 |
| Lemon juice | 2.3 | 0.1 N acetic acid | 2.0 |
| Ginger ale—apples | 3.0 | 0.1 N boric acid | 5.1 |
| Orange juice | 3.3 | 0.1 N $NaHCO_3$ | 8.4 |
| Sauerkraut | 3.5 | Milk of magnesia | 10.5 |
| Tomatoes | 4.2 | 0.1 N $NH_4OH$ | 11.1 |
| Carbonated water | 4.5 | | |
| Carrots and coffee | 5.1 | | |
| Molasses | 5.2 | | |
| Cabbage | 5.3 | | |
| Cow's milk | 6.6 | | |
| Shrimp | 7.0 | | |
| Pure water | 7.0 | | |

TABLE 15.7. COFFEE, BY-PRODUCTS AND CHEMICAL COMPONENTS

**Coffee and Coffee By-Products at Each Process Step**

Cherry fruit
Beans in hull (pergamino)
Hulls
Husks
Silver skins
Green coffee beans
Roast coffee beans: light, medium, dark, French and Italian
Grinds: regular, drip and fine; also pulverized
Instant coffee powder (spray dried and belt dried)
Extracted coffee grounds
Equivalent grounds mineral ash

**Volatiles**

Acids—acetic, propionic, butyric, valeric
Aldehydes—acet, propyl, butyl, valer
Ketones—ketone, methylethyl, penetanone, diactyl
Ketone-alcohols—acetol, acetoin
Esters—methyl formate, methyl acetate, methyl propionate, propyl formate
Sulfides—hydrogen, dimethyl, methyl mercaptan, thiophene
Cyclic—furan, methyl furan, pyridine
Amines—trimethyl, dimethyl, ammonia
Hydrocarbons—isoprene
Unsaturates—acrolein

**Non-Volatiles**

| | Soluble coffee, per cent | Roast coffee, per cent |
|---|---|---|
| Chlorogenic acid | 17 | 7 |
| Caramel | 56 | 23 |
| Mineral ash | 12 | 3.5 |
| Trigonelline | 3 | 1 |
| Caffeine | 4 | 1.2 |
| Oil | 0.1 | 10 |
| Caffeic acid | 0.8 | 0.3 |
| Quinic acid | 0.8 | 0.3 |
| Citric acid | 1.5 | 0.6 |
| Malic acid | 1.5 | 0.6 |
| Tartaric acid | 1.3 | 0.5 |
| Pyruvic acid | 0.1 | 0.06 |
| Guiacol | ... | ... |
| Nicotinic Acid | ... | ... |

**Acetic Acid.**—Acetic acid (vinegar) occurs at 0.30 percent/wt in roast coffee beans. Except for chlorogenic acid, acetic acid is the highest occurring individual acid and coffee flavor contributor. There is about 1 percent of acetic and homologous acids in percolated coffee extract solubles. After spray drying, almost no acetic acid is left in the instant coffee powder. When acetic acid (pure or from a coffee source) is added back to a cup of instant coffee, it contributes zest and smoothness to the coffee flavor. The acetic acid cannot be taken from any source for such reconstruction since minute impurities, incompatible with coffee flavor, can cause an off-flavor. The harsh, taste of some low altitude grown mild coffees is due to high acetic acid content of light colored roasts.

**Fatty Acids**.—During roasting the glyceride fats and oils are partly broken down to release several percent fatty acids. The acid content is small, but its presence affects taste texture because it reduces surface tension.

**Acidity and Roast**.—Titratable acidities of roast coffee can be reduced (to a fifth) from the just palatable light color roast to the almost burnt black roast. Acidities within the palatable bean roast range, on milds, can be reduced ½ to ⅓ from light to dark roasts.

**Acidities in Soluble Coffee**.—Acidities of instant coffees, resulting from acids formed in the hydrolysis of celluloses, often reduce beverage pH to 4.8 or 4.9; at pH 4.7 curdling of milk is highly probable. Soluble coffee has as much acidity as brewed coffee, but the powder has lost most of its volatile acids. Non volatile hydrolyzed acids formed give acidity. Volatile acids are only about one-fifth of the total acids in the commercial extract, but their loss under commercial drying conditions is measurable and noticeable in the taste.

**Holding Hot or Refluxing Coffee Brew and Extracts**.—Holding the brew and extracts at 212 F (100 C) for several hours will increase original acidities by about 25 percent. These developed acidities and associated flavors are repulsive. Oxidation of aldehydes, ketones, and alcohols form acids. In the absence of oxygen under the same conditions, undesirable taste and chemical changes occur and acids may not form.

## Off-Flavors Caused by Deterioration, Contamination, and Staling of Coffee

From the time the coffee is picked until a beverage is prepared, there are numerous points at which deterioration may occur. The following examples are illustrative.

**Green Coffee Beans**.—*During Harvesting*.—Most contamination occurs during the processing of the cherry when the green coffee bean is wet. Individual growers may harvest, ferment, and dry only a few dozen bags of coffee beans. Small, variegated lots of beans are sold to a mill, where the beans are graded, cleaned, and blended. Thus, the mill processor goes largely by his experience with bean appearance.

Ripe cherries should be picked when they are blood red. Immature cherries can be removed by floatation. If not removed, they do not roast well and are easily picked out as light colored beans with no coffee flavor, but with an off-flavor. To pick only mature cherries, repeated picking is needed as cherries mature. In fact, only a few pickings are made.

*During Processing*.—Most coffee cherry processors seldom have sufficient fermentation tank capacity, personnel, drying facilities, or water to do the jobs required on schedule. There is a real dilemma should it rain. Sunshine is needed for drying the coffee beans on the patio. At this time off-flavors may develop in the moist green coffee beans. For example, if fermentation occurs slowly, cherries are held longer than is desirable. A "wild ferment" may result in the piles of cherries with broken skins. Coffee beans that have been washed free of fermentable mucilage cannot always be dried promptly. When wet bean rotation is not fast enough, the piles of stored coffee beans can become moldy and then develop musty odors and tastes. Piles of incompletely dried coffee beans must be turned over to allow fuller bean exposure to air movement. Beans must be dried as soon as possible to 12 percent.

To process the wet coffee beans rapidly, one may use higher air temperatures. This can damage the flavor of the coffee beans. Fermented and moldy flavors usually do not develop in every coffee bean. They develop in enough beans to cause a flavor taint in the "lot." This is one reason why coffee "lots" of 50 bags or less are kept separate. One coffee lot may be tainted and another lot may not, so mixing lots must be done cautiously.

*Parchment Damage*.—The coffee bean below the fruity mucilage layer has a parchment. If it is not broken, taints cannot penetrate to the coffee bean. In practice, sufficient parchment shells are damaged during fruit pulp removal so that some taints do enter some beans. Subsequently, these beans must be removed by hand picking or electronic sorting. It is instructive to take unsorted green coffee beans and remove the discolored and damaged beans from one-half of the sample. Then all three bean fractions should be roasted and tasted. Removal of only 5 percent of damaged beans may, surprisingly, raise the quality of the 95 percent portion to exportable grade.

*Partly Dry Coffee Beans*.—Even after the beans are dried to 12 percent moisture, graded, and bagged, there are still conditions that may cause off-flavor to develop. Coffee beans after drying may, in the rainy season, reabsorb considerable moisture. This, then, can cause mold and even fermentation.

*Bean Storage*.—Holding bagged green coffee beans for export in a warm, damp shipping port causes coffee flavor loss. Under such storage conditions, the flavor of these coffee beans will deteriorate noticeably in weeks. In months, these coffee beans, once greenish blue and waxy, can become pale, straw-colored, or dark in appearance. They yield a flavorless cup of beverage at best.

*Aged Coffees*.—These used to come from Venezuela. They are good quality coffees that are held for several years in the cooler, drier, high altitudes. These coffee beans often have winey flavor. They are definitely heavy-flavored in the cup after years of aging. Such coffee beans are not often commercially available in world trade today. They appeal to gourmets who will pay the premium.

New crops of coffee beans are often grassy and thin-bodied in taste. They may reflect the grower's anxiety to get his coffee beans to market early for some price advantage. Thinness and grassiness are not desirable flavor properties.

*Naturally Dried Coffee Beans*.—In Brazil and Africa (Robustas) and some mild coffee growing areas, where water is scarce, or the harvester does not remove the skin and pulp, the cherries are dried as picked fruit. Such dried coffees are called ''naturals''. Washed coffees, as a rule, have better flavor than naturals. There are, however, some naturals that have a winey and heavy body with hardly any off-flavor. The best tasting naturals are produced when drying is prompt.

*Altitude*.—High grown coffees (5,000 ft above sea level) grow slowly and have greater and better flavor. The body of high grown Colombian coffees is rich, winey, and buttery with a characteristically appealing aroma. Mild coffees from 2,000 ft altitudes, which grow more quickly, are usually thin-bodied to harsh in flavor, and produce only fair aroma.

**Foreign Matter**.—Contaminants normally found among green coffee beans are corn kernels, kidney beans, fruit pits, twigs, nails, pebbles, sand, and stones. Buttons, strings, and personal items like coins are also common. Iron contaminants are common and when dissolved, iron and copper oxidize coffee flavors.

**Odor Absorption**.—Storage of odor-carrying products near green coffee beans may cause the coffee oils to absorb enough of the foreign odor to contaminate its flavor. Such contaminants might be paint, turpentine, solvents, asphalt, manure, and spices.

**Moisture Pickup and Storage Humidity**.—Before 1940 green coffee bean shipments from Latin America were damaged by water frequently. Moisture condensed on the coffee beans when the ships passed from the warm Gulf Stream into cold, Atlantic Ocean waters. The result was mold growth and off-flavored lots. Today, most ships have enough moisture control and air circulation to avoid this.

Where green coffee beans are stored in bags in an open shed during the rainy season, the equilibrium moisture in the green coffee beans rises sufficiently to cause the beans to swell and possibly rupture the bags. Spilled beans on the soil floor initiate conditions for mold growth. Moist coffees stored in the absence of adequate air circulation can become putrid and rank. Beans that are dried below 10 percent moisture often undergo loss in flavor. Poor sanitation conditions in the coffee processing and storage areas can cause contamination through foods, tobacco, rats, birds, and insects.

**Bag Stenciling Ink.**—Coffee contamination can occur when the bags of beans are stenciled with ink.

## Buyer of Coffee Beans

The buyer is an experienced taster of coffees from many lands and regions, and on his judgement rests decisions that can result in vast monetary differences in payments on quality, timing, and selling. The buyer, consequently, is well paid for doing a good job. He ferrets out "good" buys on a timely basis; he stocks up on a rising price market and keeps inventories down in a falling price market. He avoids Rioy cöffee and recognizes nonuniformity, age, mold, ferment, foreign tastes, grassiness, and other good/off-flavors.

## Aroma and Taste Differences Monitored in Processing (Blend, Roast, Grind and Extraction)

Taste differences in a coffee product are a result of the type of green coffees used and the processing conditions used. If the cook or superintendent of the processing plant does not taste his product, he cannot do justice to his preparation. This is true in a kitchen or in a food processing plant. If the raw material is poor, one cannot expect a good end product. On the other hand, a good raw material does not ensure a good final product. Good processing means the best recovery of available natural coffee flavors with a minimum of "off-flavors", i.e. those that are not natural to properly brewed coffee. Taste control is the key to overall process and product control. Few coffee or food processing plants have such formalized overall taste control. From an operating viewpoint, quality control data, after the product is finished, cannot rectify a poor processing condition. Tons of "out-of-standard" product can be produced. It subsequently must be released "as is" or be worked back into the normal product at low concentrations. There is loss in product quality, labor, and other costs. Quality control tests on the final product are, of course, useful, but they must be considered within the framework of quality of process, equipment (instruments), and plant personnel (knowledge, training, experience, and dependability).

The coffee plant superintendent must carry out the processing and exercise authority at a focal point. He gathers and receives all process data. He uses his eyes, nose, tongue and hands. Effective process control is then achieved by checking this data with working personnel and control instruments and by making necessary equipment adjustments. The effective process plant superintendent thus helps control product taste by controlling the processing conditions at each step of the process. If he can use taste in addition as a personal guide, that processor is exceptional and will produce an exceptional product.

The unit process operator cannot do this due to his limited responsibility, activity, and training. Without overall supervision, flavor process control is correspondingly impaired. With central supervisory authority and surveillance, continual probing and feedback, an effective control of human relations, equipment, mechanics, and cookery (food processing), and repeated tasting, one can ensure proper process and product adjustments. Tables 17.9 thru 17.12 list some taste effects at four key stages in processing.

TABLE 15.8. NON-VOLATILE AND VOLATILE ACID CONTENTS IN ROAST COFFEE BEANS

| | Weight Fraction |
|---|---|
| *Non-Volatile Acids* | |
| 1. Chlorogenic | 0.046 |
| 2. Caffeic | 0.003 |
| 3. Quinic | 0.003 |
| 4. Citric | 0.005 |
| 5. Malic | 0.005 |
| 6. Tartaric | 0.004 |
| 7. Oxalic | 0.002 |
| 8. Pyruvic | 0.0006 |
| Sub Total | 0.0686 |
| *Volatile Acids* | |
| 1. Acetic | 0.0036 |
| 2. Propionic | 0.0002 |
| 3. Butyric | 0.0001 |
| 4. Valeric | 0.0002 |
| Sub Total | 0.0041 |
| Total | 0.0727 (7.3%) |

## CONSUMERISM AND PHARMACOLOGY

Most people, in or out of the coffee industry, know little about the effects of caffeine and coffee roasting products on the workings of the body. As one looks into systematic scientific work carried out by hospitals and medical research teams, it is evident that coffee and caffeine have quite potent influences on many people.

The National Coffee Association (NCA) in 1921 did publicize a "white paper" by S. C. Prescott of M.I.T. However, in 1971 the NCA established a "Coffee

TABLE 15.9. TASTES CAUSED IN ROASTING COFFEE

| TASTE | CAUSE |
|---|---|
| Bland, pasty, nutty, green, baked | Light roasts have brought the bean short of the best flavor and aroma development. Similar tastes are derived by too slow a roast. There is drying but not the type of chemical destructive distillation (pyrolysis) that is necessary to fully develop the flavors of a good quality green coffee bean. |
| Unbalanced flavors, both burnt and green | From non-uniformly roasted beans. A mixture of beans with different physical properties such as large and small beans, old and new crop, dry and moist beans, dense and less dense beans, etc. |
| Full-bodied, acidy, aromatic | By blending after roasting, it is possible to roast each coffee type to the fullest flavor development from an assortment of green beans varying in type, density, moisture, age, size, perfection, etc. Roasting of blended green beans is done to simplify the roasting. Many green coffee blends are not suitable for such preparation. Separate roasting of single kinds of coffee takes more labor, but brings out the fullest flavor of each type. The resultant taste differences can be appreciated. |
| Thin-bodied, burnt, oily, carbon-like | Dark roasts where the roasting process has been carried beyond full flavor development and/or optimum volatiles content. Most volatiles of a desirable nature are driven out, leaving a mostly low-acid residue of solubles. Odor is fishy or amine-like, at times ammonia-like. The ruptured cell structure of the bean tends to release many oily colloids. Dark roasts are often applied to low grade coffees to obscure some undesirable aromas and flavors. |
| Heavy flavor, pyridine-like, stale | A heavy, undesirable flavor can be carried into the coffee by excessive water quench of the roasted coffee beans. Residual absorbed water on the bean releases the protective carbon dioxide gas. Oxygen enters the bean or grounds, staling volatilization occurs. |

TABLE 15.10. TASTES CAUSED IN GROUND COFFEE

| TASTE | CAUSE |
|---|---|
| Oily, smooth | Very fine grind, more colloids. |
| Harsh | Coarse grind—"Regular," few colloids. |
| Bitter | Exhaustive extraction of coarse grind. |
| Lacks aroma | Fine grind, not used promptly. |
| Stale | Very fine grind, left exposed to air, rapidly oxidizes. |

Information Institute" to rebut "misinformation" about the health influences of caffeine and chemicals in coffee.

The legend of the coffee industry is that "the coffee beverage is healthful." Not even the (caffeine added) cola companies have gone that far in their claims—with good reason. Coffee and caffeine are drunk for their physiological effects, which explains why bad tasting coffee has usually been acceptable. If the coffee beverage tastes good, that is just a bonus. Coffee is being drunk because of an acquired taste and for its "medicinal" benefits. Since people's bodies are different and the coffee they drink is different, both in quantity and quality, there results a broad range of influences on the body There are few accurate statistics in this regard. However,

TABLE 15.11. TASTES CAUSED IN PERCOLATOR EXTRACTION (SOLUBLE COFFEE)

| TASTE | CAUSE |
|---|---|
| Acid, astringent, dry, straw aroma | High (hydrolysis) solubles yield is accompanied by acidity. Furfural odor and flavor is easily identified. The cup pH may be low—4.80 or less. |
| Caramelized, harsh | Caramel flavor is distinctive and a part of both regular and instant coffee. When dominant, it is objectionable. This may occur with excessive time—temperature conditions destorying volatiles that constitute the overtone coffee flavors. Heating or evaporation of the extract can also give caramel taste. |
| Weak, flat | A loss of coffee volatiles. A high solubles (hydrolysis) yield dilutes the coffee flavor. Contamination by iron or copper also destroys coffee flavor. Exposure of extract to air (oxygen) causes flavor loss. Too large a solubles inventory within the percolator system also gives a flat cup. This increases residence time of coffee flavor constituents at high temperature. |
| Heavy, pyridine, heat treated, sickening Tobacco-like | Holding of aromatic coffees in the percolator system under long cycle times. Heating of extract produces this characterist flavor. A similar odor and flavor is developed by holding instant coffee powder for prolonged periods at elevated temperatures in air, also by allowing moistened coffee grounds to oxidize. |

TABLE 15.12. TASTES CAUSED IN SPRAY/DRYING (SOLUBLE COFFEE)

| TASTE | CAUSE |
|---|---|
| Flavorless | Much flavor content is lost when extracts are dried from low concentrations, notable at below 25 percent. Retention of flavor is most notable above 35 percent. Heated extracts lose more of their flavor in spray drying than the same extracts cold. A spray powder with small, average particle sizes also lacks flavor, as does a powder with excessively low moistures, e.g., 1 percent instead of 3.5 percent. High inlet air temperatures, 500 F or higher drive off more flavors. |
| Burnt and caramelized | Occurs with high inlet and outlet air temperatures. Wall buildup of powders where charring occurs. |
| Heat treated, sickening | Powder is exposed to hot air for hours or days. |
| Dusty taste texture | Aid adsorption on fine particle surfaces, intrinsic character of fines. This taste is foreign to brewed coffees. |
| Sulfurous, oily, heavy | Incomplete combustion in direct oil-fired spray driers allows contaminants to be adsorbed on the powder. Complete combustion of sulfur still results in adsorption. Low sulfur content fuels must be used. |

since decaffeinated coffees in Europe and the U.S.A. are 15 percent of the sales in the coffee market, this is a real indication of the magnitude of the public's evasiveness from natural coffee use: tens of millions of people. Although in recent years A.S.I.C. meetings have produced some interesting and enlightening technical papers about caffeine's role in human body effects, hardly any of this work has been financially supported by the commercial coffee industries in the U.S.A. Europe, or growing countries.

What has recently happened to the cigarette industry and its health condemnation and taxation ought to be fresh in everyone's minds. Coffee imports have been and still are taxed heavily in some countries, especially Germany. Coffee is taxed in Europe not just because it is classed as a luxury. For example, in Germany, there has been a continual public conflict as to the advisability of drinking coffee in so far as bodily and mental health is concerned.

Germany is the home of the discovery and isolation of caffeine by Runge in the late 1800's, and the decaffeination of the coffee bean by Rosclius in 1900. There is a concentration of coffee decaffeination processing plants and services in Germany and Switzerland. Decaffeinated coffee is manufactured in several places by the firm Kaffee HAG throughout the world. Decaffeinated coffee, especially decaffeinated instant coffee, is available virtually throughout the world's major cities.

Caffeine is not the only ingredient in coffee that causes physiological effects. The chemicals formed in roasted coffee beans, aromatic and not so aromatic, are direct contributors individually and collectively to physiological reactions.

These chemicals have come under scrutiny only since World War II, after which the more sophisticated analytical instruments became available and were more widely used.

## Sources of Physiological Information

The most readily available, rather thorough, and least biased sources of information about the influences of caffeine and most coffee products on the body are the pharmacological textbooks. Such books include the 1970 edition by Grollman, the 1970 edition by Goodman and Gilman, and the 1947 edition by Cushny. Their commentaries are factual and are backed up with references to the original researches. Some of this research continues now.

The Cardiology Division of the Philadelphia General Hospital conducted a series of experiments in the mid 1970's. In 1972 work was done by Kuchlnsky and Lullman (G. Thieme in Stuttgart, Germany). Scientific papers presented on this subject at the (ASIC) Association Scientific International on Coffee meetings every two years have been plentiful.

The French Institute of Coffee and Cocoa (IFCC) in Paris has done work in this area, as has G. Czok at the Pharmakologisches Institut at the University of Hamburg, Germany. Even General Foods has carried out some physiological studies with rats. German research papers have presented the effects of coffee on sportsmen and psycho-labile patients, on causes of gene mutations, as an emetic, on cardiovascular actions, in cases of arteriosclerosis, in stomach and liver disorders, in ulcer cases, in glaucoma cases, and in cases where patients are easily excitable. In 1965 at the ASIC meeting in Paris, Brazilians reported the levels of caffeine obtained with different methods of brewing. Dr. R. Ulrich of Bremen,

Germany reported on the influence of coffee in increasing the blood sugar levels with high intakes, but S. Heyden (in personal communication) states that this has been disproven; experts disagree.

Improved and systematic analytical methods for caffeine concentrations in commercial products have been put into use by several decaffeinated coffee processors. The chemical products of roasted coffee beans are very similar in chemical composition to those found in wood and tobacco smoke.

The following pages will elaborate on the individual known influences of caffeine and coffee roasting by products as is so far known and published to date.

## CAFFEINE IN COLAS, AND INCREASED COFFEE/CAFFEINE USE

Dr. Samuel Bellet, Chief of Cardiology at the Philadelphia General Hospital, has suggested that, "caffeine may be more important than smoking in setting the stage for heart attacks" (Phila. Eve. Bull. June 27, 1966). Mormons, Christian Scientists, Seventh Day Adventists, and other religious groups recommend abstaining from coffee use. Heyden (1973) and Paul (1968) did not agree with Dr. Bellett.

Coffee beverage can have 3 to 4 times as much caffeine per cup as 6 fl oz of cola drink. There is much resistance to consuming cola, especially in France, where much coffee is consumed. Coffee drinking has been repeatedly condemned from place to place in the course of time. In Europe, coffee is sold in specialty shops with only liquors, chocolates and teas, considered expensive luxuries.

Coffee roasters, who work toward improving their own businesses, are not likely to call coffee anything but a beneficial and healthful drink. After one reads of the effects coffee and caffeine have on the body functions, one gathers more respect for the religious abstentations and the remarks of medical investigators. Indeed, the subject is sorely in need of airing. Additional physiological investigations and a warning label with the caffeine content may be desirable. This is because coffee is such a commonly drunk beverage, without the user's realization of its potent physiological actions on his body . . . hour after hour, day after day, and in his lifetime.

No one has assessed the body's reactions to the hydrolyzed portion of the coffee bean that is now in instant coffees, which are called "100 percent pure coffee" by the coffee companies. What are the effects of these chemicals on the body?

Since World War II there has been a very significant rise in the use of decaffeinated coffees. Why? One reason is because with the increased use of Robustas, which have an unpleasant, irritating taste as well as double the caffeine content, consumers pass over the margin of being comfortable after drinking coffee to being uncomfortable. Interestingly enough, decaffeination removes the

Robusta's harshness (as well as flavor) and irritation, leaving a bland coffee beverage without irritants (and little flavor), whether consumed as R&G coffee or instant coffee.

## Bodily Influences

People's bodies are different. The coffees they drink are different. This suggests, therefore, that there might be a wide range of physiological reactions obtainable from food and drink, including coffee.

A body's reaction to coffee falls into many gray areas, and blanket comments like, "it's healthy for you" or "it can make you irritable" are too general. The body is in a different state under different conditions:

1. when one wakes up,
2. after a hard day's physical or mental work,
3. after dinner,
4. late in the evening,
5. after much outdoor exposure to cold and wind.

The meaningful influencing factors are probably infinite, so there are many light gray areas and dark gray areas, with few "black and white" areas.

People do suffer from sleeplessness, nervousness, intestinal discomfort, heart stimulation and other effects after drinking one or two cups of coffee. On the other hand, it is presumptive to say that this is only 1 percent of the drinkers (as Prescott said and Ukers re-echoed), although no such scientific measurement has been made. It is well known from personal experience that "excessive" coffee drinking can cause all those things mentioned. "Excessive" can be one cup for some, two cups for others, and three cups for people in general. Here again, we depend on what the people are doing and where they are. People that habitually drink coffee require a larger dose for stimulation. This is because they develop a tolerance to the drink. Whereas, a non-coffee drinker will get a "stimulating reaction" from a single cup of coffee.

Individual tolerances to dosages without obvious symptoms vary with age, sex, physical condition, environment, and other factors. Heretofore, caffeine has been considered the exclusive activating ingredient. But now the chemicals formed from roasting are known to cause body effects. Coffee oils, fatty acids, (hydrolyzed) cellulose components in coffee beverages held hot for many hours may also cause body discomforts.

## Tangibility in Coffee Drinking

In spite of (1) the coffee industry's lackadaisical attitude toward consumers, (2) the coffee drinking effects on the body, and (3) the fact that some individuals are adversely affected by caffeine and coffee, there are beneficial effects on the body and mind from coffee drinking.

These benefits take on the form of needs—"I've got to have my morning coffee" and words to that effect. The individual's body and mind react and are sensitive to the coffee although often he may not know why. Young people usually react negatively to drinking coffee. Middle-aged people find it stimulating, relaxing, and a "pickup" or lift. Older people find it a useful stimulant or avoid it due to "bad" body reactions. However, older people are usually the major drinkers of decaffeinated coffee.

The following cogent reasons for drinking coffee are genuine for many people:

1. *Stimulates the brain:*
   a. speeds thought processes
   b. increases idea association
   c. improves memory recall
   (can also "over-key" some "to climb the walls".)
2. Result: a more rapid and clearer flow of thoughts.
3. *Contributes a feeling* of "well-being":
   a. counteracts sleepiness
   b. counteracts physical and mental fatigue
   c. makes one generally more alert
4. *Relieves headaches:* (at times migraine)
   a. opens blood flow path to brain (dilates some blood vessels)
   b. stimulates cortex of brain (CNS)
   c. increases respiration—releases poisons/takes oxygen
   d. stimulates urination/or bowels
4. *Warms the body* through increased blood circulation
   *Cools the body* through increased blood circulation and respiration, accompanied by increased perspiration.
6. *Stimulates the heart* to pump more blood due to reduced arterial restrictions.
   a. ordinarily pulse rate not affected much
   b. ordinarily blood pressure not increased (due to dilation of blood vessels)
7. *Increases stomach acid secretion*—desirable to digest food (undesirable for ulcerated persons).
8. *Increases bile secretion* of gall bladder, assists fat digestion.
9. *Counteracts alcohol* and other drugs' depressing symptoms. Coffee after cocktails or after the hangover increases the respiration rate by stimulating medullary center.

## Oral and Medicinal Caffeine Dosages

One can see that the oral dosages of caffeine from coffee beverages (Table 15.14) are of the same magnitude as medicinal therapeutic dosages. Intravenous or intramuscular administration of caffeine acts more quickly than the oral dosage, but the effects on the body are similar.

The result in Table 17.14 lists common widely consumed beverages and cocoa products that contain caffeine. The following lists oral dosages as double salts.

Oral dosages used are 100 to 300 milligrams of caffeine. The amount depends on the condition and physical size of the patient. Double salts increase caffeine solubility.

Typical caffeine concentrations and solubilities are: a) 50/50 by weight with citric acid is soluble in 4 parts of water; b) 50/50 by weight with sodium benzoate is soluble in own weight of water; c) these are also used for intramuscular administration at concentrations of 250-500 mg/2 ml solution; and d) 55/45 by weight sodium salicylate and caffeine, respectively.

As per pharmacological texts, the functions of caffeine in medicine are:

**Stimulation of the central nervous system.** (CNS)

—in cases of nervous exhaustion

—relieves headaches, sometimes migraine

—prolongs wakefulness and keeps up intellectual faculties

—relieves physical and mental fatigue; increases work capacity (can contribute to insomnia with some people) (can make some people nervous)

**Activate the respiratory center.**

—in cases of collapse, narcotic and alcoholic poisoning

—in severe attacks of asthma

—in cases of falling respiration

**Relieves hypertension**

—dilates coronary arteries and reduces blood pressure

—improves blood circulation

—reported as valuable in treatment of angina pectoris

**Intestinal**

—with some people, it stimulates intestinal secretion and contributes to loss of appetite

—for this reason, coffee is not normally given to children

—mild diuretic-stimulates kidney activity and urine flow

Caffeine effects can last 1 to 3 hours and sometimes hours longer, depending on dosage. However, caffeine may not completely chemically decompose until 24 hours later. There is a psychic dependence for coffee and caffeine.

**Excessive Coffee Drinking.**—Even 5 cups of beverage a day, for many people, can result in their being high strung, nervous, irritable, and generally uncomfortable. These coffee chemicals do act on the nervous, respiratory and kidney systems. It certainly calls for a study of what each chemical and each combination of chemicals does to the body. No doubt some of this work has already been done, but has not been reported. It is generally recognized that large coffee consumption by individuals is more detrimental than beneficial.

## Commentary (on Reported Physiological Effects of Coffee and Caffeine)

The data reported here are largely from pharmacological texts. References to specific experimental data sometimes is contradictory. There are obvious variations in experimental technique through the years. The competency of the experimenters and their interpretations vary. There are also variations in test work with animals and humans. Dosages vary, sensitivity varies by animal groupings and within animal groupings. Side effects occur and qualities of coffees vary.

For example, reports on mg caffeine dosage per Kg animal weight immediately reveal the great differences that can be expected from a given cup of coffee in a child, in a young lady, in an average weight man, and in an above average weight man. Men's weights can range from 40 to 100 Kg. Metabolisms vary. Beverage use can range from one weak cup of coffee (at 70 mg caffeine) to two or three strong cups of coffee (totaling 400 mg caffeine).

In spite of variations in body responses, there is a definite pattern of reactions, e.g. the stimulating effect of the caffeine with all its ramifications. Toxicity limits of caffeine are known, but limits of the other roast coffee chemicals are not known. Whether blood pressure rises or falls often depends on the dose, the activity of the person, and the person's individual body reactions. Because of all these variables it becomes difficult at times to generalize about the body's reactions. But body reactions there are. When caffeine and coffee dosages are high, there are adverse effects.

So within this maze of facts it is not possible to flatly say that coffee drinking is healthy or unhealthy. It is wise to listen to those scientists who have measured and reported their findings, and to use coffee (with its caffeine content) in moderation and to govern use as the body of an individual reacts to it.

TABLE 15.13. CHRONOLOGICAL GROWTH IN COFFEE USE

| Year | Commerce/year Green coffee beans Millions bags (60 Kg) | Industrial Events |
|---|---|---|
| early 1800's | 0.20 | European Continental blockade |
| mid 1800's | 2.5 | Mechanized looms, wrought iron, steam engine, telegraph, steel mill |
| 1876 | 7.0 | Telephone, more coal produced |
| 1900 | 14.0 | Marconi wireless invented, Diesel engine, steam turbine |
| 1914-18 | 16.0 | Panama Canal opened to ships, airplane and auto established |
| 1920's | 24.0 | Radio and motion pictures, communications |
| 1930's | 26.0 | Dams, highways, radios, chemicals |
| 1940's | 26.0 | Nuclear energy discovered, rocketry applied, radar, petroleum |
| 1950's | 28.0 | Television manufacture, incorporation |
| 1960's | 40.0 | Jet planes, more travel, synthetics |
| 1970's | 60.0 | Nuclear power applied |

TABLE 15.14. CAFFEINE DOSAGES
(in beverages and chocolate)

| | Beverage or food item | Milligrams caffeine per 5 fl oz beverage |
|---|---|---|
| 1. | Brewed Arabica coffee (40 cups/lb) | 100 |
| | Brewed Robusta coffee (40 cups/lb) | 200 |
| 2. | Arabica-instant coffee (⅓ green yield) | 70 |
| | Robusta-instant coffee | 140 |
| 3. | Cola soft drinks | 25- 40 |
| 4. | Guarana soft drink (Brazil) | 40- 80 |
| 5. | Tea | 60- 75 |
| 6. | Cocoa (55 mg theobromine) | 1- 2 |
| 7. | 1.5 oz. milk chocolate | 60- 80 |
| | 1.5 oz. semi-sweet chocolate | 100-120 |
| | 1.5 oz. bitter chocolate | 120-150 |

CAFFEINE SOLUBILITY IN WATER FROM 0-100 C

| Weight % Caffeine | Temperatures C | F |
|---|---|---|
| 0.6 | 0 | 32 |
| 1.0 | 15 | 59 |
| 2.1 | 25 | 77 |
| 2.7 | 30 | 86 |
| 4.3 | 40 | 104 |
| 6.3 | 50 | 122 |
| 8.9 | 60 | 140 |
| 12.0 | 70 | 158 |
| 16.0 | 80 | 176 |
| 33.3 | 100 | 212 |

# 16

# Caffeine and Decaffeination

## Toxicology and Tolerance to Caffeine and Coffee

In man, a fatal dose of caffeine is estimated as 100 grams or about 50 to 100 cups of average coffee, and is dependent on age, weight, tolerance, etc. Over 1 gram of caffeine or the equivalent of 5 to 10 cups of coffee (depending on strength and volume) can cause excessive influence on the central nervous and respiratory systems, characterized by restlessness, excitement, insomnia, and possibly mild delirium. Sensory disturbances, like ringing in the ears or flashes of light, are not uncommon. Hands can tremble and muscles can become tense. Breathing is quickened and urination is urgent. At the level of several grams of caffeine, convulsion, epileptic behavior, headache, palpitation, dizziness and nausea can occur. From a fatal dose of caffeine, death occurs from respiratory failure. Persons who are habitual coffee and tea drinkers, develop a tolerance to the effects of caffeine.

## CAFFEINE AND DECAFFEINATION

## Introduction

Eight to ten percent of the world's produced and exported coffee beans are decaffeinated, e.g. if 55 million bags are exported, close to 4.4 million bags green coffee are decaffeinated. About 55 percent, or 2.4 million bags, are decaffeinated in European countries: primarily, Germany, Switzerland, France, Belgium, Italy, Holland and Spain. About 40 percent, or 1.8 million bags per year, are decaffeinated in the U.S.A., and the rest elsewhere, e.g., in Brazil and in El Salvador, Central America. Decaffeinated coffee is consumed primarily in the U.S.A. and Europe; since they consume 80 percent of world imports, these areas use an average of 10 to 12 percent decaffeinated coffees. Localized areas like Germany and Switzerland can use 15 to 20 percent decaffeinated (or treated) coffees, while

U.S. use is close to 15 percent of all green coffee imports. Most decaffeinated coffees are sold in the instant form, e.g. in the U.S.A. only 6 percent is sold as roast and ground coffee, whereas 20 percent is sold as instants. There has been a continual growth in the percentage of consumers who use decaffeinated coffees, both in the U.S. and Europe. There has been a continual growth in the percentage of consumers, who use decaffeinated coffees, both in the U.S. and Europe. In Spain and Switzerland, there is a disproportionately high use of both instants and decaffeinated coffees. Decaffeination processes originated in Germany before World War I and have been a strongly pursued industry in Germany ever since, with probably 40 percent of European production there, followed by Switzerland, France and Holland.

Secrecy has been a policy of almost all coffee bean decaffeinators, so little process data has been published. Patents have been the major source of public information. However, process "know-how" has become progressively available as the industry has spread out and more and more companies and people have become involved. Fresh thinking has entered from new processors, and it has constituted a healthy, competitive situation contributing to improved product qualities and more efficient processing in the more progressive firms.

Decaffeinated coffees have always been second class in taste and aroma properties, as well as appeal. The reasons for this lie in the severe treatment to which the green coffee beans are exposed. The steam wetting, the chlorinated solvent rinsing, and the many hours of steaming to distill off the last residues of solvent tend to destroy the factors that contribute to natural coffee bean aroma and flavor. The green beans suffer further in the final step of drying to restore a shippable green bean; this final moisture is less than its original.

The net result is that decaffeinated coffee beans change color. They are often not as green as they started, but vary in shades from dark brownish black, to dark green (at best), to reddish brown, and sometimes discolored. Most of the bean surface waxes have been removed, giving the beans a dull appearance before and after roasting. If sugars have been allowed to remain on the bean surface, these burn during roasting, lending an unsightly appearance to the roasted bean. In addition to the 1 or 2 percent caffeine extracted, another 3 to 4 percent of weight is lost. These losses are added into "contract" decaffeination work. Decaffeinated beans roast differently than natural green beans, with less bean expansion, less "popping," etc., because the sucrose content has been altered.

## Market for Decaffeinated Coffees

In the period from 1960 to 1970 decaffeinated coffee use increased 3 fold in the U.S.A., according to the PACB, rising from 0.06 to 0.18 cups/person/day; but by 1976 it was 0.31 cups/person/day, or 15 percent of cups of coffee consumed. This shows continued and startling rise in decaffeinated coffee use.

In the same period, per capita coffee use in the U.S.A. fell from 22 to 17 lbs green/person/year. This period was also one in which Robusta coffee imports into the U.S.A. increased from 3.8 to 6.7 million bags/year, or a 67 percent increase in ten years. A somewhat similar increased use of Robustas occurred in Europe. All these facts can be used to give the following interpretation.

Robustas have twice the caffeine content of Arabica mild/Brazil beans. Robustas have been used increasingly in the lower priced and commercial coffees. Hence, the consumer (unknown to him) has been taking into his body 50 percent to 100 percent more caffeine than before. Pressure roasting and steaming roast beans, in order to drive off the unpleasant Robusta flavor, also have caused the resulting roasted beans to be more thoroughly extracted in all components including caffeine.

Hence, the marginally physiologically affected consumer (by caffeine and roasted coffee products) passes over into the affected area. These consumers then switch either completely away from coffee, decrease their use (as per capita use trends show), or switch to decaffeinated coffees.

Ironically, many of the decaffeinated coffees are made from Robusta coffee beans for 3 reasons: (1) more caffeine yield for the processor; (2) Robusta coffee beans cost less; and (3) so much flavor is lost in decaffeination that good quality beans are not used. About 1968 the major U.S. brand of decaffeinated coffee was markedly improved in flavor, simply by the more thorough removal of traces of chlorinated solvent used in decaffeination.

**Quality and Prices**.—The taste and aroma of decaffeinated coffees fall far short of the non-decaffeinated coffees, as a rule. The cause of this can be green bean blend, processing, or both. Some decaffeinated coffees at the R&G stage are fairly good, but most of the instant coffees are rather flat. The person who buys decaffeinated coffee uses it out of necessity as an aid to his physical well being. He makes the flavor compromise willingly within the limited choices of decaffeinated coffee quality available to him, often at prices 20 percent higher than for non-decaffeinated types.

## CAFFEINE HISTORY AND STATUS

### Ludwig Roselius—Inventor of Decaffeinated Coffee

In Jacob's book on coffee, he tells of the origin of decaffeination. Ludwig Roselius's father, a coffee taster, died prematurely and Ludwig ascribed the death to coffee poisoning. This led him to study the deleterious effects of coffee in other persons. He found that coffee caused heart trouble, gout, and arteriosclerosis, and was forbidden by doctors to patients with diabetes and liver troubles. There was no question that coffee was a contributory factor. But life in the latter part of the 19th century had become faster. New ailments became more prevalent than before.

In 1820, Ferdinand Runge, a German chemist, isolated caffeine from coffee beans. These beans were sent to Runge by Goethe, because of his close friend's insomnia caused by drinking coffee. This discovery caused considerable excitement, since now the purified drug (also from other sources) was made available for prescription in pharmacies.

With German perseverance, Roselius about 1900 decided to produce a caffeine-free coffee left with the aroma and agreeable drinking qualities in the coffee bean. At the same time he could draw people away from drinking substitutes like acorns, wheat, and barley.

Since coffee flavor is developed in roasting the beans, it was logical to remove the caffeine from the green coffee bean. The size of the original green coffee bean could not be reduced or it would not roast properly. Steam superheating of the green coffee beans with a little alkali or acid prepared them for chloroform, benzene, or similar solvent extraction of the caffeine. Subsequently, the solvent was steam distilled away from the bean. The green coffee beans were dried and then roasted in the conventional way.

In 1906 Roselius founded the decaffeination firm Kaffee HAG in Bremen, which by 1912 challenged Hamburg as a coffee city. Similar decaffeination operations were started in the U.S.A. by Kaffee HAG before World War I, but were expropriated by the U.S. Alien Property custodian during World War I. General Foods later bought out the Sanka trademark in 1932 and the process for exclusive use in the U.S.A. Kaffee HAG did not return to the U.S.A. It was not until the mid 1950's that Nescafe and C&S introduced their decaffeinated coffees into the U.S. market to compete against Sanka, which had a virtual market monopoly.

## Commercial Caffeine Origin and Production

Caffeine was isolated in pure form by Runge and others in 1820. It was prepared by methylating theobromine by Strecker in 1961.

Some of caffeine's natural sources are as follows:

| *Name of plant* | *Part of Plant* | | *% Wt* |
|---|---|---|---|
| Coffee (bean) | seed | Arabica | 1.1 |
| | | Robusta | 2.0–2.2 |
| Tea | leaf | | 2–4 |
| Guarana (Brazil) paste | seed | | >4 |
| Cola or Kola nut | seed | | 1.5 |
| Cocoa pod | seed | | 0.1 |

Caffeine's primary commercial source is the methylation of theobromine from cocoa wastes, and its secondary commercial source is the solvent or water extraction from coffee beans; it is also extracted from tea leaf wastes.

Two-thirds to three-fourths of caffeine is used in cola beverages. The balance of the caffeine is used in headache and cold remedies.

Production imports, and price of caffeine in the U.S.A., have changed since 1945, as seen below:

| YEARS . . . . . . . . . . . . . | 1945 | 1969 | 1977 |
|---|---|---|---|
| Production—U.S.A. | 725,000 lbs. | 2,000,000 lbs. | — |
| Imports into U.S.A. | 300,000 lbs. | 2,000,000 lbs. | — |
| (mostly from Germany) | | | |
| Price in U.S.A. | | | |
| (truck load) Dollars | \$3.75–\$7.00/lb.<br>\$3.10/lb. | \$2.10/lb. | \$3.10/lb. |

The market price of caffeine is determined not only by demand for use in cola beverages and headache remedies, but by the production from synthetic sources, e.g., methylation of theobromine from cocoa wastes.

## Decaffeination Methods (Solvent and Water methods)

Many patents have been granted for producing decaffeinated coffee. There are two classes: (1) solvent extraction of the caffeine after various preliminary treatments that may include steaming, acid treatment, and alkali treatment and (2) water extraction of the caffeine. Solvent extraction has been the method used for producing almost all of the commercially decaffeinated coffee. Acid and alkali treatment of coffee beans has never produced a salable decaffeinated coffee because the flavor of the product was inferior to the coffee resulting from the normal process.

**Solvent Extraction of Caffeine.**—There are five steps: (1) the moisture content of the green coffee is increased by steaming from 10 percent to about 40 percent; (2) the coffee is treated countercurrently with an organic solvent methylene chloride, for 12 to 18 hrs—time sufficient to extract 97 percent of the caffeine; (3) the coffee is steamed with live steam to remove all of the residual solvent; (4) the excess moisture is removed by air or vacuum drying; and (5) the decaffeinated coffee is roasted. The first three operations are carried out in a rotating drum.

Solvent is recovered from the caffeine-solvent mixed solution by distillation in a continuous natural circulation evaporator. The concentrated residue is a solution of about 60 percent caffeine and 40 percent other materials, composed largely of the wax that originally constituted the protective layer on the outside of the coffee bean. This wax has never been used for any commercial purpose. The solution is sent to the caffeine refining operation for the recovery of 99.9 percent pure, pharmaceutical grade caffeine crystals. Caffeine is transferred from the solvent phase to water phase, concentrated, crystallized, and refined.

In the 1960's, some European decaffeinators used flammable solvents and had disasterous fires and explosions, subsequently resulting in more modern facilities using chlorinated solvents like methylene chloride. In the early 1970's medical research found that chloroform and TCE (tri-chlorethylene) contributed to cancer in rats and mice. Since 1976 TCE has been banned by the U.S. FDA and methylene chloride has been used. Less than 10 ppm solvent residues must be obtained in the roasted coffee beans; in fact, higher residues leave a very bad taste.

**Purification of caffeine.**—The hot water solution with dissolved caffeine is cooled to crystallize out crude caffeine. Resolution of the crude caffeine crystals in water, carbon treatment, filtration, and recrystallization by cooling produces purified caffeine crystals. Free water is centrifuged away, and the one molecule of water associated with each molecule of caffeine crystal is dried with steam heated air in a rotary drier.

**Water Extraction of caffeine.**—(A process developed to minimize solvent residues) Green coffee beans are countercurrently extracted in a percolation column battery with a caffeine depleted but saturated water solution of green coffee solubles (about 25 percent solubles). After the caffeine enriched solution leaves the green bean battery, it is liquid/liquid GATX extracted with TCE, a chlorinated solvent. The caffeine is recovered from the solvent and purified as described above.

## INDUSTRIAL ASPECTS OF DECAFFEINATION PLANT AND PROCESS

Engineering knowhow for a commercial decaffeination plant can be purchased from several firms and should include all engineering details for purchase and construction of process equipment for a 20 ton green beans/day plant. Time elapsed between contract and operation of the plant would be two years. Land, building services, and engineering on site are extra.

Some of the features that characterize a decaffeination plants performance are as follows:

1. About five times the volume of solvent must be stored relative to weight of "as received" green beans. For example, a 20 ton/day plant needs at least 100 tons of solvent, such as methylene chloride, in the main storage tank, plus solvent "in-process."
2. Since the green coffee beans are rinsed with solvent on the basis of 1.5 tons

of solvent per ton of approximately raw green beans per cycle, over 25 tons of solvent are evaporated per day. Although methylene chloride has a low boiling point (40° C) and a latent heat of vaporization one-seventh of water (142 BTU/lb), it still requires 4 tons of steam heat per ton or green beans to distill the solvent for recovery. This, however, can be reduced 2 to 3 fold by programmed reuse of solvent (last rinses used on first rinse).

3. Since steaming is used for pre-heating and wetting of the beans, as well as steam stripping solvent from the beans, we can assume at least 1.0 to 1.3 tons of steam per ton of raw green beans.

4. The moist (45 percent moisture) green beans must be dried, so the equivalent steam load is about one ton of steam to remove one ton of water associated with the green beans, and possibly much more when considering drying efficiencies and heat losses or vacuum drying.

5. Overall, a 4 ton beans/day decaffeination plant uses 5 tons of steam/hr, or a 20 ton beans/day plant will use or need 25 tons of steam/hr, or 40 gal. fuel oil/hr. Since fuel oil or heat energy is becoming more and more expensive, heat conservation is important to economy.

6. All the live steam used to strip solvent from the decaffeinated beans must be condensed to recover solvent; hence, this constitutes a great deal of cooling water. For example, for a 20 ton beans/day decaffeination plant and a 15° F water temperature rise for the cooling water, we have about 4 tons of steam to be condensed per hour, corresponds to a water flow of over 1500 gpm and would ordinarily require a sizeable cooling tower, as well as an adequate makeup (150 gpm at 10 percent and allowable source of disposal (from cooling tower 4 percent is 60 gpm).

7. Condensed steam distillate has to be cooled to recover the solvent, since methylene chloride has a 1.3 percent solubility in water at 25° C, and this solvent must be stripped from the water before it can be disposed of or returned to the boiler.

8. Condensers to condense steam and solvent during stripping are large, e.g. 16 $m^2$ surface, and are not only costly, but call for top quality engineering, shop construction and installation. Also, the solvent condensers for a 20 ton beans/day plant are huge, e.g., 60 $m^2$ surface, with an evaporator of 10 $m^2$ meters of surface, etc.

9. Electrical use is projected at 300 KwH/ton of dry decaffeinated beans in a Buss-HACO or Coffex-Nestle rotating vessel system, and would be less in the SEDA stationary vessel system.

10. Solvent losses are taken as 2 percent, which is 20 Kg of methylene chloride per ton of green beans.

11. Labor costs for the 20 ton per day plant would be six men in a 24 hour period, including lab and supervision.

12. Maintenance labor and material costs, as well as cost of spare parts, must be allowed for.

13. Material losses due to decaffeination usually mean that 3 to 5 percent by weight of incoming green beans is lost in the process above the 1 to 2 percent caffeine removal and moisture adjustment. When making instant coffee from decaffeinated coffee beans, there is invariably a 3 percent lower solubles yield under the best of circumstances and could be 4 percent lower solubles yield, depending on the roast.

14. Neither caffeine recovery costs nor the specific equipment indicated here, but assuming 2.2 percent wt caffeine is recovered from Robusta beans, then 44 lbs of caffeine will be recovered per ton.

## COMMENTS ABOUT INDUSTRIAL DECAFFEINATION

Caffeine concentrations during repeated rinsings:

The rate of diffusion of caffeine out into the methylene chloride solvent is slow and controlling, and this is why it is possible to eliminate rotary vessels or stationary vessels with internal mixing blades, as in the SEDA method. Such a SEDA modification reduces initial plant investment as well as subsequent maintenance. It is instructive to visualize how the caffeine concentration changes during 15 solvent rinses of the beans. If we assume that the green beans are Robustas with 2.3 percent caffeine (dry basis), and the beans have 40 percent moisture on initial extraction with solvent, the first solvent contact removed can have 3.0 percent caffeine.

Note from the solubility table that at 25 C, caffeine solubility in water is 2 percent and in methylene chloride 8 percent; whereas at 38 C caffeine solubility in water is 4 percent and methylene caffeine 10 percent. Each solvent contact rinse cycle can take 1 to 2 hours. Since most of the caffeine is removed in the first few contact rinses, the latter rinses are to get the caffeine level in the bean down to 97 percent removed. For example, if the solvent to bean weight ratio is taken as 1.5/1.0, we may have the following pattern of caffeine concentrations in the solvent rinses:

PERCENT CAFFEINE LEFT IN BEAN AFTER RINSES (DRY BASIS)

| | | | | | |
|---|---|---|---|---|---|
| Rinse 1 | 0.80 | Rinse 6 | 0.25 | Rinse 11 | 0.080 |
| Rinse 2 | 0.64 | Rinse 7 | 0.20 | Rinse 12 | 0.065 |
| Rinse 3 | 0.50 | Rinse 8 | 0.16 | Rinse 13 | 0.050 |
| Rinse 4 | 0.40 | Rinse 9 | 0.13 | Rinse 14 | 0.040 |
| Rinse 5 | 0.33 | Rinse 10 | 0.10 | Rinse 15 | 0.030 |

NOTE: For actual concentration of caffeine in bean, multiply percent in table by percent caffeine at start.

In commercial practice, it is not clear why 97 percent caffeine removal was chosen. In fact, there have been products with 92 percent caffeine removed, and from a practical point, even 80 percent removed ought to be marketable.

Industrially, it is not necessary to analyze the caffeine removed from each solvent rinse because from repeated experiences, it is known what is happening and only the final caffeine content is worth analyzing. The operator actually has a good guide as to the adequacy of the caffeine extraction, because he can take a few ml solvent extract, let it evaporate, and note the amount of white caffeine crystal residue on a watch glass.

It is important that during the initial wetting of the green beans they are not saturated with water; otherwise, free water saturated with green coffee solubles will result, which becomes a solubles loss. This aspect is especially important during the steam distillation of the solvent away from the decaffeinated bean, and the steam quality entering must be rigorously controlled.

Color of the final decaffeinated bean is strongly influenced by the acidity or alkalinity of the green beans. When this is not controlled, final bean colors vary from orange to red-brown to green-brown.

Drying of the decaffeinated coffee bean also influences color, especially when the steamed beans are exposed to the atmosphere and oxidation occurs, rapidly turning the yellow-green beans to dark brown. Since the public never sees these beans, its color is less important in "in-house" processing like at General Foods or Nestle. But where green decaffeinated beans are desired for resale, it is very important to try to obtain a final color approximating the original bean, with surface discolorations or dark spots (e.g. leached sugars) to a minimum. This is accomplished by polishing the dried beans to produce a smooth, clean, uniform surface. It is notable that the best preservation of the green coffee across the drying step is by vacuum drying, which does not allow the wet steamed beans to come into contact with air (oxidation), and the vacuum conditions keep the beans at near 30 C, no higher. However, vacuum drying is more expensive in initial investment and in operating costs.

It is important to realize that solvent extraction removes almost all the waxes on the bean surface, and this results not only in a dull surface, but removes a natural protective barrier. The result is that the green decaffeinated beans are dull in appearance when green and roasted. Polishing improves this situation. Further, the dried decaffeinated beans must be dried to about 6 percent moisture, compared to the 12 percent of non-decaffeinated beans. The reason for this is that decaffeinated beans are more readily subject to mold and fermentation. This additional drying degree and load further adds injury to the decaffeinated bean—double drying and drying to a lower moisture level. It is well known that the lower the moisture in dried beans, the greater is the loss in flavor and aroma, whether decaffeinated or not.

Reference to the previous page itemizing operating costs in 20 ton beans/day decaffeination plant shows an approximate breakdown: cents per lb green beans: (1) fuel 0.8¢; (2) electricity 0.8¢; (3) water 0.5¢; (4) solvent losses 0.5¢; (5) labor 1.0¢; (6) maintenance 1.0¢; (7) amortization 10 year, 2.5 to 3¢; (8) overhead 1.0¢;

(9) other taxes, waste disposal, fringe benefits, interest on investment, etc., 1¢; (10) credit for caffeine 3¢/lb. Total operating costs are about 6¢/lb green beans, plus 4 percent wt loss at $2.50/lb green in 10¢/lb green bean. Commission costs, therefore, must be 18 to 20¢/lb green bean, plus whatever transport/finance costs are to/from.

## Seda Decaffeination of Green Coffee Beans

Green coffee beans are dumped from burlap bags, passed through an air cleaner, are weighed, and are delivered to the top of the extraction vessel. Steam with a certain wetness is then delivered to the extraction vessel below for several hours until the coffee beans are wet, hot, and swollen. Steam hydrolysis liberates caffeine for subsequent extraction. Solvent rinsing is then begun by slowly filling the extraction vessel with solvent until all beans are covered. The caffeine from the wet beans, ranging in concentration from 1.0 to 3.0 percent, diffuses out into the solvent. Each soak cycle takes about 1.5 hours, after which the caffeine rich solvent passes to the solvent with caffeine storage tank, before being pumped to the evaporators. The sequence of rinses of the beans is made so that the solvent may be used up to three times, since solvent picks up very little caffeine in the last four or five contacts with beans. Steam stripping away the solvent from the decaffeinated beans then takes place for several hours. The solvent residues in the bean must be below 15 ppm and usually are taken below 5 ppm.

Condensing steam-stripped vapors is essential to solvent recovery and calls for much cooling water, preferably at low temperatures to minimize the size of the condensers. Solvent evaporators are needed to recover most of the solvent, while at the same time concentrating the residue caffeine until a saturated solvent solution is obtained. The caffeine is then transferred to an aqueous media, which has higher caffeine solubility, and the aqueous, hot, saturated solution is dumped into indicated tank, where it cools and allows crude caffeine to crystallize out. The mother liquor is drawn off and recycled. Solvent recovery systems are required to avoid solvent losses from the process system. Mineral oil or carbon bed systems may be used which automatically recover and return solvent from the vented lines. Sight glasses are provided in appropriate lines to see solvent flow and to judge when solvent residues result, e.g., as in steam stripping. Decanters are required to automatically separate the water and solvent phases, the solvent being heavier.

Drying the wet coffee beans after decaffeination may be carried out with warm air in rotary, fluid bed, and in vacuum driers, depending on the quality of end product sought. Polishing the finally dried beans may be desirable when whole bean appearance is important.

Laboratory measurements for solvent residues and caffeine residues is essential on a routine and continuing basis. High solvent residues give a very unpalatable roasted coffee and beverage. There are standard procedures for these tests and minimum standards must be maintained. Final bean moistures must be reduced to about 6 percent because decaffeinated beans, due to their loss of surface waxes, are much more susceptible to mold growth if moistures are higher, like incoming beans at 12 percent. Bean colors of decaffeinated coffees range from dark brown to light brown to orange to even green, but this depends on the processing conditions imposed, methods of drying and polishing, etc.

Taste changes resulting from properly decaffeinated coffees usually makes Robusta coffee beans much milder in taste and aroma, and often upgrades harsh Arabica coffee beans into more mild and more acceptable tasting beverages.

Costs of decaffeination on commission basis are today at least 25¢ per lb and include (the caffeine plus other substance) weight losses on a dry basis of 4 to 5 percent, plus whatever shipping to and from costs are involved. However, if one invests in the decaffeination plant, all costs ought to run (including five year amortization) less than ten cents per pound of green coffee beans, and of course, the weight loss indicated on a dry basis. Since the process is slow, only one man is required per shift in a 4 ton per day plant, and two men per shift for a 20 ton per day plant, leaving considerable slack time for cleaning, breaks, etc.

Safety in operations is axiomatic, once suitably selected and trained personnel are on the job. Only nominal features in the plant layout and building arrangement are required, since the solvent is sealed in the system.

Solvent losses for methylene chloride are experienced and projected at 30 grams maximum per Kg beans. Caffeine purifications can be made periodically as adequate amounts of crude caffeine accumulate.

The S.E.D.A. process is typical of all commercial decaffeination processes, but includes the economy of stationary extraction vessels instead of rotary drums with rotary seals, as well as knowhow gained in many years of commercial processing.

## CAFFEINE BY METHYLATION

### Caffeine Physical and Chemical Properties

Pure caffeine is a white powder. Sublimed crystals are hexagonal prisms, and crystals from saturated aqueous solutions are long, silky white needles which mat together readily. It crystallizes from water solutions with one molecule of water, which effloresces gradually in air at ambient temperatures, rapidly at 80 C, and completely at 100 C. When caffeine crystallizes from non-aqueous solvents, it is anhydrous, melts at 235-237.5 C, and sublimes without decomposition at 176 C at atmospheric pressure. It sublimes at lower temperatures when held at lower pressures. Considerable caffeine is recovered from sublimed product collected in the chimneys of coffee bean roasters. See the article on drying refined caffeine at Certified Processing in Hillside, N.J.

Caffeine is odorless, but has a bitter taste. Its aqueous solutions are neutral to litmus. However, it is considered a weak mono-acidic base; its salts dissociate when aqueous solutions are evaporated. The salts with acids are not stable. It does form relatively stable combinations with sodium benzoate and sodium salicylate which are used in oral and intramuscular medicinal applications. Caffeine is decomposed by hot alkalies and reacts with chlorine. Limewater does not affect caffeine, and so it is used in some extraction processes. Caffeine is taken up by ion exchange resins and can be separated from theobromine on fine granules in a column of Fuller's earth (analytical technique).

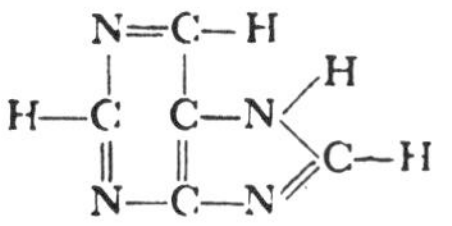

Purine

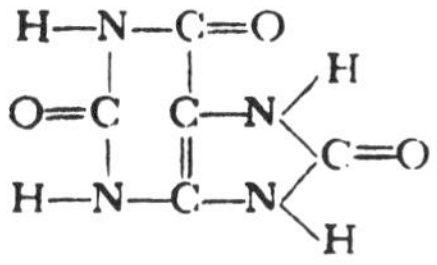

Uric Acid

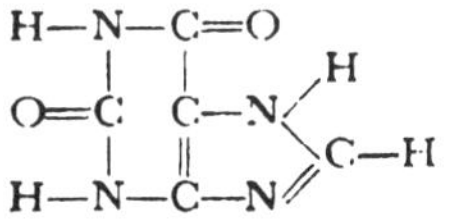

Xanthine (dioxypurine)

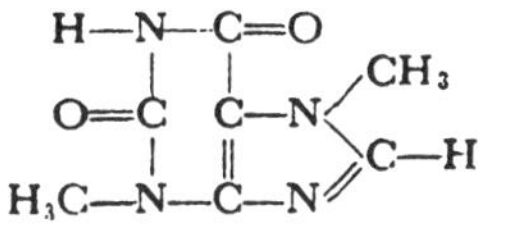

Theobromine (dimethyl xanthine)

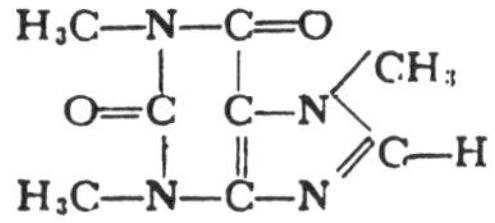

Caffeine (trimethyl xanthine)

Caffeine solubilities in various solvents is listed in Table 16.1. Solubility distributions between methylene chloride and water at 80° F is virtually linear from 0 to 1.0 percent caffeine in water in equilibrium with 0.7 percent wt caffeine in methylene chloride, e.g. 0.5 percent wt in water to 0.3 percent wt in methylene chloride. Solubilities increase at 38° C or 1 Atm pressure, and some processors operate at several atmospheres solvent pressure to speed up the caffeine transfer from bean.

There is not yet any published data on caffeine solubility or Chlorgenic acid solubilities in high pressure and liquid carbon dioxide, but since patents have been taken in this field, such data and industrial developments may evolve in the next decade.

## CAFFEINE REFINING AND DRYING

With today's fuel costs and uncertain supply, efficient drying of a product can be a decisive factor not only in the profitability of an operation but in its feasibility. At Certified Processing, this efficiency is also a matter of product quality.

The Hillside, N.J. firm processes caffeine derived from the decaffeination of coffee, and purifies it for use in soft drinks and pharmaceuticals. The purification process produces a centrifuge cake with about 35 percent moisture that must be dried to 0.25 percent moisture—uniformly and with negligible attrition to the crystals. Should their moisture content exceed 0.25 percent, they would have to be reprocessed. Any substantial breakage of the crystals would render the product unacceptably dusty and non-free flowing.

To achieve these product qualities at considerably lower processing and maintenance costs, Certified switched from vacuum batch drying to a continuous tray-type unit called a *Turbo-Dryer,* which has reduced:

A 10-12 hr drying cycle to less than 1 hr.
Fuel cost by a factor of 2 to 3.
Labor requirements to ¼ man-hour.
Product loss resulting from former handling and cleaning of trays.
Floor space requirements by a factor of 2.

**Details of Dryer**

The continuous dryer at Certified is about 9½ ft (2.9m) diam and 15 ft (4.6m) tall. Construction is Type 304 stainless, including a vertical stack of 28 circular trays and integral turbine-type fans. Auxiliaries include an external steam heater for the drying air, a separate exhaust fan, interconnecting duct-work, inlet air filter and a wet scrubber. Almost all of the dried product is discharged from the base of the unit, as normally very little material is carried over into the scrubber.

In operation, the trays rotate slowly as material to be dried enters the dryer at the top and drops onto the top tray. As the trays rotate, the product is wiped off of each tray to the one below by a stationary wiper (diagram). As material drops onto a tray, it is leveled on the tray to increase the surface for heat and mass transfer. Horizontal internal circulation of air from the fans permits zoned drying and high evaporation rates. For operating flexibility, it is possible to vary fan speed, tray speed, layer thickness and air temperature.

## DECAFFEINATION ANALYSES

Four important analyses are required to be carried out routinely in a decaffeination plant: 1) caffeine; 2) chloride solvent residues; 3) pesticide residues; and 4) moistures.

## Caffeine

Until the late 1960's, wet chemistry methods were most common. The official A.O.A.C. method used Phospho-molybdic acid to remove caramelized impurities from coffee solutions, followed by chloroform extraction of the caffeine, evaporating solvent, and weighing caffeine crystals. This method was not as popular as the MgO treatment to remove impurities from aqueous solutions, followed by chloroform extraction and measurement of spectrophotometric absorption at 273 millimicron wavelength vs a standard caffeine solution. An aluminum amalgam has been used to remove chicory impurities.

Most large decaffeinating firms have gone to automated systems for caffeine analysis, so as to reduce labor, increase accuracy, and increase numbers of samples analyzed so that production personnel have a better insight into processing circumstances. Before 1960, or even before 1965, firms depended on wet chemistry methods. Results were slow in forthcoming, not always accurate or confirmable, and often process engineers simply ran the plant by the seat of their pants, hoping that 90 percent decaffeination was obtained in the final product.

Caffeine in solution samples are set up in test tubes on a programmed, rotating turntable. The samples are, one by one, sucked up by a tiny proportioning pump, simultaneously with dilution water and methylene chloride. The three streams are injected with air, and passed through mixing coils (glass-small), wherein the caffeine transfers into the methylene chloride phase. The water-methylene chloride mixture goes to a phase separator, and the caffeine in methylene chloride flows through a spectrophotometer. The difference in the selected light wavelength, compared to the pure solvent and to control samples, is recorded as caffeine concentration on a strip chart, or can be printed out in digits corresponding to the numbered sample. Needless to say, such automated continuous processing of samples yields great accuracy and reliability in the results obtained, and production personnel can make frequent samplings to determine test variations in their processing.

HAG in Bremen, Germany has a somewhat different system. Samples are boiled with MgO, filtered, cooled, dosed with 5-amino-quinoline (internal standard), and programmed at 5 min intervals from a turntable into a gas chromatic column (Chromosorb G treated 100/200 mesh with 2 percent Carbowax 20 M, column at 220 C, with helium gas purge at 30 ml/min, and flame ionization detector). See abstract below dated 1974.

HACO near Berne, Switzerland also uses a chromatographic system. Abstract below (1973) gives some details. The procedure for natural origin and decaffeinated coffees differs. Original coffees and extracts are reflux extracted directly with chloroform, to which pyrene (internal standard) is added after filtering. Decaffeinated coffees are refluxed with 96 percent ethanol, filtered, vacuum dried recovery of solid residue, which is then water solubilized and filtered again. Then chloroform extraction is carried out and the solution is concentrated in a vacuum evaporator and dosed into the chromatograph. Three distinct peaks are sensed (flame ionization) in the following order: chloroform, caffeine, and finally, pyrene within 15 minutes.

A.S.I.C. meetings have produced a number of technical reports on caffeine analyses:

1963 Paris. R. F. Smith and F. Albanese (HAG).

1965 Paris. J.J.L. Willems (TNO); F. Verlangia (Brazil).
1967 Trieste. O.Z. Vitzthum (HAG); P. Navellier (I.F.C.C.).
1969 Amsterdam. Fazzina (G.F., U.S.A.) and P. Navellier (I.F.C.C.).
1971 Lisbon. Xabragas and Polonia (Angola); Lopez (Mozambique).
1973 Bogota. R. F. Smith; Bibliography; F. Chassevent (IFCC).
1975 Hamburg. R. F. Smith; Bibliography; Quijano (caffeine sublimation-simple micro technique).

**Direct Analyses of Caffeine**.—Virtually all the wet chemistry and GC methods cited to date require sample preparation, such as treatment with MCO, refluxing, solvent extraction, and resolutions, and even in the Quijano sublimation method, post extraction and solutions are required. The Italian paper by Carisano of STAR SpA (1972) offers direct determination without sample preparation on a GC system.

In 1971, in the Journal of Chromatography V62, Wolford, Dean and Goldstein reported a novel, accurate and direct method for caffeine analyses inadvertently discovered when headache tablets were routinely separated during drug examinations. A Bio-Rad chelating resin, 200-400 mesh is saturated with copper ions (by copper sulfate solution), and then the caffeine sample is metered through the resin by continuous flow, and the caffeine is absorbed on the resin. Subsequently, the caffeine is eluted with 3M ammonia and the resulting effluent is monitored by a Beckman DB-GT with flow cell at 260 millimicrons vs standard. Coffee, cola beverages, tea, etc., can be applied to the column directly without pre-treatment. Results of liquid chromatograph of caffeine are shown below.

SEDA in Palencia, Spain, uses the European Association of Decaffeinators' method for caffeine analysis (AOAC spectrophotometric method applicable to decaffeinated coffees). A 1 gram green or roast decaffeinated sample or 0.5 gram soluble coffee sample in a 100 ml beaker is mixed with 5 ml ammonia hydroxide (1+2) and is heated a few minutes on steam bath; then 6 grams of Celite 545 are mixed in thoroughly. An acidic and a basic column are prepared: (15.024). ACID-glass wool plug in 25x250 mm tube, followed by 2 gm Celite 545 (acid washed) and mixed with 2 ml 4N sulfuric acid tamped in above wool, then wad of glass wool above. BASIC-column is 3 gm Celite 545 plus 2 ml 2N sodium hydroxide, mixed and tamped into similar tube. The sample is tamped in above the "acid Celite" (see sketch). Pass 150 ml ethyl ether ($H_2O$) saturated through columns in the series; discard it. The tamped sample and alkaline tube is above the acid Celite tube. The caffeine has been eluted from the upper alkaline Celite to the lower acid Celite beds. Rinse the lower acid column with 50 ml diethyl ether. Then rinse the lower acid column with 50 ml ($H_2O$ sat.) chloroform, thereby eluting the caffeine; make up to 50 ml with $CHCl_3$; read at 276 mu vs $CHCl_3$ blank. Repeat spectrophotometric readings with caffeine standards ¼, ½ and ¾ mg per 50 ml.

## Analysis for Chlorinated Solvent Residues in Decaffeinated Coffees

The removal of residual amounts of chlorinated solvents from the decaffeinated coffee is done with steam for perhaps 8 to 12 hours, or until there is less than 10 ppm solvent left in the green coffee beans (dry basis). The reason this stripping of the solvent takes so long is because it is dissolved in the oils of the coffee bean, and must diffuse out of the bean and be carried away by the purging steam flow at atmospheric pressure. Stripping out TCE was much more difficult than methylene chloride because the latter is much more volatile. The 10 ppm upper limit has been set by the U.S. FDA, and 25 ppm are allowed in instant coffees. If solvent residues are 30 to 40 ppm or more, they characterize the resulting coffee beverage with a very bad taste. Such residues and bad tastes did occur in commercial coffees decades ago.

The following sketch shows the European decaffeinators' method for this analysis. Fifty grams of R&G coffee finely ground are placed in a liter flask with about 500 cc of water. This is kept at near boiling for 5 hours. During this time about 80 ml/min or 5 liters per hour of air that has been cleaned with sulfuric acid and alkaline permanganate, heated to 950 C, followed by a solution of sodium arsenate, bubbles through the coffee solution to strip out the residue solvent. The stripping air passes through the 905 C furnace in quartz tubes and bubbles through the alkaline arsenate solution. Solvent chlorides are converted to chlorine in the furnace, and oxidize the arsenate solution, forming chlorides. At the end of the stripping and absorption process, the chlorides are titrated with silver nitrate solution using potentiometric instruments and silver and mercuric sulfate electrodes. The instrument is "zeroed" with sodium arsenate solution, and standardized with sodium chloride dosages. Analyses of some commercially decaffeinated coffees in 1976 showed close to 6.0 ppm solvent residues.

Decaffeination Process, 1973—U.S. 3,671,262 (A. H. Wolfson; The Procter & Gamble Co.) Employing High Caffeine-Solvent Velocity and Exchange Rate to Reduce Decaffeination Time.

Decaffeination Process, 1972—U.S. 3,671,263 (J. M. Patel, A. B. Wolfson; The Procter & Gamble Co.) A semi-continuous, staged, countercurrent extraction process, involving high extraction temperatures, high-wetting moisture, high solvent exchange rates, and high solvent superficial velocity, is used to substantially decrease caffeine extracting processing times.

Decaffeination Process, 1974—Can. 939,962 (J. Patel, A. B. Wolfson; The Procter & Gamble Co.) A semi-continuous staged countercurrent extraction process involving high extraction temperatures, high pre-wetting moisture, high solvent exchange rates, and high solvent superficial velocity.

Decaffeination Process, 1973—U.S. 3,700,464 (J. M. Patel, A. B. Wolfson, and B. Lawrence; The Procter & Gamble Co.) Decaffeination total process and caffeine extraction times are substantially reduced by utilizing high caffeine extracting temperatures and high pre-wetting moisture ranges.

Coffee Bean Drying, 1973—U.S. 3,700,465 (B. Lawrence, A. B. Wolfson, and J. M. Patel; The Procter & Gamble Co.) Drying times for decaffeinated green coffee beans are substantially reduced by drying in a non-vacuum dryer at bean surface temperatures of 220 –300 F for 15–60 min.

Coffee Bean Processing, 1974—Ger. 1,960,694 (Coffein Cie Erich Scheele GmbH) Preheated coffee beans are washed with a chlorohydrocarbon such as 1,2-dichloroethane to extract waxlike substances and carboxylic acid 5-hydroxy) tryptamide without reducing original caffeine content.

Coffee Dewaxing, 1975—Ger. 2,031,830 (Hag AG) Waxy materials and 5-hydroxy-tryptamides are removed from coffee without appreciably affecting the coffee's caffeine content by treating the coffee beans, without any previous water or steam treatment, with methylene chloride.

Coffee Wax Removal, 1972—Br. 1,280,387 (A. G. Hag) Undesirable wax substances are removed from green coffee beans by treatment with methylene chloride.

Coffee Decaffeination, 1975—Br. 1,372,667 (Hag AG) Raw coffee is decaffeinated by adjusting its moisture content to 10–60 percent and selectively extracting the caffeine with liquid carbon dioxide containing water at a pressure above the critical pressure of carbon dioxide.

Coffee Decaffeination, 1973—Br. 1,290,117 (Studienges Kohle GmbH) Raw coffee is contacted with moist carbon dioxide in supercritical state to remove caffeine.

Caffeine Removal, 1973—Ger. 2,005,293 (Studienges Kohle GmbH) Caffeine is removed from coffee by treating with moist carbon dioxide at temperatures of 40–80 C under pressure of 120–180 atmospheres.

Dried Coffee Product, 1974—U.S. 3,809,775 (N. Ganiaris; Struthers Scientific & Int. Corp.) Process for preparation of liquid extract concentrate is described in which carcinogenic elements are eliminated by precipitation from chilled decaffeinated extract solution, which is thereafter dried. Caffeine is added to liquid extract before drying and after removal of carcinogenic elements.

Coffee Extraction—Br. 1,336,929 (Struthers Scientific & Int. Corp.) Caffeine is removed from aqueous coffee extracts to reduce solubility of carcinogenic materials, thus permitting them to be more easily removed. Caffeine is then restored to extract prior to drying.

Caffeine Recovery, 1974—U.S. 3,806,619 (K. Zosel; Studiengesellschaft KmbH) Process for obtaining caffeine from green coffee by recirculating moist carbon dioxide in supercritical state, which comprises removing caffeine from caffeine-loaded carbon dioxide by repeated treatment with water, and recovering caffeine and water from resultant dilute aqueous caffeine solution by recycling stream of air or nitrogen under superatmospheric pressure.

Decaffeinated Coffee Extract, 1975—Ger. 2,119,678 (Studienges Kohle mbH) Ground roast coffee is extracted with a dry, supercritical fluid and then moistened with water to a moisture content of 20–55 percent. The composition is then treated with moist, supercritical carbon dioxide to extract caffeine, after which it is extracted with water to remove water soluble constituents. The aqueous extract is then spray or freeze dried to yield a powder which is aromatized by the coffee oil obtained in the initial stage.

Decaffeination Process, 1973—U.S. 3,749,584 (R. H. Kurtzman Jr., S. Schwimmer; U.S. Secretary of Agriculture) Coffee, tea, and the like are decaffeinated by a procedure wherein caffeine is metabolized by a microorganism, *Penicillium crustosum*.

Caffeine Stimulation Reduction, 1969—Br. 1,157,919 (General Foods Corp.) The stimulating effects of caffeine in coffee are reduced by incorporating compounds like purine, xanthine, 1-methylxanthine, 3-methylxanthine, 7-methylxanthine, or theobromine.

Decaffeinating Process, 1973—Br. 1,313,047 (General Foods Corp.) Hydrated green coffee beans are extracted with aqueous mixture containing a solvent such as a mono-, di- or triacetic acid ester of glycerol to extract caffeine without extracting significant amounts of sugar or other coffee solubles.

Green Coffee Decaffeination, 1972—U.S. 3,682,648 (W. A. Mitchell, R. Klose; General Foods Corp.) Process for decaffeinating green coffee beans using solutions of esters of polyhydric alcohols and edible carboxylic acids.

Coffee Bean Treatment, 1973—Ger. 1,692,234 (K. Beil) Combination chemical and mechanical process for producing a non-toxic coffee by reduction of chlorogenic acid content of roasted beans.

Coffee Decaffeination, 1973—U.S. 3,740,230 (J. P. Mahlmann; General Foods Corp.) Improved decaffeinated coffee is prepared by removing green coffee flavors and aromas prior to counter-current extraction for water extracting caffeine after which flavors and aromas are returned to the decaffeinated coffee prior to discharge from extraction zone.

Coffee Decaffeination, 1975—U.S. 3,879,569 (O.V. Bremen, P. Hubert; HAG Aktiengesellschaft) Raw coffee is moistened to a moisture content of 10–60 percent, after which it is extracted with liquid carbon dioxide under conditions such that the carbon dioxide is maintained in a liquid state below 31.4 C.

Coffee Extract Powders, 1975—Ger. 2,152,793 (Hag AG) A coffee extract powder having the color of milled coffee is obtained by freezing an aqueous coffee extract while simultaneously foaming it with nitrous oxide or other inert gas, followed by freeze-drying.

Decaffeinated Coffee Extracts, 1974—Br. 1,346,134 (Hag, AG) Crushed, roasted coffee is extracted with a dry super-critical fluid to remove coffee oil containing aroma constituents. Residual material is first extracted with water, then extracted with supercritical carbon dioxide to remove caffeine. Residue is then again water extracted and extract spray dried to yield powder which is aromatized by recombination with coffee oil removed in first step.

Vegetable Oil Extraction—Br. 1,356,749 (Hag AG) Solvent extraction process for the recovery of vegetable fat and oils, employing gases under supercritical conditions.

Antioxidant from Green Coffee Beans, 1972—Br. 1,275,129 (G. Lenmann, O. Neunhoeffer, W. Roselius, O. Vitzthum; Hag AG) Caffeine is precipitated from an aqueous extract of green coffee beans. The precipitate is then made alkaline, after which an antioxidant material is extracted from the solution by solvent partition. The antioxidant may be used with fats, oils, aromatic substances, dried milk etc., and may be added to roasted coffee to improve its stability.

Coffee Bean Processing, 1974—U.S. 3,770,456 (W. Roselius, O. Vitzhum, and P. Hubert; HAG Aktiengesellschaft) Undesirable irritants are removed from raw coffee beans without reduction in caffeine content by extraction with low boiling organic solvent at temperature above boiling point of solvent but below 80 C.

Process for Removal of Irritants from Coffee, 1972—Ger. 2,031,830 (W. Roselius, O. Vitzthum, P. Hubert; HAG Aktiengesellschaft) Green, unroasted coffee beans are treated with a low-boiling organic solvent, e.g. dichloroethylene, trichloroethylene, methylene chloride, chloroform, acetic acid, ethylene esters, trichlorotrifluoroethane, dipropyl ether, petroleum ether, and/or hexane at a temperature of 50–80 C for 0.5–4.0 hours. The amount of solvent used in the process is 2–6 kg/kg coffee beans. The beans are then treated with saturated steam for 1–5 hours at 0.3 absolute atm.-3.0 atm. gauge, and finally dried to 8–13 percent moisture content. The amount of caffeine in the coffee is only slightly reduced, while the carbonic acid-5-(hydroxy)tryptamide content is reduced considerably.

Process for the Protection of Autooxidizable Materials, 1972—Ger. 3,663,581 (Gunter Saarbrucken, Otto Neunhoeffer, Homburg Saar, Wilhelm Roselius, Bremen-St. Magnus, and Otto Vitzhum; Bremen, Germany, assignors to Hag Aktiengesellschaft, Bremen, Germany) A process for protecting auto-oxidizable materials through the addition of an antioxidative substance in the form of an extract of green coffee beans.

Decaffeinated Coffee, 1975—Can. 957,894 (T. L. Fazzina, G. V. Jones, and R. P. Scelia; General Foods Corp.) Superior flavored decaffeinated coffees are produced by separately solvent decaffeinating a high grade coffee and separately water decaffeinating a low grade coffee and then blending the separately decaffeinated coffees together.

Decaffeinated Coffee, 1975—U.S. 3,840,684 (T. L. Fazzina, G. V. Jones, and R. P. Scelia; General Foods Corp.) Blends of high grade and low grade roasted and ground decaffeinated coffee having improved flavor, containing at least 20 percent high grade coffee which has been organic solvent decaffeinated, and at least 30 percent low grade coffee which has been water decaffeinated.

Coffee Extraction—U.S. 3,843,823 (R. A. Chaplow and R. A. Hodgman; General Foods, Ltd.) A filtered aqueous coffee extract is freeze concentrated and the concentrate subjected to a desludging-type of centrifuge to remove insolubles which would otherwise cause black spots to appear upon reconstitution.

Decaffeinated Coffee—U.S. 3,843,824 (W. Roselius O. Vitzthum, and P. Hubert; HAG Aktiengesellschaft) Rough ground roast coffee is extracted with a dry supercritical fluid, after which the coffee is wet with water and extracted with supercritical carbon dioxide to remove caffeine. The coffee is then extracted with water to produce an aqueous extract which is dried to yield a powder residue and combined with the coffee oil containing aroma substances from the original extraction step.

Green Bean Decaffeination, 1972—U.S. 3,669,679 (H. P. Panzer, R. S. Yare, and M. R. Forbes; General Foods Corp.) Decaffeination of green coffee is achieved by extraction with fluorinated hydrocarbons.

Green Bean Decaffeination, 1974—Can. 934,217 (M. R. Forbes, H. P. Panzer, and R. S. Yare; General Foods Corp.) Decaffeination of green coffee is achieved by extraction with fluorinated hydrocarbons.

Coffee Bean Processing, 1973— Br. 1,294,416 (General Foods Corp.) Green coffee beans, prewetted to a moisture content of 50 percent, are subdivided into particles so at least 90 percent are retained on a 16 mesh sieve, optionally decaffeinated and roasted.

Coffee Extraction, 1964—Can. 689,892 (W. V. White, M. Hamell, and E . Danielczik; General Foods Corp.) A flavor-bearing coffee fraction substantially free of caffeine and containing a minimum of nonaromatic oils is obtained from roasted coffee by wetting the coffee with an aqueous wetting liquor to form a substantially dry bed and contracting this bed with methyl chloride as the final extracting solvent.

Solvent Coffee Extract, 1967—U.S. 3,908,033 (N. Ganiaris; Struthers Scientific & Int. Corp.) Process in which caffeine is solvent extracted from green coffee beans, after which the beans are roasted and extracted with water under heat and pressure to yield an extract which is held under conditions to precipitate benzopyrenes and benzofluoranthanes. The liquid extract is then centrifuged to remove solids and mixed with the caffeine which was removed in the first step.

Process for Extracting Chlorogenic Acid From Coffee and Coffee Extracts by Chemisorption, 1974— Ger. 1,692,249 (G. Lehmann, O. Neuhoeffer, W. Roselius and O. Vitzthum; Hag AG) Chlorogenic acid is removed from extracts of green and/or roasted coffee using chemisorbents, which are either water-insoluble or water-soluble but rendered water-soluble as a result of Chlorogenic acid sorption and/or by heat denaturation. Suitable media are high molecular compounds having numerous acid amide groups, e.g. protein substances, polyamides, polyacrylamide, polyurethanes, condensation products of diamines and di-isocyanates or of formaldehyde and urea and/or dicyanodiamide and/or melamine. Graft polymers, e.g., protein post-treated with disocyanates and diamincs, may also be used. In an example, a roast coffee extract containing 10 percent (in terms of DM) Chlorogenic acid is treated with 30 percent activated polyamide powder for 30 min. The powder is then removed by filtration, and after washing with water, the coffee extract contains 60 percent of the acid.

Method for the production of Caffeine-free Coffee Extract, 1971—Ger. 3,843,824 (Vilhelm Roselius, Bremen-St. Magnus, Otto Vitzthum, Bremen, and Peter Hubert, Bremen-Lesum, assignors to HAG Aktiengesellschaft, Bremen, Germany).

Int. Cl. A23f *1/10*

U.S. Cl. 426–386 17 Claims

1. A method for the production of substantially caffeine-free coffee extract products which comprises

a. separating coffee oil containing aroma substances by extraction of rough ground roast coffee with a dry supercritical fluid having selective dissolving capacity for the coffee oil and aroma substances, at a pressure of at least 80 atmospheres excess and a temperature above 30 C.

b. wetting the thus extracted rough ground coffee with water and extracting the wetted product with wet supercritical carbon dioxide to remove the caffeine in the rough ground coffee.

c. extracting the rough ground coffee remaining from the previous step with water to produce an aqueous extract.

d. drying said aqueous extract to obtain a powder residue.

e. recovering the coffee oil containing aroma substances from the extraction of step (a) and adding at least a portion thereof to a coffee extract.

Process for Producing Low-caffeine or Caffeine-free Coffee—Ger. 1,805,391 (Valdmann, Joh. Jacobs and Co.)

Green coffee beans are steeped in water to cause expansion. The beans, which may be left in the water or drained, are then gradually cooled to −35 C at less than 0.5 C/min, preferably 0.1 C/min. This causes the water inside the beans to crystallize out. After thawing, caffeine is extracted with a solvent mixture, e.g. methylene chloride, containing methanol or ethanol. The solvent mixture is separated by vacuum steam. Adverse effects on flavor and quality are prevented.

Process for Producing a Raw Coffee with a Low Content of Stimulants—Ger. 1,960,694 (G. Kurz, H. O. Vahland; Coffeine Compagnic Dr. Erich Scheele GmbH)

Cleaned, dry, raw coffee beans are heated to about 40–48 C and are then washed, while being agitated, for 25–40 min in a chlorinated hydrocarbon to remove the surface wax layer which consists mainly of carbonic acid hydroxytryptamide. A suitable solvent is methylene chloride or 1,2 dichloroethane or trichloroethane which is maintained near its bp during the process. After washing, the beans are steam-treated and dried to the original moisture content while the bean surface is abraded. Coffee digestibility is approved.

Coffee Decaffeination, 1976—Can. 979,725 (O. Bitzhum, P. Hubert; HAG AG) Raw coffee is brought to a moisture content of 10–60 percent by the addition of water and caffeine selectively extracted with liquid carbon dioxide at a pressure above the critical pressure.

Coffee Decaffeination, 1975—U.S. 3,879,569 (O. Vitzthum, P. Hubert; Hag AG)
Raw coffee is moistened to a moisture content of 10–60 percent, after which it is extracted with liquid $CO_2$ under conditions such that the $CO_2$ is maintained in a liquid state below 31.4 C prior to recovery of the caffeine.

Decaffeinated Coffee, 1976—Ger. 2,212,171 (Kaffee-Verdedlungs-W) The hydroxytryptamide content and the ether soluble adjuvant content of raw coffee is reduced by alternately treating the coffee with hot, moist air and then cooling under reduced pressure, each treatment being carried out several times.

Decaffeination of Coffee, 1976—Ger. 2,212,281 (Hag AG) Process for the decaffeination of raw coffee by selective extraction with aqueous liquid carbon dioxide at a pressure above the critical pressure.

Coffee Decaffeination, 1975—Br. 1,372,667 (Hag AG)
Raw coffee is decaffeinated by adjusting its moisture content to 10–60 percent and selectively extracting the caffeine with liquid $CO_2$ containing water at a pressure above the critical pressure of $CO_2$.

Decaffeinated Coffee, 1975—U.S. 3,843,824 (W. Roselius, O. Vitzthum, P. Hubert; HAG AG)
Rough ground roast coffee is extracted with a dry supercritical fluid, after which the coffee is wetted with water and extracted with supercritical carbon dioxide to remove caffeine. The coffee is then extracted with water to produce an aqueous extract, which is dried to yield a powder residue and combined with the coffee oil containing aroma substances from the original extraction step.

Decaffeinated Coffee—U.S. 3,840,684 (T. L. Fazzina, G. V. Jones, R. P. Scelia; General Foods Corp.)
Blends of high- and low-grade roasted and ground decaffeinated coffee having improved flavor contain greater than or equal to 20 percent high-grade coffee which has been organic solvent decaffeinated, and greater than or equal to 30 percent low-grade coffee which has been water decaffeinated.

Rapid Method for Approximate Determination of Caffeine in Roast Coffee, 1972.

The method depends upon the precipitation of caffeine with iodine to yield the periodide. A comparison to the precipitate is made between an aqueous extract of the coffee sample and a standard 0.01 percent caffeine solution. The coffee sample must first be decaffeinated by extraction with $CHCl_3$ since Chlorogenic acid also reacts with iodine; this also applies to caffeine-free samples. Similar determinations are made on the extract from this coffee so that a correction factor can be applied. In samples of caffeine-free, Arabica, and Robusta coffees tested the correction factor was 0.2–0.3 percent (in terms of DM).

Coffee Decaffeination—Can. 926,693 (K. Zosel; Studiengessellschaft Kohle GmbH) Coffee is contacted with moist carbon dioxide in a supercritical state to effect removal of caffeine.

TABLE 16.1. CAFFEINE SOLUBILITY IN VARIOUS SOLVENTS

| Solvent | Weight % Caffeine | Temperatures C | F |
|---|---|---|---|
| 1. Tri chlor ethylene, $C_2H\ Cl_3$ | 1.5 | 29 | 84 |
| | 3.0 | 67 | 153 |
| | 3.6 | 85 | 185 |
| 2. Methylene chloride, $CH_2Cl_2$ | 9.0 | 33 | 91 |
| 3. Chloroform, $CHCl_3$ | 15.0 | 25 | 77 |
| 4. Di chlor ethylene, $C_2H_2Cl_2$ | 1.8 | 25 | 77 |
| 5. Benzene, $C_6H_6$ | 1.0 | 25 | 77 |
| | 4.8 | 100 | 212 |
| 6. Acetone, $(CH_3)_2CHO$ | 2.0 | 25 | 77 |
| 7. Ethyl Alcohol | 1.5 | 25 | 77 |
| | 4.8 | 60 | 140 |
| 8. Ethyl acetate | 4.0 | 77 | 171 |
| 9. Ethyl ether | 0.2 | 20 | 68 |

Steam Treatment of Coffee Beans—S. Gal, P. Windemann and E. Baumgartner)

The reactions occurring during steam treatment of green coffee beans were investigated. 3-Methoxy-4-hydroxystyrene (3,4-MHS) could be detected in the condensate as well as in the treated beans. Benzaldehyde, phenylacetaldehyde, isovaleric acid, furfural, and caffeine were identified in the condensed steam. The 3,4-MHS is formed from the feruloylquinic acid present in raw coffee. A method for quantitative detn of 3,4-MHS was developed (limit of detection in green coffee 2 ppm, relative SD 7 percent at concentration less than 20 ppm).

Gas-chromatographic Determination of Dichloromethane Residues in Decaffeinated Roasted Coffee, 1972—P. Schilling and S. Gal.)

Roasted coffee beans were covered with distilled water, steam distilled for 35 min, and the condensate collected in ice-cooled receivers. Hydroxylamine hydrochloride was then added to the distillate (this increases the sensitivity of the method), which was then left for 15 min before extraction with n-hexane. The hexane extract was dried over anhydrous $Na_2SO_1$ and an aliquot subjected to gas chromatography against a standard solution of dichloromethane. A logarithmic relationship was found between peak height and concentration of dichloromethane. Columbia tape coffee was analysed for dichloromethane in both the green and roasted forms (after milling and infusing), and it was found that levels of dichloromethane were always higher in green coffee than in roasted coffee.

Green Bean Decaffeination, 1974—Can. 934,217 (M. R. Forbes, H. P. Panzer and R. S. Yare; General Foods Corp.)

Decaffeination of green coffee is achieved by extraction with fluorinated hydrocarbons.

Effects of Caffeine on the Human System, 1975—(A. W. Burg)

Published data on concentration of caffeine and other methylxanthines (theobromine, theophylline) in coffee, tea, cocoa, and cola drinks are critically reviewed. New data obtained from 29 coffee, 13 tea, and 2 cocoa products (2223 individual values) are given and include: caffeine contents (mg/cup) R&G coffee (percolator) range 64-124 (average 83), (R&G decaffeinated 2-5(3); instant 40–108 (59); instant decaffeinated 2–8(3); bagged tea, no range given (42); leaf tea 30–48 (41); and instant tea 24–31 (28). Methylxanthine concentration (caffeine, theogromine, and theophylline, respectively) were (Mg/cup); instant tea -,2.4,-; black leaf tea 42.22. less than 0.3; African cocoa 6,272,-; and South American cocoa 42,232,-. Standard average caffeine contents/cup suggested are 85 mg for roast and ground coffee, 60 mg for instant coffee, 3 mg for decaffeinated coffee and 30 mg for instant tea

Coffee Decaffeination, 1975—Br. 1,372,667 (Hag AG)

Raw coffee is decaffeinated by adjusting its moisture content to 10–60 percent and selectively extracting the caffeine with liquid $CO_2$ containing water at a pressure above the critical pressure of $CO_2$.

# 17

# Brewing Technology

## THE BREWING OF COFFEE

Brewing coffee was left pretty much to the preference and ignorance of the consumer until the Pan-American Coffee Bureau (coffee growers of the Western Hemisphere) and the National Coffee Association (United States coffee roasters) formed the Coffee Brewing Institute (C.B.I.) in 1952. The objectives of the C.B.I. are to encourage improvements in the preparation of coffee beverages.

The outstanding facts that led to the establishment of this C.B.I. scientific project were: (1) common practices in brewing coffee were poor and (2) the average taste of a cup of coffee was not nearly as good as it might have been, given the available blends and roasts. This was true both in home and in large scale brewing of coffee. One of the main causes of this situation was that for reasons of economy during and after World War II, and then often merely as a matter of habit, the usual coffee brew was about 60 cups per pound of roast coffee. This brew was weak in flavor and also unnecessarily high in very harsh and undesirable flavor notes. Merely using a higher ratio of roast coffee to water was not enough. The roast coffee flavor might still be spoiled by poor brewing techniques.

By 1957, some of the C.B.I. sub-contracted research work on coffee chemistry and brew preparation became available to members and those interested in the subject in the coffee industry. The work was developed largely under Dr. Earl E. Lockhart, the C.B.I. Scientific Director. Dr. Lockhart formerly had been with the teaching staff of the food technology department of the Massachusetts Institute of Technology (M.I.T.). It was at M.I.T. in 1954 that numerous card indexes of published articles about coffee since 1925 were prepared. The nature of the C.B.I. support is such that it does not recommend coffee blends. Not recommending blends may be polite, but it certainly limits a complete presentation or judgment of coffee flavor values and even brewing. For example, some modes of brew preparation may be more suitable for Brazilian than for Colombian coffees.

Brewing, an unorganized and personal matter, still allows a wide area for useful research, consumer education, and publication. In 1961, the C.B.I. took an important step forward by establishing a brewing school in its offices in New York City to train sales and other interested personnel of coffee roaster members in the recommended ways of brewing coffee. Proper brewing of a good cup of coffee, regardless of blend and roast, requires specific education and a knowledge of principles and procedures. One main impetus came from the fact that in the United States much roast coffee was being brewed at the rate of 60 cups per pound. Brewing 40 to 50 cups per pound of roast coffee would give the consumer a better flavored beverage.

The growers and roasters of coffee stood to gain sales if the consumers used more coffee per cup, and if they drank better brewed coffee. In this way, both producers' and consumers' interests would be served. The C.B.I. has been progressively successful in attaining these goals.

Before examining current brewing practices, it is worthwhile to look at the origin of coffee beverages and coffee use. Coffee beans contain protein, carbohydrates, and oils. Coffee beans were not originally made into a beverage; coffee beans were a food. The coffee cherry was candied. The bean was used as a medicine. Coffee was first drunk as an ingredient of a wine. About 900 A.D. African warriors crushed the green beans with fat and made food ration balls. A fermented drink was made from the pulp. Although green coffee beans were extracted about 1000 A.D., roasting of the beans did not take place until about 1400 A.D. The water soluble extract of roast coffee did not evolve until about 1300 or 1400 A.D. At first, roasting was done in earthenware trays; later, metal trays were used. The roast beans were fragmented by mortar and pestle, cooked in boiling water, and then consumed as grounds and a strong water extract. This is still the custom in the Middle East. Changes in boiling and additions of cinnamon, cloves, or other spices developed untold variations of coffee flavors. These beverage preparation procedures predominated through the 17th century when the beginnings of the boiling and serving pot came into use. Orientals are reported to be the first to pour boiling water on the roast coffee grounds in the cup. About 1625 in Cairo the use of sugar in coffee was introduced to reduce bitterness. When much sugar, often brown sugar, is added to the strong brew, a syrupy cup results.

This chapter covers the history and principles of brewing, as well as the aspects and principles of different kinds of brewing machines and the characteristics of the beverages they produce. Since 1977 there has been no official trade guide to brewing as was once sponsored by the Pan American Coffee Bureau. The brewing center was abandoned in 1976.

## Brew Preparation

Boiling the beverage continued well into the 18th century, and it was usually served black in English coffee houses. About 1660 in Holland and 1685 in France, milk was first used in coffee. By 1760 boiling coffee was generally recognized as producing an objectionable tasting beverage, and steeping with or without a bag to separate the grounds was first used in France. In France, the drip method of coffee preparation was initiated about 1800, and was actually percolation (central tube lift of boiling beverage) as we know it today. Originally, iron or tin and then porcelain brewers were used. During the 19th century, many improved brewing devices were evolved, especially the vacuum and drip methods as widely used today. Although many of these devices were known in the early 1800's, they were not widely or commercially manufactured until after 1900. Automated brewing equipment was neither commercially important nor fully developed until after World War II.

Coffee was largely for the wealthy, or, in any case, a luxury beverage through the 18th century. Being exotic, it was accepted as prepared for almost a century. As coffee's commercial importance grew in the 19th century, more attention was given to its natural flavor, quality, origin, and modes of preparation. Not until after World War I was there any significant scientific work done on coffee chemistry and beverage preparation; and only since World War II has this work been reinforced by the Coffee Brewing Institute and brewing equipment fabricators and users.

In colonial America coffee grounds were boiled in tin coated copper pots and drunk with sugars and spices, sometimes with milk. In New Orleans, due to the French influence, the coffee was brewed generally by percolating the coffee grounds and allowing the water to drip slowly through. New Orleans and the southern part of the United States still produce, on the average, a relatively good and strong cup of coffee. In Europe's Scandinavian countries and in parts of the United States where Scandinavians settled, egg was often used in the hot brew to clarify it and to reduce bitterness. Since 1900 until today, the three most common methods of brewing coffee at home are the vacuum, percolator, and drip systems. Figure 17.1 shows the home brewing equipment. For large scale coffee beverage preparation, as in institutions or restaurants, the muslin bag full of coffee grounds flushed with hot water or drained-off coffee brew has been common. Since 1946, the vacuum ten-cup brewer and more recently, the automatic coffee grounds extraction system, have been more widely used.

FIG. 17.1. HOME COFFEE BREWERS: VACUUM, DRIP, AND PERCOLATOR

**Brew Taste.**—Brewing can be considered a personal taste adventure. The majority of people brew coffee to make an acceptable to pleasant brew for others. Not everyone's taste preferences can be pleased, nor to the same degree. But the majority of persons served can be given an acceptable cup of coffee.

There are some generally accepted rules about the proper way to brew a good cup of coffee. It is the lack of good quality coffee taste experience and the numerous variables in brew preparation and coffee use that confuse the consumer. This gives restaurateurs or roasters an opportunity to claim that their coffee is "best" or "good."

## Influences on Brew Quality

It is worthwhile to examine why a really good tasting cup of coffee is hard to find.

**Roast Coffee Quality.**—Competition in the United States fixes large volume sale prices. Most coffee blends must be bought to fit within this price structure. If the consumer boils his coffee brew, uses sugar and cream or milk, has poor quality water and poor brewing procedures, he might just as well buy the cheapest roast coffee. This could be a Robusta blend in a paper bag exposed to air. For the person who has some knowledge and appreciation of the better brewed coffee flavors, a premium priced roast coffee of good reputation is in order. The number of consumers buying average priced coffee far exceeds the consumers buying premium priced coffee; hence, the roast coffee market functions for the broader market. The large middle fraction coffee consumer is not so fussy about coffee flavor that he (usually the housewife) will pay a premium price. Yet he does not want to be stuck with the "bottom of the barrel" quality. So in the United States he buys the large regional or national brand of roast coffee, often at a reduced price. This brand of roast coffee is blended to suit the average competitive price. Roast coffees in the United States are of fair cup quality. Probably 75 percent of the bag and/or vacuum can brands purchased would make an acceptable to pleasing cup of coffee if brewed properly. Since almost half of the United States' imports are Brazilian coffees, the average cup is bound to be only fair tasting, depending on how much and what quality mild coffees are blended with Brazilian coffees. Only a small percentage of name brand roast coffees use Robustas in their blends, but Robustas are used in some roast coffees. In France, Robustas are about 95 percent of coffee imports, and the discussion of coffee blends within any such political boundary must take these factors into consideration.

**Watering and Staling.**—Most coffee in the home in the United States is brewed too weak. For example, 60 cups per pound of roast coffee are not uncommon. This watering ratio markedly reduces the possibility of a good tasting cup of coffee in most homes, regardless of the quality of the roast coffee. Furthermore, staling of an opened 1 lb can of R&G coffee invariably downgrades cup flavor in more than half of the cups of coffee prepared. So with water dilution and staling, perhaps one can produce one good tasting cup of coffee in four, probably fewer. But these are only two of the many coffee flavor deteriorating factors in brewing.

**Water Quality**.—In the United States, drinking water quality, as good as it is relative to the public water supplies of the rest of the world, is often poor for brewing coffee. Chlorination, organic content, hardness, softness, alkalinity, brackishness, yearly quality variations, and off-odors and tastes downgrade the water quality to varying degrees and hence, downgrade the resulting brewed coffee flavor. Water quality influences are important in developing a clean coffee flavor and will be discussed in detail in the following sections. The net result is that with all brewing conditions perfect except for water, at least half the coffee brewed in the United States will be downgraded by the water quality. The basis of this statement is Fig. 17.2, a water hardness map of the United States. Hardness in natural waters in limestone areas is usually associated with comparable alkalinity. In some areas, good brew quality is impossible due to the water quality.

**Home Percolation of Coffee**.—This boiling action of the coffee brew raises the brew up a central tube to flow over and through a shallow bed of coffee grounds. It is a common method of brewing coffee. With percolation time exceeding 10 min, most of the natural coffee aromas and flavors have been depleted and altered. The continual exposure of roast coffee grounds and diluted hot brew to air drives off aromatic volatiles, which one can easily smell, and oxidizes brew flavor solubles. The resulting brew is often heavy tasting, caramelized, nonaromatic, and oily. Vigorous percolation is a daily ritual in many homes and is, no doubt, suited to low grade, stale coffees. Automatic percolators are widely used. If percolation is limited to 5 to 10 min, the flavor damage done is not as serious; that is, the coffee beverage is still acceptable in flavor. However, considerably more natural coffee flavor will be retained in the cup if the roast coffee has been drip or vacuum brewed (extracted) in 1 to 3 min with much less exposure to oxygen.

Taking into consideration available coffee blends, water quality, water to coffee ratios used, and modes of extraction, it is probable that only a small fraction of the coffee brewed has good quality, natural coffee flavor.

**Materials of Construction**.—The lower priced percolators are made of aluminum which dissolves in acidic coffee brew. Aluminum coffee pots corrode. Aluminum also reacts with fatty acids. Even though coffee brew contact time is short and aluminum contamination is small, there may be a small deterioration of

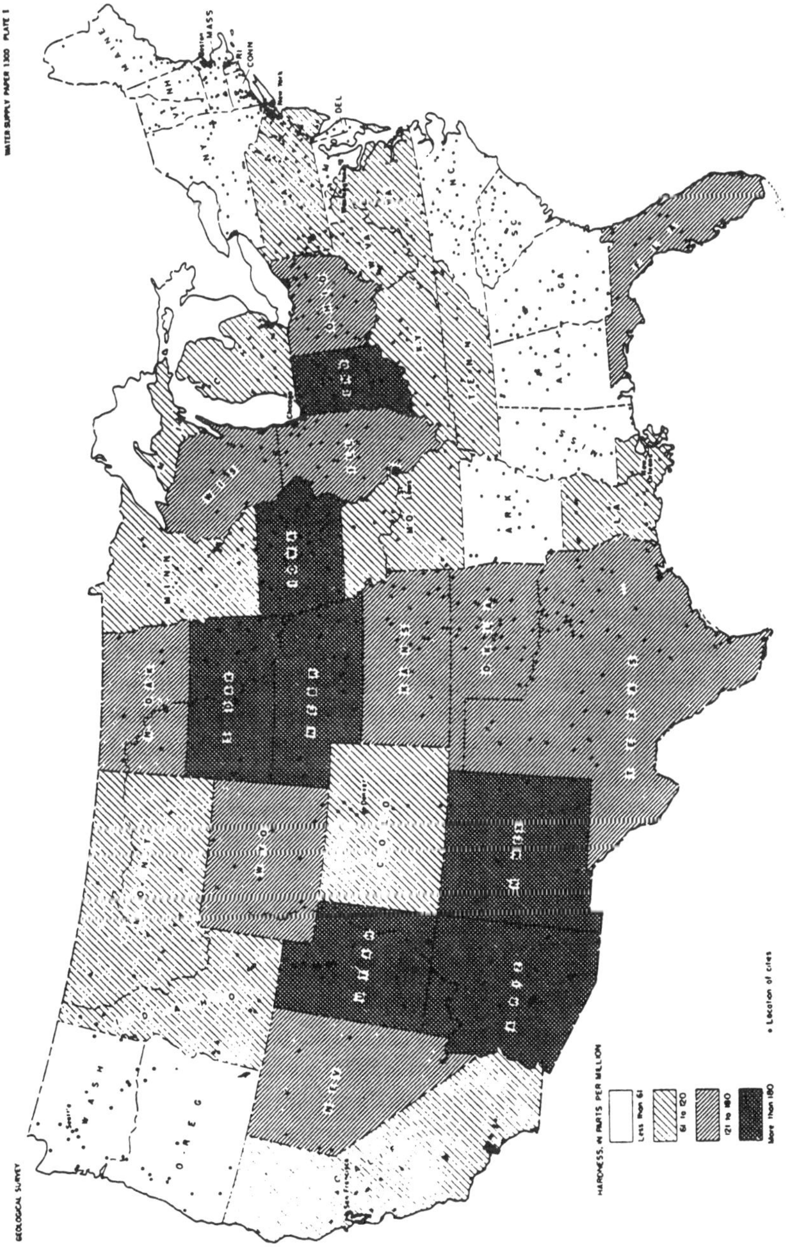


*Courtesy U.S. Geologic Survey*

FIG. 17.2. WATER HARDNESS MAP OF THE UNITED STATES

coffee flavor. Aluminum is not inert under brewing conditions and is now considered inferior to stainless steel. The materials of construction of coffee brewing equipment deserve special attention. When copper and copper alloys, as well as iron or carbon steel, are in contact with coffee brew they cause flavor deterioration. In general, glass, stainless steel, glazed ceramics, tin, nickel, and chromium plate are inert to coffee brew acids. The disadvantage of metal plating is that when the film is broken, the protective surfaces are lost.

**Cleanliness.**—The design of the brewing equipment so that it can be easily and thoroughly cleaned is as important as the materials of construction. It should be possible to inspect and wipe all surfaces of a brewing vessel. Alkali cleaners recommended for such use normally do a good job of removing coffee oils, tars, and films. Inspection after cleaning is desirable to be really sure. A sanitary brewer will have shiny, smooth, and curved surfaces; there should be no crevices for deposits to build up. A dirty brewer contributes to brew flavor alteration and, usually, degradation.

**Commercial Urn Brewing.**—Several practices which downgrade coffee flavor are common here. Muslin bags may not be kept clean; urns may not be cleaned every time they are used; extraction may not be uniform; brew may be held for hours; the brewer may be used at only half capacity, which means inefficiency; concentrated coffee extract in the lower part of the urn will be poured through the sack of grounds, losing solubles and extracting bitter flavors. Solubles may not be mixed uniformly in the urn reservoir. The residue of the urn brew will be added to a fresh batch of brew, thus downgrading all the freshly prepared coffee brew. In other words, the urn brew preparation is susceptible to poor brew management. As this has been recognized, automatic, small, fast batch brewers have been growing in use.

**Restaurant Coffee Brewing.**—This has advantages and disadvantages over modes of home brewing. In some cases, vacuum methods are used both in the home and the small restaurant. Restaurants often contract to buy their roast coffee from coffee roasters and/or suppliers in exchange for the loan of brewing equipment. Under such an arrangement, the restaurant owner often does not have much freedom to choose his coffee quality and price. The profits to this coffee roaster must cover his investment in outstanding loaned equipment. The blends supplied are, therefore, frequently high in Brazilian and sometimes even Robusta coffees. The restaurateur may pay 5¢ per lb more for comparable quality coffee. If he uses 40 lb per day for 365 days, he has paid out $730, which would amortize most of his investment in brewing equipment. Many restaurants would be better off buying name brand roast coffee and their own brewing equipment, on a bank loan if necessary. This discussion shows how coffee blends and brewing equipment are controlled in some restaurants. It is, therefore, not uncommon to have an excellent meal and service at a restaurant with a fair to good reputation, and then to be served an unpalatable cup of coffee.

To obtain a clean flavored, fresh cup of brewed coffee at the right concentration and temperature and containing most of the natural coffee aroma and flavor is not simple. This is amply demonstrated by the coffee served in private homes and in many public restaurants. Part of the attainment of such a goal is taste, brew, and coffee education for both consumer and server. Attaining better brewed coffee flavor involves the use of better brewing equipment procedures, water quality, and coffee blends.

## Five Essentials of Good Coffee Flavor

To control the five essentials for preparing a cup of really good coffee takes many talents and skills. One must know the quality of the coffee used; a brand name does not necessarily ensure coffee quality. One must also know something about keeping coffee fresh and be able to evaluate the packaging. One must determine whether the water quality is influencing the cup flavor. For the taste desired, economy must be secondary and the right water-to-coffee ratio must be used. The brewing method must be fast and thorough, yet not exhaustive. The serving must be prompt. In a fine coffee shop, the green beans are chosen, roasted on the day of use, ground just before extraction, brewed in minutes if not in seconds, and served promptly in suitable cups.

**Coffee Quality.**—Admittedly, experience can tell a person what a good tasting coffee is. However, to be absolutely sure of the basis of choice, the coffee type by variety (Arabica or Robusta) and origin (mild or Brazilian) must be known. Then the geographical (political) source, altitude of growth, quality of processing of green coffee beans (fermentation, drying, cleaning, grading, storage), age, purity (freedom from foreign matter), bean size, and moisture ought to be known. Usually most of this is not possible, so the taste and aroma must become the final basis of judgment.

Assuming that a good quality green coffee is used, aroma and flavor are not developed until roasting. The degree of roast, uniformity of roast, size of the roaster, fuel for heat, the time of roasting, use of batch or continuous roaster, heating temperatures, and mode of cooling may influence the aroma and flavor of the final roast coffee.

Depending on the brewing equipment used, the grind or fineness of final coffee particles before extraction has a great deal to do with the subsequent rate of staling, aroma loss, and flavor of brewed coffee. Further, whipping in chaff, beating out oil, and driving off volatiles wtih incorporation of air (oxygen) or even adding water can lose a great deal of flavor during the grinding operation.

**Package**.—Reports have been cited here and elsewhere that coffee can be held fresh indefinitely in a vacuum can, vacuum bag, inert gas filled container, etc. In every case, some oxygen is left in the container, and this oxygen proceeds to react with the key aroma and flavor factors in the coffee. Hence, some staling occurs within every package. Considerable staling occurs after the package is opened, so the major portion of coffee from every package is stale long before it is brewed. To put it another way, the best vacuum packed coffee is never as good in flavor as the day it was placed in the can. In fact, the process of drawing a vacuum about the coffee particles draws flavor away. The storage of coffee before it enters the can exposes it to staling. Hence, the coffee house that roasts and grinds its coffee just before use has a real flavor advantage over the commercial vacuum pack. Some evaluations state that ground coffee stays fresh up to three or more weeks. According to any coffee taster, this is absurd. The ground coffee may be acceptable to a less critical and/or a less informed consumer who uses milk and sugar and possibly not the best brewing and serving methods. Hence, the freshness of the roast coffee's volatile falvors and their abundance are important factors from the moment of grinding until brewing, and package protection can play a vital part in this time interval.

**Water Quality and Water-to-Coffee Ratio**.—Since water constitutes 99 percent of the coffee beverage, its impurities and quality can markedly downgrade coffee brew flavor. Water may be hard, alkaline, or brackish, or have organic matter, chlorine, odors, gases, or metal impurities. It is an error to make weak coffee, because this type of extraction brings out the undesirable flavors. An axiom in brewing is, in case of doubt make the cup stronger. A frequent error is not measuring water nor coffee at all, so that control is lost entirely in this variable.

**Brewing Method**.—Numerous methods have been cited; some of the most popular home and commercial methods destroy the aromatic properties of the best coffees. These include boiling, preparation of large batches that cannot be rapidly served, urn extraction, mixing an old batch with a fresh batch, and percolation, which intimately mixes coffee extract brew with air and spent coffee grounds. Essential to good brewing technique are speed, a reasonable extraction yield, and solubles concentration. Most brewing methods take five or more minutes. Channelling of rinse drip, percolate, or wash does not uniformly extract the coffee solubles. Fine grinds are seldom used. In fact, fine grind is difficult to get in many United States retail stores. Some of the newer pressurized extraction methods through granular coffee beds, not slurries, give fast, uniform extraction. Some methods such as urn, drip, and percolation cannot provide uniform flow through the granular bed.

**Serving Method**.—If all the foregoing essentials of preparation have been perfect, it is not uncommon to find the brew boiling or near boiling for many minutes on a pot heater. Admittedly, adding sugar and milk or cream covers up the bitterness of the coffee. But if really good coffee is desired, it is best enjoyed black. Cleanliness of utensils is, of course, essential for brewing and serving but is frequently neglected. Adding a held-over pot to a freshly brewed pot is common "economical" practice. The espresso method, which brews one cup at a time, offers some very desirable features.

## Rise in Instant Coffee Use

The fact that there are so many variables in preparing and serving a cup of good coffee has been responsible for a wide latitude in acceptable flavors for coffee. Further, the fact that brewing coffee produces variable results has increased the consumption of instant coffees.

Instant coffees also vary in flavor quality, so this discussion will be limited to the best flavored instant coffees currently available on the market. Instant coffees eliminate all brew processing and equipment. There is no cleaning of equipment. If the cup flavor concentration is weak, more powder is added; if flavor is too strong, more water is added. There is no wasted coffee brew in the pot and no grounds to be disposed of. There are neither filters nor accessories. A pot of boiling water or a decanter are the only preparation tools required. Instant coffee is ideal for one or two cup servings. Dissolved in a decanter, instant coffees retain more and better flavor than when made in individual cups. Some of the shortcomings mentioned earlier in brewing and brewed coffees have been eliminated by the use of instantly soluble coffees, and not always at a sacrifice in coffee flavor value. Instant coffees, cup for cup, are about half the price of regular coffee. Instant coffee has even more advantage when brewing time, brewing equipment, unused coffee brew, grounds disposal, pot cleaning, and staling of roast coffee are considered. Good instant coffees are better flavored and more uniform in cup flavor than poorly brewed coffees, even at this stage of development of the instant coffee industry. With about 18 percent of roastings going into soluble coffees, at least 30 percent of the cups of coffee consumed in the United States are instant coffee. The success of soluble coffees is in no small part due to the existence of poor quality and stale roast coffees. The best instant coffees are uniform, acceptable, and sometimes enjoyable. Thus, the growing acceptance and sales of instant coffees are a natural evolution and upgrading of available coffee beverage. Restaurants, the military, and other institutions are now using more instant coffee. Although an average roast coffee brew may be superior in flavor to an average instant coffee brew, the better flavored instant coffees are to be preferred over the poorer flavored roast coffee brews.

## Factors in Brew Preparation

One must have all beverage preparation factors under control in order to prepare a good tasting cup of brewed coffee: coffee and water, brewing equipment, and finally, the mode of serving and consuming. It is not always possible to control all preparation factors. In some cases one preparation factor may be restrictive or controlling, e.g., the quality of R&G coffee. The roast coffee quality may be dictated by price, not by consumer preference.

The brewing equipment manufacturer may not be directly concerned with the quality of roast coffee used, water quality available nor how long the coffee brew is held before it is consumed. The brewing machine should be designed for low price, utility, elimination of labor, and reproducibility of efficiently extracted, full flavored coffee brew. Commercial coffee brewing equipment, as used for large volume service in large feeding areas, is sometimes similar to mechanisms designed for use in vending machines. The most highly developed restaurant brewing equipment is completely automatic. With a coin and dispensing mechanism, the brewer might be a vending machine.

Some of the shortcomings of the urn brewing of coffee have already been mentioned. The shortcomings of brewing with many manual steps are self-evident from higher labor costs as well as from variable brew flavor results.

**Time Factor**.—The most important point regarding the processing of coffee, whether it is roasted coffee, instant coffee powder, or brewed coffee beverage, is that processing must be done quickly. Coffee aroma and flavor are very delicate and transient phenomena. Elusiveness is part of their attractiveness. They can be captured only momentarily (unless one stands downwind of the roaster or grinder) at the consumer level. The aroma and flavor are fragile and fleeting. Thus, the R&G coffee must be fresh and of good flavor. The water must be of good quality. The apparatus must brew with speed and reproducibility. The brew must be delivered and consumed promptly. Any gap in this timing sequence destroys the end result and the pleasure and satisfaction the brew ought to bring. Reducing the brewing process to a set of mechanically reproducible steps, automatically carried out without human interruption until the brewed beverage is dispensed, attains the desired goal. This already has been achieved to varying degrees in commercial brewing units which control times, temperatures, and proportions accurately.

**Grind**.—With 185 F (85 C) water, the rate of solubles extraction with fine, drip, and regular grinds is relatively low after 5 min of slurrying, at which time about 20, 18, and 16 percent solubles have been respectively extracted. Thus, the C.B.I. recommends that suitable brewing times for each grind are, respectively, 1 to 3 min for fine grind, 4 to 6 min for drip grind, and 6 to 8 min for regular grind.

**Solubles Yield and Concentration**.—Further, the C.B.I. states that an 18 to 22 percent solubles extraction is desirable for idealized brew cup flavor at a 1.15 to 1.35 percent solubles concentration in the cup, with outside limits of 1.0 to 1.5 percent solubles concentration. In practice, solubles yields of closer to 18 percent are more common.

Measurement of soluble solids content of coffee extract is, of course, an indication of solubles extraction efficiency. However, solubles concentration and yield must be considered within the whole structure of coffee brew flavor. This includes volatile aromatics and the basic quality of the roast coffee being extracted. The amount of colloids in the brew reflects blend, roast, and grind; colloids contribute mouth texture to the coffee beverage. Thus, aroma and colloid quality and quantity influence coffee brew acceptability.

The C.B.I. of necessity has evolved the solubles concentration chart for coffee brews. This shows: (1) if a watered coffee brew has been made, it is measured objectively; and (2) if a roast coffee extraction in an urn is inefficient, the low solubles yield and concentration will be seen.

The solubles yield is associated with a coffee brew solubles concentration. The two conditions of coffee solubles concentration and solubles yield must be satisfied so that they fall within the idealized cup flavor area (see Fig. 17.3). Low solubles yields do not produce a wholesome flavored cup of coffee, even in the ideal solubles concentration range. High solubles yields do not give a wholesome coffee brew flavor either, even if the brew solubles concentration is in the acceptable range.

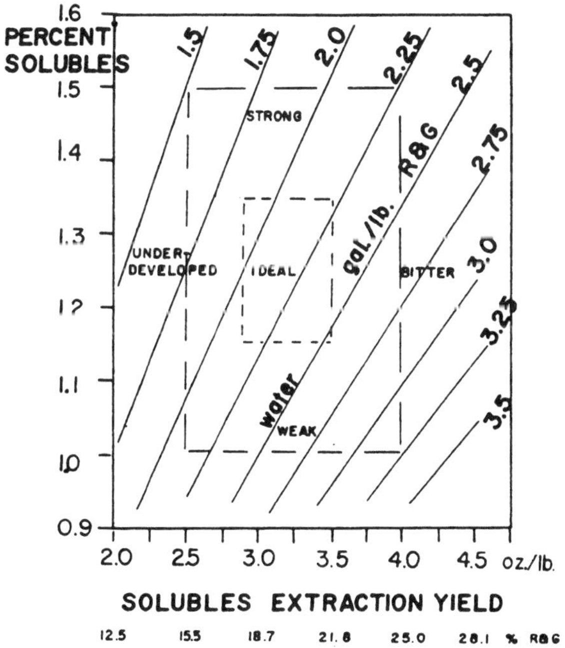


*Courtesy Coffee Brewing Institute*

FIG. 17.3. COFFEE BREW CONCENTRATION OF SOLUBLES VS YIELD ON PARAMETERS OF WATER/COFFEE RATIO

**Brew Specific Gravity.**—Variances for Brazilian and mild coffees on a specific gravity (C.B.I.) chart for brew are not likely to be as great as for Robusta coffees which give higher solubles yields and concentrations and a very different coffee brew flavor. A 1.5 percent brew at 80 F (27 C) has a specific gravity of only about 1.006, and the C.B.I. has obtained special hydrometers for this use. Flavor boundary conditions in Fig. 17.3 are not sharp because of individual taste preferences, influences of blend, roast, and grind, and time and mode of extraction, as well as pH and titratable acidity. However, the solubles yield and concentration target, as fixed, are improvements over no target at all. Thus, the Coffee Brewing Institute has given the current users of coffee urns a measuring tool for their brew extraction process.

**Water to Roast Coffee Ratio.**—In the English weights and measures system, the water to coffee ratio of about 2.00 to 2.25 gal of water per pound of roast coffee is recommended for the best brew flavor. This corresponds in the metric system to a water to coffee ratio of 16.7 to 18.8 liters per kilogram.

**Cup Sizes.**—A secondary problem enters here in that the number of cups of brewed coffee per unit weight of R&G coffee is loosely spoken of, e.g., 40 cups per pound of roast coffee, or 88 cups per kilogram of roast coffee. Figure 17.3 on coffee solubles concentration and yield fixes the volume of brew solution drawn off, hence the number of cups. However, there are various cup sizes and cup fills. Cups per unit weight of roast coffee can, therefore, vary accordingly.

The most common cup size in the United States home is 8 fl oz (240 ml) and would be filled with brew to 6 fl oz (180 ml). But smaller cups are 7 and 6 fl oz, holding 4.5 to 5.5 fl oz of brew. Restaurants usually have the smaller cup sizes. Demitasse cups, when full, may hold only 3 fl oz (90 ml) or 4 fl oz (120 ml) and actually hold 1.5 to 3 fl oz of coffee.

**Water Absorption.**—Roast coffee grounds hold twice their dry weight of water, plus whatever water does not drain off completely from the interstices. Therefore, each pound of original R&G coffee holds about 10 percent solubles after extraction and about 760 gm absorbed water plus interstitial undrained water. For convenience, the drained coffee brew can be taken as 90 percent of the original water added to the roast coffee grounds. For example, 2 gal (7,570 ml) water and 1 lb (454 gm) of grounds will yield close to 1.8 gal or 6,810 ml drained brew. If the R&G coffee was flushed so that few solubles remain in the spent grounds, (10 percent of total solubles yields) then the 91 gm of solubles (20 percent yield on R&G coffee) are dissolved in the 6,810 ml water. This is a 1.30 percent solubles solution. If 4.5 fl oz (135 ml) cup portions are used, this is equivalent to 50 cups; if a 5 fl oz cup portion is used, this is equivalent to 45 cups, and so forth.

**Urn vs Automatic Brewer**.—Urns are an obsolete way of coffee brew preparation under service circumstances where about 500 cups of coffee per hour (or 10 cups per minute) can be prepared from an automatic brewer. These can be prepared in brew batches of 60 cups; other units can prepare 12 cup portions in 3 min and 1 cup in 10 sec.

The advantage of the automatic coffee brewing units is that the human element is removed from the operation and the resulting batches are small—12 cups or less. Yet if heavy demand occurs, either the 60 cups per 6 min or the 12 cups per 3 min would handle the corresponding demands. The larger unit is for the higher demand rates, and the smaller unit may be used by the same caterer during low demand periods.

It hardly seems justifiable today to use an urn with an initial coffee brew reserve of 5 gal or more; this would be more than an hour's coffee supply, serving at about 3 cups per minute. The hold time for finished brew should be such that the longest a brew is held is less than 1 hr, preferably only a few minutes. Appropriate brew makers are readily available today. The restaurant that uses large brew batch preparations (aside from other shortcomings in urn use, which are many) will find his competitors able to deliver fresher coffee. To be able to deliver a fresher cup of brewed coffee is consistent with the transient nature of coffee aroma and flavor. Figure 17.4 shows a section through a coffee urn system. In Fig. 17.4 the urn has a

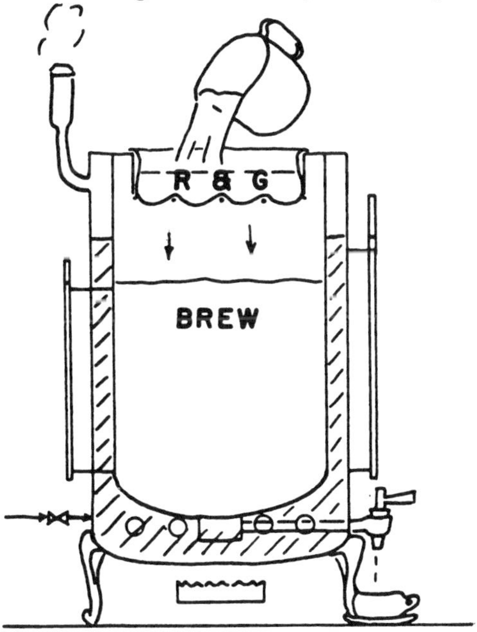


FIG. 17.4. SECTION THROUGH COFFEE URN

gas burner with vents for combustion gases, but electric heating is also very common. The outer water jacket is usually not insulated, but it should be. There are sight glasses for the inner brew reservoir and for the outer water jacket. Excessive pressure on the jacket is avoided through a safety valve. Water from the jacket is used to flush through the R&G coffee. Make-up water goes to the jacket. Figure 17.5 shows a pair of urns in one water jacket. Figure 17.6 shows a water distributor used over R&G coffee on a perforated plate. Figure 17.7 shows the grid and muslin bag. Urns of 2 or 3 gal size are most consistent with a fresh coffee brew concept, with complete consumption in less than 1 hr. The brew equipment buyer may be influenced by the lower initial cost and higher labor operating costs with urns compared to possibly higher initial cost and lower labor costs with automatic brewing equipment. Maintaining freshness of coffee brew flavor at any time means a minimum holding time for brew. This is an objective consistent with retaining the most coffee flavor, while satisfying the rate of consumer demand as it

*Courtesy Western Urn Manufacturing Company*

FIG. 17.5. PAIR OF COFFEE URNS IN ONE WATER JACKET

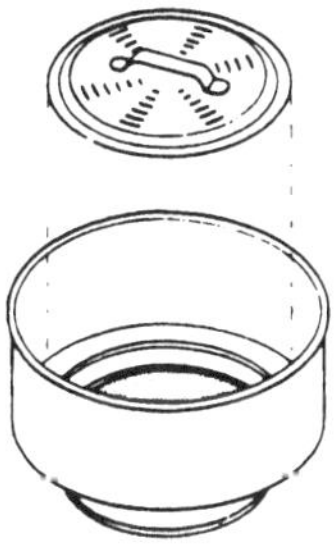

*Courtesy Western Urn Manufacturing Company*

FIG. 17.6. WATER DISTRIBUTOR OVER COFFEE GROUNDS RESTING ON PERFORATED PLATE

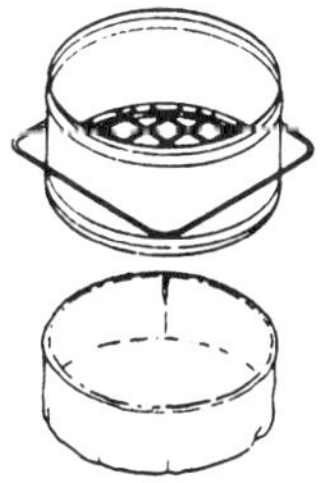

FIG. 17.7. URN SUPPORT GRID FOR MUSLIN BAG

arises. The small batch, fast brewer is well suited to this objective. During periods when the rate of coffee use is only a few cups per hour, an instant coffee, espresso, or one-cup drip brewer is best. For 12 to 200 cups per hour, the small batch, fast brewer is best. Home use drip and percolator coffee brewers for 40 cups per batch are sold, both in aluminum and stainless steel, but their usefulness is related mostly to the speed of coffee consumption.

## Brewing Principles

In principle there are several ways to prepare coffee brew. The *urn method* allows boiling water to pass slowly through the roast coffee grounds by gravity percolation (often with channelling of flow). The *espresso brewing machine* achieves the same pressure extraction effect except that the roast grounds are finer and better confined; channeling does not occur for the most part. Figure 17.8 shows an application of the espresso principle in an Italian home coffee brewing machine. Figure 17.9 shows a section of an espresso machine, and Fig. 17.10 shows espresso machines in a retail store in Bremen, Germany. Espresso extraction takes only 1 or 2 sec. Water volume is fixed usually by a volumetric piston displacement. The water heating reservoir is under 5 psig steam pressure. The volumetric water measure by piston is more usual for commercial espresso machines, while the steam pressure from a confined volume of water is more common to smaller, less expensive home use models.

**Espresso Machine.**—In principle, this machine performs good extraction for the following reasons: (1) coffee is ground just before use; (2) the coffee used is volumetrically measured, as is the amount of extraction water; (3) the water in the

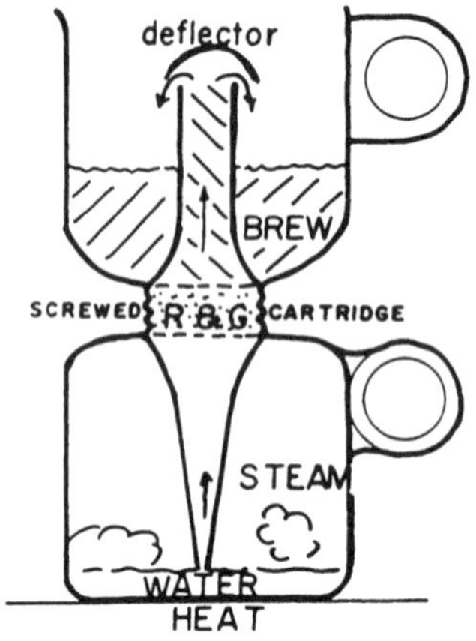


FIG. 17.8. ITALIAN HOME BREWER, UPWARD PRESSURE EXTRACTION

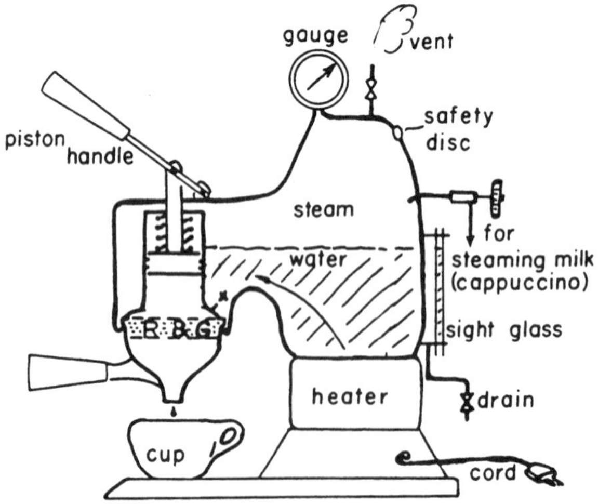


FIG. 17.9. SECTION THROUGH ESPRESSO MACHINE

reservoir comes to a full boil to expell any dissolved gases before extraction begins (see oxygen solubility in Fig. 17.11); (4) the extraction time is almost instantaneous.

The amount of espresso extraction water used is much less than in United States brews. United States methods use a water to R&G ratio of 20 to 1 and attain a 1.25 percent solubles content in the extract. The espresso method uses 6 to 8 to 1 water to R&G coffee ratio, and yields concentrations in the extract of from 3 to 5 percent solubles. Its extraction solubles yield may even be higher than with United States

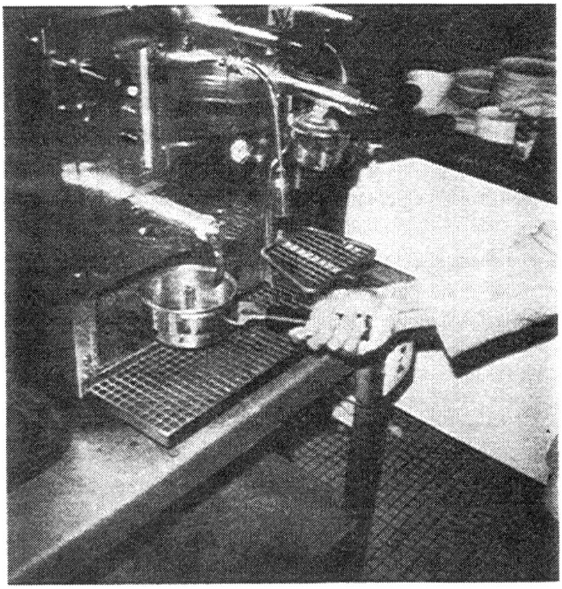

*Courtesy Probat-Werke*

FIG. 17.10. ESPRESSO MACHINES IN COFFEE SHOP

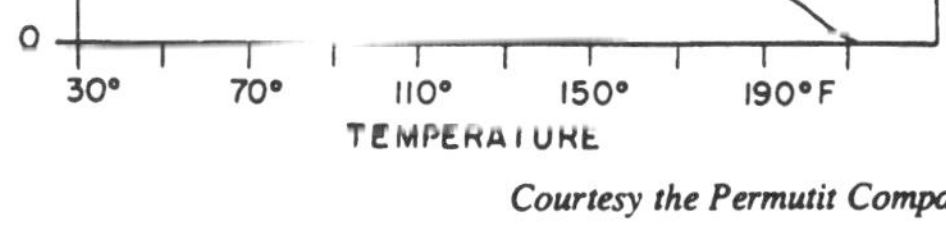


*Courtesy the Permutit Company*

FIG. 17.11. OXYGEN SOLUBILITY IN WATER VS TEMPERATURE

methods. But, due to the potentially higher solubles and oil yields available from pulverized coffee, the rapid extraction is not exhaustive. Hence, the resulting brew does not have the bitter taste that occurs with exhaustive, long time diluted solubles extraction by the United States types of brewers.

Of course, espresso coffee is quite a different beverage from the lower solubles and flavor content of United States brewed coffee. Espresso is drunk with sugar and perhaps a twist of lemon, hot milk, or liquor. The harm of exhaustive extraction is easily demonstrated by allowing perhaps 50 percent more hot water than normal through an espresso machine process. The resulting coffee flavor will be bitter and unpalatable. This phenomenon shows that extraction must not be exhaustive. It is a common procedure in some European countries and in Latin America to make the original coffee brew very strong, but not to take a high and complete solubles yield from the fine grounds. Thereafter, boiling water is added to suit personal taste.

This is one way to make a good coffee brew, but it is not customary in the United States or in many other places. Espresso machines are usually used to extract Italian (almost burnt) roasts, but the extraction machine works just as well with a French roast.

**Airline Brewer and Espresso Machines.**—This principle has been used in the coffee-brewing machine designed especially for vending coffee and for the jet airliners (see Fig. 17.12). These coffee brewers make an excellent cup of brewed coffee and operate as follows: a 3.2 oz (91 gm) polyethylene packet of fine grind roast coffee is emptied into a plastic cup of about 8 fl oz (240 ml) in volume. This has a screen of about 100 mesh in the bottom. See brewer section and Fig. 17.13, below. The filled tube cap is screwed on and is then set inside a stainless steel tube which is twist-set and tightened into an upper gasket, see Fig. 17.14 for assembly. The R&G coffee is then ready for extraction. A receiving flask (later used for pouring into cups) is the bottom stainless steel bowl of a vacuum brewer. This is set in a spring clamp which simultaneously sets an actuating micro-switch to release an interlock so that coffee brewing can begin. A button is pushed to open a valve

*Courtesy REF Manufacturing Corporation and Coffee & Tea Industries Magazine*

FIG. 17.12. AIRLINE HOSTESS DRAWING COFFEE

from the hot water supply and to actuate a metering pump. In 3 min, after the metered volume of about 1,350 ml (or ten 5.5 fl oz cups) of hot water has passed through, coffee brew is produced and the system shuts itself off. A panel light shows that the extraction is complete and the coffee brew can be served. Extract at 18 percent solubles yield is 1.26 percent solubles concentration (2 gal per lb). With

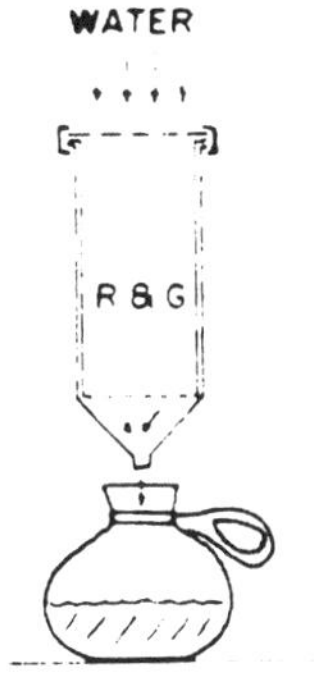


FIG. 17.13. SECTION THROUGH WATER PRESSURIZED EXTRACTOR

*Courtesy REF Manufacturing Corporation and Coffee & Tea Industries Magazine*

FIG. 17.14. PARTS OF AIRLINE BREWER

good roast coffee this machine makes a delicious cup of coffee. All processing factors are effectively controlled.

**Comparison of Extraction Systems.**—Espresso extraction is the positive diplacement of water through the granular coffee bed. It is efficient and rapid, and results in a good flavored coffee brew which somewhat resembles commercially percolated soluble coffee. The major differences are in the ratio of water to R&G coffee, the geometry of the R&G coffee bed, concentration, yield, and time. The *espresso machine* has an R&G coffee chamber that is about 1 cm high and 5 cm in diameter; this will hold about 20 ml of fine grind dark coffee, or about 8 gm. The volume of water that passes through is less than 2 fl oz (60 ml), yielding a 3 to 4 percent solubles concentration and 25 percent solubles yield. Note that the inlet water to R&G coffee espresso ratios are, by volume, 3 to 1, and by weight, 8 to 1. Effluent temperatures are low.

There is an Italian household extractor that looks like bowl over bowl, but operates on a positive displacement principle. The cartridge volume of R&G coffee lies between the bowls. When the water boils, it is forced up through the chamber of ground coffee, and the extract is deflected down into the top bowl (see Fig. 17.8).

Espresso coffee is collected in the drinking demitasse cup and is brought directly to the consumer. There is no reheating of brew to be concerned about. It is prepared cup by cup, not by the gallon. The espresso brew method is technically sound. However, poor quality coffee or water and over-extraction can make a poor tasting cup. In other words, the espresso machine is only one part of the brew process. Even though it does a good job, all the sequential brewing factors have to be correct to obtain a good cup of coffee.

In the airline cartridge of R&G coffee (about 5 cm in diameter and 10 cm high), the 200 ml volume holds 96 gm of R&G coffee. The inlet water to R&G coffee ratios are, by volume, 7.5 to 1, and by weight, 15 to 1.

In the commercial percolator for soluble coffee manufacture is a series of columns; hydrolysis yields differ from the foregoing. But due to the high extract solubles concentrations achieved, the water to R&G coffee ratio is only about 3.5 by weight.

In every case, the extraction conditions are controlled.

**Vending Machine Brewers.**—Having evolved the variable factors in the brewing of coffee, it requires only a nominal modification to install the same extraction mechanism in a coin-operated vending machine. Three general types of extractors have been developed.

The Vendo Company of Kansas City, Missouri has used (1956 to date) a system of puncturing a standard ¼ lb can of vacuum packed coffee with several slotted, tapered points. The hot water enters from one side, and the brew leaves from the same side. The machine holds 54 ¼-lb cans of regular grind roast coffee, averaging 16 cups per can or 64 cups per pound. The vacuum sealed can protects the roast coffee until it is brewed. A time mechanism may purge the batch to drain if it is held too long.

The Automatic Canteen Company of America also makes a coffee brewing vending machine, which makes 10 cups of coffee per batch; hot chocolate and soup are also dispensed. The volume of coffee is measured and washed free of solubles with pressurized water. If the brew is held too long, the old batch of brew is dumped.

Since 1958, Rudd-Melikian has made 1 cup at a time by packaging the roast coffee in 1-cup portions in a filter fabric tape. This eliminates the poor weighing mechanisms used in vending equipment with loose ground coffee. Also, by a Pliofilm liner, the tape gives limited protection from oxygen and staling until the coffee is used. This avoids piling the ground coffee in an unclean hopper. Rudd-Melikian pioneered the single-cup tape principle (see Figs. 17.17 and 17.18).

The International Bally Coffee Vending Company has a novel arrangement wherein 1 cup is brewed at a time. A compartment holds 7 lb of R&G coffee. One cup of hot water is placed in a reservoir; then about 8 gm of R&G coffee are added. A top chamber cover with built-in filter closes. Upward piston motion of the bottom of the chamber displaces the brew into the cup. The chamber then opens, and water flushes the grounds out for the next cup. This system is not unlike an espresso operation except that the coffee grounds are not fully water washed.

Another machine, called Perk-O-Fresh, used a radial design of 40 cartridges of R&G coffee each dispensing a 16 cup batch of brew.

Still another type of vending brew machine makes 1 cup at a time and operates so that the volumetric measure of coffee grounds falls into a stainless steel funnel. Six ounces of hot water flush the R&G particles about, and the brew slowly filters through a fine stainless steel screen into the cup. By having two such funneled cups, one over the other, two rinses are effected: one of a single rinsed coffee, and a second of fresh R&G coffee. The wet coffee grounds will stale rapidly.

**Restaurant Slurry Brewer.**—The single rinsed R&G coffee slurry extractor which filters out the brew has been used in the numerous 8 to 12 cup restaurant brewers. This slurrying of coffee grounds followed by drainage of brew differs from the columnar displacement of extract with water sometimes called pressure brew coffee making. The use of a paper filter takes out most of the colloidal matter as well as fine grind particles. Such paper filters are used as a means of disposing of the spent coffee grounds, which is neater than washing out the funnel chamber and refilling the metal cartridge with fresh R&G coffee. Paper absorbs stale coffee and other odors, and therefore must be carefully handled.

Positive water displacement extractors are the most efficient because finer grinds can be used, action is positive, and timing is accurate. Also, the extraction is countercurrent. The slurry extractors have solubles in batch equilibrium. Figure 17.15 shows the inside of a restaurant pressure brewer.

## Brewing Equipment Standardization

The Coffee Brewing Institute observes that there has been little attempt in the past to standardize equipment design in terms related to the quality of coffee beverage prepared. This is also true in soluble coffee plants, and, no doubt, for other types of food processors. In addition, there has been a general lack of suitable instructional material emphasizing proper use of brewing equipment to attain product flavor quality and equipment performance. This is no doubt due to the fact that the brewing equipment manufacturers are not experienced with coffee quality or really do not have an ideal brew flavor in mind when they design and build brewing machines. The user of the equipment, however, must strive for good quality brew.

The advantages of improved brewing equipment and procedures, as well as coffee quality and water quality, are increased coffee beverage consumption and increased consumer satisfaction. For example, the improved quality of pressurized brew on some of the jet airliners showed immediate passenger response by increasing consumption by more than 50 percent. The use of 2 vs 3 gal of water per pound of R&G coffee in the airlines' brew equipment reveals marked preference for the stronger brew, which is in agreement with other surveys.

# WATER PROPERTIES THAT INFLUENCE COFFEE BEVERAGE FLAVOR

Water represents about 99 percent of the coffee beverage, whether it is brewed or instant coffee. Yet it was not until about 1955 that the Coffee Brewing Institute supported the initial research for gaining understanding of the effects of water composition on coffee flavor. Since then four reports have been issued. These were based on three groups of experiments covering mineral and ion threshold concentrations as they directly and indirectly influence final coffee brew flavor. No work about the effect of water properties on instant coffee flavor has been published to date.

## Mineral Ion Taste Thresholds

Thresholds were found to vary with taster, with associated cation or anion, and whether ion is in water or coffee brew. The following ranges of ion concentrations are near taste threshold:

*Courtesy Western Urn Manufacturing Company*

FIG. 17.15. RESTAURANT WATER PRESSURIZED EXTRACTOR

1. Calcium ion about 125 ppm in water, 300 ppm in coffee brew
2. Magnesium ion about 100 ppm in water, 200 ppm in brew
3. Potassium ion about 300 ppm in water, 400 ppm in brew
4. Sodium ion about 100 ppm in water, 250 ppm in brew
5. Ferric ion about 4 ppm in water and brew, although others report less
6. Phosphate ion about 100 ppm in water
7. Chlorides about 200 ppm in water, 400 ppm in brew
8. Carbonates about 50 ppm in water, 100 ppm in brew
9. Bicarbonates about 500 ppm in water, 1,000 ppm in brew

## The United States Public Health Drinking Water Standards

These standards call for clear, colorless, odorless, and tasteless water, free of bacteria and with less than: 0.2 ppm copper, 0.3 ppm iron, 250 ppm sulfates, 100 ppm magnesium, 250 ppm chlorides, 1,000 ppm total solids, and with no caustic akalinity, more than 10 ppm alkalinity yet less than 50 ppm sodium and potassium carbonate alkalinity. It should be noted that 0.3 ppm copper sulfate will kill many types of fish.

It is beyond the scope of this text to go into water analyses and the meaning of various ion contents and water properties. The U.S. Department of Interior water supply papers are useful in giving the water hardness distribution for 1,300 major

United States cities. Figure 17.2 shows this geographically. These reports contain water analyses and many other details. For example, the water hardness maps can serve as alkalinity maps for many parts of the United States since water hardness and alkalinity are almost the same in limestone areas of the Midwest of this country. In areas of the West and Southwest this relation usually does not hold because hardness is a non-carbonate type. Analyses are provided for the waters of the Great Lakes and major rivers.

**Iron and Copper**.—When comparing brewed coffee samples containing cream and one brew has 1 ppm iron, there is a greenish cast which becomes more prominent and more evident without a control cup above 4 ppm. The iron is believed to react with phenolic compounds in coffee brew. The reaction of iron and phenols is so sensitive that it is used as a qualitative and quantitative test. Iron may enter soft (non-alkaline) water from rusting pipes after the water leaves the treating plant. Copper ion in such aggressive water areas also dissolves in water and is evident from green sink stains. Such copper content in the water may produce some coffee brew flavor deterioration at concentration levels comparable to iron. However, the color changes in the brew are not so evident.

Local water supplies may have exceptionally high contents of certain ions. Sea water intrusion into potable water supplies will result in high chlorides—for example, 400 ppm chlorides in Galveston, Texas water. Sarasota, Florida water has about 800 ppm sulfates. Some deep well waters in arid regions, parts of Michigan, and elsewhere are very brackish and have over 1,000 ppm and sometimes several thousand ppm of solids.

**Sodium Bicarbonate and Time of Drip**.—The results of brewing tests with various formulated and actual city waters in the time required for water to drip through roast coffee grounds are as follows. The test results showed that when bicarbonates were 300 ppm or higher, drip-through time was longer. For example, with de-ionized water as a 6-min control, a 300 ppm $NaHCO_3$ solution took almost 9 min to drip through, and the 400 ppm $NaHCO_3$ solution took 10 min to drip through. Similar results were obtained with natural waters having bicarbonate alkalinity. Further, removing water hardness with substitution of sodium ion for calcium or magnesium ions increased the drip-through time 50 percent. The series of tests showed that alkalinity of carbonate and bicarbonate anions, regardless of cation (Na, K, or Li), and other alkalinities increased drip-through time. This data has direct bearing for drip brew devices and percolators. Excessive pressure buildup and a definite change in the character, composition, and taste of the brew has been found to occur with vending brewing devices. This appears to be less significant in commercial percolator extractors. In any case 6 to 9 min drip-throughs were found to produce a less flavorful coffee brew. A coffee flavor deterioration also occurs because the water alkalinity neutralizes part of the coffee acids. The change in pH is sufficient to affect rates of extractability but would more directly affect the brew pH, brew titratable acidity, and brew flavor.

**Coffee Acid Neutralization by Alkaline Waters**.—This has always been a serious problem with soft beverage acids. The loss in acidity and corresponding soft beverage or coffee flavor is costly. The coffee buyer argues about ¼ ¢ per lb of green coffee. Then the coffee user, even with 100 ppm alkalinity water, goes about neutralizing roughly one-third of coffee acidity and corresponding coffee flavor. Such a coffee acid neutralization raises the pH from about 5.1 to 5.4, resulting in a flat cup taste. Figure 17.16 shows the neutralization effect of waters of various alkalinity on hydrolysate coffee solubles, instant coffee solubles and brew coffee solubles. Not only are one-third of the coffee acids neutralized at 100 ppm alkalinity, but the cup taste becomes flavorless. This is not a 33 percent quality loss on the coffee purchase price, but a 100 percent loss. What good is the fraction of 1 ¢ per lb more paid on green coffee quality when the water used in coffee brewing process spoils the brew flavor? No one brewing or using instant coffee with water

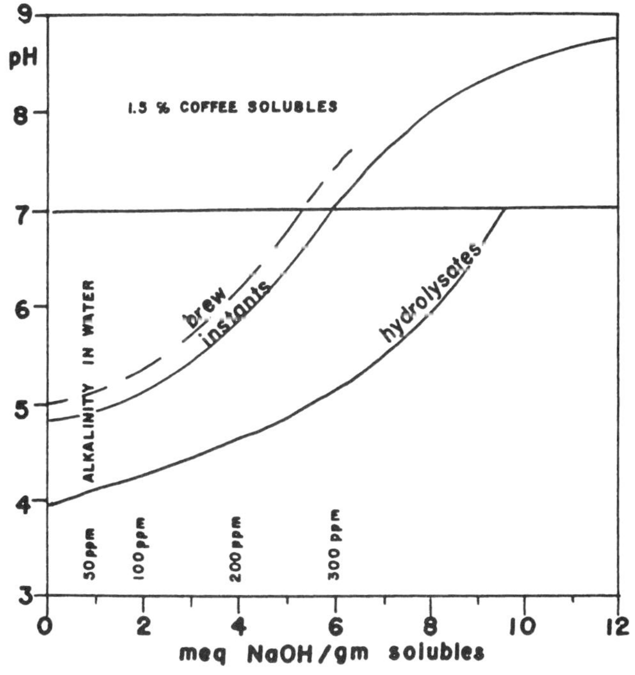


FIG. 17.16. ALKALINE WATER TITER VS pH OF COFFEE SOLUTIONS

having 100 ppm or higher alkalinity can make a good tasting cup of coffee without neutralizing the alkalinity. Some flavor advantage can be gained by using 33 percent more coffee, a higher coffee to water ratio, i.e., more acid. But this is an expensive way to neutralize the water alkalinity.

**Liberation of $CO_2$.**—The introduction of bicarbonate alkalinity into R&G or instant coffee acids still results in an acidic solution, and all the $CO_2$ gas is released. Water having 300 ppm alkalinity, for example, will liberate 44 mg of $CO_2$, which is 22 ml of $CO_2$ gas for 150 ml of water and 10 gm (20 ml) of R&G coffee. Even at 100 ppm alkalinity 7 ml of tiny $CO_2$ gas bubbles per 10 gm of R&G coffee properly distributed in particle cells may retard water flow and wetting. Ten grams of R&G coffee with 1 percent $CO_2$ (100 mg of $CO_2$) would give up about 50 ml of $CO_2$ gas into the extraction process.

**Neutralizing Alkaline Waters.**—Alkaline waters raise brew pH (or reduce the final solution acidity) and give a darker brew color (less light transmission).

Earlier it was stated that coffee roasters tend to use lighter roasts in the Midwest limestone areas of the United States where alkaline water is prevalent. The higher coffee acidity at lighter roasts neutralizes water alkalinity, leaving more coffee acidity for flavor.

Ten grams of R&G coffee have about 3.5 meq of acids per gram; 80 percent are non-volatile acids, depending on the roast and blend. The NaOH titer to completely neutralize brew or home type R&G coffee extract is only about 0.5 meq acid per gram of cup solubles. More of the coffee acidity stays within the unneutralized grounds during such a short extraction time with limited water rinse. Adding a pinch of citric acid or other suitable food acid, such as ascorbic (Vitamin C), into the boiling water just prior to extraction reduces water alkalinity but it is still not a cheap neutralization method.

The rise in brew pH from distilled to 200 ppm alkaline water in cups of coffee from commercial percolator extract may be 4.80 vs 5.00 pH; for instant coffee powder from the same percolation extract, 4.90 vs 5.20 pH. Distilled water is best for controls, but the quality of the waters used by the consumers must be evaluated.

## COFFEE ACIDS ACCOUNTABILITY

The volatile and non-volatile acids of roast coffee supply 0.345 meq acidity per gram of roast coffee on exhaustive extraction; 80 percent of the acidity is non-volatile. Brew acidity is determined chiefly by the blend and the degree of roast. For example, NaOH titer from light to dark roast varied from 1.0 down to 0.5 meq per gm of solubles, assuming that a 16 percent solubles yield from R&G coffee was obtained. A 0.2 meq per gm of solubles adjustment is made for Chicago water, which has about 110 ppm alkalinity.

For roast and ground coffee these acidities would correspond to about 0.16 to 0.08 meq acidity per gram or about half the exhaustive extraction yield of acids.

Titratable acidity is related to acidity recognized by taste and taste comments on acidic coffees, relative to aromatic, neutral, and rough coffees. Between high grown Colombian mild and Brazilian coffees there is about 0.1 meq acidity difference per gram of R&G coffee. The higher grown mild coffees are more acidic. This 0.1 meq per gm of R&G coffee is about 20 percent of total coffee acid.

There is no published data on Robusta coffee acid titer, but brewed coffee from mild and Robusta coffees may be, respectively, at pH 5.1 and 5.5; the difference in titer may be 100 percent as judged from Fig. 17.16, coffee soluble NaOH titer vs pH.

Note that 100 ppm alkalinity in the water used to prepare instant coffee will neutralize about one-third of the coffee acidity, 200 ppm alkalinity will neutralize two-thirds of the coffee acidity flavor, and 300 ppm alkaline water will neutralize all coffee acidity.

Altogether the green coffee blends, degree of roast, extraction yield, and water alkalinity markedly influence the final cup pH and titratable acidity.

The acids in coffee are weak organic acids. Figure 17.16 shows this in coffee solubles titer vs pH. There is no sharp change from acidity to alkalinity. The NaOH titer curve crosses the pH 7.0 line gradually. Further, if every carboxylic acid

*Courtesy of Rudd-Melikian, Inc.*

FIG. 17.17. TAPE OF ROAST AND GROUND COFFEE UNITS

group were neutralized with NaOH, the resulting pH would be 8.0 or higher. This is because we have a weak acid and a strong base compound. Further, many of the organic acids are only slightly dissociated and acidic hydrogen ions evolve only as neutralization takes place. It is this dissociation that causes pH buffering. This means that only slight changes in pH are accompanied by large changes in titratable acidity. As far as taste is concerned, pH changes of 0.1 are noticeable.

Thus, even though there are 0.345 meq acid per gram of light roast coffee in exhaustive water extraction, normal brewing is not exhaustive extraction. When titrating with NaOH to neutrality (pH 7.0), not all the organic acids react fully.

## Hydrolysis and Natural Acids

For instant coffees whose beans are roasted darker, the atmospheric or brew level solubles yield (about 20 percent) produce titratable acidities of 0.5 meq per gm of solubles or 0.1 meq per gm of R&G coffee. However, hydrolysis of coffee solubles in pressure percolation forms acids. In Fig. 18.16 the NaOH titer of hydrolysate coffee solubles vs pH shows 0.8 meq acidity per gram of solubles.

Assume: 1 gm of R&G coffee yields 0.15 gm of hydrolysates and 0.25 gm of free solubles; 0.40 gm is the total. The acidic contribution of: (a) hydrolysates is 0.8 meq per gm $\times$ 0.15 = 0.120 meq, and (b) free solubles is 0.5 meq per gm $\times$ 0.25 = 0.125 meq; 0.245 meq per 0.40 gm of solubles is the total, or 0.613 meq per gm of R&G coffee for the instant coffee compared with 0.1 meq per gm of R&G coffee for brewed coffee mentioned above. In other words, the instant coffee solubles are more acidic than brewed coffee solubles from the same source. This is confirmed by tasting.

Consider the following four sources and types of acidities in coffee solubles: extractable (1) non-volatile, (2) volatile; and hydrolyzable (3) non-volatile, and (4) volatile. These are depicted in the following three equations in milliequivalents per gram of solubles.

| | 1 | | 2 | | 3 | | 4 | | Total |
|---|---|---|---|---|---|---|---|---|---|
| Brew | 0.4 | + | 0.1 | + | 0 | + | 0 | = | 0.50 |
| Extract | 0.4 | + | 0.1 | + | 0.2 | + | 0.05 | = | 0.75 |
| Powder | 0.4 | + | nil | + | 0.2 | + | nil | = | 0.60 |

**Distilled Water.**—Using distilled water is no assurance that the water has no odor or taste. Not infrequently, organic volatiles come over in the distillation process, depending on original water quality, oxidation of the organics prior to distillation, and degree of entrainment. This is one reason why distilled water is sometimes observed to have a taste and odor.

**Temporary Hardness.**—Boiling water that is hard due to bicarbonate salts of calcium and magnesium will drive off $CO_2$ and precipitate insoluble carbonates. This is called temporary water hardness. Boiling to remove temporary hardness reduces total hardness and alkalinity which otherwise may noticeably harm coffee flavor.

**Chlorine.**—Most municipally treated and chlorinated waters, especially when organic matter is high, will leave chlorine residues (at several tenths up to 1.0 ppm) to destroy bacteria. A 0.3 ppm chlorine level in the coffee brew is clearly damaging to coffee flavor. Since brewed coffee contains around 1 percent solubles, this content of chlorine in the water is 30 ppm chlorine in coffee solubles. Chlorine's chemical equivalent weight relative to coffee solubles is about 0.5 equivalent per million (epm). The coffee solubles from the R&G coffee may have 1.0 to 2.0 epm. The net effect is a chlorinated coffee taste. This taste is unmistakable, undesirable, and disagreeable. At other times, chlorinated waters from natural or industrial areas carry phenols that have an objectionable flavor (especially when hot) and do great harm to natural coffee flavor. Chlorine is easily detected in water by taste and smell. Chlorine can be measured readily with an ortho-tolidine reagent and color comparator tube. Residue chlorine is removed entirely by passing water slowly over an activated carbon bed. One basic water supply problem is the seasonal variation in organic matter content in natural waters in some areas. This is accompanied by excessive chlorination by municipalities during spring periods of water runoff. The original organic matter content and the high chlorine dosage characterize the water with an odor and taste that cannot be hidden without complete water treatment. A demonstration of what less than 1 ppm chlorine can do to coffee flavor can be made by diluting standardized solutions of chlorine water and brewing coffee.

**Organic Matter.**—Organically contaminated water can usually be identified quickly by mustiness or off-odor, yellow color, and the reduction of applied chlorine. Waters in warm climates that are rich in vegetation or areas where runoff of water is high have periods of high organic content, as, for example, in Rio de Janeiro, Brazil, or Houston, Texas.

Instant coffees undergo more flavor change from impure waters than do brewed coffees. Brewed coffees have greater flavor buffering capacity due to the greater mass of grounds. It is unlikely that such water quality situations will be resolved completely in the foreseeable future. Water treatment, being a municipal function and drawing public taxes, does not progress much faster than its citizens allow. Often water treatment is not practiced until water quality becomes intolerably poor. Enlightened, progressive communities willing to pay for good water are rare. In certain geographical areas, where mountain water comes from melted snows to constitute a water reserve, there is a clean water supply.

Water quality in commercial soluble coffee percolation is less important because only two parts by weight of water leave the percolation system with one part by weight solubles. For brews, this ratio is 100 to 1. In plant percolation the feed water travels through 60 to 100 ft of coffee grounds which act as adsorbents. Use of an acidic water to feed the percolation system may accelerate hydrolysis rates at lower temperatures and will avoid scale on the feed water heaters.

**Oxygen Solubility.**—Figure 17.11 shows the solubility of oxygen in water from freezing to boiling temperatures. Cold water can dissolve 10 ml of oxygen per liter (about 14 ppm or 0.9 epm of oxygen). The solubility falls off with rising temperature. The data shown is at equilibrium. Raising water temperature does not drive off the oxygen and gases instantly; it takes time. That is why water must come to and be kept at a rolling boil before brew extraction. It must be recalled that 0.3 ppm chlorine is equivalent to 30 ppm or 0.5 epm in solubles, which is half the 1.0 epm of oxygen in cold water. Oxygen and chlorine are strong oxidizing agents.

The oxygen is removed by bringing water to a boil in a plant water system feeding the boiler so as to remove $CO_2$ and $O_2$, both of which are corrosive to carbon steel in the presence of water. The de-aerator can also be used for removing gases entering the percolator feed water system.

Oxygen in air is in contact with percolated coffee extract at 45 F (7 C) in storage tanks. Oxygen is somewhat less soluble in extract than in pure water, but oxygen still may be at 10 ppm. On a coffee solubles basis this is about 0.6 epm, whereas the oxidizables in the coffee extract are about 1.0 epm. It is for this reason that inert gas purging of fruit juices during cold storage is practiced so as to reduce the oxygen and oxidizing conditions that deteriorate flavor. Excluding oxygen is also good practice for coffee extract flavor protection.

**Hydrogen Sulfide Gas.**—When $H_2S$ is present in water supplies it is readily noticed by its rotten egg odor at 0.05 ppm in water and 0.10 ppm in coffee brew. Hydrogen sulfide gas occurs occasionally in some well waters, and its presence is sufficiently objectionable and corrosive to make it necessary to remove $H_2S$ before the water is used for food preparation or other uses.

## COFFEE BEVERAGE VENDING MACHINES

There were no coffee beverage vending machines during and prior to World War II. Melikian and Rudd, ex-servicemen, built their first coffee vending machine in a Philadelphia garage in 1946 and made their first sales to a local football audience. Both men were engineers, and their early machines were designed to use coffee extract. R&G coffee brewed through tapes was not commercially made by Melikian and Rudd until 1958. Their business grew in two ways: manufacture of frozen coffee extract and the use of extract in coffee vending machines. Kwik-Kafe at Hatboro, Pa. made percolated coffee extracts which were kept frozen until placed in the vending machine. Their sales grew from a beginning in 1946 to about 45,000 vending machines and 2,000,000 cups of coffee per day, with a nationwide franchise system by 1970. But the reported total daily cups of vended coffee in the United States in 1973 was about 15,000,000. Other vending machine manufacturers developed their own coffee brewing machines. The Vendo Company of Kansas City did not enter this market until 1956 with a 16-cup brew from a punctured ¼ lb can of R&G coffee. Vendo Company is one of the largest vending machine manufacturers in the United States.

## Vended Coffee: One-Half Percent of Total United States Coffee Use

The volume of sales of vended coffee must be placed in proper perspective in the United States green and roast coffee sales picture. One-half percent would amount to two billion cups of coffee. Calculated as instant coffee at 2 gm per cup, this is 9,000,000 lb of instant coffee or about 5 percent of United States coffee solubles consumption. In 1961, 60 percent of soluble coffees was actually vended as powder and 40 percent as extract. The United States vended use of soluble coffees in 1961 was 6,000,000 lb of solubles. The coffee extract use has been losing ground to R&G coffee, which in 1961 represented about one-third of the vended coffee sales. Thus, two-thirds of a billion cups of brewed coffee (from R&G) at 10 gm per cup requires about 100,000 bags of green coffee. This is only about ½ percent of United States imports of green coffee.

**Extracts.**—Coffee extract flavor can deteriorate easily in a few days. In many of the coffee extract vending machines, extract holding temperatures are too high. They should be about 35 F (2 C). The extract is exposed to air and storage for days, if not weeks. This is why coffee extracts have lost flavor. Extracts under these storage conditions produce a poorly flavored cup.

**R&G Coffee.**—As this is subject to staling, it also has a storage and flavor preservation problem not easily solved.

**Instant Coffee Powders.**—These may be lacking in flavor appeal, depending on the quality used, but their flavor does not deteriorate under proper storage. In current practice bulk powder flavor deterioration has been noticed in a few days. This is due mostly to moisture pick-up. With a single cup of powder or R&G coffee brew there is no waste as there is in batch brew preparation of R&G coffee.

## Factors in Flavor Quality of Vended Coffee

Inasmuch as the merchandise vending machine business is a specialized field and coffee is but a small part of this dynamic business, only the highlights of vended coffee beverage operations will be covered.

The flavor quality of vended coffee beverage is too often poor in quality. In addition to the conditions of coffee brewing that make for a low probability of a good flavored cup of coffee, other factors in vended coffee also affect this probability.

Instant coffee powders currently represent 40 percent of vended cups of coffee. Soluble coffee processors make what is called a "special" coffee for vending machines. This means that the vending machine manufacturers have adopted the use of a heavier powder bulk density, that is, 33 instead of 22 gm per 100 cu cm. It so happens that these bulk densities are usually achieved by particle breakage, which downgrades the flavor of the instant coffee. It also reduces instant solubility and makes for poor powder fluidity. Another special property of vended instant coffee is that it has to be low priced. Low price is usually achieved in two ways. The lowest priced coffees, such as Robustas or other types with off-flavor that may have heavy body, are used. Usually, a very dark roast is used—a French, or perhaps Italian, roast. Heavy body coffees at a dark roast do increase cup strength taste but not flavor quality. This has been described aptly as thinking only in terms of acceptability of the cup of coffee. The net effect is that instead of using at least 2 gm of instant coffee per cup (as recommended for home use), which gives about 227 5-oz cups per pound of instant coffee, instant coffee powders are used in quantities of 1.4 to 1.9 gm per cup. This yields, respectively, 324 and 239 cups per pound, with the emphasis on the 324 cups per pound of instant coffee. This economical policy on the instant coffee product is said to be due to the following reasons: (1) location management wants a high commission; (2) vending management wants a high return; (3) the coffee buyer is pressured to make a good showing by paying a low price; and (4) it is assumed that the consumer is not very critical and will accept a low standard of quality and strength.

The instant coffee powder may be packaged in an 8 oz (½ lb or 227 gm) laminated moisture-proof paper bag. This portion is used for filling the vending machine powder reservoir.

The R&G coffees used in beverage vending machines also have special properties and prices. Coffee beverage vending machines operate on several different principles, hence, roast coffee grounds fluidity, grind size, and several other coffee properties might have to be specified. Figure 17.17 shows the Rudd-Melikian tape with ground roast coffee chamber. Figure 17.18 shows the Rudd-Melikian tape being prepared. Figure 17.19 shows the inside of the tape used for 1-cup coffee brew vending machine.

Low cost R&G coffee blends in tapes are as common as in other fields. Cups of coffee are brewed 45 to 65 cups per pound of R&G coffee (7 to 10 gm in one tape portion). There may be some cup strength at the 65 cup per pound level, but not much flavor.

## Machine Costs for Vending Coffee Beverage

In an average year's sales, the machine can be amortized with maintenance in about 3 to 4 yr; in a good location the machine can be amortized in less than 2 yr.

*Courtesy of Coffee & Tea Industries Magazine*

FIG. 17.18. TAPE OF ROAST AND GROUND COFFEE UNITS BEING PREPARED

*Courtesy of Rudd-Melikian, Inc.*

FIG. 17.19. VENDING MACHINE FOR ONE CUP TAPE BREW

There are vending operations in which an individual owns about 10 to 25 vending machines; he maintains and services the machines and the income constitutes his whole livelihood.

## One-Cup Machine

In principle, the 1-cup coffee brew or vending machine gives the least coffee brew deterioration since there is no holding of a batch. Staling of ground coffee in brew vending machines is a problem. For this reason and because they are basically lower in cost than R&G brews, instant coffee powders hold their sales well. From a sanitation and health viewpoint and the public health codes that have to be met, the powdered instant coffee with powdered cream and sugar is the best arrangement. But it may not make the best flavored cup of coffee. Brewed coffee can be good when it is fresh, but that depends on the probability of delivering a fresh brew into the vending machine.

One of the factors influencing cup quality in vending machines is that many machines have a captive market in plants, shops, offices, amusement places, or other transient areas. Where local managements, as in company plants, insist on cup flavor quality and there is repeated machine use, day after day, quality will be offered. It takes incentive and pressure to maintain consistently high quality.

## Water Quality

Another factor that makes it difficult to vend a good flavored cup of coffee is water quality. Some machines have no water treatment equipment. Machines simply draw city water. If there is water alkalinity, chlorine, organic matter, or a noticeable odor and taste, each will downgrade the coffee flavor. The vending machine is not capable of dealing with wide variations in water qualities. However, filters can remove insoluble particles. Resins can remove brackishness, alkalinity, hardness, $H_2S$, and chlorine. Carbon also removes chlorine. A fraction of a ppm chlorine will ruin the coffee flavor. Few vending machines have ion exchanger resins in the water supply line. Further, cold beverages can taste satisfactory with a water heavy in some organic matter, but the malodorous substances become prominent and objectionable in the hot beverage.

Most coffee vending machines do not boil water, but the machine is designed to hold 208 to 210 F (98 to 99 C) water which largely obviates a flat taste.

Water heating systems in vending machines are constructed usually of stainless steel. Metal contamination from iron or copper causes serious flavor loss and change, as has been stated.

Dispensing of liquid coffee extract, liquid cream, and liquid sugar is mechanically simple. However, it has associated bacterial growth, sediment, and cleaning problems. The resultant flavor loss has caused liquid systems to lose their share of the total vending market. Patents have been issued for eliminating precipitated gels from frozen coffee extracts. Also there is a patent for adding a desiccant to instant coffee powder to retard its caking when it absorbs moisture from the air.

For quality flavor in vending coffee, the vending machine operator must be persuaded to use good quality coffee whether it is liquid extract, instant powder, or ground roast coffee. Another major technical problem is to control feed water quality which is so bad in some areas that good coffee cannot be made without water treatment. Until complete purification of water for a vending machine is developed, a "quality vending operation" could use a 5-gal. bottled water container similar to an office cooler. A 20-gal. water reservoir could be built in the vending machine and water could be transferred into the reservoir from the portable tank at each servicing. Twenty gallons is equivalent to 500 cups of coffee, which is about the vending capacity of most machines. Water storage eliminates plumbing in the vending machine installation.

In the matter of driving out all the dissolved gases from the hot water, the vending machine could be operated like an espresso machine, which keeps the water supply at about 5 psig above atmospheric pressure with intermittent gas venting. A safety disk or valve would have to be used.

Every coffee flavor quality suggestion, of course, means additional but not prohibitive cost. Also, the machine that makes a really good cup of coffee will draw more customers.

## Instant Coffee Envelopes

Another system of providing coffee beverage is the sale or give-away of 1 cup portions of instant coffee powder in laminated moisture-tight envelopes at motels, offices, shops, etc. With these, powdered cream in envelopes, powdered sugar in envelopes, and a source of near boiling 210 F (99 C) water, cup, and mixing stick must be available.

**Cups.**—Before the advent of hot coffee beverage vending machines in 1946, wax coated or paper cups were used for cold beverages. However, wax or paper cups cannot be used with hot coffee. The wax melts and heat penetrates, making these cups impractical. Non-wax type coatings were developed. About 1957 foam plastic insulated cups came into use. These cups protect the hand contact surface from the hot cup contents. A critical problem in using plastic or paper cups is to prevent the odor or flavor of the cup material from affecting the coffee aroma and flavor. It is a common observation that coffee, beer, and other beverages do not taste "right" in some paper cups, but cups are being improved rapidly.

# COFFEE FLAVOR ADDITIVES, SUBSTITUTES AND SYNTHETICS

Additives to foods, especially for flavor change, are as old as time. With laws on the labeling of packages and jars, additives vary from commonplace substances, such as sugar and salt, to pyroligneous acid and artificial flavors.

## Purpose of Additives

The National Academy of Science (1956) lists twelve groups of intentional additives used in processed foods: (1) preservatives; (2) antioxidants; (3) sequestrants; (4) surfactants; (5) stabilizers and thickeners; (6) bleaching agents; (7) maturing agents; (8) buffer acids and alkalies; (9) food colors; (10) non-nutritive sweeteners; (11) nutrient supplements; (12) flavoring agents, and miscellaneous. In regard to coffee flavoring agents there are aromatic chemicals, essential oils, and others that have been used for decades or longer, especially in candy and imitation fruit flavors. Other chemicals used in imitation flavors that are related to coffee flavor are butter, rum, butterscotch, caramel, imitation maple syrup, and others. Specifically, the traditional items that have been identified and confirmed in coffee aroma and flavors are as follows:

**Acids**.—Formic, acetic, propionic, butyric, valeric, pyruvic, and oleic are acids found. Aldehydes found are acet-, propyl, n- and iso-, butyr-, valer-, and furfur-, and all appear in natural fruits. But propyl- and valeraldehydes are not mentioned much in prepared flavor compositions. Acetoin or acetylmethylcarbinol has been used for a long time in butter flavor as has diacetyl. Ketones are used in some imitation flavors, but acetone and methylethyl ketone are not specifically noted. Imitation coffee flavor constituents appear to run more toward esters, possibly due to the instability of aldehydes and ketones. Instability is, of course, the very nature of coffee aroma and flavor. Ethyl acetate and ethyl formate have been used for a long time in many types of beverages, candy, ice cream, and baked goods. The latter is associated with rum flavor. Furfuryl mercaptan has been used in imitation coffee flavor, but it is really not representative of coffee.

Pyroligneous acid is used for smoke flavor. Dimethyl sulfide is used in imitation garlic flavor; this compound is a key natural flavor note in high grade coffees. Oddly enough, dimethyl sulfide also appears in Merory's imitation raspberry flavor.

**Imitation Flavors**.—These are the stock in trade of flavor compounders. It is difficult to find much published information in this field. Much of what has been published is outdated by current knowhow evolved from modern analytical instruments like gas, column, and paper chromatography, infrared patterns, and mass spectroscopy.

The U.S. Food and Drug Administration has been more liberal in allowing food additives that have been used for decades, especially at low levels, such as a few parts per million. Although almost every one of these flavor houses offers an imitation coffee flavor for sale, none of these offerings ever is acceptable to people who know coffee aroma and flavor. Most of the flavor makers will confidentially agree with this conclusion. This is so because coffee is made up of numerous unstable and volatile constituents. The ratio of investment cost to gain for resolving the components of coffee aroma and flavor is estimated to be too high relative to other investments that flavor makers can make. This is partly a rationalization because with a good chemical and working background in coffee aroma and flavor chemistry plus modern analytical instruments and skills, results from coffee analysis come much faster and at less cost than is realized by those who work at coffee without the requisite background.

## United States Food and Drug Law

With the change in the United States Food and Drug Law in 1958 and as amended in 1960, it is necessary to keep abreast of current progress in approved additives. This can be done through references in food magazines, especially from the continuously revised Food and Flavor Additives Directory by the Hazelton Laboratory of Falls Church, Virginia, as well as the Food Law Institute in New York City, the Federal Register, and the *Food Drug Cosmetic Law Journal*.

From time to time, articles are published in journals like the *Perfumery and Essential Oil Record (London), American Perfumer and Aromatics* and others that mention flavor additives. Publication of advances in flavor composition is slow even though considerable expansion of flavor chemistry knowledge has occurred in the last decade with the application of chromatography techniques.

## Flavor Additives and Imitation Flavors

These have been used widely (and, to some extent, almost exclusively) in the candy and soft beverage fields for decades. Imitation fruit flavors are cheaper and have a chemical stability formulated into them that the natural flavor products sometimes do not have. Successful imitation of natural flavors is variable, but some imitation flavors are very good representations of their natural counterparts. In such cases, certain flavors have become firmly and relatively permanently established in the flavor trade. In some cases a single chemical compound may dominate the flavor property; for example, vanillin from lignin for the vanilla bean extract; for licorice flavor, anethole, the major chemical portion of anise; for caffeine bitterness, methylation of theobromine; for sugar sweetness, there are saccharin and cyclamate; for butter, margarine (representing 50 percent of the market) has color and flavor additives although the fats are easy to distinguish; for cocoa butter, reasonable substitutes have been developed; benzaldehyde is classically cherry or almond flavor, depending on concentration; butyric and caproic acids have long been recognized as constituents of fermented cheeses and dairy products and are used in salad dressing mixtures; monosodium-glutamate is a distinct chemical protein and with similar compounds has evolved in the past decade into a significant flavor additive.

The purpose of the foregoing discussion of flavor additives of discreet chemical composition is to show that additives are commonplace in the food industry and that as chemical identities and compositions are evolved, the use of flavor additives also grows. There is no reason why many additives cannot be used to fortify instant coffee aroma and flavor in the foreseeable future. It has always been the dream of coffee men to upgrade and control green coffee quality, and indeed we stand at the threshold of such an application.

Instant coffee flavor is a departure from natural coffee flavor due to differences in brew composition. Flavor additives can help make instant coffee taste more like natural coffee. Appealing additives can restore some of the imbalance coffee flavor undergoes in preparing instant coffee. Introduction of discrete chemical compositions promises to exercise an element of control over the coffee aroma and flavor not yet attained.

## Technological Progress

This is man's increasing control over his environment as exemplified by synthetics: fibers, rubber and plastics, detergents for soap, dyes, pigments, paints, metals, gems such as diamonds, rubies, and emeralds, medicinals, vitamins, and insecticides. Instant foods are in some cases a distinct flavor departure from their original nature. This modification of natural flavors has been intensified since 1950. The net result has been to widen the convenience, number, and appeal of foods and flavor available to the consumer. If an acceptable tasting, reasonably priced food can be produced in a process plant, it makes little sense for each housewife to devote a disproportionate number of hours to such preparations. Unfortunately, the quality of mass processed convenience foods is sometimes inferior, and these foods cost more than natural food prepared at home. But flavor quality progress is continually being made and is gaining customer acceptance and purchasing power.

## Time for New Flavor Acceptance

The transition from the use of a natural flavor to a synthetic flavor usually does not occur overnight. There must be a process of acceptance and transition. This may take decades. Change in the properties of a food, especially a basic food flavor and appearance, cannot be accepted promptly.

After flavor additives, which only superficially fortify or perhaps only subtly change food aroma, flavor, and appearance, there are substitutes.

## Substitute Flavors

These imitate the natural flavors but usually do not produce exactly the original flavor. Coffee substitutes and additives have a long and continuing history. Chicory is an outstanding example, as are other cereal grains, caramelized carbohydrates, and fruit sugars (figs, raisins, etc.). Some are beverage extenders used in time of coffee shortage, while others are appealing and economical enough to perpetuate their use.

It is interesting to note that many of the men and women who pick coffee and have a relatively reasonable income for only a few months of the year, cannot afford to buy even coffee, in many cases. The use of soluble coffees in the coffee growing countries is a significant sales item because it offers a uniform product of acceptable quality and at a lower price than the locally roasted (burnt), adulterated (beans and corn) stale roast coffees. The coffee industry is still fluid, and new firms or new managements rise to prominence every five to ten years. The opportunities for profitable action are still present.

It is likely that coffee flavor additives and substitutes may yet enjoy their first sales in countries that are economically impoverished. Here the desire for a better standard of living will have to be made possible at first by lower priced alternates. For example, in South Africa a chicory-carbohydrate instant coffee is very popular, and in some countries control of flavor additives is practically non-existent. The circumstances are thus ripe for commercial coffee flavor developments to be pioneered.

## Coffee Flavor Additives and Fortifiers

It is most likely that development of coffee flavor additives and fortifiers will come through instant coffees. Current methods including coffee oil add-back to spray dried instant coffee and the addition of freeze-dried coffee brew still do not make a wholly balanced coffee flavor.

In the 1920's imitation and synthetic coffee flavors were investigated. Now, from time to time, news is made by someone compounding a substitute coffee beverage. Such announcements are often exaggerated, partly because of the fear among the coffee growers and coffee roasters who see their stable businesses threatened. The original report and projection for the development of synthetic coffee was made by the Stanford Research Institute. It was a study and report to the Committee on Foreign Relations of the United States Senate of September 1959. Synthetic coffee was viewed as a political problem that could result from technological advances displacing agricultural crops of some nations. The suggestion that coffee beverage could be substituted or synthesized has often been made. But the formal presentation of the idea to the United States Senate Committee caused considerable consternation in coffee quarters.

## Brewing Notes

Although this text is not concerned with the consumer's individual taste preferences and the many ways of preparing coffee beverage, the following comments are listed as being pertinent to the preparation of good tasting coffee beverage from roast and ground coffee:

(1) *The coarser the coffee grind* the less efficient is the water extraction and the yield of coffee flavor and solubles.

(2) Any brewing method that involves *holding the hot coffee beverage* will contribute to its deterioration in flavor. Household percolation is a poor method of preparing flavorful coffee beverage. In fact, household coffee percolation drives off the coffee aroma by steam distillation and leaves a strong tásting solubles residue.

(3) The drip or vacuum type of coffee preparation is superior because the contact *time* between the water and the coffee is short. Steeping roast coffee grounds for a few minutes in water that has just boiled, followed by separating (decanting or cloth filtration) the coffee grounds promptly from the beverage make a fine tasting beverage provided no additional heating is done. The use of espresso demitasse which is prepared by allowing pressurized water to pass through a finely ground coffee offers a fine tasting beverage.

(4) *Cleanliness of the equipment, purity of the water, the grind, and freshness of the roast coffee* as well as the *quality of the coffee,* the *weight ratio of coffee to water,* and the *time of contact* will bear on the flavor of the coffee beverage prepared. The Coffee Brewing Institute Standard is to prepare 40 to 45-5-fl oz (150 ml) cups of brew per pound of roast coffee. In Latin America 30 cups per pound are common with a darker roast, finer grind, heavier blend, and different modes of brew preparation.

## U.S. Government Specifications for Roast and Instant Coffees

The Federal specification for roast coffee is HHH-C-571 D Amendment 1971 and for instant coffee, it is HHH-C-575 C, 1976 . The Quartermaster Food and Container Institute for the Armed Forces oversees such specifications, evaluations of samples, and bids submitted. They are located in Natick, Mass. Roast coffee blends are specified as being Santos 4's—70 percent by weight and MAM's (Colombians—Medellin, Armenias, or Manizales)—30 percent by weight. This specification is, incidentally, better than the average restaurant or home user can procure. The instant coffee specifications state caffeine and carbohydrate levels which must not be exceeded.

# BREWING PRINCIPLES

There are many types of coffee brewing devices, but how they are used and how they perform can be more easily appreciated by understanding the few basic input and output variables involved, as follows:

(1) Quality of R&G Coffee Used. This includes botanical variety, e.g. Arabica, grade (number of defects), altitude of growth, place of origin, basic taste and aroma, how roasted, degree of roast, uniformity and fineness of grind, freshness (time and storage conditions since roasted and since ground), and so forth. Obviously, if the coffee used is of poor quality, the beverage will also be of poor quality.

(2) Quality of Water Used. This includes a consideration of any odors and tastes: organic, mineral, acidity, or alkalinity. The temperature of the water when it contacts the R&G coffee is also very important. Low temperatures will give poor yield and an unbalanced taste; high temperatures will cause a harsh taste.

(3) Quantity Ratio of Water to R&G Coffee (by weight). An appropriate example would be to add 1.152 liters of boiling water to a beaker containing 60 grams of finely ground, good quality coffee; that is input. Output will be 1,012 grams of beverage and 200 grams of wet spent coffee grounds. The beverage will have a 1.2 percent wt solubles, and the spent grounds are 75 percent water. Solubles yields from R&G coffees can vary considerably for various reasons, but under near ideal brewing conditions, about 20 percent wt solubles yield ought to be achieved from fine grinds. An exhaustive extraction of Robusta coffees may give 25 percent solubles yield, whereas a non-exhaustive extraction of fine grind or a non-exhaustive extraction of a light roast, coarse grind may give only 16 percent solubles yield. The level of solubles yield in commercial practice simply shows the efficiency of solubles removal, and reflects in the adequacy of the raw material, water, and extraction thereof. The very best quality coffee can give an unsatisfactory beverage if it gives over or under extraction yields and low solubles in cup. Rule of thumb: it is far better to make the beverage too strong or stronger, because it will have good flavor, and can always be diluted slightly to suit some drinkers; whereas, a weakly prepared beverage cannot be strengthened, and usually has the over-extracted, high yield, objectionable tastes.

(4) Method or Apparatus. This includes how R&G coffee and water are put together, and how resulting beverage and spent grounds are separated. A few examples:

(a) Vacuum flask: Boiling water rises into R&G charge, mixes, wets while steam mixes. Within 1 minute, vacuum in lower flask sucks down beverage, while leaving spent grounds in upper flask.

(b) Espresso, can or vending systems: Hot water in a measured quantity is forced down through a cartridge of R&G coffee, delivering beverage quickly.

(c) Drip type brewers: Boiled water is allowed to percolate down through a shallow bed of R&G coffee, delivering beverage below in a few minutes.

Note: A better variation of the drip system that obviates the problem of incomplete wetting or extraction is to slurry R&G coffee with hot water prior to passing it through a strainer, e.g. Brazilian Quador.

(d) Pot steeping: R&G coffee and water are held hot or even heated therein for 10 to 20 minutes until the coffee solubles are largely extracted, and the spent grounds have wetted and settled. Then, the relatively sediment-free beverage is poured from an upper spout and decanted. The Arab use of Ibrik is similar in use.

(e) Cup steeping: In many Moslem countries, it is common to pulverize the R&G coffee and to place a teaspoon full into a demitasse sized cup. Then boiling water is poured in, much as one might do when preparing instant coffee. Since the R&G coffee is very fine, it is readily wetted by the hot water, readily gives up its solubles at well over 25 percent solubles yield, and is ready to drink as is or with a bit of sugar. The fine spent grounds form a sludge in the bottom of the cup, which the drinker carefully leaves.

(5) Time. Brewing time is a very important variable. Where the granulation is fine and hot water pressure is used, as in an espresso or vending machine, brewing or extraction time is 1 to 5 seconds. When a high concentration of solubles results, as in the Italian espresso beverage, the extraction is not exhaustive; but where low solubles concentrations are taken, as in vending, the extraction is exhaustive. The vacuum flask brewing is usually less than 1 minute, so it is, in principle, a good method for preparing beverage, in spite of its diminished use. Drip brewers can perform well, but since they invariably involve use of filter papers, whether in urn or ½ gal brewer, there is the taste pickup from the paper; the paper also regulates water outflow so as to flood the R&G coffee cradled in the paper cup, and to effect steeping of 5 minutes at least. Pot and urn preparations can take up to ½ hour, depending on various operating features.

Other input factors can be classified as those causing deterioration, e.g. excessive or inadequate heating or times of brewing, steam distillation as from a beverage "pumping" percolator, contamination from corrosive metals or taste imparting plastics, cloths, paper, utensils, etc. A notoriously degrading input factor is the prolonged and excessive heating of the prepared beverage prior to serving.

Factors affecting output or the resulting beverage include (1) quality of flavor and aroma (taste appreciation and acceptance); (2) strength (intensity of solubles concentration)—quantitative and cultural; and (3) solubles yield, as related to quantity, economics, and equipment performance. The coffee drinker is most concerned with (1) and (2); the institutional user is sometimes more concerned with (3).

To obtain a clean flavored fresh cup of brewed coffee at the right concentration and temperature containing most of the natural coffee aroma and flavor is not simple. This is amply demonstrated by the poor coffee beverage served in many private homes and in many public restaurants. The attainment of such a goal is by taste, brew, and coffee education, for both consumer and restaurant. Attaining better brewed coffee flavor involves the use of better coffee blends, good brewing equipment procedures, and good water quality.

The yield of beverage in liters (or gallons) per kilogram (or pound) from a given coffee and brewing system at the desired solubles concentration is fixed. But to the institutional server, the size of the portion served and the size of the cup are pertinent. The PACB has used 150 ml (5 fl oz) of beverage as a U.S. serving, which is a good point of reference. However, the OCS and vending trade serve 7 fl oz or 6 fl oz of weak beverage. In Europe, on the other hand, 120 ml (4 fl oz) of stronger beverage (1.33 to 1.5 percent solubles) is not uncommon, as is possibly 3 percent solubles from espresso machines in 1 to 2 fl oz (30 to 60 ml) portions.

## Time

It is important to emphasize the time factor. After poor quality water, excessive times between roasting and grinding and use, are most responsible for causing staleness. Excessive times taken in brewing beverage, in separating the grounds, and in serving all individually and cumulatively contribute to beverage quality deterioration. The public in general, as well as responsible people in the trade, unwittingly contribute to such deterioration by preparing too much beverage and holding it hot for an hour or more. People actually believe that they can hold coffee beverage hot for many hours without any significant deterioration. And roaster propaganda about reheatable coffee is totally commercial and without taste for quality. It cannot be said too often: time is the essence when preparing good tasting coffee.

The most important point is that brewing must be done quickly. Coffee aroma and flavor are delicate and transient, but elusiveness is part of their attractiveness. Aroma can be captured by the consumer only momentarily. The aroma and flavor are fragile and fleeting. Thus, the R&G coffee must be fresh and of good flavor. The water must be of good quality. The apparatus must brew with speed and reproducibility. The beverage must be delivered and consumed promptly. Any delay in this timing sequence destroys the end result: the pleasure and satisfaction the beverage can bring.

Reducing the brewing process to a set of mechanically reproducible steps, automatically carried out without human interruption until the brewed beverage is dispensed, attains the desired goal. This has been achieved already to varying degrees in commercial brewing units which control times, temperatures, and water/R&G proportions accurately.

## Grind and Temperature

With 185 F water, the solubles extraction yield with fine, drip, and regular grinds after 5 min of slurrying is about 20, 18, and 16 percent respectively. If the temperature of the water and grounds is low, fast and thorough wetting will not occur, especially with light roast, coarse grounds.

C.B.I. recommends that suitable brewing times for each grind are, respectively,

1 to 3 min for fine, 4 to 6 min for drip, and 6 to 8 min for regular. Coarser grounds give less solubles and harshness.

## Deterioration (Air Addition, Heating, Holding and Evaporation)

Prompt serving of the prepared coffee beverage is necessary to present fresh flavor to the consumer. A common error also is to drink the beverage too hot, which is not conducive to taste and aroma appreciation. A burnt tongue or heated mouth is not one that can discriminate flavors and aromas.

All the foregoing steps in beverage preparation can be perfect, yet it is common to find the beverage at near the boiling temperature for ½ hour on a heater. Admittedly, adding sugar and milk or cream covers up the bitterness of the coffee. But if really good coffee flavor is obtained, it is best enjoyed black. The espresso method which brews into one cup at a time obviates this hot pot holding.

## Chances for Consistently Brewing a Flavorful Coffee Beverage

With ten or more preparation variables, an uneducated approach offers only a very small chance because of the following compounding factors. However, recognition and control of these contributing factors can raise one's chances considerably for preparing a tasty coffee beverage.

(1) Mass marketed and low priced coffees do not offer the best coffee quality or flavor. Brand name tells the buyer nothing about the true nature of the coffees used.

(2) Buying or preparing freshly roasted coffee beans is important.

(3) Vacuum packed cans of commercial coffee are only partly fresh; staling is rapid after one opens the can. Buy only 1 lb at a time; keep unused part in the freezer.

(4) The average user must acquire more knowledge about brewing and coffee to make a tasty beverage consistently.

(5) Water quality over large geographic areas noticeably downgrades coffee flavor.

(6) Guessing weights of coffee and water; hence, their ratio is unscientific and not reproducible, leading to dilute and watery preparations.

(7) Pumping percolation does not get much coffee flavor in the cup, but plenty of aroma in the kitchen. Urn coffee, even if prepared reasonably well, will taste harsh and acidy after being held hot for an hour or two before serving.

(8) Timing in any brewing process must be reproducible.

(9) Then there are the factors like cleanliness of equipment, materials of construction, flavor contributing filter papers or rubber gaskets, etc. that are important.

(10) Have a scale of values. Not recognizing these circumstances (and this is

really running blind), one cannot expect to receive more than one cup in ten that will be a pleasure, perhaps three or four cups in ten that will have acceptable taste, and the rest will not taste good. The brew may then be consumed with milk or cream and sugar, but not with much joy. However, taking precautions about these factors can give most coffee brews really acceptable to enjoyable flavor.

## YIELD

Cup size and beverage fill must be stated. Yield of satisfactory cups of beverage per pound of R&G coffee (not only water to R&G coffee ratio) are further influenced by:

(1) Blend or flavor quality level of the original green coffee beans can increase yield by over 25 percent.
(2) Degree and quality of roast. A 20 percent or more difference in yield of cups is possible.
(3) Fineness of grind. A 50 percent increase in cup yield is possible.
(4) Freshness of the R&G coffee. A 10 to 15 percent increase in yield from fresh, very freshly roasted coffee.
(5) Efficiency of flavor recovery in brewing-equipment, method, etc. This can vary 100 percent.
(6) Water quality. This can influence cup yields up to 100 percent.
(7) User's taste preferences and habits—e.g., using chicory, milk, etc. This can vary to 100 percent.

Coffee beverages have a range of taste acceptability over a range of solubles concentrations. But the many preparative and consumptive factors must be taken into consideration

The one consideration is strength. How strong do you want to brew the coffee? How strong do you like your coffee? Most Americans fall into the yield range of 50 to 80 cups, but most Latins and Europeans fall into the range of 30 to 40 cups per pound of R&G coffee. However, these yield figures are somewhat meaningless unless we cite the influencing factor of aroma and flavor, the roast bean properties. Strength is not just solubles concentration, it is a complex mixture of blend, roast, grind, botany, etc. For example: Hawaii's Kona or Colombian MAM coffees are far more flavorful than low grown centrals or old Brazilian beans, and especially Robustas.

### Solubles Yield from Roast Beans and Solubles Concentration in Beverage

The C.B.I. states that an 18 to 22 percent solubles extraction is desirable for idealized beverage flavor. This is 1.15 to 1.35 percent solubles concentration in the beverage with outside limits of 1.0 to 1.5 percent solubles concentration. In practice, solubles yields from R&G are closer to 18 percent. Measurements of solubles content of coffee beverage is only an index of extraction efficiency, nothing more.

However, solubles concentration and yield must be considered within the whole structure of coffee flavor. This includes volatile aromatics and the basic quality of the roast coffee beans being extracted. The amount of acidity in the beverage reflects blend, roast, and grind. Colloids contribute texture to the coffee beverage. Thus, aroma, acidity, and colloid content influence the coffee taste and acceptability. The C.B.I. has evolved the solubles concentration and yield chart for coffee beverage which shows: (1) if a watered coffee beverage has been prepared; it is measured objectively. (2) if the extraction in an urn is inefficient, the low solubles yield and concentration will be easily measured.

The solubles yield is related to the coffee brew solubles concentration. The two conditions of coffee solubles concentration and solubles yield must be satisfied so that they fall within the idealized cup flavor area. Low solubles yields do not give a wholesome coffee brew flavor either, even if the brew solubles concentration is in the acceptable range.

*Specific Gravity.* Variances for Brazilian and mild coffees on a specific gravity (C.B.I.) chart for beverages are not likely to be as great as for Robusta coffees (which give higher solubles yields and concentrations and a very different coffee flavor). A 1.5 percent soluble brew has a specific gravity of only about 1.006 at 80 F. The C.B.I. has obtained special hydrometers for measuring these small differences.

Flavor boundary conditions are not sharp because of individual taste preferences, influences of bean blend, roast, grind, time, and mode of extraction, as well as (pH) bean acidity.

However, the "solubles yield and concentration target" area is an improvement over "no target at all." Thus, the Coffee Brewing Institute has given the user of coffee urns a measuring tool for the extraction process.

## DOMESTIC COFFEE BREWERS IN U.S.A.

### The Beverage "Pumping" Percolator

With this brewer, all one has to do is fill it with water to the number of cups desired, set in the hollow riser tube and basket, fill the basket with regular grind R&G coffee, and plug the extension wire into a 110 volt electrical outlet; return in 15 minutes and the beverage is ready. An electrical heater at the base of the percolator below the tube riser heats the water. When the water at the base of the central tube starts to boil up the tube, after 4 or 5 minutes, hot water spills over into the basket of R&G coffee. This repetitive "spilling over" of hot water over the R&G coffee grounds quickly becomes a solution of coffee solubles. Effectiveness of rinsing out solubles depends on how well the hot water is distributed over the R&G coffee in the perforated basket and how uniformly the water trickles down over the R&G coffee bed, wetting the granules and dissolving away its solubles.

The "high-heat" element of the percolator is cut off by a thermostat when the coffee beverage reaches 205 F. This takes about 9 minutes, depending on the specific percolator design. Then the secondary "low heat" electric element maintains the coffee beverage at about 180 F. In many models a red light comes on, indicating that the percolation process is complete; that is, the end temperature has been reached and held for so many minutes.

What the public likes about this brewing apparatus is that it is automatic and does not require continual attention, as a manually heated percolator does. Hence, it leaves one free to prepare the eggs or shave while this percolator "pumps" away.

But this percolator, through the process of steam distillation, drives out the volatile coffee aromas all over the kitchen and house. This creates an ambiente that so many people have become accustomed to—the nice smell of coffee aroma. But there is rarely much aroma left in the poured cup.

Some percolators have adjustments on the thermostat to lengthen the time of percolation. Additional time drives off and destroys even more aromatics and gives only the impression that the beverage tastes stronger. The quality of taste has actually deteriorated. A sight glass on the side of the opaque percolator vessel shows how much water to add initially and how much coffee beverage is left.

## Materials of Construction

Copper and brass are the metals that have been used for cooking utensils for the longest time, and are still used today to prepare coffee beverage equipment.

Iron, due to the ease with which it rusts, is quite unsuitable for equipment used to prepare coffee. The presence of copper with acidic, hot coffee beverage for prolonged times is detrimental to the coffee flavor. If the copper or brass utensils are clean to the metal surface, and brewing or boiling, as with the Ibrik, is limited to only a few minutes, copper contamination is negligible. Minimal equipment corrosion is reflected in minimal coffee beverage contamination.

Historically, coffee pots were made in the 1700's from tin, silver, and porcelain. Porcelainized steel pots are chemically inert and satisfactory, but are subject to chipping damage. After World War I aluminum spun pots came into wider use since they were lightweight and economical.

Aluminum is slowly attacked by the coffee acids and fatty acids. But it is not so detrimental to coffee flavor as copper or iron. Hence, the use of aluminum domestic percolators is still popular. Aluminum, like copper, is one of the best conductors of heat, which glass and stainless steel are not.

In the 1920's and 30's, the vacuum glass (Mae West shape) brewers became popular, and glass carafes are still much used today in restaurants. The lower (fragile) glass bowls of vacuum coffee brewers are often replaced with stainless steel or aluminum to minimize breakage. Electroplating chromium over steel percolator pots has been done commercially for 40 years. It was preceded by hot dip and electro-tin coating steel utensils. Tin was used also to coat the inside of copper pots. Silver is used to coat copper by hot dip or electroplating processes.

Although stainless steel was available before World War II, it was too expensive for domestic brewers. However, after World War II stainless steel cooking and coffee brewing equipment was manufactured at increasing rates. Today, most coffee brewing urns, flasks and hot water tanks are stainless.

## Consumer Reports

Complaints about these units from time to time by *Consumer Reports* and *Consumer Bulletin* is that some models are difficult to clean. Some are very dangerous to clean, having sharp metal edges inside. Pumping percolators are rated for 8 to 10 cups, and often do not properly brew 3 or 4 cups. Models draw 6 to 9 amperes at 110 volts; the higher the amperage, the faster the rate of heating and percolation. Aluminum pots are the hardest to clean. That is why these pots are usually made from highly polished chrome on copper or aluminum, or highly polished aluminum. Glass is, of course, very smooth, chemically inert, and easy to clean, but it is fragile. Ceramic is opaque and less fragile than glass. Filter paper brewer use in homes has been displacing percolators slowly.

# PRESSURE BREWERS

All the brewers discussed so far function at or near atmospheric pressure. Only the force of gravity is used in their functioning. Finer grinds require water pressure.

## Napier or Vacuum

This depends on the vacuum created from the condensed steam in the lower bowl (of a Silex type of brewer). Vacuum sucks down the coffee beverage from the upper bowl compartment where it has been mixing and steeping for a minute. Actually, it is the atmospheric air pressure pushing the coffee beverage through a uniformly compacted granular bed of coffee that results in self-filtering and a uniform, downward beverage flow. This brewer has the important feature of effectively wetting each granule and thereby driving out the carbon dioxide gas and the interstitial air which can otherwise make for inefficient extraction.

The Napier vacuum brewer has the additional advantages of natural, bubbling agitation, momentarily hotter temperatures, better driving off of gases, and a natural reconsolidation of the granular coffee bed (coarsest granules at the bottom) as the positive downward pressure forces the steeped coffee beverage to filter through the fine granules, making, so to speak, the last rinse of solubles and flavor from the granules. The vacuum filtration process is positive and countercurrent. Also, the vacuum process does not require a possible contaminant, like filter paper. The beverage is filtered through the bed of coffee granules.

The vacuum brewer gives a much better tasting beverage than the "pumping-percolator." The vacuum brewer makes good tasting coffee beverage when it performs correctly.

## Espresso Type of Coffee Brewers—Domestic and Commercial

The espresso class of coffee brewers operates with pressurized hot water. Normally Italians prefer to use a dark roasted coffee bean, finely ground just before use. This results in a strong tasting (about 3 to 5 percent solubles) and oily beverage, but with good aroma and flavor.

**"Range-Type"**.—The simplest and most economical domestic brewer, it is made in Italy from cast aluminum. It consists of a bottom water chamber, a tube riser into the R&G coffee chamber (with perforated top and bottom discs), and a beverage receiving upper chamber. The steam pressure developed in the lower water compartment forces most of the hot water up through the bed of granular coffee. When steam appears at the top lip of the riser and cover, then there is no more water left to rise from the base chamber. This manually operated brewer, which does require attention so as not to overheat the coffee nor the aluminum vessel, makes beverage similar in taste to the espresso coffee beverage from larger automated commercial machines.

*Safety*: The pressurized lower compartment has a eutectic metal safety plug which will melt and then release steam pressure if the metal pot temperature (and pressure) gets too high. This type of pressure vessel is a potential bomb, and this feature, no doubt, is unattractive for the average U.S. housewife. If the heating is done too quickly, the issuing coffee beverage can spill out and make a mess. The screwing together of the R&G coffee chamber (upper to lower compartments) also is not safe or convenient for the housewife. The cheap gaskets provided often leak. Hence, this pressure brewer is used more by Italians and others who like the espresso coffee beverages. There are two or three other variants of this pressure design operating on the same principles and giving about the same quality of coffee beverage. Normally the steam pressures attained are only 1 or 2 pounds above atmospheric pressure. The ½ in. thick chamber of roast coffee granules, even with finer than "fine grind" granulation, offers small resistance to water flow. Commercial espresso machines come with several dispensing heads. Large chambers (e.g., 500 grams of R&G) deliver 1 or 2 liters of beverage from a large espresso brewing machine.

**Espresso Brewer Extraction Efficiency**.—It is not only fast, but it attains 2.5 to 4.0 percent solubles concentration beverage at 20 percent solubles yields. Whereas the common U.S. urns and domestic brewers take 10 to 15 minutes extraction time and yield only 1.2 percent solubles concentration at less than 20 percent solubles yields. The time-temperature exposure of the espresso coffee beverage is 100-fold less than U.S. brewers. The residual spent grounds are swollen tight in their chamber; clean for the next brew.

This strong tasting coffee beverage is usually drunk with hot milk in the morning, or black with a twist of lemon peel in the evening. After dinner, a bit of brandy with good espresso coffee is a delight. Most users add just a heaping spoon of sugar.

**Commercial Espresso Machines**.—Once one departs from the range-type, heated espresso brewer, the prices jump many fold. For example, a range-type 4-cup unit may cost $15 whereas a 2-cup, electrically heated, small espresso machine may cost over $200. Commercial espresso machines cost 1, 2 or 3 thousand dollars, depending on their size and make. This high pricing in itself has been a detriment to broader use of the espresso machine in the U.S.A., as much as the U.S. taste is for weak coffee beverage.

Espresso machines are fairly common in Europe and Latin America where a strong, dark, small demitasse of coffee beverage is preferred. A foam on the beverage served is considered essential by some persons. Commercial espresso machines are sized by the tank volume of pressurized hot water and the number of dispensing valves.

That espresso coffee beverage is made by passing steam through the coffee granules is an erroneous idea. The idea no doubt developed because one sees water vapor (steam) escaping out of the boiler vent and about the dispensed coffee beverage. A separate steam vent is used for heating milk for cappucino. The boiler is under about 10 psig of pressure.

The fixed volume of hot water passes down through the chamber of R&G coffee. The resulting beverage flows out into the single espresso cup and, in Italy, is not more than 1 fl oz in volume. Elsewhere, perhaps 2 fl oz of espresso beverage may be dispensed.

The hot water that contacts the bed of granular coffee has been cooled upon entering the piston walled compartment to less than 200 F. The water is cooled even further on mixing with and passing through the R&G coffee. The issuing espresso beverage delivered to the demitasse is very near drinking temperature (150 F). Some coin vending machines have been built in Italy using these same principles.

The cartridge of R&G coffee normally holds about ¼ oz or 7 grams of pulverized R&G coffee compacted between two perforated discs (1 cm apart and 5 cm diameter). A simple 90° twisting action locks the R&G coffee cartridge in the extraction position. Two to four fl oz (60 to 120 ml) of hot water are passed through the coffee granules, of which 1 fl oz of water remains with the spent coffee grounds. Extraction of solubles is not exhaustive. Extraction is efficient due to the uniform distribution of hot water flow over the full face of the fine granular coffee bed. The extraction time is about 4 seconds—hence, the name "espresso."

Espresso has another connotation, that of "removing the essential part." In this case, only flavorful and aromatic coffee solubles are removed.

## INSTITUTIONAL COFFEE BREWING IN THE U.S.A.

Before World War II, and for a few years thereafter until the early 1950's, urns were widely used in high demand restaurants. Vacuum Silex glass and metal flasks were also quite common. However, it was the CORY pressurized and metered hot water brewer that had the greatest popularity, and indeed prepared the best tasting coffee beverage. The Cory brewers lasted well into the early 1960's when the ½ gal. filter brewers were making heavy inroads into institutional equipment sales. It is not clear why the CORY Brewer lost favor in the end, but was no doubt a combination of factors: cost, grounds handling, not delivering a clear beverage as the filter units did, pressurized hot water system, higher maintenance, etc. An unfortunate application of the Cory unit to obviate grounds handling was the procedure to use "tea bag" cellulose pouches for the R&G coffee portions and to stuff them into the Cory extraction chamber. Not only was this type of pack not reliable in its extraction, but the porous bag allowed accelerated coffee staling.

From 1959 the Pan American Coffee Bureau had the Coffee Brewing Institute, which sponsored some basic research on coffee and evaluated institutional brewing equipment, sometimes making suggestions for improved design to manufacturers of equipment and had frequent schools for teaching proper brewing methods on urns, ½ gal. brewers, and vending machines mostly for the salesmen of coffee roasting firms. Dr. Ernest Lockhart developed the system for measuring the solubles content of brewed beverages with a hydrometer, as a means of determining if proper solubilization was occurring in the equipment under investigation. The PACB in the late 1960's and 1970's (until it closed in December, 1976), was receiving financial support from Brazil and the ICO in London, members of which were the Robusta producing countries. Therefore, the PACB never mentioned blend and flavor differences between producing countries. The "Gold Cup" awarded to over 4,000 eating places in the U.S.A. signified only adequate mechanical performance of the brewer machine, and not what went into it, nor what came out of it. The public never knew this. But even the Gold Cup award was fraught with problems, because it required a PACB fieldman to check and recheck that those machines were performing mechanically well. With only ten fieldmen nationally, this became an impossible task, because of revocation of Gold Cups. However, it was an on-going educational effort and did to that extent keep restaurateurs "on their toes" and conscious of coffee brewing-methods and cleanliness.

Another ongoing factor influencing coffee brewing methods and types of equipment used has been, and still is, the "tie-in". The coffee bean roaster salesman offers the restaurant the use of a brewer as long as he buys his coffee. The irony of this arrangement is that most restaurateurs accept it, knowing full well that they are paying for that brewing equipment in the price of the coffee, but they do it anyway. It is not uncommon to find the roaster's coffee to be better during the first period, when the new client has his new equipment in use; whereas later the coffee quality deteriorates. This unfortunate arrangement, which many consider a restraint of trade that the FTC has not yet challenged, requires not only that the roaster have capital to carry on such practices (vs. the roaster without capital), but is supported by the institutions because "all coffees taste the same". The coffee roaster also gives the buyer repair service at cost or less, and the coffee roaster often has several depots where urns and brewers are repaired. Since repairs can become a heavy expense, roasters do tend to offer newer, more reliable brewing equipment. Brewers that are less than 3 years old usually require little to no repairs, depending on water hardness and other water properties that can cause brewer damage. This "tie-in" of R&G coffee and brewing equipment goes so far that some roasters own brewing equipment manufacturers, e.g., Farmer Bros. owns BREWMATIC in California. Of course, the roasters are able to buy equipment cheaper, since they buy brewers by the dozens or even 100 and gain discounts that a single restaurant, or even a small chain could not obtain. There is no question that the restaurant does not save money in such "loan" arrangements; but human nature of the restaurateur, believing that he is getting "something for nothing", blinds him, as he pays and pays year after year with no control over the quality nor freshness of the R&G coffee he buys. Restaurants that do control their coffee quality invariably own their own equipment and choose, if not dictate, the blends they use.

Brewing coffee for institutional uses is fragmented into a number of parts, and urns and half gallon brewers constitute about three-quarters of the equipment used, with half gallon carafe types representing about 80 percent of located equipment, urns less than 20 percent, possibly only 15 percent in 1977, and a few percent of "instant" types.

Some small restaurants use instant coffee powder, either spray dried or freeze

TABLE 17.1. COFFEE BREWING EQUIPMENT SERVICING IN RESTAURANTS

| Age of brewers (years) | Percent of all brewers |
|---|---|
| 1 | 16 |
| 2–5 | 50–60 |
| 6–10 | 20–25 |
| over 10 | 2–10 |
| over 20 | less than 1 |

| | | |
|---|---|---|
| Roasters placing: | 1972 | 1971 |
| Urns | 23% | 23% |
| ½ gal. brewers | 67% | 71% |
| Vacuum brewers | 5% | 4% |
| Percolators | 2% | 2% |
| Other equipment | 3% | 1% |
| Total | 100% | 100% |
| Urns by type: | | |
| Automatic | 62% | 72% |
| Semi-automatic | 9% | 8% |
| Manual | 29% | 20% |
| Who does it: | 1972 | 1971 |
| Roaster personnel | 85% | 75% |
| Restaurant personnel | 23% | 25% |
| Restaurant pays for: | | |
| Parts only | 36% | 50% |
| Parts and labor | 18% | 15% |
| Neither | 46% | 35% |

dried in dispensers. An important example is the Columnware electromagnetic needle vibrator that feeds instant coffee or tea from an inverted quart jar directly into a hot water stream, dissolving the powder, and delivering it to a cup as mixed beverage. The consumer cannot tell from the outside of the apparatus what the source of the coffee beverage is, and indeed is deceived by a crown of whole roast beans covering the inverted jar of instant coffee.

## Brewing Coffee Beverage From Canned R&G Coffee

In the 1950's were built many machines that brewed coffee beverage from a ¼ lb evacuated can, which was pierced and had hot water flushed through, collecting beverage in a reservoir from which beverage was drawn cup by cup as vended. The problem that resulted was not that the beverage made was poor—it was not; but that the time the batch remained in the reservoir caused it to deteriorate until all was vended. By the early 1960's this vending machine was discontinued. However, the U.S. airlines took up this principle, and it has been used successfully ever since. On a plane, the brewed batch is used up rather rapidly and hence, the deterioration period is obviated. What has happenend is that the quality of the

R&G coffee going into the cans has deteriorated with often 100 percent Robusta blends being used and even when freshly brewed, many of these canned coffees are undrinkable. The fault lies not in the brewing method, but in the coffee materials used. With reasonably good quality coffee blends, the can method (even though R&G is not all fresh) results in good beverage. By stacking several cans with perforated ends or using columns of R&G coffee, concentrates can be prepared.

## URN BREWING

### History

The origins of the urn, no doubt, were in the late 18th century in Europe. Coffee drinking was more extensive in Europe than in colonial or the newly independent American states.

In Paris, London and Berlin the public was first introduced to coffee beverage from the street vendors, decades before coffee beverage was consumed in the home.

### Biggin

The "biggin," which was a popular method of preparation in England and France, was the origin of the urn. The bag of R&G coffee occupied the mouth of the large pot.

As the "biggin" pot became larger, it became less portable or manageable. It was difficult to manually pour hot water over the bag of R&G coffee granules. Beverage was held hot for hours, resulting in poor taste. Direct heating of urn vessels was common, and reflected in the development of heat treated beverages.

The direct heating of the urn was a primary development. Wood and charcoal heat was most common. Heating allowed the water to be boiled in the beverage preparation vessel (urn) and also allowed one to keep the coffee beverage warm thereafter.

The manual withdrawal of hot water from the urn vessel and pouring it through the bag of R&G coffee granules was very common in the U.S.A. until 1960. Carry-through of fines of spent grounds contributed tars to the urn walls and beverage, a poorer taste, and the time-temperature degradation.

Even though automatic coffee brewing urns are now available, hand pouring of the hot water is still not uncommon on the older urns. These "pour-over" urns can still be purchased for those who do not want to automate or make the higher initial cost for automated urn equipment. Hand pouring of water is menial, laborious, and dangerous.

The faucet to withdraw coffee beverage from the base of the urn vessel was necessary on the first urn. Wood and charcoal fuels gave way to oil and then gas in the early 1900's. A sight glass on the side of the urn vessel indicated the amount of coffee beverage therein.

## Bain Marie

The direct heating of the beverage vessel was realized to be detrimental to taste in the early 1900's, so the "Bain Marie" or water-bath principle was introduced into urn design. The use of the hot water bath had the additional advantage of a hot water reserve for the next batch of beverage.

The inner beverage compartment, called the liner, and the outer water bath compartment, both needed liquid level sight glasses. The water volume in the bath area needed to be several times greater than what was required for beverage preparation. Then as cold water was replenished into the jacketed area, the resulting cooling of the outer bath water did not unduly cool the prepared coffee beverage. A several fold increase in water bath volume made the whole urn more costly. Hence, the insulation of the inner liner (to buffer excessive cooling or heating) was introduced by simply making a double walled beverage liner.

**Siphon**.—The hand pouring of the boiled water over the bag of R&G coffee was a continually undesirable operating feature. By the early 1900's the water bath area was pressured by heating the water to slightly above atmospheric pressure in a sealed but safety vented jacket. This slight pressure increase was not sufficient to add cost to the water boiler, but was sufficient to cause the hot bath water to rise a foot above the normal coffee beverage level. This technique allowed hot bath water to pass through a spray head over the bag of R&G coffee, thereby obviating the manual repouring. The adequate water pressure would be indicated by steam issuing from the jacket relief valve. The operator would then open the pressure/siphon valve to the spray head. Once the siphon flow was started, the heat could be turned down on the jacketed water bath. The volume of siphoned water flow over the R&G coffee was evidenced by the resulting beverage level in the liner. When the correct beverage level was reached the hot water siphon valve was manually shut.

## Urn Preparation of Coffee Beverage

Brewing of coffee in urns is an outgrowth of steeping and straining or placing ground coffee in a cloth bag and swirling it in a pot of hot water. Urn brewing techniques have been refined over the years. Even the cloth bag filled with coffee grounds fixed at the top of the urn has been replaced by filters. Near boiling water is now pumped and sprayed over the R&G coffee instead of manual "pour-over".

Urn preparation of coffee beverage has the following disadvantages:

1. Coffee brew falls through air, allowing oxidation and flavor loss.
2. Urn investment is high.
3. Large urn vessels are unwieldy to keep clean.
4. Coarser granulations (regular) produce lower solubles and flavor yields.
5. Flow of water through the R&G coffee bed is not uniform.
6. The time for preparation of gallons of beverage and the time for dispensing gallons of beverage (except for peak demand periods) result in hours of storage with flavor deterioration. The beverage can become heavy, bitter, and acidy.

On the other hand, urns can produce 200 cups per hour (3 gal. pair) or 400 cups per hour (6 gal. pair) of beverage for high demand, short consuming period situations with 20 to 30 minute urn cycles. Urn vessels are paired: one urn is being drawn from while the other urn is brewing. When an urn holds hot beverage too long, especially Robusta blends, tars build up on walls, making cleaning more difficult.

The use of metered and pressurized hot water sprayed over the bed of R&G coffee makes the 6 gallon urn almost as productive as the manually operated 12 gallon urn. The manual labor has now been removed both in mixing finished beverage (now done with air bubbles), and in supplying hot water. The once virgin beverage no longer is recycled through the spent coffee grounds. The automation of the urns with metered water and hot water shower head rinsing R&G coffee has both eliminated labor and has improved the coffee beverage flavor, but holding time still deteriorates flavor.

The faster brewing cycles have allowed the use of smaller sized urns for a given cup productivity, thereby reducing initial investment and reducing beverage holding time. Electrically heated water and water temperatures controlled with thermostats provide better temperature (hence extraction) control over brewing than ever before.

## Support Grids

Even after World War II muslin bags were used in urns without under supports. This situation caused the R&G coffee to hang centrally, and good hot water distribution through the uneven parabolic shaped R&G coffee bed was not obtained. A 2 in. grid at the base of a stainless steel cyclinder was then used to support the sagging bag, which resulted in a more uniform and more level R&G bed. In addition, a perforated cover plate was used which gave more uniform hot water distribution across the top of the R&G coffee being extracted. The top perforated plate also prevented disturbance of the R&G coffee, as had occurred with ordinary pour-over systems.

In the mid 1960's, the muslin bags virtually disappeared from use as filter papers cradled in wire baskets came into use with automated hot water spray heads. The paper filters found favor because they passed no particulate matter, so this allowed easier cleaning of the urns, but it also removed good colloidal flavor and contributed a papery taste (latter points were ignored commercially). The use of filter papers facilitated the disposal of spent coffee grounds, also reduced labor and reduced tars easily deposited from Robusta coffee blends.

## Urn Cleanliness and Sanitation Procedures

Cleanliness starts with the adequacy of urn construction as designed and built and as approved by the National Sanitation Federation. Inaccessible crevices are avoided, and corners are smooth and round. Materials of construction that are inert to coffee beverage and hot water are selected: glass, stainless steel, chrome plated brass, tinned copper, etc. The stainless steel used is of a highly polished finish (4B or better). Then insolubles do not adhere; smoothness helps in periodic cleaning with alkaline phosphates and bristle brushes.

Each residue batch of coffee beverage should be drained off. The faucet lines should be rinsed and be brushed with hot water before preparing the next fresh batch of coffee beverage. Muslin filter bags, when not in use, must remain in cold water to rinse out coffee residues and to prevent souring. Build-up of any caramelized coffee solids in the beverage liner that would subsequently cause a valve to leak or deposit in the customer's cup needs to be conscientiously removed through systematic cleaning.

Coffee oils at near 200 F quickly turn into objectionable tasting substances. Oils also contribute to oily droplets on the surface of the coffee beverage. Alkaline cleaners effectively remove these oils. It is easier to clean off these oily deposits at short intervals (such as daily), than after several weeks. Then such built-up deposits caramelize, and become tarry and quite difficult to remove. Hard water can deposit lime on spray head perforations, subsequently obstructing normal spray patterns. Water hardness, if not controlled, will deposit unsightly whitish deposits on the sight glasses as well as on heated surfaces, causing resistance to heat transfer and eventual heater failures.

The use of paper filters has reduced the quantity of grit and oil that carries over into the beverage, making urn cleaning easier and less frequent. A few gallons of cold water are usually left in the urn overnight and drained in the morning before brewing. Urn cleaning chemicals are now conveniently marketed in weighed portions in sealed pouches, making their use safer, more effective, and easier.

## Selection of Urn Size

Size of the urn is governed by the peak cups of coffee consumed per hour. Bottle or carafe filter brewers can produce 12 5-fl oz cups of coffee beverage per cycle. Four cycles per hour is feasible with 48 cups per hour. Six cycles per hour are possible with close attention to yield 72 cups per hour. From there on up, two or three ½-gallon brewers are needed, and rates up to 200 cups per hour are possible. However, at this level of use the urn becomes more practical because it requires less labor. Also, use of beverage is fast enough to not allow the beverage to deteriorate by remaining in the urn over one hour. Two 3-gallon urns would nicely yield 250 cups per hour.

**Time for Cup Draft**.—Two 5-gallon liners filled with beverage receive the peak load. The second liner is brewed when nearly empty; the first liner may be about half empty. Often two liners of beverage are prepared so that customers can draw the cups simultaneously.

*Drawing the cups becomes the controlling rate factor*.—Normally 5 fl oz of beverage runs out into the cup in 4 to 6 seconds, depending on the head of coffee beverage in the liner. But the customer can take 10 seconds fussing around before and after drawing the cup of coffee. Also, he or she may not open the draw valve fully, thereby making a longer draw time. Even at a 15 second draw time per cup, one after the other, a 5 gallon urn would take at least 30 minutes to drain just from the pure mechanics.

The restaurateur could buy two 3-gallon (75 cups each) liners to brew twice per hour on each, giving a total of 300 cups available in that hour. This leaves only one faucet to draw from during the 10 to 15 minute brew period on each liner. This method becomes marginal in service and requires close attention by the person brewing. A 10 to 15 minute delay in starting an automatic brewing cycle can be ruinous at the peak demand period.

**Excessive Brewing Capacity**.—Generally restaurants provide excessive capacity in coffee beverage supply. In cases where 12 cup carafes are filled from the urn, it is desirable to specify larger, faster flowing draw valves. At 600 cups per hour use, two 9-gallon liners will perform adequately. In exceptionally high demand areas, two 20-gallon urn liners for 1,000 cups per hour are needed. When these larger urn sizes are used, it is often with the use of circulatory pumps to deliver coffee beverage to several stanchions. Then the customer or employee drawoff is not concentrated at one urn.

## WATER QUALITY AND COFFEE FLAVOR

Good quality, fresh flavored coffee beans are more frequently ruined by incompatible water quality than from all other factors combined. World and local water supplies are being progressively polluted, contaminated, and chlorinated. These chemicals react with aromatic coffee flavor components resulting in some very bad tastes. Bad water tastes may not be noticeable in cold water, but the same water can be very objectionable-tasting when hot. Municipal chlorination of water supplies that still contain organic matter, especially phenolic industrial wastes, can produce very repulsive tasting water and coffee beverage. Hard and alkaline waters, as well as brackish waters, can neutralize the natural coffee acids and otherwise "flatten" the coffee beverage taste. Softened hard water should never be used for preparing coffee because it gives a bad tasting beverage. Each of these influencing factors will be discussed in detail.

To attain better water:

1. Use bottled, spring, or demineralized water.
2. Contact your local water department and find out the cause of the poor taste and its remedy.
3. If chlorine or small amounts of organic matter are the problems, install a carbon or charcoal cartridge on your cold water line to a separate drinking water faucet.
4. If the water has excessive salts (brackish), you may pass the tap water through a demineralizer cartridge. The same resin types are used to demineralize water for steam irons. These are readily available.
5. Make sure the water used for instant or R&G coffee is well boiled. Temporary bicarbonate hardness can thereby be removed. In some localities this boiling will markedly improve the water quality for coffee use. Boiling water also drives off dissolved air, which if not removed, gives coffee beverage a flat taste.
6. Sometimes, just using more R&G coffee and/or better quality coffee and/or finer grind will introduce sufficient additional coffee flavor to offset the detriments of the water impurities. Simply making the coffee beverage stronger (less water per unit of R&G coffee) is sufficient to upgrade off-flavors of some waters.

In view of these circumstances, it is incredible how many restaurants and public catering places continue to use bad tasting tap water, year after year. They destroy coffee's flavor as they brew and serve it.

## Coffee Beverage is 99 percent Water (in the U.S.A.)

Restaurant chains and similar operations that ignore geographical differences in water quality relegate the coffee beverages they prepare from good R&G coffee into a checkerboard pattern of flavors.

Good quality water is a real economy when preparing coffee beverage and reconstitutable soups and beverages. In warmer climates, in the southern part of the U.S.A. and especially in downstream locations, organic matter from nature and/or industry often contributes a bad taste to water. Municipalities often do not remove all such organic matter. Also, alkaline, brackish, hard, chlorinated, chemically contaminated, and otherwise impure waters reduce, alter, and degrade the flavor of coffee beverages. There are rental services that offer reverse-osmosis cartridges and demineralizers to purify such poor quality waters. Fortunate are those who live in areas that are serviced with clean, pure mountain water, relatively free of minerals, e.g., New York City, San Francisco, Seattle, Portland, and other mountain localities.

Coffee beverage is 99 percent water (in the U.S.A.). This means that for every part of water impurity (in parts per million parts of water), there will result 100 ppm impurity in relation to instant coffee solubles or even 500 ppm impurity to the coffee solubles from R&G coffee (that is 0.05 percent). When considering influence on taste, this level of impurities is very high.

Take the case of 100 ppm alkalinity in tap water from Lake Michigan. This is not very high alkalinity by water purity standards. But this becomes 10,000 ppm alkalinity relative to the coffee solubles, or 1 percent weight. Since the coffee water solubles have only the equivalent of a few percent acidity, this kind of water alkalinity is highly detrimental to the natural coffee acid flavors. A few parts per million of impurity in the water that is reactive with the coffee flavor and aroma components, based on 600 ppm aroma constituents in the R&G coffee, can cause marked changes in taste.

## Alkaline Water Neutralizes Coffee Flavor Acidity

About half the coffee beverage prepared with tap waters will be downgraded only by the alkalinity. Downgrading occurs also due to chlorine, phenols, organic matter, etc. The ppm hardness in water in the mid-western U.S.A. is usually equivalent to that level of ppm alkalinity. Hardly any municipalities reduce hardness, alkalinity or other impurities in all city water. In the Midwest and in alkaline water regions, it is common for coffee roasters to use poorer quality coffees. This is because the water ruins the coffee flavor and who can tell after brewing? A light bean roast also leaves more coffee acidity.

Lightly roasted coffee beans are more acidic and good beans stand up in flavor better with mildly alkaline waters. Continental Coffee Co. in Chicago has a U.S. Patent for adding acids to R&G coffee for use in such alkaline water areas. The pH (which is the exponent of the digit ten of the hydrogen ion concentration per liter of solution) is an index of acidity in water. This pH is very close to 5.0 for coffee beverages prepared from mild coffee beans. A pH of 4.80 is not uncommon with instant coffees available today, when dissolved in pure water. These same coffee beverages prepared with alkaline water can have pH's of up to 5.5 and over. The layman can differentiate pH differences of 0.10 pH units. The acidity of the coffee beverage is reduced by alkaline waters. The beverage color in the cup becomes darker brown with higher alkalinities. Robusta coffees, when prepared with pure water, yield pH's of about 5.50. These beverages are dark brown to black in color.

Dark color in the cup gives one the impression that the coffee beverage is strong; in fact, the beverage is only lacking in acidity. For those people brewing coffee beverage in such alkaline water areas, another solution is to add the proper quantity of weak fruit acid (e.g. citric acid) to the water used before brewing so as to neutralize the excessive water alkalinity. Since dark roast coffee beans are less acidic than lighter roast beans, the taste of beverages made from dark roast beans are more influenced by alkaline waters. More acidic high grown coffees like Colombian can be used to help neutralize the water alkalinity, but that is an expensive way to correct the situation.

## Softened (Hard) Water Downgrades Coffee Beverage Taste and Impedes Coffee Brewing

About 1962 the PACB sponsored some systematic studies on the influence of softened water on coffee brewing. This work clearly showed that the "drip through" time in brewers was increased 50 percent and was accompanied by a disagreeable bitter taste with softened water. For example, deionized water took 6 minutes to drip through while water with 300 ppm $NaHCO_3$ took 9 minutes.

Softened hard water is simply water in which there has been an exchange of sodium ions for naturally occurring calcium or magnesium ions. The resulting blockage in water flow rate through the bed of R&G coffee is attributed to sodium stearate or soaps forming with the fatty acids in the roast coffee granules. The R&G coffee bed becomes slimy or soapy. In a coffee beverage vending machine, when the softened water is forced through the bed of R&G coffee by pressure, the resistance to water flow becomes so great that the R&G coffee extraction compartment often ruptures. Even as recently as the early 1960's, firms selling water softening equipment and services were recommending softened water for use in coffee brewing. Fortunately, after painstaking and persistent education of these firms by the PACB, this promotional matter was brought under better control. The slimy and soapy nature of the R&G coffee beds help to trap carbon dioxide gas liberated from the coffee granules which retards water extraction. The acids in coffee are neutralized by the sodium bicarbonate of the soft water. Five fl oz of 300 ppm alkaline water can liberate about 22 ml of $CO_2$ gas on neutralization. With ten grams of R&G coffee per 5 fl oz cup of beverage, this is a release of 50 ml or 2 fl oz of $CO_2$ gas.

The mineral softening of water by complete removal of the calcium and magnesium ions as well as bicarbonate and carbonate ions, as is done by deionization or demineralization of the water, benefits brewing, and is not to be confused with the sodium substitution softening.

## Minerals in Water that Influence Coffee Beverage Taste

The U.S. Public Health drinking standards for water call for clear, odorless, and tasteless water, free of bacteria and with less than: 0.2 ppm of copper, 0.3 ppm of iron, 250 ppm of sulfates, 250 ppm of chlorides, 100 ppm of magnesium, 1,000 ppm of total dissolved solids, but more than 10 ppm alkalinity (but no caustic alkalinity), less than 50 ppm of sodium and potassium alkalinity.

Note that 0.3 ppm of copper will kill many types of fish. Copper is also very detrimental to coffee flavor. Iron at 1 ppm easily colors (gray-green) a coffee beverage with milk. Iron reacts with the phenolics in coffee to give a gray color, and it also destroys coffee flavor by catalyzing aldehyde oxidation.

Some local water supplies have high levels of certain mineral ions. For example, salt water intrusion into Galveston, Texas water supplies gives over 400 ppm chlorides. Sarasota, Florida has 800 ppm sulfates. Deep wells in Michigan and other arid areas have over 1,000 ppm minerals; their waters are called brackish.

In 1955, the CBI (Coffee Brewing Institute) sponsored some taste tests on water and coffee beverages to measure the taste thresholds for some common mineral ions that occur in water. The results in ppm for water and beverage respectively, were: calcium—125/300; magnesium—100/200; potassium—300/400; sodium—100/250; iron 4/2 or less; chlorine—200/400; phosphate—100/100; and carbonate—50/100.

Chlorine usually is added to treated municipal waters to destroy bacteria and organic solubles. Insoluble organic matter is usually coagulated and filtered out. A few tenths ppm of chlorine residual is left in the water going to consumers, and this is readily detected by odor and taste. A 0.3 ppm chlorine residual is equivalent to 30 ppm of chlorine reacting with the coffee solubles in the beverage, giving a very foul phenolic taste which is easily recognized and very disagreeable.

The organic matter in rivers varies in concentration with season, being greater during the rainy seasons. A yellow color and mustiness indicates organics in water.

Oxygen occurs at 14 ppm in cold water, but boiling drives off most of the oxygen. Unboiled water gives coffee beverages a flat taste.

**Steeping R&G Coffee in Water**.—Steeping R&G coffee in boiled (not boiling) water is probably the simplest, cheapest and one of the more effective ways of drawing off into the water phase the solubles portion of the R&G coffee granules. Steeping allows time of contact for (first) wetting the granules, and then for solubles to diffuse out of the granules. Steeping R&G coffee is an intrinsic part of almost every brewing and/or extraction technique. In the simplest application, granulated roast coffee is added to just-boiled water, and these are mixed. They are allowed to steep in a pan for several minutes with intermittent mixing.

In Brazil, for example, this is a standard brewing technique. The stirred mixture is decanted into a half pint sized flannel sack, called a Quador. The spent coffee grounds mostly are left in the pan. Grit-free coffee beverage passes through the flannel bag. There is no boiling. The flavor of this coffee beverage is faithful to its bean ingredients.

## TASTING

Commercial coffee tasters and buyers use the steeping method, but without filtration. From every green coffee bean sample, about 100 grams of beans are roasted and then ground just prior to tasting. Ten grams (two 5-cent coins) of drip grind roast coffee are placed in each of two or three cups for each bean sample. The sameness in taste of the two or three cups relates to the uniformity of the coffee beans in the sample. When all cups are filled with 10 grams of R&G coffee, the boiling water is added. The lightly roasted grounds are allowed to steep for 5 or more minutes. Since a few percent of the coarser grounds may float (because they do not wet), these grains are skimmed off with a perforated spoon. The mixture in the cup is then given a quick stir with a spoon to make sure the solubles are uniformly in solution. A few more minutes are allowed to let the wetted granules settle to the bottom of the cup, and for the coffee beverage to cool to tasting temperature. Then, with a silver (or silver plated) spoon, often private to each taster, the coffee beverage is spooned up, without disturbing the grounds at the bottom of the cup. The beverage is slurped with a suction of air into the mouth (with accompanying noise). This slurping action cools the coffee beverage, while allowing aroma and flavor to be thoroughly wafted into the nostrils.

Steeping is, therefore, a very honorable and traditional way of preparing coffee beverage for critical evaluation as well as for enjoyment. Normally the professional taster, at his swivel stool and table, does not swallow the coffee beverage, but spits it out immediately.

Properties of Good Tasting Coffee Beverage
(Using Good Quality Water and Methods)

1. Recognizable quality and quantity of desirable aroma and flavor (a control sample is essential to orient and guide taster by comparison, since taste/aroma memory is short).
2. An acidity (pH between 4.90 and 5.10) taste characteristic of fine quality coffees.
3. Dissolved coffee solubles of at least 1.0 percent and to 1.5 percent in U.S.A. (higher in Europe, espresso, etc.).
4. Moderately dark reddish color at meniscus of cup (not brown nor black).
5. Relatively little sediment (U.S.A. cultural standard) or suspension, as contrasted to espresso, which is high in sediment (Arabic preparation) and high in oils.

## FLAVOR STRENGTH OF FREEZE AND SPRAY DRIED INSTANT COFFEES

With all things equal, freeze dried instant coffees are much stronger in coffee aromatics and flavor than are spray dried coffees, as measured by objective taste testing and gas chromatographic analyses. One can use one-quarter to one-third less freeze dried coffee per cup for a satisfactory tasting coffee beverage, e.g. 1.50 grams of freeze dried vs 2.25 grams of spray dried instant coffee.

Solubles by themselves do not characterize acceptable taste levels. The content of volatiles contributing to aroma and taste sensations are very important in what is a satisfactory taste. For this reason, one can have lower solubles concentrations in freshly prepared brews or concentrates, yet they satisfy the drinker.

Soluble concentration levels diminish with freeze dried instants, then brewed beverages, and then liquid extracts relative to spray dried solubles. This sequence of ingredient strengths is logical since:

1. Commercial instants have high cellulose, (hydrolysis sugars) non-coffee content, and lose much or almost all volatile aromas in spray drying.
2. Commercial freeze dried instants lose less volatiles than in spray drying, but more than in brewing.
3. Commercial or domestic brewing is less efficient than scientifically designed extractors for 200 F.

This difference in solubles strengths between spray and freeze dried instants (from the same extract) have not been adequately communicated to the public using these products.

Where a low grade freeze dried coffee is compared with a high grade spray dried instant coffee these solubles differences in taste may not hold.

Since spray dried solubles have 3.0 to 3.5 percent moisture, while freeze dried instants usually have less than 2 percent moisture, this is a small but positive factor that contributes to the solubles concentration use differences. Obviously, the flavor of a good brewed coffee would be a point of reference against current commercial R&G instant coffees. But since the latter are invariably accepted by the consumer as another coffee-like drink and are consumed with dairy additives, a direct taste comparison is not possible. It is well known that better quality, higher grown, freshly roasted coffees "go further" when preparing coffee beverages.

In 1976 and 1977 there was a strong movement to use instant coffees, mainly freeze dried instant coffees in the institutional outlets as well as in some office coffee service outlets and in vending machines. The circumstances conducive to this move were the fast increasing coffee prices, the existing poor quality of freshly brewed coffees, especially when held hot for a period, and lower cost of freeze dried dispensing equipment.

Until the problem of freeze dried instant coffees picking up moisture (they are very hygroscopic) and forming gummy deposits, inactivating the brewing machines, was placed under better control, the use of freeze-dried instant coffees was hindered.

## Quality

Many of the major coffee roasting firms have been using either pressure roasting or steaming of the roasted Robusta coffee beans from Africa and Indonesia to drive out the objectionable Robusta tastes. These techniques simultaneously have increased the solubles yields from R&G coffee to 30 percent.

A high grown, good quality, mild Arabica coffee, like those from Colombia or Guatemala, can give a better tasting cup of coffee with 1.1 percent solubles than can a cheaper, lower grade Robusta coffee with 1.35 percent solubles. Unfortunately, once the original coffee beans are blended by the roaster, it becomes difficult for the layman to discriminate between the original blend and influencing taste quality.

## PAPER FILTERS

In the chemical laboratory, it was no effort to pour the steeped, decanted coffee beverage through a paper filter in a laboratory glass funnel. It was about 1936 when Peter Schlumbohm, a German immigrant to the U.S.A.with a Ph.D. in chemistry, devised the Pyrex hour-glass shaped paper filter coffee brewer called Chemex. This single glass apparatus had the advantage that the steeping could be done in the top portion filter cone simultaneously while slowly filtering coffee beverage. This did not require a separate steeping pot (that would get dirty with grounds), plus it had the throwaway feature for the spent grounds with filter.

The filter paper was so fine in porosity (in order to control a slow outflow of coffee beverage) that high clarity of coffee beverage was achieved and hardly any oil passed through with the beverage. With a hot water rinsed filter paper (that did not contribute taste), fine tasting coffee beverage could be made which was representative of the coffee beans of origin by rinsing foreign tastes from the paper.

This brewing device was exhibited in the Museum of Art in New York City in 1943 and won an I.I.T. award in 1958 for one of the best designed products of modern times.

Strangely enough, shortly after Dr. Schlumbohm's death, this filter principle of coffee brewing began to dominate the domestic ½ gallon and multi-gallon urn brewers in the U.S.A. There were simple variations of increasing batch sizes, metering hot water, fluting the filters and/or providing filter paper grids to facilitate flow-through of coffee beverage.

Similar to the Chemex hour-glass flask is the Melitta German brewer, which uses a separate conical fluted, porcelain funnel to hold the filter paper. The funnel piece rests on a porcelain serving pot. After the coffee beverage has filtered through, the top funnel piece is removed and a cover is put on the coffee server pot. The paper filter, with spent grounds, is thrown away.

The use of filter paper was, of course, only an extension of the drip pot coffee brewer. The paper replaced the perforated metal that retained the roast coffee granules, and controlled the rate of water flow out of the funnel.

These filter paper, beverage retaining brewers allow steeping of the R&G coffee. Steeping improves the filter brewer performance over the pumping-percolator, which normally has little steeping action but only a rinsing action on the granules. That is, the perforated basket that holds the R&G coffee normally does not hold beverage, because basket porosity and bed porosity (with regular grind) allow the watery beverage to quickly trickle through.

On the other hand, the filter paper brewers hold back water among the granules, causing steeping before the coffee beverage trickles out. This temporary steeping period improves solubles extraction without resorting to the repetitive beverage rinsing of the pumping-percolator.

A further advantage of the fluted filter paper unit is that it has more filtration area than the limited holes in a drip pot. Hence, the fluted filter paper brewers ordinarily do not plug or take excessive time to allow coffee beverage through, as can occur in an uncontrolled drip pot. In other words, the rate of beverage flow through the filter, as related to steeping extraction time, is being controlled. And if R&G fines do seal off bottom flow, beverage can seep through the side walls.

The technology of filter papers for the chemical laboratory has been highly developed for over 100 years. But the filter paper application (without taste contribution) to coffee brewing has been broadly applied only recently. The taste contribution of filter papers to the coffee beverage is a serious shortcoming. Packaging of small numbers of filter papers in sealed polyethylene bags is an important step toward minimizing and eliminating additional odor and taste pickup by these papers.

The filter paper media was the principal factor in the commercialization and widespread use of the ½ gal brewer, which is so common in the U.S.A. and now in Europe. Until about 1960 the Cory slurry-type brewer, with cloth/screen retainer, was one of the best coffee brewers in the restaurant industry. Afterwards, the Bunn brewer, with its non-pressurized hot water spraying over the R&G coffee (lying in a paper filter), became one of the most popular mechanisms for coffee beverage preparation. The ease of filling the filter cup with R&G coffee and its ease of total disposal were major factors in the broad acceptance and use of these principles and methods.

Brewing coffee beverage consists essentially of two steps: (1) extracting the soluble portion of the coffee grounds with hot water, and then (2) separating the spent or extracted coffee grounds from the desired beverage.

The second step is to separate the unwanted, solubles exhausted coffee granules from the wanted beverage. Traditionally, this separation is by decanting the coffee beverage away from the spent grounds at the bottom of the pot. Cleaner beverage recovery (with no grit) is obtained by using a metal strainer, woven cloth, or filter paper. These separators progressively allow less oil and grit through into the potable beverage. Filter papers, therefore, remove natural and desirable coffee flavor. However, when the two steps are combined on the filter paper, the paper must control outflow rates.

Since before recorded times, this separation of coffee grounds from beverage has been done in several ways and with varying effectiveness. In Brazil and other parts of Latin America, the QUADOR or flannel or muslin sack is used to filter or strain the steeped coffee grounds to recover the beverage.

For example, Arabs pour the coffee beverage from their Ibriks. About 90 percent of the grounds remain in the bottom of the Ibrik. Any floating granules are removed with palm fibers in the mouth of the Ibrik. Similarly, when R&G coffee is steeped in hot water, careful decanting can leave over 95 percent of the spent coffee grounds in the pot. What grounds are not left behind are the coarse coffee granules that never were wetted and remained as "floaters". The finest coffee granules are easily decanted over; also the oils and oily particles tend to float and are carried over with the beverage. The clarities of paper filtering did not come into widespread use in the U.S.A. until the 1950's and intensively institutionally until the 1960's.

Pouring the slurry of beverage and coffee grounds through a strainer or perforated metal disc eliminates over 99 percent of the grounds and reduces the oiliness of the coffee beverage. Hence, perforated R&G coffee retaining discs became an integral part of the drip-maker, "pumping" percolator, and vacuum brewer. Also, for beverage grounds separations, a muslin bag was used ("muslim" is an Arabic term).

In all of these separation methods, it is not only the chosen filter media that performs the separation, but the built-up spent granular coffee cake on the filter media, that affects the beverage filtration. Indeed, some of the best filtrations of beverage are achieved on a fine granular bed of coffee without any cloth or other fibrous separation media.

The carryover of spent coffee grounds and oil is, of course, a lack of perfection in the separation or beverage preparation. It is annoying to the consumer to taste the oil and the grit except where this is culturally normal and acceptable.

## Filter Paper: Properties and Specifications

Table 17.2 compares several properties of commercial grades of filter papers and pouches.

*Thickness* of media: Porosity is not directly related to thickness. Most fibrous media are 2 mils thick; the thickest are 8 mils.

*Color:* This is usually white or whitish and can have a brightness standard.

*Permeability* to water (distilled): This is usually based on a 1 in. diameter flow area under a head of 2 in. of water and is about a pint per min. The fines of R&G coffee embed themselves into the pores of the paper, thereby considerably reducing the throughput flow rate.

There is also an air permeability test where a 2 oz metal cylinder falls by gravity in a tube pushing the 100 $cm^3$ air down through the 1 sq in. area dry test paper (1.2 in. diameter orifice) in 0.1 sec.

TABLE 17.2. COMPARISON OF FILTER PAPER PROPERTIES

Several commercial filter pouches and papers were studied to compare certain relative properties. The rate of through-flow when using a slurry of R&G and water, the area per unit weight on a square inches/gram basis, film thickness of thousandths of an inch, capacity, and taste contribution are compared in the table below.

| | Filter | Rate | Area Weight | Thickness | Porosity | Taste |
|---|---|---|---|---|---|---|
| 1. | Star | Fast | 125 | 2 | veil-like | nil |
| 2. | Tea bags | Fast | 100 | 2 | less porous than (1) | nil |
| 3. | Hill-Shaw pouch | Fast | 62 | 3½ | less porous than (2) | nil |
| 4. | Eaton Dikeman Grade ED #980 | Fast | 32 | 6 | translucent | nil |
| 5. | Georgia Pacific Coronet toweling | Fast | 33 | 8 | less translucent than (4) | nil |
| 6. | N.W. Coffee Co. semi-circle sewn | Med. | 30 | 5 | like (4) | high |
| 7. | Chemex | Slow | 13 | 8 | like (4) | noted |
| 8. | Fairfield Tomlinson | Fast | 27–40 | 5½–7 | translucent | nil |
| 9. | Dexter | Fast | 100 | 2 | see through | nil |

*Star filter of tea bag texture, sold for percolator use, filter fast.
A cellulose fiber matte; taste not notable. R&G coffee sealed in pouches have ruptured in service.
Chemex paper was least permeable and most expensive.

*Density* of paper or media: It is usually expressed as the weight of 500 sheets (ream) of paper 24 in. x 36 in., e.g. 4–6 mil is 30 to 35 lb stock.

*Wet burst* (or dry burst) strength: This can be run on a standard tester wherein a rubber diaphragm under air pressure expands until the paper bursts.

*Visual porosity:* This is simply a "feel" test. The 2 mil thick Star media (which is like a tea bag) is quite transparent, whereas the 6 to 8 mil papers are barely transluscent when held up 1 foot away from a 60 watt bulb.

*Taste contribution:* This is barely mentioned. It is, in fact, the most important specification for coffee beverage preparation, equal to porosity.

Filter papers for commercial restaurant or institutional urns cost about 4¢ each. Domestic retailed smaller filter papers cost 2 to 3¢ each. These filter papers are often packaged with minimal protection against outside odor pickup. Wet strength can be designed into the filter paper by use of resin.

If a ⅛ in. support grid or strainer is used to hold these paper media, they are all surprisingly fast filtering, rather thorough in particle retention, and have no more taste contribution than the commercial filter papers used for coffee. It must be re-emphasized that speed of beverage filtration has a direct bearing on taste contribution of the paper. Slower filtering papers tend to give more of their taste.

The cheese-cloth pouch used for Max-pax R&G coffee is adequate evidence that oil particles are not being retained, as is claimed. The filter-media also gives a definitely noticeable and objectionable taste.

Filter paper is a good adsorbent of foreign odors and tastes. Filter papers pick up foreign odors, and these often are readily transmitted to the consumer's coffee beverage. Even if the fluted filter papers arrive at the restaurant in odor-free condition, the restaurant kitchens are moist and have spicy, meaty, fishy, and doughy odors in the air. This is hardly the ideal environment for filter paper storage. Further, these kitchen volatile odors, after adsorption on the filter papers, are oxidized and chemically modified to give objectionable resulting odors and tastes.

Astonishingly, these prepared paper filters are often shipped in cardboard cartons in lots of 250 or more, and are only unitized in smaller groups and in polyethylene liners on special request. This unprotective type of packaging reflects on the crudeness of the supply industry. For restaurant use, the paper filters ought to be sealed in polyethylene bags in units not exceeding 25 or 50. Also, there are less odor permeable plastics than polyethylene that can be used to protect the paper.

Most retail paper filters are sold in units of 100, and more often than not, in ordinary cardboard boxes. In the supermarket, these papers pick up some exotic odors and tastes, like detergents and perfumes.

## Modes of Filter Paper Use

Modes of use can influence the filter paper performance. For example, taking a circle of filter paper and folding it into a 90° quadrant does the following:

1. The half of the filter paper with 3 layers in the funnel is not going to filter much coffee beverage.
2. The single layer of filter paper against the smooth funnel wall is wetted to the wall. This offers a small area for coffee beverage permeability.

There is a class of filter papers that overcome these shortcomings. They have sewn semicircles, which when opened are cupped quadrants. However, this has not solved the problem of the paper wetted and sealed to the smooth, solid funnel wall. Thick crepe paper does help to keep its shape and keeps paper away from wall, but is of very low permeability usually. In Mellita funnels, an undulated inner funnel wall keeps the filter paper from touching all of the funnel walls. This helps to increase filter area.

The solution is to place the filter paper conic cup into a conic strainer, with ½ in. mesh openings. Then all filter paper surfaces are free to allow flow of beverage through (with only 10 percent or less screen contacted to paper). The ½ gal. coffee brewer and the sprinkler urns use a coarse grid of 1½ in. to 2 in. spacing. But this lacks adequate support for the paper. A ¼ in. grid strainer, for example, would support thinner, less rigid, and more porous filter paper, hence, would give more faithful coffee beverage preparation and flavor. Highly porous filter paper has the additional advantage that the coffee beverage residence time or contact time with the filter media is reduced to less than a minute and preferably to tens of seconds. The importance of this contact time is that most filter paper media are not taste free, and hence, impart their flavor to the coffee beverage. Reduced contact time reduces the taste contribution of the media to a lower or negligible level. Time is very important when trying to not contaminate the original fresh coffee flavor.

## Metal Support Grid

The metal support grid shapes the resulting bed of wetted R&G coffee. A dramatic illustration of the importance of this grid support for the paper filter media, can be obtained by simply leaving out the grid in a Bunn ½ gal. coffee brewer. Then the filter paper will seal itself to the funnel, and hardly any coffee beverage will drip through.

The fluted paper cup design is primarily to aid the user to have a container or cup to receive the R&G coffee before it is placed into its gridded cradle. After the filter paper is wet, the fluted paper support is largely lost and retention of shape is by the metal grid or ring.

## Paper Filters—Urn Use

Paper filters in urn use were, until about 1960, more the exception than the rule. However, since then filter papers have enjoyed greater use both in urns and in the ½ gal. brewer. An estimate in 1972 was that at least three-fourths of urn users used filter paper.

The muslin bag required emptying the R&G coffee and washing; the bag never was completely rid of stale, oily grounds even when it was water-soaked overnight. Soaking was often forgotten because it depended on someone to do it. There is labor saved with filter paper use. Costs are 4¢ to 3¢ per lb of R&G.

Some of the coffee brewing equipment manufacturers sell filter papers under their brand names. They often are not the prime paper fabricators nor converters. They do this for a profit as well as a service. There is some assurance that the proper filter media for their urn system is being used.

Filter paper shapes are standardized for urns as truncated cones with fluted sides. This allows for the R&G coffee to be poured into the "cup cake" structure. It is common to see dozens of these cupped papers filled with R&G coffee on a restaurant counter. The attendant, in getting ready to brew coffee, does not realize that the open R&G coffee is rapidly staling. The fluted filter paper cup should first be set in its wire grid support, then filled with R&G, and then transferred to the urn. The paper flutes give the paper more structural strength while dry than wet. Once wetted, the filter paper is supported by the wire grids. Filter papers are usually white in color. Papers must be free of any taste or odor. Papers should be in several dozen units, preferably sealed in polyethylene envelopes to protect them from moisture, dirt and foreign odors that papers can readily absorb.

Aside from some routine specifications already indicated, there are two important properties that the filter paper must have: (1) air and water permeability, and (2) adequate wet burst strength.

The coffee urn equipment manufacturers and/or filter paper suppliers order the correct specification filter paper from a prime supplier. They have the paper cut to the diameters required by shear dies. Then dozens of paper discs with a thermal setting resin are moistened and steam pressed into a fluted metal mold for the desired fluted cup structure.

## Fineness of Filter Media

A 60 mesh stainless steel strainer quite effectively removes all coarse floaters and almost all particles larger than what might make a fine sludge at the bottom of the cup. Oil is not removed.

Some cultural coffee beverage consumers, like Italians, French, Turks, Arabs, and Greeks, have acquired a liking for the dark roast, oily coffee beverage with sediment in the cup. Passing their strong coffee beverage through a paper filter would be ruining it.

Indeed, thorough beverage filtration does remove considerable coffee flavor and aroma which are carried by the oily particles. On the other hand, if the original coffee beans were Robusta (of bad taste and aroma), the thorough removal of the oily particles would reduce the objectionable tastes.

With one-third of consumed cups of coffee in the U.S.A. being instants and of high clarity, this fastidiousness in cup clarity has influenced the consumer's standards and expectations. The consumer now expects to see that kind of beverage clarity and freedom from oil, without considering what has happened.

How thorough should the filtration be? Most commercial paper filters are too thorough in their clarification, when one uses good quality blends of fresh R&G coffee. There is a per capita reduction in U.S. coffee use, largely due to increased use of cheaper Robusta blends. The commercial, thorough paper filtration is consistent with the Robusta type of coffee beans being used.

With a good quality R&G coffee, the filtration separation should not be so thorough. Oily, richly aromatic coffee particles (still below 100 to 200 mesh) ought to be allowed through into the potable beverage. This "leakage" can be achieved by the tea bag type of texture of filter media. The close cropped fibers and thick filter papers now being used can be replaced by thinner, more porous papers or fiber mats (where paper structure is not self-supporting).

Most users of filter papers have never given these concepts a thought. The consumer simply uses what papers are commercially available, and lets it go at that. The sophistication in filter paper use for the coffee beverage preparation has hardly begun. The user at present has no choice of the porosities of filter paper he can buy. In the chemical laboratory there is a broad assortment of filter papers that the chemist can choose from, depending on his objectives. The filter paper user in the coffee brewing industry ought to have some choices of porosity, such as "fine," "medium" and "coarse."

## SUMMARY

Good tasting coffees are prepared from good quality beans—wholesome beans that are fresh, cleanly and properly roasted, finely ground, and promptly brewed and served. Any claims made that deviations from these methods can give satisfying coffee are subject to serious criticism.

The reader should note that in the U.S.A. the per capita coffee consumption has declined one-third in the period from 1960, while per capita coffee use has increased practically everywhere else. This is real testimony that the policies, equipment, methods, and raw material use in the U.S.A. are generally wrong.

It should be noted that steeping of finely ground, freshly roasted coffees of Arabica types with few defects is the best way to prepare a good tasting cup of coffee. The strength of the beverage must be based on at least the use of 60 grams of finely ground R&G coffee to produce a liter of beverage, and that solubles content must be at least 1.2 percent weight. Time is of the essence. The time between when the green beans are processed in the growing country and when the beans are roasted ought to be less than one year and preferably less than six months with proper storage and handling in the interim. The roasted beans must be consumed within one week preferably; grinding ought to be just before beverage preparation. Storage of frozen roasted beans in sealed containers may extend bean freshness to several weeks and ground coffee freshness to several days. Vacuum can packaging of R&G coffee will always yield a partly stale and aroma-depleted product similar to gas flushed pouches. Spent coffee grounds must not be filter separated from the beverage; the colloidal aromatic parts should be allowed to pass into beverage. Separation of spent grounds from beverage ought to be within a minute or two. Once beverage is prepared, it should be served promptly and not be held on heaters for hours.

Interestingly enough, the espresso machine fulfills most of the requirements indicated above, whereas the procedures used in the U.S.A. violate most of them. Improper brewing is a waste of a natural resource—the good, green coffee bean.

With increased world population and increased world demand for coffee, hence, also good coffee quality, we should see coffee prices progressively rise and discrimination in purchasing and quality increase where it is demanded.

Most of the bad features in handling coffee in the U.S.A. (and elsewhere too) are due to mass production and centralized marketing and sales. The indicated particular treatment of green, roast ground coffees and their brewing is in conflict with corporate mass production and selling policies. The mass brewing in urns, in restaurants, etc. contributes to use of poor coffees and downgrades all coffees used. Resolution of this direct conflict in the near future is not foreseeable nor anticipated without a discriminating consumer.

Commercial aspects, that is profit-making, invariably override technical considerations and process technology. To this extent, it is to the self-interest of each commercial group to put forth its concepts and cost savings in production, and not to proclaim real quality advantages unless they coincide (they seldom do). It is the chemist and engineer that can understand what is involved in coffee processing technology.

Automatic drip filter brewers have made great inroads into consumer use in Europe and in the U.S.A. but the beverage cannot be better than what goes into the brewer, and automation and filter papers detract from what quality is potentially possible.

Finally, real quality progress in coffee beverages must be supported by the consumer himself. So long as his expectations are low, his standards are low. The acceptance of poor tasting coffees is perpetuated by tacit acceptance, without question and motivation to improve this single aspect of his existence, even if he drinks a coffee beverage several times every day.

# Useful Tables

TABLE A-1. EXHIBITS

**Coffee and Coffee By-Products at Each Process Step**

Cherry fruit
Beans in hull (pergamino)
Hulls
Husks
Silver skins
Green coffee beans
Roast coffee beans: light, medium, dark, French and Italian
Grinds: regular, drip and fine; also pulverized
Instant coffee powder (spray dried and belt dried)
Extracted coffee grounds
Equivalent grounds mineral ash

TABLE A-2. CHEMICALS IN COFFEE

**Volatiles**

Acids—acetic, propionic, butyric, valeric
Aldehydes—acet, propyl, butyl, valer
Ketones—ketone, methylethyl, penetanone, diactyl
Ketone-alcohols—acetol, acetoin
Esters—methyl formate, methyl acetate, methyl propionate, propyl formate
Sulfides—hydrogen, dimethyl, methyl mercaptan, thiophene
Cyclic—furan, methyl furan, pyridine
Amines—trimethyl, dimethyl, ammonia
Hydrocarbons—isoprene
Unsaturates—acrolein

**Non-Volatiles**

| | Soluble coffee, per cent | Roast coffee, per cent |
|---|---|---|
| Chlorogenic acid | 17 | 7 |
| Caramel | 56 | 23 |
| Mineral ash | 12 | 3.5 |
| Trigonelline | 3 | 1 |
| Caffeine | 4 | 1.2 |
| Oil | 0.1 | 10 |
| Caffeic acid | 0.8 | 0.3 |
| Quinic acid | 0.8 | 0.3 |
| Citric acid | 1.5 | 0.6 |
| Malic acid | 1.5 | 0.6 |
| Tartaric acid | 1.3 | 0.5 |
| Pyruvic acid | 0.1 | 0.06 |
| Guiacol | ... | ... |
| Nicotinic Acid | ... | ... |

TABLE A-3. TYPES OF COFFEE BREWING APPARATUS

Steeping and straining or decanting
Boiling (driving off volatile flavors and aromas)
Percolation (recycling brew over spent grounds in air)
Drip (water once through shallow grounds bed)
Vacuum (slurry with straining)
Cold extractor (similar to drip but regulated cold water flow)
Pressure entraction with hot water (commercial soluble coffee)
Urn (batch rinsing of solubles from grounds in bag)
Tape (rinsing solubles from 1 cup equivalent of R & G coffee)
Espresso—domestic (upward or downward hot water/steam flow). One cup per cycle.

TABLE A-4. UNIT CONVERSION FACTORS

| | |
|---|---|
| Length | One centimeter (cm) = $10^{10}$ micromicrons (mu mu) = $10^{8}$ Angstroms (A) = $10^{7}$ millimicrons (m mu) = $10^{4}$ microns (mu) = 10 millimeters (mm) = 0.3937000 inch (in.). |
| | One meter (m) = 100 cm = 39.37000 in. = 3.280833 ft = 1.09361 yard (yd). |
| | One kilometer (km) = 1,000 m = 0.62137 statute mile. |
| | One statute mile = 5280 feet (ft) = 1,760 yd = 1.60935 km. |
| | One nautical mile (1 min of arc at equator) = 1.1516 statute mile = 6080.2 ft. |
| | One ft = 0.3048006 m. One in. = 2.540005 cm. One rod = 16.5 ft. |
| Area | One sq mile = 640 acres = 2.59000 sq km. |
| | One acre = 160 sq rods = 4,840 sq yd = 43,560 sq ft = 0.404687 hectares (ha) = 4046.9 sq m = 208.71 ft squared. |
| | One ha = 100 ares = 2.471044 acres = 100 m squared. |
| | One sq ft = 929.0341 sq cm. One sq in. = 6.4516258 sq cm. |
| Volume | One U.S. gallon (gal) = 3.78533 liters (l) = 231.00 cu in. = 0.83268 British gal. |
| | One gal water at 59 F (15 C) weighs 8.337 pound (lbs). |
| | One British gal = 1.20094 U.S. gal. One U.S. quart (qt) = 0.946 l. |
| | One l = 1,000 milliliters (ml) = 1.056710 qt = 0.26418 U.S. gal. |
| | One U.S. qt = 32 fluid ounces (fl oz). One fl oz = 29.6 ml. |
| | One cu ft = 28.316 l = 7.481 U.S. gal. |
| | One cu m = 35.314445 cu ft = 1.307943 cu yd. |
| Weight | One hundredweight (cwt) = 100 lb avoirdupois (avdp) U.S. = 112 lb avdp U.K. |
| | One oz avdp = 28.349527 grams (gm) = 437.5 grains (gr) avdp. |
| | One gm = 1,000 milligrams (mg) = 15.4324 gr. One lb = 7,000 gr. |
| | One Spanish lb = 1.014 lb avdp. One arroba = 12.5 kilogram (kg). |
| | One quintal = 100 lb avdp or 100 Spanish lb according to local custom. |
| Velocity | One ft per second (sec) = 0.681818 miles per hour (hr) = 1.09828 km per hr. |
| Density | One gm per ml = 62.426 lb per cu ft = 1,685.50 lb per cu yd = 2,204.55 lb per cu m = 8.345 lb per gal = 1000 kg per cu m. |
| Pressure | One atmosphere (atm) = 14.697 lb per sq in. (psi) = 1.0332 kg per sq cm = 760.00 mm = 29.921 in. of mercury (Hg) at 32 F (0 C) = 10.295 m = 33.899 ft of water at 39.1 F (3.94 C). |
| | One kg per sq cm = 14.223 psi. One psi = 0.070307 kg per sq cm. |
| Power | One horsepower (hp) = 745.70 watts (w) = 1.0139 metric hp = 178.130 gm calories (gm-cal) per sec = 0.7070 British thermal unit (Btu) per sec. |
| | One boiler horsepower (bhp) = 33,500 Btu per hr = 34.5 lb water per hr evaporated to steam from and at 212 F (100 C). |

| | |
|---|---|
| | One ton of refrigeration = 200 Btu per min = 12,000 Btu per hr = 288,000 Btu per 24 hr = latent heat of freezing of 1 ton of water frozen in 24 hr. |
| Heat | One Btu (1 lb water 1 deg F, mean value 32 to 212 F) = 252.00 gm-cal (1 gm water 1 deg C mean value) = 0.25200 kg-cal = 778 foot-pounds (ft-lb) |
| | One kilowatthour (kwhr) = 3413 Btu. |
| | One Btu per lb = 0.5556 kg-cal per kg = 0.5556 gm-cal per gm. |
| Heat Flow | One gm-cal per sec per sq cm = 13,272 Btu per hr per sq ft = 4.186 w per sq cm. |
| Heat Conductivity | One gm-cal per sec per sq cm per cm thickness per deg C = 4.186 w per sq cm per cm per deg C = 2,903 Btu per hr per sq ft per in. per deg F. |

TABLE A-5. RELATIONS BETWEEN TEMPERATURE, VAPOR PRESSURE, AND LATENT HEAT OF VAPORIZATION OF WATER[1]

| Temperature | | Vapor Pressure | | | Latent Heat | |
|---|---|---|---|---|---|---|
| F | C | mm Hg | psia[2] | kg/cm[2] | Btu/lb | kg-cal/kg |
| 32 | 0 | 4.6 | 0.0886 | 0.00623 | 1072 | 595 |
| 77 | 25 | 23.7 | 0.4581 | 0.03221 | 1048 | 582 |
| 122 | 50 | 92.3 | 1.7849 | 0.12549 | 1023 | 568 |
| 167 | 75 | 289.0 | 5.589 | 0.3329 | 997 | 554 |
| 212 | 100 | 760.0 | 14.697 | 1.0333 | 970 | 539 |
| 338 | 170 | 5937.0 | 114.79 | 8.0705 | 880 | 489 |

[1] From steam tables.
[2] Absolute pressure in pounds per square inch.

TABLE A-6. BOILING TEMPERATURES OF WATER AND BAROMETRIC PRESSURES AT VARIOUS ALTITUDES

| Altitude | | Temperature | | Pressure | | |
|---|---|---|---|---|---|---|
| ft | m | F | C | mm Hg | in. Hg | psia |
| 0 | 000 | 212.0 | 100.0 | 760 | 29.92 | 14.70 |
| 3,000 | 914 | 206.4 | 96.9 | 677 | 26.65 | 13.09 |
| 5,000 | 1524 | 202.6 | 94.8 | 629 | 24.76 | 12.16 |
| 7,000 | 2134 | 198.8 | 92.7 | 582 | 22.91 | 11.25 |

TABLE A-7. STEAM GAUGE PRESSURES AT VARIOUS ALTITUDES

| Altitude | | Pounds Per Square Inch | | | |
|---|---|---|---|---|---|
| ft | m | 212 F/100 C | 230 F/110 C | 248 F/120 C | 260 F/127 C |
| 0 | 000 | 0 | 6.1 | 14.1 | 20.7 |
| 2,000 | 610 | 1.0 | 7.1 | 15.1 | 21.7 |
| 4,000 | 1219 | 2.0 | 8.1 | 16.1 | 22.7 |
| 6,000 | 1829 | 2.9 | 9.0 | 17.0 | 23.6 |

TABLE A-8. ALKALINITY AND HARDNESS OF VARIOUS UNITED STATES WATER SUPPLIES[1]

| Lake Waters From 8 Large Lakes, results in ppm | Lake Superior, Sault Ste. Marie, Mich. | Lake Huron, Port Huron, Mich. | Lake Erie, Buffalo, N.Y. | Lake Ontario, Toronto, Ont., Canada | Lake Michigan, Chicago, Ill. | Yellowstone Lake, Nevada | Lake Champlain, New York & Vermont | Lake Okechobee, Florida |
|---|---|---|---|---|---|---|---|---|
| Total hardness as $CaCO_3$ | 46 | 89 | 109 | 123 | 125 | 23 | 48 | 107 |
| Calcium hardness as $CaCO_3$ | 33 | 60 | 78 | 89 | 80 | 22 | 36 | 78 |
| Magnesium hardness as $CaCO_3$ | 13 | 29 | 31 | 34 | 45 | 1 | 12 | 29 |
| Alkalinity as $CaCO_3$ | 46 | 82 | 94 | 94 | 122 | 42 | 51 | 93 |
| Sodium + potassium as Na | 3 | 4 | 7 | 8 | 3 | 20 | 6 | 18 |
| Chlorides as Cl | 1 | 3 | 9 | 16 | 2 | 9 | 1 | 28 |
| Sulfates as $SO_4$ | 2 | 6 | 13 | 22 | 7 | 8 | 7 | 7 |
| Nitrates as $NO_3$ | .5 | .4 | .3 | 1.3 | ... | .. | .. | .1 |
| Iron as Fe | .06 | .04 | .07 | .05 | .03 | .. | .. | .1 |
| Silica as $SiO_2$ | 7 | 12 | 6 | 8 | 5 | 42 | 1 | 8 |

| Hardness of Nine Rivers | Average | Maximum | Minimum |
|---|---|---|---|
| Willamette at Salem | 19 | 27 | 12 |
| Raritan at Bound Brook | 46 | 56 | 29 |
| Hudson at Hudson | 69 | 89 | 45 |
| Potomac at Cumberland | 79 | 149 | 41 |
| Iowa at Iowa City | 203 | 269 | 82 |
| Missouri near Florence | 246 | 412 | 154 |
| Smoky Hill near Lindsborg | 376 | 510 | 209 |
| Brazos near Waco | 481 | 652 | 176 |
| Arkansas near Deerfield | 721 | 1008 | 292 |

| High Sodium Alkalinity Waters, results in ppm | Bryan, Tex., Well | Lufkin, Tex., Private Well | Waverly, Kan., Well | Cristfield, Md., Private Well |
|---|---|---|---|---|
| Sodium alkalinity as $CaCO_3$ | 121 | 170 | 280 | 460 |
| Total hardness as $CaCO_3$ | 12 | 7 | 24 | 0 |
| Calcium hardness as $CaCO_3$ | 8 | 5 | 15 | 0 |
| Magnesium hardness as $CaCO_3$ | 4 | 2 | 9 | 0 |
| Total alkalinity as $CaCO_3$ | 133 | 177 | 304 | 460 |
| Free carbon-dioxide as $CO_2$ | 0 | 0 | 0 | 0 |
| Chlorides as Cl | 18 | 6 | 14 | 54 |
| Sulfates as $SO_4$ | 2 | 26 | 74 | 42 |
| Iron as Fe | .5 | .2 | .1 | .1 |
| Silica as $SiO_2$ | 35 | 12 | 7 | 0 |

[1] From Permuti Water Conditioning Handbook (1949).

TABLE A-8. (*Continued*)
COMPOSITION OF TYPICAL U.S. WATER SUPPLIES[2]

| Location | Composition, Parts per Million | | | | | | | | | | | |
|---|---|---|---|---|---|---|---|---|---|---|---|---|
| | Total Dissolved Solids | $SiO_2$ | Fe | Ca | Mg | Na | K | $HCO_3$ | $SO_4$ | Cl | $NO_3$ | Total Hardness $CaCO_3$ |
| Augusta, Me. | 31 | 3.8 | 0.01 | 5.0 | 1.0 | 2.4 | 0.8 | 15 | 5.8 | 1.4 | 0.50 | 17 |
| Boston, Mass. | 43 | 2.1 | 0.12 | 4.4 | 1.0 | 3.9 | | 16 | 9.8 | 3.2 | 0.31 | 15 |
| Providence, R.I. | 54 | 11 | .05 | 12 | 0.5 | <5 | | 17 | 12 | 4.4 | 0.53 | 32 |
| New York, N.Y | 31 | 2.4 | 0.14 | 5.8 | 1.4 | 1.7 | 0.7 | 14 | 9.7 | 2.0 | .54 | 20 |
| Trenton, N.J. | 70 | 9.0 | .07 | 12 | 3.3 | 5.4 | | 46 | 12 | 2.9 | 1.1 | 44 |
| Philadelphia, Pa. | 99 | 2.4 | .01 | 17 | 6.4 | 6.6 | 2.1 | 54 | 25 | 10 | 4.5 | 69 |
| District of Columbia | 130 | 7.0 | 2.7 | 34 | 1.9 | 6.8 | | 79 | 34 | 5.2 | ... | 93 |
| Miami, Fla. | 191 | 9.0 | .02 | 35 | 4.8 | 12 | 1.5 | 47 | 61 | 20 | 4.5 | 107 |
| Chicago, Ill. | 157 | 6.4 | 0.2 | 34 | 9.7 | 5.1 | | 146 | 12 | 4.5 | ... | 125 |
| St. Louis, Mo. | 228 | 8.0 | 2.5 | 21 | 7.5 | 36 | | 48 | 94 | 18 | 4.9 | 83 |
| New Orleans, La. | 158 | 7.8 | ... | 17 | 7.1 | 23 | | 45 | 43 | 28 | ... | 72 |
| Minneapolis, Minn. | 210 | 9.7 | 0.07 | 44 | 15 | 4.0 | | 167 | 36 | 3.8 | .20 | 172 |
| Omaha, Neb. | 334 | 12 | 2.2 | 44 | 12 | 50 | | 130 | 138 | 12 | ... | 159 |
| Tulsa, Okla. | 98 | 4.0 | 0.1 | 30 | 1.3 | 4.7 | | 93 | 7.5 | 4.5 | Trace | 80 |
| Ponca City, Okla. | 1005 | 8.5 | 0.1 | 143 | 20 | 157 | | 306 | 133 | 276 | ... | 440 |
| Dallas, Tex. | 1119 | 15 | 13 | 14 | 3.0 | 367 | | 451 | 382 | 88 | ... | 47 |
| Los Angeles, Calif. | 421 | ... | 0 | 75 | 21 | 46 | | 223 | 130 | 39 | ... | 274 |
| Tacoma, Wash. | 76 | 25 | .29 | 7.9 | 4.4 | 4.1 | 1.4 | 38 | 4.9 | 4.0 | 5.3 | 38 |

[2] From Kunin, R. 1958. Ion Exchange Resins. John Wiley and Sons. New York.

TABLE A-9. WATER ANALYSES CONVERSION UNITS (EQUIVALENTS OF $CACO_3$)

| | ppm | gr/gal U.S. | gr/gal English |
|---|---|---|---|
| Parts per million or mg/liter | 1 | 0.058 | 0.07 |
| Grains per U.S. gallon or German degrees (practically) | 17.1 | 1.0 | 1.2 |
| Grains per English gallon or Clark degrees | 14.3 | 0.83 | 1.0 |

One French degree = one part per 100,000.

TABLE A-10. SCREEN EQUIVALENTS

| Screen No. | Sieve Openings: U.S. Standard Screens in. | U.S. Standard Screens mm | Tyler Screens mm | Tyler Screens in. |
|---|---|---|---|---|
| 8 | 0.0937 | 2.38 | 2.362 | 0.0929 |
| 10 | 0.0787 | 2.00 | 1.651 | 0.065 |
| 12 | 0.0661 | 1.68 | 1.397 | 0.055 |
| 14 | 0.0555 | 1.41 | 1.168 | 0.0459 |
| 16 | 0.0469 | 1.19 | 0.991 | 0.039 |
| 18 | 0.0394 | 1.00 | ... | ... |
| 20 | 0.0331 | 0.84 | 0.833 | 0.0328 |
| 25 | 0.0280 | 0.71 (24-mesh) | 0.701 | 0.0276 |
| 30 | 0.0232 | 0.59 (28-mesh) | 0.589 | 0.0232 |
| 35 | 0.0197 | 0.50 | 0.417 | 0.0164 |
| 40 | 0.0165 | 0.42 (42-mesh) | 0.351 | 0.0138 |
| 45 | 0.0138 | 0.35 | ... | ... |
| 50 | 0.0117 | 0.297 (48-mesh) | 0.295 | 0.0116 |
| 60 | 0.0098 | 0.250 | 0.246 | 0.0097 |
| 70 | 0.0083 | 0.210 | ... | ... |
| 80 | 0.0070 | 0.177 | 0.175 | 0.0069 |
| 100 | 0.0059 | 0.149 | 0.147 | 0.0058 |
| 200 | 0.0029 | 0.074 | 0.047 | 0.00185 |
| 325 | 0.0017 | 0.044 | ... | ... |

TABLE A-11. ABBREVIATED STEAM TABLE

| Absolute Pressure, lb per sq in. | Temperature F | Temperature C | Specific Volume, cu ft per lb-steam | Latent Heat of Evaporation, Btu per lb |
|---|---|---|---|---|
| 1 | 102 | 39 | 334 | 1036 |
| 14.7 | 212 | 100 | 26.8 | 970 |
| 65 | 298 | 148 | 6.7 | 912 |
| 115 | 338 | 170 | 3.9 | 880 |
| 165 | 366 | 186 | 2.75 | 857 |
| 215 | 388 | 198 | 2.13 | 837 |
| 265 | 406 | 208 | 1.75 | 822 |

TABLE A-12. TEMPERATURE CONVERSION CENTIGRADE/FAHRENHEIT

| F | C | F | C | F | C | F | C |
|---|---|---|---|---|---|---|---|
| −459 | −273 | −166 | −110 | 115 | 46 | 421 | 216 |
| −454 | −270 | −148 | −100 | 133 | 56 | 482 | 250 |
| −436 | −260 | −130 | −90 | 151 | 66 | 572 | 300 |
| −418 | −250 | −112 | −80 | 169 | 76 | 662 | 350 |
| −400 | −240 | −94 | −70 | 187 | 86 | 752 | 400 |
| −382 | −230 | −76 | −60 | 205 | 96 | 842 | 450 |
| −364 | −220 | −58 | −50 | 223 | 106 | 932 | 500 |
| −346 | −210 | −40 | −40 | 241 | 116 | 1022 | 550 |
| −328 | −200 | −22 | −30 | 259 | 126 | 1112 | 600 |
| −310 | −190 | −4 | −20 | 277 | 136 | 1202 | 650 |
| −292 | −180 | 14 | −10 | 295 | 146 | 1292 | 700 |
| −274 | −170 | 25 | −4 | 313 | 156 | 1382 | 750 |
| −256 | −160 | 34 | 1 | 331 | 166 | 1472 | 800 |
| −238 | −150 | 43 | 6 | 349 | 176 | 1562 | 850 |
| −220 | −140 | 61 | 16 | 367 | 186 | 1652 | 900 |
| −202 | −130 | 79 | 26 | 385 | 196 | 1742 | 950 |
| −184 | −120 | 97 | 36 | 403 | 206 | 1832 | 1000 |

Conversion Formulas:

$$C = \frac{5}{9}(F - 32) \qquad F = 1.8C + 32$$

TABLE A-13. FREEZING TEMPERATURES OF SALT (NACL) BRINES

| Approximate Salt, Per Cent | Freezing Temperature F | Freezing Temperature C | Specific Gravity | Grams NaCl per Liter Solution |
|---|---|---|---|---|
| 5 | 24.8 | −4 | 1.034 | 51.6 |
| 10 | 19.4 | −7 | 1.071 | 107.1 |
| 15 | 12.2 | −11 | 1.108 | 166.0 |
| 20 | 6.8 | −14 | 1.148 | 230 |
| 25 | −0.4 | −18 | 1.189 | 292 |

TABLE A-14. SPECIFIC GRAVITY OF AMMONIA 20 C/4 C

| Specific Gravity | °Baumé | Per Cent $NH_3$ | Grams $NH_3$ per Liter | Normality |
|---|---|---|---|---|
| 0.977 | 13.1 | 5 | 49 | 2.75 |
| 0.958 | 16.2 | 10 | 96 | 5.62 |
| 0.923 | 21.7 | 20 | 185 | 10.84 |
| 0.892 | 27.0 | 30 | 268 | 15.7 |

TABLE A-15. GAUGE AND THICKNESS U.S. STANDARD SHEET METAL

| Gauge No. | Inches | Millimeters |
|---|---|---|
| 3 | 0.250 | 6.35 |
| 6 | 0.203 | 5.16 |
| 8 | 0.172 | 4.37 |
| 10 | 0.141 | 3.58 |
| 12 | 0.109 | 2.77 |
| 14 | 0.078 | 1.98 |
| 16 | 0.062 | 1.57 |
| 18 | 0.050 | 1.27 |
| 20 | 0.0375 | 0.95 |
| 22 | 0.0312 | 0.79 |
| 24 | 0.0250 | 0.64 |

TABLE A-16. SPECIFIC GRAVITY OF SOLUTIONS HEAVIER THAN WATER

| °Baumé | Twadell | Specific Gravity | Per Cent by Weight | | | | |
|---|---|---|---|---|---|---|---|
| | | | $H_2SO_4$ | HCl | $HNO_3$ | NaOH | $Na_2CO_3$ |
| 5 | 7.2 | 1.036 | 5.3 | 7.2 | ... | 3.2 | 3.4 |
| 10 | 14.8 | 1.074 | 10.8 | 14.8 | 12.9 | 6.6 | 7.0 |
| 15 | 23.0 | 1.115 | 16.4 | 22.9 | 19.4 | 10.3 | 10.7 |
| 20 | 32.0 | 1.160 | 22.3 | 31.5 | 26.2 | 14.4 | ... |
| 25 | 41.6 | 1.208 | 28.3 | 41.7 | 33.4 | 18.7 | ... |
| 30 | 52.2 | 1.261 | 34.6 | ... | 41.3 | 23.5 | ... |
| 35 | 63.6 | 1.318 | 41.3 | ... | 50.3 | 28.8 | ... |
| 40 | 76.2 | 1.381 | 48.1 | ... | 61.4 | 35.0 | ... |
| 45 | 90.0 | 1.450 | 55.1 | ... | 77.2 | 42.0 | ... |
| 50 | 105.2 | 1.526 | 62.2 | ... | ... | 50.1 | ... |
| 55 | 122.2 | 1.611 | 69.7 | ... | ... | ... | ... |
| 60 | 141.2 | 1.706 | 77.7 | ... | ... | ... | ... |
| 65 | 162.6 | 1.813 | 88.7 | ... | ... | ... | ... |

Conversion Formulas

Baumé heavier than water

$$\text{Bé} = 145 - \frac{145}{\text{sp gr}} \qquad \text{sp gr} = \frac{145}{145 - \text{Bé}}$$

Baumé lighter than water

$$\text{Bé} = \frac{140}{\text{sp gr}} - 130 \qquad \text{sp gr} = \frac{140}{130 + \text{Bé}}$$

Twadell

$$\text{Tw} = 200(\text{sp gr} - 1.000) \qquad \text{sp gr} = 0.005\text{Tw} + 1.000$$

# Index